Wilhelm Krelle (Editor)

# The Future of the World Economy

## Economic Growth and Structural Change

With 124 Figures

Springer-Verlag
Berlin Heidelberg GmbH

Professor Dr. Wilhelm Krelle
Institute of Social Sciences
and Economics
Bonn University
Adenauerallee 24-42
D-5300 Bonn 1, FRG

IIASA
International Institute for Applied
Systems Analysis
A-2361 Laxenburg, Austria

ISBN 978-3-662-02588-8 ISBN 978-3-662-02586-4 (eBook)
DOI 10.1007/978-3-662-02586-4


Originally published by Springer-Verlag Berlin Heidelberg New York Tokyo in 1989
Softcover reprint of the hardcover 1st edition 1989



2142/7130-543210

# Contents

# Foreword

It would be difficult to find more important or controversial economic variables than economic growth and structural change. Growth and change are generally related to each other in many disciplines. However, it is sometimes argued that, in economic theory and practice, the interdependence of growth and change is less well understood than in biology, for example. It was precisely this fact that presented a challenge to Professor Krelle and his collaborators and led them to select this topic for the joint IIASA–Bonn University research project.

The merging of economic growth and structural change *per se* may be considered an exciting approach. Here, even more interesting is the fact that the approach was applied to economies on a global scale, fully integrating of socialist countries. The researchers who comprised the IIASA–Bonn University core were able to stimulate their colleagues in several different countries and, as a result of the ensuing collaborative effort, it was possible to overcome many methodological difficulties. In the course of the research, detailed economic data were collected and consistent data bases created that permitted comparative analyses to be carried out.

This book contains the basic findings of this international study, which were discussed at the final conference, held at IIASA in November 1986.

These findings, relevant to medium-term forecasting, are application-oriented. They describe the likely path of the forces behind economic growth and structural change and their causal relationships. As a consequence, they are important not only to researchers in econometrics, but also to those who must make important decisions relating to this time frame.

It is most gratifying to be able to note that IIASA was instrumental in initiating this complex and important project. It is an illustration of how much can be accomplished on a collaborative basis. Nonetheless, it must be recognized that none of this work would have been possible without the personal efforts, involvement, and dedication of Professor Wilhelm Krelle.

*Robert H. Pry*
Director
International Institute for
Applied Systems Analysis

# Preface

In this book the results of the Bonn–IIASA Research Project on Economic Growth and Structural Change are presented to the public. The project was initiated in June 1983, when the former director of IIASA, Professor Holling, asked me to draft a research plan on this subject and to act as program leader. It was his intention that this project should become one of IIASA's central research programs. I agreed to draft such a research plan, to start the project and to act as a program leader, if I could find sufficient support from collaborating research groups in the most important countries. The research was to cover the whole world, and it was clear that a project such as this one could not succeed without the whole-hearted support of outstanding scholars in the most important countries and from national and international institutions as well.

The original research program was outlined in the "IIASA Research Plan 1985" (April 17, 1984, ECO I, pp. 1–7). The plan provided that a central group at IIASA would build a highly aggregated world model to relate the results of parallel research in a consistent way. Then collaborating groups in different countries would prepare detailed forecasts of economic development and structural changes in their countries. Similarly, the most important commodity markets would be analyzed as well. Finally, there would be related special studies on the driving forces of economic development, on the statistical and econometric identification and treatment of structural change, on linking input–output models to the world model, and on the world steel industry.

From February to June 1984, I stayed at IIASA to organize the project, to draft a detailed research plan and to see whether enough international support could be obtained. In March 1984, I presented the plan (and some related research of my own) to several Institutes of the Academy of Sciences in Moscow [All-Union Research Institute for Systems Studies (VNIISI), Central Economic–Mathematical Institute, Institute of World Economic and International Affairs and others], and again, in May 1984, at an IIASA workshop at Albena (Bulgaria), and finally at the IIASA Conference on the Analysis and Forecasting of Economic Structural Change, May 14–16, 1984, in Laxenburg. This conference was organized to bring together the leaders of research groups willing to cooperate in order to discuss the research plan. If we could find enough support, the project would start. We found it, but unfortunately the director of IIASA informed me shortly afterward that IIASA could not finance the project owing to unforeseen financial constraints. It would have been a pity to bury the project after all this preparation and given the enthusiasm from so many sides that became evident at the conference. Thus, I sought the necessary additional support from German funds. I succeeded, and the *Bonn-IIASA Research Project* was born.

My first duty is to thank all those institutions that supported the project. Of course, these thanks belong first and foremost to IIASA, but also to a large extent the Deutsche Forschungsgemeinschaft (German National Science Foundation), the Sonderforschungsbereich 303 (Special Research Unit 303) at Bonn University, the Rechts- und Staatswissenschaftliche Fakultät (Faculty of Law and Economics) of Bonn University, and Bonn University itself. For financial reasons, it was natural for the central group to become established at Bonn University. It worked intensively during 1985 and 1986 (with five scholars) and during the first half of 1987 in reduced form (with two scholars). Thus, so far as Part I of the book is concerned, it presents the work of two and a half years of a small, very efficient and highly motivated group of scholars.

The links to IIASA were maintained by regular seminars at the Institute, where some of the results of the project were presented and discussed. Anatoli Smyshlyaev and, later, Tibor Vasko represented the project in Laxenburg and were most helpful in organizing the workshops in which the central group and the collaborating country groups met and coordinated their work. These workshops were held in Lodz, December 9–10, 1985, and in Sofia, June 24–25, 1986. The final conference, where earlier versions of the chapters of this book were read and discussed, took place in Laxenburg, November 24–25, 1986.

The members of the central group are the authors of the chapters of Part I of the book. In addition, Professor Dubovsky, Dr. Eismont and Dr. Vasilyev [from the Institute for Systems Studies (VNIISI), Moscow] and Dr. Gajda, Dr. Sztaudynger and Dr. Czyzewski (from the University of Lodz) have been working for some time in the central group at Bonn University. Of course, we also had many visitors from all over the world. I wish to thank all the collaborators in the central group for their full commitment to the project and for their successful work. I extend these thanks to all collaborating scholars and groups all over the world. Most of them appear as authors in Parts II to IV of the book. There are some exceptions, notably Lawrence Klein, who fully backed the project and was present at almost all the meetings and conferences so that the project could profit from his experience and advice. The same applies to Jean Waelbroeck.

As may be seen from the contents, the Bonn–IIASA Research Project covers only part of the originally planned research. Some parts had to be dropped; but one other part has survived: the research of the statistical and econometric identification of structural change, headed by Professor Peter Hackl of the University of Vienna. The results of this research should have been used in this project, and the experiences of this project might well have inspired statisticians and econometricians to look for more appropriate estimation procedures. But this was not possible. As a substitute, there was a parallel Bonn research program on time-dependent parameters and latent variables in econometric models and on estimation procedures in models with errors in the variables. The results are in part reflected in our research. It was fortunate that Peter Hackl reported on the work of his group at the final conference, in November 1986, and Johannes Ledolter presented a paper on adaptive estimation and structural change in regression and the time series models, so that at least part of their message would inform this volume. The results of the statistics group will be published separately, as Professor Hackl mentions in his "Methodological Note" to this book.

I wish to thank the present director of IIASA, Professor Robert H. Pry; the former director of IIASA, Professor Thomas Lee; and especially the former deputy director, Vitali Kaftanov, and the IIASA Council representative of the Federal Republic of Germany at IIASA, Wolf Häfele – all of whom encouraged the project from start to finish, visited the central group at Bonn and furthered our work through their sympathy and interest. Thanks, too, to the following members of the Publications Department at IIASA – Robert A. Duis (Head), Betsy Schmidt (Senior Editor), and Ewa Delpos (Graphic Artist) – all of whom helped to make a unified book out of a complex, disparate set of manuscripts.

I hope that the scientific community as well as the political authorities find the methods and the results stimulating and useful.

*Wilhelm Krelle*
Institut für Gesellschafts- u.
Wirtschaftswissenschaften
Universität Bonn

# Methodological Note

At the planning stage of the University of Bonn–IIASA Research Project on Economic Growth and Structural Change, it became clear that the methodological side of the task would be crucial for the strength and credibility of its results. As a consequence, it was decided to create an informal Methodological Research Group at IIASA on the statistical analysis of structural change. Experienced statisticians and econometricians in this field were invited to contribute to the aims of this group according to a research plan.

As mentioned in Professor Krelle's preface, the idea of transferring knowledge or results of the IIASA Methodological Research Group into the central and local groups of the University of Bonn–IIASA Project could not be realized. Reasons – among others – were limitations of financial resources, and difficulties in organizing communication with the various research groups in East and West.

Nevertheless, some 50 statisticians and econometricians contributed to the results of the IIASA Methodological Research Group. Meetings took place in Lodz, Poland (May 1985), Berlin, GDR (June 1986), and Sulejov, Poland (September 1986). A selection of the papers presented at these meetings, together with some invited papers, is to be published by Springer-Verlag as a multi-authored volume entitled *Statistical Analysis and Forecasting of Economic Structural Change*; it is scheduled to appear in the 1988–1989 academic year. In 1985, IIASA published Collaborative Paper 85-31, *Statistical Analysis of "Structural Change" – An Annotated Bibliography*, by P. Hackl and A. Westlund, which contains about 300 references on formal aspects of the analysis of structural change.

The volume *Statistical Analysis and Forecasting of Economic Structural Change* documents the status of the art in this field, which has recently attracted the interest of both statisticians and economists. The authors give insights into available methods and highlight new developments and trends. It is hoped that the volume will further stimulate the interest of statisticians and make economists aware of available methods and their relevance and importance. The four sections of the book are as follows:

(1) As an introduction, a chapter on "What can statistics contribute to the analysis of economic structural change?", tries to bridge the gap between the ways of thinking and approaches of economists and statisticians. The chapter deals with the terminology of "structural change", discusses the process of model building for nonexperimental data and particularly the role of significance testing.
(2) "Identification of structural change" encompasses chapters that are concerned with the detection of parameter nonconstancy. The procedures discussed cover a broad spectrum of techniques from classical methods, such as the CUSUM test, to new concepts, e.g., tests based on nonparametric statistics. Several chapters treat the robustness of the procedures with respect to such conditions. Many of the chapters are illustrated by numerical data analyses.
(3) "Model building in the presence of structural change" discusses various generalizations of constant-parameter models. Specification of models, estimation of parameters and forecasting in the presence of structural change are treated.
(4) "Data analysis and modeling" deals with "real-life" structural change situations, such as analysis of the poverty structure in a society, the notion of technical progress, the dynamics of the IS–LM concept, Lucas's critique, and changing causality. The application of suitable statistical methods and the interpretation of the results exemplify the relevance and difficulties.

The authors are: Y. Abrahamsen (St. Gallen, Switzerland), L.D. Broemeling (Washington, USA), M. Deistler (Vienna, Austria), F.X. Diebold (Washington, USA), J.-M. Dufour (Montreal, Canada), J. Dziechciarz (Bremen, FRG), V. Fedorov (Laxenburg, Austria), A.C. Harvey (London, UK), M. Huskova (Prague, Czechoslovakia), W. Katzenbeisser (Vienna, Austria), A. Keller (Paris, France), J. Kleffe (Berlin, GDR), W. Krämer, (Hannover, FRG), J. Ledolter (Iowa City, USA), H.-J. Lenz (Berlin, FRG), J. Lukashin (Moscow, USSR), G.E. Mizon (Southampton, UK), T. Ozaki (Tokyo, Japan), V. Ozaki (Tokyo, Japan), P. Pauly (Philadelphia, USA), G.D.A. Phillips (Manchester, UK), W. Ploberger (Vienna, Austria), W. Polasek (Vienna, Austria), J. Praagman (Eindhoven, Netherlands), P. Robinson (London, UK), B. Schips (St. Gallen, Switzerland), P.K. Sen (Chapel Hill, USA), P.H. Tong (Canterbury, UK), B. Törnkvist (Umeå, Sweden), H. Tsurumi (New Brunswick, USA), Z. Wasilewski (Lodz, Poland), and A. Westlund (Stockholm, Sweden).

Although collaboration between the University of Bonn–IIASA Research Project on Economic Growth and Structural Change and the IIASA Methodological Research Group could not be established as closely as was planned, the results of the whole project will surely interest both economists and statisticians. I hope that many economists find our volume *Statistical Analysis and Forecasting of Economic Structural Change* useful and stimulating. The present book, which reports on a very ambitious and rather unique modeling task, can inform statisticians of the needs and problems of economists. For the authors of the book, particularly Professor Wilhelm Krelle, I hope that statisticians as well as economists will invest their efforts and profit from reading it.

*Peter Hackl*
Institut für Statistik,
Wirtschaftsuniversität Wien
and IIASA, Laxenburg

# PART I

# Analysis and Forecast of Economic Growth and Structural Change on the Basis of a World Model: Results of the Central Research Group

CHAPTER 1

# Growth, Decay and Structural Change

*Wilhelm Krelle*

## Summary

We attempt to show the historically unique situation in which mankind now finds itself, which forces are working and what may be expected in the future. These general considerations are then used to study economic development up to the year 2000. The driving forces are identified and their development explained by a latent variable called "degree of economic activity". This chapter provides the philosophical background for the understanding of the following chapters.

## 1.1. Introduction: The Background of the Bonn–IIASA Research Project

We are lucky and privileged to live in an extraordinary period of the 2 million or so years of the history of man. Mankind experienced only two real revolutions of its way of life in its history. The first happened about 4000 to 6000 years ago when mankind succeeded in taming domestic animals, growing corn, and invented spinning, weaving, pottery, smelting and casting copper, bronze and iron, wheeled vehicles and sailing boats. This was accomplished at different places, but in the short span of about 2000 years. It took some 1000 years for this knowledge to spread from its places of origin over the world. This first agricultural and artisan revolution radically changed the way of life of mankind. Instead of living in small nomadic tribes, gathering food and hunting, man settled in permanent houses, cities came into being, large empires developed, writing, reading, calculating, and all kinds of cultural activities emerged. The size of the world population increased by a factor of six between 4000 and 2000 B.C.

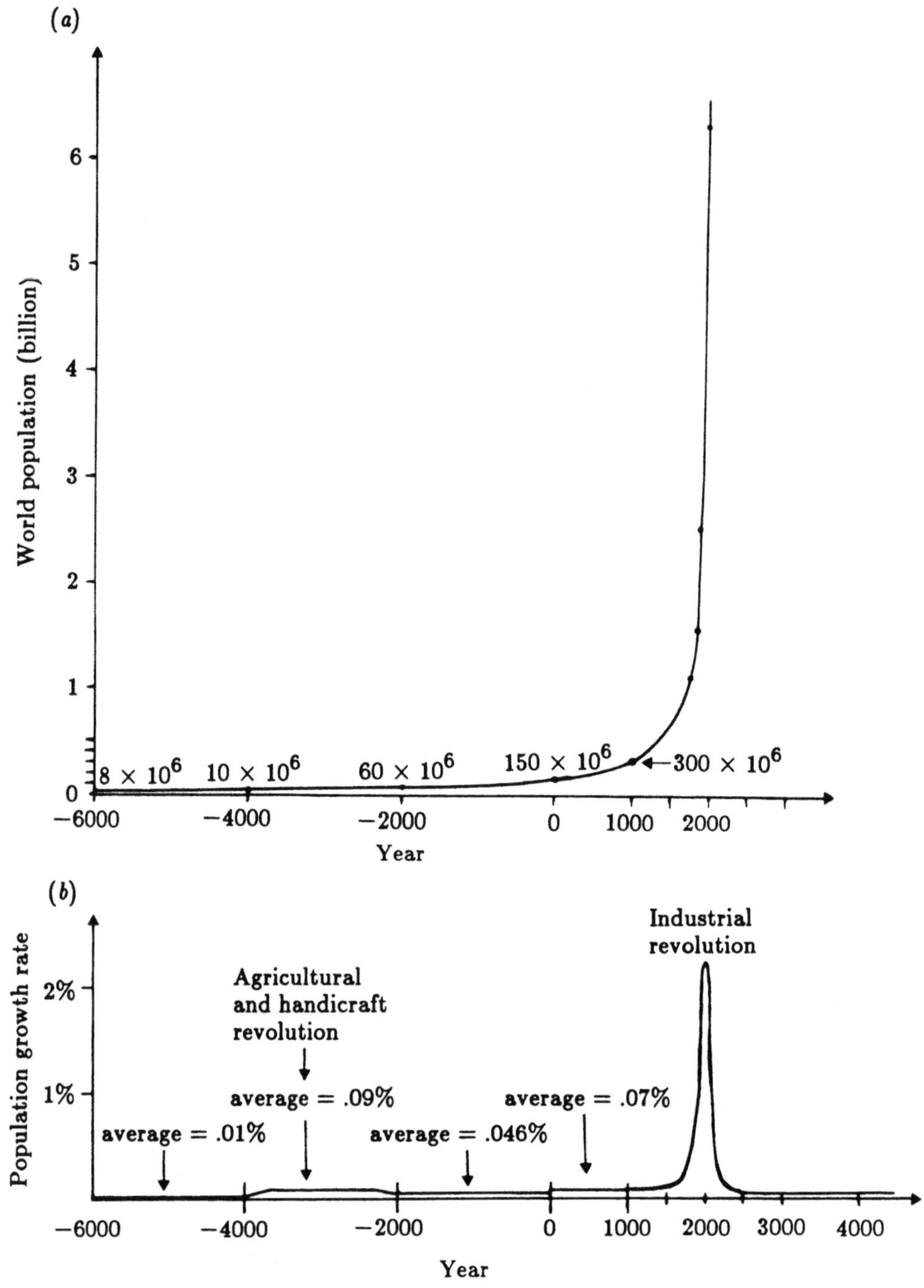


*Figure 1.1.* World population. Sources for (*a*): (1) W. Fucks, Über die Zahl der Menschen, die bisher gelebt haben, Z.f.d. ges. St. W. 107 (1951), pp. 440–450. (2) Carr-Saunders, *World Population*, Oxford (1936), Tab. 8, p. 42. (3) UN, *Demographic Yearbook* (1959), Tab. 2, p. 127. (4) UN, *The Future Growth of World Population*, New York (1958), p. 23. Sources for (*b*): (1) See above. (2) Forecasts from R. Freeman and B. Berelson, The Human Population, *Scientific American* (Sept. 1974), pp. 36–37.

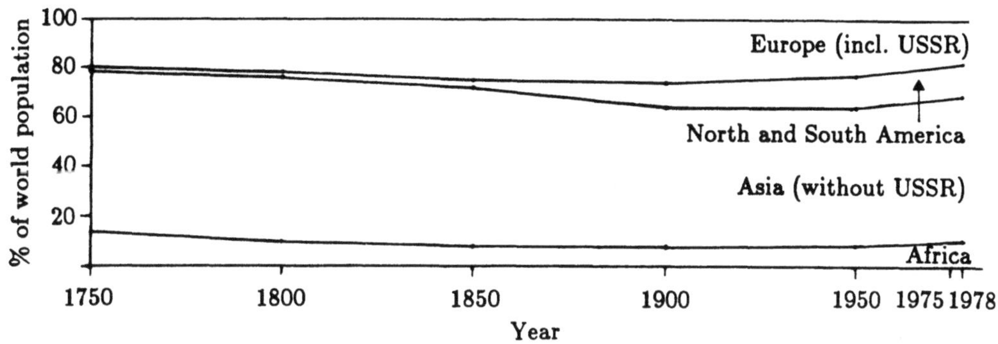


*Figure 1.2.* Regional distribution of world population. Sources: (1) Q.H. Stanford, *The World's Population, Problems of Growth*, Oxford (1972), p. 19. (2) UN, *Demographic Yearbook* (1956 and 1978).

As far as technology is concerned nothing much happened after this until the industrial revolution started in the 18th century in Great Britain. Living in Goethe's time in Germany was not much different from living in emperor Augustus' time. But our current way of life is simply incomparable to that. The industrial revolution is now spreading very fast around the world – fast indeed if one compares this with the hundreds or thousands of years it took for the use of a new tool or knowledge to spread over the world in former times; cf. the details given in the *Oxford History of Technology*. The world is now capable of supporting a much larger population. *Figure 1.1* shows the singularity of our period graphically. *Figure 1.2* illustrates that the population explosion followed the path of the industrial revolution: first the European population expanded substantially, increasing from 20% to 36% of the world population. This trend is now being reversed. Of course, world GDP per capita increased substantially from \$200 in 1775 to \$1800 in 1975 (see *Figure 1.3*) and will increase further if the growth process can be continued. This assumption is not unrealistic. If the growth process continues with a per capita growth rate of 1%, the world average income per capita in about 200 years would be that of people in the Federal Republic of Germany, or France, or Japan today. If this increases by only 0.5%, world average income in 200 years would be comparable to that of the citizens of Portugal and Greece today. This, of course, is not a forecast but simply a trend extrapolation. We do not need it further. But it is interesting that it does not yield unreasonable and unacceptable results.

The industrial revolution not only increased the average world per capita income enormously (it grew by a factor of 9 in 200 years), it also substantially widened the difference in per capita income between nations. This is intuitively clear: if some nations succeed in extending their GDP per capita and others stay at the old level, world income distribution must become more unequal. The imbalance is now substantial, as may be seen from *Table 1.1*. Average GDP per capita was \$9440 in the OECD countries and \$230 in the low income countries in

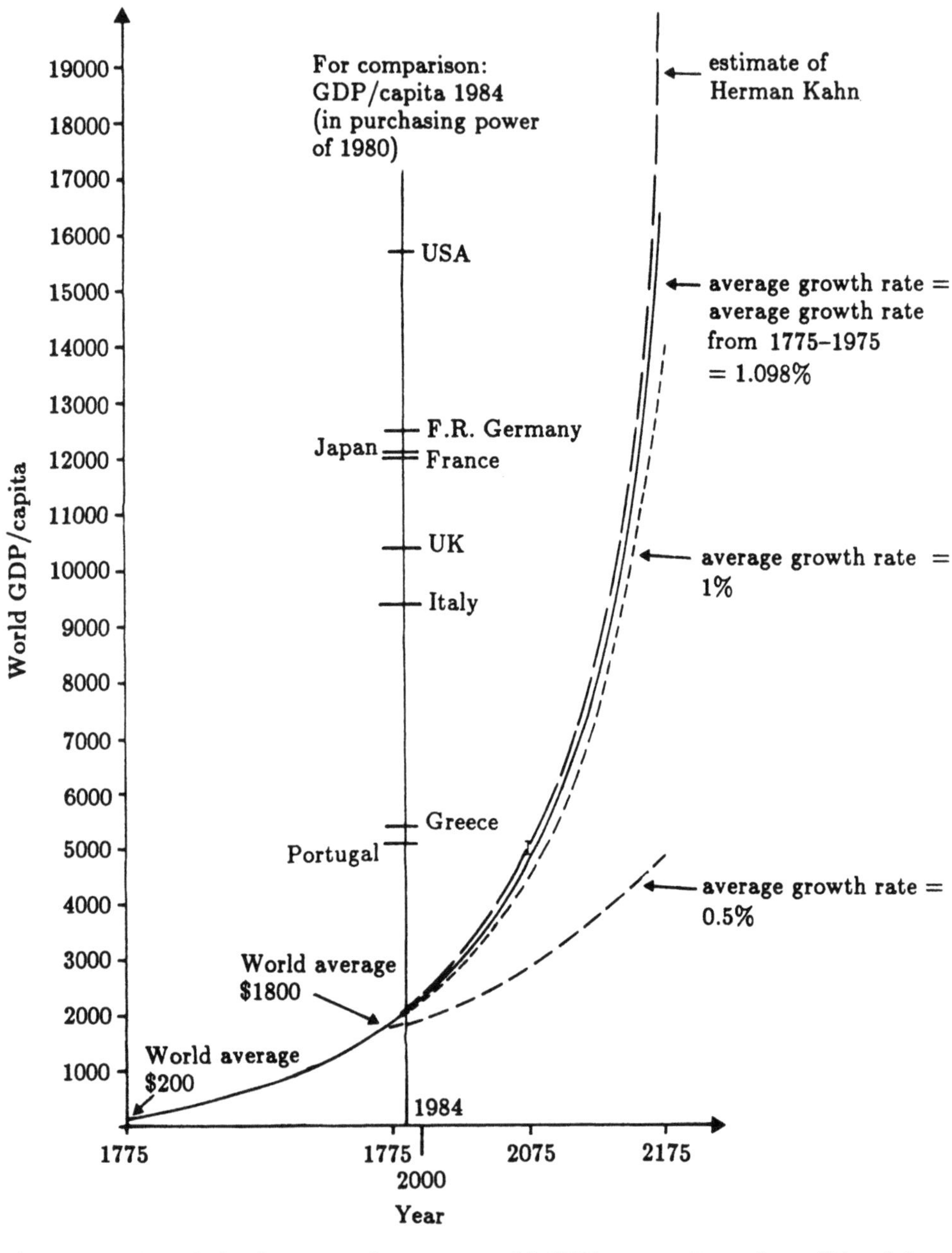


*Figure 1.3.* Actual development of average world GDP per capita and conditional forecasts. Sources: (1) ***Statistisches Jahrbuch 1986***, for the Federal Republic of Germany, p. 717; a DM/$ exchange rate of 2 has been assumed. (2) Herman Kahn, ***World Economic Outlook, 1979 and Beyond***, New York (1979), p. 66. (3) Our calculations.

1979. The OECD countries comprise about 16% of the world population, the low income countries about 53%. The Lorenz curve of world income distribution (with respect to groups of countries in *Table 1.1*) is given in *Figure 1.4*. The degree of inequality is comparable to that of Great Britain in 1801, though the British income distribution at that time was more unequal in the higher income brackets and less unequal in the lower ones than the world distribution today. It is also apparent from *Figure 1.4* that world income distribution became more unequal from 1960 to 1984. This unfortunate trend will continue until birth rates are reduced in the low income countries and the wave of industrialization reaches them fully. But after that the developing countries will catch up to the average level of the presently industrialized ones. This will reduce the inequality of world income distribution. It may become comparable to the domestic income distribution of an industrialized nation now; see the Lorenz curve for the Federal Republic of Germany in *Figure 1.4*.

There are people who think that the developing nations do not have the chance to reach the level of the developed ones since there are not enough natural resources left to them. This argument takes no consideration of the working of the price system with respect to exhaustible resources. The price of a commodity rises and its consumption per unit of output is reduced if its supply declines. *Figure 1.5* shows this for crude oil. The actual figures of total supply of crude oil, of crude oil price and of the depletion rate (= ratio of extraction to known extractable reserves) are represented for the years 1900–1985 or 1950–1985, respectively [1]. Since about 1965 the depletion rate has stayed constant in spite of rising extraction because prospecting for new sources has stayed in fixed relation to the exhaustion of the old ones. This will continue until all extractable resources are known, say in the year $T$. Up to that year the real price of crude oil will increase at an average rate approximately equal to the rate of growth of world GDP, and total deliveries will stay approximately constant. After the year $T$ the Hotelling rule applies: the real oil price will increase at the real rate of interest (about 5%), and total supply will decline at the rate of depletion (about 3%). The year $T$ might lie well beyond the year 2000.

With increasing prices of a resource, substitutes become economically feasible, e.g., for crude oil: atomic energy, wind, geothermal energy, solar energy, perhaps later fission energy. Moreover, the demand is reduced. Resource saving is also enforced by the declining proportion of manufacturing in GDP. This follows from the changing demand structure if real income per capita increases (Fourastie's law). *Figure 1.6* shows this trend for the USA in the past. Of course, total energy and raw material consumption on the world scale will still increase until the developing countries are fully industrialized. The decline starts after that.

The spearhead of scientific, economic and cultural development which is connected with the center of political power will change as well, as it did in the past: see *Figure 1.7*. Between 4000 and 100 B.C. the most advanced centers of human culture were in China, India, Mesopotamia and Egypt. These centers developed more or less independently of each other. Nevertheless, a slow flow of information from one center to the other took place. Europe was touched very

*Table 1.1.* GDP per capita, world regions.

| *Indicator* | *Industrialized market economies (OECD countries)* | *Nonmarket industrial economies (CMEA countries)* | *Mostly developing countries* | | |
|---|---|---|---|---|---|
| | | | *Middle income countries* | *Low income countries* | *High income countries* |
| Population, 1984 (million) | 733 | 389 | 1,189 | 2,390 | 19 |
| GDP/capita, 1984 ($) | 11,430 | 4,230[a] | 1,250 | 260 | 11,250 |
| GDP/capita, average growth rate, 1965–84, (%) | 2.4 | 4.3[a] | 3.1 | 2.8 | 3.2 |
| GDP, average growth rate (%): | | | | | |
| 1965–73 | 4.7 | 4.8 | 7.4 | 5.6 | 9.0 |
| 1973–84 | 2.4 | 5.2 | 4.4 | 5.3 | 4.5 |
| Population, average growth rate (%) | | | | | |
| 1965–73 | 1.0 | 0.8 | 2.5 | 2.6 | 4.5 |
| 1973–84 | 0.7 | 0.8 | 2.4 | 2.0 | 5.1 |

Source: *Weltentwicklungsbericht* [World Development Report] *1986*, The World Bank, Washington, D.C. (1986), pp. 206–209, 254–255.
[a]These figures are missing from the *World Development Report 1986*. The figures reproduced above are taken from the *World Development Report 1981*, pp. 135, 166–167, and refer to the year 1979 or to the periods 1960–70, 1970–79, respectively.

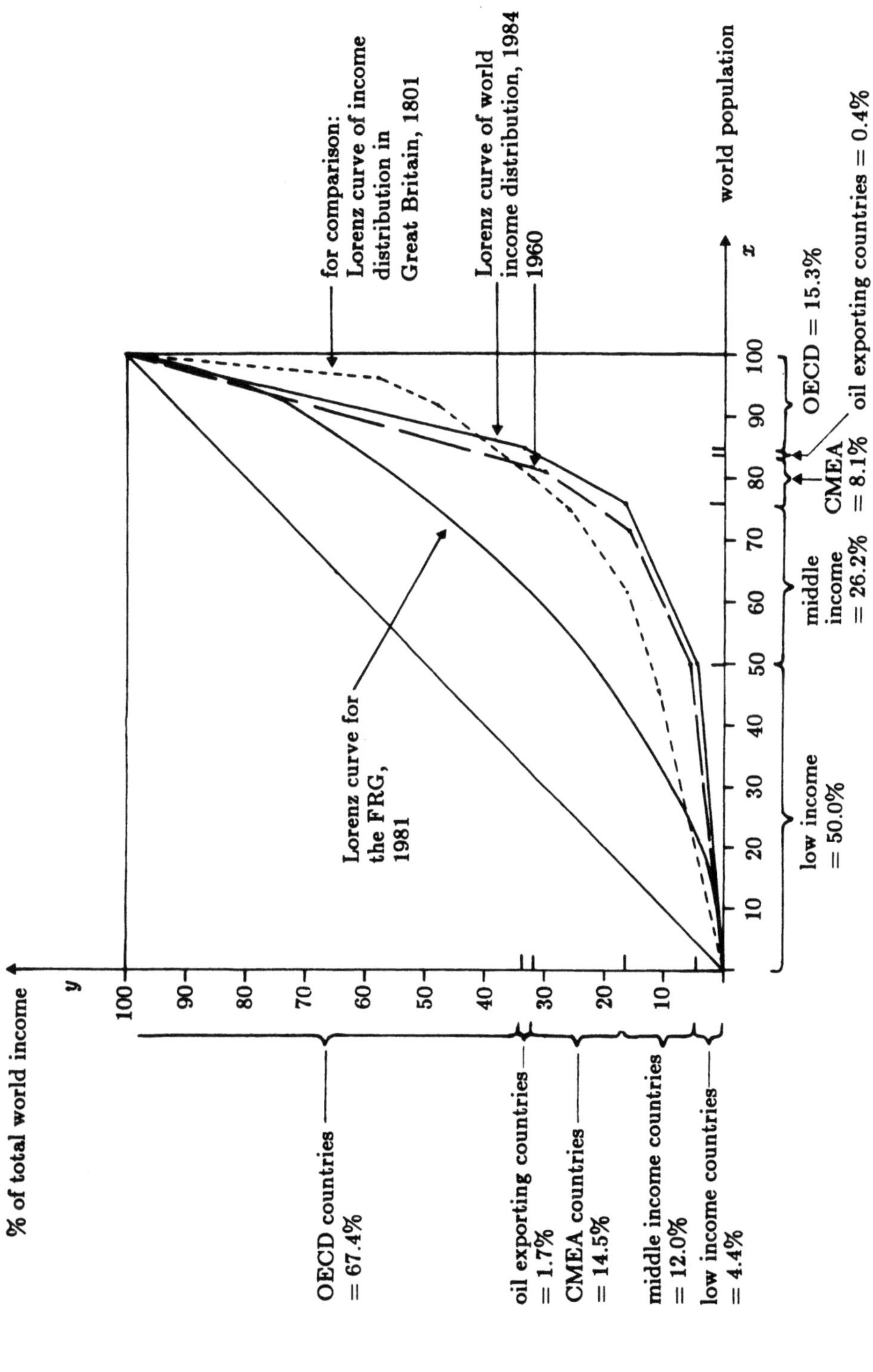


*Figure 1.4.* Lorenz curve of world income distribution, 1960 and 1984.

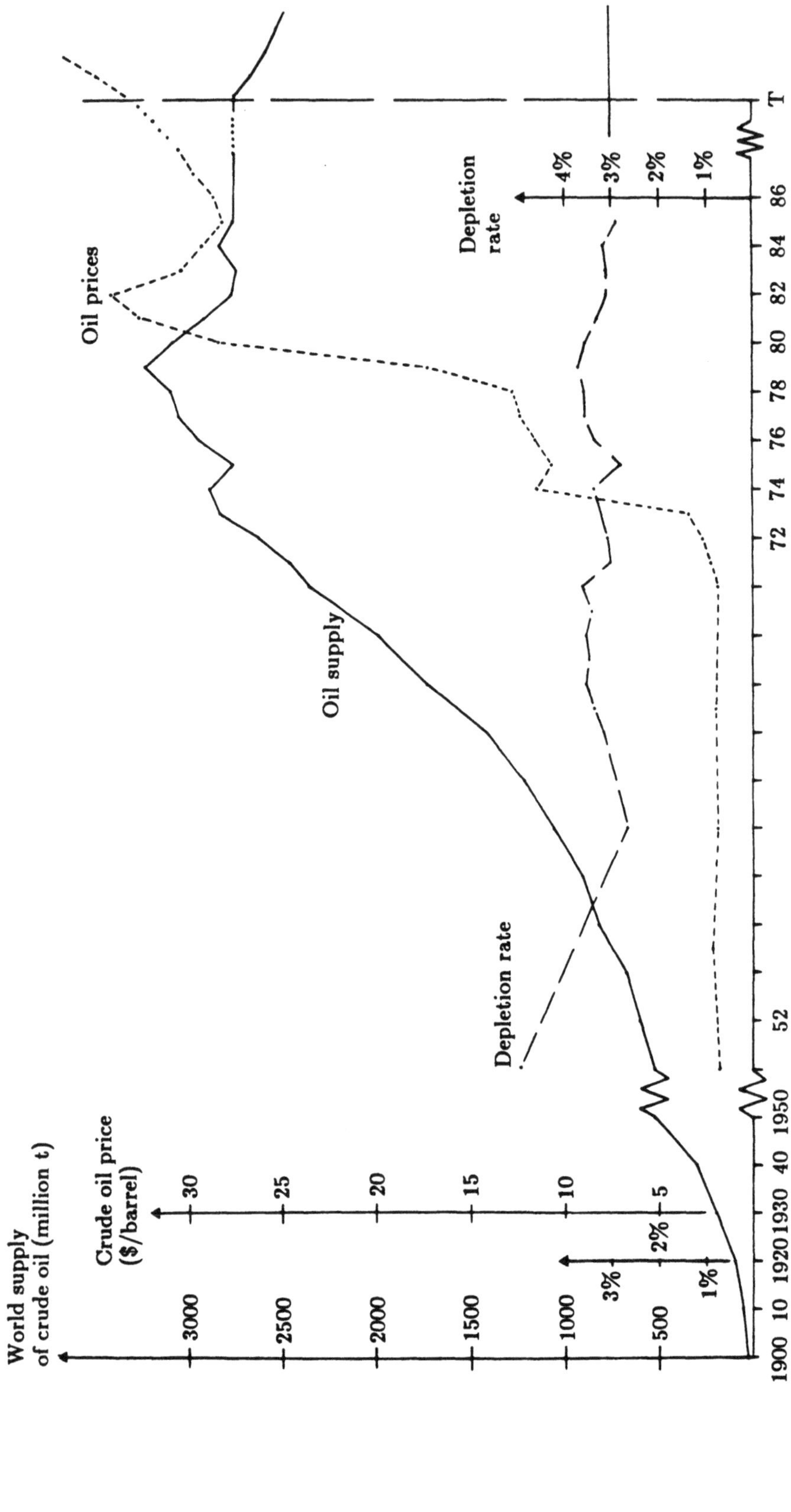


*Figure 1.5.* Oil supply and oil prices.

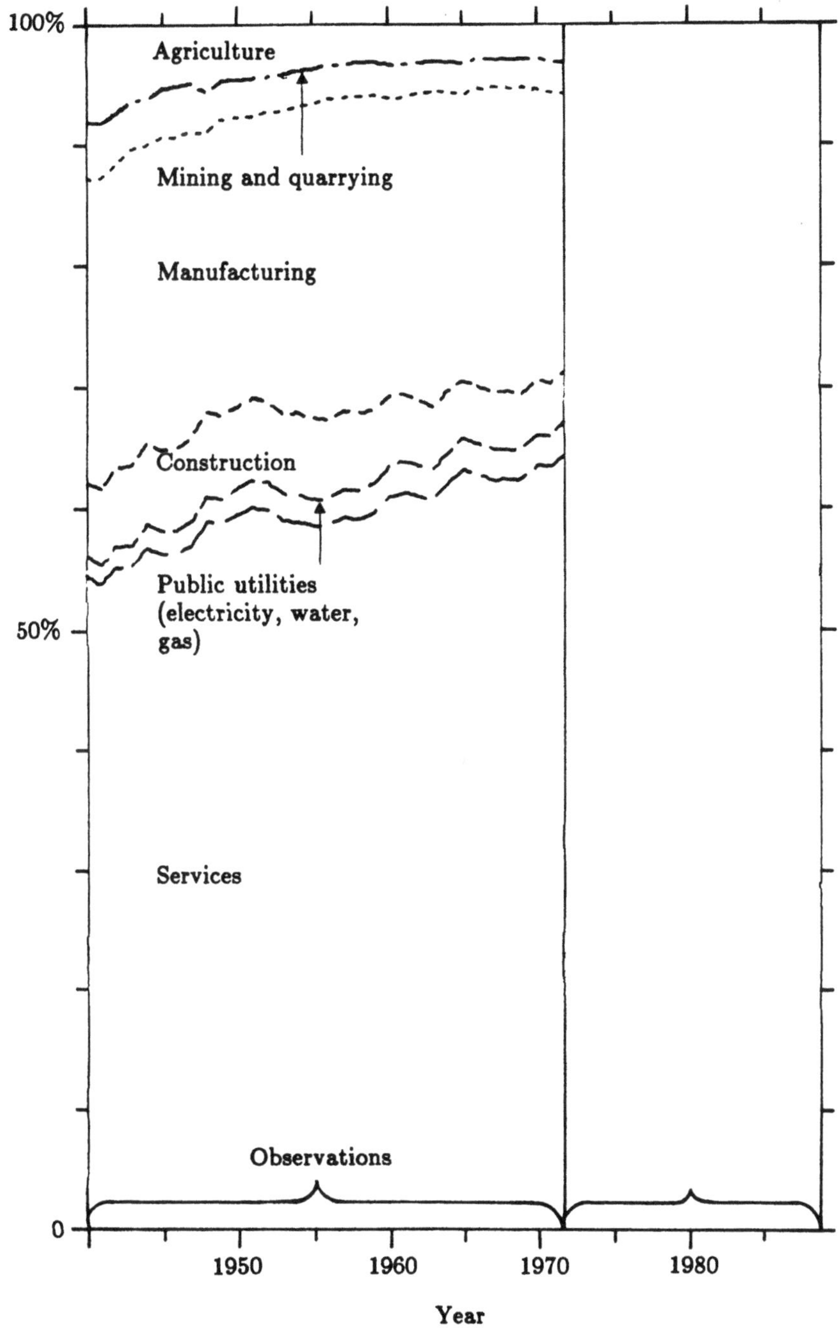


*Figure 1.6.* Shares of value added, USA.

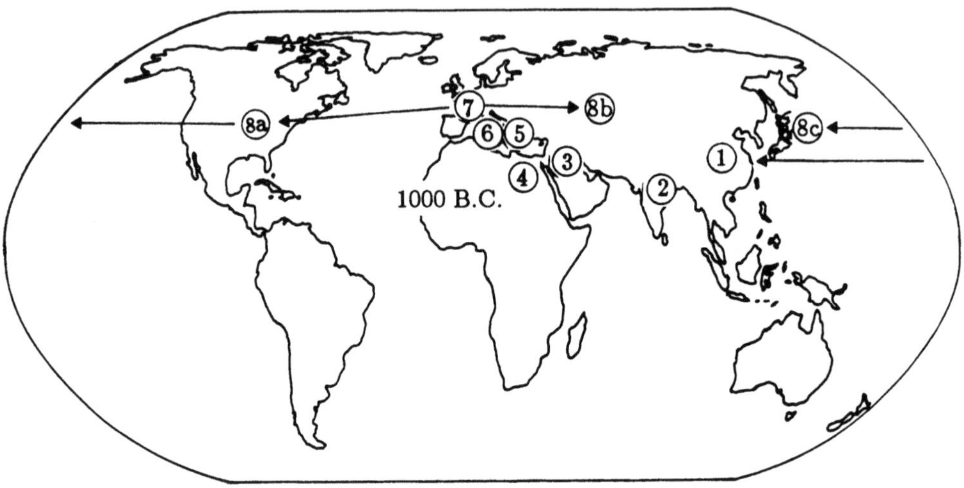


*Figure 1.7.* Changes of the centers of culture in the world.

late. First Greece became a center (600–200 B.C.), then Rome (200 B.C.–500 A.D.), Western and Northern Europe (Spain, Portugal, France, Germany, Great Britain, 500–1900), and finally North America and Russia (1900–?). Now the wave is turning back to Japan and China (1980–?). Of course the old centers (as a rule) kept part of their cultural inheritance and influenced other countries. But they lost their superiority. It is interesting to note that the centers of culture and power have existed only in the northern hemisphere and have moved predominantly from East to West (*ex oriente lux*), with the notable exception of Russia, where Czar Peter the Great opened the country to western influence. In these centers of culture new knowledge was created. It spread with greater or lesser speed (in fact very slowly in the past, but now rather faster) to other countries through scholars coming to learn in these centers, by trade, traveling, books, and other means of information. It takes time for new insights into forms of social organization and behavior to be taken up by other nations.

Assume that the rate of change $\dot{x}$ in the number of $x$ of people who know of a certain new discovery (or who conform to a new behavior or take up a new production process which has proved to be more advantageous to society) is proportional to $x$ (since the more people that are informed, the more likely it is that they "infect" the uninformed ones) and to the number of those who are not informed (because the more people that are not informed, the more likely it is that one of the set of uninformed people will be informed). Thus we have

$$\dot{x} = \alpha x(\bar{x} - x), \quad \alpha > 0$$

a proportionality factor, where $\bar{x}$ is the total population. This is the differential equation of the logistic curve which describes the spread of an epidemic disease. The solution of this differential equation is

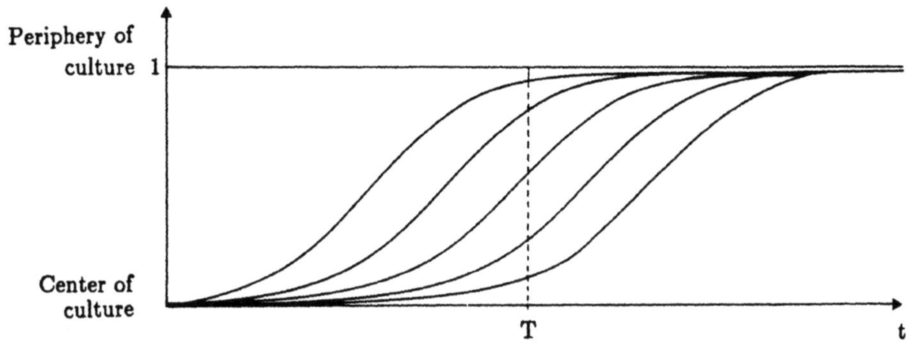


*Figure 1.8.* Spread of knowledge or behavior.

$$x_t = \frac{a/b}{1 + \frac{c}{b}\, e^{-dt}}$$

where $\bar{x} = a/b$, $\alpha\bar{x} = d$, and $c =$ a constant determined by the initial condition.

*Figure 1.8* shows the graph of a family of these functions. It shows that there is a hierarchy of regions. It takes about $T$ years for the new knowledge (or behavior or production process) to spread to the periphery of the development area. Thus there will always be a difference between the degree of development of nations though the center of gravity of culture may (and will) change [2].

## 1.2. Explaining Long-Term Development with the Latent Variable "Degree of Activity"

Mankind is organized in societies which are now mostly equivalent to nations. There are "active" societies which are striving for achievements in different fields (political dominance, economic efficiency, cultural and scientific accomplishments) and "passive" ones which do not move. They are trapped in a repetitive behavior pattern and from outside they seem to be petrified. It is clear that active societies will dominate in one form or another over those passive ones which are "near" to them, if they are not too small and if their state of technology in relevant fields is superior enough. Thus other societies are incorporated into or affiliated with or subdued by the active ones, and large empires arise. They may survive for centuries if they provide some advantages to the dependent societies as well. This superposition of one active society over one or several passive ones may well be one reason for the formation of a class structure within the society, as Alexander Rüstow (1950) suggests, but there are others.

After a while the "active" society becomes saturated. Its success seems to indicate that no further change is necessary. The degree of activity declines and other societies take over the leadership.

This picture has to be modified today. There are no isolated regions of the earth anymore. The world will become an interdependent network of centers rather than a hierarchical order of nations as far as science and technology is concerned. One nation may be leading in one field, others in other fields. The technological "distance" between them may become a matter of weighing different fields. Of course, larger nations will have the chance of being in the leading position in more fields than smaller ones. The fast flow of information allows for only small and temporary advantages of one nation with respect to others. Thus the industrialized centers of the world will move nearer together and the developing countries will catch up after a while, though not in the time span we are considering [3]. Of course, there will always be an uneven distribution of income and wealth between countries as well as within countries. The gap in GDP per capita between developed and developing countries will widen further in the next 20 years. Later one may expect a tendency to a more even distribution on the world scale.

A similar development is to be expected as far as political power is concerned. There will be several centers, but the predominances and alliances will change. As far as one can see now, China may become one of the leading centers in the next century.

We explain these secular changes by relative movements of the degree of activity of a society. This is a latent variable comparable to "intelligence" or "willpower" on the individual level or to "entropy" in thermodynamics. These concepts describe the state of a complicated system in an aggregated way without going into detail and without (by themselves) explaining the reasons for this state. The explanation is left to a second stage of research which, of course, must follow. But first the latent variable "degree of activity" of a society should be filtered out from a set of appropriate indicators, just as, for example, the "intelligence" of a person may be measured by a system of tests comprising his (or her) ability to read, to write, to find out patterns etc.

Degrees of activity may be measured in different fields: political, military, economic, research, the arts and others. These are interrelated. Political and military activities yield power which (as with economic activities) in turn provides financial means for cultural and scientific activities. Thus in the long run the centers of political power will coincide with the centers of cultural and scientific activities. In the following chapters we only consider the degree of economic activity, but here we reflect on the degree of activity "in toto".

When does a society have momentum? The answer is: when a large number of individuals in the population are willing to work hard and to make sacrifices to achieve the same goal, and if these efforts are coordinated. Coordination may be achieved by a strong leadership but also by an appropriate incentive system, e.g., an appropriate religion or an ideology which provides a common value system and psychological rewards for activities in this direction. Of course, there are also "passive" religions and ideologies which put brakes on all or many secular activities. In modern societies the coordination of economic activities is largely performed by the price system. But there are limits to both approaches: too many and too stringent a set of regulations hinder or even destroy individual activity. Conversely, if the price system is the only regulator of economic activities, there will be monopolies and an unequal and unjust distribution of income and wealth and, as a consequence, class struggle and internal social unrest which will also reduce the degree of activity. There is an interdependence between the "ideological superstructure" of a society (i.e., its religious and other ideas and ideologies) and the actual situation: the value system determines the actions which in turn codetermine the actual situation, and the value system itself is heavily influenced by the actual situation. This is Hegel and Marx united. There is nothing mystical about such interdependencies.

It is difficult to keep a society on the track of a high degree of activity for a long time. There is always a nonzero probability of falling off one of two precipices: too many regulations and overly rigid organizations which yield a petrification of the society; or too few regulations and organizations which result in a disintegration of society. In both cases competing societies with a higher degree of activity will eventually take over the leadership. This was the case in the past and will happen in the future. The transition need not come by force. The old leadership may simply fade away, as the British Empire did. This seems to be more likely than World War III. There are no economic reasons for a long-term breakdown of human society.

This is the (somewhat optimistic) background for the Bonn–IIASA Research Project. It deals with economic growth and structural change over the relatively small time span of about 20 years. This allows a much more detailed use of the available economic theories, and may give more specific hints as to the appropriate economic policy in each country. But the results should fit into the broader picture of historical development and of very long-term projections.

It is a control theoretical problem to keep a society on the path of a relatively high degree of economic activity. If the society shows signs of weakness because it moves too much in one of the two dangerous directions, efforts will be made to correct the aberration. Thus we may expect cyclical movements of the degree of activity. We shall show later that these cycles may be identified with the long-term Kondratieff cycle. Our own projections are based on this approach. But before coming to this we shall briefly review other long-term projections.

## 1.3. Some Other Long-Term Projections

There are other long-term projections which come to similar conclusions and those which come to quite different ones from ours. We shall start with the more "philosophical" approaches.

Marx (1894, Ch. 15) projects the "end of the historical epoch of capitalism" (which may be interpreted as a breakdown of the market economies) whereas the communist economies would fully enjoy the blessings of technical progress and growth. Spengler (1919) finds a "Decay of the West" (which may be interpreted as a decline of European and American economies in absolute or relative terms compared to the Asian or African countries). He thinks of a culture as behaving in an analogous way to an individual: it passes through a period of youth, full development of its abilities and of old age and death. Toynbee (1947) comes to similar conclusions, but for different reasons: it is a lack of response to the requirements of the natural, social, and political environment that leads to the ultimate decay. Olson (1982) specifies this idea by pointing to institutional rigidities as the main reason for this lack of vitality. Sorokin (1957) believes in a cyclical change between three types of culture (ideational, idealistic, and sensational) and thinks that we are now in a transition from the sensate (= positivistic, science-oriented, agnostic) phase of culture to an idealistic or ideational (= philosophical or religious) phase where the center of creativity shifts from Europe "to the larger area of the Pacific-Atlantic".

There are also theories which do not recognize a rise and decline of a certain culture but identify changing stages or cycles of performance of a society. Aristotle may be first named here. In his book *Politics* he distinguishes three "good" types of constitutions: kingship (the rule of the best to the advantage of society), aristocracy (the rule of a few excellent citizens to the advantage of society) and politie (the rule of all citizens to the advantage of society). Each type may degenerate: kingship to tyranny, aristocracy to oligarchy and politie to democracy. The degenerate forms are all defined as despotism to the advantage of those in power without rule of law. The degenerate forms of constitutions lead to decay of the community, to revolution, and finally to a change of the constitution. There is no strict order of change. Pareto (1916) sees history as the result of a circulation of the elites. Elite is a name for the ruling group. It imposes its rule on the majority and identifies its own perception of utility with the utility of the population. If the elite deteriorates and loses the willingness to use force, there will be revolution, disorder and decay until a new elite takes over and organizes society according to its perception.

Berdiajew (1927) denies any progress in history. Each generation lives in its own right. Cultures come and go, but there is no trend, no goal in history. This is close to Alfred Weber (1951), who distinguishes between culture and civilization (this conforms to an old tradition in German philosophy). Civilization comprises "practical knowledge" as provided by, for instance, the natural sciences and engineering. Culture means the value side of human life and comprises religion, philosophy, literature, art, and the like. There is no progress in culture, though there may be cyclical movements, whereas progress is possible in civilization.

On the economic side Kondratieff (1926) must be mentioned: he supported the existence of long-term waves without presenting a theoretical background. Schumpeter (1939) and others explained them on the basis of great tasks which confronted society: the introduction of the steam engine in industry, the use of it in transportation (railways and steam boats), the electrification, later the motorization of society, etc. But this does not explain the regularity of these waves, nor the low tides in between, since the basic knowledge for each new wave was available long before the wave actually started. We shall suggest an explanation later.

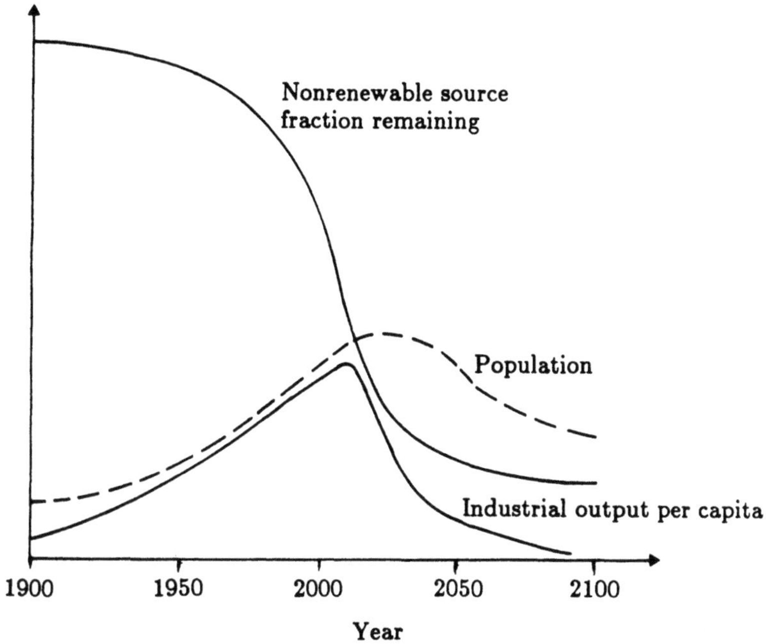


*Figure 1.9.* Forecast of world economic development by Meadows *et al.* (1972); see *Figure 1.10.*

In the present computer age *world models* have been constructed to forecast the *long-run development* of the world economy. One of the first (and the most spectacular one) was that of Meadows *et al.*, *The Limits to Growth* (1972). It was built in order to show that the world is on the edge of an abyss and will collapse in a relatively short time if basic changes are not initiated at once. These changes are: a total stop to further industrialization, at least in the already industrialized countries (zero growth), substantial savings in the use of energy and minerals, conservation of nature, reduction of pollution, and basic redistribution of income from North to South. The most striking results of Meadows *et al.*, can be seen in *Figure 1.9* where a collapse of industrial production and of population was forecast relatively soon after the year 2000. The forecasts of Mesarovic and Pestel (1974) (see *Figures 1.10* and *1.11*) are more "normal". The socialist world and the Middle East are the winners in these forecasts. The *Bariloche Model* [see Bruckmann (1976) and Herrera, A.D., Scolnik, H.D. *et al.*, (1976)] has quite another character. It is a planning model constructed from the point of view of the developing regions of the world. It allocates capital and labor worldwide in such a way that life expectancy becomes maximal at birth in all regions of the world. This, of course, amounts to a huge redistribution of capital and wealth to the advantage of the developing regions. In this case "basic needs" could be fulfilled in Latin America in the early 1990s, in Africa in 2008 and in Asia in 2040. In developed countries the growth rate of GDP will be

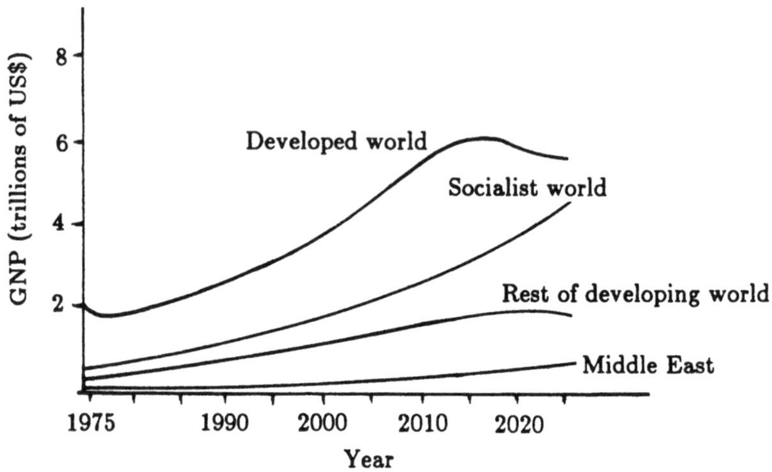


*Figure 1.10.* Results of low-level oil price scenario. Forecast I of world economic development by Mesarovic and Pestel (1974); reproduced from Meadows *et al.* (1982), pp. 19, 41.

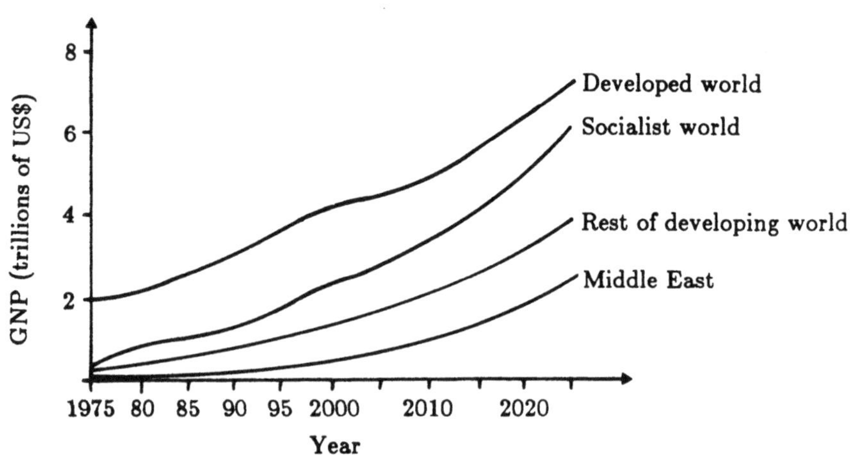


*Figure 1.11.* Results of "optimal" oil price scenario. Forecast II of world economic development by Mesarovic and Pestel (1974); reproduced from Meadows *et al.* (1982), p. 41.

drastically reduced. The authors do not see any physical limits to growth such as Meadows does.

The *UN Model* constructed by Leontief *et al.*, (1977) is an input–output model for the whole world. Different target growth rates of GDP per capita are assumed for developed and developing regions of the world and the implications for structural change, international trade, the balance of payments, and the income gap between developed and developing countries are estimated. The results are that there are no physical barriers to growth (neither with respect to natural resources nor with respect to abatement), but far-reaching changes in the social and political order of developing countries are necessary.

The most recent and most comprehensive world model is the *Globus Model* developed by Karl W. Deutsch and collaborators at the Wissenschaftszentrum Berlin and published recently by Stuart A. Bremer (1987). It is not constrained to the economic sphere but covers also demographic and political processes (domestic as well as international). Globus is a huge system of some 40,000 equations; it is dynamic, continuous and recursive, noneconometric, constructed in the spirit of control theory and systems analysis. The results are optimistic: the growth process will continue and the north–south GDP per capita differential will decline. Our result is more pessimistic as far as the north–south GDP per capita ratio is concerned.

Of course, these are only crude outlines of the methods and results reached, and not all existing long-term world models are covered. For more details and an overview of this whole area see Meadows *et al.* (1982), which is a report on the Sixth IIASA Symposium on Global Modeling.

The pessimistic outlook of some of these results and especially the consequences drawn by some of their authors (that the economic order of the world must be changed as soon as possible to avoid disaster) are to be explained by the time of their origin when, at the end of more than 20 years of unprecedentedly rapid economic growth, the widening gap of GDP per capita between most developing and developed countries and the necessity of structural changes became apparent. If there is enough flexibility in the price system and in organizational forms or (on the side of planned economies) in the planning system, and if the developing nations can change their social and political order fast enough and reduce their birth rates sufficiently, the necessary changes will be accomplished more or less smoothly and without much ado. But otherwise there will be a crisis accompanied by a feeling of being caught in an inappropriate system which has to be radically changed. But the process of development takes time since it depends on education, transfer of knowledge, change of behavior and of institutions. There is no way of jumping from one stage to the next. Impatience and force will only prolong the state of transition that the world is in.

There is a series of studies of the Hudson Institute, especially by Herman Kahn and others, which lie somewhere in between the "philosophical" and the formal approaches, see, e.g., Kahn (1979). Kahn and his collaborators use partly formal and partly informal methods of forecasting different features of the world to come. One of his forecasts is reproduced in *Figure 1.3*. These projections are mostly more optimistic than those mentioned earlier as far as averages are concerned. As to the distribution of world income they are more pessimistic.

There are several *world models* in operation now which are constructed and used for *short-term forecasting* (say from 1 to 5 years) and for other purposes, e.g., the *Link Model* of Lawrence Klein, the *Wharton Econometrics Model, and the EPA Model* of the Economic Planning Agency of Japan [4]. They are much more detailed than the long-term world models. But since they follow the Keynesian line in so far as they are demand driven, and since the long-term supply conditions are usually taken as exogenous, they cannot be used for real long-term purposes.

The Bonn–IIASA world model can be placed between the "futurologist" and the short-term world models. Its results are consistent with those long-term projections which do not forecast total disaster, at least not in the foreseeable future. This implies that mankind is able to manage its own problems: first, that devastating wars can be avoided; second, that the rising prices of exhaustible resources will induce substitution by other resources and savings such that these resources will always be available (though in decreasing amounts); third, that pollution can be controlled and the natural environment preserved such that this planet stays inhabitable; fourth, that all nations now lagging behind in economic development will catch up eventually; some of them may even reach a leading position. But this will take time.

## 1.4. The Bonn–IIASA Research Project

The Bonn–IIASA Research Project is concerned with economic growth and structural change "in the medium term" of about 20 years. Economic growth and structural change are phenomena which are barely perceptible in the short run but which exert a decisive influence in the long run. On the other hand, 15–20 years are short enough to take the existing states and their basic political and economic order as given and to forecast labor supply and technological development. The existing economic theory is applicable and allows forecasts. These forecasts are not prophecies. They show future lines of economic development of nations conditional upon assumptions on the trends of the driving forces of this development. Hopefully, those nations which find themselves on the dark side of development will counteract and reverse the trend. This would, of course, invalidate our forecasts. We would be happy were this to happen.

Mankind lives in societies that are organized as nations or countries. Thus the project considers the most important developed countries separately. The analysis and the projections comprise:

(1) The growth of GDP or NMP in total and per capita of the most important countries. This will show the change in the relative economic importance of nations.
(2) The sectoral composition of GDP or NMP, i.e., its subdivision into agricultural production, manufacturing, services, etc.
(3) The commodity flows between countries (exports and imports).

(4) The commodity composition of these flows.
(5) The capital flows between countries and the cumulation of international debts and assets.

Of course, in order to analyze and forecast these variables we shall need other variables (such as prices, exchange rates, capital stocks) which have to be explained in turn. A large amount of data had to be gathered, made consistent, and then analyzed. This would not have been possible without the wholehearted support of collaborating research groups and scholars in many countries. Details of the system and the forecasts will be given in the following chapters. Here we shall present the basic approach.

## 1.5. The Main Driving Forces of Economic Growth

In order to understand economic growth and to forecast possible future developments it is necessary to identify the main *driving forces* of the economic development which in turn are governed by the latent variable "degree of activity". Our approach is best understood by assuming a production function for each country where the production $Y^*$ (measured in a specific way) is a linear homogeneous function of labor input $L$, of the "state of technology" $\tau$, of the capital equipment $K$ and of the access to foreign resources $IM_R$ approximated by real imports of raw materials and energy:

$$Y^* = f(L, \tau, K, IM_R) \tag{1.1}$$

Domestic secondary inputs are netted out. This allows us to group the direct *real driving forces* of economic growth in four categories, namely in those which influence $L, \tau, K$ or $IM_R$, respectively.

(1) The first driving force is growth of labor input $L$, measured in working hours. We have by definition

$$L = POP \cdot \lambda \cdot h \cdot e \tag{1.2}$$

where $POP$ = number of population, $\lambda$ = labor participation rate, $h$ = working hours per employed person per year, $e$ = rate of employment (including self-employment). For OECD countries we used forecasts for $POP$ from UN statistics and estimated the development of $\lambda$, $h$ and $e$. For CMEA countries $L$ is approximated by the number of persons employed in the material sphere. Forecasts were taken partly from estimates of the UN, partly from estimates of the countries themselves. For developing countries $L$ is approximated by the labor force as forecast by the UN. Thus we took $L$ as a predetermined known variable which is not influenced by the latent variable "degree of activity".

(2) The next (and most important) driving force is ***technical progress***, i.e., the growth rate of $\tau$. We use this term in a very broad sense to include all influences which come from research, development, education, organization of society, natural environment and from qualitative changes in capital which are not reflected by the price system. We define

$$\tau = \tau_a \cdot \tau_w \cdot \tau_o \cdot \tau_e \cdot \tau_k \tag{1.3}$$

where $\tau_a$ is an index of ***ability*** (or knowledge) of the members of the society, $\tau_w$ an index of their ***willingness*** to work, $\tau_o$ an index of the ***organizational state*** of the society, $\tau_e$ an index of the state of the ***environment*** (in the sense of the easiness of access to natural resources within the country) and $\tau_k$ is an index of the quality of ***capital***. $\tau_a$, $\tau_w$ and $\tau_o$ relate to labor and may be put together to form an index $\tau_l$ of the "quality" of labor [5]. All constituents of $\tau$ with the possible exception of $\tau_e$ depend on the latent variable "degree of activity" $u$. Thus we get

$$\tau \approx \tau(u), \quad \tau' > O \tag{1.3a}$$

We estimated and forecast $\tau$ on the base of possible future values of $u$ [6].

(3) The third driving force is ***capital accumulation***. The capital $K$ of a country is an index for the production power of domestic capital goods $K^{dom}$ and imported capital goods $K^{imp}$:

$$K = K^{dom} + K^{imp}$$

where a possibly higher productive power of imported capital goods is, as a rule, taken care of by their higher real prices in domestic currency. Capital accumulation is defined by

$$K = K_{-1}\,(1-d) + I \tag{1.4}$$

where $d$ is the rate of depreciation and $I = \Delta K^{dom} + \Delta K^{imp}$ is determined by the investment ratio $s$:

$$I = s \cdot Y \tag{1.5}$$

$Y$ = real GDP or NMP. Thus capital accumulation is determined by the real shape of $d$ and $s$. We kept $d$ constant at the average historical rate [7]. The investment ratio $s$ depends on GDP or NMP per capita (= $Y/POP$), on the willingness $\tau_w$ to work for the future, on organizational abilities $\tau_o$ of the leading groups of society and on the future discount rate $\rho$:

$$s = f(Y/POP, \tau_w, \tau_o, \rho) \tag{1.6}$$

$\tau_w$ and $\tau_o$ depend positively on $u$. The influence of $Y/POP$ is small for the ranges of $Y/POP$ to be expected in the forecasting period. The more uncertain the future, i.e., the higher the future discount rate, the smaller the propensity to invest. Thus we have

$$s \approx \varphi\left(\overset{+}{u}, \overset{-}{\rho}\right) \tag{1.6a}$$

where the signs above the arguments indicate the signs of the partial derivatives. But the future discount rate $\rho$ also depends on the latent variable $u$: a higher degree of activity is correlated with a more optimistic outlook into the future, i.e., with a smaller discount rate. Thus we have

$$\rho \approx \rho(u), \quad \rho' < 0 \tag{1.6b}$$

and therefore

$$s \approx s(u), \quad s' > 0 \tag{1.6c}$$

The relations (1.3a), (1.6b) and (1.6c) are used to estimate the latent variable $u$; see Section 1.6.

From (1.6b) and (1.6c) we see that $s$ and $\rho$ are inversely related. This may also be derived by optimizing consumption over an infinite time horizon under the constraint of a production function for a given rate of time discount; see Krelle (1987).

(4) The last (but not least) driving force in this category is determined by *greater international division of labor* which in our approach is measured by an increasing amount of imports $IM_R$ of secondary factors of production, mostly raw materials, intermediate goods and energy. This is where the influence of the international economic order and of international capital flows comes in (the other driving forces are of domestic origin). In the current approach we took the secondary imports to be endogenous variables which depend directly only on other economic variables, not on the latent variable $u$. But since the other economic variables depend on $u$ via $\tau$ and $s$, the imports do likewise.

We now turn to the *monetary side* of the economy. Here the exogenous "driving force" is the money supply $M$ (for OECD countries and developing market economies) and the monetary wage level $l$ (for CMEA countries). We estimated functions for the velocity of money (for OECD and developing countries) from which the domestic price level is determined. For CMEA countries the domestic price level follows from the ratio of the nominal wage rate to the real wage rate which is related to labor productivity.

Summing up, we have for each country the main exogenous driving forces $L$, $\tau$, $s$, $d$ and $M$ (or $l$) where $\tau$ and $s$ depend in turn on the "degree of activity" $u$. The driving forces of all countries determine simultaneously the economic performance of all countries. Details are given in the following chapters.

## 1.6. Some Empirical Results: Development of the Rate of Technical Progress, of the Savings Ratio, and of the Time Discount Rate

These three variables are used as indicators for the latent variable "degree of activity" $u$; but they are also (along with the rate of growth of the labor force) the most important driving forces of economic growth. Therefore it is interesting to look at their time shape in the past.

We start with the *rate of technical progress* $w_\tau = (\tau - \tau_{-1})/\tau_{-1}$. This variable is not directly observable. We use the following method for the identification of $w_\tau$ [8]. As pointed out earlier, the major part of $\tau$ is associated with the quality of labor. Therefore we specify the technical progress as Harrod-neutral. Instead of equation (1.1) we use

$$Y^* = f(L\tau, K, IM_R) \tag{1.1a}$$

This equation may be rewritten in terms of growth rates [9] and solved for the latent variable: $w_\tau$:

$$\begin{aligned} w_\tau &= 1/\alpha\ (w_Y - \alpha w_L - \beta w_K - \gamma w_M) \\ \alpha &+ \beta + \gamma = 1 \end{aligned} \tag{1.7}$$

where $\alpha$, $\beta$, $\gamma$ denote the production elasticities of labor, capital and imported factors of production, respectively, and $w_M = w_{IM_R}$. *In equilibrium growth:*

$$w_Y = w_K = w_M \tag{1.8}$$

and equation (1.7) becomes simply

$$w_\tau = w_{Y/L} = w_Y - w_L \tag{1.9}$$

In equilibrium growth, the rate of technical progress coincides with the growth rate of labor productivity. For *disequilibrium growth* we define

$$d_K = w_K - w_Y, \quad d_M = w_M - w_Y \tag{1.10}$$

and obtain from equation (1.7):

$$w_\tau = w_Y - w_L - (\beta/\alpha)d_K - (\gamma/\alpha)d_M \tag{1.11}$$

In disequilibrium the growth rate of technical progress is determined only partly by the rate of growth $w_Y - w_L$ of labor productivity $Y/L$. The term $(\beta/\alpha)d_K$ measures the influence of a changing capital coefficient $K/Y$, the term $(\gamma/\alpha)d_M$ the influence of a changing import coefficient $IM_R/Y$, i.e., of a changing international division of labor. Both influences have to be deducted from the growth rate of labor productivity to get the rate $w_\tau$ of technical progress proper.

In order to make the influence of $d_K$ and $d_M$ visible, we also define an adjusted growth rate of technical progress:

$$\tilde{w}_\tau = w_{Y/L} - (\beta/\alpha)d_K = (1/\tilde{\alpha})(w_Y - \tilde{\alpha}w_L - \tilde{\beta}w_K) \tag{1.12}$$

$$\tilde{\alpha} = \alpha/(1-\gamma), \quad \tilde{\beta} = \beta/(1-\gamma), \quad \tilde{\alpha} + \tilde{\beta} = 1$$

Now we get

$$w_\tau = \tilde{w}_\tau - (\gamma/\alpha)d_M \tag{1.13}$$

We estimated $\alpha$, $\beta$, $\gamma$ for the reference period 1955–1982 (see the following chapter) and thus arrived at three different measures for the rate of technical progress:

- labor productivity $w_{Y/L}$,
- the adjusted growth rate $\tilde{w}_\tau$ of technical progress, and
- the growth rate $w_\tau$ of technical progress proper.

*Figures 1.12–1.15* show the graphs of these three functions for the USA, the FRG, Japan and the USSR. The figure for the other countries look similar, with the exception of the GDR. The vertical distance between the graphs of $w_{Y/L}$ and $\tilde{w}_\tau$ shows the influence of a rising capital coefficient, the vertical distance between $\tilde{w}_\tau$ and $w_\tau$ the influence of a rising import coefficient, i.e., the influence of an increasing international division of labor. In equilibrium growth all three graphs would coincide.

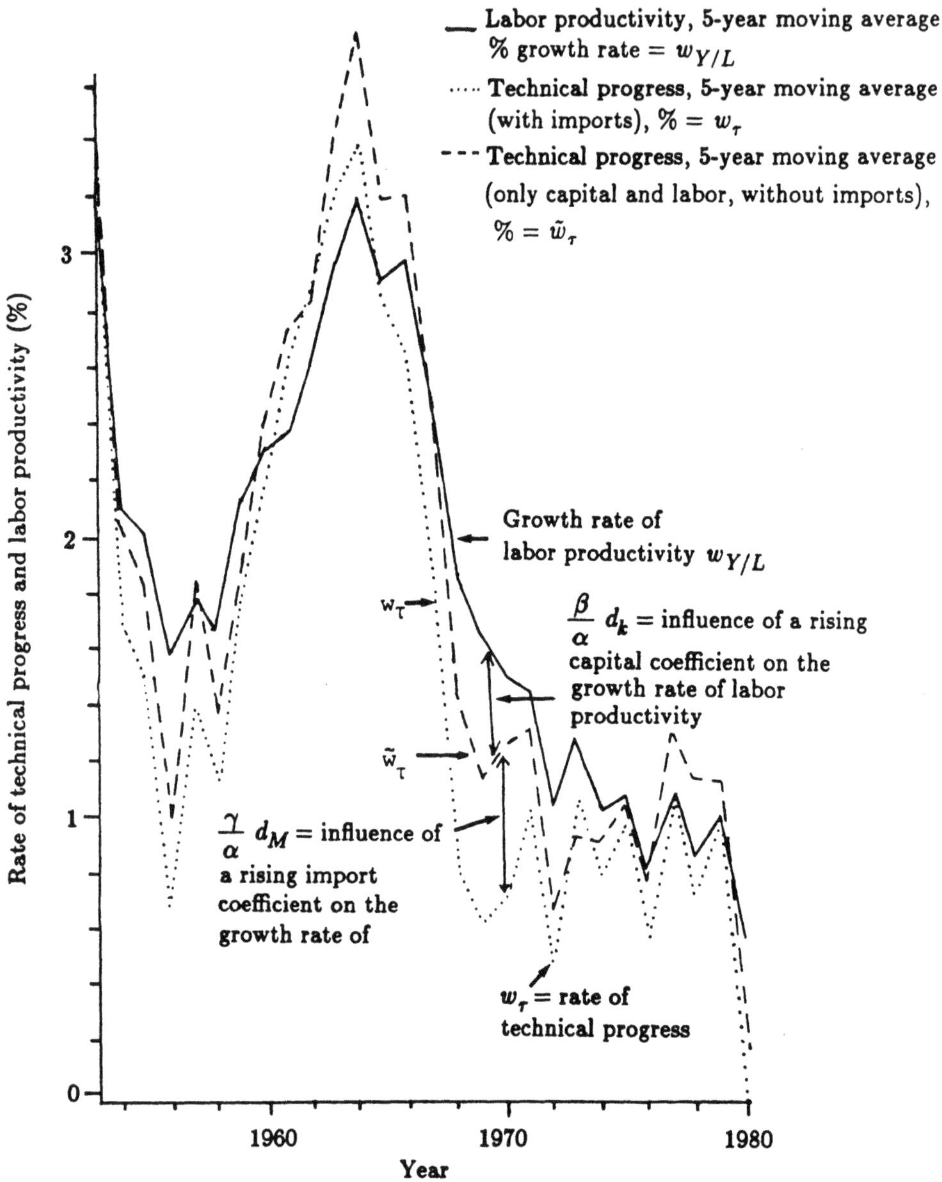


*Figure 1.12.* USA: actual development of labor productivity and technical progress. Sources: (1) OECD, *National Accounts*, Vol. I: Main Aggregates, Paris (various years). (2) OECD Labor Force Statistics, Paris (various years). (3) ILO, *Yearbook of Labor Statistics*, Geneva (various years). (4) Our calculations and estimates.

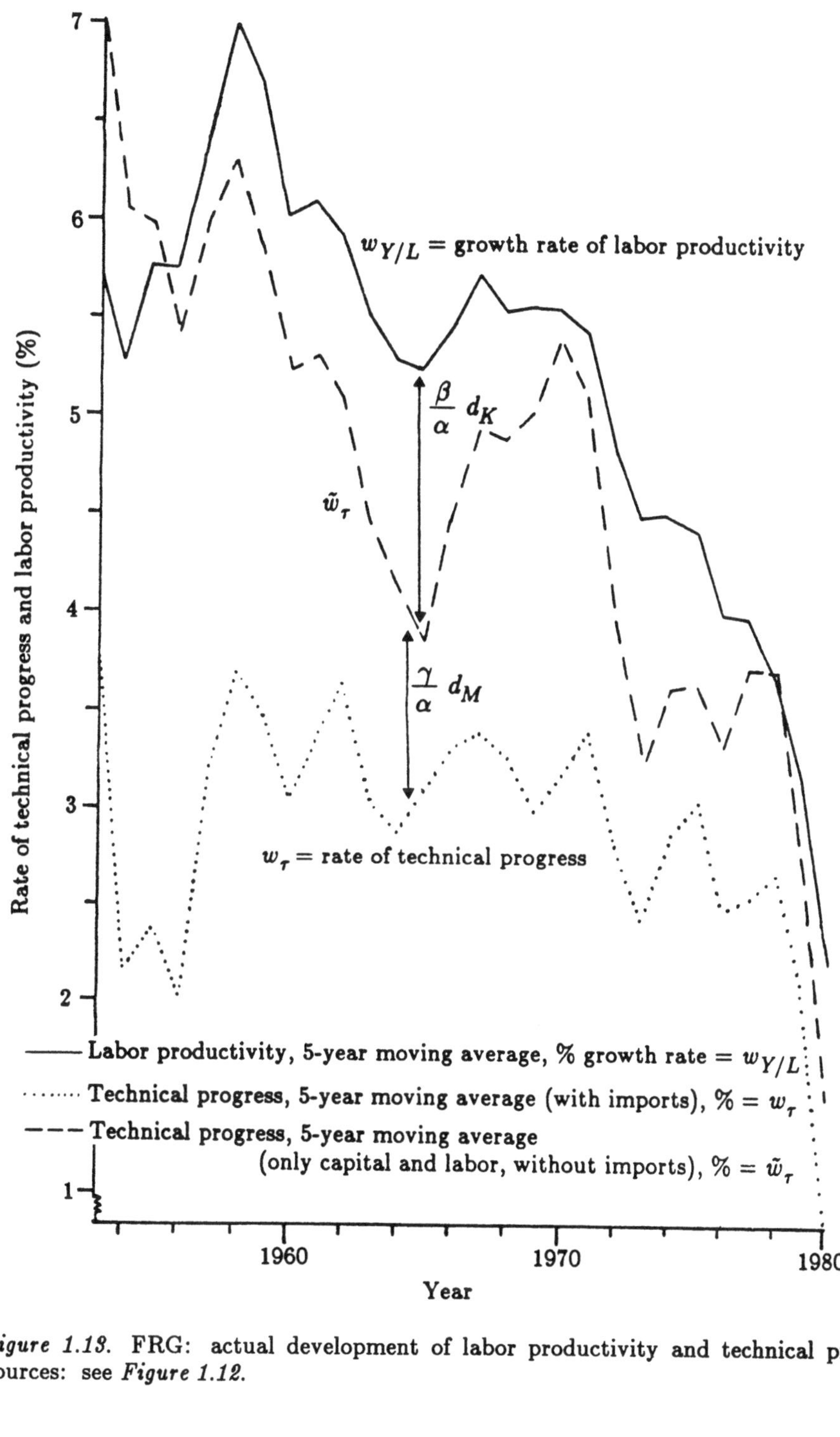


*Figure 1.13.* FRG: actual development of labor productivity and technical progress. Sources: see *Figure 1.12*.

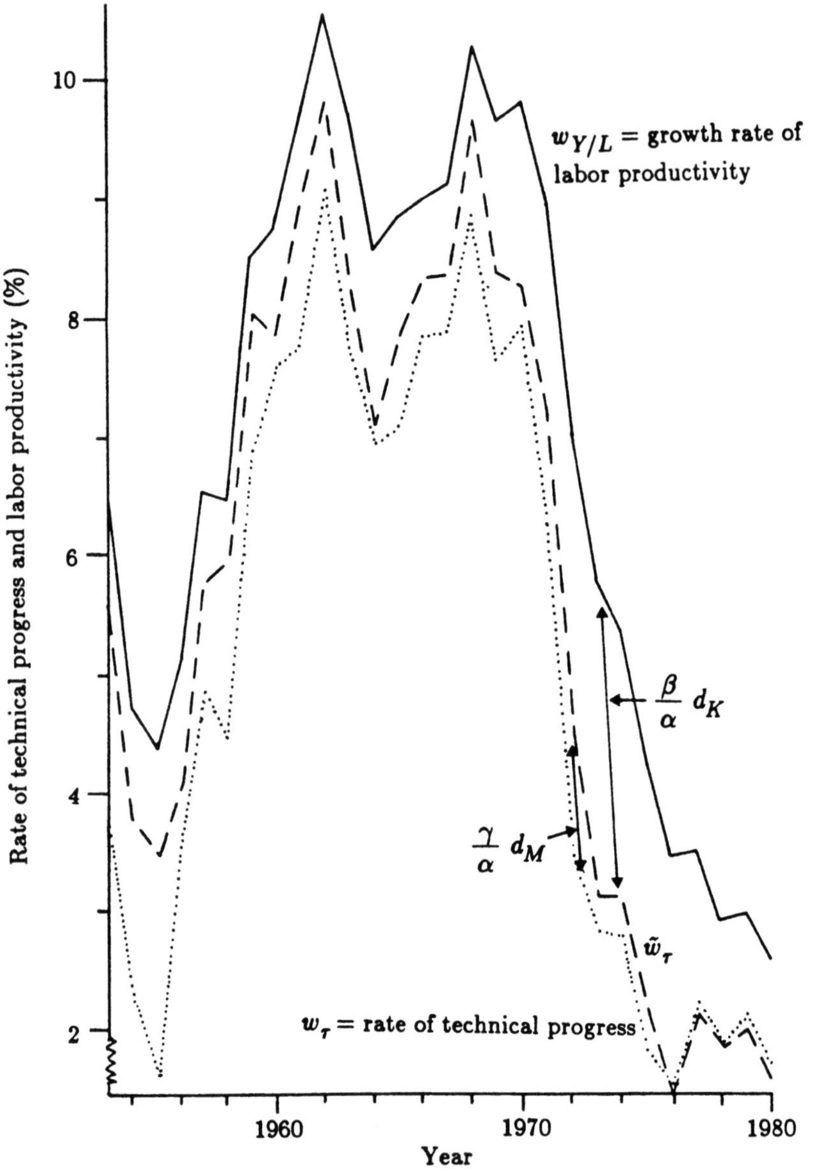


—— Labor productivity, 5-year moving average, % growth rate = $w_{Y/L}$

······· Technical progress, 5-year moving average (with imports), % = $w_\tau$

– – – Technical progress, 5-year moving average (only capital and labor, without imports), % = $\tilde{w}_\tau$

*Figure 1.14.* Japan: actual development of labor productivity and technical progress. Sources: see *Figure 1.12.*

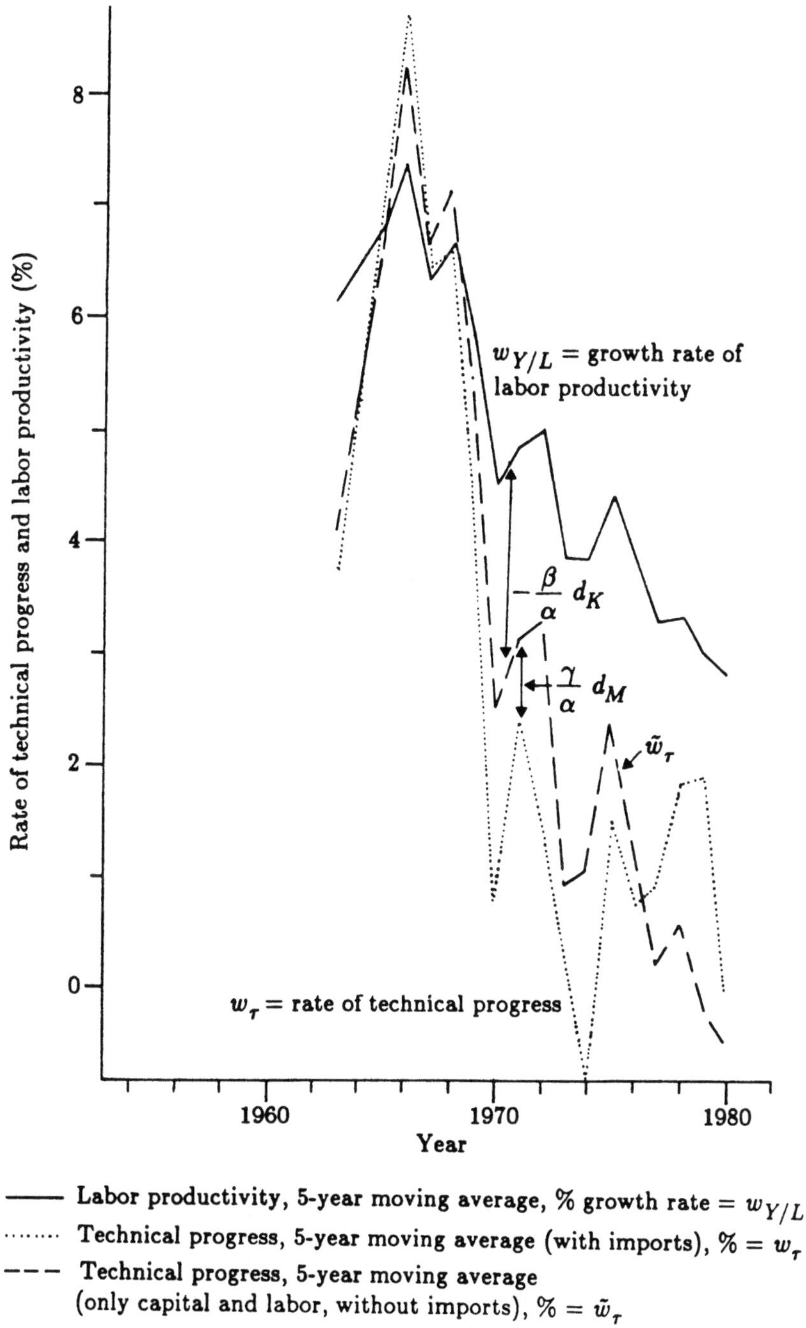


*Figure 1.15.* USSR: actual development of labor productivity and technical progress. Data sources for the CMEA countries: (1) UN *Yearbook of National Accounts Statistics*, 1970–1982. (2) *Statistical Yearbook of the CMEA* (Statisticheskij Ezhegodnik Stran Chlenov SEV), 1971–1983. (3) UN, *Yearbook of International Trade Statistics*, 1977, 1982. (4) National Statistical Yearbooks of USSR, Bulgaria, CSSR, GDR, Hungary and Poland for various years.

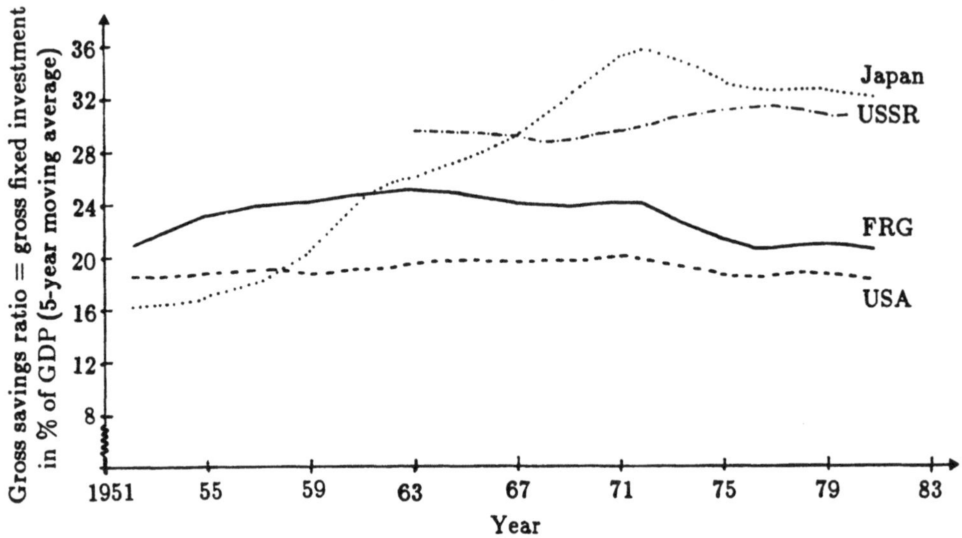


*Figure 1.16.* Investment ratios. Sources: (1) OECD, *National Accounts*, Vol. I: Main Aggregates, Paris (various years). (2) Our own calculations. (3) For USSR, see *Figure 1.15*.

The most striking feature of these figures is the fundamental decline of the rate of technical progress, starting already in the 1960s. This is true for all countries in East and West with the notable exception of the GDR where this rate stays almost constant.

The same unfortunate development can be seen in the *investment ratio* (see *Figure 1.16*) and also in the *future discount rate.* We approximate this rate by the government bond yield. *Figure 1.17* shows the time shape of the negative of the bond yield for the USA, the FRG and for Japan.

## 1.7. The Identification of the Latent Variable "Degree of Activity"

Here we wish to forecast technical progress and the investment ratio. Both depend on the latent variable "degree of activity" $u$ [see equations (1.3a) and (1.6c)]. We estimated $u$ from the three indicators $w_T$, $s$ and $\rho$ by the method of principal components and fitted a sinusoidal curve $\bar{u}$ through these points. *Figure 1.18* shows the results for the FRG. The curves for the other OECD countries look similar. For the USA, FRG, Japan, France, UK and Italy they are reproduced in *Figure 1.19* and extended to the year 2000.

*Table 1.2* shows the cycle length in years. Details are given in Krelle (1987). This looks rather like a Kondratieff cycle. In our perception it is the reaction of society to unfortunate developments which induces the change. The fight against stagnation and decline may be successful, and often

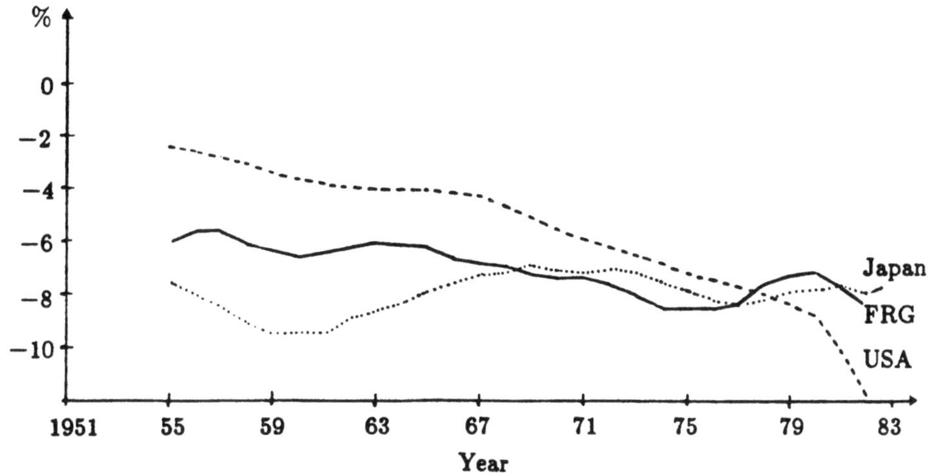


*Figure 1.17.* The negative of government bond yield. Source: IMF, *International Financial Statistics Yearbook*, Washington, D.C. (1983).

will be, but not always. The extrapolations in *Figure 1.19* must not be mistaken for forecasts. Moreover, the interrelation between the growth paths of different countries as caused by foreign trade and capital flows are disregarded. But the curves show that we may expect a turn of the tide: the rate of technical progress and the investment ratio will not continue to fall but will increase again, though very likely not to the same level as in the 1960s. This assumption underlies our forecasts.

*Table 1.2.*

| | *USA* | *FRG* | *Japan* | *France* | *UK* | *Italy* | *NL* | *Bel/Lux* |
|---|---|---|---|---|---|---|---|---|
| Cycle length in years | 33.6 | 42.7 | 30.8 | 33.2 | 28.8 | 31.1 | 28.8 | 29.5 |

## 1.8. A Simplified Theory for the Long-Term Cycle of Degree of Activity

How can the Kondratieff-type long-term wave of economic activity be explained? In Krelle (1973, 1982, 1984) it was shown that the process of transfer of information and valuation from person to person may (and as a rule will) lead to cycles of this type. If one treats a society as a person the idea may be formulated as follows. A deviation of the degree of activity $u$ from its normal zero value intensifies this deviation in the next period, i.e., there is a law of "psychological infection" in a society. But after a while the limitations of this new attitude and

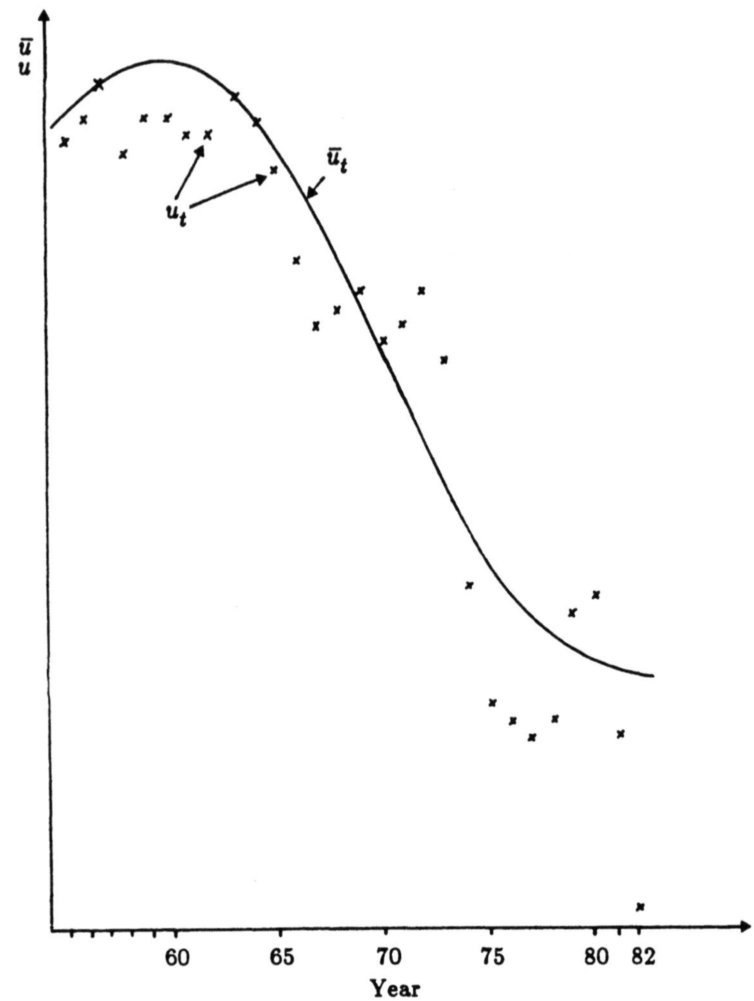


*Figure 1.18.* FRG: a sinus curve fitted to the estimated "degrees of activity".

its drawbacks become apparent. Thus there is a retracting, equilibrating force from the previous period but one. Therefore we arrive at the equation

$$\bar{u}_t = a_1\bar{u}_{t-1} + a_2\bar{u}_{t-2} \tag{1.14a}$$

where $a_1 > 0, \quad a_2 < 0$ . If $|a_2| > (a_1/2)^2$, the solution is

$$\bar{u}_t = b_1 \sin(b_2 t + b_3) \tag{1.14b}$$

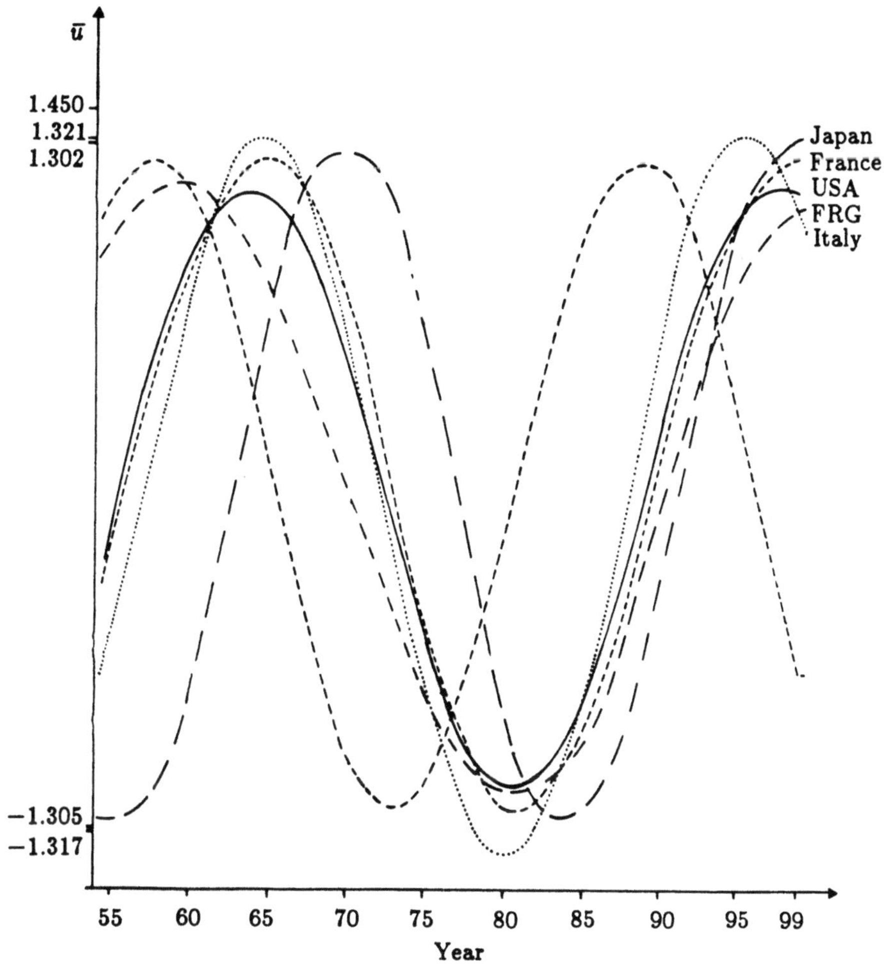


*Figure 1.19.* Trends of the degrees of activity in USA, FRG, Japan, France, UK and Italy.

Of course, this is the most simplified version of a more elaborate theory; see the Krelle references above. If this is basically an appropriate explanation for the long-term sinusoidal movements of the driving forces of economic growth, we may expect that they will continue in the future.

## 1.9. Growth and Structural Change

The growth process is always connected with structural change. If structural change is inhibited, e.g., for social reasons (because people want to keep their old working places), growth will stop. The demand structure changes if people become richer and if technology advances and the international division of labor increases. There is a whole literature on this subject: Dennison (1974), Giersch (1981), Nelson (1981), Kendrick (1981), Stoneman (1983), Conrad (1985), Jorgenson, Gollop, Fraumeni (1986), to name only a few. We use the "upside down" approach: growth determines structural change. Hindrances in structural change manifest themselves as a decline of the degree of economic activity and therefore result in a decline of the rate of growth. Details will be given in the next chapters.

## 1.10. Conclusion

This chapter provides the philosophical background of the Bonn–IIASA Research Project. The industrial revolution which we are experiencing is a unique period in human history and will lift mankind to a new, better life. Some nations have led the way, the others will follow with some time lag. Thus income distribution on a world scale will inevitably become more uneven for quite a while. But in the historical perspective this will be a transitory phase.

Superimposed on this very long-term there are long-term Kondratieff-type cycles of economic activity. We identified these cycles by using indicators and looking for the principal components, and found that the main driving forces of economic growth (the rate of technical progress and the investment ratio) show sinusoidal fluctuations of a time length between 30 and 40 years. We are now a short way behind the trough of a Kondratieff wave.

Since our project is concerned with the economic development over the next 20 years, we may assume that the driving forces of economic growth will not continue to decrease as in the last 15 years but increase again, though to different degrees. This assumption underlies our forecasts.

### Notes

[1] I wish to thank C.C. v. Weizsäcker, director of the Institute for Energy Economics at the University of Cologne, for providing these data.

[2] Marchetti used this type of law of extension of knowledge and behavior extensively and showed that it is applicable in very many fields; see, e.g., Marchetti (1986), who used a graphical representation of a transformation of the logistic: $\log F/(1-F) = \log b/c + \mathrm{d}t$, where $F = x_{\hat{t}}/\bar{x})$. Thus, the logistic curve becomes linear in time space.

[3] Since there does not seem to be a basic difference in natural gifts between nations, a law of "regression to the mean" will work, which is called "Stein's law"; cf. Efson and Morris (1977).

[4] For a complete list of world models see Uebe *et al.* (various dates), and regular printouts of the latest state of the bibliography by Uebe.

[5] There are similar, but not identical, classifications of the constituent parts of technical progress; see, e.g., Dennison (1967, 1974) and Kendrick (1981, p. 117). Since they do not consider imported factors of production explicitly, they cannot deal explicitly with the consequences of an increasing international division of labor.

[6] In fact we measured the rate of growth $w_\tau = (\tau - \tau_{-1})/\tau_{-1}$ of the state of technology, but did not break this index down into constituent parts as indicated in equation (1.3). This remains a task for future research. It can be done in the following way. The index $\tau_a$ of "ability" of the population may be measured by the average schooling and education of the population, by the amount of research and development and by the transfer of knowledge from other nations. The index $\tau_w$ of the "willingness to work" indicates the inclination of the population to work hard, to show discipline, to make sacrifices for the wellbeing of others and for the future, to take risks, to try something new and to adjust to new requirements of the economy. It is an index of "entrepreneurship" in the sense of Schumpeter and may be measured by several indicators. The index $\tau_o$ measures the ability of the leading groups to organize the society, to overcome old religious or philosophical taboos and to change organizations and institutions if they become a hindrance to economic progress. In a market economy it means keeping the price system flexible and total demand and income distribution at acceptable levels; in a planned economy it means improving the efficiency of planning as much as possible. In both systems it measures the capacity to realize the necessary structural changes. Thus $\tau_o$ is also an index of "economic activity" or "entrepreneurship" but related to the leading groups of a country. The index $\tau_e$ measures the ease of access to natural resources, e.g., by the relative amount of labor and capital used in the extraction of natural resources.

[7] For OECD countries, $d$ has also been varied; see Chapter 2.

[8] For details, see Krelle (1987). Section 1.6 follows this article.

[9] We always write $w_x = (\dot{x}/x) \approx (x - x_{-1})/x_{-1} =$ growth rate of a variable $x$.

## References


Berdiajev, N. (1927), *Der Sinn der Geschichte.* (Tübingen: Reidel).

Bremer, St. A. (ed.) (1987), *The Globus-Model. Computer Simulation of Worldwide Political and Economic Developments.* (Frankfurt and Boulder, Colorado: Campus/Westview).

Bruckmann, G. (ed.) (1976), *Bariloche Group: The Latin American World Model, Proceedings of the Second IIASA Conference on Global Modeling*, Collaborative Paper CP-76-8. (Laxenburg, Austria: International Institute for Applied Systems Analysis).

Conrad, K. (1985), *Produktivitätslücken nach Wirtschaftszweigen im internationalen Vergleich.* (Berlin/Heidelberg/New York/Tokyo: Springer).

Dennison, E.F. (1967), *Why Growth Rates Differ.* (Washington, D.C.: The Brookings Institution).

Dennison, E.F. (1974), *Accounting for United States Economic Growth 1929–1969.* (Washington, D.C.: The Brookings Institution).

Efron, B. and Morris, C. (1977), Stein's Paradox in Statistics, *Scientific American*, **236**, May, pp. 19–127.

Gerstenberger, W. (ed.) (1985), *Ansätze und Methoden zur Strukturanalyse.* (Göttingen: Vandenhoede & Ruprecht).

Giersch, H. (ed.) (1981), *Towards an Explanation of Economic Growth.* Symposium 1980. (Tübingen: Mohr).

Hansen, A. (1938), *Full Recovery or Stagnation.* (New York).

Herrera, A.D., Scolnik, H.D. *et al.* (1976), *Catastrophe or New Society? A Latin American World Model.* (Ottawa: International Development Research Centre). German edition (1977): *Grenzen des Elends.* (Frankfurt: S. Fischer).

Jorgenson, D.W., Gollop, F.M., and Fraumeni, B.M. (1986), *Productivity and Growth of Sectoral Output in the United States 1948–1979*, Discussion Paper 1217. (Cambridge, Mass.: Harvard Institute of Economic Research).

Kahn, H. (1979), *World Economic Development 1979 and Beyond.* (New York: Morrow Quill Paperbacks).

Kendrick, J.N. (1981), Why Productivity Growth Rates Change and Differ, pp. 111–148 in: H. Giersch (ed.), *Towards an Explanation of Economic Growth* (Tübingen: Mohr).

Kleinknecht, A. (1984), *Innovationsschübe und lange Wellen: Was bringen "Neo-Schumpeterianische" Kriseninterpretationen?* Research Memorandum No. RM-84-009. (Maastrich, Netherlands: Limburg University, Faculty of Economics).

Kondratieff, N.D. (1926), Die langen Wellen der Konjunktur, *Archiv für Sozialwissenschaft*, **56**, p. 573ff.

Krelle, W. (1973), Dynamics of the Utility Function, pp. 92–128 in: J.R. Hicks and W. Weber (eds.), *Carl Menger and the Austrian School of Economics.* (Oxford: Clarendon Press).

Krelle, W. (1982), The Dynamic Interdependence between Information Valuation and Production in a Society, pp. 97–106 in: G.R. Feiwel (ed.), *Samuelson and Neoclassical Economies*, (Boston/The Hague/London: Klüwer–Nijhoff).

Krelle, W. (1984), Waves of Entrepreneurial Activity Induced by Transfer of Information and Valuation, *Zeitschrift f. Nat. Ök., Journal of Economics*, Suppl. 4, pp. 71–92.

Krelle, W. (1985), *Theorie des wirtschaftlichen Wachstums.* (Berlin/Heidelberg/New York/Tokyo: Springer).

Krelle, W. (1987), *Long-Term Fluctuations in Technical Progress and Growth*, Discussion Paper, Bonn, May 1986, to be published in the *Journal of Institutional Economics (Zeitschr. f. d. ges. Staarswiss)*.

Kuczynski, T. (1985), Das Problem der "langen Wellen" – einige Überlegungen, pp. 89–120, Band 18, in: *Wirtschaftsgeschichte und Mathematik* (Forschungen zur Wirtschaftsgeschichte), (Berlin-Ost: Akademie-Verlag).

Leontief, W., Carter, A., and Petri, P. (1977), *The Future of the World Economy, A United Nations Study.* (Oxford and New York: Oxford University Press).

Marchetti, C. (1986), *The Future of Hydrogen: An Analysis at World Level, with Special Look at Air Transport*, Working Paper WP-86-25. (Laxenburg, Austria: International Institute for Applied Systems Analysis).

Marx, K. (1984), *Das Kapital*, Buch III.

McClelland, D.C. (1961), *The Achieving Society.* (Princeton/Toronto/New York/London: van Nostrand).

Meadows, D., Meadows, D.L., Randers, J., and Behrens, W. (1972), *The Limits to Growth.* (New York: Universe Books, A Potomac Associates Book).

Meadows, D., Richardson, J., and Bruckmann, G. (1982), *Groping in the Dark.* (Chichester/New York/Brisbane/Toronto/ Singapore: John Wiley).

Mesarovic, M.D. and Pestel, E. (1974), *Mankind at the Turning Point.* (New York: Dutton).

Mensch, G. (1975), *Das technologische Patt.* (Frankfurt/Main: Umschau Verlag).

Nelson, R.R. (1981), Research on Productivity Growth and Differences, *Journal of Economic Literature*, **XIX**, No. 3, pp. 1029–1064.
OECD (1985), *Employment, Growth and Structural Change.* (Paris: OECD).
Olson, M. (1982), *The Rise and Decline of Nations.* (New Haven and London: Yale University Press).
Pareto, V. (1916), *Trattato di Sociologia generale.* (Firenze). German translation by G. Eisermann (1962), *Vilfredo Paretos System der allgemeinen Soziologie.* (Stuttgart: Enke).
Rostow, W.W. (1975), Kondratieff, Schumpeter, Kuznets: Trends Periods Revisited, *Journal of Economic History*, **35**, p. 729ff.
Rostow, W.W. (1986), *The Stages of Economic Growth.* (Cambridge: University Press).
Rüstow, A. (1950), *Ortsbestimmung der Gegenwart*, Band 1–3. (Zürich/Stuttgart: Erlenbach).
Schüller, A. (ed.) (1984), Wachstumsverlangsamung und Konjunkturzyklen in unterschiedlichen Wirtschaftssystemen, in *Schriften d. Vereins f. Sozialpolitik*, N.F. Bank 142. (Berlin: Duncker & Humblot).
Schumpeter, J. (1912), *Theorie der wirtschaftlichen Entwicklung.* (Leipzig).
Schumpeter, J. (1939), *Business Cycles.* (New York and London: McGraw-Hill).
Sorokin, P. (1957), *Social and Cultural Dynamics.* (Boston: Porter Sargent).
Spengler, O. (1919), *Der Untergang des Abendlandes.* (München: Beck).
Stoneman, P. (1983), *The Economic Analysis of Technical Change.* (Oxford: Oxford University Press).
Toynbee, A.J. (1947), *A Study of History.* (New York and London: Oxford University Press).
Wachter, M.L. and Wachter, S.M. (eds.) (1984), *Removing Obstacles to Economic Growth.* (Philadelphia: University of Pennsylvania Press).
Weber, A. (1951), *Prinzipien der Geschichts- und Kultursoziologie.* (München: Piper).

CHAPTER 2

# Main Results of the Bonn–IIASA Research Project

*Wilhelm Krelle*

## Summary

This chapter presents the main results of the Bonn–IIASA Research Project on economic growth and structural change as far as the work of the central group is concerned. First, the model is outlined. The basic assumptions about the main driving forces of the economic development (technical progress and capital accumulation) are stated, and finally the results (given in the Annexes) are discussed. They may be summarized as follows. The growth process will continue, but at a slower rate. The CMEA countries will come nearer to the OECD countries, but not much. Within the group of the OECD countries, some smaller ones will pass the USA as far as labor productivity and standard of life is concerned. Some developing countries (especially Black Africa and India) will stay behind so that the world income distribution will further deteriorate. The developing countries (with the exception of the oil-exporting countries), the USA and some European countries will keep their trade deficit, though (in the case of the USA) it will decline.

## 2.1. Introduction: The Model

We are concerned with medium-term forecasts of the trend of economic growth and structural change on the world level, i.e., with trend forecasts until the year 2000. These are conditional forecasts, of course; we are not prophets. The conditions are assumptions on the future time paths of the driving forces of economic growth. They depend on the success of the different governments in creating favorable economic conditions within their countries (or, in the case of

CMEA countries, in organizing their economies efficiently) and in taking advantage of world trade and international division of labor. Given the size and development of these driving forces, we may estimate the future development of GDP or NMP, its sectoral composition, exports and imports and their commodity composition, the general price level, the price levels of groups of commodities, and the exchange rates by means of a world model which consists of linked national models (or, in the case of developing countries, models for groups of countries). The countries and groups of countries considered in the project are given in Annex 1 at the end of the book. The model is outlined in Annex 2 at the end of the book. Here we only present the general construction principles of the model and the underlying motivations.

One basic theoretical construct of the model is a *production function* for each country of the Cobb–Douglas type:

$$Y^* = \tau L^{\alpha_1} K^{\alpha_2} (IM_R)^{\alpha_3}, \quad \alpha_i > 0, \quad \sum_i \alpha_i = 1 \tag{2.1}$$

where $Y^*$ is total production after netting out domestic secondary inputs, $\tau$ is an index for the state of technology, $L$ = labor employed, $K$ = capital, $IM_R$ = imported secondary inputs (energy, raw materials, half-finished goods, etc.). $w_\tau = (\tau - \tau_{-1})/\tau - 1$ is the rate of technical progress.[1]

It is necessary to defend the choice of this approach (which uses a very simple functional form) in the light of other much more sophisticated approaches. We have to consider the aim of our research. We want to forecast the medium-term trend of economic development, not the short-term economic results, say, GDP or NMP in the next or next but one year. In this case we would have to use much more complicated functions. But for medium- and long-run trends, this type of function is most appropriate.[2]

Another basic assumption is the *monetaristic approach*: money supply (we choose M2 as the money concept) and the velocity of money determine the *domestic price level*, given real production. The velocity of money is explained by a function, the most important arguments of which are the capital output ratio, labor productivity and the foreign debt ratio with respect to production. The *nominal rate of interest* is explained by the rate of inflation, the rate of interest abroad (we took the US rate as a proxy), the savings ratio, the foreign debt ratio and the rate of change of the exchange rate. The *exchange rates* of all countries should be explained simultaneously. This approach is presented by Dr. Welsch in Chapter 8. Unfortunately, we could not solve the total system numerically with this complicated exchange rate subsystem. Thus we had to simplify the approach. We explained the trend of the exchange rate by variants of the purchasing power parity theory. This should be improved later.

For *CMEA countries* we estimated the price level by the ratio of the nominal wage rate (which is a decision variable of the government or of the planning office and forecast exogenously) and the real wage rate which follows (among others) from labor productivity. This is an endogenous variable.

*Foreign trade* and *capital flows* form the links between all economies. We explain *real imports* of each country by a dynamized version of the linear expenditure system.[3]

The *import prices* are determined by the export prices of other countries and by the exchange rate. They are scaled in such a way that real imports equal real exports for the whole world. *Nominal exports* are also explained by dynamized versions of the linear expenditure system, where the total monetary demand for exports equals the value of imports. The *export prices* are explained by the domestic price level and by import prices, basically. This approach guarantees consistency of the model (total imports = total exports in current as well as in constant $) without using a trade matrix and thus without throwing away the export functions. This approach works well in the reference period.

Of course, the current account *balance of trade* and therefore the development of international indebtedness follow immediately. Finally the commodity structure of production, exports and imports is estimated by structural equations.

## 2.2. The Driving Forces of Economic Development

The driving forces of economic development are those which increase production $Y^*$ in equation (2.1). These are directly:

(1) The rate $w_\tau$ of technical progress.
(2) The growth rate $w_L$ of labor input.
(3) The growth rate $w_K$ of invested capital. This rate follows from the gross investment ratio $s$ and the depreciation rate $d$ by the definitional equation

$$K = K_{-1}(1-d) + I_{-1} \tag{2.2}$$

where $I$ = gross investment, $Y$ = GDP or NMP and $I = sY$.

The exact definitions of $L$, $K$, $I$, and $Y$ differ between OECD countries, CMEA countries and developing countries. Imports $IM_R$ are related to total imports $IM$. Thus we consider as main driving forces $w_\tau$, $w_L$, $s$ and $d$. They are estimated for the reference period (mostly 1960–1982) and forecast till the year 2000. For reasons which were pointed out in Chapter 1, we assume a "turn of the tide" with respect to the negative development of $w_\tau$ and $s$ in the OECD and CMEA countries in the past. *Figure 2.1* shows the general principle. Details are given in the next section.

(4) Since commodity prices and interest rates influence imports and exports, and therefore production, the *money supply* (or for CMEA countries: the monetary wage rate) comes up as another more indirect driving force. It affects GDP at least transitorily. Finally, we forecast the price index of fuels exogenously because of the political character of this price. Thus it may also be counted as a driving (or, better, a restraining) force of economic development.

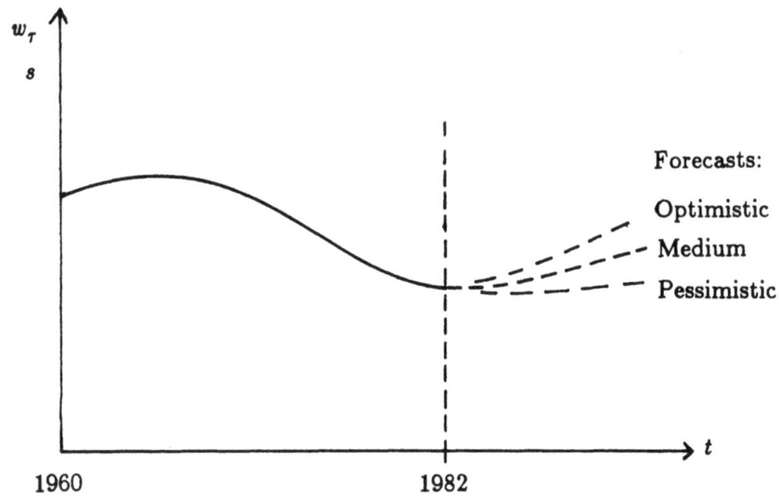


*Figure 2.1.* The general principle of forecasting the exogenous variables.

The driving forces of all countries simultaneously determine the rate of economic growth of each country since the economies of all countries are linked by foreign trade and capital flows. But the predominant cause of economic growth of each country is still the development of its own driving forces.

## 2.3. Assumptions about the Future Development of the Driving Forces

We start with the *growth rate* $w_L$ *of employed labor.* There is no great uncertainty on the figures for the future population of working age – the number of those who may join the labor force till the year 2000 or may drop out of it is approximately known. But immigration or emigration may change this figure, and the labor participation rate, the average working hours and the employment rate are rather uncertain. For *OECD countries* all these figures are known for the past, i.e., for the reference period 1961–1984. Averages are presented in Appendix 2A. On the basis of these observations, of some outside information and of subjective judgments of the author, the growth rate $w_L$ of employed labor (in working hours) has been estimated, see Appendix 2A.1 and *Table 2.1.* For *CMEA countries* we estimated the growth rates of persons employed in the material sphere from the available resources as indicated in Appendix 2A.2 and *Table 2.2.* For *developing countries* the number of the economically active population has been forecast on the base of UN, UNCTAD and other data. We reproduce the growth rates, in Appendix 2A.3 and *Table 2.3.* We keep these forecasts constant for all three scenarios (see below).

The same applies for the *money supply* (for OECD and developing countries) and for the *growth rate of monetary wages* for CMEA countries. These forecasts are based on actual behavior of the past; see Appendix 2B and *Tables*

*Table 2.1.* Definition of scenarios for the driving forces of economic development, 1985–2000: OECD countries.

| *Indicator* (%) | *Scenario* | *USA* | *FRG* | *Japan* | *France* | *UK* | *Italy* | *Neth.* | *B/L* | *Can* | *Rest of OECD* |
|---|---|---|---|---|---|---|---|---|---|---|---|
| Growth rate of labor input (yearly working hours), $w_L$ | constant | 1.4 | –0.4 | 0.4 | –0.7 | 0.5 | 0.3 | 0.2 | –0.7 | 1.7 | 0.3 |
| Growth rate of money supply, M2 | constant | 9.0 | 6.5 | 9.0 | 11.0 | 11.0 | 16.0 | 9.0 | 7.5 | 12.0 | 13.0 |
| Depreciation rate, average[a] | | 5.95 | 3.90 | 5.86 | 4.51 | 4.81 | 3.38 | 2.69 | 2.77 | 5.00 | 2.73 |
| Rate of (Hicks-neutral) technical progres[b] | optimistic | 0.85 | 2.80 | 3.20 | 3.00 | 1.20 | 3.15 | 1.75 | 2.65 | 1.00 | 2.25 |
| | medium | 0.55 | 2.25 | 2.45 | 2.45 | 0.95 | 2.00 | 0.95 | 2.20 | 0.45 | 1.60 |
| | pessimistic | 0.30 | 1.75 | 1.75 | 2.00 | 0.65 | 1.20 | 0.25 | 1.90 | 0.00 | 1.00 |
| Gross investment ratio (1985–1999 average)[c] | optimistic | 21.13 | 22.74 | 35.11 | 23.82 | 20.49 | 21.60 | 20.98 | 21.05 | 24.09 | 26.15 |
| | medium | 18.78 | 20.08 | 30.99 | 20.95 | 18.14 | 19.05 | 18.48 | 18.49 | 21.17 | 22.60 |
| | pessimistic | 16.44 | 17.42 | 26.86 | 18.13 | 15.79 | 16.54 | 15.98 | 15.83 | 18.25 | 19.11 |

[a]This rate is not constant; see Appendix 2C.
[b]In Appendix 2E, these rates are also given for the Harrod-neutral concept.
[c]The rates are not constant; see Appendix 2F.

*Table 2.2.* Definition of scenarios for the driving forces of economic development, 1986–2000: CMEA countries.

| *Indicator* (%) | *Scenario* | *USSR* | *Bulgaria* | *CSSR* | *GDR* | *Hungary* | *Poland* | *Romania* |
|---|---|---|---|---|---|---|---|---|
| Growth rate of employment in the material sphere | – | 0.0 | 0.05 | 0.00–0.36[a] | 0.33 | 0.00–−0.17[a] | 0.00–0.48[a] | 0.00–−0.08[a] |
| Growth rate of nominal wages | – | 5.5 | 5.0 | 4.0 | 4.0 | 8.5 | 12[b] 10[c] | 7.5 |
| Depreciation rate[d] | constant | 1.35 | 2.35 | 1.40 | 0.90 | 1.20 | 0.90 | 1.70 |
| Rate of (Hicks-neutral) technical progress | optimistic | 1.8 | 2.1 | 1.7 | 2.1 | 1.8 | 2.0 | 2.3 |
| | medium | 1.1 | 1.5 | 0.9 | 1.6 | 0.9 | 1.3 | 1.5 |
| | pessimistic | 0.4 | 0.9 | 0.2 | 1.2 | 0.2 | 0.6 | 0.6 |
| Gross investment ratio (1985–1999 average | optimistic | 29 | 32 | 29 | 27 | 28 | 30 | 31 |
| | medium | 27 | 30 | 27 | 25 | 26 | 28 | 29 |
| | pessimistic | 25 | 28 | 25 | 23 | 24 | 26 | 27 |

[a]These rates change over time; see Appendix 2A.
[b]For 1986–1989.
[c]For 1990–1999.
[d]Defined as the rate of "sorting out" capital.

*Table 2.3.* Definition of scenarios for the driving forces of economic development, 1982–2000: developing countries.

| *Indicator* (%) | *Scenario* | *Oil-exporting countries* | *Asian countries without India* | *India* | *Developing African countries* | *Developing Latin Amer. countries*[a] | *Mexico, Brazil, Argentina* | *North African, Middle East countries* |
|---|---|---|---|---|---|---|---|---|
| Growth rate of economically active population (1982–1999 average)[b] | - | 2.46 | 2.16 | 1.77 | 2.72 | 2.72 | 2.66 | 2.85 |
| Growth rate of money supply (M2) | - | est.[c] | | 18 | 15 | 20 | 45 | 45 |
| Depreciation rate | constant | 2.5 | 3.3 | 1.7 | 2.3 | 3.2 | 1.4 | 1.5 |
| Rate of Hicks-neutral technical progress (1985–1999 average)[d] | optimistic | 1.79 | 1.91 | 0.97 | 0.90 | 1.19 | 1.97 | 2.22 |
| | medium | 0.75 | 1.26 | 0.50 | 0.38 | 1.00 | 1.68 | 2.00 |
| | pessimistic | –0.23 | 0.74 | 0.50 | –0.10 | 1.00 | 1.44 | 2.00 |
| Gross investment ratio (1985–1999 average) | optimistic | 29.77 | 28.56 | 25.35 | 23.78 | 18.63 | 23.91 | 25.82 |
| | medium | 24.77 | 24.78 | 22.20 | 21.84 | 15.89 | 20.61 | 21.75 |
| | pessimistic | 20.21 | 21.25 | 19.31 | 16.54 | 11.49 | 17.54 | 18.07 |

[a] Excluding Mexico, Brazil and Argentina.
[b] These growth rates are not constant; see Appendix 2A.
[c] However, the rate of change of OECD dollar import prices of mineral oil and fuels is 12.8% (1982–1987), 2.3% (1988–1993), and 4.4% (1994–1999). See also Appendix 2B.
[d] Most of these rates are not constant; see Appendix 2E.
[e] These ratios are not constant. For details, see Appendix 2F. For illustration, see *Figure 2.9.*

*2.2–2.3*. Of course, the behavior of the government and of the monetary authorities may change. This is a political problem. Our forecasts are made on the "behavior as usual" assumption.[4] For the oil-exporting countries the *dollar price of mineral oil and fuels* is of special importance. We estimated it as the dollar import price of mineral oil and fuels of OECD countries (see Appendix 2B), in accordance with forecasts of UNCTAD. For the *CMEA countries* the rate of growth of the nominal wage rate plays a role similar to the money supply in OECD countries. It was forecast on the base of past experience; see Appendix 2B and *Table 2.2*.

The *rate of depreciation* for OECD countries follows a "back to normality" formula (see Annex 2, Appendix 2C and *Table 2.1*). For the other countries this rate was estimated exogenously on the base of past observations; see *Tables 2.2* and *2.3*.

For the other most important driving forces of economic development,[5] the rate of technical progress and the investment ratio, *three different scenarios* have been defined which span the most likely range of possible future developments. Which scenario will be realized in each country depends on whether there are able governments and other social forces which motivate the society such that the economic activity increases or whether a spirit of laziness or fear spreads or even social unrest, revolution or war. Of course, this will be different in different countries.

We shall present estimations of the future development based upon an "optimistic", a "medium" and a "pessimistic" scenario with respect to the future development of these two driving forces. Following Murphy's law ("What can go wrong will go wrong") we should perhaps concentrate on the pessimistic scenario. Of course, this scenario is not really as "pessimistic" as it might be, since revolutions, wars, and other political disturbances or catastrophes are not considered.

The general assumption underlying all scenarios is demonstrated in *Figure 2.1*: there will be a turn of the tide. The negative trend experienced in the 1970s will not continue. The three scenarios differ in the *rate of technical progress* and in the *investment ratio*. As to the *rate of technical progress* we assume for the optimistic scenario that its average in the future will be approximately the same as the average from 1961–1984, for the medium scenario approximately the same as the average from 1971–1984 and for the pessimistic scenarios approximately as the average from 1976–1984.[6] *Tables 2.1–2.3* show these rates in the Hicks-neutral concept. In Appendix 2E we present the average of technical progress in the past (in the Hicks-neutral as well as in the Harrod-neutral concept) for OECD countries. For CMEA and developing countries the same procedure has been used. As to the *investment ratios*, a "back to normal" formula was used for the *OECD countries*; see *Annex 2*. The results are shown in *Table 2.1* and Appendix 2F. For *CMEA countries* the average performance in the past was taken as the medium forecast. In the optimistic scenario this ratio was increased by two percentage points, in the pessimistic scenario it was lowered by the same percentage. For *developing countries* a variant of the method used for the OECD countries was applied. (For details, see Chapter 6.)

## 2.4. Assumptions to Make the Productive Performance of OECD and CMEA Countries Comparable

All figures used in this project are based on the official statistics. They may not be comparable from country to country. This is especially important in comparing figures for CMEA countries with those of OECD countries. We made the results approximately comparable, first, by using conversion factors (shown in *Table 2.4*) to change the NMP figures of CMEA countries in domestic currency to those in dollars.

*Table 2.4.* Conversion factors for CMEA currencies.

| *Country* | *Currency* | *To 1975 US $* | *GDP/NMP 1975* | *NMP/growth rate GDP* |
|---|---|---|---|---|
| USSR | Ruble (1973) | 0.518 | 1.29 | 1.3 |
| Bulgaria | Leva (1980) | 0.689 | 1.32 | 1.8 |
| CSSR | Koruna (1977) | 7.86 | 1.19 | 1.5 |
| GDR | Mark (1980) | 2.78 | 1.26 | 1.5 |
| Hungary | Forint (1976) | 14.0 | 1.24 | 1.5 |
| Poland | Zloty (1982) | 55.8 | 1.28 | 1.3 |
| Romania | Leu (1981) | 8.99 | 1.20 | 1.8 |

But this was not sufficient. Since the national accounting system used in CMEA countries (which is based on the NMP concept) is rather different from that of OECD countries (which is based on the GDP concept), we had to use GDP/NMP conversion factors as well as conversion factors for NMP/GDP growth rates to make the growth paths of OECD and CMEA countries comparable. It is now generally acknowledged that the differences in level and in growth rates between OECD and CMEA countries partially follow from the different national accounting methods (GDP *versus* NMP) and from different statistical procedures. Such differences must exist because otherwise (taking into account the much higher growth rates of NMP in the CMEA countries compared to the generally much lower growth rates of GDP of the OECD countries) the CMEA countries should long have passed the OECD countries in standard of living, which has not happened. There have been different suggestions from CMEA and OECD economists for solving this problem. The World Bank established a research group which carefully examined the problem on the basis of available information. The results were published in Marer (1985). Basically, we rely on these results, but combine them with other results published in CMEA countries; for details, see the Appendix to Chapter 30.

To compare the economic development of OECD and CMEA countries, we first converted the NMP figures to GDP figures for one base year (1975) by using the conversion factors suggested by Marer (1985, pp. 18–19) for the year 1980 (see *Table 2.4*).

Thus, we calculated GDP for the year 1975 for all CMEA countries. By using the conversion factors for the currencies (see *Table 2.4*) and applying the official growth rates of NMP to these GDP figures we got GDP "type 1" (in $) for all CMEA countries. Thus the growth rates of NMP and GDP "type 1" are equal. This may constitute an upper limit to an estimate of economic performance of the CMEA countries.

But, as pointed out above, the NMP growth rates are too high to be applied to GDP. Therefore we reduced the official NMP growth rates to equivalent GDP growth rates on the basis of Marer's estimate (see Marer, 1985, pp. 184–185); *Table 2.4* and the Appendix to Chapter 30.

Application of these conversion factors to the official NMP growth rates and taking the GDP figure (in $) for 1975 as estimated above, we arrive at time series for GDP "type 2". This may be considered as a lower bound to CMEA economic performance.

## 2.5. Graphical Representation of the Main Assumptions

In order to show how our main assumptions are related to other observations in the past we represent some of these figures graphically. *Figures 2.2 and 2.3* represent employment figures (working hours for OECD countries, persons employed for CMEA countries) and the trend forecasts. They remain unchanged in all three scenarios. *Figures 2.4–2.6* show the optimistic, medium and pessimistic trend forecasts for the rate of technical progress of some selected OECD, CMEA and developing countries. For the other countries the graphs look similar. *Figures 2.7–2.9* show the same for the investment ratio. In our opinion, these developments of the main driving forces of economic growth cover the range of what might be possible in the future under the general assumption that there is some "turn of the tide", as was indicated in *Figure 2.1*.

## 2.6. The Main Results of the Project

The main results are given in *Tables 2–38* of Annex 3. In *Table* 1(*a–c*) of that annex some basic figures from other sources are reproduced. We comment on all of them. The following discussion refers consistently to tables appearing in Annex 3 of this volume.

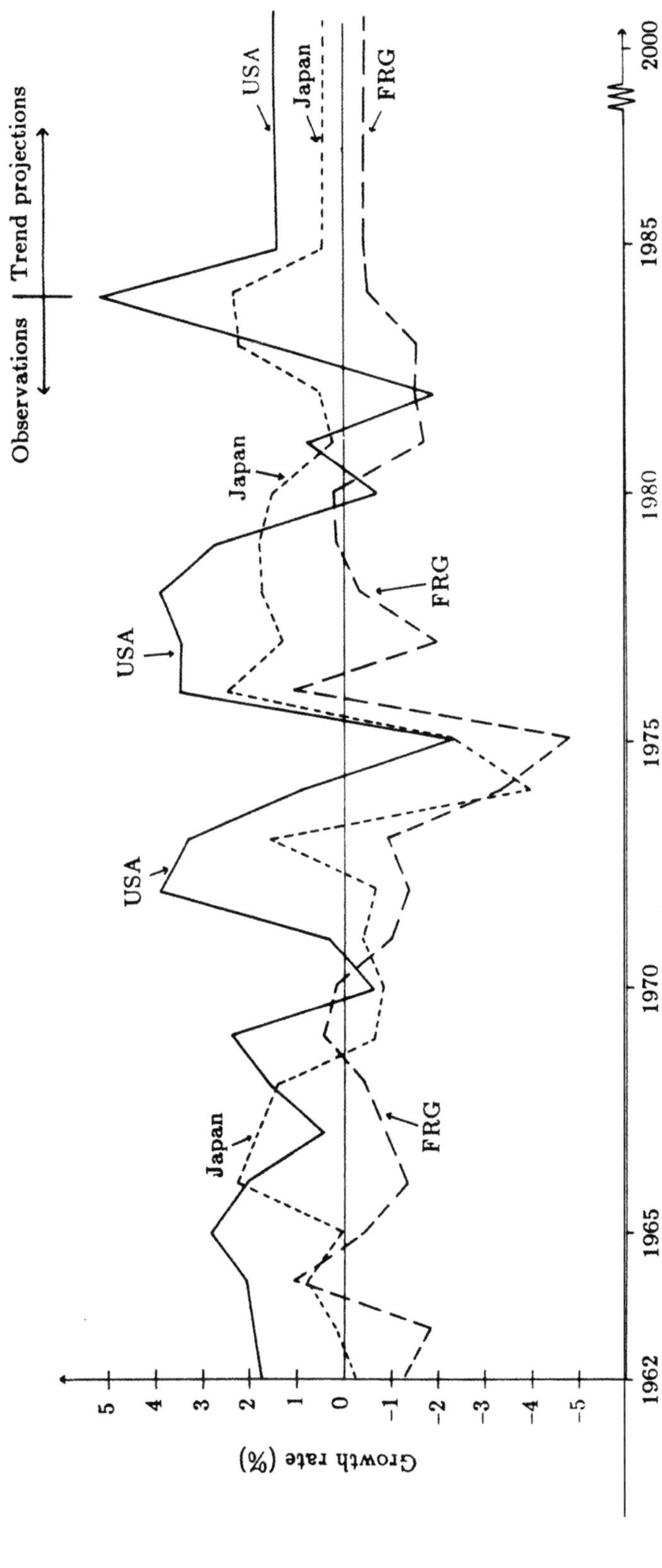


*Figure 2.2.* Growth rates of working hours: OECD countries.

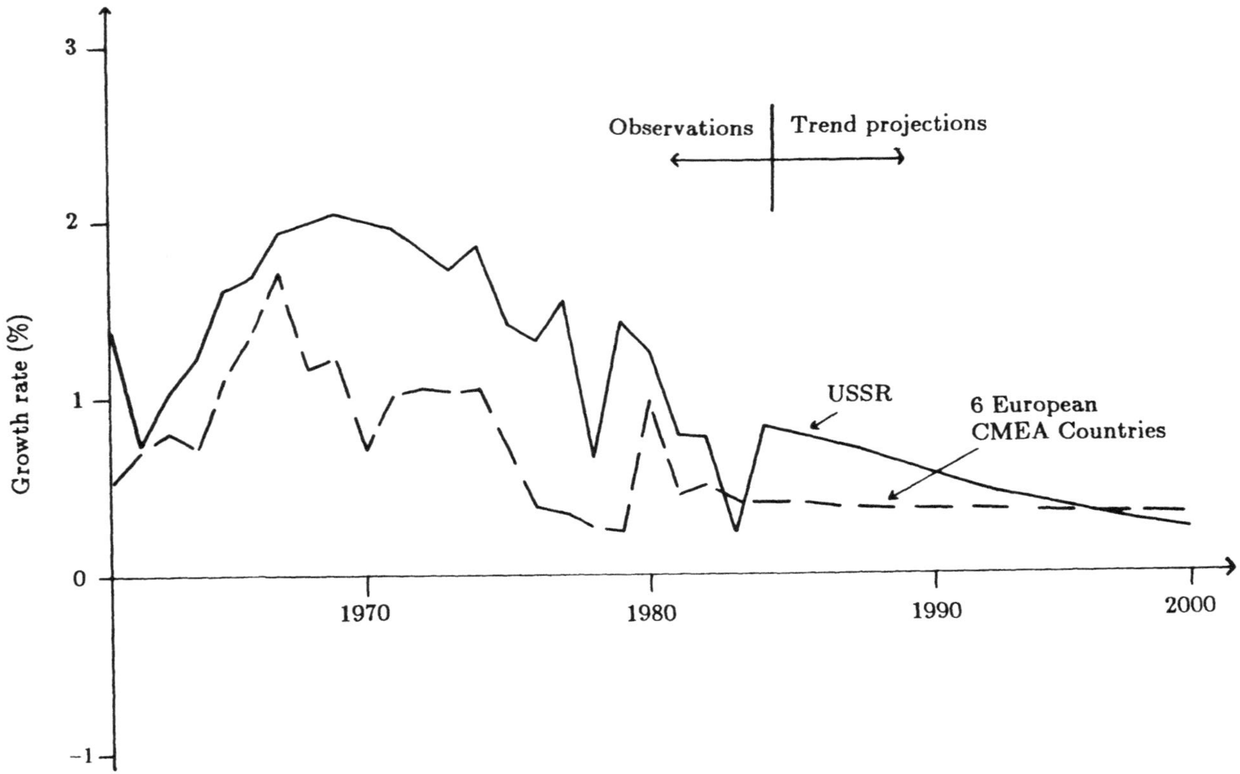


*Figure 2.3.* Growth rate of employed labor force: CMEA countries. Note that this graph shows the growth rates of *total* employment (in number of persons), not only of those employed in the material sphere. The latter are reproduced in Appendix 2A. The forecasts are based on trend extrapolations and outside information.

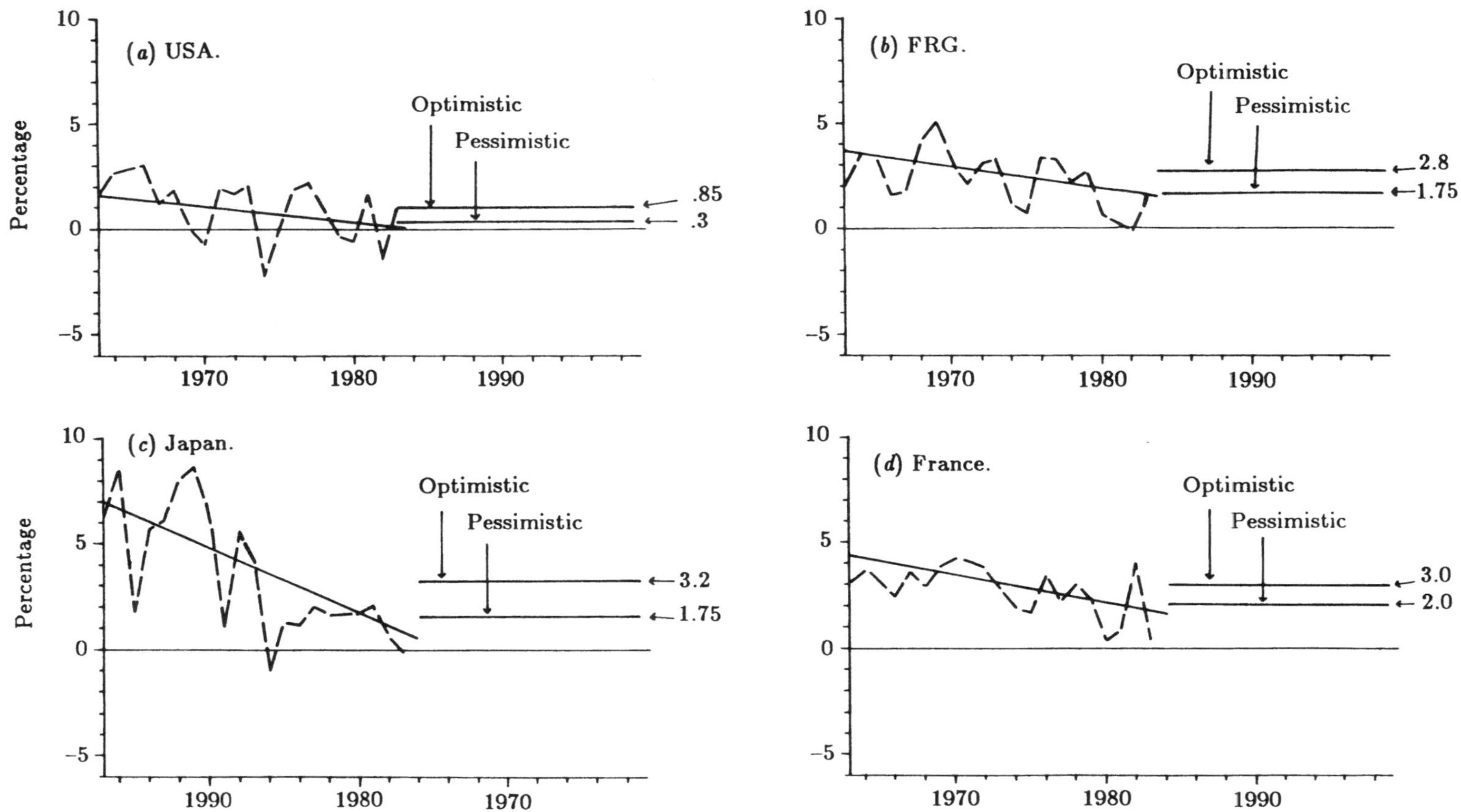


*Figure 2.4.* Rate of (Hicks-neutral) technical progress: selected OECD countries. Linear trends are shown by the solid lines; actual rates of technical progress are represented by hatched lines. The forecasts for the medium scenario lie between the optimistic and pessimistic scenarios and are not reproduced here for graphical reasons.

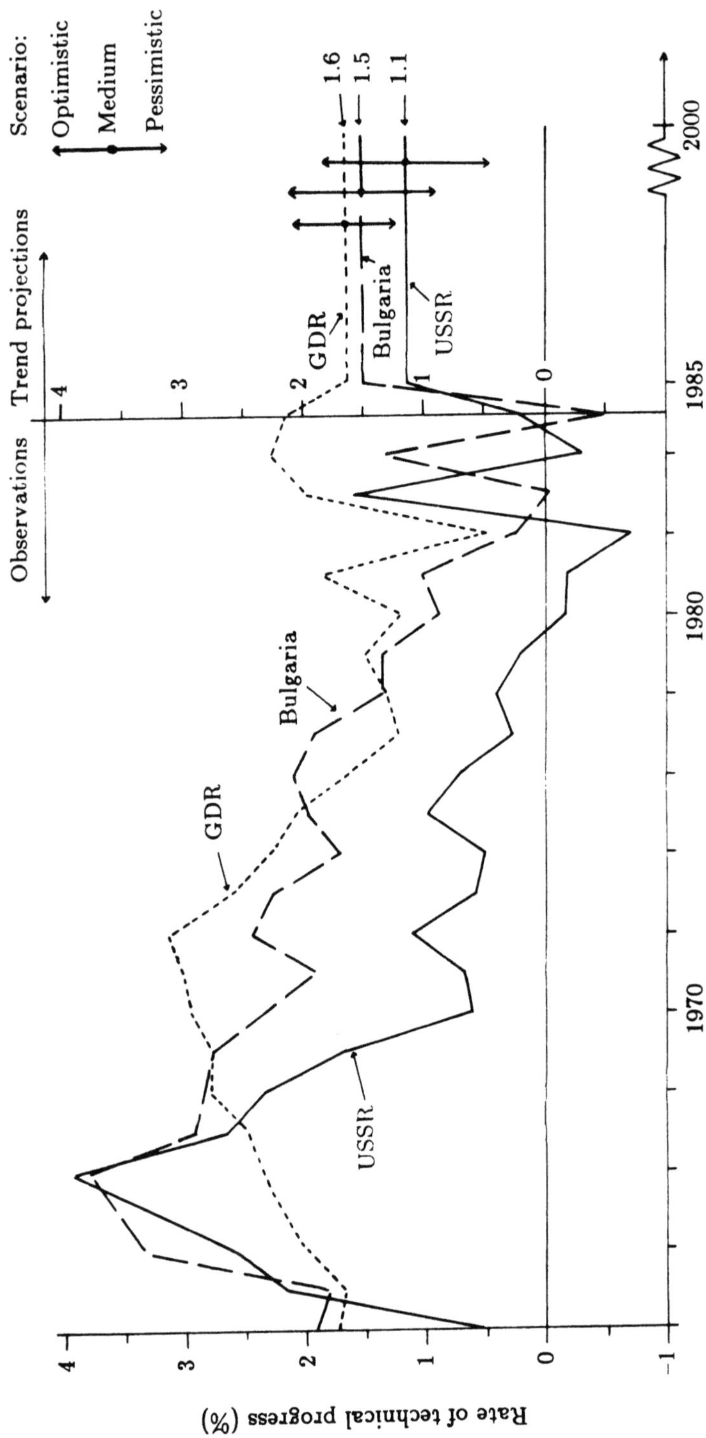


***Figure 2.5.*** Rate of (Hicks-neutral) technical progress: CMEA countries.

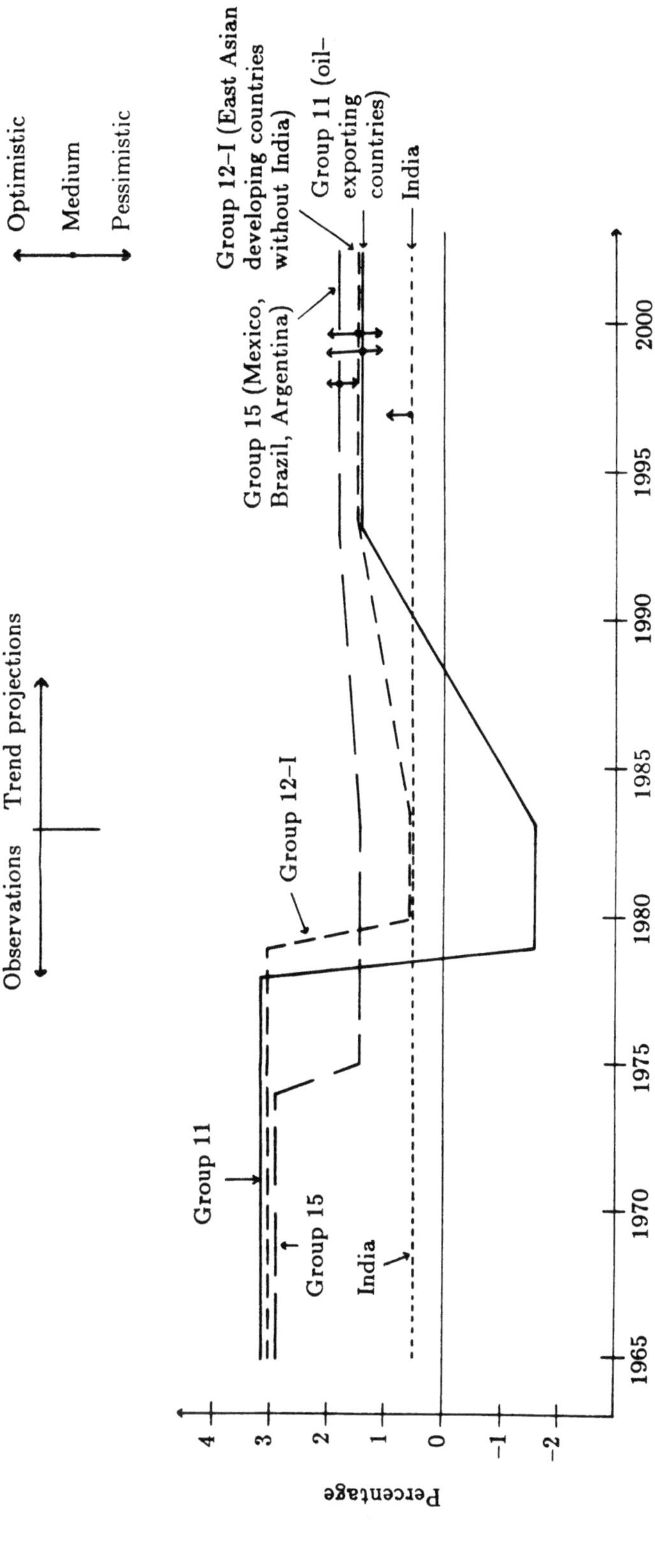


***Figure 2.6.*** **Growth rates of technical progress (averages): developing countries.**

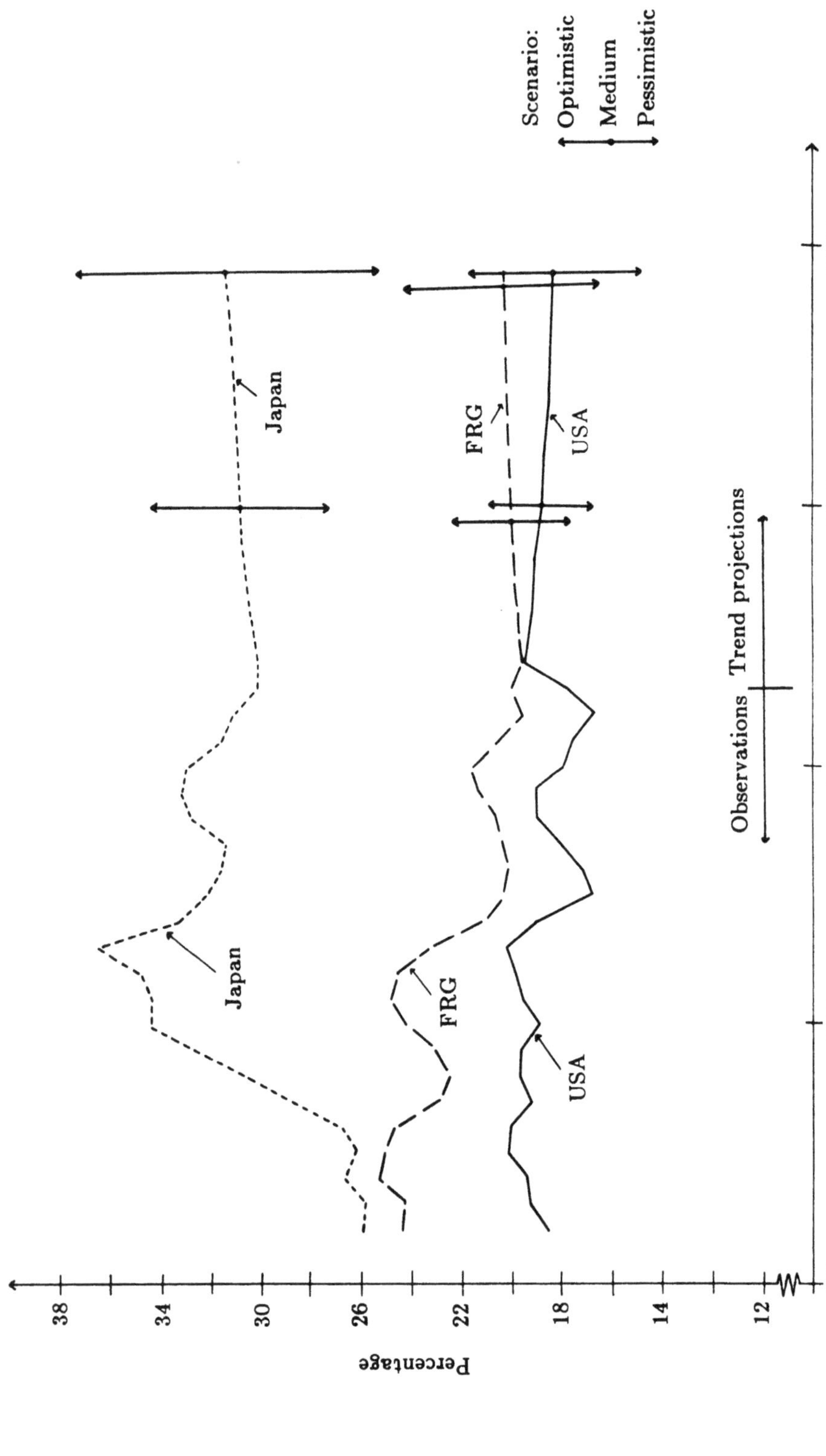


*Figure 2.7.* Gross investment ratio: selected OECD countries.

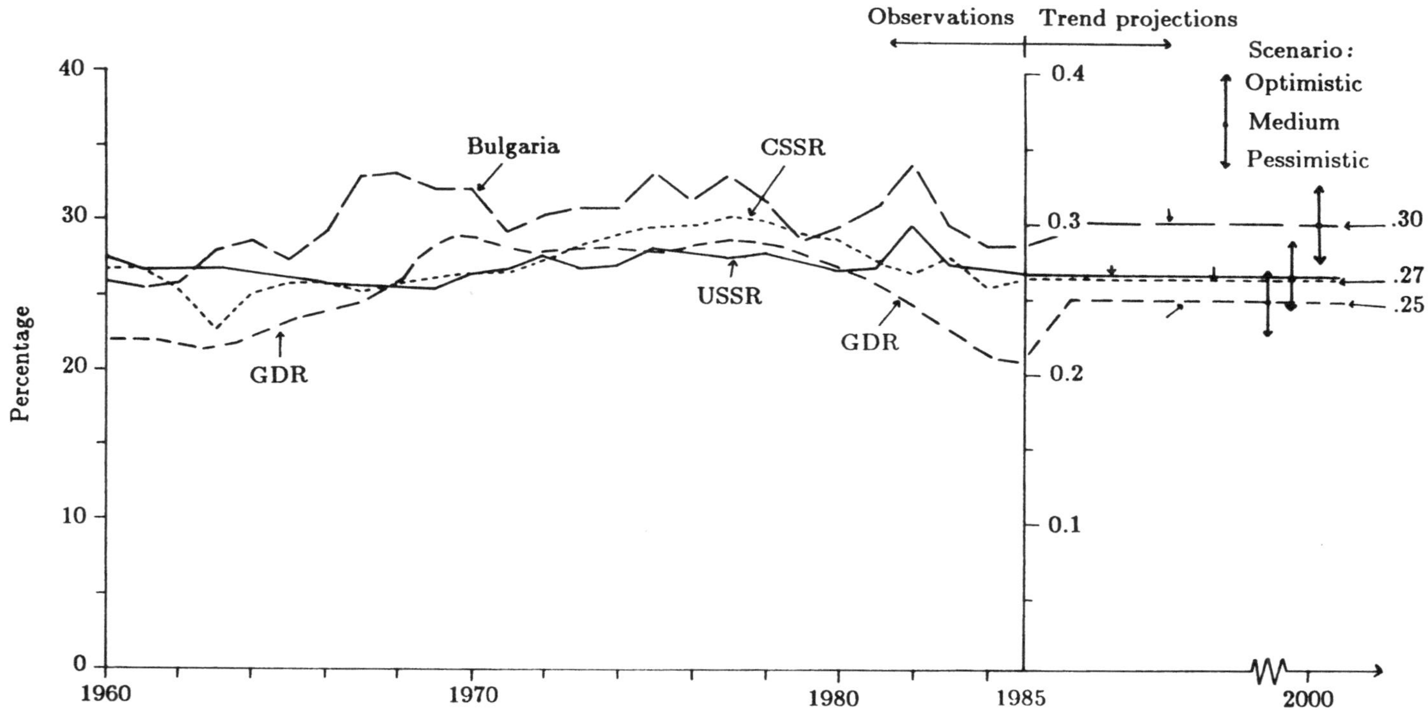


*Figure 2.8.* Gross investment ratio: selected CMEA countries.

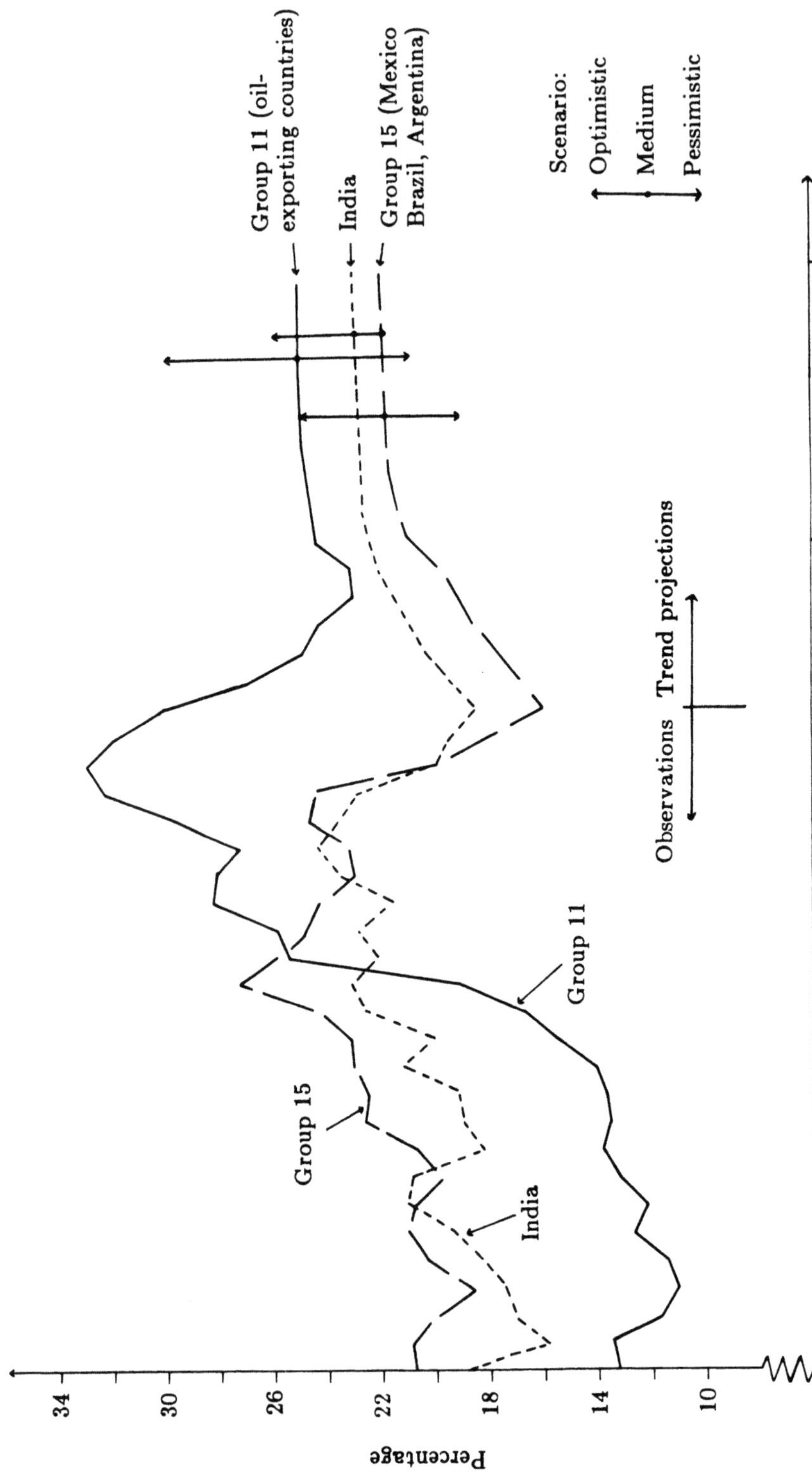


*Figure 2.9.* Gross investment ratio: selected developing countries.

### 2.6.1. Population and employment

Until the year 2000 the proportion of world population living in industrialized market economies will shrink from 20.9% in 1960 to 14.7%. The proportion of world population living in European CMEA countries (including the USSR) will shrink from 10.7% to 7.1% in this time span; see *Table 1(a)*. This means that almost 80% of the world population in the year 2000 will live in what is now called developing countries, whereas in 1960 only about 70% were living there. It also means that the pressure on redistribution of income of the world level will increase.

The displacements of employed labor forces are a bit smaller, but not fundamentally; see *Table 1(b)*. Looking at the average growth rates of the employed labor force is more informative; see *Table 1(c)*. It is striking that the OECD countries will experience a modest increase in employment of about 1.3%, whereas the labor force in the European CMEA countries stays almost constant. The largest increases (of 2 to 2.4%) are in the developing countries, of course.

### 2.6.2. Growth rates of real GDP

Annex *Table 2* shows the real GDP growth rates for the past and the forecasts until 1999 for *OECD countries* for the *medium scenario*. The trend is almost constant between 2 and 3% with the exceptions of Japan, where the growth rate lies above 4% but is declining, and Great Britain (UK), where this rate is only just above 1%. The FRG, France, Italy and Belgium/Luxembourg are close together (with growth rates between 2.8 and 2.9%). The US growth rate is around 2.2%. In the *optimistic scenario* the growth rates of the continental European OECD countries are almost 1% higher, for Japan even 1.3–1.5%, but for the USA and UK only around 0.5%. In the *pessimistic scenario* the growth rates for the continental European OECD countries and for Japan are 1–1.3% lower, for the USA and UK only around 0.6%. Thus the variance of the growth paths is lower for the slower-growing than for the faster-growing countries. All scenarios show that the US economy advances at a lower rate than almost all other OECD economies. The relative importance of the US economy declines. We shall come back to that point later. This, of course, does not mean the most advanced technologies might not be developed in the USA, though the chances that the other OECD countries will equal or pass the USA in this respect are rising. *Figure 2.10* shows the growth rates of some OECD countries graphically.

Annex *Table 3* presents the equivalent figures for *CMEA countries* with respect to the growth rates of real NMP (which is identical with the growth rates of real GDP "type 1"). In the medium scenario these growth rates lie mostly between 2.9 and 3.7%. They are slightly decreasing. For Bulgaria, the GDR and (astonishingly) for Romania they lie between 4.5 and 5%. Hungary and the CSSR stay behind. But these growth rates are based on the domestic NMP figures, which are not directly comparable with GDP figures and might not be fully comparable between the CMEA countries themselves. *Table 4* shows the growth rates of GDP "type 2" for CMEA countries where adjustments have been

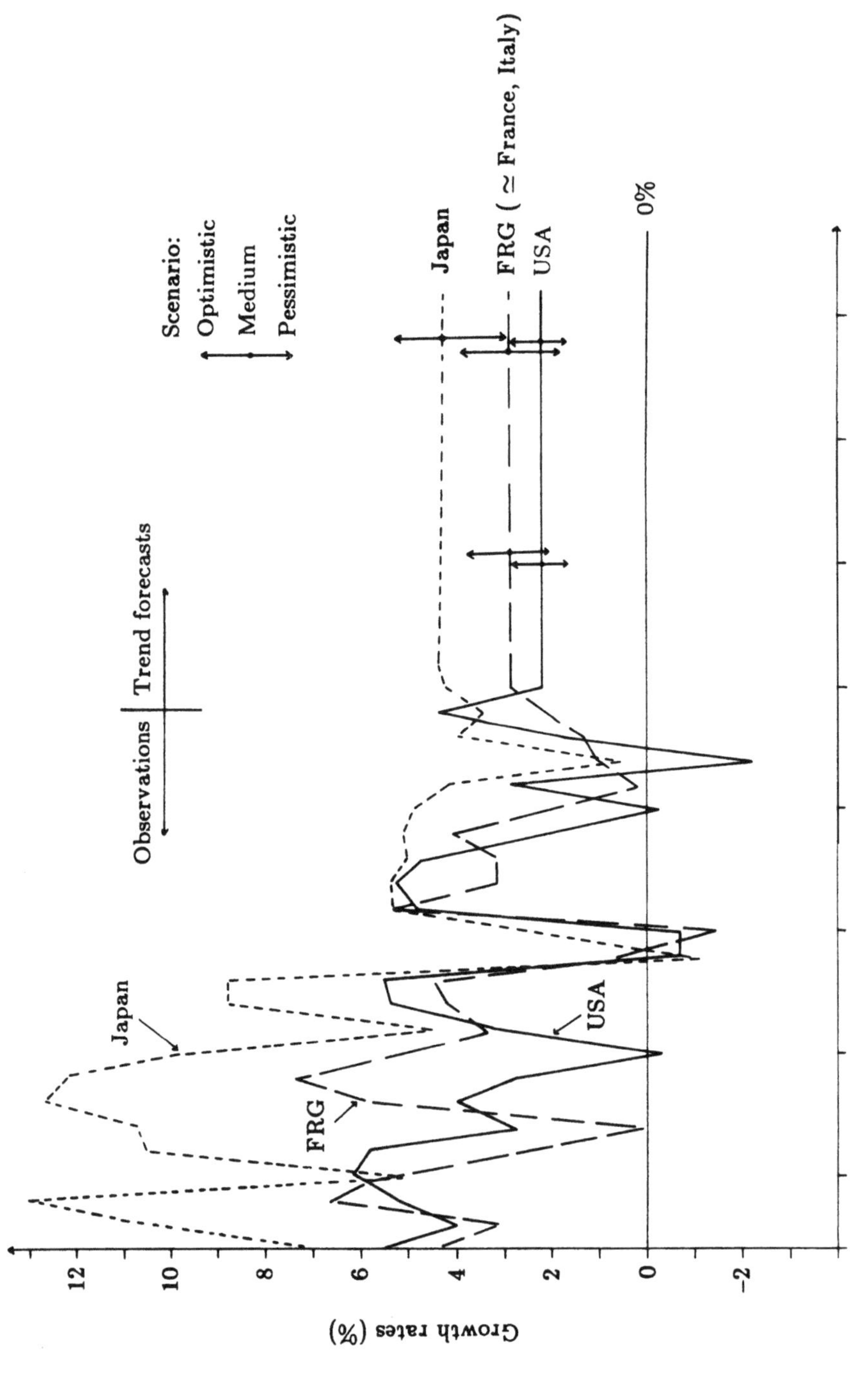


*Figure 2.10.* Growth rates of real GDP: selected OECD countries.

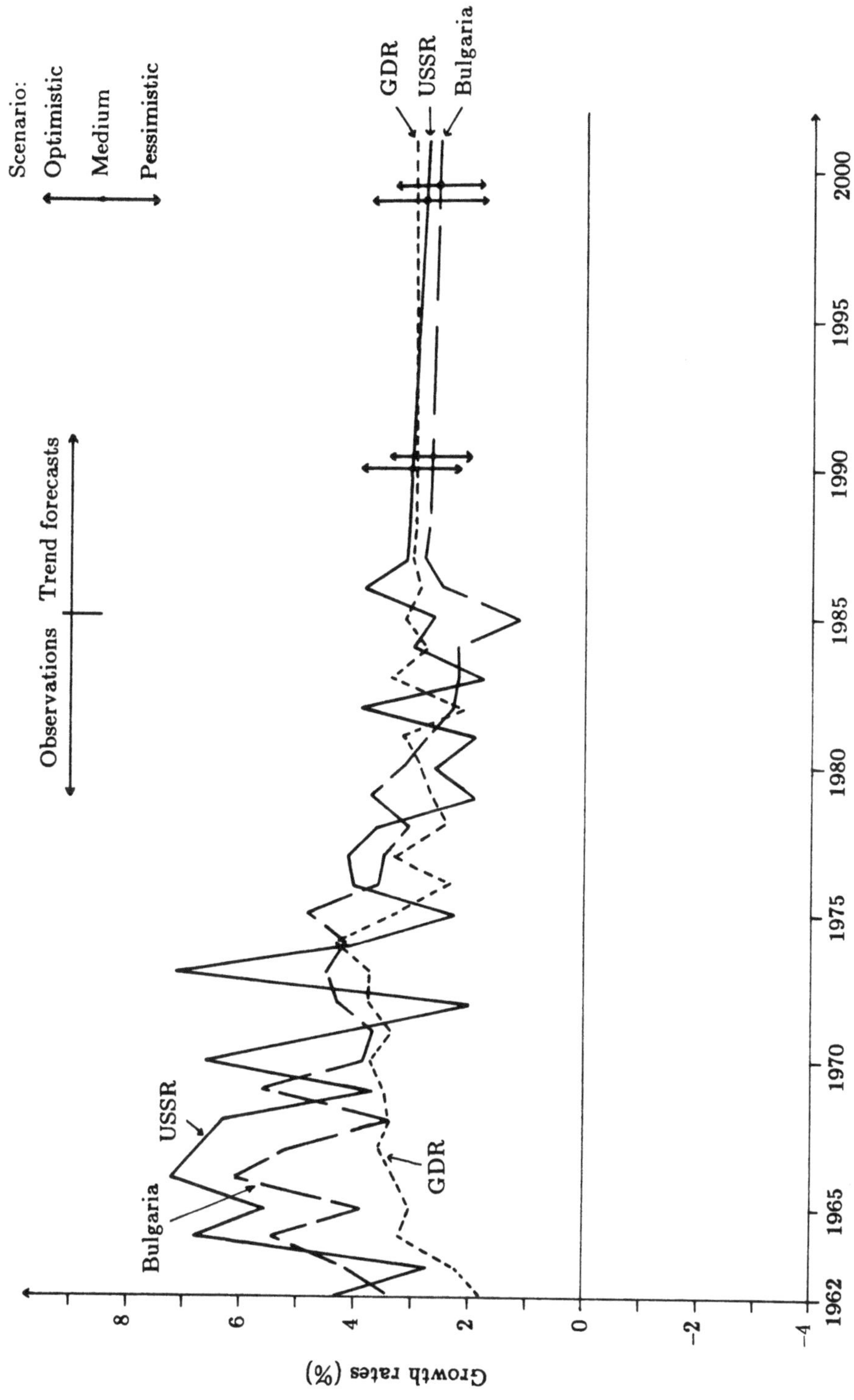


*Figure 2.11.* Growth rates of real GDP "type 2": selected CMEA countries.

made in order to come nearer to comparability with the OECD countries. Now the GDR comes out best. All CMEA countries lie rather near together (growth rates of about 2.6–2.9%, which are similar to those of the OECD countries). Also, after this adaptation, the CSSR and Hungary stay behind. *Figure 2.11* shows these results for some CMEA countries.

The growth rates of *developing countries* are higher; see Annex *Table 5*. They lie between 3.4 and 6.5% in the medium scenario. But this is mostly due to the growth of the labor force and does not indicate an adjustment of the standard of living. Mexico, Brazil, Argentina, the Middle East and North Africa are doing well whereas India and Black Africa stay behind. The pessimistic scenario does not reduce this growth too much since the main driving force, population growth, is not influenced by it.

It is interesting to compare the growth rates of the GDP of the world regions – OECD, CMEA and developing countries – and to estimate the growth rate of world GDP; see Annex *Table 6* and *Figure 2.12*. In the medium scenario, the growth rates of the OECD and the CMEA group are approximately equal. This means that the comparative standards of living would stay approximately the same as they are now. The roughly 2% higher growth rates of developing countries must be seen against the background of roughly 2% higher growth rates of population and labor force in these countries; see *Table 1(c)*. Thus, the standard of living of the developing countries as a whole will not improve more than that of the developed countries, if all regions follow the medium scenario. Of course, this need not be the case. But if the pessimistic scenario were to come true for the developing countries and the medium scenario for the developed ones, there will be a fundamental deterioration of the distribution of world income.

### 2.6.3. GDP per capita

Annex *Table 7* shows the forecasts of GDP per capita (representing standard of living) for the *OECD countries*. The USA will be passed by several OECD countries, notably Japan, the FRG, France, Belgium/Luxembourg. The UK and Italy will stay behind. *Figure 2.13* illustrates this for some OECD countries. Of course, for one country the pessimistic scenario may come true (e.g., for the FRG), whereas for others (e.g., France) the optimistic one may prevail.

Annex *Table 8* shows *GDP "type 1" per capita for the CMEA countries*. Bulgaria as well as the GDR will enjoy a much higher standard of living than the USSR. Poland stands at the end. Its standard of living will be only half of that of the GDR and about two-thirds that of the USSR.

If one takes GDP "type 1" as a comparable measure of income in the CMEA countries, Bulgaria and the GDR will pass the standard of living of the USA before the year 2000, whereas the USSR will stay at about three-quarters of it. Hungary passes the level of the UK. This seems unlikely.

If one takes GDP "type 2" as the appropriate measure (see Annex *Table 9*), the GDR comes out best among the CMEA countries. In the medium scenario it passes the level of the UK and reaches about 60% of the standard of living of the

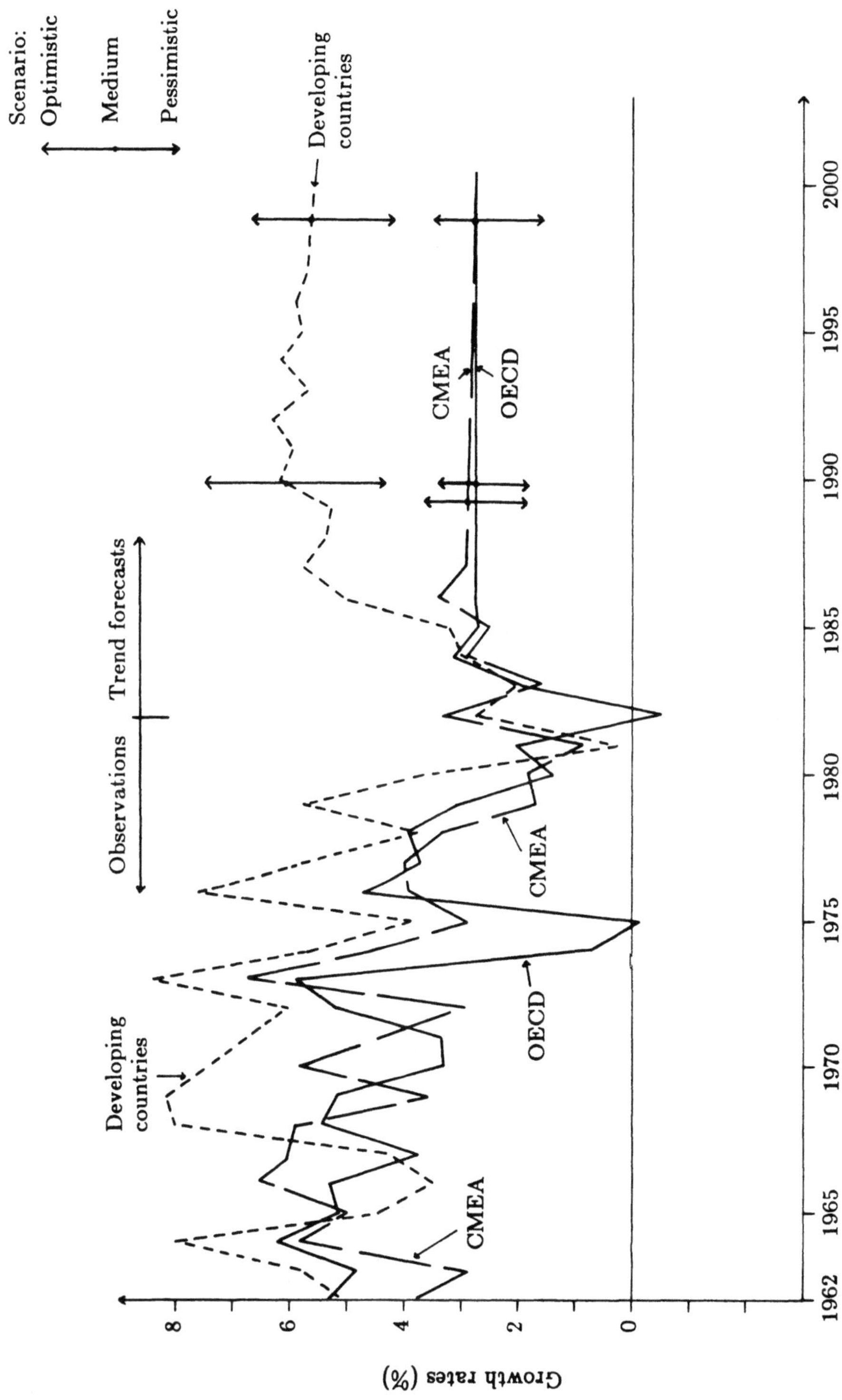


*Figure 2.12.* Growth rates of real GDP (for CMEA countries – GDP "type 2"): world regions.

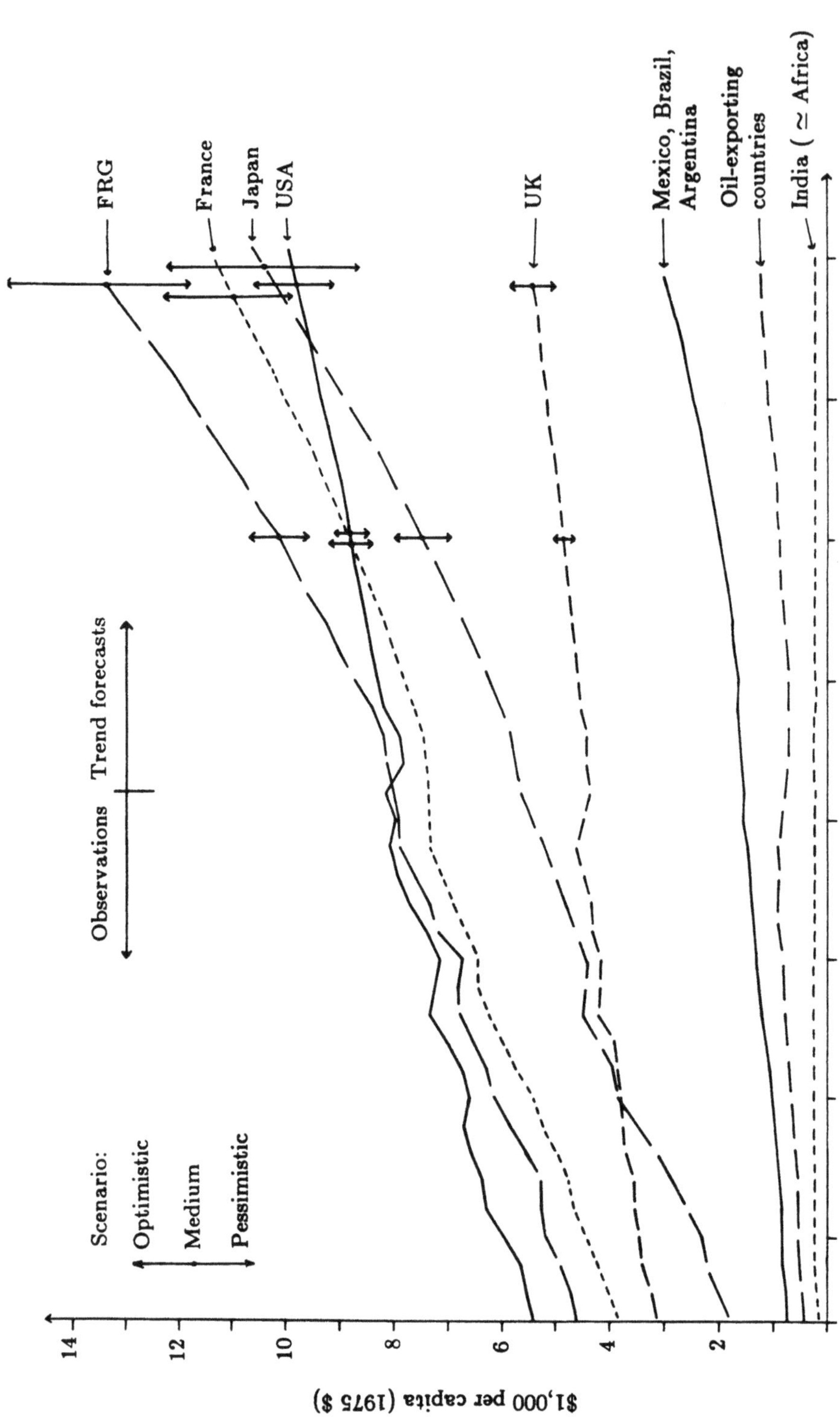


*Figure 2.13.* GDP per capita: selected OECD and developing countries.

FRG. The standard of living in the USSR will also pass that of the UK, but reaches only about 60% of that of the USA. ***Figure 2.14*** shows the development of GDP "type 2" per capita for some CMEA countries. For comparison's sake the GDP per capita of the USA is also reproduced.

The ***developing countries*** are also making headway, but stay far behind the OECD and the CMEA countries; see Annex ***Table 10*** and ***Figure 2.13***. In the year 2000 Mexico, Brazil and Argentina will almost reach the standard of living enjoyed by the UK at the beginning of the 1960s. India and Black Africa continue to be the poorhouses of the world. The oil-exporting countries are also on the lower side. Their special position is gone. Thus income distribution on the world scale becomes more unequal in the period considered. This is also visible in Annex *Table 11*, where the GDP per capita of the world regions is reproduced. In the medium scenario in 1999 the OECD countries reach a GDP per capita of around $9,000, the CMEA countries (in GDP "type 2") around $5,800 and the developing countries around $1,000.

### 2.6.4. GDP per employed person or labor productivity

Labor productivity should be measured by GDP per working hour. Unfortunately, these figures are available only for OECD countries. For comparison's sake, we approximate that measure by GDP per employed person. Since the average working hours per employed person and per year are declining in OECD countries, this measure understates the "true" labor productivity. Annex *Table 12* shows the labor productivity in this sense. The USA will be passed by many OECD countries, with the notable exceptions of the UK and Italy. Japan's labor productivity increases fast, but is still lower than those of the FRG and France. This, of course, refers to the average. In the most advanced industries it may very well be the other way round. ***Figure 2.15*** illustrates these results.

For *CMEA countries* we again have to differ between GDP "type 1" and GDP "type 2". Measured in terms of GDP "type 1" (see Annex *Table 13*) labor productivity of Bulgaria and the GDR will reach or pass that of the USA around the year 2000, whereas the USSR will stay at about three-quarters of the labor productivity in the USA in the medium scenario. If we take GDP "type 2" as the appropriate measure (see Annex *Table 14*), the GDR (with the highest labor productivity in the CMEA countries) will stay at about half of the labor productivity of the FRG and about three-quarters of the labor productivity of the USA around the year 2000. The USSR will then have about 65% of the labor productivity of the USA. ***Figure 2.16*** illustrates this development.

For *developing countries* we take GDP per economically active person as a measure of labor productivity (see Annex *Table 15*). This is not directly comparable to GDP per employed person. But if we make this comparison nevertheless, we see that the most advanced countries in this group, namely, Mexico, Brazil and Argentina, will in the year 2000 in the medium scenario reach the labor productivity of Japan at the beginning of the 1970s. This is about 40% of the labor productivity of the USA at the end of this century. India and Black Africa are also far behind in this respect. The other countries are catching up

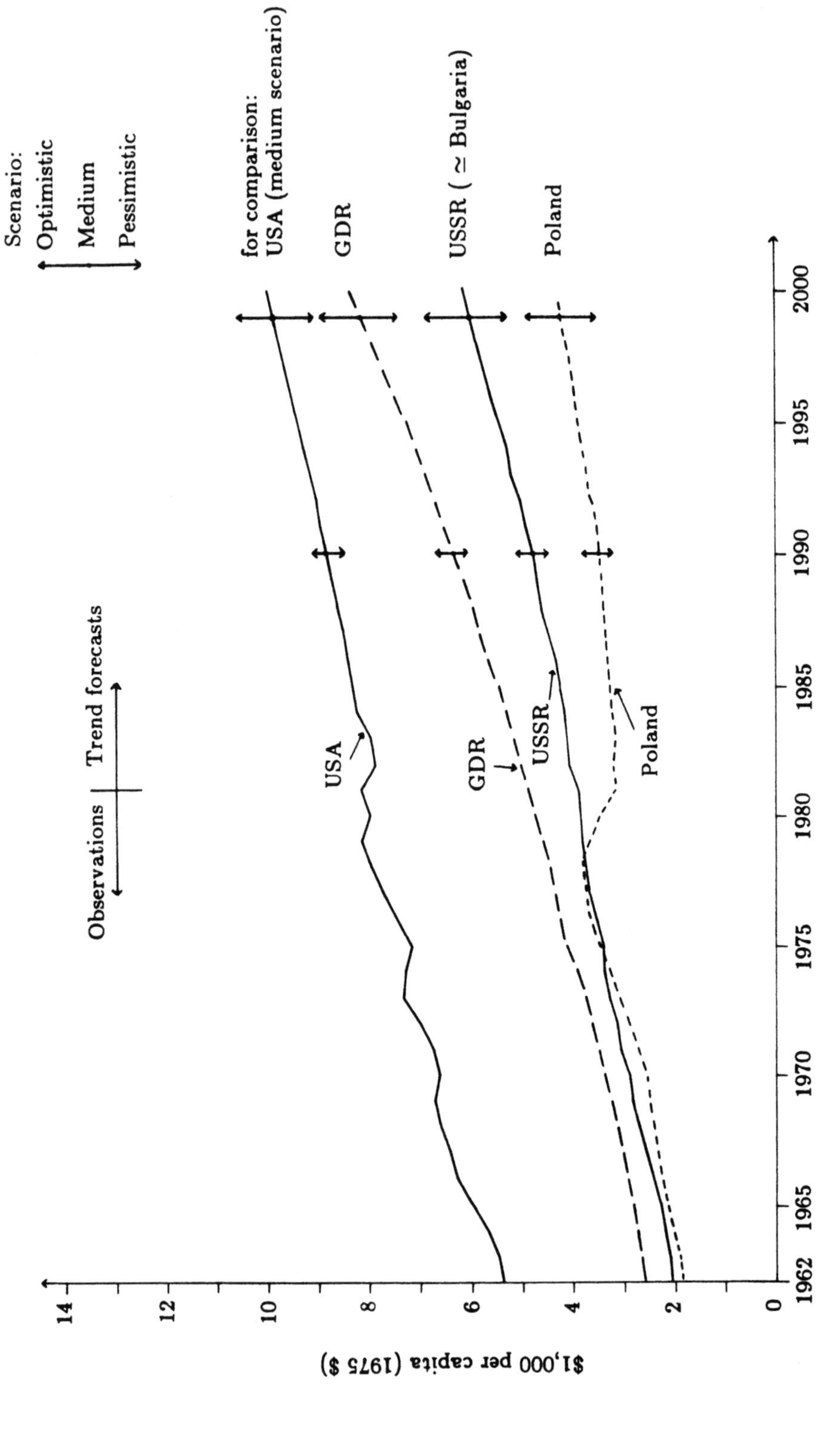


*Figure 2.14.* GDP "type 2": selected CMEA countries.

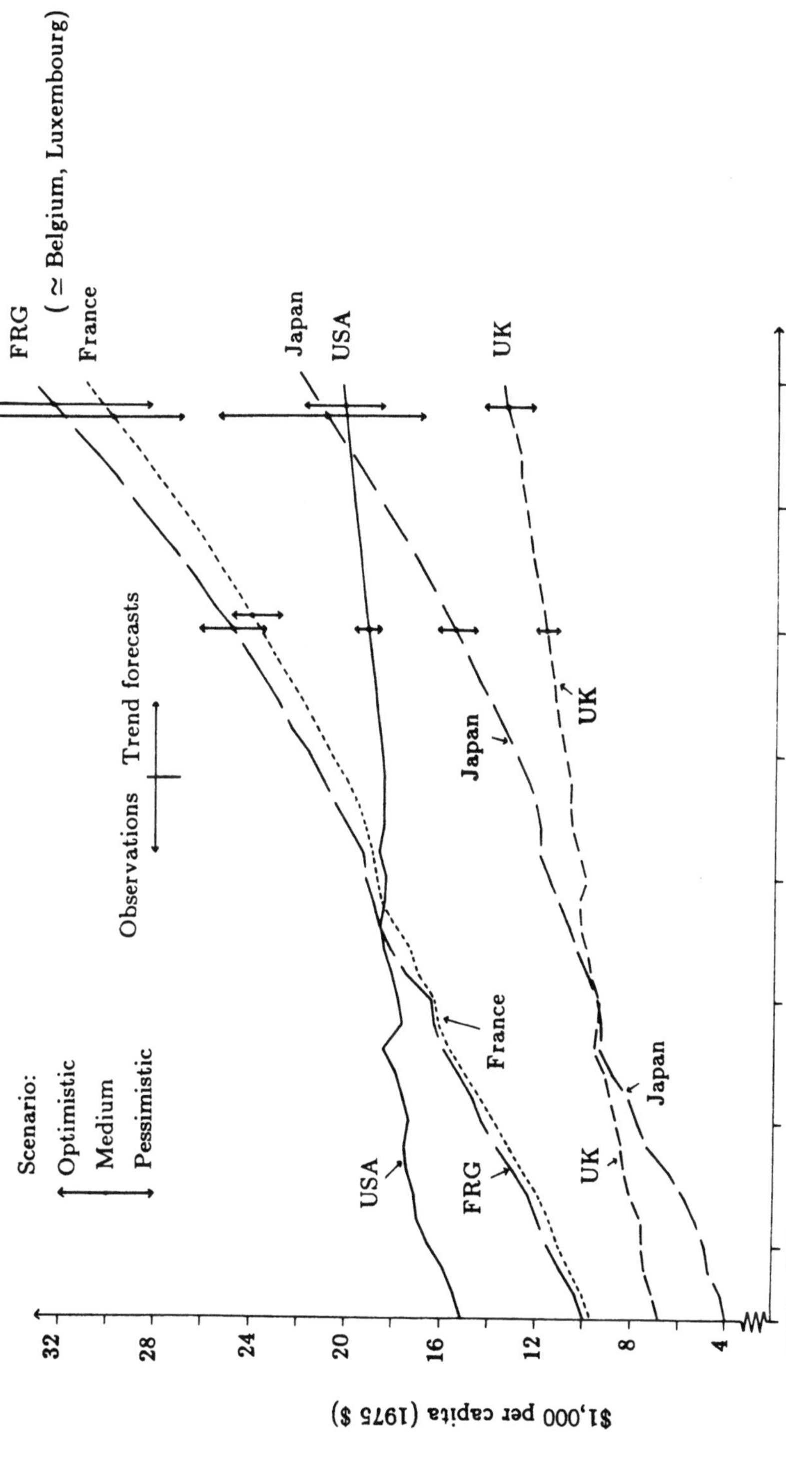


*Figure 2.15.* GDP per employed person ≈ labor productivity: selected OECD countries.

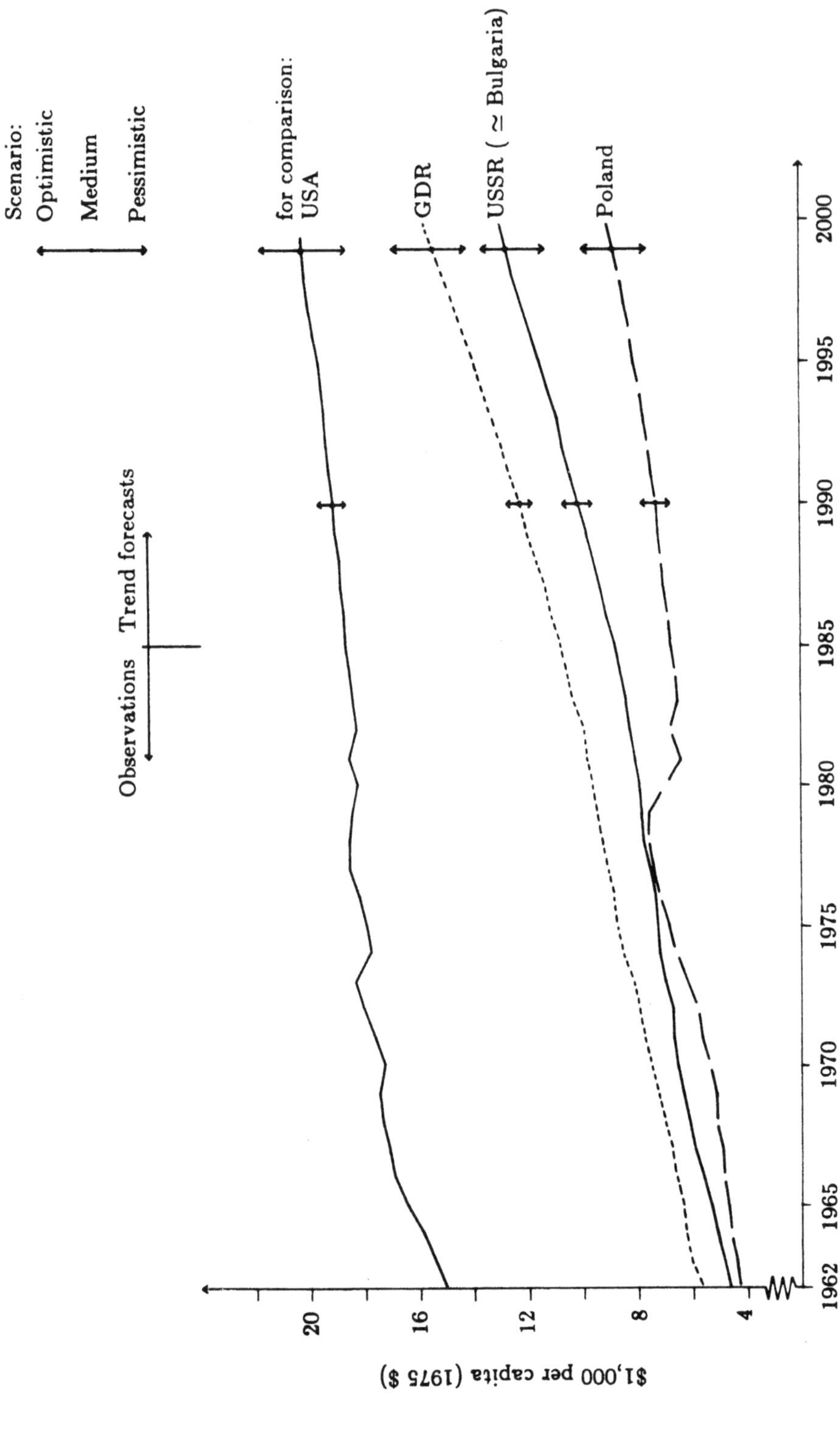


*Figure 2.16.* GDP "type 2" per employed person ≈ labor productivity: selected CMEA countries.

more or less slowly. In Annex *Table 16* the labor productivity of the three world regions (OECD, CMEA and developing countries) is compared. If one sets the average labor productivity of the OECD countries = 100%, the labor productivity of the CMEA countries (with respect to GDP "type 2") was 51% in 1981 and will be 58% at the end of this century. For developing countries these figures are 11% and 13%, respectively. Thus, on the average the gaps are closing, but very slowly.

### 2.6.5. Foreign trade

Real imports and exports of OECD, CMEA and developing countries are shown in Annex *Tables 17–23*. In *Table 24* the development of total world trade may be seen. The following list summarizes this result in form of growth rates (%) of world trade i.e., total real exports, observed until 1981 and forecast thereafter, under the medium scenario:

| | | | | | |
|---|---|---|---|---|---|
| 1964 | 9.1 | 1976 | 10.0 | 1988 | 4.3 |
| 1965 | 7.5 | 1977 | 6.1 | 1989 | 4.3 |
| 1966 | 8.0 | 1978 | 4.1 | 1990 | 4.6 |
| 1967 | 5.5 | 1979 | 6.3 | 1991 | 4.4 |
| 1968 | 11.5 | 1980 | 3.7 | 1992 | 4.5 |
| 1969 | 10.5 | 1981 | 3.1 | 1993 | 4.4 |
| 1970 | 8.8 | 1982 | −1.8 | 1994 | 4.4 |
| 1971 | 6.2 | 1983 | 2.6 | 1995 | 4.3 |
| 1972 | 7.8 | 1984 | 7.5 | 1996 | 4.4 |
| 1973 | 10.7 | 1985 | 0.3 | 1997 | 4.4 |
| 1974 | 5.7 | 1986 | 4.0 | 1998 | 4.4 |
| 1975 | −2.5 | 1987 | 4.3 | 1999 | 4.4 |

In the pessimistic scenario the growth rates decline from around 2% at the end of the 1980s to 1% at the end of the century. In the optimistic scenario these rates stay at about 6.6% from the end of the 1980s to the end of the century. Judging from these figures, the most likely growth path would seem to lie between the medium and the pessimistic scenarios.

*Table 2.5* shows the shares of OECD, CMEA and developing countries in total world trade (calculated from *Table 24* in the Annex).

*Table 2.5.* Shares (%) of total world trade, medium scenario.

| | *1962* | *1970* | *1980* | *1990* | *1999* |
|---|---|---|---|---|---|
| OECD countries | 71 | 71 | 74 | 75 | 72 |
| CMEA countries | 7 | 8 | 7 | 7 | 7 |
| Developing countries | 22 | 21 | 19 | 18 | 21 |

The share of CMEA trade stays constant at about 7%. The share of OECD trade rises until the 1990s and falls thereafter, when the more advanced developing countries turn the tide. As to the *balance of trade* (including services, i.e., the current account balance without transfer payments), *Tables 25–27* in the Annex show that the USA will reduce its trade deficit from $84 billion in 1984 to $29 billion in 1999 in the medium scenario and to about $14 billion in the optimistic scenario. But the deficit will remain. The surpluses of the FRG and Japan will slowly increase in the medium scenario (to about $22 billion for the FRG and to about $35 billion for Japan). But both countries will end up with trade deficits of about $9 billion if the pessimistic scenario comes true. France will have a large deficit, and Italy and Belgium/Luxemburg small ones, whereas the UK and the Netherlands come up with relatively large trade surpluses. The CMEA countries are able to earn small surpluses. The oil-exporting developing countries enjoy large surpluses, whereas all other developing countries have relatively large deficits. This refers to the medium scenario. In the optimistic scenario these deficits are even larger.

These results depend on the *exchange rates*; see *Table 29* in the Annex. As was pointed out earlier, our exchange rate equations are based on purchasing power parity theory and do not take into account speculative waves. The latest devaluation of the dollar with respect to the DM and the yen is not considered. Thus, the growth path of the European and the Japanese economies might be too high compared to that of the USA. Therefore we should concentrate more on the pessimistic scenario for these countries.

Of course, the exchange rates are related to the *general price levels*, which in turn depend on money supply or on the development of the nominal wage rates. In Annex *Tables 32–34* the resulting *rates of inflation* are reproduced. The USA ends up with a rate of inflation of about 7%, the FRG 2–3%, Japan 4–5%, France 7–8%, the UK 10%, Italy 12%–13%. The majority of the CMEA countries keep their inflation rate at 1–2%, with the exception of Hungary ( $\simeq$ 5%) and Poland (6–7%). Of course, the monetary regime may change. If such information becomes available, the system should be recalculated under these new assumptions.

### 2.6.6. Relative size of economies

Labor productivity and standard of living are important indicators of economic performance, but the sheer size of an economy, measured in GDP, is important as well. It is one measure of political "power".

*Table 35* in the Annex gives the ratios of GDP of different OECD countries with respect to the "dominant" economy of the USA and the ratios of GDP "type 1" and "type 2" of different CMEA countries with respect to the "dominant" economy of the USSR. One sees that the relative importance of the USA declines with respect to the other OECD countries and that these other countries, taken together, will have about 50% more productive power than the USA at the end of this century. Within the CMEA countries the USSR keeps its rank. The other CMEA countries taken together stay at about 35% of the productive power of the USSR. If the USSR as well as the USA perform according to the same scenario (optimistic, medium, or pessimistic), the USSR would improve its GDP ratio relative to the USA from 58.8% in 1984 to between 63 and 69%; see *Table 35(d)* in the Annex.

### 2.6.7. Sectoral composition of GDP or NMP

Annex *Table 36* shows the sectoral composition of GDP for *OECD countries*. In the USA the service sector will expand further, but slowly. There is no sign of "deindustrialization". In Japan and Italy manufacturing expands; in the other countries it stays more or less constant. In the *CMEA countries* the industrial sector still expands substantially, mostly at the expense of the agricultural sector. Only part of the services is included in NMP, and this part stays more or less constant. In *developing countries* the manufacturing sector is expanding, but is still smaller than the agricultural sector in many groups of countries. Services keep a surprisingly high level and continue to expand.

## 2.7. Which Growth Path Will Be Realized?

It should be emphasized once again that the results presented in Annex 3 and commented upon in the foregoing section are not predictions, but indicate possible consistent growth paths. Which path will be realized in each country depends heavily on the domestic performance of the economic agents, including governments. This is reflected in the two main driving forces of economic development: the investment ratio and the rate of technical progress. Of course, it cannot be assumed that they follow the same scenario in each country – the medium scenario, for instance. Especially countries with an unsatisfactory economic performance in the past (such as the USSR and Great Britain) will have a high incentive to turn the tide, whereas those with a better record in the past (such as the FRG and Japan) will not do as much to facilitate the necessary structural changes. This conforms to the general theory of economic development presented in Chapter 1 and illustrated in *Figure 2.1* of this chapter. It is especially evident in the FRG.

It seems to be likely that the USA, UK and the USSR will be able to realize the optimistic scenario; the FRG and Japan will follow the pessimistic scenario; and all other OECD, CMEA and developing countries will stay in the medium scenario. The world model could be solved under these assumptions without difficulties, but the limited time of the research project did not allow this. Thus, *Figure 2.17* presents only a rough approximation of the economic performance of some countries under these assumptions, measured by GDP per capita (or GDP "type 2" for CMEA countries). It is drawn by putting together the solutions for different scenarios and therefore does not constitute a consistent solution of the world model; but it may be used as an approximation.

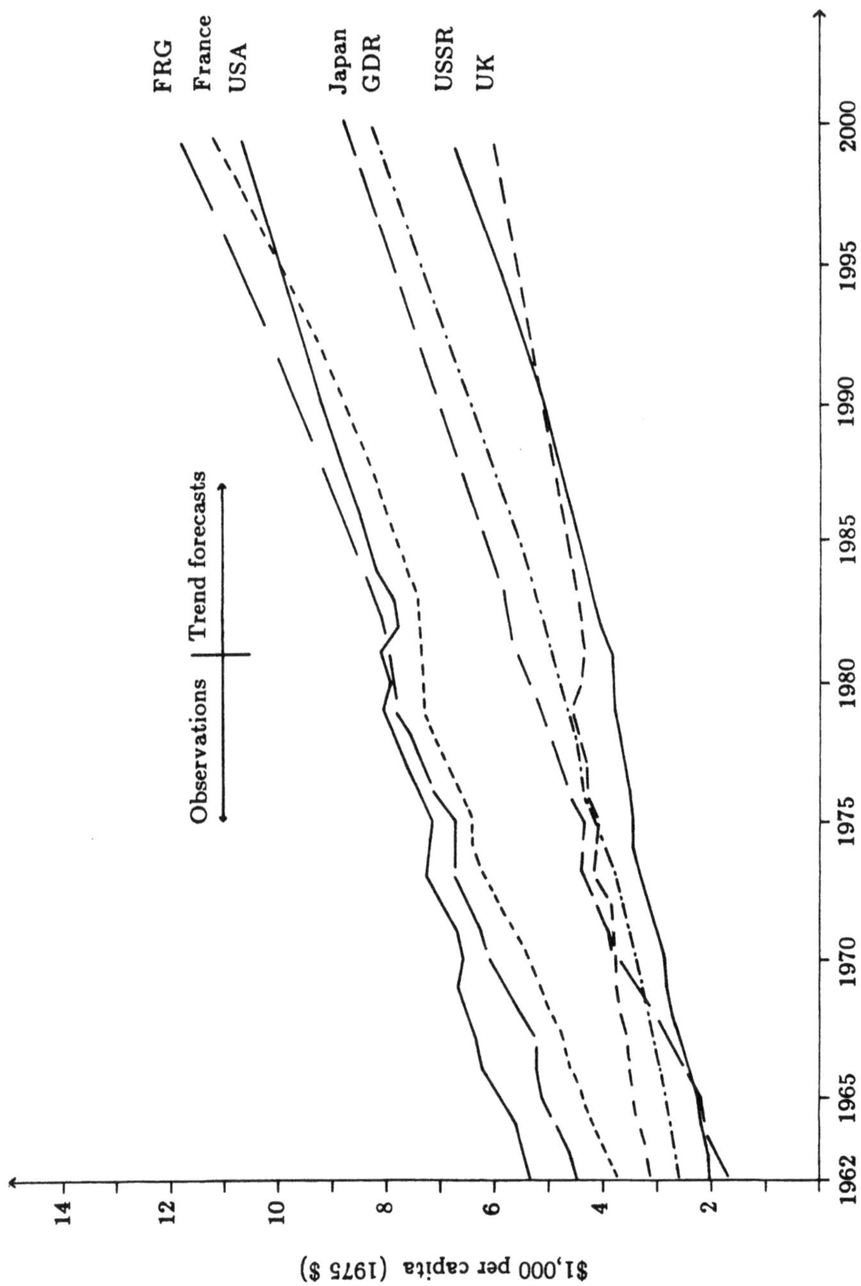


*Figure 2.17.* GDP per capita (OECD) or GDP "type 2" (CMEA) under the assumptions that the optimistic scenario will be realized for the USA, USSR and UK; the pessimistic scenario for the FRG and Japan; and the medium scenario for all other countries.

## 2.8. Final Remarks

This is only a part of the results of the Bonn–IIASA research project. Other results will be presented in the following chapters. All of them are estimates of the consequences of basic behavioral assumptions, which we have tried to make explicit at the outset. If these results are unsatisfactory for a country, that nation should make efforts to change its basic economic behavior. As a rule, this is not easy.

## Notes

[1] We use the notation $w_x$ for the growth rate of a variable $x$, i.e., $w_x = \dot{x}/x$ if $x$ is continuous in time, $w_x = (x - x_{-1})/x_{-1}$ if $x$ is discrete.

[2] There are many reasons for this. We mention here only the the most important ones.

(a) Given the technology $\tau$, constant returns to scale $(\sum_i \alpha_i = 1)$ are almost self-evident in the long run: doubling labor, capital and imported factors of production must double the product. Limitations coming from exhaustible or nonaugmentable factors (such as land) are taken care of by changes in $\tau$.

(b) The production elasticities $\alpha_i$ must stay within certain limits $0 < \underline{\alpha}_i \leq \alpha_i \leq \bar{\alpha}_i < 1$. If we are only concerned with the trend, we take the mean of $\alpha_i$ such that $\sum_i \alpha_i = 1$. If the unknown functions determining $\alpha_i$ within these limits do not change in the medium term (as we assume), we take this mean as expectation for the future.

Arguments (a) and (b) alone give us the Cobb–Douglas function. But there are other arguments in favor of it as well.

(c) Jorgenson (1971, 1972) in defending his approach on investment functions (which implies the Cobb–Douglas production function for a firm) showed that many reliable estimations of production functions did not yield significant deviations of the elasticity of substitution from one which implies a Cobb–Douglas production function.

(d) Krelle and Pallaschke (1981a) derived the integrability conditions for a linear and a quadratic approximation of a general demand system. One term which always appears in this approximation to any production function is a Cobb–Douglas term of the form above. Empirical research showed that the deviations from this term are small, i.e., that the other terms do not contribute much to the explanation of production; see Krelle (1981b).

(e) Houthakker (1955–1956) assumed fixed production coefficients $a_i$ for all firms and assumed a Pareto distribution

$$\phi(a_{1,\dots,}a_n) = A \cdot a_1^{\alpha_1 - 1} \dots . a_n^{\alpha_n - 1}, \qquad \alpha_i \geq 0$$

for these coefficients. He showed that, by aggregation, this yields a Cobb–Douglas function for the economy. Steindl (1965) proved that the Pareto distribution follows from a very general stochastic process.

Thus there are many *a priori* arguments in favor of our approach. The results support our approach, too.

[3] The linear expenditure system is:

$$x_i = \bar{x}_i + \frac{a_i}{p_i}\,(Y'N - \sum_j p_j\bar{x}_j), \qquad a_i \geq 0, \qquad \sum_i a_i = 1, \qquad i = 1,\ldots,n$$

where $x_i =$ total demand of commodity $i$, $p_i =$ price of commodity $i$, $\bar{x}_i =$ predetermined demand of commodity $i$, and $Y'N =$ total expenditure for all commodities. Assume $\bar{x}_i = \lambda_i x_{i-1}$ and rewrite the system as

$$x_i = b_i x_{i,-1} + c_i\,\frac{Y'N}{p_i} - d_i \sum_{j\neq i} \frac{p_j}{p_i}\,\mu_{ji}\, x_{j,-1}$$

where $b_i = \lambda_i(1-a_i)$, $c_i = a_i$, $d_i = a_i\lambda_i$, $\mu_{ji} = \lambda_j/\lambda_i$. This is the demand system we use, basically. For details, see Chapter 8.

[4] Since we use the asymptotic purchasing power parity theory to explain the exchange rates, the influence of the monetary side is limited, as far as the trend is concerned.

[5] We do not consider the ratio $\mu_R$ of imports of secondary factors to total imports as a driving force. It is endogenous for OECD countries, but exogenous for CMEA and developing countries. This treatment is only preliminary. The ratios used in the forecasts are:

| *For OECD countries* | *Average of $\mu_R$ (in %)** |
|---|---|
| USA | 14.25 |
| FRG | 16.32 |
| Japan | 42.72 |
| France | 13.43 |
| UK | 11.36 |
| Italy | 22.13 |
| Netherlands | 17.21 |
| Belgium/Luxembourg | 15.26 |
| Canada | 7.30 |
| Rest of OECD | 13.23 |
| *For CMEA countries:* | |
| USSR | 52.0 |
| Bulgaria | 78.0 |
| CSSR | 77.0 |
| GDR | 74.0 |
| Hungary | 74.0 |
| Poland | 72.0 |
| Romania | 80.0 |
| *For developing countries:* | |
| Oil-exporting countries | 5.74 |
| Asian countries without India | 11.04 |
| India | 59.59 |
| Africa | 10.28 |
| Latin America without Mexico, Brazil and Argentina | 14.63 |
| Mexico, Brazil and Argentina | 18.87 |
| North Africa and Middle East | 19.27 |

*Note that the rates for the OECD countries vary over time; see Appendix 2D. Moreover, the rates for the OECD countries are not comparable with those for CMEA, because the latter include machinery parts, tools, and so on.

[6] These averages have sometimes been modified according to other information and judgments; see Appendix 2E for OECD countries.

# Appendix 2A. Forecast of Employed Labor Force (in Working Hours)

By definition:

$$w_L = w_{POP} + w_{LPR} + w_{ER} + w_{WH}$$

where $w_L$ = growth rate of employed labor (in working hours), $w_{POP}$ = growth rate of population, $w_{LPR}$ = growth rate of the labor participation rate, $w_{ER}$ = growth rate of employment rate, and $w_{WH}$ = growth rate of average working hours.

**2A.1.** Labor supply in OECD countries.

| | $w_{POP}$ | | | | $w_{LPR}$ | | | |
|---|---|---|---|---|---|---|---|---|
| | *Observed means* | | | *Forecast* | *Observed means* | | | *Forecast* |
| *Country* | *61–84* | *71–84* | *76–84* | *1985–2000* | *61–84* | *71–84* | *76–84* | *1985–2000* |
| 01 USA | 1.1 | 1.0 | 1.0 | 1.0 | 0.9 | 1.2 | 1.1 | 0.8 |
| 02 FRG | 0.4 | 0.06 | -0.1 | -0.2 | -0.3 | 0.1 | 0.4 | 0.4 |
| 03 Japan | 1.0 | 1.0 | 0.8 | 0.6 | 0.1 | 0.0 | 0.4 | 0.1 |
| 04 France | 0.8 | 0.6 | 0.4 | 0.3 | 0.1 | 0.2 | 0.4 | 0.3 |
| 05 UK | 0.3 | 0.1 | 0.1 | -0.1 | 0.1 | 0.4 | 0.4 | 0.3 |
| 06 Italy | 0.6 | 0.4 | 0.2 | 0.3 | -0.3 | 0.4 | 0.8 | 0.6 |
| 07 NL | 1.0 | 0.7 | 0.6 | 0.4 | 0.4 | 0.7 | 1.3 | 0.6 |
| 08 B/L | 0.3 | 0.2 | 0.1 | 0.1 | 0.3 | 0.6 | 0.6 | 0.4 |
| 09 Canada | 1.4 | 1.2 | 1.1 | 1.4 | 1.4 | 1.6 | 1.3 | 0.9 |
| 10 Rest of OECD[a] | 1.4 | 1.3 | 1.2 | 1.2 | -0.1 | 0.0 | 0.0 | 0.0 |

| | $w_{ER}$ | | | | $w_{WH}$ | | | | $w_L$[b] | | | |
|---|---|---|---|---|---|---|---|---|---|---|---|---|
| | *Observed means* | | | *Forecast* | *Observed means* | | | *Forecast* | *Observed means* | | | *Forecast* |
| | *61–84* | *71–84* | *76–84* | *1985–2000* | *61–84* | *71–84* | *76–84* | *1985–2000* | *61–84* | *71–84* | *76–84* | *1985–2000* |
| 01 | –0.1 | –0.2 | –0.2 | –0.1 | –0.4 | –0.4 | –0.2 | –0.3 | 1.6 | 1.7 | 2.0 | 1.4 |
| 02 | –0.3 | –0.6 | –0.5 | –0.2 | –0.9 | –0.8 | –0.4 | –0.4 | –1.1 | –1.2 | –0.7 | –0.4 |
| 03 | –0.04 | –0.1 | –0.1 | –0.1 | 0.0 | –0.3 | 0.5 | –0.2 | 0.6 | 0.6 | 1.6 | 0.4 |
| 04 | –0.4 | –0.6 | –0.7 | –0.5 | –0.7 | –1.0 | –1.0 | –0.8 | –0.25 | –0.8 | –0.9 | –0.7 |
| 05 | –0.4 | –0.7 | –1.0 | –0.5 | –0.4 | –0.3 | –0.03 | –0.2 | –0.4 | –0.5 | –0.5 | –0.5 |
| 06 | –0.2 | –0.4 | –0.6 | –0.4 | –1.0 | –1.0 | 0.3 | –0.2 | –10 | –0.6 | 0.8 | 0.3 |
| 07 | –0.6 | –1.0 | –1.1 | –0.5 | –0.8 | –0.6 | –0.2 | –0.3 | –0.04 | –0.2 | 0.6 | 0.2 |
| 08 | –0.5 | –0.9 | –1.2 | –0.7 | –0.9 | –1.2 | –0.3 | –0.5 | –0.7 | –1.3 | –0.8 | –0.7 |
| 09 | –0.2 | –0.4 | –0.5 | –0.4 | –0.2 | –0.2 | –0.02 | –0.2 | 2.4 | 2.2 | 1.9 | 1.7 |
| 10 | –0.2 | –0.4 | –0.6 | –0.4 | –0.7 | –0.7 | –0.1 | –0.5 | 0.3 | 0.1 | 0.4 | 0.3 |

[a]The means for country 10 are calculated only up to 1982 (instead of 1984).
[b]With precision up to the rounding errors.

**2A.2.** Labor supply in CMEA countries: Growth rate $w_{LM}$ of employed persons $LM$ in the material sphere, 1986–2000 (in %).

| *USSR* (16) | *Bulgaria* (71) | *CSSR* (72) | | *GDR* (73) | *Hungary* (74) | | *Pol* (75) | | *Romania* (76) | |
|---|---|---|---|---|---|---|---|---|---|---|
| 0 | 0.05 | 1986: | 0.6 | 0.33 | 1986–87: | −0.17 | 1986–89: | ~0.50 | 1986–87: | 0.06 |
| | | 1987: | 0.34 | | 1988–93: | −0.16 | 1990–94: | ~0.49 | 1988: | 0.03 |
| | | 1988–90: | 0.33 | | 1994–99: | ~−.0165 | 1995–99: | ~0.47 | 1989: | 0 |
| | | 1991: | 0.40 | | | | | | 1990: | −0.03 |
| | | 1992–97: | 0.33 | | ø: | −0.165 | ø: | ~ 0.49 | 1991: | −0.05 |
| | | 1998: | 0.39 | | | | | | 1992: | −0.07 |
| | | 1999: | 0.32 | | | | | | 1993: | −0.10 |
| | | | | | | | | | 1994: | −0.12 |
| | | ø: | 0.36 | | | | | | 1995: | 0.14 |
| | | | | | | | | | 1996: | −0.16 |
| | | | | | | | | | 1997: | −0.17 |
| | | | | | | | | | 1998: | −0.19 |
| | | | | | | | | | 1999: | −0.21 |
| | | | | | | | | | ø: | −0.15 |

**2A.3.** Labor supply in developing countries: growth rate $w_L$ of economic active population $L$, 1982–2000 (in %).

| *Year* | *Oil-exporters* (11) | *Asia without India* (12I) | *India* (34) | *Africa* (13) | *Latin America without M,B,A* (14) | *Mexico, Brazil, Argentina* (15) | *N. Africa, Middle East* (18) |
|---|---|---|---|---|---|---|---|
| 1982 | 2.14 | 2.09 | 1.67 | 2.67 | 3.35 | 2.78 | 3.27 |
| 1983 | 2.36 | 2.37 | 1.66 | 2.60 | 2.78 | 2.57 | 2.71 |
| 1984 | 2.30 | 2.16 | 1.63 | 2.28 | 2.70 | 2.77 | 2.64 |
| 1985 | 2.16 | 2.41 | 1.61 | 2.48 | 2.63 | 2.56 | 3.00 |
| 1986 | 2.63 | 1.91 | 1.69 | 2.42 | 2.56 | 2.75 | 2.92 |
| 1987 | 2.39 | 2.31 | 1.62 | 3.07 | 2.92 | 2.68 | 3.24 |
| 1988 | 2.42 | 1.98 | 1.63 | 2.52 | 2.83 | 2.73 | 2.75 |
| 1989 | 2.20 | 2.22 | 1.60 | 2.68 | 2.36 | 2.54 | 2.29 |
| 1990 | 2.31 | 1.76 | 1.58 | 2.40 | 2.31 | 2.59 | 3.36 |
| 1991 | 3.01 | 2.53 | 2.07 | 3.40 | 3.38 | 2.96 | 2.53 |
| 1992 | 2.56 | 2.34 | 2.03 | 3.09 | 2.91 | 3.09 | 3.52 |
| 1993 | 2.71 | 2.28 | 2.02 | 3.19 | 3.18 | 2.79 | 3.40 |
| 1994 | 2.71 | 2.23 | 1.95 | 2.90 | 3.08 | 3.02 | 2.30 |
| 1995 | 2.77 | 2.06 | 1.91 | 2.82 | 2.99 | 2.63 | 3.22 |
| 1996 | 2.37 | 2.14 | 1.88 | 2.74 | 2.26 | 2.76 | 2.49 |
| 1997 | 2.38 | 2.10 | 1.87 | 2.67 | 2.84 | 2.50 | 2.74 |
| 1998 | 2.57 | 2.05 | 1.81 | 2.43 | 2.15 | 2.62 | 2.37 |
| 1999 | 2.32 | 2.01 | 1.78 | 2.54 | 2.70 | 2.46 | 2.60 |
| ø | 2.46 | 2.16 | 1.77 | 2.72 | 2.77 | 2.66 | 2.85 |

# Appendix 2B. Forecast of Money Supply

**2B.1.** OECD countries: rate of growth of money supply, M2, (%).

| | *Observed means* | | | *Forecast* |
|---|---|---|---|---|
| *Country* | *1961–84* | *1971–84* | *1976–84* | *1985–2000* |
| 01 USA | 8.1 | 9.0 | 9.5 | 9.0 |
| 02 FRG | 9.5 | 8.2 | 6.6 | 6.5 |
| 03 Japan | 14.8 | 12.6 | 9.4 | 9.0 |
| 04 France | 13.1 | 13.5 | 11.4 | 11.0 |
| 05 UK | 11.5 | 15.7 | 14.5 | 11.0 |
| 06 Italy | 16.2 | 17.8 | 16.8 | 16.0 |
| 07 NL | 11.2 | 11.3 | 9.4 | 9.0 |
| 08 B/L | 9.3 | 10.0 | 7.8 | 7.5 |
| 09 Canada | 12.0 | 13.5 | 12.2 | 12.0 |
| 10 Rest of OECD | 12.8 | 14.4 | 13.9 | 13.0 |

**2B.2.** Developing countries rate of growth of money supply, M2, (%): forecast 1982–2000. [For oil exporting countries (Group 11) we do not estimate M2 since we determine the price level in another way; see Annex 2.]

| *Asia without India* (12 I) | *India* (34) | *Africa* (13) | *Latin America without M,B,A* (14) | *Mexico, Brazil, Argentina* (15) | *N. Africa, Middle East* (18) |
|---|---|---|---|---|---|
| 18 | 15 | 20 | 45 | 45 | 22 |

**2B.3.** Rate of change of OECD import price (in \$) of mineral oil and fuel (P\$MFL W1, see Annex 2) (%): forecast 1987–1999.

| *1982* | *83* | *84* | *85* | *86* | *87* | *88* | *89* | *90* | *91* | *92* | *93* | *94* | *95* | *96* | *97* | *98* | *99* |
|---|---|---|---|---|---|---|---|---|---|---|---|---|---|---|---|---|---|
| −0.7 | −10.9 | −5.0 | −15.0 | −40.0 | −5.0 | 0.0 | 1.0 | 2.0 | 3.0 | 3.5 | 4.0 | 4.2 | 4.4 | 4.5 | 4.5 | 4.5 | 4.5 |

**2B.4.** Rate of growth of the nominal wage rate, CMEA countries (%): forecast 1986–1999.

| *USSR* (16) | *Bulgaria* (71) | *CSSR* (72) | *GDR* (73) | *Hungary* (74) | *Poland* (75) | *Romania* (76) |
|---|---|---|---|---|---|---|
| 5.5 | 5.0 | 4.0 | 4.0 | 8.5 | 1986–89:12<br>1990–99:10 | 7.5 |

# Appendix 2C. Forecast of the Rate of Depreciation of Capital (%), OECD Countries, 1985–2000

We estimated the rate of depreciation by a "back to normal" formula; see Annex 2. Therefore, there is a smooth change of this rate, and we only give the rates for some years.

| *Year* | *USA* (01) | *FRG* (02) | *Japan* (03) | *France* (04) | *UK* (05) | *Italy* (06) | *NL* (07) | *B/L* (08) | *Canada* (09) | *Rest of OECD* (10) |
|---|---|---|---|---|---|---|---|---|---|---|
| 1982 | 5.89 | 3.91 | 5.66 | 4.68 | 4.77 | 3.39 | 2.62 | 2.68 | 4.80 | 2.47 |
| 1985 | 6.26 | 3.91 | 5.82 | 4.53 | 4.91 | 3.36 | 2.67 | 2.73 | 4.99 | 2.58 |
| 1988 | 6.08 | 3.91 | 5.84 | 4.52 | 4.85 | 3.37 | 2.68 | 2.75 | 5.00 | 2.67 |
| 1991 | 5.95 | 3.91 | 5.86 | 4.52 | 4.81 | 3.38 | 2.69 | 2.76 | 5.00 | 2.73 |
| 1994 | 5.86 | 3.90 | 5.87 | 4.51 | 4.78 | 3.39 | 2.69 | 2.77 | 5.00 | 2.78 |
| 1997 | 5.79 | 3.90 | 5.88 | 4.51 | 4.76 | 3.39 | 2.69 | 2.78 | 5.00 | 2.81 |
| 1999 | 5.75 | 3.90 | 5.88 | 4.51 | 4.75 | 3.39 | 2.69 | 2.78 | 5.00 | 2.83 |
| ø1985–1999 | 5.95 | 3.90 | 5.86 | 4.51 | 4.81 | 3.38 | 2.69 | 2.77 | 5.00 | 2.73 |

The rates of depreciation assumed for the CMEA and the developing countries are held constant for the forecasting period; see *Tables 2.2* and *2.3* in the text.

# Appendix 2D. Forecast of the Ratio (%) of Secondary Factor Imports to Total Imports, 1985–2000 (= $\mu_R$, see Annex 2)

**2D.1.** OECD countries.

| *Year*[a] | *USA* (01) | *FRG* (02) | *Japan* (03) | *France* (04) | *UK* (05) | *Italy* (06) | *NL* (07) | *B/L* (08) | *Canada* (09) | *Rest of OECD* (10) |
|---|---|---|---|---|---|---|---|---|---|---|
| 1982 | 15.79 | 16.56 | 42.15 | 14.47 | 12.20 | 22.44 | 16.97 | 15.66 | 9.18 | 13.60 |
| 1985 | 14.47 | 16.44 | 43.37 | 13.82 | 11.71 | 22.59 | 17.39 | 15.50 | 7.73 | 13.44 |
| 1988 | 14.34 | 16.32 | 43.00 | 13.60 | 11.52 | 22.43 | 17.29 | 15.36 | 7.53 | 13.22 |
| 1991 | 14.25 | 16.23 | 42.73 | 13.44 | 11.38 | 22.32 | 17.21 | 15.26 | 7.39 | 13.23 |
| 1994 | 14.18 | 16.17 | 42.53 | 13.32 | 11.28 | 22.23 | 17.15 | 15.19 | 7.28 | 13.17 |
| 1997 | 14.13 | 16.12 | 42.39 | 13.23 | 11.20 | 22.17 | 17.11 | 15.14 | 7.21 | 13.12 |
| 1999 | 14.11 | 16.10 | 42.31 | 13.19 | 11.16 | 22.14 | 17.09 | 15.11 | 7.17 | 13.10 |
| ø1985–1999 | 14.25 | 16.32 | 42.72 | 13.43 | 11.36 | 22.13 | 17.21 | 15.26 | 7.39 | 13.23 |

[a]The introductory explanation of Appendix 2C applies here analogously.

**2D.2.** CMEA countries. (This ratio is kept constant; see text note 5.)

**2D.3.** Developing countries. (These ratios result from a solution of a submodel; see Chapter 6.)

| *Year* | *Oil-exporters* (11) | *Asia without India* (12I) | *India* (34) | *Africa* (13) | *Latin America without M,B,A* (14) | *Mexico, Brazil, Argen-tina* (15) | *N. Africa Middle East* (18) |
|---|---|---|---|---|---|---|---|
| 1982 | 3.56 | 18.01 | 41.54 | 11.39 | 11.72 | 17.76 | 16.68 |
| 1985 | 4.40 | 16.23 | 42.19 | 10.26 | 12.23 | 17.95 | 15.67 |
| 1988 | 6.02 | 14.28 | 56.28 | 10.26 | 16.21 | 17.64 | 18.06 |
| 1991 | 5.84 | 12.22 | 60.18 | 10.22 | 16.62 | 18.22 | 19.71 |
| 1994 | 5.83 | 10.23 | 62.58 | 10.32 | 16.18 | 19.59 | 20.34 |
| 1997 | 5.94 | 8.36 | 64.67 | 10.33 | 15.58 | 20.04 | 20.61 |
| 1999 | 6.04 | 7.14 | 65.85 | 10.32 | 15.20 | 20.41 | 20.74 |
| ø1985–1999 | 5.74 | 11.04 | 59.59 | 10.28 | 14.63 | 18.87 | 19.27 |

# Appendix 2E. Observations and Forecasts of the Rate of Technical Progress (%) for Different Scenarios

**2E.1.** OECD countries.

| Country | Hi or Ha[a] | Observed means[b] 1961–84 | 1971–84 | 1976–84 | Scenarios[b] Optimistic | Medium | Pessimistic |
|---|---|---|---|---|---|---|---|
| 01 USA | Hi | 0.85 | 0.45 | 0.30 | 0.85 | 0.55 | 0.30 |
| | Ha | 1.20 | 0.70 | 0.40 | 1.20 | 0.80 | 0.40 |
| 02 FRG | Hi | 2.90 | 2.30 | 2.05 | 2.80 | 2.25 | 1.75 |
| | Ha | 4.10 | 3.30 | 2.90 | 4.00L | 3.20 | 2.50 |
| 03 Japan | Hi | 4.00 | 2.50 | 1.75 | 3.20 | 2.45 | 1.75 |
| | Ha | 5.60 | 3.50 | 2.40 | 4.50 | 3.40 | 2.40 |
| 04 France | Hi | 3.10 | 2.45 | 2.10 | 3.00 | 2.45 | 2.00 |
| | Ha | 4.30 | 3.40 | 3.00 | 4.20 | 3.40 | 2.80 |
| 05 UK | Hi | 1.20 | 0.85 | 0.65 | 1.20 | 0.95 | 0.65 |
| | Ha | 1.80 | 1.30 | 1.00 | 1.80 | 1.40 | 1.00 |
| 06 Italy | Hi | 3.85 | 2.05 | 1.20 | 3.15 | 2.00 | 1.20 |
| | Ha | 4.80 | 2.60 | 1.50 | 4.00 | 2.50 | 1.50 |
| 07 NL | Hi | 2.30 | 0.95 | 0.25 | 1.75 | 0.95 | 0.25 |
| | Ha | 4.20 | 1.70 | 0.50 | 3.20 | 1.70 | 0.50 |
| 08 B/L | Hi | 3.05 | 2.50 | 2.20 | 2.65 | 2.20 | 1.90 |
| | Ha | 4.80 | 3.90 | 3.50 | 4.20 | 3.50 | 3.00 |
| 09 Canada | Hi | 1.10 | 0.00 | –0.60 | 1.00 | 0.45 | 0.00 |
| | Ha | 1.70 | 0.00 | –0.90 | 1.50 | 0.70 | 0.00 |
| 10 Rest of | Hi | 2.75 | 1.70 | 1.15 | 2.25 | 1.60 | 1.00 |
| Canada | Ha | 4.20 | 2.60 | 1.80 | 3.50 | 2.50 | 1.50 |

[a] Hi = Hicks neutral; Ha = Harrod neutral. If $w_\tau$ = rate of technical progress in the Hicks-neutral sense and $w^*_\tau$ the same in the Harrod-neutral sense and $\alpha$ = production elasticity of labor, we have the relation: $w^*_\tau = w_\tau / \alpha$.

[b] All figures are rounded to 0 or 5 in the second decimal place.

**2E.2.** CMEA countries: rate of (Hicks-neutral) technical progress. See *Table 2.2* in the text.

**2E.3.** Developing countries: rate of (Hicks-neutral) technical progress (medium scenario, %).[a]

| Year | Oil-exporters (11) | Asia without India (12I) | India (34) | Africa (13) | Latin America without M,B,A (14) | Mexico, Brazil, Argentina (15) | N. Africa, Middle East (18) |
|---|---|---|---|---|---|---|---|
| 1982 | -1.643 | 0.50 | 0.50 | -0.793 | 1.0 | 1.443 | 2.0 |
| 1985 | -1.015 | 0.70 | 0.50 | -0.484 | 1.0 | 1.504 | 2.0 |
| 1988 | -0.072 | 1.00 | 0.50 | -0.021 | 1.0 | 1.596 | 2.0 |
| 1991 | 0.871 | 1.30 | 0.50 | 0.441 | 1.0 | 1.689 | 2.0 |
| 1994 | 1.500 | 1.50 | 0.50 | 0.750 | 1.0 | 1.750 | 2.0 |
| 1997 | 1.500 | 1.50 | 0.50 | 0.750 | 1.0 | 1.750 | 2.0 |
| 1999 | 1.500 | 1.50 | 0.50 | 0.750 | 1.0 | 1.750 | 2.0 |
| ⌀1985–99 | 0.750 | 1.26 | 0.501 | 0.380 | 1.005 | 1.680 | 2.0 |

[a] For the groups of developing countries we estimated an average rate of technical progress for the period 1960–1974 (or 1975, ..., 1979 for some groups) and an average rate from the end of the first period until the end of the reference period (1983); see *Figure 2.6*. This latter rate was substantially lower than the first. The reason for this procedure was that we could not get reliable estimates of a continuous time trend for this rate, whereas a break in the development of this rate yields significant estimates. For the forecast we assumed that a rate in between will be reached again after a while; see *Figure 2.6* for the special linearity assumptions made in this respect. For India, some Latin American countries and North Africa and the Middle East, there was no significant difference between the two periods.

For the other scenarios similar tables exist. We reproduce only the average values 1985–99:

| *Scenario* | (11) | (12I) | (34) | (13) | (14) | (15) | (18) |
|---|---|---|---|---|---|---|---|
| Optimistic | 1.79 | 1.91 | 0.97 | 0.90 | 1.19 | 1.97 | 2.22 |
| Pessimistic | -0.23 | 0.74 | 0.50 | -0.10 | 1.00 | 1.44 | 2.00 |

# Appendix 2F. Forecast of the Investment Ratio, Medium Scenario (%)

**2F.1.** OECD countries.

| Year[a] | USA (01) | FRG (02) | Japan (03) | France (04) | UK (05) | Italy (06) | NL (07) | B/L (08) | Canada (09) | Rest of OECD (10) |
|---|---|---|---|---|---|---|---|---|---|---|
| 1982 | 16.90 | 19.62 | 31.17 | 21.24 | 17.02 | 18.76 | 17.36 | 17.59 | 21.86 | 21.97 |
| 1985 | 19.48 | 19.71 | 30.36 | 20.19 | 18.27 | 18.74 | 17.84 | 17.07 | 20.00 | 22.25 |
| 1988 | 19.08 | 19.92 | 30.72 | 20.63 | 18.19 | 18.92 | 18.21 | 17.86 | 20.67 | 22.45 |
| 1991 | 18.79 | 20.08 | 30.99 | 20.95 | 18.14 | 19.05 | 18.48 | 18.44 | 21.17 | 22.60 |
| 1994 | 18.57 | 20.19 | 31.179 | 21.18 | 18.10 | 19.14 | 18.67 | 18.86 | 21.53 | 22.71 |
| 1997 | 18.42 | 20.28 | 31.32 | 21.35 | 18.08 | 19.21 | 18.82 | 19.17 | 21.79 | 22.79 |
| 1999 | 18.34 | 20.32 | 31.39 | 21.43 | 18.06 | 19.25 | 18.89 | 19.33 | 21.93 | 22.83 |
| ø1985–99 | 18.78 | 20.08 | 30.99 | 20.95 | 18.14 | 19.05 | 18.48 | 18.49 | 21.17 | 22.60 |

[a] The forecasting equation is reproduced in Annex 2. We give the figures for only some years; the others may be interpolated.

**2F.2.** CMEA countries: the investment ratios are kept constant in the forecasts; see *Table 2.2* in the text.

**2F.3.** Developing countries.[a]

| Year | Oil-exporters (11) | Asia without India (12I) | India (34) | Africa (13) | Latin America without M,B,A (14) | Mexico, Brazil, Argen-tina (15) | N. Africa, Middle East (18) |
|---|---|---|---|---|---|---|---|
| 1982 | 33.00 | 25.00 | 20.00 | 22.00 | 16.00 | 20.00 | 25.00 |
| 1985 | 27.00 | 22.00 | 29.50 | 18.00 | 12.00 | 17.00 | 18.00 |
| 1988 | 23.06 | 24.35 | 21.65 | 20.93 | 15.05 | 19.29 | 20.53 |
| 1991 | 24.72 | 25.24 | 22.69 | 22.54 | 16.59 | 21.37 | 22.52 |
| 1994 | 24.93 | 25.40 | 22.88 | 22.81 | 16.84 | 21.78 | 22.84 |
| 1997 | 24.98 | 25.46 | 22.96 | 22.93 | 16.94 | 21.93 | 22.95 |
| 1999 | 25.00 | 25.49 | 22.99 | 22.98 | 16.98 | 21.98 | 22.99 |
| ø1985–99 | 24.77 | 24.78 | 22.20 | 212.84 | 15.89 | 20.61 | 21.75 |

[a] The investment ratio of the developing countries declined substantially since the middle of the 1970s. We assumed that this decline will continue for a while (until the debt problem has been settled), but that afterwards a certain fraction of the old investment ratio will again be reached asymptotically. For details, see Chapter 5; for an illustration, see *Figure 2.9* in Chapter 2.

NOTE: For the optimistic and the pessimistic scenarios, similar tables exist. The average values are shown in *Table 2.3* in the text.

## References

Efron, B. and Morris, C. (1977), Stein's Paradox in Statistics, *Scientific American*, **236** (5), May, pp. 119–127.

Houthakker, H.S. (1955–1956), The Pareto Distribution and the Cobb–Douglas Production Function in Activity Analysis, *Review of Economic Studies*, **23**, pp. 27ff.

Jorgenson, D.W. (1971), Econometric Studies of Investment Behavior: A Survey, *Journal of Economic Literature*, **9**, pp. 111ff.

Jorgenson, D.W. (1972), Investment Behavior and the Production Function, *Bell Journal of Economics and Management Science*, **3**, pp. 220ff.

Krelle, W. and Pallaschke, D. (1981), A General Demand System, *Zeitschrift für Nationalökonomie*, **41**, pp. 223ff.

Krelle, W. (1981), Die repräsentative Firma - Ein sinnvolles Konzept in Volks– und Betriebswirtschaftslehre? pp. 81ff. in: K. Brockhoff and W. Krelle (eds.), *Unternehmensplanung.* (Berlin, Heidelberg, New York: Springer).

Krelle, W. and Welsch, H. (1986), *Determination of Exchange Rates and Capital Flows for OECD Countries*, Discussion paper B 46, Sonderforschungsbereich 303, University of Bonn.

Marer, P. (1985), *Dollar GNPs of the USSR and Eastern Europe.* (Baltimore and London: Johns Hopkins University Press).

Steindl, J. (1965), *Random Process and the Growth of Firms. A Study of the Pareto Law.* (London: Griffin).

CHAPTER 3

# Economic Growth and Structural Change of OECD Countries

*Hermann Ross*

## Summary

In this chapter we present the estimates of production functions for OECD countries and define the different scenarios for the driving forces of economic growth. They form the basis for the solution of the world model and determine our forecasts. In addition we estimate other functions (to explain the velocity of money, the rate of interest and structure of production) and present forecasts for these variables.

## 3.1. Introduction: The Group of Industrialized Market Economies

The industrialized market economies considered in the model include 20 countries (with Belgium and Luxembourg treated as one country; see Annex 1 at the end of this volume). With the exception of South Africa all of them are members of the OECD. The countries were selected with respect to their share in world trade at the end of the 1970s. The 20 countries covered 60.6% of world exports and 72.3% of world imports (without intra-CMEA trade) reported for 1980 in the Direction of Trade Statistics (DOTS) of the International Monetary Fund. Nine countries which covered 56.5% of all exports and 59.8% of all imports in 1980 are treated separately. The other 11 countries are aggregated and treated as one economic unit [1].

## 3.2. The Production Function and Technical Progress

As was explained in Chapter 2 we use a linear homogenous Cobb–Douglas production function which we write as

$$Y = (\tau \cdot L)^{\alpha} \cdot K^{\beta} \cdot R^{\gamma} \cdot a, \quad \alpha + \beta + \gamma = 1 \tag{3.1}$$

where $Y$ = production, $\tau$ = index of technical knowledge, $L$ = labor input, $K$ = stock of fixed capital and $R$ = imported raw materials. $\tau$ is used here in the notation of Harrod-neutral technical progress. This ensures that in equilibrium growth (with the rate of growth of $K$ and $R$ being equal to the rate of growth of $Y$) the rate of technical progress is equal to the rate of growth of labor productivity. If $\alpha = 2/3$, the rate of technical progress, $\tilde{\tau}$, in Harrod-neutral notation is 50% higher than the rate of technical progress in Hicks-neutral notation ($\tilde{\tau} = \tau^{\alpha} \rightarrow w_{\tau} = w_{\tilde{\tau}}/\alpha$, where $w_x$ is the growth rate $\dot{x}/x$ of a variable $x$).

We measure imported raw materials as the imports of goods in the SITC-Sections 2, 3 and 4 (i.e., inedible crude materials, except fuels; mineral fuels, lubricants and related materials; animal and vegetable oils, fats and waxes).

As was pointed out in Chapter 2 our conception of production is not gross domestic product, but total production minus domestic secondary inputs. In accordance with statistical practice we use a Paasche price index for total production.

Labor input is the product of the number of people actually employed and the number of average working hours per year ("actually worked" if such figures are available, "paid for" otherwise). Capital stock series have been calculated as the accumulated net fixed investment at constant prices starting with some estimated initial value for 1950.

Technical progress is not directly observable. It must be identified from other measurable variables. First we used a method suggested by Solow. We shall present these results. For the Bonn–IIASA model we use another, related approach.

Solow (1957) estimated technical progress as a discrete approximation to a Divisia index. The derivative of the logarithm of the production function with respect to time yields

$$w_{\tau} = (w_Y - \alpha w_L - \beta w_K - \gamma w_R)/\alpha \tag{3.2}$$

If marginal productivity theory holds, the income shares (or cost shares) of the factors should be equal to the elasticities of production with respect to the corresponding factors of production.

We calculated the income shares for all countries and found that they are in most cases nearly constant. As an example, see *Figures 3B.1(a)–(h)* in Appendix 3B (where $MRAW = R$ in the production function). They exhibit small cyclic fluctuations which may be explained by the business cycle and an increase of the cost share for raw materials in 1973/74 and 1979/80. Mean values for 1962–1983 are listed in *Table 3A.1* of Appendix 3A.

We used income shares corrected for changes in the number of self-employed persons (see also Eismont and Ross (1985)]. Our labor data include employed and self-employed persons whereas the statistical figures on wages do not include labor income of self-employed persons. This income is included in profits and entrepreneurial income. To correct this we assumed the average labor income per working hour to be the same for wage earners and self-employed persons. The series for labor cost $LCC^n$ and capital stock $KCC^n$ which we used are therefore defined by

$$LCC^n = WG^n / WES \tag{3.3}$$

$$KCC^n = Q^n - WG^n \cdot \frac{1 - WES}{WES} + KC^n \tag{3.4}$$

where $WG^n$ = nominal wage, $WES$ = wage earners' share, $Q^n$ = nominal profit and entrepreneurial income and $KC^n$ = nominal depreciation.

Technical progress (or the increase of total factor productivity, as it is also called), if calculated by Solow's method, shows a similar trend for all countries in the 1970s and early 1980s, especially if one looks at five-year moving averages of the growth rates instead of yearly rates. In *Figures 3B.2(a)–(h)* of Appendix 3B yearly rates of (Harrod-neutral) technical progress and five-year moving averages are shown [2]. It is interesting to observe that the moving average of $w_T$ reaches a local minimum around the year 1958. After that a period of high growth of technical progress starts. There is a first interruption around the years 1965/67 in most countries followed by a new peak around 1970. In the United States this peak is missing. The rate of technical progress in the USA was already in decline since the late 1960s. All other countries followed this trend during the 1970s.

We fitted a linear trend through the values of technical progress obtained in this way (but with imported raw materials included) for the years 1962–1983. The results are given in *Table 3A.1* of Appendix 3A. The trend is always negative, but there are differences between countries. The USA and the UK showed a rather low level of technical progress (3.3 and 2.6% in 1960) but only a slight decline on the average (−0.15 and −0.05% per year, respectively). Japan, Italy and the Netherlands realized the highest level of growth rate in 1960 (11.2%, 9.2% and 7.7%, respectively) but also the highest decline on the average (−0.49, −0.42 and −0.35% per year). The Federal Republic of Germany, France, Belgium/Luxembourg, Canada and the group of other industrialized market economies showed a medium growth of technical progress in the beginning (about 6% in 1960) and a medium decline (about −0.2% per year).

*Table 3.1* shows the average yearly growth rates of production, labor productivity and technical progress between 1962 and 1977. The extent to which the increase in labor productivity is explained by technical progress for the period is smaller than the 88% value obtained by Solow (1957, p. 316) for the United States for 1909–1949, but similar to his result for 1909–1929. Our results for the USA are very close to those to be expected for equilibrium growth.

For the Bonn–IIASA world model we estimated technical progress and the elasticities of production in a slightly different way. In this project we are only interested in the trend. Therefore we estimated the system:

$$Y_t = [\exp(w_\tau + \frac{\delta}{2}\, t) \cdot t \cdot L_t]^\alpha \cdot K_t^\beta \cdot R_t^\gamma \cdot a \tag{3.5a}$$

$$LCC_t^n = LCC_{t-1}^n + \alpha \cdot [(LCC_t^n + KCC_t^n + R_t^n) - (LCC_{t-1}^n + K_{t-1}^n + R_{t-1}^n)] \tag{3.5b}$$

$$KCC_t^n = KCC_{t-1}^n + \beta \cdot [(LCC_t^n + KCC_t^n + R_t^n) - (LCC_{t-1}^n + KCC_{t-1}^n + R_{t-1}^n)] \tag{3.5c}$$

simultaneously under the restriction $\alpha + \beta + \gamma = 1$. The reference period is 1963–1983. This specification of the index of technical knowledge yields a linear time trend of the rate of technical progress: $w_\tau = \bar{w}_\tau + \delta \cdot t$, which conforms to the observations for 1962–1983 (see *Figure 3B.2* in Appendix 3B). The results of the estimate are listed in *Table 3A.3* of Appendix 3A and (for comparison) also in *Table 3.2*. The differences from the first approach are small, *cf. Table 3.1.* For the elasticity of production with respect to raw materials we get values which are larger than the average factor shares. This is due to a slight increase of this share in the reference period and the greater weight given to the last observations when series with a trend are estimated with an additive error term.

## 3.3. The Development of the Driving Forces of Economic Growth

The driving forces for the development of an economy are the rate of growth of labor input $w_L$, the investment ratio $s$, the rate of depreciation $d$ and the rate of technical progress $w_\tau$. For reasons which were explained in Chapter 1, we assume the following development of these driving forces.

### 3.3.1. The investment rate

The ratio of real gross fixed investment to real gross domestic product declined in most countries since the late 1960s. It differed from country to country. As a rule, this ratio lay between 15% and 25%, with the exception of Japan where the ratio is exceptionally high [up to 36% in 1973, see *Table 3A.4* in Appendix 3A

*Table 3.1.* Average yearly growth rates in production, labor productivity and technical progress, 1962–1977.

| *Growth rates* | *USA* | *FRG* | *Japan* | *France* | *UK* | *Italy* | *Neth.* | *B/L* | *Canada* | *Other* |
|---|---|---|---|---|---|---|---|---|---|---|
| 1. Production | 3.4 | 3.4 | 7.5 | 4.3 | 2.2 | 3.9 | 4.0 | 3.6 | 4.7 | 4.0 |
| 2. Labor productivity | 1.8 | 4.7 | 7.3 | 4.4 | 2.5 | 5.1 | 4.3 | 4.2 | 2.3 | 3.7 |
| 3. Technical progress[a] | 1.2 | 3.0 | 4.2 | 2.8 | 1.4 | 3.8 | 2.7 | 2.8 | 1.5 | 2.8 |
| Row 3/Row 2 (%) | 69.7 | 64.7 | 57.6 | 64.8 | 55.7 | 73.7 | 64.0 | 65.8 | 67.1 | 75.9 |

[a]Hicks-neutral notation.

*Table 3.2.* Asymptotic value $\bar{s}$ of the investment rate (%).

| *Investment rate* $\bar{s}$ | *USA* | *FRG* | *Japan* | *France* | *UK* | *Italy* | *Neth.* | *B/L* | *Canada* | *Other* |
|---|---|---|---|---|---|---|---|---|---|---|
| Pessimistic | 13.5 | 15.4 | 23.8 | 16.4 | 13.5 | 14.6 | 14.4 | 15.0 | 16.9 | 17.3 |
| Medium | 18.0 | 20.5 | 31.7 | 21.8 | 18.0 | 19.4 | 19.2 | 20.0 | 22.5 | 23.0 |
| Optimistic | 22.5 | 25.6 | 39.6 | 27.3 | 22.5 | 24.3 | 24.0 | 25.0 | 28.1 | 28.8 |

*Table 3.3.* Asymptotic value $\bar{d}$ of the rate of depreciation (%).

| | *USA* | *FRG* | *Japan* | *France* | *UK* | *Italy* | *Neth.* | *B/L* | *Canada* | *Other* |
|---|---|---|---|---|---|---|---|---|---|---|
| Depreciation rate $\bar{d}$ | 5.6 | 3.9 | 5.9 | 4.5 | 4.7 | 3.4 | 2.7 | 2.8 | 5.0 | 2.9 |

and *Figures 3B.3(a)–(h)* in Appendix 3B]. In almost all countries we observe a similar pattern: the trend of the rate of investment increases during the 1950s and the 1960s and decreases in the late 1960s or the early 1970s. The upward trend was broken before the oil price increase of 1973/74 in several countries (e.g., UK, Italy, Netherlands and Belgium/Luxembourg).

For the medium scenario the rate of gross fixed investment was assumed to approach the mean value of the decade 1975–84, $\bar{s}$:

$$s_t = \lambda \cdot s_{t-1} + (1-\lambda) \cdot \bar{s} \tag{3.6}$$

The adjustment speed parameter $\lambda$ was set to 0.9. This implies that an initial difference $(s_t - \bar{s})$ in 1984 will be reduced to roughly 50% of that value in 1990 and 20% in 2000.

For the optimistic and pessimistic scenarios we fixed the final values $\bar{s}$ at 25% above and below the value for the medium scenario. The values are listed in *Table 3.2.*

This method may be slightly favorable for Japan and somewhat pessimistic for the USA and the UK. The results are given in *Table 3A.4* of Appendix 3A for some selected years and graphically demonstrated in *Figure 3B.3* of Appendix 3B.

### 3.3.2. The rate of depreciation

The rate of depreciation $d$ was rather constant for most countries [see *Table 3A.5* of Appendix 3A and *Figures 3B(a)–(h)* of Appendix 3B]. There are some exceptions to the rule: in the Japanese economy $d$ increased from 6.7% in 1950 to approximately 10% in 1964–1969 and decreased afterwards. The US rate of depreciation increased from about 5% until the early 1970s to above 6% in 1983/84. For Canada we observe a decline from about 6% in the early 1960s to about 5% in 1984. A higher rate of depreciation slows down the accumulation of capital, but it increases technical progress by sorting out older and less efficient capital goods. This is not made explicit in our model.

For all scenarios we assumed the depreciation rate of each country to approach its mean value of 1975–1984, $\bar{d}$, asymptotically:

$$d_t = \lambda \cdot d_{t-1} + (1-\lambda) \cdot \bar{d}, \quad \lambda = 0.9 \tag{3.7}$$

Asymptotic results are shown in *Table 3.3*; figures for the rate of depreciation are given in *Table 3A.5* of Appendix A and graphically represented in *Figures 3B.3(a)–(h)* in Appendix 3B.

Since the USA and the UK have a relatively low rate of investment but a relatively high rate of depreciation, their capital stock will grow relatively slowly. The accumulation equation

$$K_t = K_{t-1} \cdot (1-d) + s \cdot Y_{t-1} \tag{3.8}$$

implies a growth rate of the capital stock of $w_{Kt} = s \cdot Y_{t-1}/K_{t-1} - d$. A growth rate of capital stock of 3% (4%) in the UK in the next few years would require a 20.8% (23.5%) rate of investment with a depreciation rate unchanged at 4.7%, or a depreciation rate of 3.7% (2.7%) with the investment rate unchanged at 18%.

### 3.3.3. Labor input

Only in four countries was total labor input (in billion working hours) higher in 1984 than in 1960 (see *Table 3.4*).

*Table 3.4(a)*. Growth rate of labor input, 1960–1964, for selected industrialized market economies.

| *Labor input* | *1960* | *1984* | *Average growth rate (%)* |
|---|---|---|---|
| USA | 132.0 | 192.7 | 1.6 |
| Japan | 108.0 | 124.1 | 0.6 |
| Canada | 12.5 | 22.0 | 2.4 |
| Other industrialized market economies | 87.5 | 94.0[a] | 0.3 |

[a]1982 value.

*Table 3.4(b)*. Yearly change in labor input, 1960–1984, for selected industrialized market economies.

| *Labor input* | *1960* | *1984* | *Yearly change (%)* |
|---|---|---|---|
| FRG | 55.6 | 42.8 | –1.1 |
| France | 45.2 | 42.6 | –0.25 |
| UK | 56.0 | 50.8 | –0.4 |
| Italy | 41.2 | 32.1 | –1.0 |
| Netherlands | 10.6 | 10.5 | –0.0 |
| Belgium/Luxembourg | 7.8 | 6.6 | –0.7 |

All other countries experienced a decrease of labor input on the average [see *Table 3.4(b)*]. But also in these countries labor input showed a transitory increase in the middle of the 1960s and in the beginning of the 1970s.

Labor input ($LAB$) can be factorized into

$$LAB = POP \cdot LP'POP \cdot ER \cdot HWW \cdot 52/1000 \tag{3.9}$$

where $POP$ is the population, $LP'POP$ is the share of labor force in the population (potential labor participation rate), $ER$ is the employment rate, and $HWW$ is the average number of working hours per week. $LABF = POP \cdot LP'POP$ is the labor force and $LABF \cdot ER$ is the number of employed persons [see *Table 3A.6* in Appendix 3A and *Figures 3B.4(a–h)* in Appendix 3B].

These factors contributed to the rate of change of labor input in the following way:

(1) The population increased in all countries in the period 1960–1984.
(2) The potential labor participation rate increased in all countries, on average, with the exception of the FRG and Italy.
(3) The employment ratio decreased sharply after the beginning of the 1970s, with the exception of Japan.
(4) The average number of working hours per week decreased continuously in all countries.

*Table 3.5* shows the determinants of the average growth rates of labor input (in percentages) for all countries in the period 1961–1984. We estimated growth rates for the four determinants of labor input for the years 1985–2000, or we took them from other sources. Growth rates of *population* were taken from UN forecasts (DRPA statistics).

The potential *labor participation* rate $LP'POP$ was assumed in the long run to approach the high level of about 49%, similar to that of Japan and Canada. We restricted the labor participation growth rate to no more than 1% a year. The results are given in *Table 3.6*.

We assumed that the *employment rate ER* will increase slowly until the year 2000. *Table 3.7* shows the "required" growth rate of *ER* which would yield an employment rate of 100% in the year 2000. We did not take this growth rate of *ER* but some lower value, called the "assumed" growth rate. By this method we restricted the employment growth rate to less than or equal to 0.5%.

We assumed that the general trend of declining *working hours per week* will continue. In detail, we assumed that the average rates of decline during the years 1961–1984 will stay unchanged in the future.

These assumptions on future labor input are summarized in *Table 3.8*. A comparison of the second and the fourth row shows that the increase of the labor participation rate is to a large extent cancelled by the decrease in hours worked per week. But considering the increasing population and the reduction of unemployment, we arrive at rather high growth rates of labor input. These assumptions on future trends of labor supply are represented in *Figures 3B.4(a–h)* in Appendix 3B.

### 3.3.4. The rate of technical progress

For reasons explained in Chapter 1, we assume that the declining trend of technical progress will be reversed in the future. For the optimistic scenario, this is modeled as a return to approximately the average level of 1961–1984; for the medium scenario, as a return to approximately the average level of 1971–1984; and for the pessimistic scenario, as a return to approximately the average of 1976–1984. If outside information was available or our own judgment pointed in

*Table 3.5.* Average growth rates of the determinants of labor input, 1961–1984 (%).

| *Determinants* | *USA* | *FRG* | *Japan* | *France* | *UK* | *Italy* | *Neth.* | *B/L* | *Canada* | *Other*[a] |
|---|---|---|---|---|---|---|---|---|---|---|
| Population | 1.1 | 0.4 | 1.0 | 0.8 | 0.3 | 0.6 | 1.0 | 0.3 | 1.4 | 1.4 |
| Labor participation | 0.9 | −0.3 | 0.1 | 0.1 | 0.1 | −0.3 | 0.4 | 0.3 | 1.4 | −0.1 |
| Employment | −0.1 | −0.3 | −0.0 | −0.4 | −0.4 | −0.2 | −0.6 | −0.5 | −0.2 | −0.2 |
| Working time | −0.4 | −0.9 | −0.5 | −0.7 | −0.4 | −1.0 | −0.8 | −0.9 | −0.2 | −0.7 |
| Labor input[b] | 1.6 | −1.1 | 0.6 | −0.2 | −0.4 | −1.0 | 0.0 | −0.7 | 2.4 | 0.3 |

[a]Average 1961–1982 only.
[b]Due to rounding, the first four figures in the columns do not always add up to the last figure.

*Table 3.6.* Assumed development of the labor participation rate.

| | *USA* | *FRG* | *Japan* | *France* | *UK* | *Italy* | *Neth.* | *B/L* | *Canada* | *Other* |
|---|---|---|---|---|---|---|---|---|---|---|
| $LP'POP$ (1984) | 48.0 | 44.2 | 49.5 | 42.3 | 47.2 | 40.0 | 40.2 | 42.3 | 49.5 | 40.7[a] |
| Assumed growth rate (%) | 0.2 | 0.8 | 0.0 | 1.0 | 0.3 | 1.0 | 1.0 | 1.0 | 0.0 | 1.0 |

[a]1982 value.

*Table 3.7.* Estimate of the employment rate (*ER.*)

| *ER* | *USA* | *FRG* | *Japan* | *France* | *UK* | *Italy* | *Neth.* | *B/L* | *Canada* | *Other* |
|---|---|---|---|---|---|---|---|---|---|---|
| Actual (1984) | 92.5 | 91.3 | 97.3 | 90.1 | 89.1 | 89.5 | 85.8 | 86.3 | 88.7 | 93.7[a] |
| 100% (1984) | 7.5 | 8.7 | 2.7 | 9.9 | 10.9 | 10.5 | 14.2 | 13.7 | 11.3 | 6.3 |
| "Required" growth rate (%) | 0.45 | 0.52 | 0.17 | 0.59 | 0.65 | 0.63 | 0.83 | 0.81 | 0.67 | 0.38 |
| "Assumed" growth rate (%) | 0.3 | 0.4 | 0.1 | 0.5 | 0.5 | 0.5 | 0.5 | 0.5 | 0.5 | 0.3 |
| Result in 2000 | 97.0 | 97.3 | 98.9 | 97.6 | 96.5 | 96.9 | 92.9 | 93.5 | 96.1 | 98.3 |

[a]1982 value.

*Table 3.8.* Assumptions regarding labor input.

| *% growth rate of* | *USA* | *FRG* | *Japan* | *France* | *UK* | *Italy* | *Neth.* | *B/L* | *Canada* | *Other* |
|---|---|---|---|---|---|---|---|---|---|---|
| Population | 1.0 | −0.2 | 0.6 | 0.3 | −0.1 | 0.3 | 0.4 | 0.1 | 1.4 | 1.2 |
| Labor participation | 0.2 | 0.8 | 0.0 | 1.0 | 0.3 | 1.0 | 1.0 | 1.0 | 0.0 | 1.0 |
| Employment rate | 0.3 | 0.4 | 0.1 | 0.5 | 0.5 | 0.5 | 0.5 | 0.5 | 0.5 | 0.3 |
| Working time | −0.4 | −0.5 | −0.5 | −0.7 | −0.4 | −1.0 | −0.8 | −0.9 | −0.2 | −0.7 |
| Labor input | 1.1 | 0.5 | 0.2 | 1.1 | 0.3 | 0.8 | 1.1 | 0.7 | 1.7 | 1.8 |

another direction, we adapted these figures a little. This may be seen from Appendix 2E in Chapter 2. We do not reproduce these figures here.

## 3.4. Solution of Isolated Country Models: Future Growth Rates

The forecasts for OECD countries which follow from the solution of the total world model were reviewed in Chapter 2. They are reproduced in Annex 3. We also solved the country models separately by making some simplifying assumptions on total imports and imports of raw materials for each country [3]. The average growth rates of GDP for the years 1983–1999 which result from these solutions of isolated country models are reproduced in *Table 3.9.*

*Table 3.9(a).* Average growth rates derived from individual country models and from the solution of the interdependent world model. Average growth rates of GDP, 1983–1999, were estimated by individual country models.[a]

| *Scenario* | *USA* | *FRG* | *Jap* | *Fra* | *UK* | *Ita* | *NL* | *B/L* | *Can* | *Other* |
|---|---|---|---|---|---|---|---|---|---|---|
| Optimistic | 3.1 | 4.4 | 6.9 | 4.8 | 2.4 | 5.1 | 4.2 | 4.6 | 3.5 | 5.5 |
| Medium | 2.7 | 3.9 | 5.0 | 4.1 | 2.0 | 3.7 | 3.0 | 3.8 | 3.0 | 4.5 |
| Pessimistic | 2.1 | 2.7 | 2.6 | 3.1 | 1.5 | 1.7 | 1.5 | 2.6 | 1.5 | 2.8 |

[a]Estimated by $100 \cdot [(x_{99}/x_{82})^{1/17} - 1]$.

*Table 3.9(b).* Average growth rates of GDP, 1985–1999, estimated by the interdependent world model.[a]

| *Scenario* | *USA* | *FRG* | *Jap* | *Fra* | *UK* | *Ita* | *NL* | *B/L* | *Can* | *Other* |
|---|---|---|---|---|---|---|---|---|---|---|
| Optimistic | 2.8 | 3.9 | 5.5 | 3.7 | 1.6 | 4.4 | 3.5 | 3.8 | 3.5 | 4.0 |
| Medium | 2.2 | 2.8 | 4.1 | 2.8 | 1.1 | 2.9 | 2.1 | 2.8 | 2.5 | 2.9 |
| Pessimistic | 1.6 | 1.9 | 2.9 | 1.9 | 0.5 | 1.8 | 0.9 | 1.9 | 1.7 | 1.7 |

[a]Estimated as the arithmetic mean of the growth rates 1985–1999.

Of course, these growth rates differ from those of the solution of the interdependent world model, cf. *Table 2* in Annex 3. Averages are given in *Table 3.9(b)*. It is surprising that the growth rates of all OECD countries are reduced by considering the international interdependence of all countries (including the developing ones). This seems to be due to the special way in which we isolated the countries from each other; see Ross (1986). We put an upper limit on the import ratio which does not depend on price ratios. But there might also be other reasons for the upward bias of isolated country models of the type used in the Bonn–IIASA research project.

## 3.5. The Rate of Inflation

The price level $p$ of GDP is explained by a generalization of the classical quantity theory:

$$p = v \cdot \frac{M2}{GDP} \tag{3.10}$$

where $v$ = velocity of money M2 = money supply (in our analysis we use the money concept usually referred to as M2, but include saving deposits, as is done by the IMF in the IFS). If $v$ is constant, the rate of inflation is equal to the difference between the growth rate of money supply and the growth rate of GDP.

The velocity of money in the industrialized market economies was not constant in the years 1960–1984, with the exception of Belgium/Luxembourg and the group of other industrialized market economies; see *Table 3A.7* of Appendix 3A and *Figures 3B.5(a)–(h)* of Appendix 3B. In the UK [*Figure 3B.5(e)*] the velocity of money started to decrease in 1970 but turned back to an even higher level in the high-inflation period of 1974–1981. For the USA there is an increasing trend of $v$. In all other countries we observe a decrease of the velocity of money which was stopped or even reversed in the early 1980s.

The velocity of money reflects short-term effects as well as the long-term changes in the marginal propensity to hold money. This propensity depends on business customs, techniques of payments, the level and distribution of wealth, the rate of interest, the capital coefficient, the foreign debt ratio and other variable. We specified the equation for the velocity of money as:

$$\tilde{v}_t = a_1 \cdot \tilde{v}_{t-1} + a_2 \cdot (\tilde{k}_t - \tilde{k}_{t-3}) + a_3 \cdot (\tilde{\pi}_t - \tilde{\pi}_{t-3}) + a_4 \cdot \frac{A_t^n}{\mathrm{GDP}_t^n} + a_5 \tag{3.11}$$

where $\tilde{x}$ means the logarithm of $x$, $k$ = capital coefficient $K/Y$, $\pi$ = labor productivity $Y/L$ and $A^n$ = accumulated foreign trade surplus. The estimated results are reproduced elsewhere [4]. Some comments must suffice here.

In general the approach worked quite well. The coefficients $a_2$ is always negative, if it is significantly different from zero, and the same is true for $a_4$. As the capital coefficient was increasing in all countries during this period, the coefficient $a_2$ partly explains the declining velocity of money. If the capital coefficient can be interpreted as a measure of relative wealth, then a higher wealth in a country will imply a higher propensity to hold money. The negative coefficient $a_4$ indicates that trade surpluses (accumulated) slow inflation and trade deficits speed up inflation. The dynamic behavior of the equation is satisfactory. If we assume constancy of the capital coefficients, balanced trade and 3% growth of labor productivity, we get the asymptotic values for the velocity of money shown in *Table 3.10*.

*Table 3.10.* Asymptotic velocity of money.

| | *USA* | *FRG* | *Jap* | *Fra* | *UK* | *Ita* | *NL* | *B/L* | *Can* | *Other* |
|---|---|---|---|---|---|---|---|---|---|---|
| $\bar{v}$ | 2.14 | 1.93 | 1.29 | 2.14 | 2.72 | 1.30 | 1.95 | 2.02 | 2.23 | 1.50 |

This holds a reasonable relationship with the observed values (see *Table 3A.7* in Appendix 3A).

The forecasts of the growth rates of money supply (M2) are related to the historical averages (see *Table 2.1* in Chapter 2).

The resulting "rates of inflation" are reproduced in *Table 32* of Annex 3 for the past as well as for the future. *Table 3.11* shows forecasts of the rate of inflation for the years 1990 and 1999 for the OECD countries for the different scenarios.

*Table 3.11.* Forecasts of the rate of inflation.

| *Scenario* | *USA* | *FRG* | *Jap* | *Fra* | *UK* | *Ita* | *NL* | *B/L* | *Can* | *Other* |
|---|---|---|---|---|---|---|---|---|---|---|
| Optimistic | | | | | | | | | | |
| 1990 | 5.3 | 2.2 | 4.6 | 6.4 | 10.5 | 11.5 | 7.3 | 3.8 | 8.3 | 7.7 |
| 1999 | 5.8 | 1.0 | 3.7 | 6.7 | 9.8 | 10.5 | 6.1 | 3.4 | 8.0 | 8.3 |
| Medium | | | | | | | | | | |
| 1990 | 6.3 | 3.4 | 6.0 | 8.0 | 11.0 | 13.0 | 8.5 | 4.6 | 9.3 | 9.1 |
| 1999 | 7.7 | 2.5 | 4.9 | 7.9 | 10.4 | 12.6 | 7.7 | 4.4 | 9.2 | 9.7 |
| Pessimistic | | | | | | | | | | |
| 1990 | 7.3 | 4.5 | 7.3 | 9.4 | 11.6 | 14.2 | 9.5 | 5.4 | 10.2 | 10.5 |
| 1999 | 10.4 | 4.1 | 6.2 | 9.3 | 11.1 | 14.1 | 9.3 | 5.6 | 10.3 | 11.2 |

These results indicate that inflation is not a matter of the past if the rate of growth of money supply stays within the order of magnitude of the past (see *Table 2.1* in Chapter 2). Money has to grow at a lower rate if inflation is to stay at the low levels which are nowadays current.

## 3.6. The Nominal Rate of Interest

Interest rates influence the exchange rates and capital flows between countries. This in turn influences foreign trade and the foreign debt situation and therefore the domestic price level. Unfortunately, this part of the research program could not be connected with the world model for reasons explained in Chapter 2. We had to use a simpler approach for exchange rate determination. Nevertheless, rates of interest are explained by our model.

We do not differ between different asset markets and their special rates of interest. Only one average rate of interest, $z$, is considered. It is measured as an average of the short-term money market rate, the government bond yield and the discount rate.

It is well known that the nominal interest rates of all countries are highly correlated. The interest rate of the USA plays a leading role. Therefore we estimated, for all other countries, the equation

$$z_t = a_1 \cdot z_{t-1} + a_2 \cdot w_{p,t} + a_3 \cdot z_t^{\mathrm{USA}} + a_4 \cdot \left(\frac{s_t}{s_{t-4}} - 1\right) + a_5 \cdot \left(\frac{A_t^n}{\mathrm{GDP}_t^n}\right) + a_6 \tag{3.12}$$

The term $s_t/s_{t-4}$ is an indicator of the change of investment activity which will influence interest rates. $A_t^n/\mathrm{GDP}_t^n$ is the ratio of cumulated trade surplus to nominal gross domestic product which should decrease domestic interest rates (deficits should increase $z_t$). Indeed the coefficient $a_5$ turns out to be always negative and such that the rate of interest $z_t$ increases about 0.2 to 0.5 percentage points if the ratio of foreign debts to GDP increases by one percentage point. The coefficient $a_2$ connected with the rate of inflation $w_p$ is rather low whereas the coefficient $a_3$ for the US interest rate $z^{\mathrm{USA}}$ is always very high.

The figures for $RC^2$ and for $DW$ indicate that this approach takes into account the most important determinants [5]. One of them is the interest rate of the USA. For the USA we used the same approach, but set $a_3$ to zero. The estimated results are quite good. The dynamic behavior of the equation is acceptable, too. If, in the long run, the inflation rate is constant at $w_p$, the investment rate $s$ is also constant and the foreign debts are zero, then the US rate stabilizes at

$$\bar{z}^{\mathrm{USA}} = \frac{a_2}{1-a_1} \cdot \bar{w}^p + \frac{a_6}{1-a_1} = 0.73 \cdot \bar{w}^p + 4.48 \tag{3.13}$$

With an inflation rate of 5% the nominal rate of interest would stabilize at 8.13% and the real rate of interest at 8.13% – 5% = 3.13%, which is considerably smaller than the rate of profit which follows from the marginal productivity theory (for $K/Y = 3$ and $\beta = 0.25$, it follows that $z = 0.25/3 = 8.3\%$).

Under the same assumptions the nominal rates of interest for the other countries would stabilize at the values shown in *Table 3.12*.

Interest rate forecasts are given in *Table 3A.8* of Appendix 3A. The optimistic scenario yields lower interest rates eventually than the medium scenario, which in turn produces lower rates than the pessimistic scenario with the notable exception of France, where the opposite is true. This needs further research.

*Table 3.12.* Asymptotic nominal rates of interest.

| | *FRG* | *Jap* | *Fra* | *UK* | *Ita* | *NL* | *B/L* | *Can* | *Other* |
|---|---|---|---|---|---|---|---|---|---|
| $\bar{z}$ | 7.1 | 8.5 | 7.4[a] | 9.8 | 8.6[a] | 7.7 | 8.5 | 9.6 | 6.8 |

[a]Additional assumptions: $s_t = 20.0$ for France; no change in the exchange rate for Italy.

## 3.7. The Structure of Production

There are a number of structural variables which are determined by the model: the capital coefficient, the import and export share in GDP, the share of consumption in GDP, labor productivity, the capital/labor ratio, and others. These variables are related to the structure of production and may be used as explanatory variables to forecast structural change. By "sectoral structure of production" we mean the share in GDP of real value added to different sectors of the economy. *Tables 3A.9(a)* and *(b)* in Appendix 3A show the size and the stability of the shares of real value added and nominal value added.

We specified a log-linear function to explain the shares of real value added:

$$\beta = \pi^{a_1} \cdot (K/L)^{a_2} \cdot (X/Y)^{a_3} \cdot (M/Y)^{a_4} \cdot s^{a_5} \cdot a_6 \tag{3.14}$$

where $\beta$ = real value added share of the sector, $\pi$ = labor productivity, $K/L$ = capital/labor ratio, $X/Y$ = export share, $M/Y$ = import share, and $s$ = gross investment ratio. The estimated results are reported elsewhere [6]. They are quite satisfactory for the great majority of cases, but certainly need some more work in some cases. The results are reproduced in *Table 36* of Annex 3 and commented upon in Chapter 2.

## 3.8. Conclusion

We think that the models developed for OECD countries are able to simulate the actual behavior of the most important aggregates of these economies in a satisfactory way, if the driving forces are estimated correctly. This, of course, is the crux.

### Notes

[1] The method of aggregation is reported in Ross (1986). The data bank used is described in detail in Ross (1985). National accounts data have been taken from OECD publications, employment figures from ILO and OECD publications and

monetary items from the International Financial Statistics (IFS) of the International Monetary Fund.

[2] For details, see Eismont and Ross (1985).

[3] For details, see Ross (1986). Basically, we related real imports to real GDP. Thus the international price competition has been neglected.

[4] See Ross (1986), Appendix A, *Table 9*.

[5] Details may be seen in Ross (1986, Appendix A, *Table 10*).

[6] See Ross (1986, Appendix A, *Table 12*).

# Appendix 3A. Model Parameters (tabulated)

Country codes:

01 = USA
02 = Federal Republic of Germany
03 = Japan
04 = France
05 = UK
06 = Italy
07 = Netherlands
08 = Belgium and Luxembourg
09 = Canada
10 = Other industrialized market economies

*Table 3A.1.* Average cost structure and trend of the Divisia Index of (Harrod-neutral) technical progress (%).

| | 1[a] | 2[b] | 3[c] | 4[d] | 5[e] |
|---|---|---|---|---|---|
| *Country* | $\alpha_3$ | $\beta_3$ | $\gamma_3$ | $W'T$ "60" | $\delta_3$ |
| 01 | 0.731 | 0.252 | 0.017 | 3.3 | −0.15 |
| 02 | 0.687 | 0.262 | 0.051 | 5.9 | −0.18 |
| 03 | 0.725 | 0.213 | 0.062 | 11.2 | −0.49 |
| 04 | 0.705 | 0.246 | 0.049 | 5.7 | −0.15 |
| 05 | 0.698 | 0.247 | 0.054 | 2.6 | −0.05 |
| 06 | 0.774 | 0.158 | 0.068 | 9.2 | −0.42 |
| 07 | 0.647 | 0.253 | 0.100 | 7.7 | −0.35 |
| 08 | 0.657 | 0.234 | 0.109 | 6.2 | −0.18 |
| 09 | 0.688 | 0.284 | 0.028 | 5.1 | −0.28 |
| 10 | 0.699 | 0.248 | 0.053 | 6.8 | −0.29 |

[a]Labor share using $K,L,R$ 1962–1983 average. [b]Capital share using $K,L,R$ 1962–1983 average. [c]Raw imports share using $K,L,R$ 1962–1983 average. [d]Intercept: growth rate of technical progress in 1960. [e]Slope: yearly change of growth rate of technical progress.

*Table 3A.2.* Parameters of the production function (%).

| | 1[a] | 2[b] | 3[c] | 4[d] | 5[e] |
|---|---|---|---|---|---|
| *Country* | $\alpha_3$ | $\beta_3$ | $\gamma_3$ | $W'\tau$ "60" | $\delta_3$ |
| 01 | 0.706 | 0.259 | 0.035 | 2.5 | −0.11 |
| 02 | 0.644 | 0.240 | 0.115 | 5.9 | −0.15 |
| 03 | 0.717 | 0.143 | 0.140 | 10.3 | −0.40 |
| 04 | 0.717 | 0.209 | 0.073 | 6.3 | −0.17 |
| 05 | 0.684 | 0.255 | 0.061 | 3.0 | −0.10 |
| 06 | 0.791 | 0.109 | 0.100 | 9.8 | −0.43 |
| 07 | 0.553 | 0.317 | 0.131 | 9.7 | −0.47 |
| 08 | 0.634 | 0.187 | 0.179 | 6.8 | −0.17 |
| 09 | 0.648 | 0.320 | 0.032 | 6.8 | −0.35 |
| 10 | 0.647 | 0.251 | 0.102 | 7.6 | −0.32 |

[a]Elasticity of production with respect to labor. [b]Elasticity of production with respect to capital. [c]Elasticity of production with respect to imported goods. [d]Rate of technical progress in 1960. [e]Yearly change of the rate of technical progress.

*Table 3A.3.* Parameters of production functions (values in parentheses are $t$-values).

| *Country* | $\alpha$ | $\beta$ | $\gamma$ | $\bar{W}_T$ | $\delta$ |
|---|---|---|---|---|---|
| 01 | 0.706 | 0.259 | 0.035 | 2.465 | −0.106 |
| | (33.88) | (13.20) | (3.58) | (7.15) | (4.40) |
| 02 | 0.644 | 0.240 | 0.115 | 5.866 | −0.154 |
| | (26.92) | (8.93) | (7.38) | (17.81) | (6.41) |
| 03 | 0.717 | 0.143 | 0.140 | 10.284 | −0.404 |
| | (18.47) | (5.01) | (4.90) | (12.41) | (7.30) |
| 04 | 0.717 | 0.209 | 0.073 | 6.327 | −0.174 |
| | (66.35) | (18.61) | (5.68) | (27.95) | (10.05) |
| 05 | 0.684 | 0.255 | 0.061 | 3.001 | −0.103 |
| | (27.88) | (9.59) | (4.65) | (8.02) | (3.55) |
| 06 | 0.791 | 0.109 | 0.100 | 9.807 | −0.427 |
| | (27.14) | (4.62) | (7.83) | (16.57) | (10.03) |
| 07 | 0.553 | 0.317 | 0.131 | 9.701 | −0.474 |
| | (16.74) | (10.60) | (7.36) | (19.17) | (12.08) |
| 08 | 0.634 | 0.187 | 0.179 | 6.794 | −0.173 |
| | (20.33) | (7.45) | (6.11) | (10.93) | (3.70) |
| 09 | 0.648 | 0.320 | 0.032 | 5.834 | −0.347 |
| | (20.58) | (11.70) | (2.57) | (17.61) | (12.97) |
| 10 | 0.647 | 0.251 | 0.102 | 7.614 | −0.316 |
| | (39.11) | (16.62) | (8.80) | (19.48) | (11.24) |

*Table 3A.3.* Continued.

| *Country* | *Constant* | *DW* | *R2C* | *SEE* | *Est.* |
|---|---|---|---|---|---|
| 01 | 1.370 | 1.17 | 0.99 | 18.054 | 63–83 |
| | (21.57) | | | | FIML |
| 02 | 1.702 | 0.80 | 1.00 | 10.106 | 63–83 |
| | (18.80) | | | | FIML |
| 03 | −0.217 | 0.65 | 1.00 | 2.747 | 63–83 |
| | (1.79) | | | | FIML |
| 04 | 2.019 | 1.32 | 1.00 | 12.415 | 63–83 |
| | (50.85) | | | | FIML |
| 05 | 0.248 | 1.37 | 0.97 | 1.590 | 63–83 |
| | (5.81) | | | | FIML |
| 06 | 0.548 | 0.90 | 0.98 | 2.400 | 63–83 |
| | (11.50) | | | | FIML |
| 07 | 1.213 | 0.71 | 0.99 | 3.122 | 63–82 |
| | (12.04) | | | | FIML |
| 08 | 3.413 | 0.78 | 0.98 | 53.761 | 63–83 |
| | (22.81) | | | | FIML |
| 09 | 1.015 | 1.29 | 1.00 | 1.855 | 63–83 |
| | (14.40) | | | | FIML |
| 10 | 0.620 | 1.39 | 0.99 | 5.055 | 65–82 |
| | (12.72) | | | | FIML |

*Table 3A.4.* Ratio of gross fixed investment to GDP (%).

| | | | | | Forecast | |
|---|---|---|---|---|---|---|
| *Country* | *1960* | *1968* | *1976* | *1984* | *1992* | *1999* |
| 01 | 18.528 | 19.700 | 17.229 | 19.642 | 18.707 | 18.338 |
| 02 | 24.422 | 22.603 | 20.260 | 19.620 | 20.121 | 20.319 |
| 03 | 22.411 | 30.656 | 31.717 | 30.206 | 31.057 | 31.392 |
| 04 | 19.222 | 23.500 | 23.025 | 20.014 | 21.031 | 21.432 |
| 05 | 16.793 | 20.989 | 19.020 | 18.295 | 18.127 | 18.061 |
| 06 | 25.192 | 24.235 | 19.765 | 18.663 | 19.083 | 19.248 |
| 07 | 21.760 | 26.865 | 19.320 | 17.693 | 18.551 | 18.890 |
| 08 | 21.428 | 23.041 | 22.150 | 16.740 | 18.597 | 19.329 |
| 09 | 22.727 | 22.282 | 23.573 | 19.717 | 21.302 | 21.927 |
| 10 | 22.117 | 24.558 | 24.085 | 21.968 | 22.640 | 22.828 |

Note: 1960–1984 observed, 1992–1999 assumptions for the medium scenario.

*Table 3A.5.* Rate of depreciation (%).

| | | | | | Forecast | |
|---|---|---|---|---|---|---|
| *Country* | *1960* | *1968* | *1976* | *1984* | *1992* | *1999* |
| 01 | 5.023 | 5.039 | 5.291 | 6.331 | 5.915 | 5.751 |
| 02 | 3.899 | 4.041 | 3.956 | 3.912 | 3.905 | 3.902 |
| 03 | 8.807 | 10.083 | 6.195 | 5.814 | 5.863 | 5.882 |
| 04 | 4.444 | 4.367 | 4.724 | 4.531 | 4.513 | 4.506 |
| 05 | 4.537 | 4.561 | 4.446 | 4.933 | 4.800 | 4.748 |
| 06 | 3.757 | 3.656 | 3.442 | 3.358 | 3.382 | 3.391 |
| 07 | 2.581 | 2.661 | 2.684 | 2.669 | 2.687 | 2.694 |
| 08 | 3.420 | 3.336 | 2.909 | 2.719 | 2.765 | 2.783 |
| 09 | 5.968 | 5.775 | 5.254 | 4.994 | 4.998 | 4.999 |
| 10 | 2.232 | 2.659 | 2.547 | 2.466 | 2.749 | 2.828 |

Note: 1960–1984 observed, 1992–1999 assumptions for the medium scenario.

*Table 3A.6.* Determinants of labor input.[a,b]

| *Country/Determinant* | *1960* | *1968* | *1976* | *1984* | *1992* | *1999* |
|---|---|---|---|---|---|---|
| *USA*: | | | | | | |
| labor input (bill. h/year) | 132.030 | 149.228 | 166.605 | 192.747 | 210.376 | 227.119 |
| population (mill.) | 180.680 | 200.710 | 218.040 | 236.680 | 256.289 | 274.774 |
| potential labor participation | 38.538 | 39.229 | 44.101 | 47.969 | 48.742 | 49.428 |
| employment rate | 94.468 | 96.422 | 92.298 | 92.489 | 94.731 | 96.738 |
| hours of work per week | 38.600 | 37.800 | 36.100 | 35.300 | 34.186 | 33.240 |
| *Federal Republic of Germany*: | | | | | | |
| labor input (bill. h/year) | 55.644 | 50.556 | 45.946 | 42.831 | 44.574 | 46.157 |
| population (mill.) | 55.430 | 60.170 | 61.510 | 61.180 | 60.208 | 59.370 |
| potential labor participation | 47.312 | 42.902 | 42.463 | 44.157 | 47.063 | 49.762 |
| employment rate | 98.967 | 98.749 | 95.942 | 91.312 | 94.275 | 96.947 |
| hours of work per week | 41.230 | 38.140 | 35.260 | 33.390 | 32.077 | 30.971 |
| *Japan*: | | | | | | |
| labor input (bill. h/year) | 107.954 | 115.746 | 110.459 | 124.130 | 126.130 | 127.907 |
| population (mill.) | 94.100 | 101.960 | 112.770 | 120.020 | 125.903 | 131.287 |
| potential labor participation | 47.938 | 49.637 | 47.699 | 49.375 | 49.375 | 49.375 |
| employment rate | 98.337 | 98.834 | 97.992 | 97.300 | 98.081 | 98.769 |
| hours of work per week | 46.800 | 44.500 | 40.300 | 41.400 | 39.773 | 38.401 |
| *France*: | | | | | | |
| labor input (bill. h/year) | 45.246 | 47.194 | 45.983 | 42.577 | 46.471 | 50.170 |
| population (mill.) | 45.680 | 49.910 | 52.890 | 54.950 | 56.282 | 57.475 |
| potential labor participation | 41.486 | 40.387 | 41.310 | 42.317 | 45.823 | 49.128 |
| employment rate | 98.739 | 97.882 | 95.455 | 90.057 | 93.722 | 97.052 |
| hours of work per week | 46.500 | 46.000 | 42.400 | 39.100 | 36.963 | 35.190 |
| *United Kingdom*: | | | | | | |
| labor input (bill. h/year) | 55.980 | 55.528 | 53.236 | 50.806 | 52.038 | 53.140 |
| population (mill.) | 52.350 | 55.050 | 55.890 | 56.490 | 56.040 | 55.648 |
| potential labor participation | 45.818 | 45.373 | 46.085 | 47.233 | 48.378 | 49.403 |
| employment rate | 98.641 | 97.830 | 95.089 | 88.663 | 92.271 | 95.549 |
| hours of work per week | 45.500 | 43.700 | 41.800 | 41.300 | 39.997 | 38.890 |

| | | | | | | |
|---|---|---|---|---|---|---|
| *Italy*: | | | | | | |
| labor input (bill. h/year) | 41.211 | 36.650 | 31.270 | 32.060 | 34.170 | 36.130 |
| population (mill.) | 49.640 | 52.990 | 56.170 | 56.980 | 58.362 | 59.598 |
| potential labor participation | 43.280 | 38.622 | 37.686 | 40.032 | 43.348 | 46.475 |
| employment rate | 94.346 | 94.273 | 93.263 | 89.531 | 93.175 | 96.485 |
| hours of work per week | 39.100 | 36.530 | 30.460 | 30.190 | 27.858 | 25.965 |
| *Netherlands*: | | | | | | |
| labor input (bill. h/year) | 10.612 | 10.753 | 9.995 | 10.482 | 11.441 | 12.351 |
| population (mill.) | 11.480 | 12.720 | 13.770 | 14.420 | 14.888 | 15.310 |
| potential labor participation | 36.690 | 36.454 | 35.817 | 40.153 | 43.479 | 46.615 |
| employment rate | 99.288 | 98.447 | 94.363 | 85.751 | 89.241 | 92.411 |
| hours of work per week | 48.800 | 45.300 | 41.300 | 40.600 | 38.073 | 35.992 |
| *Belgium and Luxembourg*: | | | | | | |
| labor input (bill. h/year) | 7.784 | 7.886 | 7.212 | 6.568 | 6.945 | 7.293 |
| population (mill.) | 9.464 | 9.957 | 10.180 | 10.246 | 10.328 | 10.401 |
| potential labor participation | 39.074 | 37.893 | 40.265 | 42.260 | 45.762 | 49.062 |
| employment rate | 96.836 | 97.085 | 94.511 | 86.305 | 89.817 | 93.008 |
| hours of work per week | 41.800 | 41.400 | 35.800 | 33.800 | 31.442 | 29.513 |
| *Canada*: | | | | | | |
| labor input (bill. h/year) | 12.531 | 15.912 | 19.076 | 22.022 | 25.201 | 28.357 |
| population (mill.) | 17.910 | 20.730 | 23.030 | 25.130 | 28.086 | 30.957 |
| potential labor participation | 35.796 | 38.471 | 44.316 | 49.331 | 49.331 | 49.331 |
| employment rate | 93.043 | 95.210 | 92.877 | 88.731 | 92.342 | 95.623 |
| hours of work per week | 40.400 | 40.300 | 38.700 | 38.500 | 37.888 | 37.361 |
| *Other Ind. Market Economies*: | | | | | | |
| labor input (bill. h/year) | 87.367 | 92.126 | 92.544 | 97.382 | 112.321 | 127.261 |
| population (mill.) | 94.340 | 106.690 | 118.360 | 130.250 | 143.291 | 155.769 |
| potential labor participation | 41.758 | 40.593 | 40.534 | 41.529 | 44.970 | 48.213 |
| employment rate | 98.743 | 98.528 | 97.557 | 94.249 | 96.534 | 98.579 |
| hours of work per week | 43.192 | 41.519 | 38.024 | 36.735 | 34.727 | 33.061 |

[a] 1960–1984 values are observed, 1992–1999 values are forecasts for the medium scenario.
[b] Sources: OECD, ILC, UNO, author's calculations.

*Table 3A.7.* Velocity of money (M2) for selected years.[a]

| Country | 1960 | 1968 | 1976 | 1984 | 1992 | 1999 |
|---|---|---|---|---|---|---|
| 01 | 2.429 | 2.374 | 2.550 | 2.694 | 2.756 | 2.651 |
| 02 | 2.644 | 1.929 | 1.863 | 1.753 | 1.854 | 1.792 |
| 03 | 1.495 | 1.350 | 1.166 | 1.039 | 1.278 | 1.326 |
| 04 | 2.802 | 2.164 | 1.974 | 2.133 | 2.222 | 2.165 |
| 05 | 2.465 | 2.746 | 2.800 | 2.336 | 2.415 | 2.545 |
| 06 | 1.791 | 1.361 | 1.106 | 1.304 | 1.358 | 1.407 |
| 07 | 2.213 | 2.107 | 1.907 | 1.572 | 1.762 | 1.970 |
| 08 | 2.088 | 1.952 | 1.974 | 1.947 | 1.990 | 2.024 |
| 09 | 2.948 | 2.539 | 2.307 | 2.207 | 2.145 | 2.207 |
| 10 | 1.797 | 1.654 | 1.631 | 1.563 | 1.587 | 1.572 |

[a]1960–1984 observed, 1992–1999 assumptions for the medium scenario.

*Table 3A.8.* Nominal interest rates (%) for some selected years: 1980 and 1984 observed; 1990, 1995, and 1999 forecast.

| | Observations | | Optimistic scenario | | | Medium scenario | | | Pessimistic scenario | | |
|---|---|---|---|---|---|---|---|---|---|---|---|
| Country | 1980 | 1984 | 1990 | 1995 | 1999 | 1990 | 1995 | 1999 | 1990 | 1995 | 1999 |
| 01 | 12.0 | 11.6 | 10.3 | 9.7 | 9.3 | 9.8 | 10.1 | 10.2 | 9.2 | 10.5 | 11.4 |
| 02 | 8.5 | 7.0 | 6.7 | 5.7 | 5.3 | 6.7 | 6.6 | 6.4 | 6.5 | 7.3 | 7.6 |
| 03 | 9.1 | 6.1 | 8.3 | 8.4 | 8.5 | 8.4 | 8.8 | 9.0 | 8.5 | 9.2 | 9.6 |
| 04 | 11.4 | 11.1 | 11.2 | 11.5 | 11.7 | 9.9 | 9.8 | 9.8 | 8.7 | 8.2 | 8.1 |
| 05 | 14.3 | 11.5 | 12.0 | 11.7 | 11.4 | 11.6 | 12.2 | 12.3 | 11.1 | 12.6 | 13.5 |
| 06 | 16.6 | 17.9 | 14.5 | 12.9 | 12.5 | 14.4 | 13.8 | 13.8 | 14.1 | 14.4 | 15.0 |
| 07 | 9.4 | 7.7 | 9.0 | 8.9 | 8.7 | 8.7 | 9.4 | 9.6 | 8.4 | 9.9 | 10.7 |
| 08 | 11.8 | 11.4 | 10.2 | 9.6 | 9.3 | 9.9 | 10.2 | 10.3 | 9.5 | 10.7 | 11.5 |
| 09 | 14.2 | 12.5 | 12.5 | 11.8 | 11.4 | 11.8 | 12.4 | 12.7 | 11.0 | 12.9 | 14.2 |
| 10 | 9.5 | 9.1 | 8.2 | 7.9 | 7.8 | 8.2 | 8.3 | 8.3 | 8.1 | 8.6 | 9.0 |

*Table 3A.9(a)*. GDP: Structural composition of real value added in 1960–1982.[a,b]

| *Country* | *AGR* | *MIN* | *MAN* | *EGW* | *CON* | *IND* | *SRV* |
|---|---|---|---|---|---|---|---|
| 01 | 3.5 | 2.7 | 24.8 | 2.4 | 5.6 | 35.5 | 61.0 |
| USA | 0.5 | 0.2 | 1.0 | 0.1 | 0.9 | 1.6 | 1.8 |
| | 14.3% | 7.4% | 4.0% | 4.2% | 16.1% | 4.5% | 3.0% |
| 02 | 3.2 | 1.8 | 37.1 | 2.3 | 7.5 | 48.7 | 48.1 |
| FRG | 0.5 | 0.7 | 1.8 | 0.4 | 1.0 | 2.6 | 2.9 |
| | 15.6% | 38.9% | 4.9% | 17.4% | 13.3% | 5.3% | 6.0% |
| 03 | 7.3 | 0.8 | 28.7 | 2.0 | 8.9 | 40.4 | 52.4 |
| Japan | 3.0 | 0.2 | 4.0 | 0.1 | 0.7 | 4.3 | 1.5 |
| | 4.1% | 25.0% | 13.9% | 5.0% | 7.9% | 10.6% | 2.9% |
| 04 | 6.6 | 1.4 | 29.0 | 1.7 | 7.9 | 40.0 | 53.4 |
| France | 1.5 | 0.5 | 1.3 | 0.3 | 1.0 | 4.3 | 1.8 |
| | 22.7% | 35.7% | 4.5% | 17.6% | 12.7% | 3.8% | 3.4% |
| 05 | 2.7 | 2.8 | 27.4 | 2.7 | 7.7 | 40.5 | 56.8 |
| UK | 0.1 | 1.0 | 1.9 | 0.4 | 1.0 | 1.7 | 1.6 |
| | 3.7% | 35.7% | 6.9% | 14.8% | 13.0% | 4.2% | 2.8% |
| 06 | 8.9 | 2.4 | 28.2 | 4.8 | 9.5 | 44.9 | 46.2 |
| Italy | 1.3 | 0.2 | 2.3 | 0.4 | 1.7 | 1.2 | 0.7 |
| | 14.6% | 8.3% | 8.2% | 8.3% | 17.9% | 2.7% | 1.5% |
| 07 | 5.2 | 0.1 | 28.2 | 2.1 | 7.8 | 38.2 | 56.6 |
| Neth | 0.4 | 0.0 | 1.7 | 0.5 | 1.4 | 1.7 | 1.5 |
| | 7.7% | 0.0% | 6.0% | 32.8% | 17.9% | 4.5% | 2.7% |
| 08 | 3.9 | 1.3 | 26.5 | 2.5 | 7.6 | 37.9 | 58.3 |
| B/L | 1.0 | 0.7 | 1.8 | 0.7 | 0.8 | 1.5 | 1.0 |
| | 25.6% | 53.8% | 6.8% | 28.0% | 10.5% | 4.0% | 1.7% |
| 09 | 5.9 | 4.1 | 21.8 | 2.4 | 8.0 | 36.3 | 57.8 |
| Canada | 1.0 | 0.4 | 1.2 | 0.5 | 0.5 | 1.4 | 2.0 |
| | 16.9% | 9.8% | 5.5% | 20.8% | 6.3% | 3.9% | 3.5% |
| 10 | 8.1 | 2.7 | 26.2 | 2.3 | 9.3 | 40.6 | 51.3 |
| Other | 1.3 | 0.1 | 0.8 | 0.3 | 0.9 | 1.2 | 1.5 |
| ind. | 16.0% | 3.7% | 3.1% | 13.0% | 9.7% | 3.0% | 2.9% |

[a] Explanation: In 1960–1982, 3.5% = the average share, 0.5 = the standard deviation of the share, and 14.3% = the variation coefficient.

[b] AGR = agriculture, MIN = mining and quarrying, MAN = manufacturing, EGW = electricity, gas, water, CON = construction, IND = total industry = MIN+MAN+EGW+CON, SRV = services.

*Table 3A.9(b)*. GDP: Structural composition of nominal value added in 1960–1982.[a,b]

| *Country* | *AGR* | *MIN* | *MAN* | *EGW* | *CON* | *IND* | *SRV* |
|---|---|---|---|---|---|---|---|
| 01 | 3.1 | 2.4 | 26.0 | 2.5 | 4.8 | 35.6 | 61.3 |
| USA | 0.4 | 0.6 | 2.3 | 0.2 | 0.2 | 1.9 | 2.1 |
| | 12.9% | 25.0% | 8.0% | 8.0% | 4.2% | 5.3% | 3.4% |
| 02 | 3.7 | 1.5 | 39.6 | 2.5 | 7.8 | 51.4 | 44.9 |
| FRG | 1.1 | 0.5 | 1.8 | 0.2 | 0.8 | 2.7 | 3.7 |
| | 29.7% | 33.3% | 4.5% | 8.0% | 10.3% | 5.3% | 8.2% |
| 03 | 7.1 | 0.9 | 33.0 | 2.5 | 8.0 | 44.4 | 48.5 |
| Japan | 2.7 | 0.4 | 2.3 | 0.4 | 1.0 | 1.8 | 4.1 |
| | 38.0% | 44.4% | 7.0% | 16.0% | 12.5% | 4.1% | 8.5% |
| 04 | 7.1 | 1.1 | 30.4 | 1.9 | 7.5 | 41.0 | 52.0 |
| France | 2.1 | 0.4 | 2.1 | 0.1 | 0.8 | 2.7 | 4.5 |
| | 29.6% | 36.4% | 6.9% | 5.3% | 10.7% | 6.6% | 8.7% |
| 05 | 2.8 | 2.7 | 29.7 | 3.0 | 6.5 | 41.9 | 55.3 |
| UK | 0.5 | 1.5 | 3.5 | 0.2 | 0.6 | 2.9 | 3.3 |
| | 17.9% | 55.6% | 18.5% | 6.7% | 9.2% | 6.9% | 6.0% |
| 06 | 9.0 | 2.4 | 27.7 | 5.4 | 8.0 | 43.5 | 47.5 |
| Italy | 2.2 | 0.2 | 1.4 | 0.6 | 0.3 | 0.9 | 1.9 |
| | 24.4% | 8.3% | 5.1% | 11.1% | 3.8% | 2.1% | 4.0% |
| 07 | 6.1 | 0.8 | 30.9 | 2.4 | 7.6 | 41.6 | 52.3 |
| Neth | 1.5 | 0.6 | 2.9 | 0.1 | 0.6 | 3.4 | 4.8 |
| | 24.6% | 75.0% | 9.4% | 4.2% | 7.9% | 8.2% | 9.2% |
| 08 | 4.2 | 1.2 | 30.6 | 2.7 | 7.2 | 41.7 | 54.1 |
| B/L | 1.5 | 0.7 | 2.7 | 0.4 | 0.6 | 2.6 | 3.8 |
| | 35.7% | 58.3% | 8.8% | 14.8% | 8.8% | 6.2% | 7.0% |
| 09 | 5.0 | 4.4 | 23.5 | 2.9 | 6.4 | 37.2 | 57.8 |
| Canada | 0.9 | 0.8 | 2.4 | 0.4 | 0.6 | 1.5 | 2.3 |
| | 18.0% | 18.2% | 10.2% | 13.8% | 9.4% | 4.0% | 4.0% |
| 10 | 8.9 | 2.6 | 26.8 | 2.6 | 8.9 | 41.0 | 50.0 |
| Other | 2.3 | 0.8 | 1.1 | 0.1 | 0.6 | 0.9 | 2.9 |
| ind. | 22.8% | 30.8% | 4.1% | 3.8% | 6.7% | 2.2% | 5.8% |

[a] Explanation: In 1960–1982, 3.1% = the average share, 0.4 = the standard deviation of the share, and 12.9% = the variation coefficient.
[b] AGR = agriculture, MIN = mining and quarrying, MAN = manufacturing, EGW = electricity, gas, water, CON = construction, IND = total industry = MIN+MAN+EGW+CON, SRV = services.

# Appendix 3B. Model Parameters (graphed)

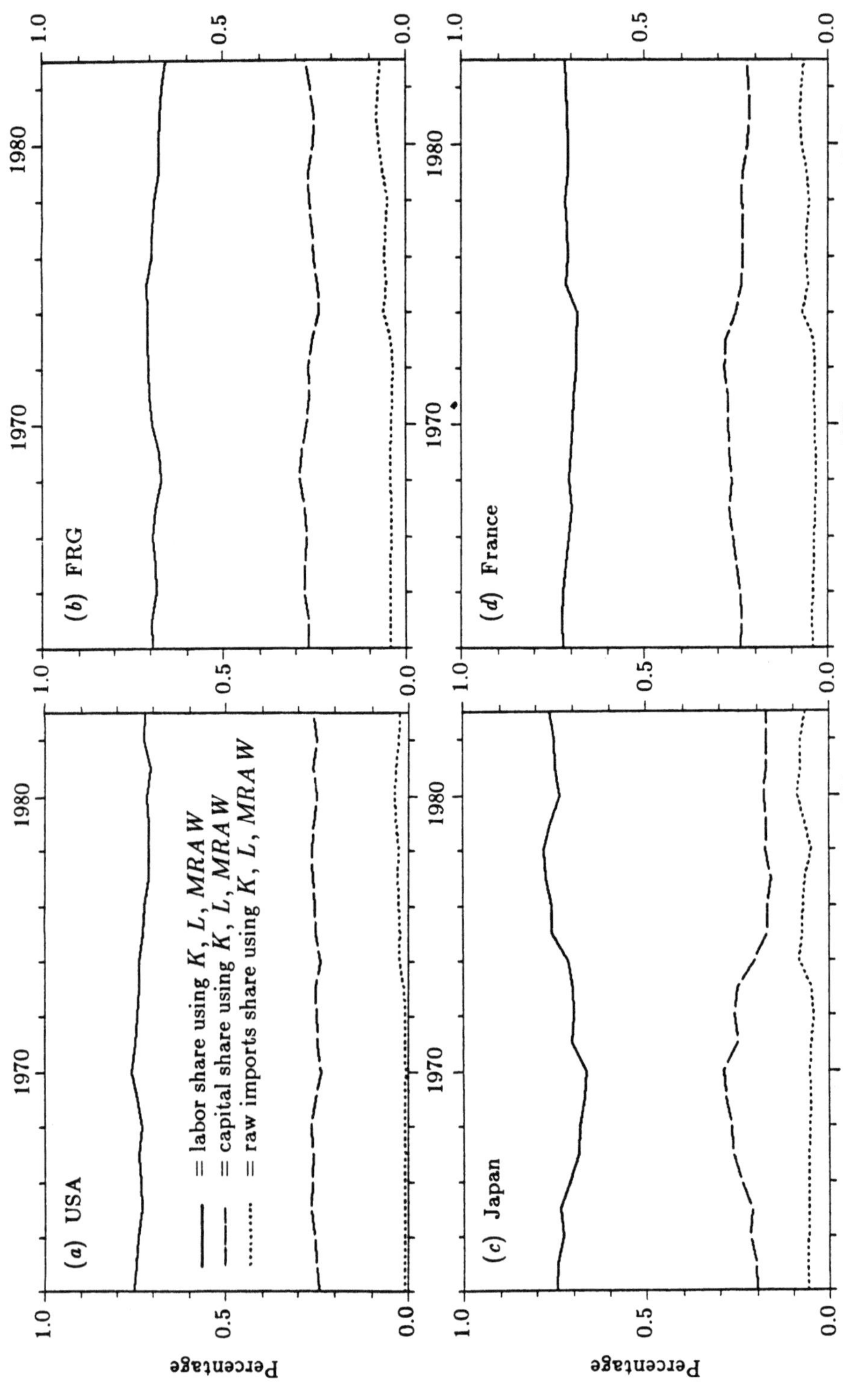


*Figure 3B.1.* Cost shares in production.

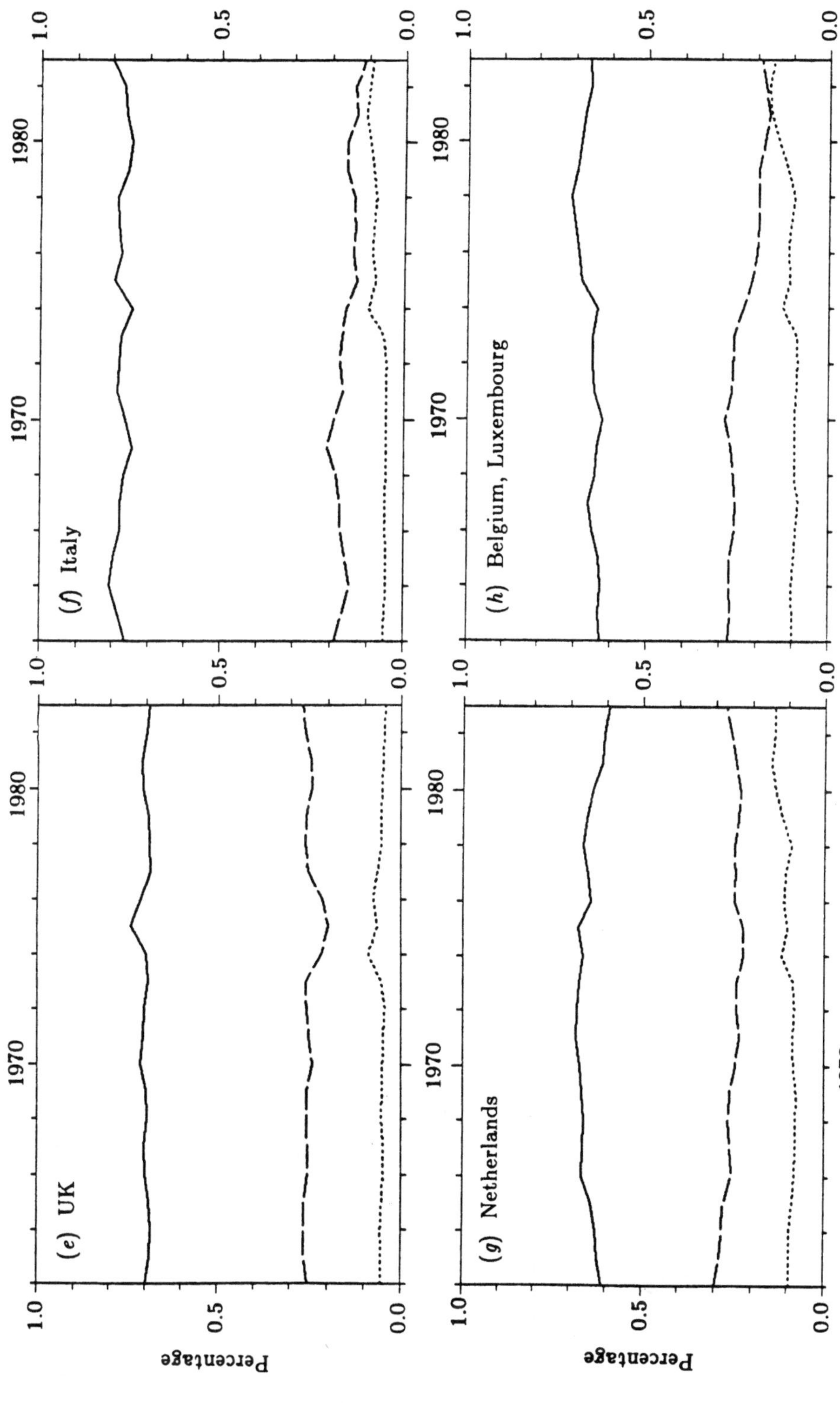


***Figure 9B.1.*** Continued.

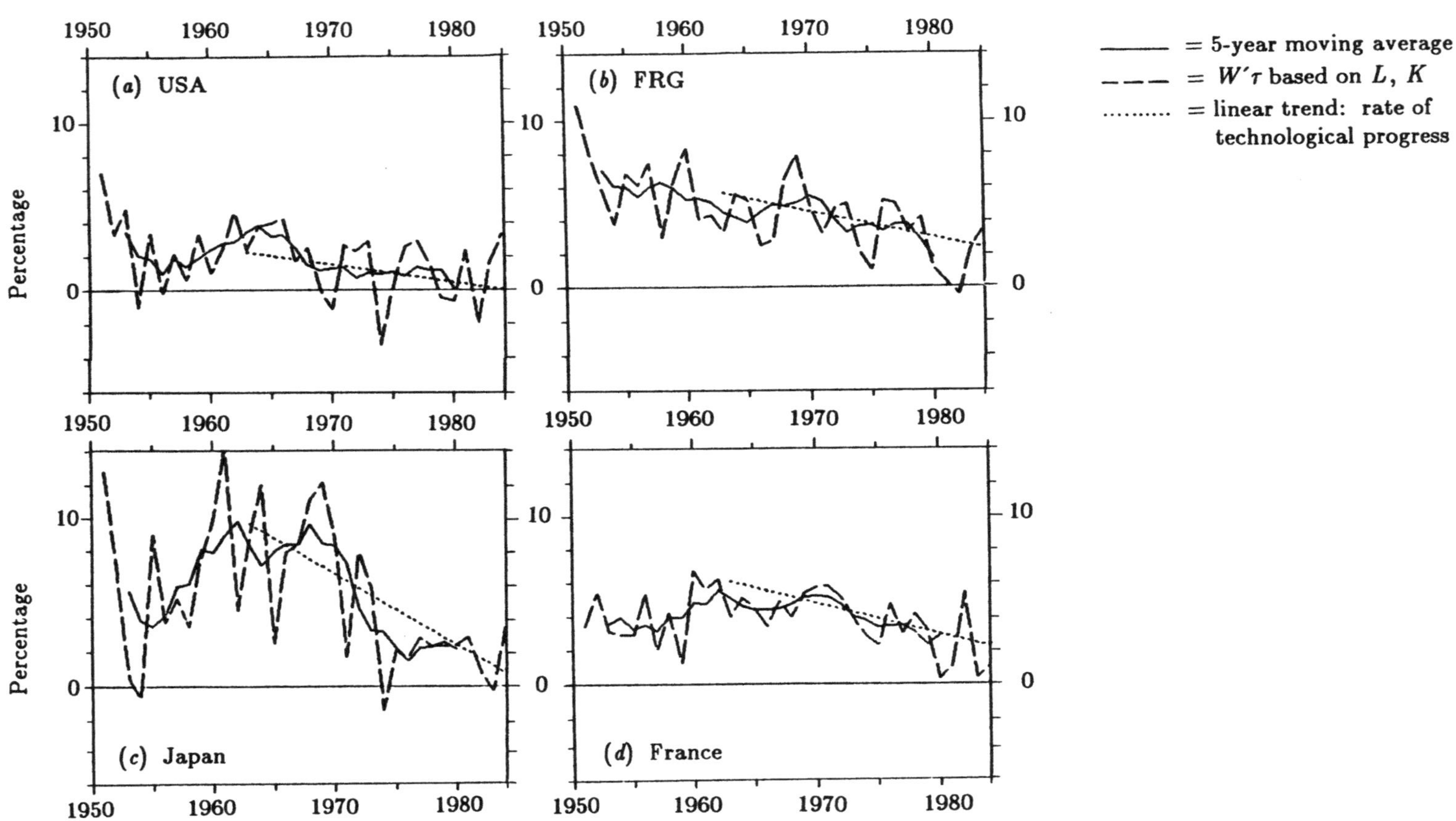


*Figure 3B.2.* Rates of (Harrod neutral) technical progress in different countries. (We show the results based on a production function without imported raw materials in order to cover a longer time period. The inclusion of imports does not change the results substantially. $w_\tau = W'\tau$ = rate of technical progress.)

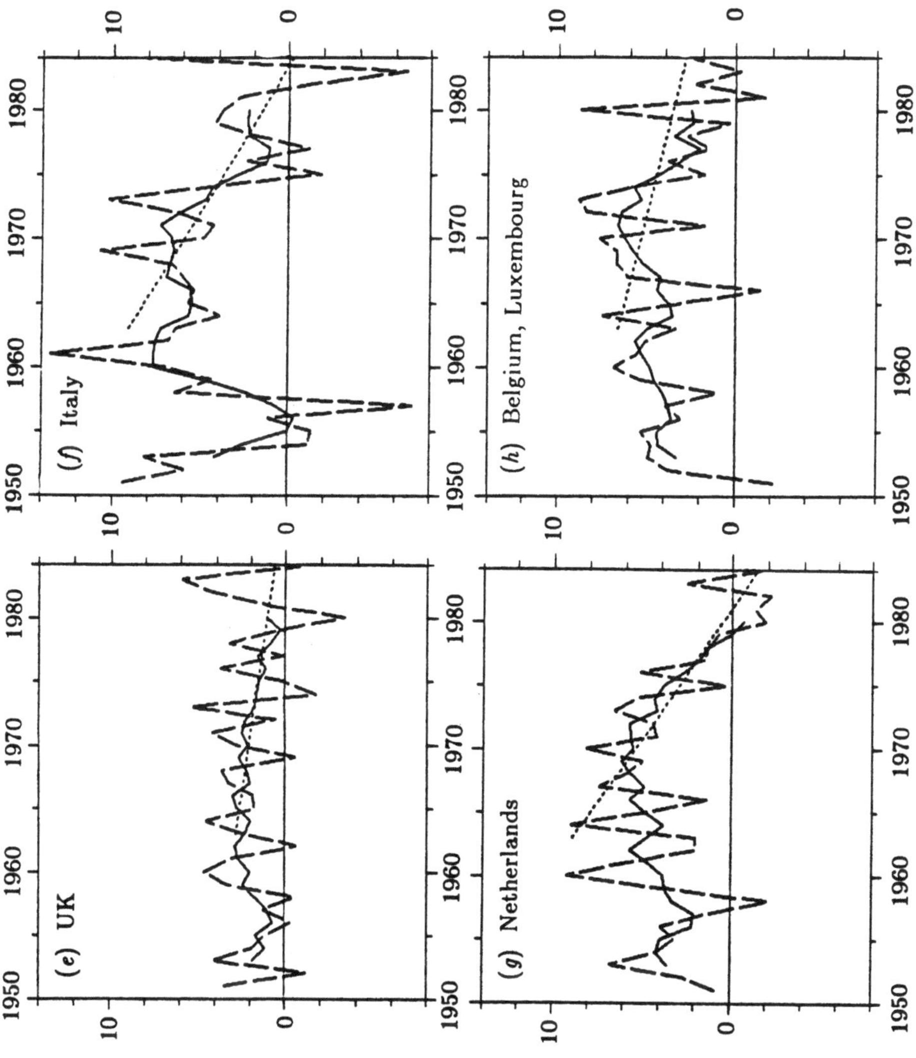


***Figure 3B.2.*** Continued.

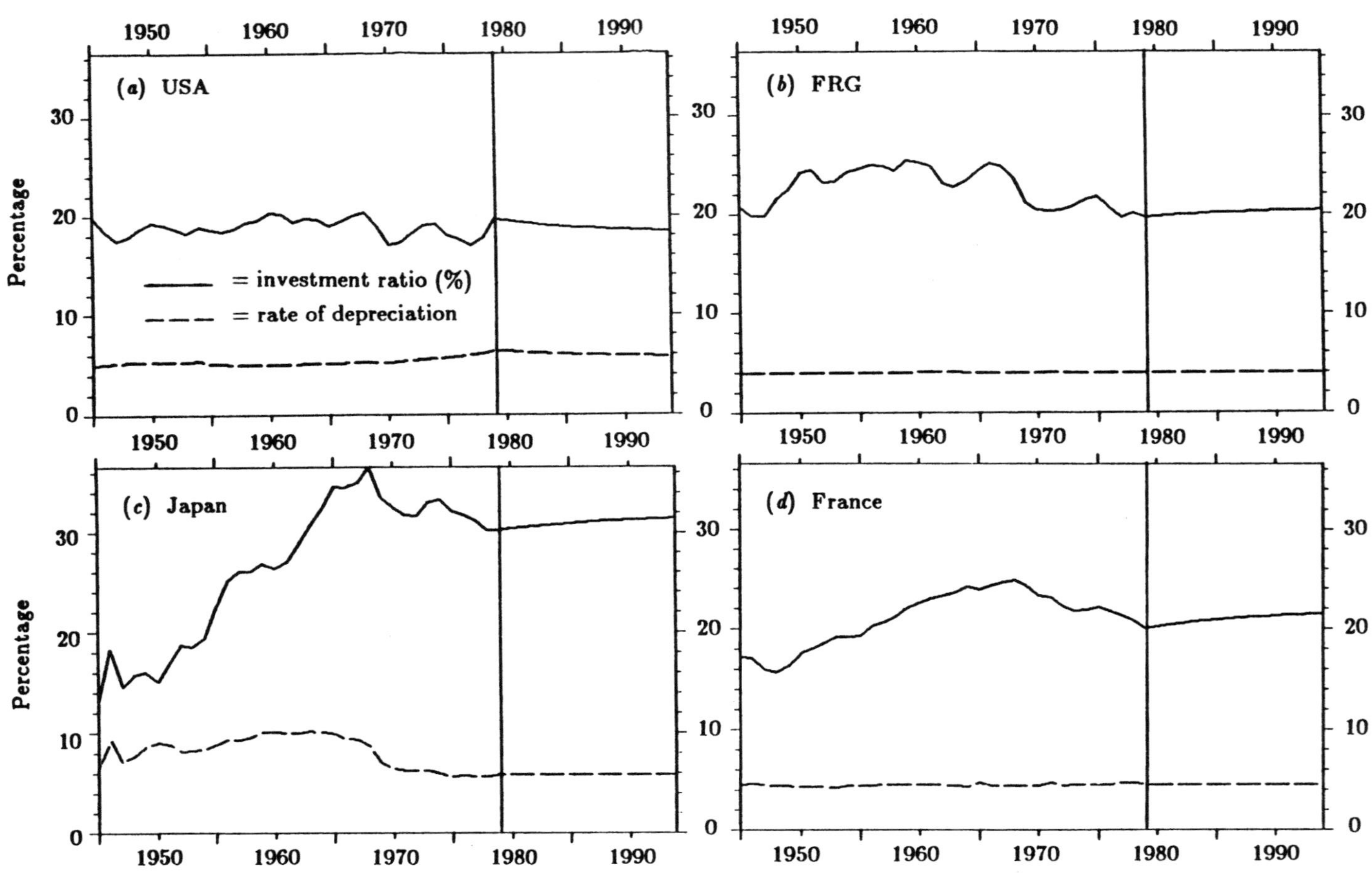


*Figure 3B.3.* Investment ratio and rate of depreciation. Figures from 1950–84 = observations, 1985–1999 = forecasts, medium scenario.

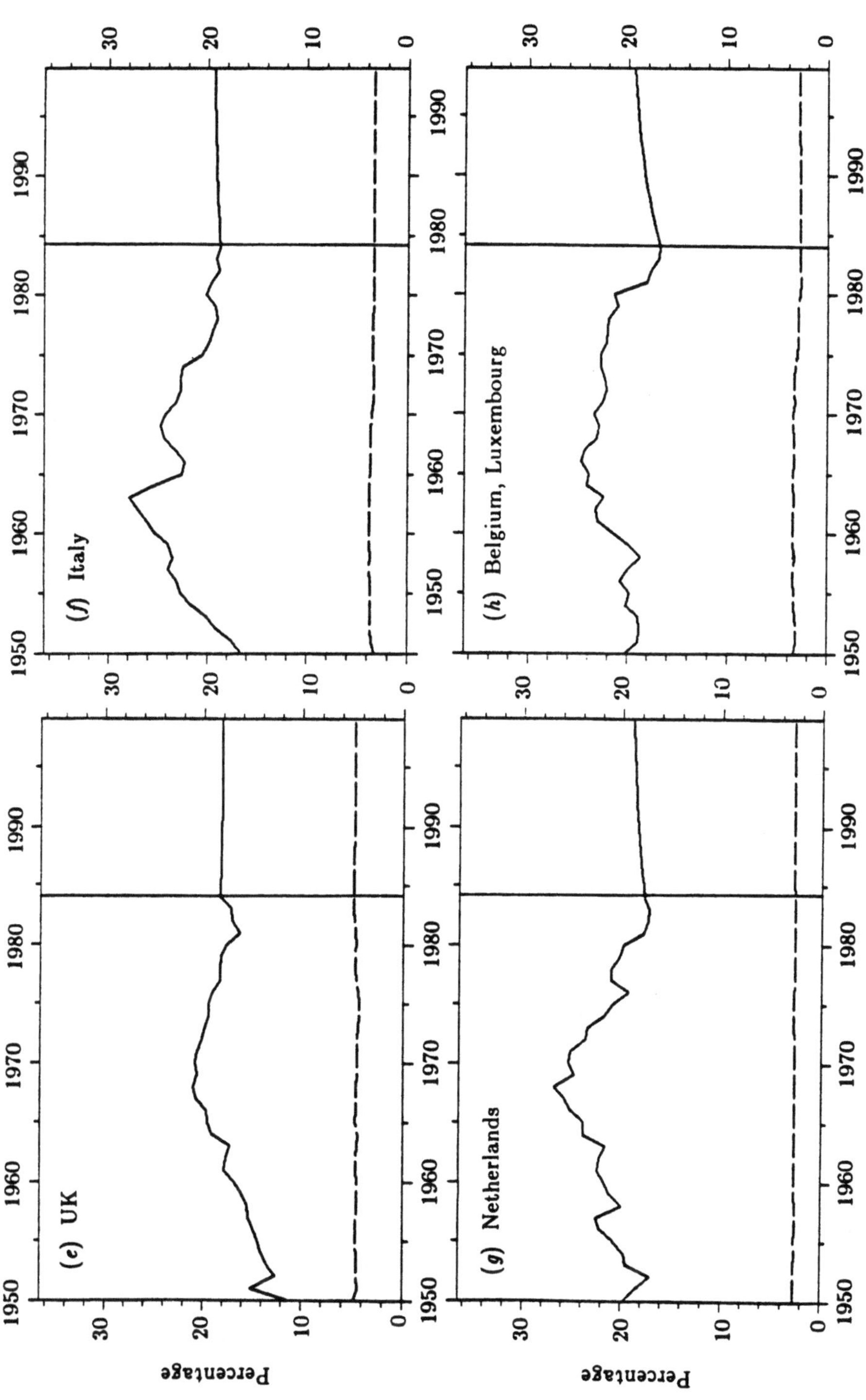


*Figure 3B.3.* Continued.

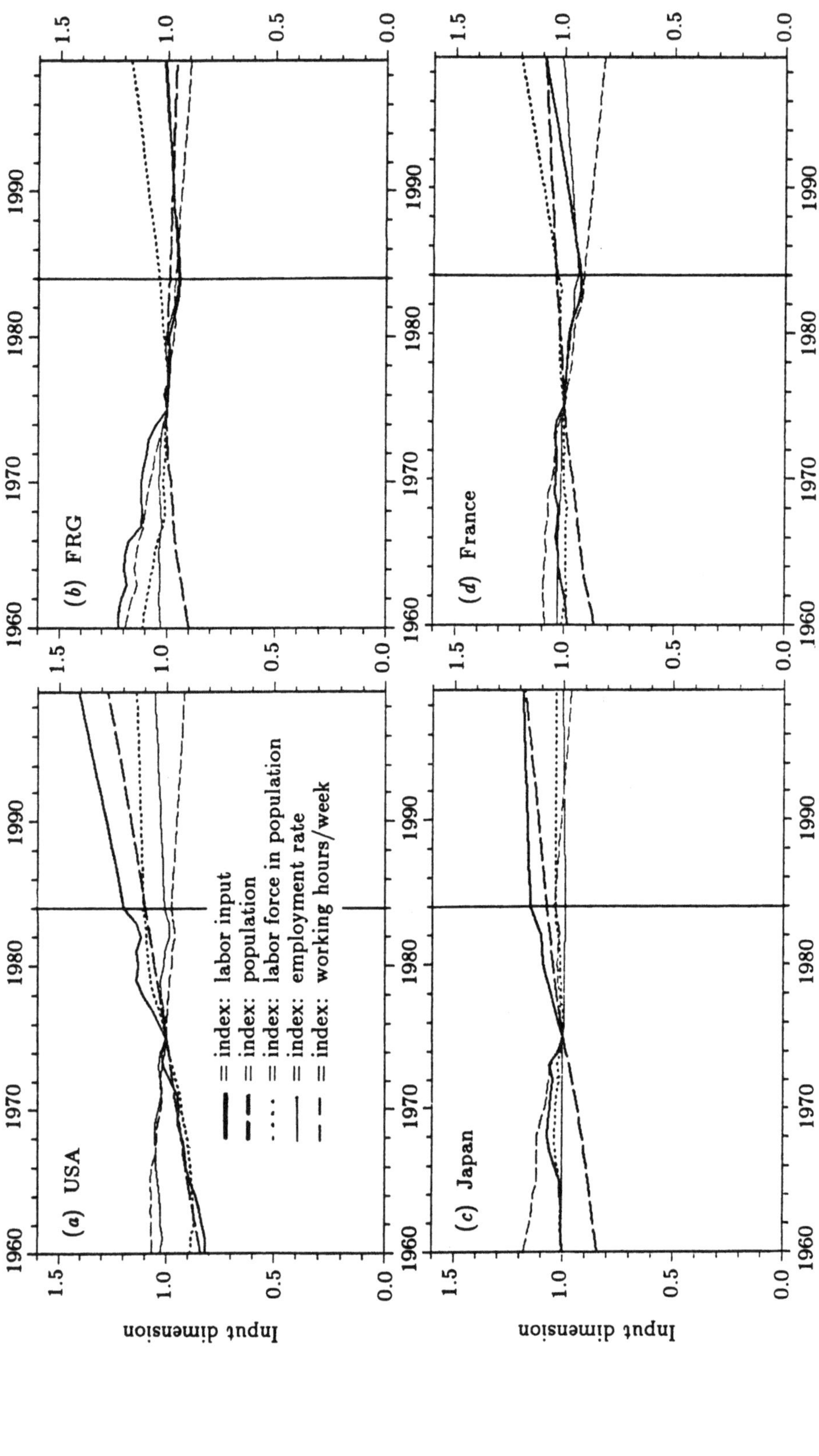


*Figure 9B.4.* Components of labor input (for all indexes, 1975 = 1). Figures from 1960–1984 = observations; 1985–1999 = forecasts.

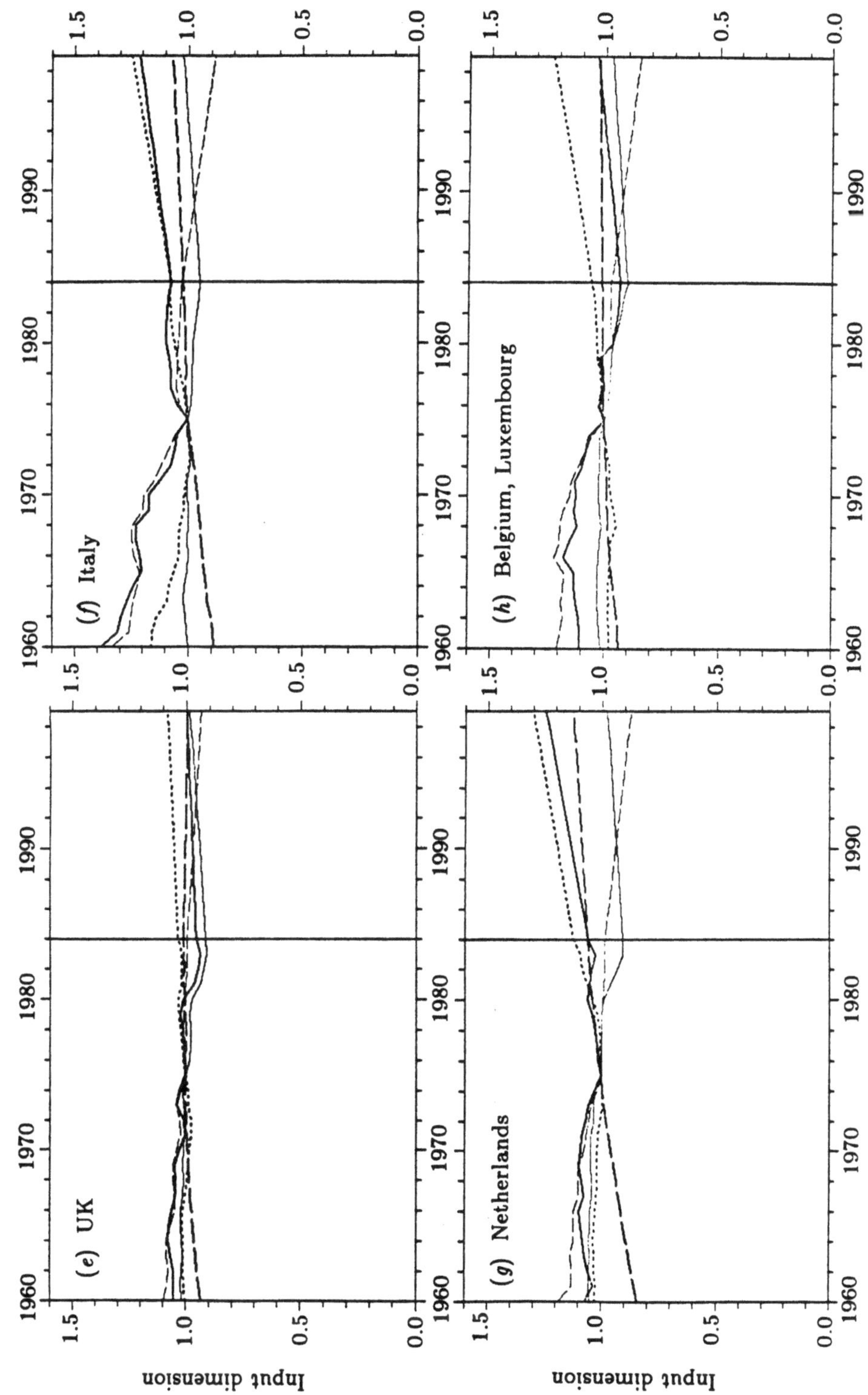


*Figure 3B.4.* Continued.

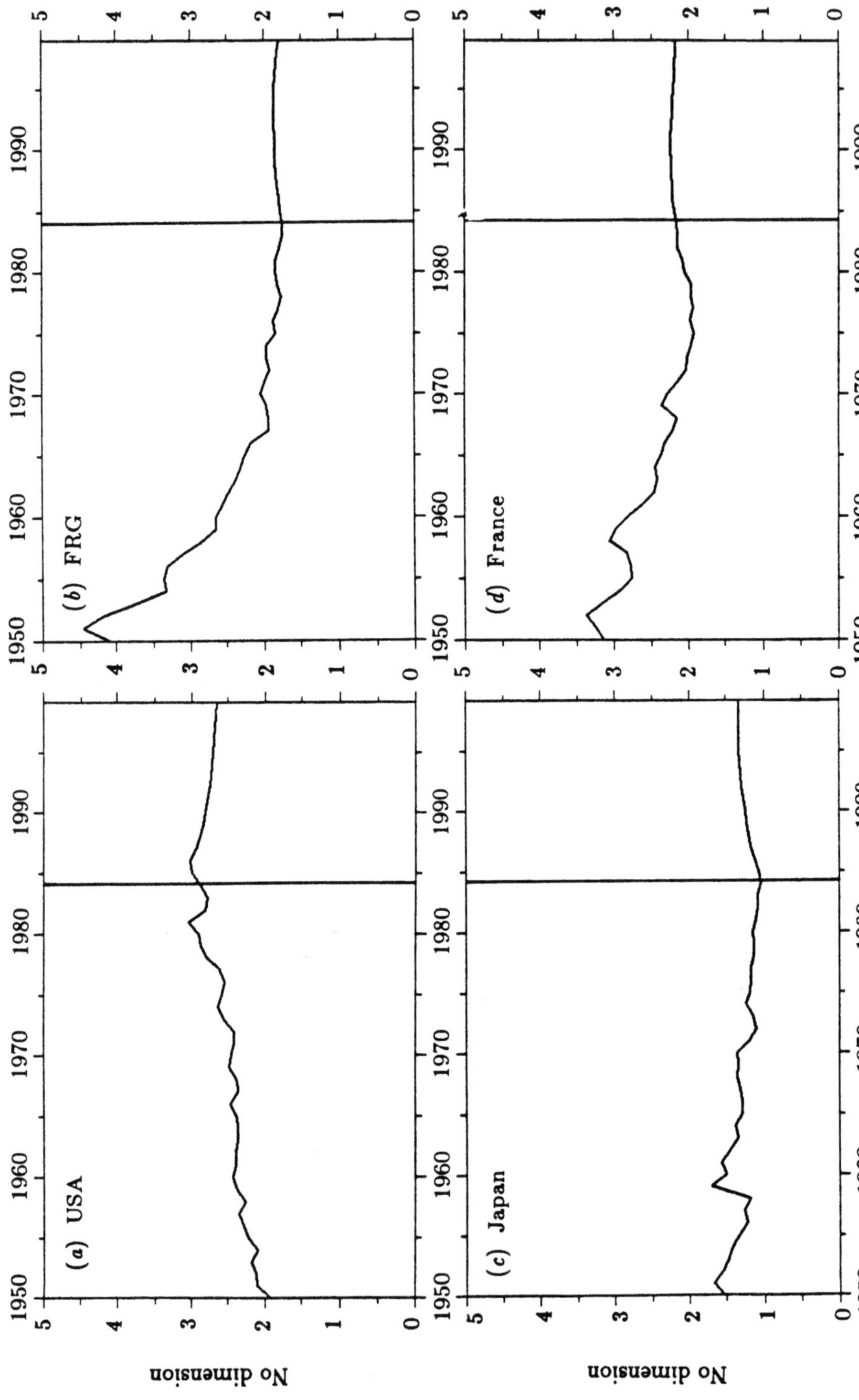


*Figure 9B.5.* Velocity of money (M2). Figures from 1950–1984 = observations; 1985–1999 = forecasts.

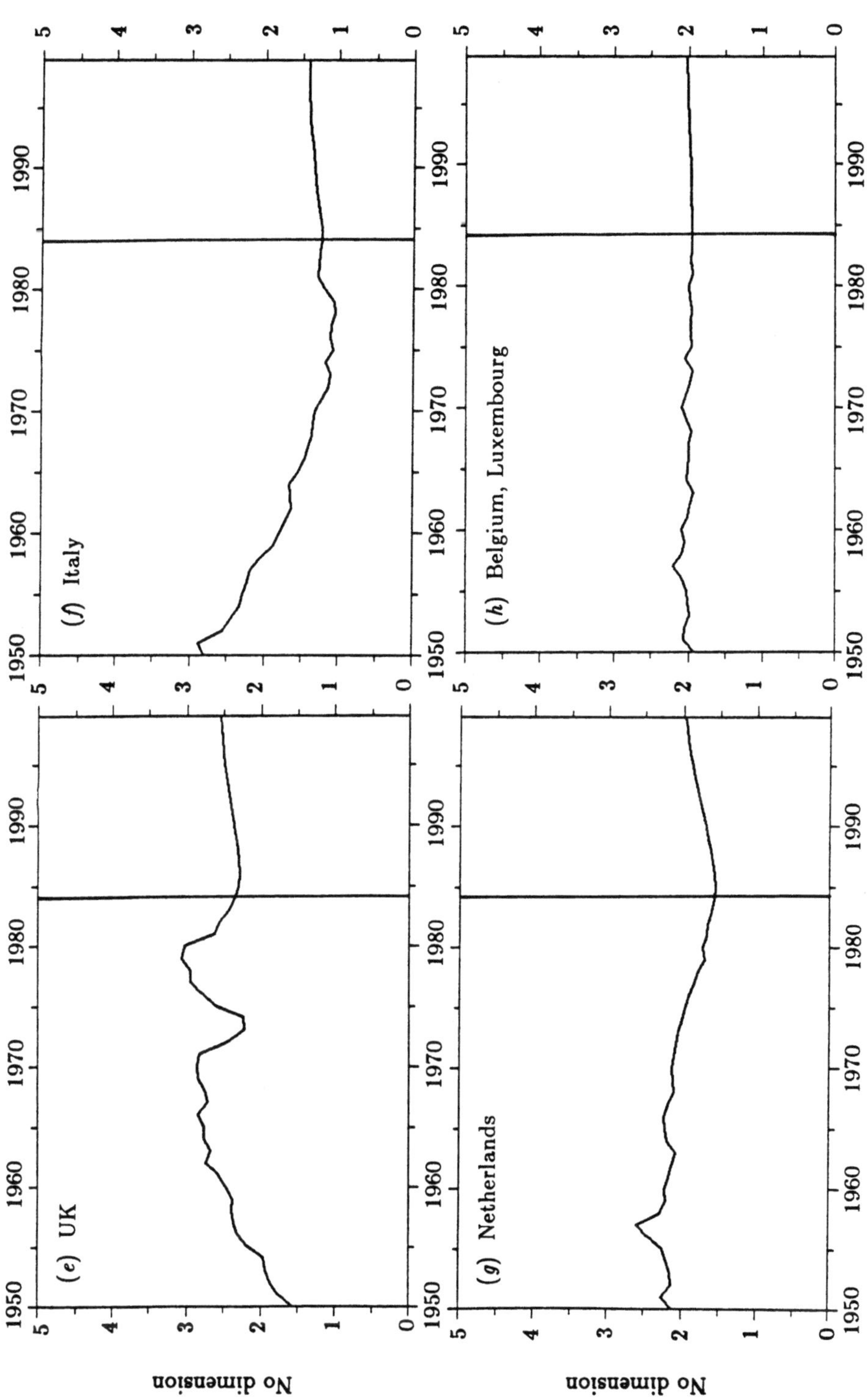


*Figure 3B.5.* Continued.

## References

Eismont, O. and Ross, H. (1985), *The Measurement of the Rate of Technical Progress*, Working Paper WP-85-95. (Laxenburg, Austria: International Institute for Applied Systems Analysis).

Ross, H. (1985), *A Database for Nine OECD Countries.* Working Paper. (Bonn).

Ross, H. (1986), *Economic Growth and Structural Change of OECD Countries*, Discussion Paper B-63, Sonderforschungsbereich 303, Bonn University. (Paper read at the IIASA Conference on Economic Growth and Structural Change in Laxenburg, Austria, Nov. 24–25, 1986).

Solow, R.M. (1957), Technical Change and the Aggregate Production Function, *Review of Economics and Statistics*, **39**, pp. 312–320.

CHAPTER 4

# Economic Growth and Structural Change of CMEA Countries

*Rumen Dobrinsky*


## Summary

This chapter deals with two main topics: first, an historical perspective on economic growth and structural change in the European CMEA countries in the period 1960 to 1985 is given; second, different scenarios for the future are presented and analyzed. The driving forces of economic growth are analyzed in greater detail. Three scenarios of the driving forces are designed on the basis of the observed past trends and the actual long-term policies of the countries. Some *ex ante* simulation results are also reported and discussed.


## 4.1. Introduction

One part of the Bonn–IIASA Research Project is devoted to the analysis and forecast of economic growth and structural change in the seven European CMEA countries: Bulgaria, CSSR, GDR, Hungary, Poland, Romania and the USSR. (Throughout this chapter, the countries are ordered alphabetically.) Some of the main results of this study are reported in this chapter, concentrating on the factual aspects of the analysis and forecast.

The specific methodological aspects of the approach applied with respect to the CMEA countries are described in more detail in Chapter 6. It should be mentioned that the CMEA country models were constructed in accordance with the System of Material Product Balance accounting system (MPS), and there are some differences in the macroeconomic indicators used for the analysis and forecast of CMEA countries as compared to the market economies. The actual data base for the CMEA countries which was used in the project was compiled on the basis of contributions from many collaborating groups. This data base is described in more detail in Dobrinsky (1986a).

Several comments should be made in order to clarify our attitide with respect to the results reported in this chapter and, especially, with respect to the forecasts. Although this chapter concentrates on the numerical results of our studies, we do not think that the figures, taken at face value, should be considered as the main outcome of this research. The models produce forecasts only in combination with the expert knowledge of the analyst working with them, and the result is a conditional forecast of the future which depends on the vision of the driving forces of long-term development, as introduced by the analyst. This is how we suggest that the three forecast scenarios (which are discussed later in the chapter) be interpreted. We shall try to illustrate the potential of the models as analytical tools rather than pretend to a strong predictive power in the absolute sense.

## 4.2. Economic Growth and Structural Change in the CMEA Countries, 1960 to 1985

### 4.2.1. An overview of the main developments

Economic growth has always been a problem of major importance in the economic policy of the socialist countries. Historically, most of them started from a rather low level of economic development, and their main goal for quite a long period was to achieve fast growth through rapid industrialization, high investment rates and full utilization of domestic resources,. In the period under consideration, the European CMEA countries in general successfully managed to follow these policies and achieved rates of economic growth which were among the highest in the world (*Table 4.1*). This was especially manifested in the 1960s and in the first half of the 1970s, when several countries hit the 10% level of annual NMP growth rates in some years.

*Table 4.1.* Growth rates of NMP (%).

| | *Period* | | | |
|---|---|---|---|---|
| *Country* | *1961–1985* | *1961–1970* | *1971–1980* | *1981–1985* |
| Bulgaria | 6.6 | 7.7 | 7.0 | 3.7 |
| CSSR | 4.0 | 4.4 | 4.6 | 1.8 |
| GDR | 4.5 | 4.4 | 4.8 | 4.4 |
| Hungary | 4.5 | 5.5 | 5.0 | 1.6 |
| Poland | 4.3 | 6.1 | 5.2 | −0.8 |
| Romania | 7.9 | 8.4 | 9.2 | 4.3 |
| USSR | 5.5 | 7.3 | 4.7 | 3.5 |

However, in the second half of the 1970s the rates of economic growth started to decline. This tendency continued and was even more accentuated in the first half of the 1980s. A more detailed analysis of the driving forces of

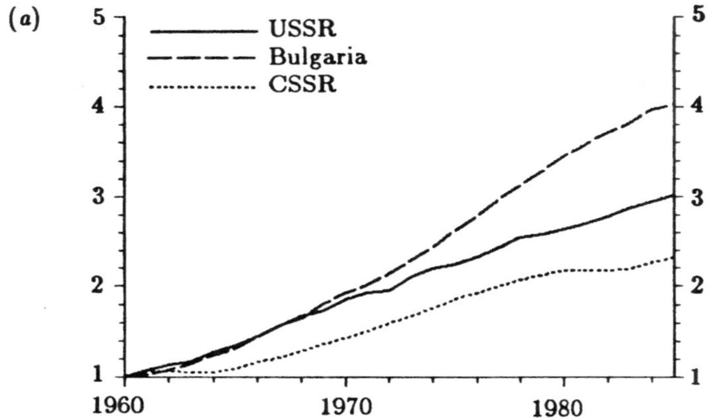


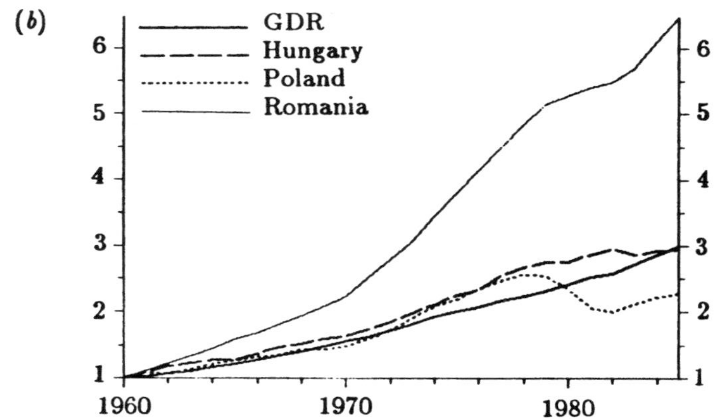


*Figure 4.1.* Index of labor productivity in the CMEA countries (1960 = 1).

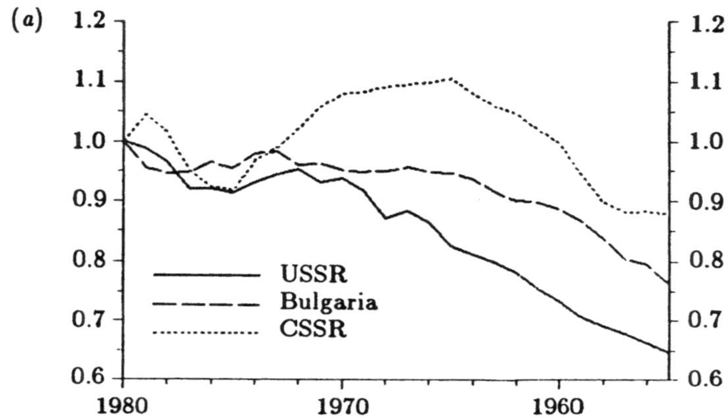


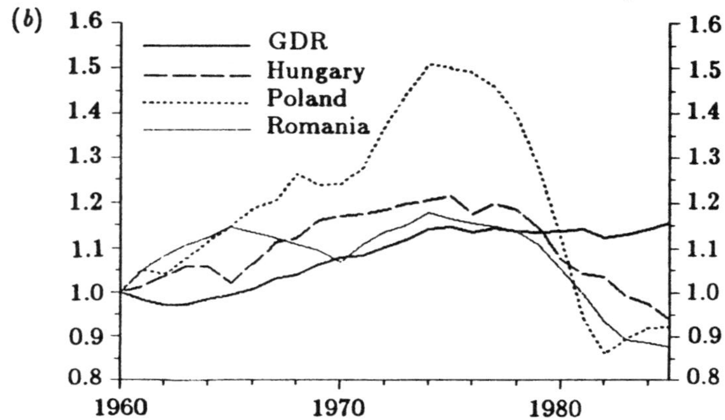


*Figure 4.2.* Index of NMP/fixed assets ratio in the CMEA countries (1960 = 1).

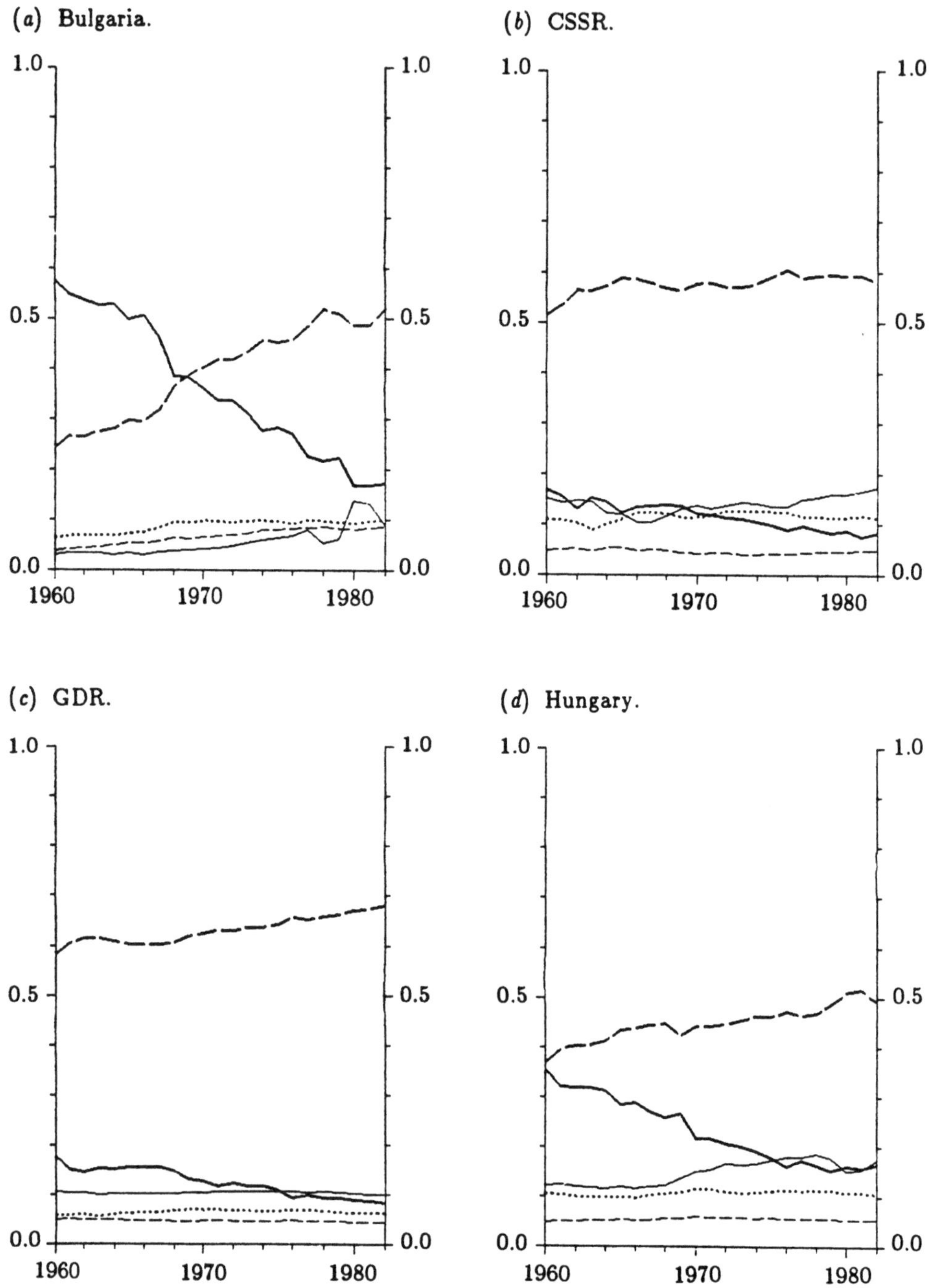


***Figure 4.3.*** **Sectoral structure of NMP in the CMEA countries, 1960–1982.**

(*e*) Poland.

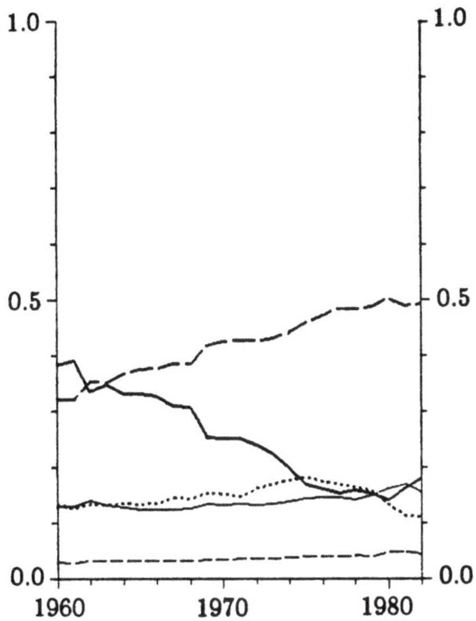


(*f*) Romania.

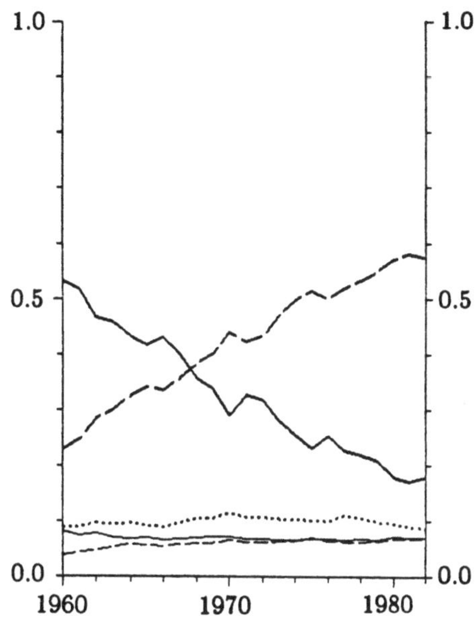


(*g*) USSR.

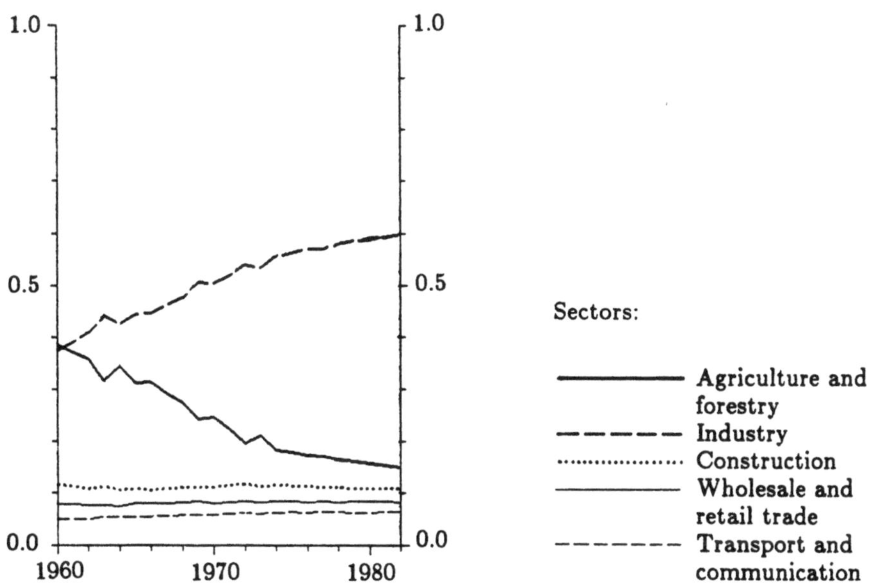


*Figure 4.3.* Continued.

economic growth will follow in the next section. Here we would only like to mention some of the factors which brought about this decline in the rates of economic growth.

First of all, the sources of extensive growth which had played an important role in the previous subperiod were nearly exhausted: there was almost no growth in the labor supply (a consequence of the demographic development) and, as a result, subsequent expansion of production became more difficult. Growth in the utilization of domestic natural resources could not be sustained, either because of depletion or by rising extraction costs due to unfavorable natural conditions. Other unfavorable factors in the most recent years were the deteriorating terms of trade and large foreign debts in some countries.

The overall effect of these developments can be traced in the long-term behavior of such synthetic indicators as "labor productivity" and "output/capital ratio". The trends in these indicators in the seven countries in the period 1960–1985 are given in *Figures 4.1* and *4.2*. In these plots labor productivity is measured by NMP produced by one person employed in the material sphere, and the output/capital ratio is measured by NMP produced per unit of fixed assets. Both indicators are presented in index form, with the value of 1960 taken as 1.0. The labor productivity index, as may be expected, follows the NMP growth almost identically. More interesting is the development of the output/capital ratio. In most countries, until about the middle of the 1970s, this ratio showed an increasing trend which, on a macro level, suggests increasing efficiency of the fixed assets. Only in the case of Bulgaria and the USSR was the trend decreasing, but at a rather moderate rate. However, starting from the middle of the 1970s this trend changed and the output/capital ratio started to decline in all countries with the exception of the GDR, where it remained more or less at a constant level.

The development of the production structure in the CMEA countries measured in constant prices by sectors of economic activity in the period 1960–1982 is shown in *Figure 4.3* [1]. In accordance with the MPS, six (macro-) sectors of economic activity in the material sphere are treated separately: Agriculture and Forestry, Industry, Construction, Wholesale and Retail Trade, Transport and Communication, and Other activities of the material sphere. The residential sixth sector, which is usually insignificant in value terms, is not shown in the plots (according to statistical practice in the USSR, this sector is not separated at all). In the plots the names of the sectors appear in abbreviated form. Although there were differences in the patterns of economic development followed by the countries, there was one common feature in the dynamics of structural change of all countries: the increasing share of industry at the expense mainly of a declining share of agriculture. This feature is revealed to a different extent in the various countries, depending on the starting development level, the natural conditions and the economic tradition, as well as on the economic policies pursued.

Thus, by the beginning of the 1960s, the GDR and the CSSR had already reached a relatively high level of industrialization reflected in the high level of the industrial sector's share and continued this line quite smoothly. In both countries the share of industry in NMP during 1960–1982 increased by about

10% [2] while the share of agriculture decreased by about half. In the USSR, Hungary and Poland the shares of agriculture for this period also decreased by about one half, but the starting level at the beginning of the 1960s was higher. This change in the shares of agriculture in the three countries was accompanied by a more considerable increase in the shares of industry. In the USSR the shares of industry increased by about 20%, in Hungary by about 15%, and in Poland by about 18%.

Most substantial structural change took place in Bulgaria and Romania. Historically, these countries were industrially underdeveloped, their economies relying strongly on agriculture. In the postwar years they followed a policy of rapid industrialization which, in the period under consideration, brought about major shifts in their production structure. The shares of industry in the NMP of the two countries doubled while the shares of agriculture decreased to about one third of the initial level.

In all countries the other three sectors did not undergo such substantial changes and in most cases manifested a slowly increasing trend. Some exceptions were the trade sectors in Bulgaria, the CSSR and Hungary, which experienced considerable fluctuation, and the construction sectors in Poland and Romania, which started to decline in the second half of the 1970s.

The dynamics of the sectoral shares highlights only one side of the process of structural change with respect to production. This picture can be completed by the dynamics of the real NMP produced in the different sectors of the economy. Average growth rates in the production sectors of the European CMEA countries for the period 1961–1982 are presented in *Table 4.2.*

*Table 4.2.* Average growth rates of NMP (%), by sectors of economic activity, 1961–1982.

| | *Sector* | | | | |
|---|---|---|---|---|---|
| *Country* | *Agriculture and forestry* | *Industry (total)* | *Construction* | *Wholesale and retail trade* | *Transport and communication* |
| Bulgaria | 1.4 | 10.9 | 9.3 | 12.3 | 11.0 |
| CSSR | 0.7 | 4.7 | 4.2 | 4.8 | 4.1 |
| GDR | 1.6 | 5.2 | 4.7 | 4.2 | 3.8 |
| Hungary | 1.4 | 6.4 | 5.0 | 6.7 | 5.5 |
| Poland[a] | 1.2 | 6.2 | 3.7 | 5.0 | 6.5 |
| Romania | 3.0 | 12.8 | 8.1 | 7.5 | 10.9 |
| USSR | 1.2 | 8.0 | 5.5 | 6.0 | 6.9 |

[a]The growth rates for Poland refer to the period 1961–1985.

As can be seen from the figures, during the period under consideration the countries that experienced the highest overall NMP growth rates underwent the most substantial economic structural change as well, and vice versa. Throughout this period all countries completed the process of industrialization of their economies, and those which had gone through that stage earlier stabilized this development. By the beginning of the 1980s all European CMEA countries were approaching a similar economic structure, which is in line with the CMEA policy for equalizing the levels of economic development in the member countries.

### 4.2.2. The driving forces of economic growth and structural change

In this section we present a brief analysis of the main driving forces of economic growth and structural change in the CMEA countries during the reference period, following the general approach of the Bonn–IIASA Project. Here we concentrate only on the numerical results of this analysis. The underlying general methodology is given in Chapters 1 and 2, and the specific methodological aspects concerning the CMEA countries in Chapter 5.

We start with a general analysis of the factors of growth.

We describe the production technology by a Cobb–Douglas production function, which in the case of the CMEA countries relates total domestic material output $Y^*$ [3] to the production factor: $L$ (number of employed persons in the material sphere), $K$ (fixed assets in the material sphere) [4], $IM_R$ (imports of raw materials and intermediate products). The level of technical progress (or, of total factor productivity) $\tau$ is taken in its explicit Hicks-neutral form ($w_\tau$ is the growth rate of $\tau$).

$$Y^* = \alpha_0\, \tau_{-1}\, (1 + w_\tau)\, L^{\alpha_1}\, K^{\alpha_2}\, IM_R^{\alpha_2}\,, \quad \sum_{i=1}^{3} \alpha_i = 1 \tag{4.1}$$

This assumption about the production technology implies that the growth rates of the variables are related as:

$$w_{Y^*} = w_\tau + \alpha_1\, w_L + \alpha_2\, w_K + \alpha_3 w_{IM_R} \tag{4.2}$$

where $w_{Y^*}$, $w_L$, and $w_{IM_R}$ are the growth rates of $Y^*$, $L$, $K$ and $IM_R$, respectively.

Each of our four terms on the right-hand side of (4.2) can be interpreted as the contribution of each of the four main "factors of growth" to the growth rate of the total domestic material output $w_{Y^*}$. Such an analysis for the seven European CMEA countries for the period 1961–1985 is provided in *Table 4.3*. Besides the values for the whole period, figures for three subperiods are given (1961–70, 1971–80, 1981–85).

For each period the values of $w_{Y^*}$, $w_\tau$, $\alpha_1 w_L$, $\alpha_2 w_K$ and $\alpha_3 w_{IM_R}$ are given in the average annual percentage growth rates for the period and also the "share" of each of the four components of $w_{Y^*}$ is shown (as a percentage of $w_{Y^*}$). These "shares", or "weights" indicate the actual contribution of each the four "factors of growth" to $w_{Y^*}$ [5].

Several main tendencies can be traced from the figures in *Table 4.3*. In most countries the growth of employment has played an almost negligible role in the growth of total output both in absolute ($\alpha_1 w_L$) and in relative (percentage of $w_{Y*}$) terms. The general trend of the absolute term is declining, due to the declining (sometimes zero, or negative, in the most recent years) growth rates of employment in the material sphere.

*Table 4.3.* Analysis of the factors of growth.

| | | Contribution to $w_Y$ | | | | | | | | |
|---|---|---|---|---|---|---|---|---|---|---|
| | | $\tau$ | | | $L$ | | $K$ | | $IM_R$ | |
| *Country* | *Period* | $w_{Y*}$ | $w_\tau$ | *% of* $w_{Y*}$ | $\alpha_1 w_L$ | *% of* $w_{Y*}$ | $\alpha_2 w_K$ | *% of* $w_{Y*}$ | $w_{IM_R}$ | *% of* $w_{Y*}$ |
| Bulgaria | 1961–85 | 7.2 | 1.7 | 24 | 0.3 | 4 | 3.1 | 43 | 2.1 | 29 |
| CSSR | | 4.3 | 1.3 | 30 | 0.2 | 5 | 1.9 | 44 | 0.9 | 21 |
| GDR | | 4.4 | 2.0 | 45 | 0.0 | 0 | 1.7 | 39 | 0.7 | 16 |
| Hungary | | 5.1 | 1.4 | 28 | 0.0 | 0 | 2.0 | 39 | 1.7 | 33 |
| Poland | | 4.5 | 1.0 | 22 | 0.4 | 9 | 1.9 | 42 | 1.2 | 27 |
| Romania | | 7.9 | 2.4 | 30 | 0.0 | 0 | 4.2 | 53 | 1.3 | 17 |
| USSR | | 5.7 | 1.1 | 20 | 0.4 | 7 | 3.7 | 65 | 0.5 | 8 |
| Bulgaria | 1961–70 | 9.0 | 2.3 | 26 | 0.3 | 3 | 3.3 | 37 | 3.1 | 34 |
| CSSR | | 5.0 | 1.9 | 38 | 0.3 | 6 | 1.5 | 30 | 1.3 | 26 |
| GDR | | 4.8 | 2.2 | 46 | 0.0 | 0 | 1.5 | 31 | 1.1 | 23 |
| Hungary | | 6.6 | 2.3 | 35 | 0.1 | 2 | 1.6 | 24 | 2.6 | 39 |
| Poland | | 6.5 | 2.5 | 39 | 0.8 | 12 | 1.5 | 23 | 1.7 | 26 |
| Romania | | 9.1 | 3.2 | 35 | 0.0 | 0 | 3.8 | 42 | 2.1 | 23 |
| USSR | | 7.3 | 2.3 | 32 | 0.4 | 5 | 4.1 | 56 | 0.5 | 7 |
| CSSR | | 4.9 | 1.5 | 30 | 0.1 | 2 | 2.4 | 50 | 0.9 | 18 |
| GDR | | 4.5 | 2.0 | 44 | 0.1 | 2 | 1.8 | 41 | 0.6 | 13 |
| Hungary | | 5.5 | 1.6 | 29 | 0.0 | 0 | 2.4 | 44 | 1.5 | 27 |
| Poland | | 5.0 | 0.7 | 14 | 0.2 | 4 | 2.6 | 52 | 1.5 | 30 |
| Romania | | 8.9 | 2.6 | 29 | 0.1 | 1 | 4.8 | 54 | 1.4 | 16 |
| USSR | | 5.0 | 0.5 | 10 | 0.4 | 8 | 3.8 | 76 | 0.3 | 6 |
| Bulgaria | 1981–85 | 4.7 | 0.4 | | 0.2 | | 2.8 | | 1.3 | |
| CSSR | | 1.8 | −0.4 | | 0.2 | | 1.8 | | 0.2 | |
| GDR | | 3.7 | 1.8 | | 0.0 | | 1.7 | | 0.2 | |
| Hungary | | 1.5 | −0.7 | | 0.1 | | 1.8 | | 0.3 | |
| Poland | | −0.6 | −1.5 | | 0.0 | | 1.2 | | −0.3 | |
| Romania | | 3.6 | 0.2 | | 0.0 | | 4.0 | | −0.6 | |
| USSR | | 4.1 | 0.1 | | 0.3 | | 3.0 | | 0.7 | |

The contribution of "fixed assets accumulation" ($\alpha_2 w_K$) to the growth of $Y^*$ in all countries is quite high both in absolute and in relative terms. The declining tendency in the absolute value of $\alpha_2 w_L$ is not so pronounced as the declining tendency in the growth rates of output (starting from the middle of the 1970s) which even results in an increasing relative contribution of $\alpha_2 w_K$ in $w_{Y*}$. However, this development is accompanied by a decline in the absolute efficiency of fixed assets in most of the countries, as indicated in *Figure 4.2*.

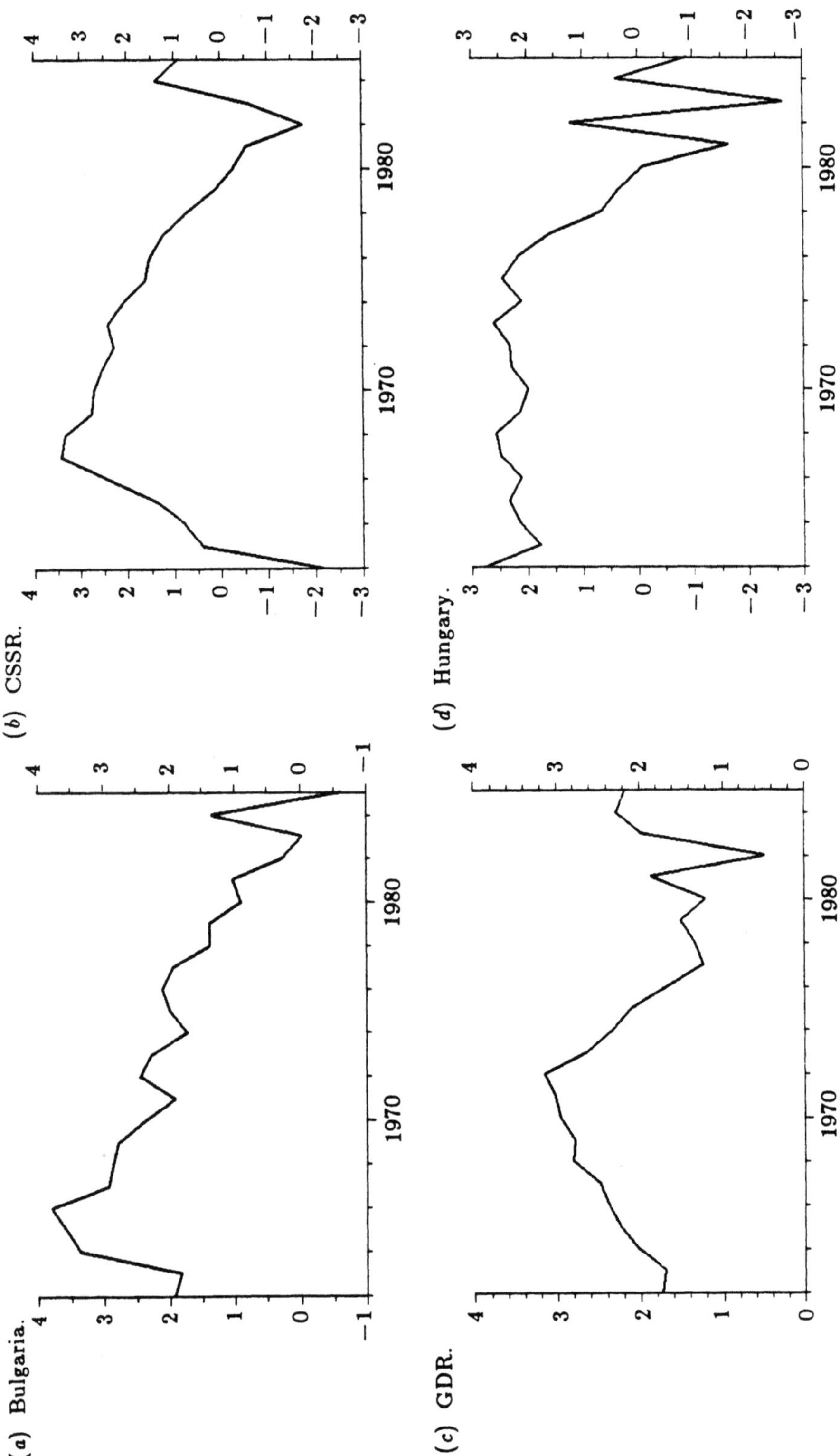


*Figure 4.4.* **Rates of technical progress ($w_\tau$) in the CMEA countries, 1961–1985.**

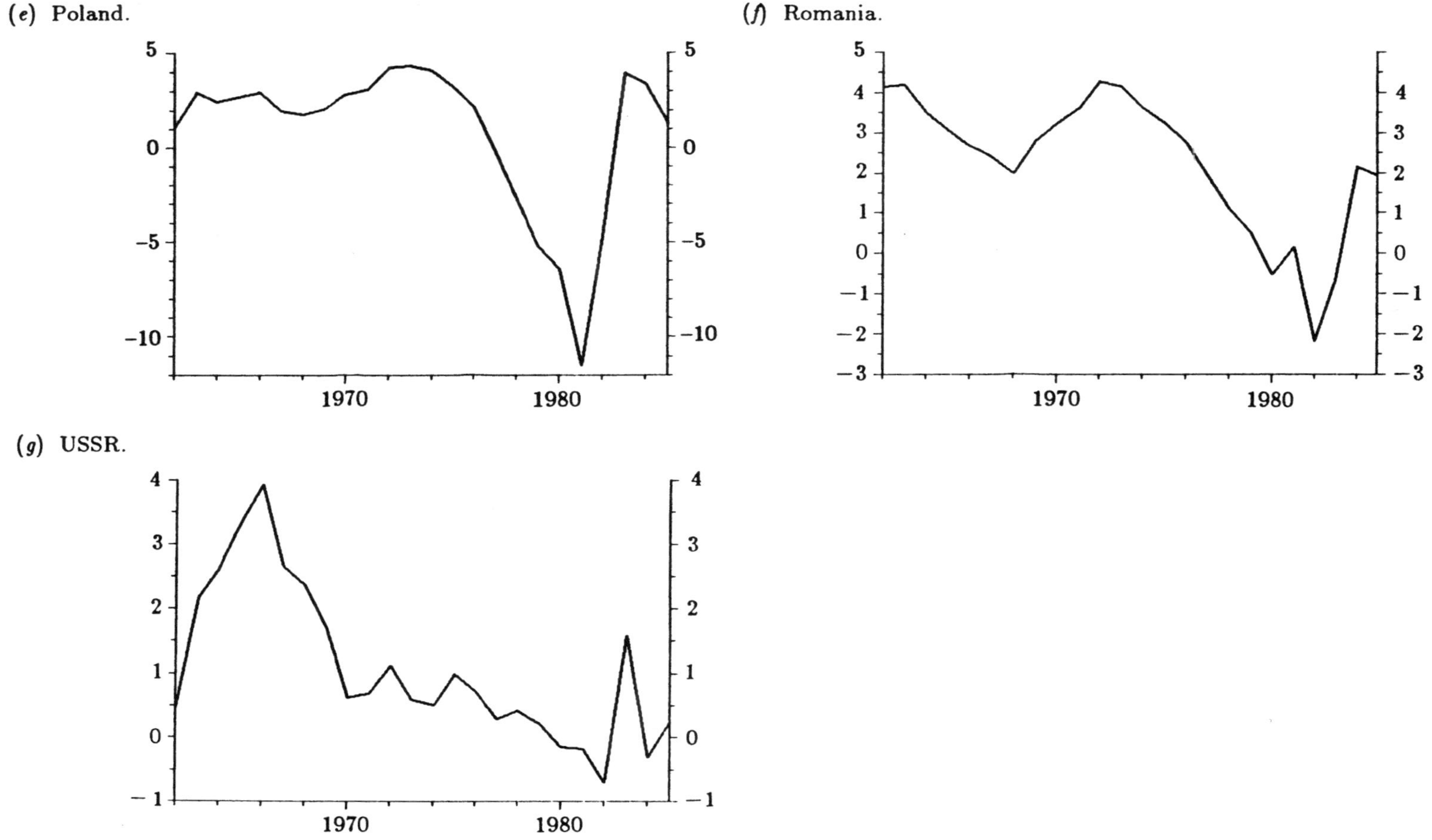


*Figure 4.4.* Continued.

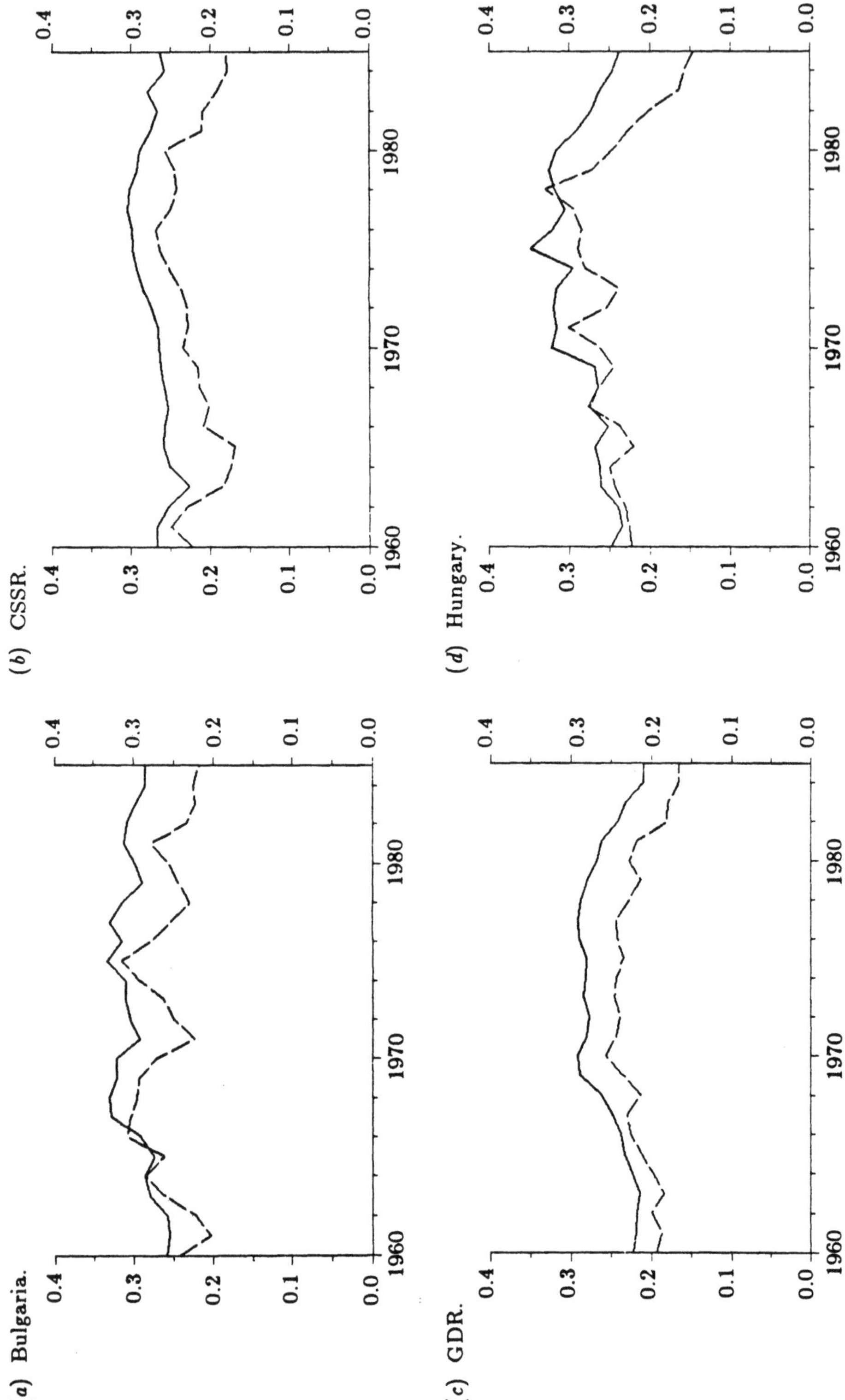


*Figure 4.5.* Savings ratios in the CMEA countries, 1960–1985.

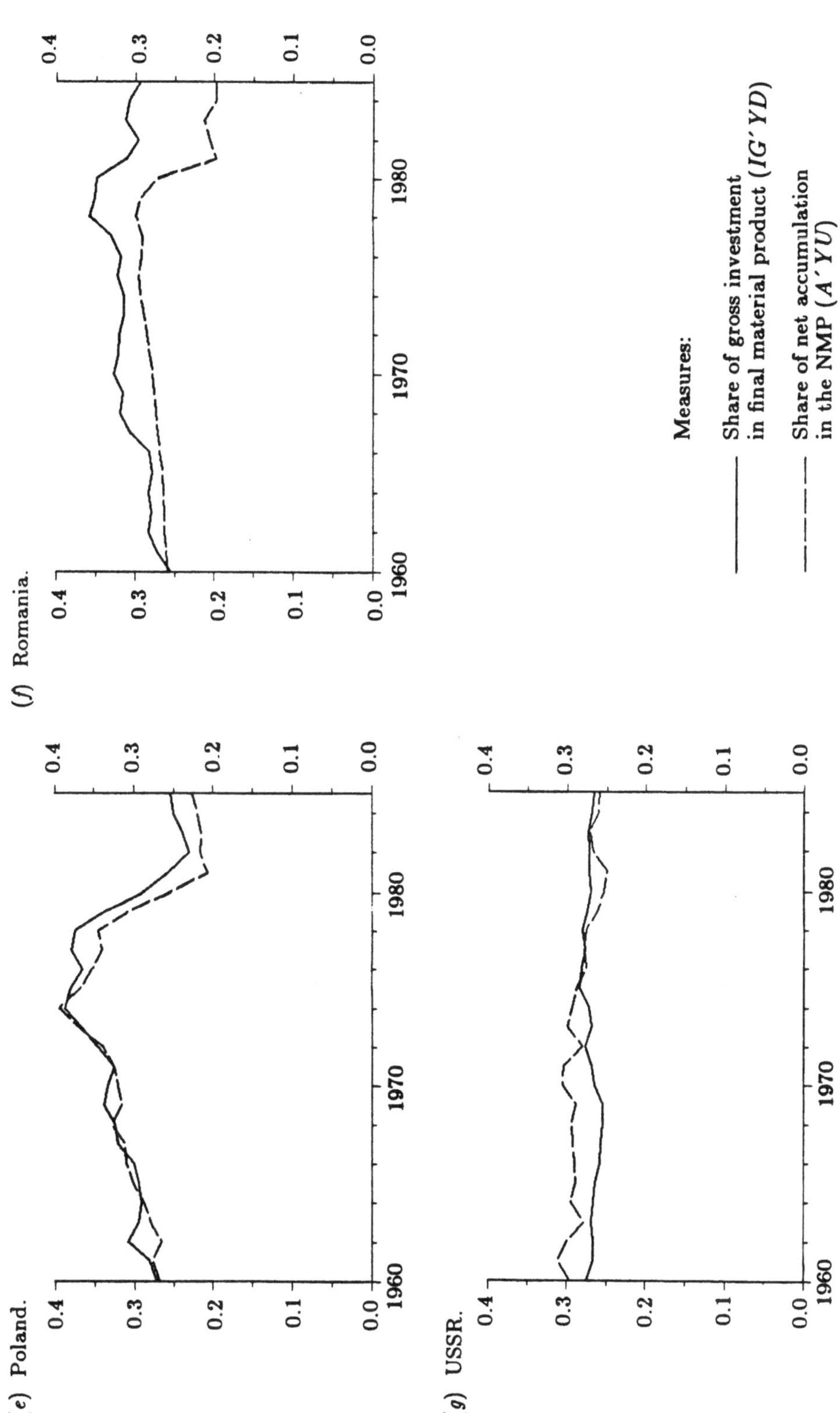


***Figure 4.5.*** Continued.

In many countries the periods of high economic growth were accompanied by even higher growth of foreign trade. During such periods (especially in the 1960s) the "imports component" ($\alpha_3 w_{IM}$) had quite a high value in both absolute and relative terms. However, when the slowdown of the economic growth started, just the reverse tendency occurred. This was especially manifested in countries with higher level of foreign indebtness.

It can be seen from *Table 4.3* that the factor which contributes most significantly to *changes* in rates of economic growth is the rate of technical progress, or total factor productivity $w_T$ [6]. The slowdown of economic growth at the end of the 1970s and in the beginning of the 1980s in most countries is linked to a significant decline in the value of $w_T$. The only exception is the GDR, where there was almost no drop in the level of $w_T$ in both absolute and relative terms and, accordingly, no slowdown of the general rates of economic growth (some drop in the value of $w_{Y*}$ in the first half of the 1980s can be related to decline in $w_{IM_R}$.

The actual figures for $w_T$, one of the main driving forces of economic growth, during the period 1961–1985, are presented in graphical form in *Figure 4.4*. Note that the values until 1980 are the five-year moving averages, whereas from 1981 they are annual values (for this reason there are more fluctuations in this subperiod). As can be seen, in all countries during the 1970s, the rate of the total factor productivity started to decline. Some of the causes for this development have already been mentioned. We shall add another possible interpretation. The exhaustion of the factors of extensive growth requires adaptation of the planning and management systems of the countries and re-orientation of the main emphasis toward the intensive factors of growth. This process is quite complex, and its actual results can be assessed only with some time lag. However, the values of $w_T$ in the first half of the 1980s indicate that the declining trend is probably over, and it is likely that $w_T$ might increase in the future.

Another main driving force of growth is the investment (savings) ratio. This determines, to a great extent, the dynamics of the fixed assets. As has been shown, their contribution to the overall rates of economic growth is quite significant.

In our models we use as a "savings ratio" the share of gross investment in the final material product (shown as *IG′YD* on the plots). Another "savings ratio" often used in the CMEA countries is the "accumulation rate" – the share of net accumulation in the NMP used (shown as *A′YU*). Both these indicators are shown graphically in *Figure 4.5.*

During the slowdown period the investment ratio in most cases decreased too. In some countries this was partly due to the decline of imports; partly this was related to the general policy of increasing (or, at least, preventing a decrease in) the level of real consumption.

There is, however, one interesting development in the most recent years in many of the countries that may be traced on the plots. As can be seen, the decline in the *IG′YD* was not so pronounced as that in *A′YU*. This can be explained by a change in the investment policy of these countries, namely, to finance a larger share of the investment outlays from the so-called "amortization fund" which is formed on the basis of the norm of depreciation of the fixed assets.

Some more data about the driving forces of economic growth will be given when the forecast scenarios are discussed.

## 4.3. *Ex Ante* Scenarios and Simulations

### 4.3.1. Medium- and long-term economic strategies and objectives of the CMEA countries

A central objective in the long-term economic strategies of the CMEA countries at the present time is the intensification of production by a massive injection of scientific and technical acumen in order to achieve higher rates of economic growth. These are also the key issues of the "Program for science and technical progress of the CMEA countries till the year 2000", which was approved by the 41st session of the CMEA in December 1985 (PROGRAM, 1985). This Program outlines the main directions of cooperation and joint work in the fields of science and technology: electronics, complex automation, nuclear energy, new materials and biotechnology. At the same time the Program sets the goal of doubling the labor productivity in the CMEA as a whole up to the year 2000.

In the period 1985–1986 almost all countries set their five-year economic goals and outlined some long-term economic policies. Here is a brief summary of the general strategies, quoted in abridged form from the Economic Survey of Europe (ESE, 1986, pp. 133–137):

(1) Transition to an intensive, resource-efficient path of economic development, with relatively high average output growth rates.
(2) Far-reaching structural change, as well as re-equipment of the productive apparatus.
(3) Reconstruction and re-equipment of many sectors of the nonmaterial sphere, especially science and research.
(4) Higher investment growth.
(5) Faster growth of accumulation, rather than of total domestic consumption.
(6) Further increases in real income and in overall living standards
(7) Changes in management and planning systems with corresponding adjustment in economic policies.
(8) Increasing participation in the international division of labor.

"Altogether, the underlying strategies contained in the five-year plans, draft guidelines and related documents open the way for accelerated, more balanced and also more efficient economic growth" (ESE, 1986, p. 137).

### 4.3.2. Scenarios for the driving forces of economic growth and structural change

The analysis of the possible paths of future development on the basis of our macroeconomic models is performed by constructing scenarios for the "driving forces" of economic growth and structural change. So the first task is to "translate" the strategy we want to study into the language of the "driving force" variables and solve the models for these scenarios.

Three main scenarios have been constructed and analyzed:

(1) A "high" scenario, scenario "A", which is in line with the strategies for accelerated growth outlined in 3.1.
(2) A "low" scenario, scenario "C", which assumes a continuation of the unfavorable "slowdown" trends of the last ten years.
(3) A "medium" scenario, scenario "B", which is a "middle-of-the-road" development between "A" and "C".

The main driving forces of economic growth which we consider (within the research scope of the project) are:

(1) The labor input.
(2) The rate of technical progress.
(3) Fixed assets accumulation.
(4) The international division of labor.

We shall briefly analyze the underlying assumptions for the driving forces in our three scenarios.

*Labor input (L)*

The labor input in the CMEA country models is given by the number of employed persons in the material sphere $L$. As has already been mentioned, this value did not change significantly over the last 25 years, even with a declining, and sometimes negative, growth rate.

For our scenarios we used forecasts of two variables: the total population in the countries (whenever available national forecasts were used) and forecast of the share of $L$ in the total population (usually as a time trend). The resulting values of $L$ were checked with national forecasts, when such were available.

*Table 4.4* shows the average historical growth rates of $L$ in 1961–1985 and the resulting *average* growth rate from the forecast for 1986–2000 (note that in the actual forecasts for $L$ the growth rate is not constant!). Only one version of the labor input $L$ was used in the three scenarios.

*Table 4.4.* Average growth rates of $L$ (%).

| *Period* | *Bulgaria* | *CSSR* | *GDR* | *Hungary* | *Poland* | *Romania* | *USSR* |
|---|---|---|---|---|---|---|---|
| 1961–1985 | 0.8 | 0.5 | 0.1 | 0.1 | 0.9 | 0.1 | 0.9 |
| 1986–2000 | 0.05 | 0.36 | 0.33 | –0.17 | 0.48 | –0.08 | 0.0 |

*Table 4.5.* Average growth rates of $w_T$ (%).

| | *Mean value for:* | | | | *Scenarios* | | |
|---|---|---|---|---|---|---|---|
| | | | | | *(Mean value for 1986–2000)* | | |
| *Country* | *1961–1985* | *1961–1980* | *1971–1985* | *1976–1985* | *"A"* | *"B"* | *"C"* |
| Bulgaria | 1.7 | 2.1 | 1.3 | 0.9 | 2.1 | 1.5 | 0.9 |
| CSSR | 1.3 | 1.7 | 0.8 | 0.2 | 1.7 | 0.9 | 0.2 |
| GDR | 2.0 | 2.1 | 2.0 | 1.6 | 2.1 | 1.6 | 1.2 |
| Hungary | 1.4 | 2.0 | 0.9 | 0.1 | 1.8 | 0.9 | 0.2 |
| Poland | 1.0 | 1.6 | 0.0 | –1.9 | 2.0 | 1.3 | 0.6 |
| Romania | 2.4 | 2.6 | 1.7 | 0.6 | 2.3 | 1.5 | 0.6 |
| USSR | 1.1 | 1.4 | 0.5 | 0.3 | 1.8 | 1.1 | 0.4 |

### *Rate of technical progress ($w_T$)*

The historical trends of $w_T$ were already analyzed in Section 4.2.2. In *Table 4.5* we give the average values of $w_T$ for some subperiods.

As can be seen, during the last decade (1976–1985) the values of $w_T$ have fallen to a rather low level. The strategies for "accelerated growth" described in Section 4.3.1. assume that there would be a (significant) rise of $w_T$ in all countries. It is more reasonable to expect that $w_T$ might rise to some level, which has already been experienced by the country. So, for our scenario "A" we assume for 1986–2000 an average level of $w_T$ which is approximately equal to the mean value in the period 1961–1980. The "low" scenario "C" is based on the mean values for 1976–1985 [7]. Scenario "B" is approximately the mean of "A" and "C". However, as it can be seen, in most cases "B" is very close to the mean historical figures for the period 1971–1985.

### *Fixed Assets Accumulation*

Two main variables determine the dynamics of fixed assets accumulation: the gross savings ratio $s$ ($IG'YD$ on the plots in *Figure 4.5*) and the rate of "sorting out" of fixed assets $d$ [8]. The higher rates of technical progress discussed above require modernization and re-equipment of the productive assets and, accordingly, larger amounts of investment outlays. So three scenarios for the "savings ratio " have been developed, in line with the three scenarios for $w_T$ (*Table 4.6*).

The "medium" scenario "B" is taken as the average historical value for 1961–1985 (excluding the investment "boom" years in some countries in the 1970s). The "high" and "low" scenarios differ from the "medium" one by two percentage points ("+" and "–", respectively).

*Table 4.6.* Gross savings ratio $s$ $(IG'YD)$.

| *Country* | *Mean value for 1960–1985* | *Scenarios: mean value for 1986–2000* "A" | "B" | "C" |
|---|---|---|---|---|
| Bulgaria | 0.30 | 0.32 | 0.30 | 0.28 |
| CSSR | 0.27 | 0.29 | 0.27 | 0.25 |
| GDR | 0.26 | 0.27 | 0.25 | 0.23 |
| Hungary | 0.29 | 0.28 | 0.26 | 0.24 |
| Poland | 0.32 | 0.30 | 0.28 | 0.26 |
| Romania | 0.31 | 0.31 | 0.29 | 0.27 |
| USSR | 0.27 | 0.29 | 0.27 | 0.25 |

The rates of "sorting out" $d$ which are used in the model are calculated on the basis of a "zero-lag" assumption of the implementation of the investment outlays (all investments of a given year are put into operation within the same year). The mean values of $d$ for 1961–1982 and the forecast values (one version for all scenarios) are given in *Table 4.7.*

In general, the scenario values are based on the mean values in the sample period; in some cases recent trends have been reflected as well.

### *The international division of labor*

The impact of this factor is modeled by the variable $IM_R$ (imports of raw materials and intermediate goods), which directly enters the production function. In the case of the CMEA countries $IM_R$ is considered as a share $b_r$ of the total imports. The historical and forecast values of $b_r$ are shown in *Table 4.8* (one version for all scenarios).

The scenarios for $b_r$ are based on the more recent development of the import structure of the countries.

### *Other factors*

Besides the main "driving forces" in the case of the CMEA countries there are two other "driving" variables which determine some of the future developments.

The "norm of depreciation of fixed assets" $a_d$ defines the share of the gross investment which is aimed at replacement of the old fixed assets, the so-called "amortization fund" $AD$.

$$I^{\text{gross}} = AD + I^{\text{net}} \tag{4.3}$$

$$AD = a_d \cdot K_{-1} \tag{4.4}$$

where $I^{\text{gross}}$ and $I^{\text{net}}$ are gross and net investment, respectively; $K$ is total fixed assets. Thus $a_d$ defines the proportion by which $AD$ is formed in relation to $K$. In the models it is used to determine NMP (by subtracting $AD$ from the value of final material product $Y^*$).

*Table 4.7.* Rate of "sorting out" of fixed assets $d$ (%).

| *Mean value of d during period:* | *Bulgaria* | *CSSR* | *GDR* | *Hungary* | *Poland* | *Romania* | *USSR* |
|---|---|---|---|---|---|---|---|
| 1961–1982 | 2.2 | 1.3 | 0.9 | 1.2 | 1.7 | 1.9 | 1.5 |
| 1986–2000 | 2.3 | 1.4 | 0.9 | 1.2 | 0.9 | 1.7 | 1.4 |

*Table 4.8.* Share of raw materials and intermediate goods in total imports $b_r$.

| *Mean value of* $b_r$ *during period:* | *Bulgaria* | *CSSR* | *GDR* | *Hungary* | *Poland* | *Romania* | *USSR* |
|---|---|---|---|---|---|---|---|
| 1961–1975 | 0.75 | 0.71 | 0.75 | 0.74 | 0.73 | 0.76 | 0.54 |
| 1976–1985 | 0.78 | 0.72 | 0.76 | 0.74 | 0.72 | 0.81 | 0.52 |
| 1986–2000 | 0.78 | 0.72 | 0.77 | 0.74 | 0.72 | 0.80 | 0.52 |

*Table 4.9.* Norm of depreciation of fixed assets $a_d$ (%).

| *Mean value of* $a_d$ *during period:* | *Bulgaria* | *CSSR* | *GDR* | *Hungary* | *Poland* | *Romania* | *USSR* |
|---|---|---|---|---|---|---|---|
| 1961–1982 | 5.3 | 2.7 | 1.3 | 3.0 | 2.4 | 4.8 | 4.1 |
| 1986–2000 | 5.3 | 2.7 | 1.7 | 2.9 | 1.6 | 5.0 | 3.8 |

The historical and forecast values of $a_d$ are given in *Table 4.9* (one version for all scenarios). As in the case of $d$, the scenario values are based on the mean values in the sample period; in some cases recent trends have been reflected as well.

Finally, the *index of nominal wages* of the employed in the material sphere is the "driving factor" in the dynamics of the general price level in the countries (together with the index of real wages, which is endogenous). The historical development and the forecast values of the growth rate of nominal wages $w_l$ are given in *Table 4.10* (one version for all scenarios).

*Table 4.10.* Growth rate of nominal wages, $w_l$ (%).

| | *Mean value for:* | | | | *Scenario mean value for:* | |
|---|---|---|---|---|---|---|
| *Country* | *1961–1985* | *1961–1970* | *1971–1980* | *1981–1985* | *1986–1990* | *1991–2000* |
| Bulgaria | 6.2 | 7.1 | 5.4 | 4.5 | 5.0 | 5.0 |
| CSSR | 3.8 | 4.5 | 3.3 | 3.0 | 4.0 | 4.0 |
| GDR | 4.1 | 3.8 | 4.4 | 4.3 | 4.0 | 4.0 |
| Hungary | 7.4 | 6.5 | 8.0 | 8.5 | 8.5 | 8.5 |
| Poland | 12.3 | 4.9 | 13.3 | 33.8 | 12.0 | 10.0 |
| Romania | 8.0 | 7.7 | 8.2 | 8.2 | 7.0 | 7.0 |
| USSR | 4.2 | 5.8 | 2.6 | 3.7 | 4.0 | 4.0 |

*Table 4.11. Ex ante* simulation results: average growth rates of important variables, 1986–1999 (%).

| *Indicator* | *Bulgaria* "A" | "B" | "C" | *CSSR* "A" | "B" | "C" | *GDR* "A" | "B" | "C" |
|---|---|---|---|---|---|---|---|---|---|
| NMP produced | 6.1 | 4.8 | 3.5 | 4.4 | 2.9 | 1.7 | 5.4 | 4.4 | 3.6 |
| NMP used | 6.4 | 5.0 | 3.6 | 4.0 | 2.7 | 1.6 | 4.3 | 3.4 | 2.7 |
| Labor productivity[a] | 6.1 | 4.8 | 3.5 | 4.0 | 2.6 | 1.3 | 5.1 | 4.1 | 3.3 |
| Fixed assets, total | 6.4 | 5.7 | 4.9 | 4.2 | 3.6 | 3.0 | 4.8 | 4.3 | 3.8 |
| NMP produced per fixed assets, total | −0.3 | −0.8 | −1.4 | −0.2 | −0.6 | −1.3 | 0.6 | 0.2 | −0.2 |
| Consumption, total | 5.9 | 4.9 | 4.0 | 3.2 | 2.4 | 1.7 | 3.1 | 2.6 | 2.2 |
| Real imports, total | 7.1 | 5.4 | 3.7 | 5.7 | 4.2 | 2.9 | 4.4 | 3.0 | 1.6 |
| Real exports, total | 6.6 | 5.1 | 3.6 | 5.9 | 4.3 | 2.8 | 7.1 | 5.7 | 4.3 |
| Real primary incomes[b] | 4.2 | 3.7 | 3.1 | 3.3 | 2.8 | 2.3 | 3.6 | 2.7 | 1.9 |
| Price deflator of NMP produced | 0.7 | 1.3 | 1.8 | 0.7 | 1.2 | 1.6 | 0.3 | 1.3 | 2.1 |
| Price deflator of total consumption | 1.1 | 1.6 | 2.1 | 0.8 | 1.1 | 1.5 | 0.4 | 1.3 | 2.0 |

[a]Defined as NMP produced per employed person in material sphere.
[b]Defined as income of one employed person in material sphere.

### 4.3.3. An overview of the main results from the *ex ante* simulations

A number of *ex ante* simulations were run on the models of the CMEA countries both in a "detached" mode (each model running separately) and in a "linked" mode (linked with the models of the other countries and groups of countries). Due to space limitations we present here only a selection of the most general and important results of these simulations. More results concerning the CMEA countries are available in Chapter 7 and in Annex 3 as well as in Dobrinsky (1986b). Here we focus our attention on the three basic scenarios described in the previous section. The simulations reported here were performed on the linked world model described in Chapter 7.

The most important findings from the three basic simulation runs are summarized in *Tables 4.11* and *4.12*. *Table 4.11* reports the major macroeconomic indicators presented in the form of the average rates of growth for the simulation period (1986–1999). *Table 4.12* reflects the changes in the structure of production in the three scenarios. Since these results are basically self-explanatory we comment only briefly on them.

The variables selected in *Table 4.11* are those indicators whose performance was analyzed in this chapter and those which were a subject of the special treatment detailed in Chapter 5. In addition, we have included total imports and exports in order to have a more complete picture of the overall economic performance. However, it should be pointed out that, since the scenarios were run on

*Table 4.11.* Continued.

| *Hungary* | | | *Poland* | | | *Romania* | | | *USSR* | | |
|---|---|---|---|---|---|---|---|---|---|---|---|
| *"A"* | *"B"* | *"C"* | *"A"* | *"B"* | *"C"* | *"A"* | *"B"* | *"C"* | *"A"* | *"B"* | *"C"* |
| 4.4 | 2.7 | 1.4 | 4.3 | 3.1 | 1.9 | 7.0 | 5.2 | 3.2 | 5.2 | 4.0 | 2.7 |
| 4.7 | 2.1 | 0.5 | 4.3 | 2.9 | 1.4 | 7.0 | 5.0 | 2.9 | 5.2 | 4.0 | 2.7 |
| 4.5 | 2.9 | 1.6 | 3.8 | 2.6 | 1.4 | 7.1 | 5.3 | 3.2 | 5.2 | 4.0 | 2.7 |
| 3.9 | 3.3 | 2.8 | 3.6 | 3.2 | 2.7 | 6.9 | 6.0 | 5.1 | 6.0 | 5.4 | 4.7 |
| 0.5 | −0.5 | −1.3 | 0.7 | −0.1 | −0.8 | 0.0 | −0.8 | −1.9 | −0.8 | −1.3 | −1.9 |
| 4.0 | 1.6 | 0.4 | 3.9 | 2.7 | 1.5 | 6.4 | 4.9 | 3.2 | 5.1 | 4.2 | 3.3 |
| 7.4 | 5.0 | 2.5 | 6.7 | 4.3 | 1.7 | 5.0 | 2.6 | 0.2 | 5.5 | 3.7 | 1.9 |
| 6.6 | 5.3 | 4.1 | 6.8 | 5.4 | 3.9 | 5.0 | 3.1 | 1.0 | 5.5 | 3.3 | 1.0 |
| 4.5 | 3.2 | 2.1 | 3.8 | 3.1 | 2.0 | 7.1 | 6.0 | 4.5 | 4.9 | 4.0 | 3.2 |
| 3.8 | 5.2 | 6.3 | 6.4 | 7.5 | 8.7 | 0.0 | 1.4 | 2.9 | 0.6 | 1.4 | 2.2 |
| 4.2 | 6.0 | 7.3 | 6.4 | 7.5 | 8.7 | 0.1 | 1.7 | 3.3 | 0.8 | 1.7 | 2.7 |

the linked world model, the foreign trade indicators reported here depend not only on the domestic development in the countries under consideration, but also on the general world economic performance. In our case all results in scenario "A" correspond to the "high" or "optimistic" scenario for the world economy; "B" to the "medium scenario" and "C" to the "low" or "pessimistic" scenario.

In general, scenarios "A" and "B" outline a stable long-term development in the countries with relatively high rates of economic growth. The principal long-term economic goals of the CMEA countries discussed in Section 3.1. are met in scenario "A". As can be seen, some important economic ratios which are subject to special attention in the economic policy are preserved in this scenario. Thus, the real wages of those employed in the material sphere do not grow faster than labor productivity; and real consumption grows more slowly than NMP used (this means an increasing share of accumulation in NMP used, which is in line with the long-term objectives mentioned in Section 4.3.1). The latter also means a higher investment growth than in the previous several years (also in line with the targets of Section 4.3.1.). Foreign trade (imports and exports) also in general grows at higher rates than NMP (increasing imports to NMP ratio), which corresponds to increasing participation in the international division of labor (another goal mentioned in Section 4.3.1).

In this scenario there are also positive tendencies with respect to the output/capital ratio, measured by NMP per total fixed assets. In three countries (GDR, Hungary and Poland) it increases; in one (Romania) it does not change; and in the rest of the countries it continues to decline, but at much lower rates than have been experienced in the last few years.

*Table 4.12. Ex ante* simulation results: sectoral structure of net material product.

| | | | Scenarios | | | | | |
|---|---|---|---|---|---|---|---|---|
| | | | "A" | | "B" | | "C" | |
| *Country* | *Sector*[a] | *1982* | *1990* | *1999* | *1990* | *1999* | *1990* | *1999* |
| Bulgaria | (1) | 17.3 | 18.6 | 12.4 | 19.9 | 14.3 | 21.3 | 16.2 |
| | (2) | 51.9 | 49.9 | 54.1 | 48.8 | 52.6 | 47.7 | 51.1 |
| | (3) | 10.0 | 10.0 | 10.4 | 9.5 | 9.8 | 9.0 | 9.3 |
| | (4) | 8.8 | 9.5 | 11.0 | 9.8 | 11.2 | 10.1 | 11.4 |
| | (5) | 8.6 | 8.7 | 9.2 | 8.6 | 9.0 | 8.4 | 8.9 |
| CSSR | (1) | 8.2 | 8.7 | 8.2 | 9.2 | 8.6 | 9.7 | 9.0 |
| | (2) | 58.0 | 55.0 | 54.3 | 55.4 | 55.5 | 55.8 | 56.7 |
| | (3) | 11.2 | 12.7 | 12.9 | 12.7 | 13.0 | 12.7 | 13.0 |
| | (4) | 17.4 | 18.4 | 19.7 | 17.3 | 17.6 | 16.3 | 15.6 |
| | (5) | 4.9 | 4.6 | 4.5 | 4.6 | 4.6 | 4.6 | 4.7 |
| GDR | (1) | 8.3 | 7.0 | 5.2 | 6.9 | 5.2 | 6.7 | 5.3 |
| | (2) | 67.9 | 68.7 | 70.7 | 69.5 | 71.0 | 69.8 | 71.3 |
| | (3) | 6.1 | 6.4 | 6.2 | 6.1 | 5.9 | 5.8 | 5.7 |
| | (4) | 10.0 | 10.1 | 10.1 | 10.1 | 10.1 | 10.0 | 10.0 |
| | (5) | 4.4 | 4.4 | 4.3 | 4.4 | 4.3 | 4.4 | 4.3 |
| Hungary | (1) | 16.4 | 14.3 | 14.1 | 15.8 | 15.4 | 17.4 | 17.4 |
| | (2) | 49.1 | 50.6 | 51.6 | 51.5 | 52.8 | 52.3 | 54.2 |
| | (3) | 10.5 | 11.3 | 11.6 | 11.0 | 11.3 | 10.8 | 11.0 |
| | (4) | 17.5 | 16.1 | 14.6 | 14.0 | 12.3 | 11.9 | 9.2 |
| | (5) | 5.4 | 6.1 | 6.3 | 6.2 | 6.6 | 6.3 | 6.7 |
| Poland | (1) | 18.1 | 16.1 | 14.0 | 15.4 | 11.9 | 14.7 | 9.6 |
| | (2) | 49.2 | 50.0 | 52.1 | 50.5 | 53.1 | 51.2 | 54.4 |
| | (3) | 11.0 | 13.0 | 12.3 | 12.8 | 12.7 | 12.7 | 13.1 |
| | (4) | 15.7 | 15.2 | 15.4 | 15.8 | 16.0 | 15.5 | 16.5 |
| | (5) | 4.5 | 4.5 | 4.8 | 4.5 | 4.8 | 4.5 | 4.8 |
| Romania | (1) | 17.9 | 16.9 | 16.6 | 18.2 | 16.0 | 19.2 | 14.0 |
| | (2) | 57.1 | 58.7 | 59.9 | 57.7 | 60.1 | 56.8 | 61.4 |
| | (3) | 8.6 | 8.6 | 8.2 | 8.0 | 7.5 | 7.3 | 6.5 |
| | (4) | 7.1 | 6.5 | 6.3 | 6.3 | 6.2 | 6.1 | 6.4 |
| | (5) | 6.8 | 7.3 | 7.2 | 7.6 | 7.8 | 7.9 | 8.6 |
| USSR | (1) | 14.8 | 9.4 | 6.3 | 11.4 | 6.9 | 13.6 | 7.9 |
| | (2) | 59.6 | 63.7 | 66.9 | 63.1 | 67.7 | 62.3 | 68.3 |
| | (3) | 10.9 | 11.5 | 11.3 | 10.6 | 10.3 | 9.6 | 9.3 |
| | (4) | 8.3 | 8.6 | 8.6 | 8.3 | 8.3 | 8.1 | 7.9 |
| | (5) | 6.4 | 6.8 | 6.9 | 6.6 | 6.8 | 6.4 | 6.6 |

[a]Sectors are as follows: (1) Agriculture and forestry, (2) Industry, (3) Construction, (4) Wholesale and retail trade, and (5) Transport and communications.

Scenario "C" should be regarded as a kind of "warning" scenario. It indicates the likely long-term consequences of a development in which the unfavorable trends observed in recent years are prolonged into the future. In many of the countries in this scenario the unfavorable tendencies deepen and some of the proportions mentioned above do not match. So scenario "C" is intended rather to indicate the necessity for a change in the economic policies of the countries (and, as was pointed out in Section 3.1, such a change has already been initiated!) than to project a realistic and sustainable development path.

As for the structure of production in the countries (*Table 4.12)* the general trend of increasing the industrial share is present in almost all scenarios. However, the development in the three scenarios differs from country to country. Thus, in Bulgaria the higher rates of economic growth imply a relatively higher share of industry and lower share of agriculture; in Poland it is the other way round; in the CSSR and Hungary both these sectors "shrink" at higher growth rates (at the expense of the other sectors); in the GDR, Romania and the USSR these results differ in the different subperiods.

## Notes

[1] It should be noted that in accordance with the national data available in the CMEA data base (Dobrinsky, 1986a), the base years for constant prices differ from country to country, namely: Bulgaria – 1980, CSSR – 1977, GDR – 1980, Hungary – 1976, Poland – 1982, Romania – 1981, USSR – 1973. The production structure shown in *Figures 4.11–17* refers to sectoral shares measured in the prices of these years.

[2] The figures reported here refer to real production shares measured in constant prices of the indicated years.

[3] For the definitional identities, see Annex 2 of this volume.

[4] Total fixed assets were used as a proxy, owing to the lack of more detailed data.

[5] However, in the presence of negative growth rates, as was the case from 1981 to 1985, these "shares" do not have such a clear interpretation, so we do not calculate them for that subperiod.

[6] In terms of the assumed production technology, changes in the level of $\tau$ mean shifts of the isoquants of the production function in the direction of the original (positive $w_\tau$) or in the opposite direction (negative $w_\tau$). These shifts are due to the combined effect of different factors, technical progress being only one of them. Technical progress *per se* can be accepted as the main driving force for positive changes in the level of $\tau$ ("improvement" in the production technology), but negative changes can be due to general deterioration of the production environment or lower degree of capacity utilization. For this reason we prefer to interpret $\tau$ as "level of total factor productivity". Whenever "technical progress" is used, it is also in the sense of "total factor productivity".

[7] There are some exceptions. The "high" and "low" values for Poland are calculated as the averages for the other CMEA countries (besides the USSR). The "high" value of $w_\tau$ for the USSR is based on the mean for 1961–1975.

[8] Note that "fixed assets" and "sorting out of fixed assets" which are used in the CMEA country models are not directly comparable with "real capital" and "depreciation of real capital" which are used in the models of market economies. "Fixed assets" denote assets which are physically available for operation. "Sorting out" denotes the process of physical liquidation of old assets, and the rate of sorting out refers only to the portion of fixed assets which have been liquidated during a specific year.

## References

Dobrinsky, R. (1986a), *The Macroeconomic Data Base for the CMEA Countries*, Working Paper WP-86-54. (Laxenburg, Austria: International Institute for Applied Systems Analysis).

Dobrinsky, R. (1986b), *Economic Growth and Structural Change in the CMEA Countries*, Discussion paper B-59, Bonn University, Sonderforschungsbereich 303.

ESE (1986), *Economic Survey of Europe in 1985–1986*, prepared by the Secretariat of the Economic Commission for Europe, Sales No. E.86.II.E.1. (Geneva and New York: United Nations).

PROGRAM (1985), *Program for Science and Technical Progress of the CMEA Countries till 2000* (Rabotnichesko delo, 19.12.1985).

CHAPTER 5

# The Macroeconomic Models of the European CMEA Countries

*Rumen Dobrinsky*


## Summary

Some specific features of the models of the European CMEA countries are described in this chapter. Three main aspects of the models are discussed: the production technology, the structure of production, and the distribution of the final product. The empirical findings concerning these aspects of the models are presented, as well as an analysis of the dynamic properties of the model as a whole.


## 5.1. Introduction

This chapter reports on the models of the seven European CMEA countries [1] and outlines some specific methodological aspects of the approach to the modeling of the economies of these countries. The factual aspects of the analysis and forecast are presented in a separate chapter.

The models have been developed following general lines of the Bonn–IIASA Project. However, the approach to the CMEA country models differs in some important details which are due to the different behavior of economic agents in a socialist economy as compared to a market-type economy. Another cause of difference is the different accounting system in the CMEA countries whose statistics are based on the System of Material Product Balances.

The actual data base for the CMEA countries which was used in the project was compiled on the basis of contributions from many collaborating groups. This data base is described in Dobrinsky (1986).

## 5.2. The Production Technology

The central part of our study is the production technology which, in our case, following the general approach of the Bonn–IIASA Project, is described by a linear homogeneous production function of Cobb–Douglas type.

The application of the neoclassical productions, and in particular, the Cobb–Douglas form, to modeling of a socialist economy has been a subject of long debate in the economic literature. Also subject to discussion and criticism have been different aspects of this approach, such as the linear homogenity assumption (Anchishkin, 1973), the restrictive assumption with respect to the elasticities of substitution between factor inputs (Klacek and Nešporová, 1983), and the necessity to distinguish between a production function for potential product and a production function for actual product (Klacek and Nešporová, 1983). These aspects of the debate are of more general character and refer to the general properties of the neoclassical approach. However, the most controversial issue has been the "marginal productivity assumption" relating the marginal productivities of the input factors to their "prices" (or factor costs). This assumption, which is the foundation of most theoretical and empirical studies in the western countries has been widely criticized by Marxist economists both in terms of its theoretical basis (Anchishkin, 1973) and of the possibilities for its empirical implementation (e.g., Pappe and Ryvkin, 1977).

The general outcome of the debate seems to be positive with respect to practical implementation; however, in most cases some modification of the traditionally used neoclassical forms have been suggested. Thus, Anchishkin (1973) suggests modeling the Soviet economy with a Cobb–Douglas production function which is not homegeneous to degree one. Simon and Samoval (1981) propose a "function of economic growth", in which the production elasticities are functions of some secondary factors. Klacek and Nešporová (1983) advocate the translog and other flexible functional forms and test them on CSSR data. Ershov and Sadykov (1986) suggest a model of the "limits of the production potential", which assumes "switching" of the production technology to different modes of substitutability of the inputs.

At the same time more traditional functional forms have also been analyzed [2]. Probably the first attempt to estimate a Cobb–Douglas production function for the Soviet economy was the works of Mikhalevski and Solov'ev (1966). Weitzman (1970) estimated a CES production function for the Soviet industry. Toda (1976) analyzed a CES (and a translog) form with respect to Soviet manufacturing. Desai (1976, 1985) used Cobb–Douglas and CES forms to analyze the factor productivity and the elasticity of factor substitution in Soviet industry. Similar approaches were applied by Toms and Hajek (1970) for Czechoslovakia; by Brown *et al.* (1974, 1976) for Hungary; and by Kemme (1984) for Poland. A Cobb–Douglas production function was used in a macromodel of the Bulgarian economy (Econometric Macromodels, 1984).

The central issue and, at the same time, the main problem of concern in all empirical studies remains the estimation of the production elasticities of the input factors, and all scholars acknowledge the serious difficulties in the estimation of these parameters [3]. Many of these problems have an objective character; but a simple and straightforward solution does not exist. Thus, the common approach of unrestricted (with respect to the estimated parameters) direct estimation of a production function with an explicitly defined shape of the technical progress variable, using time series data, is actually an attempt to "fix" the production isoquant curves only on the basis of one point (since it is assumed that the production isoquants are "shifted" in every observation, only one observation is available per isoquant). Combined with the usually strong multicollinearity of the variables in the regression, this provides an almost insoluble task, both theoretically and practically [4]. Another problem (and source of estimation difficulties) may be that the actual economy – as depicted by the empirical data – can diverge significantly from the assumptions of the theoretical model, which may be very restrictive (e.g., constancy of the production elasticities, constant, even unit, elasticities of substitution, etc.). A further difficulty is caused by the quality of the data available which is usually not very high.

In our judgment, the selection of the theoretical model and the methods for its estimation should be based first of all on the goal of the research and on the assessment of the ability of the theoretical model to serve the goal. This was the main criterion for selecting the functional form for our technology. Our goal is to study the *long-term* growth trends, and for *this goal* the Cobb–Douglas production function seems to be the most appropriate tool, in spite of its implicit restricitive assumptions with respect to the production technology. Other functional forms (e.g., the translog, or the other flexible forms) may better fit the goals of analyzing past performance, but can scarcely be used for long-term forecasts. (Unfortunately, almost all flexible functional forms are very unstable in *ex ante* simulations.)

The production function which we use in the models is specified as

$$Y^* = \alpha_0 \, \tau L^{\alpha_1} K^{\alpha_2} IM_R^{\alpha_3}$$
$$\sum_{i=1}^{3} \alpha_i = 1, \quad 0 < \alpha_i < 1, \quad i = 1,2,3 \tag{5.1}$$

where $Y^*$ = total material output; $\tau$ = the level of technical progress (total factor productivity); $L$ = the number of employed persons in the material sphere; $K$ = fixed assets in the material sphere (in our case, approximated by total fixed assets); and $IM_R$ = imports of raw materials and intermediate products.

We assume that the total material output $Y^*$ is separable in $Y$ (final material product) and $IM_R$:

$$Y^* = \beta_0 \, Y^\beta \, IM_R^{1-\beta} \tag{5.2}$$

and that $Y$ is also defined by a Cobb–Douglas production function over $L$ and $K$:

$$Y = \gamma_0 \tau^* L^\gamma K^{1-\gamma}. \tag{5.3}$$

The parameters of (5.1), (5.2) and (5.3) are related as follows:

$$\beta = \alpha_1 + \alpha_2; \quad 1-\beta = \alpha_3; \quad \gamma = \frac{\alpha_1}{1-\alpha_3} \tag{5.4}$$

Further, we assume that the planning board of the economy pursues different policies on the different levels of the production technology. First, we assume that on the level defined by equation (5.2) the planning agency follows a cost-minimization policy with respect to the utilization of domestic resources $Y$ and imported resources $IM_R$. Taking into account that in nominal terms we have

$$Y^* \cdot P'Y^* = Y \cdot P'Y + IM_R \cdot P'IM_R \tag{5.5}$$

where $P'Y^*$, $P'Y$ and $P'IM_R$ are price deflators of the corresponding variables. This assumption means that we can identify $1-\beta = \alpha_3$ by the share of $IM_R \cdot P'IM_R$ in $Y^* \cdot P'Y^*$. This leaves us only with the estimation of $\gamma$.

Next we assume that the policy with respect to the final material product $Y$ [equation (5.3)] is maximization of $Y$ and at the same time provision of a certain level of real primary income (real wages) of those employed in the material sphere $l_r$ :

$$l_r \geq \bar{l}_r \tag{5.6}$$

where $\bar{l}_r$ is the desired level of income.

In Appendix 5A, we present the solution to this problem in the case when the distribution function for the real wages is defined as

$$l_r = a_1 l_{r,-1} + a_2 \frac{Y}{L} - a_3 \frac{q_{r,-1} \cdot K}{L} \tag{5.7}$$

where $q_r$ denotes real primary incomes of enterprises per unit of fixed assets (see Appendix 5A for the complete definitions).

The solution to this problem provides the following relation between the parameters of the production and distribution functions:

$$\frac{l_r - a_1 \cdot l_{r,-1}}{a_3 \cdot q_{r,-1}} = \frac{\gamma}{1-\gamma} \cdot \frac{K}{L} \tag{5.8}$$

The economic interpretation of equation (5.8) is that there is a specific relation between the process of production and primary distribution of $Y$. From a practical point of view equation (5.8) can be used to derive additional constraints on the parameters of the estimated equations which would allow us to obtain more reliable estimates.

Another problem is the estimation of the technical progress variable $\tau$. It is a latent variable which, by its nature, changes smoothly over time (see Chapter 1). In estimation the shape of the trend is usually set exogenously and only its parameters are estimated. However, in the presence of second (and higher) order variations in the trend of $\tau$, the results of such an estimate can be highly unstable; and this was our experience in estimating the production functions for the CMEA countries. For this reason we used another approach to the estimation of $\tau$.

In the observations for the output $Y^*$ and for the production factors $\tau$, $L$, $K$, $IM_R$ (which are yearly figures) one can usually distinguish two components: a "trend" component $F(X)$, $X = \{Y^*, \tau, L, K, IM_R\}$ which defines the long-term growth path of the variables and a "fluctuation" component which in log form can be defined as $U_X = \log X/F(X)$, $X = \{Y^*, \tau, L, K, IM_R\}$ and which can be due to the presence of short-term cycles (such as the investment cycles), variation on the capacity utilization, etc. Taking into account the nature of the technical progress variable $\tau$, we can consider that the fluctuations in $Y$ are only determined by the fluctuations of $L$, $K$ and $IM_R$, which is equivalent to the assumption that $\tau = F(\tau)$ and $U_\tau = 0$.

This enables us to eliminate $\tau$ from the estimation equation which, after the substitutions, reads:

$$U_{Y*} = \alpha_1 U_2 + \alpha_2 U_K + \alpha_3 U_{IM_R} \tag{5.9}$$

The level of $\tau$ can be calculated residually from (5.1) where $Y^*$, $L$, $K$, $IM_R$ are substituted by $F(Y^*)$, $F(L)$, $F(K)$, $F(IM_R)$. In the actual estimations the trend functions $F(X)$, $X = \{Y^*, L, K, IM_R\}$ were specified as five-year moving averages.

The estimation of the production function was performed in the form (5.9) jointly with the distribution equation for real wages (5.7), imposing (5.8) as parameter restrictions. The full information maximum likelihood method was used for the estimation. The estimation period was 1960–1982 [actually 1962–1980, because five-year moving averages were used for equation (5.9)]. The estimated production function and demand function for unit income are given in *Tables 5.1* and *5.2*.

We would like to make some points concerning the estimation results. First of all, due to the specified restrictions the degrees of freedom in the system are reduced considerably (actually we estimate only three of the six coefficients). Due to this, the calculated standard errors of the estimated parameters are very low (and, accordingly, the $t$-values are high). But since we consider this as an indication of low degrees of freedom and not as a high precision of the estimate, we do not reproduce these values in the tables.

*Table 5.1.* Parameters of the estimated production functions.[a]

| | *Estimated results* | | | | | *Calculated parameters* | | |
|---|---|---|---|---|---|---|---|---|
| *Country* | $\alpha_1$ | $\alpha_2$ | $\alpha_3$ | *DW* | $R_1^2$ | $R_2^2$ | *MAPE* | $\alpha_0$ |
| Bulgaria | 0.361 | 0.405 | 0.234 | 1.53 | 0.42 | 0.998 | 0.78 | 1.476 |
| CSSR | 0.408 | 0.411 | 0.181 | 1.17 | 0.21 | 0.997 | 0.80 | 3.926 |
| GDR | 0.383 | 0.423 | 0.194 | 1.54 | 0.89 | 0.998 | 0.39 | 2.072 |
| Hungary | 0.341 | 0.416 | 0.243 | 1.58 | 0.55 | 0.994 | 0.85 | 3.486 |
| Poland | 0.424 | 0.401 | 0.175 | 0.84 | 0.22 | 0.997 | 1.24 | 7.972 |
| Romania | 0.321 | 0.501 | 0.178 | 1.03 | 0.67 | 0.998 | 0.67 | 2.168 |
| USSR | 0.421 | 0.514 | 0.065 | 1.81 | 0.10 | 0.996 | 0.81 | 1.048 |

[a]Estimated equation: $u_{Y^*} = \alpha_1 u_L + \alpha_2 u_K + \alpha_3 u_{IM_R} + \varepsilon$ (see text).

*Table 5.2.* Equations for real wages in the material sphere.[a]

| *Country* | $a_1$ | $a_2$ | $a_3$ | *DW* | $R^2$ | *MAPE* |
|---|---|---|---|---|---|---|
| Bulgaria | 0.866 | 0.161 | 0.147 | 1.49 | 0.996 | 1.67 |
| CSSR | 0.811 | 0.128 | 0.072 | 2.28 | 0.970 | 2.84 |
| GDR | 0.538 | 0.448 | 0.423 | 2.21 | 0.995 | 1.12 |
| Hungary | 0.392 | 0.361 | 0.177 | 2.02 | 0.994 | 1.83 |
| Poland | 0.484 | 0.358 | 0.208 | 2.14 | 0.992 | 1.50 |
| Romania | 0.315 | 0.574 | 0.470 | 2.41 | 0.999 | 1.09 |
| USSR | 0.507 | 0.519 | 0.519 | 1.69 | 0.999 | 0.56 |

[a]Estimated equation: $l_r = \alpha_1 l_{r-1} + \alpha_2 (Y/L) - \alpha_3 (q_{r-1} K/L) + \varepsilon$ (see text).

The next point is that two values of the goodness of fit are shown in *Table 5.1.* The first one ($R_1^2$) refers to the actual $R^2$ of the estimated equation (5.9). At first glance $R_1^2$ seems to be low in general, but we have to take into consideration that (5.9) is estimated on the residual series $u_X$ and in this sense in most cases it can even be regarded as rather high. $R_2^2$ is the equivalent (recalculated) goodness of fit of equation (5.1) on the basis of the estimated parameters. The mean absolute percentage error (MAPE) is also calculated with respect to (5.1). Both these indicators are quite satisfactory.

The values of $\tau$ which result from this estimation are given in Chapter 4 on "Economic Growth and Structural Change in the CMEA Countries" in the form of growth rates ($w_\tau$).

All parameters of the estimated equation for $l_r$ have the correct sign and order of magnitude. From the estimated equation (5.7) we can calculate the structural parameters and other characteristics of the distribution system (see Appendix 5A). In *Table 5.3* we present the most important of them: the production elasticities and the own short-run quantity elasticities of the factor remunerations $l_r$ and $q_r$ calculated as sample averages [5].

*Table 5.3.* Implied production and quantity elasticities of real factor remunerations.

| | *Production elasticity* | | *Own short-run quantity elasticity (uncompensated)* | |
|---|---|---|---|---|
| *Country* | $l_r$ | $q_r$ | $l_r$ | $q_r$ |
| Bulgaria | 0.37 | 1.54 | −0.17 | −0.85 |
| CSSR | 0.31 | 1.50 | −0.21 | −0.93 |
| GDR | 0.91 | 1.10 | −0.48 | −0.58 |
| Hungary | 0.87 | 1.09 | −0.63 | −0.82 |
| Poland | 0.76 | 1.23 | −0.53 | −0.79 |
| Romania | 1.27 | 0.78 | −0.71 | −0.53 |
| USSR | 1.12 | 0.90 | −0.51 | −0.49 |

## 5.3. The Structure of Production

The structure of production in the CMEA countries is defined by the shares of output produced in the sectors of the material sphere. Below we propose a possible approach to the derivation of a specification for explaining the production sector shares. As our output concept is based on the final material product $Y$, we derive the equations explaining the composition of $Y$:

$$Y = \sum_{i=1}^{n} y_i; \quad y_i = \beta_i \cdot Y; \quad \sum_{i=1}^{n} \beta_{i=1} \tag{5.10}$$

where $\beta_i$ denotes the share of sector $i$ in the final material product.

However, the available database (Dobrinsky, 1986) contains only data about the structure of the Net Material Product (NMP). But, as we show later, the equation derived can be applied to the NMP sector shares under some not very restrictive assumptions.

We start our derivation of the functions $\beta_i$ from the input–output system in real terms. Let $x_j$ be real total production (gross output) of sector $j$, $Y_j$ = real final product of sector $j$; $|a_{ij}|$ = the input–output matrix; $C_j$ = total consumption of the product of sector $j$; $A_j$ = gross accumulation of product $j$; $E_j$ = exports of $j$; $M_j$ = imports of $j$; $X$ = total gross output in the economy; $Y$ = total final product; $C$ = total consumption; $A$ = total gross accumulation; $E$ = total exports; $M$ = total imports; $j = 1,...,n$.

From this system we can express the final product of sector $i$ as:

$$y_i = \sum_j a_{ij} \cdot x_j - \sum_j a_{ji} \cdot x_i + C_i + A_i + E_i + M_i. \tag{5.11}$$

Denote:

$$\Omega_i = \sum_j a_{ij} \cdot x_j - \sum_j a_{ji} \cdot x_i$$

$$\beta_i^C = \frac{C_i}{C}; \quad \beta_i^A = \frac{A_i}{A}; \quad \beta_i^M = \frac{M_i}{M}; \quad \beta_i^E = \frac{E_i}{E} \tag{5.12}$$

where $\Omega_i$ sum up to zero; $\beta_i^C$, $\beta_i^A$, $\beta_i^M$, and $\beta_i^E$ sum up to 1.

From (5.10) and (5.12) we can write:

$$\beta_i = \Omega_i \cdot \frac{1}{Y} + \beta_i^C \cdot \frac{C}{Y} + \beta_i^A \cdot \frac{A}{Y} + \beta_i^M \cdot \frac{M}{Y} + \beta_i^E \cdot \frac{E}{Y}. \tag{5.13}$$

We shall now analyze the coefficients introduced in (5.12). Consider for example, $\beta_i^C$. We can assume that approximately $C_i \approx a \cdot y_i$, $C \approx b \cdot Y$ and, as a consequence,

$$\beta_i^C \approx d \cdot \beta_i \tag{5.14}$$

where $a$, $b$, and $d$ = const, which is quite natural for an equilibrium growth path. A similar assumption can also be made for $\beta_i^A$, $\beta_i^M$ and $\beta_i^E$ . As for $\Omega_i$, it accounts for the difference between the amount of sector $j$'s gross output used as intermediate product in the economy, and all intermediate products used in sector $i$. It depends of the structure of final demand. Therefore

$$\Omega_i = \Omega_i \left( \frac{C}{Y}, \frac{A}{Y}, \frac{M}{Y}, \frac{E}{Y} \right) \tag{5.15}$$

Now substitute (5.14) and (5.15) into (5.13). If we solve this nonlinear equation for $\beta_i$ we can present the linearized (by a Taylor expansion) solution as:

$$\beta_i = a_1 \cdot \frac{1}{Y} + a_2 \cdot \frac{C}{Y} + a_3 \cdot \frac{A}{Y} + a_4 \cdot \frac{M}{Y} + a_5 \cdot \frac{E}{Y} + a_6 \tag{5.16}$$

This is the form which served as basis for our estimations.

Two more points remain to be made. We use this equation to explain the structure of NMP, whereas it is derived for the final product. However, this is only equivalent to the assumption that the amortization fund is distributed among the sectors proportionally to their final product. Obviously, this is not a very restrictive assumption. The next point concerns the estimation. If we estimate (5.16) directly, we may run into problems because of the linear dependence of ($Y$, $C$, $A$, $M$, $E$). In order to avoid this we substitute $IG$ (gross investment) for $A$ (gross accumulations). From the point of view of the economic interpretation we even gain from this substitution because the difference between $A$ and $IG$ is the so-called "change in material assets and stocks" which is subject to large fluctuations and is not a significant factor of structural change.

The actual estimated results for the production sector functions for the seven CMEA countries are given in *Table 5.4*. Some of the estimated parameters may seem unsatisfactory from the point of view of the precision of the estimation (the $t$-values are very low). However, we have not excluded any variable for this reason, because in the full specification the set of sector functions for a given country preserve the valuable property of summing up to 1 [6], and this is one of the main requirements with respect to these functions. The relatively low value of $R^2$ occurs mainly in the cases of low variation of the observed values of $\beta_i$. As for the signs of the estimated parameters which indicate the direction of influence of the explanatory variables, no theoretical considerations indicating a positive or negative sign stem from our derivation. So it is difficult to make a comprehensive intercountry comparison, because the signs reflect the different processes of structural change in the countries.

## 5.4. Consumption and the Price Index for Consumer Goods

The problems of specification and estimation of a macroeconomic function explaining aggregated consumption in a socialist economy have not been studied so intensively as, for instance, the production side, especially as far as the empirical implementation is concerned. Among the well-known works we should mention the attempts of Portes and Winter to estimate an aggregated "consumption supply" function (1977) and a "consumption demand" function (1978), as well as their theoretical work on disequilibrium estimation (1980); see also Charemza and Quandt (1982). As for the price index for consumer goods, probably the most common is the cost-push approach (W. Welfe, 1985), sometimes incorporating disequilibrium techniques (A. Welfe, 1985).

The general approach to the modeling of consumption within the Bonn–IIASA Project is basically supply-driven, total consumption being determined residually from the national accounts identity (see Annex 2). However, we also considered it to be of interest to estimate "consumption demand" functions for the CMEA countries and to test their performance against the performance of consumption supply as an indicator of possible disequilibrium tendencies on the consumer goods markets.

We based our approach with respect to consumption demand on the dynamic version of the linear expenditure system, discussed in Appendix 5A. In this case we consider the distribution of final material product *used* [7] – $Y_u$ – among the final demand categories: personal consumption, other consumption, gross investment, change in material assets and stocks. Or, in a more aggregated form: total consumption $C$ and gross accumulation $A$. When we consider only two items, the theoretical setting and derivations are identical to those given in

*Table 5.4.* Sectoral production functions, by CMEA country ($\tau$ values given in parentheses).[a] All estimates cover the period 1960–1982 (OLS).

| *Country* | *Sector*[b] | $a_1$ | $a_2$ | $a_3$ | $a_4$ | $a_5$ | $a_6$ | *DW* | *R2C* | *MAPE* |
|---|---|---|---|---|---|---|---|---|---|---|
| Bulgaria | (1) | 2456.470 | −0.457 | −0.549 | 0.801 | −0.763 | 0.521 | 1.50 | 0.969 | 5.35 |
| | | (6.29) | (1.45) | (1.15) | (4.01) | (2.31) | (1.65) | | | |
| | (2) | −2251.930 | 0.664 | 0.329 | −0.578 | 0.386 | 0.157 | 1.59 | 0.976 | 2.77 |
| | | (9.39) | (3.43) | (1.12) | (4.71) | (1.90) | (0.81) | | | |
| | (3) | −256.367 | 0.092 | 0.228 | −0.056 | 0.025 | −0.006 | 1.02 | 0.895 | 3.47 |
| | | (3.75) | (1.65) | (2.73) | (1.59) | (0.42) | (0.11) | | | |
| | (4) | 83.604 | −0.291 | −0.023 | −0.153 | 0.236 | 0.256 | 1.60 | 0.648 | 22.50 |
| | | (0.29) | (1.25) | (0.07) | (1.04) | (0.97) | (1.10) | | | |
| | (5) | −396.836 | 0.051 | 0.022 | −0.039 | 0.002 | 0.077 | 0.79 | 0.966 | 3.22 |
| | | (8.29) | (1.33) | (0.37) | (1.58) | (0.04) | (1.99) | | | |
| | (6) | 364.939 | −0.059 | −0.007 | 0.024 | 0.116 | −0.004 | 2.57 | 0.785 | 6.68 |
| | | (5.50) | (1.11) | (0.09) | (0.72) | (2.06) | (0.07) | | | |
| CSSR | (1) | 9.530 | −0.152 | −0.202 | 0.040 | −0.359 | 0.145 | 2.04 | 0.882 | 5.27 |
| | | (1.01) | (0.63) | (0.92) | (0.19) | (2.11) | (0.94) | | | |
| | (2) | 12.224 | −0.468 | −0.302 | 0.542 | 0.300 | 0.653 | 1.34 | 0.668 | 1.49 |
| | | (1.04) | (1.56) | (1.11) | (2.11) | (1.42) | (3.41) | | | |
| | (3) | −14.030 | 0.160 | −0.015 | 0.092 | −0.270 | 0.112 | 1.00 | 0.464 | 3.88 |
| | | (1.97) | (0.88) | (0.09) | (0.59) | (2.11) | (0.97) | | | |
| | (4) | −7.632 | 0.043 | 0.516 | −0.583 | 0.278 | 0.101 | 1.06 | 0.543 | 5.50 |
| | | (0.60) | (0.13) | (1.74) | (2.10) | (1.21) | (0.49) | | | |
| | (5) | 2.439 | 0.072 | 0.085 | −0.151 | 0.137 | −0.028 | 1.58 | 0.624 | 3.49 |
| | | (1.01) | (1.16) | (1.50) | (2.85) | (3.12) | (0.70) | | | |
| | (6) | −2.531 | 0.041 | −0.082 | 0.060 | −0.085 | 0.016 | 1.73 | 0.647 | 14.60 |
| | | (1.83) | (1.16) | (2.53) | (1.98) | (3.39) | (0.73) | | | |

*Table 5.4.* Continued.

| Country | Sector[b] | $a_1$ | $a_2$ | $a_3$ | $a_4$ | $a_5$ | $a_6$ | $DW$ | $R2C$ | $MAPE$ |
|---|---|---|---|---|---|---|---|---|---|---|
| GDR | (1) | 14.629 | −0.325 | 0.108 | 0.067 | −0.019 | 0.215 | 1.30 | 0.918 | 3.37 |
| | | (5.72) | (1.44) | (0.54) | (0.50) | (0.11) | (1.17) | | | |
| | (2) | −15.267 | 0.280 | −0.312 | −0.074 | 0.006 | 0.639 | 1.09 | 0.911 | 0.95 |
| | | (5.89) | (1.22) | (1.55) | (0.54) | (0.04) | (3.44) | | | |
| | (3) | 1.245 | −0.070 | 0.147 | 0.018 | 0.006 | 0.062 | 2.28 | 0.901 | 1.23 |
| | | (3.16) | (2.00) | (4.83) | (0.88) | (0.24) | (2.20) | | | |
| | (4) | −1.025 | 0.088 | −0.003 | 0.024 | 0.003 | 0.031 | 1.97 | 0.818 | 0.59 |
| | | (3.73) | (3.61) | (0.13) | (1.67) | (0.18) | (1.59) | | | |
| | (5) | 0.151 | 0.050 | −0.011 | 0.003 | −0.001 | 0.009 | 2.10 | 0.752 | 1.72 |
| | | (0.40) | (1.49) | (0.39) | (0.16) | (0.04) | (0.32) | | | |
| | (6) | 0.239 | −0.021 | 0.071 | −0.040 | 0.005 | 0.041 | 2.05 | 0.847 | 1.76 |
| | | (1.01) | (0.99) | (3.89) | (3.17) | (0.32) | (2.42) | | | |
| Hungary | (1) | 27.661 | 0.126 | −0.170 | −0.137 | −0.216 | 0.245 | 2.40 | 0.979 | 2.78 |
| | | (2.23) | (0.75) | (1.19) | (1.29) | (2.11) | (2.51) | | | |
| | (2) | −41.159 | 0.169 | −0.224 | −0.110 | 0.084 | 0.531 | 1.43 | 0.852 | 2.03 |
| | | (2.28) | (0.69) | (1.07) | (0.71) | (0.56) | (3.73) | | | |
| | (3) | −2.852 | 0.011 | 0.080 | 0.033 | −0.040 | 0.085 | 0.71 | 0.623 | 2.30 |
| | | (0.65) | (0.19) | (1.57) | (0.88) | (1.10) | (2.45) | | | |
| | (4) | 23.452 | −0.373 | 0.196 | 0.207 | 0.224 | 0.115 | 0.94 | 0.842 | 5.12 |
| | | (1.80) | (2.11) | (1.30) | (1.85) | (2.07) | (1.12) | | | |
| | (5) | −5.511 | 0.068 | 0.050 | −0.010 | −0.033 | 0.026 | 1.54 | 0.748 | 1.85 |
| | | (2.91) | (2.65) | (2.28) | (0.61) | (2.11) | (1.76) | | | |
| | (6) | −1.557 | −0.001 | 0.067 | 0.018 | −0.018 | −0.002 | 2.03 | 0.833 | 11.40 |
| | | (0.74) | (0.05) | (2.76) | (1.02) | (1.04) | (0.14) | | | |

*Table 5.4.* Continued.

| *Country* | *Sector*[b] | $a_1$ | $a_2$ | $a_3$ | $a_4$ | $a_5$ | $a_6$ | *DW* | *R2C* | *MAPE* |
|---|---|---|---|---|---|---|---|---|---|---|
| Poland | (1) | 387.474 | −0.786 | −0.964 | 0.023 | −0.842 | 1.062 | 1.41 | 0.982 | 3.50 |
| | | (3.40) | (5.14) | (4.16) | (0.08) | (2.15) | (5.78) | | | |
| | (2) | −423.529 | 0.473 | 0.368 | −0.382 | 0.668 | 0.084 | 1.61 | 0.981 | 1.39 |
| | | (5.28) | (4.41) | (2.26) | (1.88) | (2.43) | (0.65) | | | |
| | (3) | 56.479 | −0.125 | 0.318 | 0.283 | 0.017 | 0.060 | 2.02 | 0.957 | 1.94 |
| | | (1.33) | (2.19) | (3.68) | (2.62) | (0.12) | (0.87) | | | |
| | (4) | 30.952 | 0.308 | 0.235 | 0.077 | 0.136 | −0.169 | 1.55 | 0.949 | 1.44 |
| | | (1.07) | (7.92) | (3.99) | (1.04) | (1.37) | (3.62) | | | |
| | (5) | −33.008 | 0.074 | 0.022 | 0.023 | −0.004 | −0.011 | 2.52 | 0.975 | 1.54 |
| | | (3.71) | (6.24) | (1.19) | (1.02) | (0.13) | (0.75) | | | |
| | (6) | −18.379 | 0.055 | 0.022 | −0.024 | 0.024 | −0.027 | 1.58 | 0.952 | 5.26 |
| | | (2.66) | (6.00) | (1.55) | (1.36) | (1.03) | (2.40) | | | |
| Romania | (1) | 33.222 | −0.630 | −1.260 | 1.123 | −1.088 | 0.933 | 1.58 | 0.971 | 4.31 |
| | | (4.90) | (2.08) | (2.48) | (2.86) | (2.07) | (2.86) | | | |
| | (2) | −35.186 | 0.517 | 0.916 | −0.943 | 0.708 | 0.053 | 1.43 | 0.978 | 2.75 |
| | | (6.17) | (2.03) | (2.14) | (2.85) | (1.60) | (0.19) | | | |
| | (3) | 2.183 | −0.115 | 0.251 | 0.011 | 0.173 | 0.038 | 1.69 | 0.823 | 2.02 |
| | | (1.90) | (2.25) | (2.90) | (0.17) | (1.93) | (0.68) | | | |
| | (4) | 3.742 | 0.061 | 0.205 | −0.175 | 0.176 | −0.040 | 1.91 | 0.631 | 2.61 |
| | | (4.44) | (1.62) | (3.24) | (3.58) | (2.69) | (1.00) | | | |
| | (5) | −5.146 | 0.064 | −0.101 | 0.051 | −0.016 | 0.058 | 2.20 | 0.878 | 3.15 |
| | | (5.20) | (1.44) | (1.36) | (0.90) | (0.21) | (1.21) | | | |
| | (6) | 1.236 | 0.103 | −0.007 | −0.071 | 0.052 | −0.043 | 1.50 | 0.795 | 7.17 |
| | | (1.53) | (2.86) | (0.11) | (1.50) | (0.82) | (1.09) | | | |

*Table 5.4.* Continued.

| *Country* | *Sector*[b] | $a_1$ | $a_2$ | $a_3$ | $a_4$ | $a_5$ | $a_6$ | *DW* | *R2C* | *MAPE* |
|---|---|---|---|---|---|---|---|---|---|---|
| USSR | (1) | 42.222 | −0.929 | −1.379 | 0.728 | −2.186 | 1.129 | 2.69 | 0.988 | 2.51 |
| | | (14.87) | (3.72) | (3.36) | (1.36) | (3.24) | (5.89) | | | |
| | (2) | −38.376 | 0.914 | 0.690 | −0.350 | 1.480 | −0.196 | 2.53 | 0.992 | 0.92 |
| | | (17.99) | (4.87) | (2.24) | (0.87) | (2.92) | (1.36) | | | |
| | (3) | 0.526 | 0.006 | 0.487 | −0.270 | 0.487 | −0.027 | 1.87 | 0.531 | 1.37 |
| | | (0.77) | (0.11) | (4.93) | (2.10) | (3.00) | (0.59) | | | |
| | (4) | −1.671 | 0.005 | 0.099 | −0.100 | 0.112 | 0.061 | 2.38 | 0.705 | 1.27 |
| | | (3.34) | (0.12) | (1.37) | (1.07) | (0.94) | (1.81) | | | |
| | (5) | −2.714 | 0.004 | 0.104 | −0.009 | 0.109 | 0.033 | 2.13 | 0.977 | 0.89 |
| | | (11.17) | (0.17) | (2.96) | (0.20) | (1.88) | (2.03) | | | |

[a] Estimated equation:

$$\beta_1 = a_i \cdot \frac{1}{Y} + a_2 \cdot \frac{C}{Y} + a_3 \cdot \frac{IG}{Y} + a_4 \cdot \frac{IMD}{Y} + a_4 \cdot \frac{EXD}{Y} + a_6 + \varepsilon$$

where $\beta_i$= share of sector $i$ in NMP produced. For definitions of the other variables, see text.

[b] Sectors are as follows: (1) Agriculture and forestry, (2) Industry, (3) Construction, (4) Wholesale and retail trade, (5) Transport and communications, and (6) Other.

the Appendix. There is a difference only in the interpretation of the system. In this case the demand system which we consider reflects the behavior of the economic agents who are responsible for the distribution of the final product in a socialist economy. We assume that their goal is the maximization of a social utility function. It is also assumed that the agents represent or act in the interest of the actual final users in the economy. By aggregation over these agents we derive the aggregated demand system for the whole economy in which the total expenditure term is represented by $Y_u$.

The derived estimable equation for the aggregate demand for consumption goods [8] is specified in the form

$$C = a_1 \cdot C_{-1} + a_2 \cdot \frac{Y_u P'Y_u}{P'C} - a_3 \cdot \frac{A_{-1} P'A}{P'C} \tag{5.17}$$

where $P'Y_u$, $P'C$ and $P'A$ are the price deflators of $Y_u$, $C$ and $A$, respectively. The parameters of the estimation $a_1$, $a_2$, $a_3$ can be interpreted in terms of the structural parameters of the demand system (see Appendix 5A). In this case again we have a system of two equations with a budget constraint so the estimation of one equation is sufficient to determine the whole system.

The results of the estimations of the aggregated consumption demand equations (5.17) for the seven CMEA countries are shown in *Table 5.5*. From a statistical point of view they can be accepted as satisfactory (except for some low $t$-values). The check of the order of magnitude indicates that all signs are correct and all parameters are in the plausible range of magnitude.

The expenditure and own price elasticities corresponding to the estimated demand systems are shown in *Table 5.6* (calculated as sample averages). One comment is necessary here. Equation (5.17) corresponds to a direct demand function with respect to price–quantity relations. However, as follows from our theoretical model, it reflects the actual demand of the final users indirectly, through the performance of the agents who are assumed to act in their interest. This fact must be taken into consideration in the interpretation of the results in *Table 5.6*.

One striking feature of the calculated elasticities is the similarity of these results across the countries, which might be interpreted as a similarity of the underlying economic policies. In all cases the expenditure elasticity of gross accumulation is much higher than that of total consumption, which indicates that the level of accumulation is affected to a much larger degree by fluctuations in the level of the final product used. Accumulation is also more price-elastic.

Another interesting feature is that the expenditure elasticities of the final use products in most of the cases (with the exception of Romania and the USSR) compare quite favorably with the production elasticities of the real factor remuneration (*Table 5.3*). Although indirectly, this might indicate a link between the primary and final distribution of output in the economy.

In order to derive an equation for the price index of consumer goods, we use a theoretical model which is similar to the one discussed in the beginning of

*Table 5.5.* Equations for the total consumption demand ($\tau$ values given in parentheses).[a] All estimates cover the period 1961–1985 (OLS).

| *Country* | $a_1$ | $a_2$ | $a_3$ | *DW* | *R2C* | *MAPE* |
|---|---|---|---|---|---|---|
| Bulgaria | 0.504 | 0.358 | 0.092 | 1.98 | 0.998 | 1.52 |
| | (6.80) | (6.27) | (1.02) | | | |
| CSSR | 0.669 | 0.340 | 0.324 | 2.36 | 0.997 | 1.19 |
| | (8.83) | (4.75) | (3.37) | | | |
| GDR | 0.616 | 0.377 | 0.305 | 1.97 | 0.999 | 0.62 |
| | (10.66) | (7.42) | (4.59) | | | |
| Hungary | 0.783 | 0.209 | 0.148 | 1.68 | 0.999 | 1.07 |
| | (20.44) | (5.60) | (3.05) | | | |
| Poland | 0.589 | 0.368 | 0.242 | 2.08 | 0.995 | 1.31 |
| | (11.42) | (6.71) | (2.98) | | | |
| Romania | 0.574 | 0.326 | 0.090 | 2.42 | 0.999 | 1.47 |
| | (8.80) | (4.41) | (0.75) | | | |
| USSR | 0.397 | 0.493 | 0.282 | 1.50 | 0.999 | 0.87 |
| | (3.47) | (5.57) | (2.21) | | | |

[a] Estimated equation:

$$C = a_1 \cdot C_{-1} + a_2 \cdot \frac{Y_u \cdot P'Y_u}{C} - a_3 \cdot \frac{A_{-1} \cdot P'A}{P'C} + \varepsilon$$

For further details, see text.

*Table 5.6.* Implied expenditure and price elasticities for final use products ($C$ = consumption; $A$ = accumulation).

| | *Expenditure elasticity* | | *Short-run price elasticity (uncompensated)* | |
|---|---|---|---|---|
| *Country* | $C$ | $A$ | $C$ | $A$ |
| Bulgaria | 0.58 | 1.70 | −0.47 | −0.90 |
| CSSR | 0.50 | 2.11 | −0.31 | −0.66 |
| GDR | 0.51 | 2.44 | −0.36 | −0.68 |
| Hungary | 0.32 | 2.40 | −0.18 | −0.85 |
| Poland | 0.56 | 1.96 | −0.38 | −0.75 |
| Romania | 0.51 | 1.90 | −0.38 | −0.90 |
| USSR | 0.77 | 1.40 | −0.58 | −0.70 |

this section [9]. Consider again our production function in the form (5.2), but formulated with respect to the total final product $X$ [10]. More generally, assume that $X$ is a linear homogenous function of "domestic inputs" $Y$ and imported inputs $M$ [11]:

$$X = f(Y, M) \tag{5.18}$$

Assume also that there exists a cost function $K = K(X)$ which is derived from cost-minimization conditions.

The product $X$ is used for consumption (total), accumulation (gross) and exports:

$$X = C + A + E \tag{5.19}$$

The planning agency distributes the output $X$ so as to maximize a social utility function, which is defined over $C$ and $A$.

$$\max U = U(C, A) \tag{5.20}$$

subject to the constraints:

(1) Revenue–cost constraint:

$$P'C \cdot C + P'A \cdot A + P'E \cdot E - K(X) \geq \bar{R} \tag{5.21}$$

where $P'E$ is the price deflator of exports; $\bar{R}$ is a certain desired level.

(2) Balance of trade constraint:

$$P'E \cdot E - P'M \cdot M \geq \bar{B} \tag{5.22}$$

where $P'M$ is the price deflator of imports; $\bar{B}$ is also a certain target level.

The desired price levels $P'C$, $P'A$ and $P'E$ will be those which equilibrate the three markets to a certain degree $b_C$, $b_A$, $b_E$, respectively.

$$\begin{aligned} C &= b_C \cdot C(P'C) \\ A &= b_A \cdot A(P'A) \\ E &= b_E \cdot E(P'E) \end{aligned} \tag{5.23}$$

where the right-hand parts are the corresponding demand functions (for simplicity of notation, the interdependences over the prices are omitted). For simplification of the further derivation, we also assume $b_C = b_A = b_E = 1$.

Imports $M$ are also considered to be defined by an import demand function

$$M = M(X) = M[C(P'C) + A(P'A) + E(P'E)] \tag{5.24}$$

If we perform the necessary substitutions we can formulate the following optimization problem:

$$\max U_{P'C,\,P'A,\,P'E}\,[C(P'C),\,A(P'A)] \tag{5.25}$$

subject to:

$$P'C \cdot C(P'C) + P'A \cdot A(P'A) + P'E \cdot E(P'E) \\ - K[C(P'C) + A(P'A) + E(P'E)] \geq \bar{R} \tag{5.25a}$$

$$P'E \cdot E(P'E) - P'M \cdot M[C(P'C) + A(P'A) + E(P'E)] \geq \bar{B} \tag{5.25b}$$

If we solve this problem of constrained optimization using the usual Lagrangean technique, after several transformations (and making some assumptions) we can arrive at an estimation equation for $P'C$, which in our derivation had the following form (in a linearized version):

$$P'C = a_1 \cdot P'Y + a_2 \cdot \frac{C}{C_{-1}} + a_3 \cdot \frac{A}{A_{-1}} \\ + a_4 \cdot \frac{C}{Y} + a_5 \cdot \frac{A}{Y} + a_6 \cdot P'M + a_7 \cdot \frac{1}{t} + a_8 + \varepsilon \tag{5.26}$$

where $t$ is a time-variable.

The expected signs of the coefficients are (+) for $a_1$, $a_3$ and $a_6$; (−) for $a_2$, $a_4$ and $a_5$; $a_7$ is not determined.

The estimation results for equation (5.26) for the seven CMEA countries for the period 1961–1982 are shown in *Table 5.7.* Some variables have been removed from the regression because of wrong signs or negligible value. We did not remove variables only because of high standard error of estimation, if they had the right sign.

Finally, we would like to make some comments on the use of the consumption function in the simulation model. As was pointed out, our general approach is "supply driven" and the investment ratio is one of the main driving factors. So, with respect to the *ex ante* forecasts, the consumption demand function plays a secondary role. The value of the expected "consumption demand" is compared to the generated value of "consumption supply". Since our production technology actually reflects the production boundary, the value of "consumption supply" can be treated as the maximum level available at this level of production. On the other hand the value of "consumption demand" reflects the normally expected level of consumption. So a resulting discrepancy between these two levels might be regarded as an indicator of possible disequilibrium in the consumer goods market.

*Table 5.7.* Equations for the price index of total consumption ($\tau$ values given in parentheses).[a] All estimates cover the period 1961–1982 (OLS).

| *Country* | $a_1$ | $a_2$ | $a_3$ | $a_4$ | $a_5$ | $a_6$ | $a_7$ | $a_8$ | *DW* | *R2C* | *MAPE* |
|---|---|---|---|---|---|---|---|---|---|---|---|
| Bulgaria | 1.001 | −0.101 | | −0.466 | −0.203 | 0.145 | −5.333 | 0.388 | 2.09 | 0.979 | 0.92 |
| | (11.94) | (0.99) | | (2.89) | (2.46) | (4.24) | (0.84) | (2.80) | | | |
| CSSR | 0.427 | −0.117 | | −0.214 | −0.238 | 0.131 | −3.508 | 0.833 | 1.99 | 0.991 | 0.41 |
| | (11.41) | (1.10) | | (1.30) | (2.30) | (6.42) | (0.80) | (4.27) | | | |
| GDR | 0.899 | −0.416 | 0.075 | −0.144 | −0.508 | | −15.995 | 0.918 | 1.67 | 0.983 | 0.54 |
| | (7.18) | (1.87) | (2.12) | (0.84) | (3.31) | | (3.19) | (2.40) | | | |
| Hungary | 1.375 | −0.445 | 0.088 | −0.977 | −0.331 | | 45.486 | 0.150 | 0.86 | 0.966 | 2.62 |
| | (7.90) | (0.94) | (0.76) | (1.84) | (0.84) | | (1.75) | (0.20) | | | |
| Poland | 0.801 | −0.183 | 0.048 | −0.252 | −0.258 | 0.136 | −4.446 | 0.488 | 1.71 | 0.999 | 1.09 |
| | (15.37) | (3.10) | (1.34) | (1.87) | (1.54) | (2.58) | (1.72) | (3.61) | | | |
| Romania | 0.935 | −0.133 | 0.195 | −0.025 | −0.790 | 0.105 | | 0.184 | 0.99 | 0.993 | 0.39 |
| | (16.76) | (2.06) | (4.26) | (0.33) | (7.14) | (11.66) | | (1.37) | | | |
| USSR | 0.899 | | | −0.340 | −0.238 | 0.050 | 15.039 | 0.156 | 1.65 | 0.915 | 0.39 |
| | (11.06) | | | (1.81) | (0.91) | (6.28) | (7.53) | (0.68) | | | |

[a] Estimated equation:

$$P'C = a_1 \cdot P'Y + a_2 \cdot \frac{C}{C_{-1}} + a_3 \cdot \frac{A}{A_{-1}} + a_4 \cdot \frac{C}{Y} + a_5 \cdot \frac{A}{Y} + a_6 \cdot P'M + a_7 \cdot \frac{1}{t} + a_8 + \varepsilon$$

For further details, see text.

## 5.5. The Complete Models and Their Dynamic Characteristics

A simplified picture of the structure of the complete macroeconomic models of the CMEA countries is shown in *Figure 5.1.* The boxes on the figure correspond to endogenous variables whereas the circles stand for exogenous ones. Although the foreign trade sector is also partly reflected on *Figure 5.1*, we shall mainly focus our attention on the domestic part. The foreign trade equations for the CMEA countries which were used in the linked world model are presented in Chapter 7 on "Growth is an Interdependent World Economy: Linking of National Models with an Integrated System of International Trade".

Within the domestic part of each country model, there are 10 main behavior equations:

(1) Production technology which determines total material output $Y^*$ [equation (5.1)].
(2) Real wages of the employed in the material sphere $l_r$ [equation (5.7)].
(3) Demand for consumption goods, denoted as $C_d$ on *Figure 5.1* [equation (5.17)].
(4) Price index for consumption goods $P'C$ [equation (5.26)].
(5) Structure of domestic production $\beta_i$ [six equations, specified as (5.16)].

The other endogenous variables including fixed assets $K$, final material product $Y$, net material product $NMP$, gross accumulation $A$, gross investment $IG$, "consumption supply" $C$, price deflator of $NMP$-$P'NMP$ and some others which are not shown on *Figure 5.1* are determined from identities. All important equations are given in Annex 2.

The main exogenous driving forces are: the rate of technical progress (or, total factor productivity) $w_{\tau}$; the number of employed persons in the material sphere $L$, the rate of sorting out of fixed assets $d$, the investment (savings) ratio $s$, the level of nominal wages in the material sphere $L$. They were analyzed in more detail in Chapter 4.

The performance of the models was tested in dynamic *ex post* simulation for 1965–1982. The mean absolute percentage errors (MAPE) for the four main behavioral variables resulting from these simulations are shown in *Table 5.8* (consumption $C$ in this table is evaluated by "consumption demand"). For a dynamic simulation these results can be regarded as satisfactory.

The dynamic performance of the models was also checked on the basis of the dynamic elasticity multipliers. They were calculated as the dynamic response of the models to a single positive shock of 10% in the level of some important exogenous driving forces (for the exact formulas, see Chapter 7).

A selection of some of the main results is given in *Table 5.9.* The table reports some elasticity multipliers calculated for a single shock in the level of technical progress $\tau$ and for the level of nominal wages in the material sphere $l$, taking place in 1986. The medium scenario (described in other chapters) was used as the control solution.

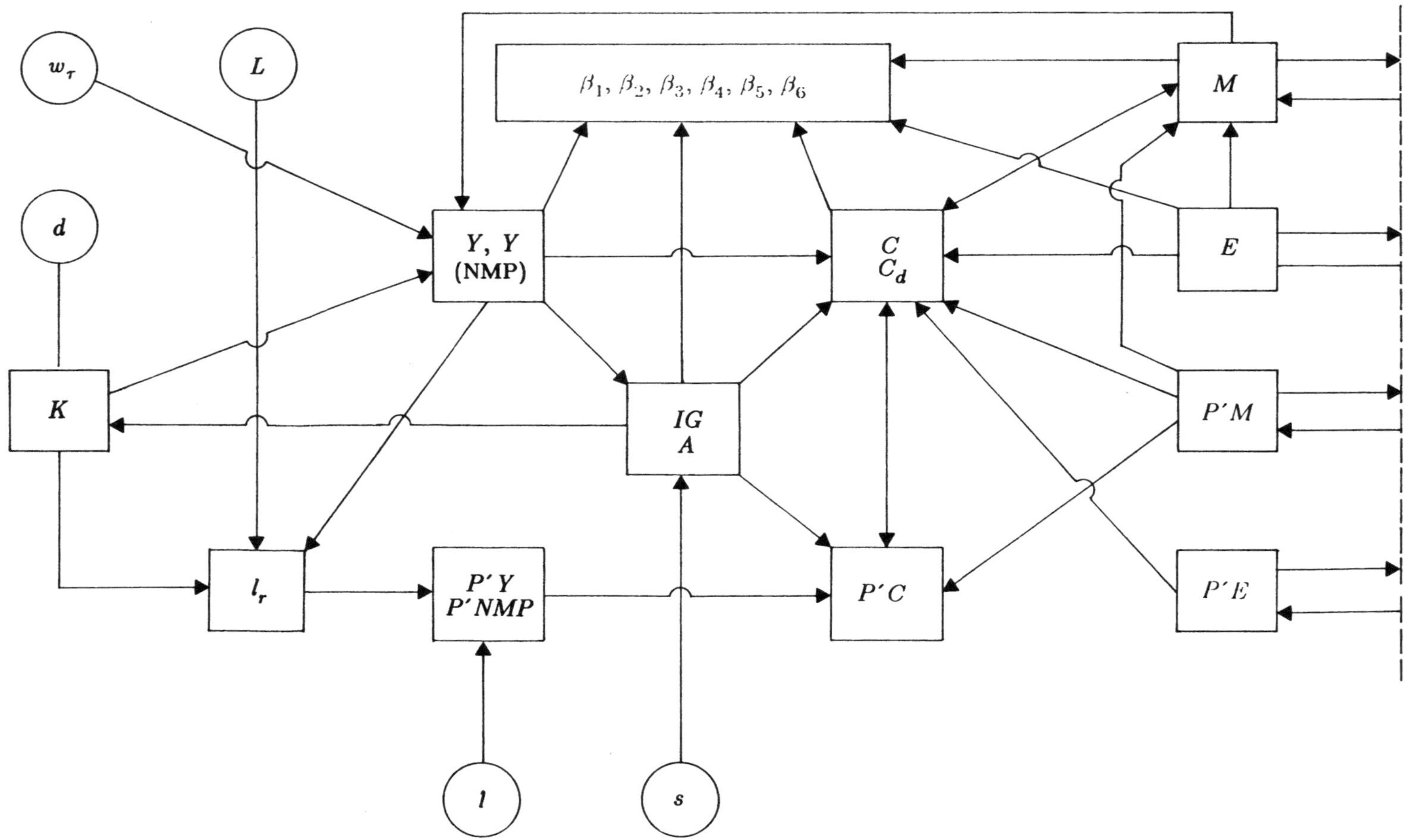


*Figure 5.1.* General structure of the models: boxes = endogenous variables, circles = exogenous variables.

*Table 5.8. Ex post* simulation test: mean absolute percentage errors of selected variables.

| | Country | | | | | | |
|---|---|---|---|---|---|---|---|
| *Variable* | *Bulgaria* | *CSSR* | *GDR* | *Hungary* | *Poland* | *Romania* | *USSR* |
| $Y^*$ | 1.02 | 0.87 | 0.41 | 1.54 | 2.55 | 1.26 | 1.20 |
| $C$ | 1.86 | 1.05 | 0.70 | 1.78 | 3.01 | 1.17 | 1.07 |
| $I_r$ | 2.19 | 2.32 | 1.63 | 1.45 | 3.58 | 1.93 | 1.21 |
| $P'C$ | 3.03 | 1.27 | 2.13 | 3.81 | 4.13 | 1.79 | 1.42 |

*Table 5.9.* Elasticity multipliers: percentage level response of selected variables to a single, positive 10% shock of $\tau$ and $l$ in 1986.

| | $\tau$ | | | | | | | $l$ |
|---|---|---|---|---|---|---|---|---|
| *Country* | $Y$ | *NMP* | $M$ | $C_d$ | $C$ | $P'Y$ | $P'C$ | $P'C$ |
| Bulgaria | 1.28 | 1.56 | 2.72 | 1.97 | 3.35 | -0.85 | -1.55 | 0.91 |
| CSSR | 1.23 | 1.43 | 1.02 | 0.86 | 1.94 | -0.27 | -0.38 | 0.44 |
| GDR | 1.25 | 1.36 | 1.01 | 0.94 | 1.74 | -1.59 | -1.97 | 0.90 |
| Hungary | 1.27 | 1.51 | 3.69 | 1.22 | 4.15 | -0.99 | -2.90 | 1.20 |
| Poland | 1.26 | 1.37 | 2.13 | 0.85 | 1.86 | -0.90 | -0.91 | 0.95 |
| Romania | 1.25 | 1.50 | 1.88 | 0.67 | 1.81 | -1.31 | -1.07 | 1.02 |
| USSR | 1.06 | 1.23 | 3.45 | 0.87 | 1.03 | -0.91 | -0.98 | 1.09 |

The calculated multipliers for $\tau$ show a similar response of $Y$ and NMP in the countries. However, the responses of the other variables are more diversified. Thus, the response of the import demand to the shock in $\tau$ (actually, the latter can be interpreted as a single shift to a higher rate of economic growth) is quite high in Bulgaria, Hungary and USSR; it is moderate in Poland and Romania and low in CSSR and GDR.

It is interesting to trace the induced response of the "consumption demand" $C_d$ and "consumption supply" $C$. Although the margins are different, the effect is similar in all the countries: in all cases the induced response of $C$ is higher than that of $C_d$. This can be interpreted as a positive effect with respect to the situation on the consumer goods markets.

## 5.6. Concluding Remarks

The macroeconomic models of the CMEA countries discussed in this chapter were used for a number of simulations. First, the models were tested as separate units; later they were linked to the other models of the Bonn–IIASA Project. The actual results of these simulations are discussed in the other chapters of this volume.

## Notes

[1] Bulgaria, CSSR, GDR, Hungary, Poland, Romania and USSR (the ordering is alphabetical).

[2] Our brief review concentrates only on some studies with empirical implementation.

[3] See, for example, the recent survey and report on measuring total factor productivity in CMEA countries in the *Economic Survey of Europe* (ESE, 1986).

[4] Because of these difficulties some of the authors of the empirical studies quoted above, although rejecting the marginal productivity assumptions theoretically, use them in empirical estimates.

[5] Formulas can be found in Welsch (1987).

[6] It is easy to show that the OLS estimation of the production sector functions in the form of (5.16) will guarantee that

$$\sum_i a_j^i = 0, \quad j = 1,2,\ldots,5; \quad \sum_j a_6^i = 1$$

which, in turn, provides that $\sum_i \beta_i = 1$.

[7] It equals final material product produced *less* the balance of exports and imports of goods and material services.

[8] Note that, in accordance with MPS, we deal here only with consumption of material goods and services.

[9] The "theoretical" setup for the derivation which follows is based on ideas suggested by Dr. H. Welsch.

[10] $X$ exceeds $Y^*$ by the value of imported investment and consumption goods which are not included in the production technology (5.2).

[11] Note that in this case total imports $M$ enter the production function.

# Appendix 5A. A Theoretical Macromodel of Production and Primary Distribution in a Socialist Economy

Consider a centrally planned economy where output $Y$ is the "final material product" (*NMP* plus the amortization fund), and where it is defined by a linear homogeneous function of the two factor inputs $L$ (number of employed persons in the material sphere) and $K$ (fixed assets in the material sphere):

$$Y = f(L, K). \tag{5A.1a}$$

In a more general formulation which we use later:

$$Y = f(x_1 x_2) \tag{5A.1b}$$

From the primary distribution of income (the third quadrant of the input–output table) we have the following definitional identity (in nominal terms)

$$Y \cdot P'Y = W_n + Q_n \tag{5A.2}$$

where $W_n$ = total primary income of the employed in the material sphere, $Q_n$ = gross primary income of the enterprises (including the net profit of the enterprises, the sum of indirect taxes collected for the state budget and, the amortization fund, since we regard $Y$ as final material product); $P'Y$ is the price deflator of $Y$.

Further, we denote

$$W_n = l \cdot L \tag{5A.3}$$

$$Q_n = q \cdot K \tag{5A.4}$$

where $l$ is the nominal wage (primary income) of one person employed in the material sphere; $q$ is the primary income per unit of fixed assets. Whereas (5A.3) is a straightforward relation, (5A.4) may seem a bit dubious, especially because of the presence of indirect taxes in $Q_n$ which are a matter of state regulation and can hardly be associated with $K$. In a theoretical setting it would be more correct to exclude taxes from $Y$. However, the relevant data for the empirical implementation of this division were missing. However, the other two terms, and most of all the amortization fund, are directly linked to the value of $K$.

Furthermore, we can write (5A.2) as

$$Y \cdot P'Y = l \cdot L + q \cdot K \tag{5A.5}$$

and

$$Y = l_r \cdot L + q_r \cdot K \tag{5A.6a}$$

In a more general formulation which we use later, and in accordance with (5A.1b) we can rewrite (5A.6a) as

$$Y^* = x_1 \cdot q_1 + x_2 \cdot q_2 \tag{5A.6b}$$

where $q_1$ and $q_2$ denote the unit primary incomes (factor remuneration). In (5A.6a)

$$l_r = \frac{l}{P'Y} \tag{5.A7}$$

is the real wage (primary income) of one employed person in the material sphere (throughout the chapter we use the term "real wage" only in this sense). Moreover,

$$q_r = \frac{q}{P'Y} \tag{5.A8}$$

is the real primary income per unit of fixed assets.

We now make a slight digression. From the setting of equations (5A.1)–(5A.8) there is only one step to the standard marginal productivity assumption. Actually, if we interpret $l_R$ and $q_r$ as "factor prices" of $L$ and $K$, and assuming profit-maximizing (or cost-minimizing) behavior of the producer, from the constrained optimization of (5A.1) we would get $\partial f/\partial L = l_r$; $\partial f/\partial K = q_r$ which is the usual neoclassical approach.

However, we consider that a step in *this direction* should not be made when modeling a socialist economy. The main argument against it is the fact that $l_r$ and $q_r$ are not and cannot be regarded as "factor prices", because (especially in a centrally planned economy) the primary division of income in (5A.2)–(5A.6) only indicates a distribution relation which is not an element of the process of creation of the income (5A.1). This fact, which is one of the main arguments against the marginal productivity theory, has been pointed out by many of its critics. However little, if anything at all, has been suggested toward a constructive continuation of this critique in the direction of workable estimation procedures for the production function (5A.1), based on alternative assumptions.

In what follows we try to suggest a possible approach which could provide an alternative to the standard marginal productivity assumptions.

We start from the fact that the primary division of income reflects an income distribution relation. In allocating the available income the planning agency follows certain criteria: it considers social justice and incentives to work by determining the relation between wage income and residual income of the enterprises (which is mostly used for accumulation). We represent these criteria in the form of a social utility function defined over the factor remunerations, i.e., $U(q_1,q_2)$. The decision-making process connected with the income distribution can be modeled as a process of maximizing social utility $U(q_1,q_2)$ in the presence of the budget constraint (5A.6). Formally, this is equivalent to the derivation of demand equations from a utility maximization problem.

As follows from demand theory (see, e.g., Phlips, 1983) the constrained utility maximization in such a system would yield the distribution functions in the form

$$\begin{aligned} q_1 &= \varphi_1(x_1,x_2,Y) \\ q_2 &= \varphi_2(x_1,x_2,Y) \end{aligned} \tag{5A.9}$$

[In these distribution functions the places of "quantities" and "prices" are inverted as compared to demand functions. However, they are not to be mixed with "inverse" demand functions in the sense of Anderson (1980). Besides, we shall speak about "production elasticities" and "quantity elasticities" instead of the usual "expenditure elasticities" and "price elasticities".]

Next we consider that the planning agency plans the production (5A.1a) (the allocation of $x_1$ and $x_2$) *taking into account* that the primary distribution will be carried out in accordance with (5A.9) and (5A.6a) and on the basis of other criteria as well. Taking into consideration the existing planning practice, one possible way to model this decision-making process is the following optimization problem. The planning agency maximizes output $Y$ and at the same time sets the goal to guarantee a certain level of the real wages of the employed person: $\bar{q}_1$.

$$\begin{aligned} &\max_{x_1,x_2} \; Y = f(x_1,x_2) \quad \text{such that} \\ &q_1 = \varphi_1(x_1,x_2,Y) \geq \bar{q}_1; \quad x_1,\, x_2 \geq 0 \end{aligned} \tag{5A.10}$$

Note that in the case of distribution functions derived from constrained maximization of a duly specified utility function (as we assume) the budget constraint will be guaranteed automatically. Since we already have a restriction on one of the two distribution functions of the system we do not have to include a budget constraint in the optimization problem and this simplifies the solution significantly. The Lagrangean function in this case can be defined as:

$$L = f(x_1,x_2) + \lambda \cdot \{\bar{q}_1 - \varphi_1\,[x_1,\, x_2,\, f(x_1,\, x_2)]\} \tag{5A.11}$$

If we take the Kuhn–Tucker conditions and make all the necessary transformations, we get the solution of this optimization problem in the form:

$$\frac{\partial \varphi_1}{\partial x_1} / \frac{\partial \varphi_1}{\partial x_2} = \frac{\partial f}{\partial x_1} / \frac{\partial f}{\partial x_2} \tag{5A.12}$$

This general relation can be regarded in a way as an analogue to the standard marginal productivity assumption, applied to a socialist economy. Obviously (5A.12) is less restrictive because it does not establish the relation between the marginal productivities as a static proportion, but as a proportion of incremental changes in the distribution function.

In order to make practical use of this derivation we have to specify the production function and the distribution system and analyze the implied relationships between the parameters. For the production function, as we have already discussed, we assume a linear homogeneous Cobb–Douglas form:

$$Y = \gamma_0 \cdot \tau^* \, L^\gamma \, K^{1-\gamma} \tag{5.A13}$$

If we also assume a Cobb–Douglas specification for the social utility function, we arrive at a distribution system which is formally equivalent to the Linear Expenditure System. We used the dynamic version of this system as formulated by Pollak (1970) and Phlips (1972) and further elaborated by Klevmarken (1981) and Welsch (1986, 1987). Without going into detail we shall just mention that the main assumption of this model (interpreted in terms of our system) is that the actual level of factor remuneration $q_i$ has two components: a "base" level $q_i'$ which is proportional to the lagged actual level:

$$q_i' = \lambda_i \cdot q_{i,-1} \, , \quad i = 1,2 \tag{5A.14}$$

where $\lambda_i$ denotes a habit formation coefficient; and an "excess" component $(q_i - q_i')$ which depends on the actual income–quantity situation $(Y, x_1, x_2)$.

The resulting distribution equation is:

$$q_i = \lambda_i q_{i,-1} + \beta_i \cdot (Y - \sum_{i=1}^{2} x_k \lambda_k q_{k,-1}) / x_i \tag{5A.15}$$

where $\beta_i$ is the marginal share in $Y$ of unit income $i$.

After transformation (5A.15) can be rearranged in the following estimable form with respect to our variables $l_r$ and $q_r$:

$$l_r = a_1 \cdot l_{r,-1} + a_2 \cdot \frac{Y}{L} - a_3 \cdot \frac{q_{r,-1} \cdot K}{L}$$

$$q_r = b_1 \cdot r_{r,-1} + b_2 \cdot \frac{Y}{K} - b_3 \cdot \frac{l_{r,-1} \cdot L}{K} \tag{5A.16}$$

where the estimable parameters $a_j$, $b_j$, $j = 1,2,3$ are related to the parameters of the structural equation (5A.15) and among themselves as follows:

$$\begin{aligned} a_1 &= \lambda_1 \cdot (1\ \beta_1) = \lambda_1 \cdot \beta_q = b_3 \\ a_2 &= \beta_1 = 1 - \beta_q = 1 - b_2 \\ a_3 &= \lambda_q \cdot \beta_1 = \lambda_q \cdot (1 - \beta_q) = b_1 \end{aligned} \tag{5A.17}$$

Obviously, since the distribution system is derived with a budget constraint it is sufficient to estimate only one equation of (5A.16) – the first one, for instance – in order to determine the whole system.

Taking the marginal productivities from (5A.13) and the quantity derivatives of $l_r$ from (5A.16) and substituting them into (5A.12) yields:

$$\frac{l_r - a_1 l_{r,-1}}{a_3 q_{r,-1}} = \frac{\gamma}{1-\gamma} \cdot \frac{K}{L} \tag{5A.18}$$

Equation (5A.18) defines the relation between the parameters of the production and the distribution functions which can be used during the estimation.

## References

Anchishkin, A. (1973), *Forecasting the Growth of the Socialist Economy.* (Moscow: Ekonomika).

Anderson, R. (1980), Some Theory of Inverse Demand for Applied Demand Analysis. *European Economic Review*, **14**, pp. 281–290.

Brown, A., Licari, A., and Neuberger, E. (1974), A Dynamic CES Production Function of Hungarian Growth, *Acta Oeconomica* [Budapest], **11**, 1973,4, pp. 305–324.

Brown, A., Licari, A., and Neuberger, E. (1976), Alternate Measures of the Productivity Growth Hierarchy in Hungarian Industry, in: *On the Measurement of Factor Productivities.* (Göttingen: Vandenhoeck & Ruprecht).

Charemza, W., and Ouandt, R. (1982), Models and Estimation of Disequilibrium for Centrally Planned Economies, *Review of Economic Studies*, **49**, pp. 109–116.

Desai, P. (1976), Technical Change, Factor Elasticity of Substitution and Returns to Scale in Branches of Soviet Industry in the Post-War Period, in: *On the Measurement of Factor Productivities.* (Göttingen: Vandenhoeck & Ruprecht).

Desai, P. (1985), Total Factor Productivity in Post-War Soviet Industry and Its Branches, *Journal of Comparative Economics*, **9**, pp. 1–23.
Dobrinsky, R. (1986), *The Macroeconomic Data Base for the CMEA Countries.* Working Paper WP-86-54. (Laxenburg, Austria: International Institute for Applied Systems Analysis).
Econometric Macromodels (1984), *Econometric Macromodels for the National Economy Development.* (Sofia: Bulgarian Academy of Sciences).
Ershov, E. and Sadykov, I. (1986), Study of the Resources Substitutability and Their Dynamic Proportionality in the Industrial Branches of the USSR, *Ekonomika i Matematicheskie Metody*, **22**, pp. 426–440.
ESE (1986), *Economic Survey of Europe in 1985–1986*, prepared by the Secretariat of the Economic Commission for Europe, Sales No. E.86.II.E.1. (Geneva and New York: United Nations).
Kemme, D. (1984), Discriminating Among Alternative Production Functions in Polish Industry, *Empirical Economics*, **9**, pp. 59–66.
Klacek, J., and Nešporová, A. (1983), *Production Functions and Economic Growth in Czechoslovakia.* (Prague: Academia).
Kelin, L. and Welfe, W. (1983), *Lectures in Econometrics.* (Amsterdam: North-Holland).
Klevmarken, N.A. (1981), *On the Complete Systems Approach to Demand Analysis.* (Stockholm: Almquist & Wiksell International).
Mikhalevski, V. and Solov'ev, Iu. (1966), Production Function of the USSR National Economy for 1951–63, *Ekonomika i Matematicheskie Metody*, **2**, pp. 823–840.
Pappe, Ja. and Pyvkin, A. (1977), About the Attempts for a Model Proof of the Outcomes of the Marginal Productivity Theory. *Ekonomika i Matematicheskie Metody*, **13**, pp. 1175–1195.
Phlips, L. (1972), A Dynamic Version of the Linear Expenditure Model. *Review of Economics and Statistics*, **64**, pp. 450–458.
Phlips, L. (1983), *Applied Consumption Analysis.* (Amsterdam: North-Holland).
Pollack, R. (1970), Habit Formation and Dynamic Demand Functions. *Journal of Political Economy*, **78**, pp. 745–763.
Portes, R. and Winter, D. (1977), The Supply of Consumption Goods in Centrally Planned Economies. *Journal of Comparative Economics and Statistics*, **1**, pp. 351–365.
Portes, R. and Winter, D. (1978), The Demand for Money and for Comsumption Goods in Centrally Planned Economies. *Review of Economics and Statistics*, **60**, pp. 8–18.
Portes, R. and Winter, D. (1980), Disequilibrium Estimates for Consumption Goods Markets in Centrally Planned Economies. *Review of Economic Studies*, **47**, pp. 137–159.
Simon, D. and Samovol, V. (1981), About the Functional of Economic Growth. *Ekonomika i Matematicheskie Metody*, **17**, pp. 148–259.
Toda, Y. (1976), Capital-Labour Substitution in Production Function: The Case of Soviet Manufacturing for 1950–1971, in: *On the Measurement of Factor Productivities.* (Göttingen: Vandenhoeck & Ruprecht).
Toms, M. and Hajek, M. (1970), Technological Change, Production Functions and Economic Growth in Czechoslovakia. *Czechoslovak Economic Papers*, **12**, pp. 51–91.

Weitzman, M. (1970), Soviet Postwar Economic Growth and Capital–Labor Substitution. *American Economic Review*, **60**, pp. 676–692.

Welfe, A. (1985), Modeling Consumers' Behaviour Under Inflation and Disequilibrium: WA-2 Model. University of Lodz.

Welfe, A. (1985), Econometric Macromodels of Unbalanced Growth. University of Lodz.

Welsch, H. (1986), *Habit Persistence and the Structure of International Commodity Trade. A Demand System Analysis for Nine Industrialized Countries*. Discussion Paper B-47. (Bonn University, Sonderforschungsbereich 303).

Welsch, H. (1987), An Aggregate Import Demand Model for Long-Term Projections, *Jahrbuch für Nationalökonomie und Statistik* **203** (forthcoming).

# CHAPTER 6

# Economic Growth and Structural Change in Developing Countries

*István Székely*

## Summary

Macroeconometric submodels of different groups of developing countries have been constructed and integrated into the Bonn–IIASA World Model. The submodels are designed to reflect the specific characteristics of these economies. The results of the medium-term scenario analysis indicate the contours of economic development in the different regions through the end of this century. A detailed analysis arrives at the conclusion that, if no far-reaching measures are undertaken, the limit of feasible growth rate differences between the developed market economies and the developing countries considered in the study lies between 1.5–2%. This pace implies scarcely any reduction of the existing income gap.

## 6.1. Introduction

Previous chapters discussed the different aspects of economic growth and structural change in developed market and centrally planned economies. In the present chapter, we continue this analysis by focusing on the specific characteristics of *developing economies*. The framework we apply to these economies is similar to the one generally used in the Bonn–IIASA Research Project. However, there are some special features of this analysis which reflect the different nature of economic growth and structural change in these economies.

Due to the limitations of our project, the developing countries are modeled as a group. The only exception is India, which is modeled separately because of its dominant size and different growth pattern. The composition of the groups differs from those of well-known world models. To achieve a reasonable degree of homogeneity and a manageable size, representative countries were selected in the different regions and considered as groups. Countries were selected based on their common cultural and historical backgrounds and on similarities in their growth patterns. The description of the groups is given in Annex 1 to this volume.

The chapter is organized as follows. In Section 6.2, the macroeconometric model used in our analysis is described in detail. In Section 6.3, attention is directed to the future development of developing countries. Using the results of a medium-term scenario analysis, the growth prospects of different regions of developing countries are investigated. Finally, in the Appendix, the estimation results are presented together with the results of an *ex post* simulation.

## 6.2. Models for Groups of Developing Countries in the Bonn–IIASA World Model

### 6.2.1. General description of the models

The structure of the models of the different groups of developing countries is essentially no different from that of the models specified for the developed market economies and described in the previous chapters. However, there are some differences reflecting the specific characteristics of the developing economies under study. In the discussion that follows, these differences will be indicated.

A full-scale model of a group consists of 83 equations, of which 29 are stochastic behavioral equations and 54 are identities. It contains 98 variables, of which 15 are exogenous. The model as a whole is highly nonlinear. In the Linked World Model, described in Chapter 7, a somewhat tailored form of these models is used. The simplifications mainly concern the commodity structure of foreign trade and were necessary to obtain a manageable size for the world model.

Since there is no natural unit of domestic currency for groups of countries and since an important economic indicator would have been lost by using dollar terms, artificial currencies were defined for these groups of developing countries. The exchange rate of an artificial currency against the US dollar is a weighted geometric average of the individual exchange rate indices of the national currencies of the group, where the 1975 values of these indices were normalized to 1, and the nominal GDPs were used for weighting. The variables in the model are expressed in terms of these artificial currencies unless otherwise stated.

Stemming from the basically supply-oriented nature of our models, the output is determined by a *production function*. To reflect the interdependencies in the production process, besides the usual factors of production, imports of raw and intermediate materials are also incorporated into the specification. In contrast to the developed market economies, marginal productivity theory is not applied in the estimation of the input elasticities. *Technical progress* is one of the main driving forces in the model.

Total output is, then, disaggregated into six sectors by *sector share functions*. This way of determining the sectoral outputs reflects the generally adopted *top-down* approach of our model.

The *investment ratio* determining the level of investment in real terms is another important exogenous driving force of the model.

The volume of imports is determined by an *import demand* function, which is based on a dynamic version of the Linear Expenditure System. The total import demand, then, is disaggregated into four commodity groups by a demand system based on the same approach. In the Linked World Model, only the aggregate import demand is determined.

In explaining export flows, the general idea of the model is that exports of the different countries or groups are determined by the pooled import demand of the others. This, together with the world market prices in the different commodity groups, determines the nominal demand in each commodity group. Interpreting these nominal demands as incomes allocated to imports of the different commodity groups and using again the framework of the Linear Expenditure System, *export demand functions* are specified for the different commodity groups. Exports of services are assumed to be proportional to exports of goods, and total exports are then determined as the sum of the different export items. In the Linked World Model, again, only aggregate exports are modeled. Specification of the export demand functions designed for this purpose is given in Chapter 7.

Effective demand meets the level of production by adjusting the level of consumption.

Turning to prices, the *general price level* is determined from the Fisher equation. Money supply is assumed to be exogenous, and the income velocity of money is endogenized.

The investment deflator is explained by the general price level and by the import deflator.

As to foreign trade prices, in contrast to the approach for the developed market economies, price-taking behavior is assumed. This means that both export and the import prices in the different commodity groups are linked to corresponding world market prices.

The consumption deflator is left to be determined from the GDP expenditure identity.

Finally, determination of the exchange rate is based on purchasing power parity theory. In spite of the widely emphasized problems of this approach, we felt that it best suited the behavior of financial markets in developing countries and the medium-term orientation of our model.

### 6.2.2. Specification of selected equations and the estimation results

*The production function*

The production function is of Cobb–Douglas type with constant returns to scale. Technical progress is assumed to be Hicks-neutral with constant exponential rate of growth, but with a specification that allows different rates of technical progress for different time periods. The estimated form of this function can be written

$$Y^* = a_1 \, e[a_2 t + a_3(t-t_c)d_c] \, K^{a_4} \, L^{a_5} \, IM_R^{a_6} + u, \quad a_4 + a_5 + a_6 = 1 \qquad (6.1)$$

where $Y^*$, $K$, $L$ and $IM_R$ are output, capital and labor input and input of imported raw materials and energy, respectively. $Y^*$ is the sum of real GDP and real imports of raw materials and energy (approximated by the sum of SITC categories 2, 3 and 4). Since no reliable indicator of capacity utilization is available, $K$ is taken as gross capital stock. $L$ represents the number of the economically active population. As to the specification of technical progress, $t$ is the time trend, $t_c$ indicates the year when the change in the rate of technical progress takes place, and $d_c$ is a dummy variable (0 until $t_c$, and 1 thereafter). Following from this specification, the rate of technical progress equals $a_2$ in the first subperiod and $a_2 + a_3$ in the second.

*Table 6A.1* in the Appendix reports the estimation results of the production function. The equations were estimated in their original form by the MIDIS (MInimum DIStance) method. In the case of group 11 (oil-exporting countries), the real value added of the sector mining and quarrying was deducted from $Y^*$. This reflects the fact that the production of this sector could not be explained by a supply-driven approach.

In order to identify the switching point in the rate of technical progress, the production function was estimated for different time periods but with a constant and uniform rate of technical progress. The switching point ($t_c$) shown in the second line of *Table 6A.1* was set where the estimated rate of technical progress changed substantially. In the estimation for the whole period, the value of $a_2$ was set at the estimated rate of technical progress in the first subperiod. In the case of India, groups 14 and 18, there was no indication of change in the rate of technical progress.

Regarding the estimated values of the output elasticity of capital ($a_4$), it can be observed that, with the exception of groups 11 and 12, they are quite near to each other, falling in the range of 0.42 and 0.46. The estimated value for group 12 is in line with the results of other studies on countries of this group (see, e.g., Chen, 1979, pp. 62–63). With respect to the output elasticity of imported raw materials and energy ($a_6$), the results clearly show the importance of this factor. Except for India, the estimated parameters are larger than 0.1 and, with the exception of groups 11 and 18, they are strongly significant. It is important to note that the real imports of raw materials and energy are determined endogenously in the model.

*The structure of production*

In the model, six producing sectors are distinguished. Agriculture comprises ISIC major division 1, mining and quarrying, manufacturing, utilities and construction refer to ISIC major divisions 2 through 5 and services cover major divisions 6 through 9.

The structure of production is explained by a Cobb–Douglas type function, having the following form:

$$y_i = a_1 \left(\frac{Y}{L}\right)^{a_2} \left(\frac{K}{L}\right)^{a_3} \left(\frac{IM}{Y}\right)^{a_4} \left(\frac{EX}{Y}\right)^{a_5} \left(\frac{I}{Y}\right)^{a_6} + u \tag{6.2}$$

where $y_i$ is the share of real value added produced by sector $i$ in real GDP; $Y$, $K$, $IM$, $EX$ and $I$ refer to real GDP, real capital stock, real imports, real exports, and gross capital formation in real terms, respectively; and $L$ represents the number of the economically active population.

Since this specification does not guarantee the fulfillment of the adding-up constraint, the estimated sector shares are scaled to add up to 1.

*The general price level*

The general price level is determined from the Fisher equation by assuming that the money supply is exogenous:

$$p = \frac{Mv}{Y} \tag{6.3}$$

where $M$ refers to money measured as end-of-period M2 money stock, and it is expressed in terms of the artificial currency defined for the group. The income velocity of money is determined by the following equation:

$$v = \exp\left[a_1 v_{-1} + a_2(1 + w_{p,-1}) + a_3 \frac{L}{Y} + a_4\right] + u \tag{6.4}$$

where $v$ is the income velocity of money (M2), $Y$ and $L$ are real GDP and economically active population and $w_p$ refers to the rate of inflation.

The original specification of equation (6.4) also included the interest rate, but finally it was left out for different reasons. First, reliable figures on interest rates were available only for some of the countries in our model. Second, in most of the countries where these figures were available, the interest rates were institutionally pegged and did not in general reflect financial market conditions. This character of the interest rates in developing countries was pointed out by several empirical studies investigating the money markets in these countries [see, e.g., Wong (1977) and Driscoll and Lahiri (1982)]. It was also pointed out in these studies that the speculative demand for money was negligible in these economies.

The estimation results are shown in *Table 6A.2* in the Appendix.

For group 11, the general price level is determined by the following function:

$$p = a_1 \, p_{IM}{}^{a_2} \, p_{-1}{}^{a_3} \, p_{EX}{}^{a_4} + u \tag{6.4}$$

where $p$, $p_{IM}$, and $p_{EX}$ are GDP, import, and export deflators, respectively. The estimation results of this equation are also shown in *Table 6A.2*. The parameters are plausible, and the specification permits correlating the short- and long-run adjustment of the domestic price level with the export and import prices.

*Real imports*

The equations describing the import demand of the different groups of developing countries are based on a dynamic version of the Linear Expenditure System. The theoretical derivation of the specification can be found in Welsch (1987).

The estimated form of the import demand function can be written

$$IM = a_1 IM_{-1} + a_2 \frac{I_{p_I}}{p_{IM}} + a_3 \frac{C_{p_C}}{p_{IM}} + a_4 \frac{EX_{p_{EX}}}{p_{IM}} + a_5 \frac{Y_{-1} p_{Y_{-1}}}{p_{IM}} + u \tag{6.6}$$

where $IM$, $I$, $C$, $EX$, and $Y$ are real imports of goods and services, gross capital formation in real terms, real consumption, real exports of goods and services, and real GDP, respectively, and the price terms refer to the corresponding deflators. One of the advantages of this specification is that it can reflect the changes in the expenditure structure of GDP. The estimation results show that, in the case of developing countries this effect is of great importance since the increase of the share of gross capital formation has a rather strong influence on their import demand.

For this specification, the short-run import price elasticity is given by

$$E^{Sr}_{IM,p_{IM}} = a_1 \frac{IM_{-1}}{IM} - 1$$

and the long-run elasticity by

$$E^{Lr}_{IM,p_{IM}} = \left( a_1 \frac{IM_{-1}}{IM} - 1 \right) / (1 - a_1)$$

*Table 6A.3* in the Appendix shows the regression results of equation (6.6). In the last two columns, the estimated values of the price elasticites can be found. The equations were estimated by the MIDIS method. In all cases, the import propensity of consumption (parameter $a_3$) was found to be insignificant. This may be due to the strong multicollinearity among the explanatory variables and to the rigidity of the import demand for consumption. The other parameters bear the signs required by the theoretical approach and are in general significant. The widely varying estimated values of the propensities of the different expenditure items seem to justify our approach of disaggregating the income term in the import demand function.

*Commodity structure of real imports*

In the foreign trade part of the model, the structure of exports and imports of goods is described by four commodity groups: agricultural goods (commodity group 1) comprise SITC major groups 0 and 1; crude materials (commodity group 2) make up SITC major groups 2 and 4; mineral fuels (commodity group 3) refer to SITC major group 3; and manufactured goods (commodity group 4) cover SITC major groups 5 through 9. In the presentation of the model, we shall always refer to these commodity groups. The commodity structure of the imports of goods is explained by a demand system also based on a dynamic version of the Linear Expenditure System:

$$IM^k = a_{1k} IM^k_{-1} + a_{2k} \frac{IM^G \, p^G_{IM}}{p^k_{IM}} - \sum_{\substack{l=1 \\ l \neq k}}^{4} a_{l+2,k} \frac{IM^l_{-1} \, p^l_{IM}}{p^k_{IM}} + u$$

$$\sum_{k=1}^{4} a_{2k} = 1 \text{ and } \frac{a_{1k}}{1-a_{2k}} = \frac{a_{k+2,l}}{a_{2l}} \quad k, l = 1,\ldots,4 \quad k \neq l \tag{6.7}$$

where superscripts $k$ and $l$ refer to the commodity groups and superscript $G$ indicates that $IM$ refers only to the imports of goods. Each demand system consists of four equations, but only three of them are estimated. The parameters of the fourth equation can be calculated from those of the first three.

In the case of group 11 (oil-exporting countries), imports of mineral fuels could not be explained by this demand system. Since the share of this commodity group in total imports of goods is negligible and of no importance in the explanation of the production, it is taken as exogenous in the model. In the case of India, imports of agricultural goods could not be explained by this specification, either. This is probably due to the fact that the import demand of India in this commodity group is strongly influenced by the harvest conditions, and this factor is not reflected in the specification. In the case of group 15, the lagged effect of the agricultural imports could not be identified and, therefore, this parameter was set to zero.

When looking at the results of the estimations in *Table 6A.4* in the Appendix, it can be found that the parameters in all cases bear the correct sign from the point of view of theory and they are in general significant. It is interesting to compare the different groups of developing countries with respect to the marginal expenditure shares ($a_2$). In doing so, it can be observed that these parameters are quite similar for all groups. The only notable exception is India, which has a rather different structure of these parameters. Comparing the structures of marginal expenditure shares of the developed countries in question to those of the developed market economy (see Welsch, 1986, p. 13), the developing countries are characterized by very high shares of manufactured goods. When making the same comparison with respect to the persistence coefficient $[a_1/(1 - a_2)]$, it can be found that the values of these parameters are considerably lower in developing countries. The only exception is group 12, for which the coefficients are rather similar to those of the developed market economies.

*Real exports*

The export demand functions are specified for the different commodity groups by:

$$EX^{\$,k} = a_1 \, EX^{\$,k}_{-1} + a_2 \frac{WT^{\$,k} \, p^{\$,k}_{WT}}{p^{\$,k}_{EX}} - a_3 \frac{EX^{\$,Fk}_{-1} \, p^{\$,Fk}_{EX}}{p^{\$,k}_{EX}} + u \tag{6.8}$$

where $k$ ($k = 1,...,4$) refers to the commodity groups, $EX$ is the real exports of the country under consideration, $WT$ is the volume of world trade in the commodity group, and $p_{WT}$ is the corresponding price deflator. Superscript $F$ refers to the countries different from the one under consideration. Superscript \$ indicates that the variables are expressed in US\$ terms.

The results are given in *Table 6A.5* in the Appendix. In the case of India and group 15, the exports of mineral fuels (SITC 3) could not be explained in this way. This is probably due to the fact that their exports were determined not only by the demand side but by the supply side as well. For group 12, it was necessary to distinquish between the import demand of the developed market economies and that of the rest of the world because of the considerable differences in the marginal expenditure shares. (The corresponding parameters in *Table 6A.5* are $a_2$ and $a_3$; $a_4$ stands for the lagged effect originated in the countries different from the ones in group 12.) The specification of the export demand function used in the Linked World Model is given in Chapter 7.

*Export and import prices*

The export and import prices of the developing countries in the different commodity groups are linked to the world market prices by the following equations:

$$p_{EX}^{\$,k} = a_1 p_{WT}^{\$,k^{a_2}} p_{EX_{-1}}^{\$,k^{a_3}} + u \quad k = 1,...,4 \tag{6.9}$$

$$p_{IM}^{\$,k} = a_1 p_{WT}^{\$,k^{a_2}} p_{IM_{-1}}^{\$,k^{a_3}} + u \quad k = 1,...,4 \tag{6.10}$$

where $p_{IM}^k$ and $p_{EX}^k$ are the export and import prices of the group under consideration and $p_{WT}^k$ stands for the world market prices. Superscript $k$ refers to the commodity groups. Superscript \$ indicates that the variables are expressed in US\$ terms.

In the Linked World Model, a similar approach was applied to the aggregate export and import prices. The design of these equations is given in Chapter 7.

### 6.2.3. *Ex post* simulations

As the models discussed above were developed for *ex ante* simulations, it is of basic importance to evaluate their ability to forecast. In the case of dynamic simultaneous econometric models, as it is widely pointed out, it is not sufficent to perform statistical tests only on the individual equations. Therefore, we made *ex post* simulations to investigate the models in this respect. Since in the next section the analysis will be based on *ex ante* simulations with the Linked World Model, the results of the *ex post* simulation presented in *Table 6A.6* in the Appendix refer to the same model [1]. The simulation period was 1966–1981. Taking into consideration the fact that the simulation period covers the two oil shocks, the values of the mean absolute percentage error reported in this table can be rated as satisfactory.

## 6.3. Growth Prospects of the Developing Countries: Results of a Medium-term Scenario Analysis

The ultimate goal of the Bonn–IIASA Research Project was to indicate the possible growth paths of the different regions of the world economy. To trace these paths, the world model developed in the project was used for a medium-term scenario analysis. Chapter 2 provided a general description of the design of this analysis and discussed the main results of the medium scenario. Here, we concentrate on the developing countries.

### 6.3.1. Assumptions of the scenarios

Similar to the procedure applied to other regions, three different sets of assumptions were used to define the scenarios. *Table 6.1* summarizes these assumptions.

*Table 6.1.* The definition of scenarios for developing countries.

| Variable | Scenario and period | Country group[a] | | | | | | |
|---|---|---|---|---|---|---|---|---|
| | | 11 | 12 | India | 13 | 14 | 15 | 18 |
| Rate of growth of labor input (economically active population) | all: | | | | | | | |
| | 1986–1990 | 2.39 | 2.04 | 1.63 | 2.62 | 2.60 | 2.66 | 2.91 |
| | 1991–1995 | 2.75 | 2.29 | 2.00 | 3.08 | 3.11 | 2.90 | 2.99 |
| | 1996–2000 | 2.37 | 2.08 | 1.82 | 2.60 | 2.46 | 2.55 | 2.54 |
| Rate of (Hicks-neutral) technical progress | high: | | | | | | | |
| | 1986–1990 | 1.69 | 1.87 | 0.96 | 0.85 | 1.18 | 1.95 | 2.22 |
| | 1991–1995 | 2.00 | 2.00 | 1.00 | 1.00 | 1.20 | 2.00 | 2.22 |
| | 1996–2000 | 2.00 | 2.00 | 1.00 | 1.00 | 1.20 | 2.00 | 2.22 |
| | medium: | | | | | | | |
| | 1986–1990 | 0.00 | 1.00 | 0.50 | 0.00 | 1.00 | 1.60 | 2.00 |
| | 1991–1995 | 1.31 | 1.44 | 0.50 | 0.66 | 1.00 | 1.73 | 2.00 |
| | 1996–2000 | 1.50 | 1.50 | 0.50 | 0.75 | 1.00 | 1.75 | 2.00 |
| | low: | | | | | | | |
| | 1986–1990 | -1.03 | 0.62 | 0.50 | -0.49 | 1.00 | 1.44 | 2.00 |
| | 1991–1995 | -0.02 | 0.81 | 0.50 | 0.03 | 1.00 | 1.44 | 2.00 |
| | 1996–2000 | 0.88 | 0.98 | 0.50 | 0.44 | 1.00 | 1.44 | 2.00 |
| Gross investment ratio | high: | | | | | | | |
| | 1986–1990 | 29.05 | 28.29 | 25.84 | 22.65 | 17.45 | 23.04 | 24.73 |
| | 1991–1995 | 29.87 | 28.89 | 25.89 | 24.69 | 19.70 | 24.82 | 26.76 |
| | 1996–2000 | 30.00 | 29.00 | 26.00 | 25.00 | 20.00 | 25.00 | 27.00 |
| | medium: | | | | | | | |
| | 1986–1990 | 24.05 | 25.25 | 21.50 | 20.86 | 14.95 | 19.22 | 20.54 |
| | 1991–1995 | 24.87 | 25.35 | 22.82 | 22.72 | 16.76 | 21.65 | 22.74 |
| | 1996–2000 | 25.00 | 25.50 | 23.00 | 23.00 | 17.00 | 22.00 | 23.00 |
| | low: | | | | | | | |
| | 1986–1990 | 19.00 | 20.73 | 18.62 | 14.89 | 10.79 | 16.13 | 16.43 |
| | 1991–1995 | 20.32 | 21.85 | 19.91 | 17.58 | 11.89 | 18.53 | 19.21 |
| | 1996–2000 | 21.00 | 22.00 | 20.00 | 18.00 | 12.00 | 19.00 | 19.50 |
| Rate of depreciation | all: | | | | | | | |
| | 1986–2000 | 2.50 | 3.3 | 1.7 | 2.3 | 3.2 | 1.4 | 1.5 |
| Rate of change of money supply (M2) | all: | | | | | | | |
| | 1986–2000 | – | 18.0 | 15.0 | 20.0 | 45.0 | 45.0 | 22.0 |

[a]Group code numbers refer to the following: group 11 = oil-exporting countries; group 12 = Asian countries excluding India; group 13 = developing African countries; group 14 = Latin American countries excluding group 15; group 15 = Argentina, Brazil, Mexico; and group 18 = Middle Eastern and North African countries excluding groups 11 and 13.

The scenarios (called high, medium, and low) differ in their rates of technical progress and in the gross investment ratios.

As to the gross investment ratio, the assumptions were designed in the following way. In the case of (oil-exporting countries), in the low scenario, the level reached right after the first oil shock; in the medium scenario, the average level between the two oil shocks; and finally, in the high scenario, the level reached after the second oil shock were assumed. For groups 12 and 15, in the low scenario, the average of the historically observed lowest values; in the high scenario, that of the historically observed highest ones; and finally, in the medium scenario, the average of the remaining observations was assumed. In the case of groups 13, 14 and 18, the same principle was used for the domestic savings ratio, but not for the external sources of investment. A substantial decrease in the latter was assumed. The resulting gross investment ratios (which are consequently rather moderate as compared with the historically observed values, but still quite high as compared with those of the developed market economies) reflect the outcome of these two assumptions. For India (where domestic saving was the dominant factor in the determination of investment), in the medium scenario, it was assumed that the level of the investment ratio reached during the second half of 1970s will be kept up in the long run. In the high scenario, the observed tendency for the investment ratio to increase was assumed to continue leveling off at 26%. In the low scenario, the average of the historically observed values was assumed.

As to the rate of technical progress, in the case of India, groups 14 and 18 (regions where there was no indication of change in this parameter), the estimated rates indicated in *Table 6A.1* in the Appendix were used in the medium and in the low scenarios; while in the high scenario, a moderate increase in these rates was assumed. For the other groups (where the estimation results indicated a substantial decrease in the rate of technical progress after the first or second oil shock), a gradual increase was assumed, leveling off differently in the three scenarios. But it was never assumed that any of these countries could reach those very high levels maintained until the switching points. (These levels are given by the estimated values of $a_2$ in *Table 6A.1* in the Appendix.)

The main results of the solutions of the Linked World Model for the developing countries under the scenarios described above are presented in *Tables 6.2* to *6.4* [2].

### 6.3.2. Growth prospects

One of the most important questions of social and economic development in the developing countries is the feasibility of accelerated growth. Such accelerated growth would be necessary in order to narrow the income gap [3] between developed and developing countries. Narrowing the income gap, it is hoped, would bring more social justice and equality of opportunity.

The feasibility of such a goal can be investigated from several points of view. Certainly, there are physical limits to faster growth in the long run, especially if it is not accompanied by the necessary structural changes. The rapidly increasing pollution of the environment could also lead to constraints on any accelerated growth, especially in the developing countries, where the new technologies necessary to reduce pollution substantially are not available. Moreover, recent structural changes in the world economy have tended to relocate the most polluting industries and technologies from the industrialized countries to the developing countries. The availability of mineral resources could also become a factor limiting economic growth. Changes in the social environment are also essential conditions for faster economic development in the developing countries.

*Table 6.2.* Results of the analysis: high scenario.

| *Variable* | *Period* | Country group[a] *11* | *12* | *India* | *13* | *14* | *15* | *18* |
|---|---|---|---|---|---|---|---|---|
| Growth rate of real GDP | 1986–1990 | 10.3 | 5.9 | 4.5 | 3.8 | 5.9 | 7.4 | 7.1 |
| | 1991–1995 | 8.8 | 5.9 | 4.5 | 4.7 | 6.2 | 7.6 | 7.8 |
| | 1996–2000 | 7.7 | 5.5 | 4.4 | 4.5 | 5.8 | 7.5 | 7.6 |
| Growth rate of real imports | 1986–1990 | 3.7 | 6.6 | 10.5 | 4.5 | 10.1 | 12.0 | 3.9 |
| | 1991–1995 | 8.4 | 6.8 | 6.0 | 4.9 | 8.2 | 7.2 | 8.1 |
| | 1996–2000 | 6.9 | 6.9 | 5.4 | 4.9 | 7.0 | 6.8 | 7.9 |
| Rate of change of import prices (in US$) | 1986–1990 | 1.3 | 0.6 | 0.8 | 2.6 | 0.2 | 1.7 | 2.5 |
| | 1991–1995 | 4.2 | 3.8 | 4.0 | 4.2 | 3.6 | 4.0 | 3.8 |
| | 1996–2000 | 4.7 | 4.3 | 4.3 | 4.4 | 4.3 | 4.4 | 4.3 |
| Growth rate of real exports | 1986–1990 | 11.2 | 5.4 | 5.8 | 2.9 | 3.4 | 5.0 | 3.7 |
| | 1991–1995 | 10.6 | 6.9 | 10.0 | 6.0 | 6.4 | 6.7 | 7.1 |
| | 1996–2000 | 7.9 | 7.3 | 8.9 | 7.2 | 6.7 | 7.3 | 7.0 |
| Rate of change of export prices (in US$) | 1986–1990 | -8.7 | 1.2 | 0.1 | 1.0 | 1.4 | 1.0 | 0.9 |
| | 1991–1995 | 2.8 | 3.8 | 2.6 | 3.2 | 4.7 | 3.5 | 4.5 |
| | 1996–2000 | 4.1 | 4.2 | 3.0 | 3.5 | 5.1 | 3.8 | 4.9 |
| Trade balance to GDP | 1986–1990 | -0.2 | -3.6 | -4.8 | -2.9 | -3.7 | -0.4 | -7.8 |
| | 1991–1995 | 2.6 | -4.0 | -5.5 | -4.8 | -6.3 | -1.5 | -8.3 |
| | 1996–2000 | 3.1 | -3.8 | -5.1 | -3.7 | -6.4 | -1.6 | -8.4 |
| Trade balance to exports | 1986–1990 | -1.0 | -6.7 | -78.5 | -10.6 | -27.1 | -5.4 | -55.1 |
| | 1991–1995 | 9.0 | -7.3 | -74.7 | -17.2 | -45.3 | -21.6 | -62.6 |
| | 1996–2000 | 10.9 | -6.4 | -57.7 | -12.5 | -44.0 | -23.5 | -63.8 |
| Imports to GDP (in real terms) | 1986–1990 | 28.6 | 60.9 | 8.1 | 36.3 | 20.3 | 9.6 | 16.3 |
| | 1991–1995 | 26.6 | 62.7 | 9.2 | 37.0 | 23.9 | 10.1 | 16.1 |
| | 1996–2000 | 25.9 | 66.0 | 9.6 | 37.5 | 25.3 | 9.8 | 16.4 |
| Exports to GDP (in real terms) | 1986–1990 | 34.3 | 55.0 | 6.2 | 29.2 | 14.1 | 9.1 | 14.8 |
| | 1991–1995 | 40.1 | 56.0 | 7.9 | 29.8 | 13.6 | 8.5 | 13.9 |
| | 1996–2000 | 41.3 | 59.7 | 9.7 | 32.8 | 14.0 | 8.3 | 13.6 |

[a] For definitions of group code numbers, see note to *Table 6.1*.

All of these aspects are very important, especially in the longer run, but the recent development of the world economy suggests that the most serious factors limiting accelerated growth in the developing countries probably originate from the international economic environment, which these countries must face. The scarcity of external financial resources necessary to increase investment and, through this, achieve a faster rate of technical progress seems to be the most acute problem in this respect.

Viewing the results of the scenario analysis from this angle, we can associate the different growth paths of the world economy described by the different scenarios with the corresponding trade balance indicators. Since we concentrate on the feasibility of accelerated growth that may ultimately narrow the income gap, *Table 6.5* gives the differences, in terms of annual growth rates, between the developed market economies as a whole and the different groups of developing countries with respect to real GDP and real GDP per capita.

*Table 6.3.* Results of the analysis: medium scenario.

| *Variable* | *Period* | *Country group*[a] | | | | | | |
|---|---|---|---|---|---|---|---|---|
| | | *11* | *12* | *India* | *13* | *14* | *15* | *18* |
| Growth rate of real GDP | 1986–1990 | 6.6 | 4.2 | 3.7 | 2.5 | 5.1 | 6.2 | 6.0 |
| | 1991–1995 | 6.8 | 4.6 | 3.6 | 3.8 | 5.3 | 6.7 | 6.8 |
| | 1996–2000 | 6.2 | 4.2 | 3.5 | 3.8 | 5.0 | 6.6 | 6.6 |
| Growth rate of real imports | 1986–1990 | -0.9 | 5.1 | 11.3 | 3.2 | 8.8 | 10.2 | 1.0 |
| | 1991–1995 | 5.8 | 4.8 | 6.2 | 3.4 | 6.9 | 6.5 | 6.1 |
| | 1996–2000 | 4.7 | 4.8 | 5.0 | 3.5 | 5.8 | 5.5 | 6.1 |
| Rate of change of import prices (in US$) | 1986–1990 | 1.7 | 0.9 | 1.2 | 3.0 | 0.5 | 2.1 | 2.8 |
| | 1991–1995 | 5.1 | 4.7 | 4.8 | 5.2 | 4.4 | 4.9 | 4.5 |
| | 1996–2000 | 5.9 | 5.4 | 5.4 | 5.6 | 5.4 | 5.5 | 5.2 |
| Growth rate of real exports | 1986–1990 | 8.5 | 3.6 | 2.3 | 1.1 | 0.8 | 3.3 | 0.7 |
| | 1991–1995 | 8.7 | 4.6 | 7.6 | 3.5 | 3.4 | 4.5 | 4.1 |
| | 1996–2000 | 6.7 | 4.9 | 6.9 | 4.7 | 4.0 | 5.0 | 4.4 |
| Rate of change of export prices (in US$) | 1986–1990 | -8.7 | 1.5 | 0.3 | 1.2 | 1.8 | 1.4 | 1.3 |
| | 1991–1995 | 2.8 | 4.7 | 3.1 | 3.9 | 5.7 | 4.3 | 5.4 |
| | 1996–2000 | 4.1 | 5.3 | 3.8 | 4.4 | 6.4 | 4.8 | 6.2 |
| Trade balance to GDP | 1986–1990 | 1.8 | -2.2 | -3.5 | -1.9 | -2.0 | 1.3 | -5.7 |
| | 1991–1995 | 4.8 | -3.4 | -5.4 | -4.8 | -4.9 | -0.3 | -6.2 |
| | 1996–2000 | 5.0 | -3.6 | -5.9 | -5.4 | -5.6 | -0.8 | -6.3 |
| Trade balance to exports | 1986–1990 | 6.5 | -4.0 | -68.2 | -6.8 | -15.8 | 16.8 | -43.4 |
| | 1991–1995 | 17.5 | -6.2 | -94.0 | -18.6 | -41.1 | -5.2 | -55.0 |
| | 1996–2000 | 18.3 | -6.5 | -91.7 | -21.0 | -48.0 | -12.6 | -59.9 |
| Imports to GDP (in real terms) | 1986–1990 | 25.9 | 60.0 | 6.5 | 34.7 | 17.5 | 7.7 | 14.0 |
| | 1991–1995 | 23.2 | 60.9 | 7.9 | 34.6 | 20.0 | 8.4 | 13.1 |
| | 1996–2000 | 21.9 | 62.0 | 8.5 | 34.1 | 20.9 | 8.1 | 12.8 |
| Exports to GDP (in real terms) | 1986–1990 | 34.0 | 55.6 | 5.3 | 29.1 | 13.3 | 9.3 | 13.8 |
| | 1991–1995 | 40.3 | 55.0 | 6.2 | 28.0 | 11.7 | 8.2 | 11.7 |
| | 1996–2000 | 41.8 | 56.0 | 7.2 | 28.5 | 11.0 | 7.6 | 10.6 |

[a] For definitions of group code numbers, see note to *Table 6.1.*

In interpreting the results for group 11 (oil-exporting countries), it should be borne in mind that, in the model of this group, the production of the oil sector (the contribution of which to total GDP is extremly high, and the production of which is almost entirely exported) is demand-driven. As a consequence, the manageable extent of growth difference is mainly determined by the overall pace of economic growth in the other regions. Under the most favorable scenario analyzed in the project, even a very high growth difference of 5–6% seems to be feasible. In terms of GDP per capita, this would mean a growth difference of 3.4–4.1%. Under the medium scenario, it decreases to 4–5% with respect to GDP and to 2.4–3.1% with respect to GDP per capita, which should also be considered high by all standards. Under the low scenario, the corresponding ranges are 2.5–3.5% and 0.9–1.6% [4]. According to the results, growth differences larger than these values tend to lead to a deterioration of the foreign trade balance.

*Table 6.4.* Results of the analysis: low scenario.

| | | Country group[a] | | | | | | |
|---|---|---|---|---|---|---|---|---|
| *Variable* | *Period* | *11* | *12* | *India* | *13* | *14* | *15* | *18* |
| Growth rate of real GDP | 1986–1990 | 3.4 | 3.2 | 3.5 | 1.2 | 3.8 | 5.4 | 5.1 |
| | 1991–1995 | 4.0 | 3.3 | 3.4 | 2.4 | 4.2 | 5.8 | 6.1 |
| | 1996–2000 | 3.7 | 2.9 | 3.2 | 2.7 | 3.9 | 5.5 | 5.9 |
| Growth rate of real imports | 1986–1990 | -5.6 | 3.7 | 10.8 | 0.7 | 2.3 | 8.7 | -2.8 |
| | 1991–1995 | 2.9 | 2.7 | 6.5 | 2.0 | 4.9 | 6.2 | 4.3 |
| | 1996–2000 | 1.3 | 2.3 | 5.3 | 1.6 | 4.0 | 3.5 | 3.3 |
| Rate of change of import prices (in US$) | 1986–1990 | 2.0 | 1.4 | 1.7 | 3.5 | 0.8 | 2.6 | 3.1 |
| | 1991–1995 | 6.2 | 5.7 | 5.9 | 6.4 | 5.4 | 6.0 | 5.3 |
| | 1996–2000 | 7.7 | 7.0 | 7.0 | 7.5 | 7.0 | 7.3 | 6.6 |
| Growth rate of real exports | 1986–1990 | 4.6 | 2.1 | -1.6 | -0.6 | -1.6 | 1.7 | -2.3 |
| | 1991–1995 | 5.9 | 2.2 | 4.1 | 0.8 | 0.0 | 2.1 | 0.4 |
| | 1996–2000 | 3.9 | 2.0 | 3.1 | 1.3 | 0.0 | 2.1 | 0.3 |
| Rate of change of export prices (in US$) | 1986–1990 | -8.7 | 1.9 | 0.5 | 1.6 | 2.3 | 1.7 | 1.7 |
| | 1991–1995 | 2.8 | 5.7 | 3.7 | 4.7 | 6.9 | 5.2 | 6.6 |
| | 1996–2000 | 4.1 | 6.9 | 4.9 | 5.7 | 8.4 | 6.3 | 8.1 |
| Trade balance to GDP | 1986–1990 | 3.3 | -1.2 | -2.6 | 2.0 | 1.0 | 2.6 | -3.8 |
| | 1991–1995 | 5.8 | -3.0 | -5.1 | -2.3 | -1.3 | 0.7 | -4.3 |
| | 1996–2000 | 4.7 | -3.9 | -6.4 | -4.4 | -2.5 | -0.0 | -4.3 |
| Trade balance to exports | 1986–1990 | 13.3 | -2.3 | -60.7 | 7.2 | 7.7 | 33.7 | -30.4 |
| | 1991–1995 | 22.8 | -5.9 | -129.8 | -9.5 | -13.1 | 11.5 | -46.8 |
| | 1996–2000 | 20.0 | -7.9 | -173.5 | -20.0 | -27.5 | -0.8 | -58.1 |
| Imports to GDP (in real terms) | 1986–1990 | 22.5 | 57.8 | 6.1 | 29.8 | 13.3 | 6.2 | 11.8 |
| | 1991–1995 | 19.5 | 57.0 | 6.5 | 29.8 | 14.2 | 6.6 | 10.1 |
| | 1996–2000 | 18.0 | 55.5 | 7.2 | 28.6 | 14.3 | 6.3 | 9.1 |
| Exports to GDP (in real terms) | 1986–1990 | 32.1 | 54.4 | 4.5 | 28.9 | 12.7 | 9.3 | 12.7 |
| | 1991–1995 | 37.9 | 51.6 | 4.4 | 26.5 | 10.1 | 7.8 | 9.4 |
| | 1996–2000 | 39.0 | 49.5 | 4.4 | 24.9 | 8.5 | 6.7 | 7.3 |

[a] For definitions of group code numbers, see note to *Table 6.1.*

In the case of the other groups, although the limits of feasible growth differences correlate positively with the overall rate of growth in the world economy, this relation does not seem to be so strong.

As to group 12 (Asian countries) the results tend to suggest that under the medium scenario, this region would be able to keep up a growth difference of 1.5–2% with respect to GDP. Although this growth path would be associated with a negative balance of trade, it does not seem unmanageable. This pace, in turn, would mean a growth difference with respect to real GDP per capita leveling off at 0.6–1.1% by the end of this century. Under the high scenario, these values would be increased by about 0.5%.

In the case of India, the alarmingly high foreign trade deficits are clear signs of the fact that even a gradually declining growth difference of 1.0–0.7% with respect to GDP (which would actually mean a slowly widening income gap between India and the developed market economies as a whole) would be hard to

*Table 6.5.* Differences in annual growth rates (%) between developed market economies and groups of developing countries.

| Variable | Scenario and period | Country group[a] | | | | | | |
|---|---|---|---|---|---|---|---|---|
| | | 11 | 12 | India | 13 | 14 | 15 | 18 |
| Real GDP | high: | | | | | | | |
| | 1986–1990 | 6.8 | 2.4 | 1.0 | 0.3 | 2.4 | 3.9 | 3.6 |
| | 1991–1995 | 5.1 | 2.2 | 0.8 | 1.0 | 2.5 | 3.9 | 4.1 |
| | 1996–2000 | 4.0 | 1.8 | 0.7 | 0.8 | 2.1 | 3.8 | 3.9 |
| | medium: | | | | | | | |
| | 1986–1990 | 3.9 | 1.5 | 1.0 | –0.2 | 2.4 | 3.5 | 3.3 |
| | 1991–1995 | 4.0 | 1.8 | 0.8 | 1.0 | 2.5 | 3.9 | 4.0 |
| | 1996–2000 | 3.4 | 1.4 | 0.7 | 1.0 | 2.2 | 3.8 | 3.8 |
| | low: | | | | | | | |
| | 1986–1990 | 1.5 | 1.1 | 1.4 | –0.9 | 1.7 | 3.3 | 3.0 |
| | 1991–1995 | 2.1 | 1.4 | 1.5 | 0.5 | 2.3 | 3.9 | 4.1 |
| | 1996–2000 | 1.8 | 1.0 | 1.3 | 0.8 | 2.0 | 3.6 | 4.0 |
| Real GDP per capita | high: | | | | | | | |
| | 1986–1990 | 4.9 | 1.1 | –0.1 | –2.2 | 0.7 | 2.2 | 1.6 |
| | 1991–1995 | 3.3 | 1.0 | –0.2 | –1.6 | 0.9 | 2.4 | 2.2 |
| | 1996–2000 | 2.4 | 0.9 | –0.1 | –1.8 | 0.7 | 2.4 | 2.1 |
| | medium: | | | | | | | |
| | 1986–1990 | 2.0 | 0.2 | –0.1 | –2.7 | 0.7 | 1.8 | 1.3 |
| | 1991–1995 | 2.2 | 0.6 | –0.2 | –1.6 | 0.9 | 2.4 | 2.1 |
| | 1996–2000 | 1.8 | 0.5 | –0.1 | –1.6 | 0.8 | 2.4 | 2.0 |
| | low: | | | | | | | |
| | 1986–1990 | –0.4 | –0.2 | 0.3 | –3.4 | 0.0 | 1.6 | 1.0 |
| | 1991–1995 | 0.3 | 0.2 | 0.5 | –2.1 | 0.7 | 2.4 | 2.1 |
| | 1996–2000 | 0.2 | 0.1 | 0.5 | –1.8 | 0.6 | 2.2 | 2.2 |

[a]For definitions of group code numbers, see note to *Table 6.1*.

maintain in the longer run. In interpreting the results of this scenario analysis, especially in the case of India, a note of caution is necessary. The aggregate level of the analysis and the demand-oriented modeling export flows prevented our considering favorable changes in the commodity structure of export supply different from the historically observed pattern. In reality, development projects in India are mostly aimed at bringing about such favorable structural changes. Successful implementations of these projects could, consequently, make at least a modest beginning toward narrowing the existing income gap. But according to the results, without these favorable structural changes in export supply (and, of course, without changes in the pattern of import demand), the income gap rather tends to widen.

The results for group 13 (African countries) are shocking. Under the most optimistic scenario analyzed in the study (high scenario), the feasible growth difference seems to be not higher than 0.5–0.8%. Under the medium scenario (which is still rather optimistic with respect to the overall rate of economic growth in the developed market economies), even this growth rate difference tends to produce a rapidly deteriorating foreign trade balance. The *shocking* fact is that this growth path is associated with a very fast increase in the already *alarmingly wide* income gap. The difference in the yearly growth rate of real GDP per capita between this group and the developed market economies as a whole would be about -2.1% to -1.8%.

In the case of group 14 (Latin American countries), the results indicate that even a very moderate breakaway from the growth path keeping up the existing level of the income gap between this group and the developed market economies as a whole might produce a rapidly deteriorating foreign trade balance.

In the longer run, group 15 (Argentina, Brazil, Mexico) seems to be in a more favorable position. Of course, in view of the present debt situation of these countries, these favorable growth paths in the longer run are feasible only if the countries in question can manage to bring down their accumulated debts to a tolerable level during the next few years – a task easy to assign to a country but very hard to achieve.

According to the model, it seems unlikely that group 18 (North African and Middle Eastern countries) can maintain their very high growth rates of the years after the oil crisis.

To sum up, taking into consideration the results of the three main scenarios and those of the alternative scenarios described in Chapter 7 and aimed at investigating similar aspects, a difference of 1.5–2.0% in the growth rates of real GDP between developed market economies and developing countries [5] seems to be the limit of feasibility [6]. In terms of our scenarios, this means that if the developed market economies were to grow along the path described by the medium scenario, the developing countries as a whole could rather be expected to follow along the growth path envisaged by the pessimistic scenario, which would hardly narrow the existing income gap at all. The results also suggest that some groups of developing countries might be expected to sustain a faster pace of economic development. Among the regions considered here, group 12 (Asian countries) seems to be in the most favorable position in this respect.

### 6.3.3. Structural change

Another important aspect of economic development in developing countries is the pattern of *structural change*. In Chapter 2, several aspects of this issue were investigated, among them the structure of production. We extend this examination by comparing the different groups of developing countries and the results of the different scenarios. *Table 6.6* summarizes the projections of the model under the different scenarios [7].

According to the model, there seems to be a general tendency for the share of agricultural production to continue its decline. The only exception is group 13 (African countries), where, under the low scenario, this share is expected to increase. Another general feature, suggested by these results, is that accelerated

*Table 6.6.* The structure of production in different groups of developing countries: share of real value added to GDP by different sectors (%).

| Sector/ Country group[a] | High scenario 1990 | High scenario 1995 | High scenario 2000 | Medium scenario 1990 | Medium scenario 1995 | Medium scenario 2000 | Low scenario 1990 | Low scenario 1995 | Low scenario 2000 |
|---|---|---|---|---|---|---|---|---|---|
| Agriculture: | | | | | | | | | |
| 11 | 20.8 | 20.0 | 18.9 | 21.9 | 21.4 | 20.5 | 24.1 | 23.0 | 22.2 |
| 12 | 14.8 | 12.9 | 11.4 | 15.8 | 14.1 | 12.8 | 16.4 | 15.1 | 13.9 |
| India | 33.9 | 32.8 | 31.6 | 37.0 | 35.9 | 34.8 | 40.3 | 39.3 | 38.3 |
| 13 | 25.9 | 25.0 | 23.9 | 27.9 | 27.7 | 27.1 | 30.1 | 31.0 | 31.1 |
| 14 | 12.8 | 11.7 | 10.6 | 13.4 | 12.6 | 11.7 | 14.2 | 13.8 | 13.1 |
| 15 | 8.2 | 6.9 | 5.7 | 9.0 | 7.7 | 6.6 | 9.7 | 8.5 | 7.5 |
| 18 | 18.9 | 17.0 | 14.9 | 20.5 | 19.0 | 17.2 | 22.3 | 21.0 | 19.4 |
| Manufacturing: | | | | | | | | | |
| 11 | 11.5 | 11.3 | 11.1 | 12.2 | 11.9 | 11.7 | 12.5 | 12.3 | 12.1 |
| 12 | 31.9 | 34.6 | 37.0 | 30.0 | 32.0 | 33.8 | 28.9 | 30.2 | 31.3 |
| India | 18.2 | 18.3 | 18.5 | 17.5 | 17.8 | 18.1 | 16.6 | 16.9 | 17.2 |
| 13 | 11.9 | 12.2 | 12.7 | 11.0 | 10.9 | 11.0 | 10.2 | 9.5 | 9.2 |
| 14 | 22.1 | 22.2 | 22.3 | 22.9 | 23.0 | 23.2 | 24.7 | 24.8 | 25.0 |
| 15 | 29.1 | 29.6 | 29.7 | 27.8 | 28.4 | 28.5 | 26.5 | 27.2 | 27.2 |
| 18 | 16.0 | 16.0 | 15.8 | 16.2 | 16.1 | 16.0 | 16.3 | 16.2 | 16.1 |
| Services: | | | | | | | | | |
| 11 | 51.8 | 51.8 | 51.7 | 51.3 | 51.3 | 51.2 | 50.0 | 50.5 | 50.6 |
| 12 | 45.1 | 44.7 | 44.1 | 45.7 | 45.6 | 45.3 | 46.3 | 46.3 | 46.3 |
| India | 40.5 | 41.5 | 42.4 | 38.3 | 39.3 | 40.2 | 36.2 | 37.0 | 37.7 |
| 13 | 45.0 | 45.0 | 45.0 | 44.6 | 44.2 | 44.0 | 44.8 | 43.5 | 42.7 |
| 14 | 52.5 | 53.5 | 54.4 | 51.5 | 52.4 | 53.2 | 49.5 | 50.2 | 50.9 |
| 15 | 52.4 | 52.6 | 52.6 | 53.1 | 53.2 | 53.4 | 53.8 | 53.9 | 54.2 |
| 18 | 50.6 | 51.8 | 53.1 | 49.6 | 50.5 | 51.5 | 48.5 | 49.1 | 50.0 |

[a]For definitions of group code numbers, see note to *Table 6.1.*

overall economic development would accentuate the changes. In the case of agriculture, this means that the high scenario would bring about a faster decrease in the share of the agricultural sector.

In spite of these and other common tendencies, the directions in which the different regions of developing countries are expected to develop seem to be somewhat diverse. In the case of group 12 (Asian countries) and group 15 (Argentina, Brazil, Mexico), manufacturing seems to be the leading sector of the economy; while in the other regions, services tend to play this role.

In the case of group 13 (African countries), these results also seem to support the finding that the growth prospects for this group are rather limited. The very high share of agriculture tends to decrease rather slowly (or actually, under the low scenario, it even increases). On the other hand, the share of manufacturing seems to remain very low. All this means that this group is likely to continue to be an exporter of primary commodities and, consequently, to face depressingly low prices and virtually no growth in demand for its exports.

## 6.4. Concluding Remarks

The results of the scenario analysis presented here tend to suggest that the growth prospects of the developing countries as a whole are rather limited. The availability of external financial sources seems to be the most serious limiting factor in the medium run. The growth paths of the world economy envisaged by the projections would hardly narrow the income gap between developed and developing countries at all.

The situations of the different regions of developing countries seem to be rather different. The African countries (as represented by group 13 in our model) are felt to be in the most critical situation, while some of the Asian countries (those in group 12) are likely to continue to catch up with the developed market economies. The projections on the structure of production in the different regions also seem to support these findings.

### Acknowledgment

We wish to express our gratitude to Jürgen Richtering and Paul Robertson (UNCTAD) for their help in the construction of our data bank and for many valuable discussions. We would also like to thank Anna Gelei (UNIDO) and Gyózó Gábor (Economic Comission for Africa, UN) for contributing important data to our data bank.

### Notes

[1] A general description of the Linked World Model can be found in Chapter 7 and in Annex 2 to the volume. During the *ex post* simulation, the exchange rates were taken as exogenous.

[2] Assumptions and results for the other regions can be found in Chapter 2.

[3] In the discussion that follows, the term *income gap* refers to the difference between the levels of real GDP *per capita* in the countries in question.

[4] One reservation should, however, be made with reference to this group. Owing to the aggregate nature of the foreign trade subsystem and to the features of the Linear Expenditure System, the framework of which is extensively used in the design of this subsystem, the projected market share of this group might very well be overstated. If this is the case, the corresponding figures on the feasible growth differences are probably considerably lower than those expressed in *Table 6.5.*

[5] Developed market economies and developing countries as represented by the countries considered in our model. A study covering the poorest African and Asian countries not considered here would probably find this difference to be smaller.

[6] A scenario analysis based on the SIGMA system developed at UNCTAD (for a description of this system, see Chapter 26 in this volume), and described in the Appendix of Chapter 7, seems to support this finding, too.

[7] It should be mentioned that the shares for group 11 (oil-exporting countries) refer only to that part of the GDP which is not produced in the oil sector (mining and quarrying).

## Appendix 6A. Sources, Abbreviations and Model Parameters

Main data sources:

*Handbook of World Development Statistics*, UN, 1983.
*International Financial Statistics Yearbook*, IMF, various years.

List of abbreviations used in the tables:

DW = Durbin–Watson *d* statistics
FIML = Full Information Maximum Likelihood method
MAPE = mean absolute percentage error
MIDIS = MInimum DIStance method
R2C = corrected $R$-squares
SEE = standard error of estimate

Country group codes:

11 = Oil-exporting countries
12 = Asian countries excluding India
13 = Developing African countries
14 = Latin American countries excluding group 15
15 = Argentina, Brazil, Mexico
18 = Middle Eastern and North African countries excluding groups 11 and 13

*Table 6A.1.* Production functions for groups of developing countries ($t$-values in parentheses).[a]

| | Country group | | | | | | |
|---|---|---|---|---|---|---|---|
| *Variable* | *11* | *12*[b] | *India* | *13*[b] | *14* | *15*[b] | *18* |
| $t_c$ | 1978 | 1979 | | 1974 | | 1974 | |
| $a_1$ | 0.215 | 0.147 | 0.312 | 0.108 | 0.749 | 0.241 | 0.226 |
| | (1.43) | (14.97) | (10.79) | (17.05) | (19.75) | (2.58) | (4.38) |
| $a_2$ | 0.031 | 0.030 | 0.005 | 0.025 | 0.010 | 0.029 | 0.022 |
| | (3.97) | | (1.99) | | (2.05) | | (2.44) |
| $a_3$ | –0.048 | –0.025 | | –0.033 | | –0.015 | |
| | (3.75) | (5.23) | | (14.34) | | (2.86) | |
| $a_4$ | 0.245 | 0.344 | 0.445 | 0.458 | 0.457 | 0.426 | 0.423 |
| | (1.54) | (18.11) | (16.67) | (9.13) | (10.62) | (3.84) | (7.73) |
| $a_5$ | 0.535 | 0.491 | 0.489 | 0.436 | 0.439 | 0.409 | 0.457 |
| | (3.42) | (22.81) | (21.52) | (7.22) | (7.86) | (5.70) | (9.66) |
| $a_6$ | 0.220 | 0.165 | 0.066 | 0.106 | 0.105 | 0.165 | 0.120 |
| | (2.37) | (5.56) | (3.51) | (5.62) | (7.34) | (3.23) | (1.90) |
| DW | 1.45 | 1.95 | 1.53 | 2.23 | 1.46 | 2.34 | 1.21 |
| R2C | 0.99 | 1.00 | 0.99 | 1.00 | 1.00 | 1.00 | 0.99 |
| SEE | 4.154 | 0.914 | 1.125 | 0.215 | 0.472 | 3.178 | 0.911 |

[a]For the specification see equation (6.1) in the text. All calculations performed using the MIDIS method for the period 1961–1981.

[b]For groups 12, 13, and 15 $a_2$ was set to the estimated rate of technical progress until the switching point ($t_c$). For a more detailed description of the estimation procedure see the text.

*Table 6A.2.* Estimation results of the equation for velocity of money ($t$-values in parentheses).[a]

| *Country group*[b] | $a_1$ | $a_2$ | $a_3$ | $a_4$ | *DW* | *R2C* | *SEE* | *Est.* |
|---|---|---|---|---|---|---|---|---|
| 12 | 0.177 | | 0.185 | 0.459 | 1.60 | 0.93 | 0.165 | 62–81 |
| | (4.61) | | (2.08) | (5.69) | | | | MIDIS |
| India | 0.223 | 0.004 | 0.190 | –0.141 | 2.21 | 0.94 | 0.181 | 63–81 |
| | (6.02) | (2.10) | (2.73) | (1.11) | | | | MIDIS |
| 13 | 0.129 | | 0.325 | 0.296 | 1.86 | 0.89 | 0.175 | 65–81 |
| | (2.68) | | (2.69) | (3.35) | | | | MIDIS |
| 14 | 0.203 | 0.002 | 1.020 | 0.111 | 2.00 | 0.79 | 0.260 | 62–81 |
| | (7.45) | (4.24) | (3.21) | (0.58) | | | | MIDIS |
| 15 | 0.110 | | 0.174 | 1.011 | 1.84 | 0.76 | 0.349 | 62–81 |
| | (2.62) | | (2.56) | (4.02) | | | | MIDIS |
| 18 | 0.322 | | 0.283 | –0.093 | 2.02 | 0.96 | 0.091 | 62–81 |
| | (6.69) | | (2.03) | (1.39) | | | | MIDIS |

[a]For the specification see equation (6.4) in the text.

[b]Estimation results for group 11 (R2C = 1.00, SE = 0.0198, DW = 199):

$$p = \underset{(83.30)}{1.081} \cdot p_{IM\,(5.79)}^{0.554}\; p_{-1\,(4.14)}^{0.1966}\; p_{EX\,(8.32)}^{0.297}$$

*Table 6A.3.* Import demand functions for groups of developing countries ($t$-values in parentheses).[a]

| Variable[b] | Country group 11 | 12 | India | 13 | 15 | 15 | 18 |
|---|---|---|---|---|---|---|---|
| $a_1$ | 0.567 | 0.485 | 0.531 | 0.240 | 0.577 | 0.665 | 0.519 |
| | (5.67) | (8.58) | (4.63) | (2.99) | (7.26) | (5.15) | (6.94) |
| $a_2$ | 0.326 | 0.268 | 0.279 | 0.587 | 0.425 | 0.198 | 0.329 |
| | (1.89) | (3.27) | (3.59) | (6.43) | (5.80) | (2.97) | (5.48) |
| $a_4$ | 0.165 | 0.459 | | 0.374 | 0.233 | 0.400 | 0.579 |
| | (3.69) | (6.50) | | (7.38) | (3.44) | (2.70) | (6.76) |
| $a_5$ | 0.014 | 0.011 | 0.026 | | 0.041 | 0.052 | 0.048 |
| | (0.50) | (0.96) | (1.81) | | (4.39) | (4.40) | (4.84) |
| DW | 1.66 | 1.88 | 1.39 | 2.26 | 2.35 | 1.71 | 1.50 |
| Durbin's $m$ | 0.29 | 0.01 | 1.56 | 0.76 | 0.98 | 0.57 | |
| R2C | 0.99 | 1.00 | 0.82 | 0.99 | 0.99 | 0.98 | 1.00 |
| MAPE | 4.53 | 1.61 | 10.45 | 3.25 | 2.41 | 4.66 | 2.13 |
| APE:[c] | | | | | | | |
| short run | −0.485 | −0.556 | −0.481 | −0.771 | −0.456 | −0.374 | −0.511 |
| long-run | −1.119 | −1.079 | −1.026 | −1.014 | −1.078 | −1.115 | −1.056 |

[a] For the definition of the variables see the text. Specification:

$$IM = a_1 IM_{-1} + a_2 \frac{I_{p_I}}{p_{IM}} + a_4 \frac{EX_{p_{EX}}}{p_{IM}} - a_5 Y_{-1} \frac{p_{Y_{-1}}}{p_{IM}} + u$$

[b] Due to insignificance $a_3$ in equation (6.6) in the text was set to zero.
[c] APE = average price elasticity.

*Table 6A.4.* Estimation results for the structure of imports ($t$-values in parentheses).[a]

| *Country group* | *Commodity*[b] | $a_1$ | $a_2$ | $a_3$ | $a_4$ | $a_5$ |
|---|---|---|---|---|---|---|
| 11[c] | 1 | 0.569 | 0.120 | 0.093 | | 0.068 |
| | | (5.77) | (9.98) | (8.91) | | (5.99) |
| | 2 | 0.760 | 0.020 | 0.013 | | 0.011 |
| | | (7.82) | (3.47) | (3.21) | | (2.90) |
| 12 | 1 | 0.941 | 0.095 | 0.097 | 0.097 | 0.103 |
| | | (29.14) | (6.39) | (6.24) | (6.34) | (6.19) |
| | 2 | 0.975 | 0.050 | 0.052 | 0.051 | 0.054 |
| | | (25.87) | (2.09) | (2.10) | (2.10) | (2.10) |
| | 3 | 0.844 | 0.170 | 0.177 | 0.175 | 0.185 |
| | | (21..66) | (4.90) | (4.77) | (4.82) | (4.70) |
| India[c] | 2 | 0.340 | 0.214 | | 0.107 | 0.199 |
| | | (2.62) | (4.06) | | (1.93) | (3.67) |
| | 3 | 0.232 | 0.536 | | 0.232 | 0.498 |
| | | (2.00) | (5.55) | | (2.40) | (4.68) |
| 13 | 1 | 0.673 | 0.080 | 0.053 | 0.065 | 0.039 |
| | | (5.81) | (2.72) | (2.19) | (2.59) | (1.38) |
| | 2 | 0.658 | 0.021 | 0.016 | 0.017 | 0.011 |
| | | (4.82) | (3.73) | (3.38) | (3.71) | (1.66) |
| | 3 | 0.772 | 0.058 | 0.042 | 0.039 | 0.029 |
| | | (11.81) | (2.05) | (1.76) | (1.67) | (1.17) |
| 14 | 1 | 0.767 | 0.079 | 0.074 | 0.048 | 0.060 |
| | | (7.80) | (3.32) | (3.44) | (2.79) | (2.91) |
| | 2 | 0.922 | 0.018 | 0.015 | 0.011 | 0.014 |
| | | (19.05) | (3.48) | (3.61) | (2.81) | (3.07) |
| | 3 | 0.494 | 0.190 | 0.158 | 0.178 | 0.144 |
| | | (4.32) | (5.64) | (5.69) | (5.83) | (5.16) |
| 15 | 1 | | 0.120 | 0.050 | 0.105 | 0.019 |
| | | | (10.88) | (3.17) | (7.97) | (1.56) |
| | 2 | 0.413 | 0.047 | | 0.041 | 0.007 |
| | | (3.65) | (5.63) | | (5.49) | (1.65) |
| | 3 | 0.845 | 0.031 | | 0.013 | 0.005 |
| | | (13.98) | (1.85) | | (1.80) | (1.35) |
| 18 | 1 | 0.287 | 0.284 | 0.165 | 0.167 | 0.192 |
| | | (2.46) | (10.93) | (4.39) | (4.59) | (7.16) |
| | 2 | 0.532 | 0.085 | 0.034 | 0.050 | 0.057 |
| | | (4.40) | (5.03) | (2.25) | (3.65) | (4.12) |
| | 3 | 0.516 | 0.121 | 0.0499 | 0.070 | 0.082 |
| | | (4.47) | (3.38) | (1.86) | (2.60) | (2.76) |

[a] For specification see equation (6.7) in the text. All calculations performed using the FIML method for the period 1964–1983.

[b] Commodity groups are as follows: 1 = agriculture, 2 = manufacturing, and 3 = services.

[c] In the case of group 11 and India, the missing commodity group was left out of the demand system.

*Table 6A.4.* Continued.

| *Country group* | *Commodity*[b] | *DW* | *Durbin's m* | *R2C* | *SEE* |
|---|---|---|---|---|---|
| 11[c] | 1 | 1.63 | 0.30 | 0.99 | 0.394 |
| | 2 | (1.93) | 0.05 | 0.98 | 0.189 |
| 12 | 1 | 2.33 | 0.78 | 0.99 | 0.175 |
| | 2 | 1.38 | 1.56 | 0.99 | 0.271 |
| | 3 | 2.10 | 0.42 | 1.00 | 0.409 |
| India[c] | 2 | 1.22 | 1.62 | 0.80 | 0.211 |
| | 3 | 2.18 | 0.56 | 0.95 | 0.366 |
| 13 | 1 | 2.09 | 0.43 | 0.95 | 0.069 |
| | 2 | 1.91 | 0.04 | 0.98 | 0.011 |
| | 3 | 2.30 | 0.75 | 0.97 | 0.102 |
| 14 | 1 | 2.90 | 1.89 | 0.95 | 0.130 |
| | 2 | 2.20 | 0.38 | 0.98 | 0.028 |
| | 3 | 2.11 | 0.43 | 0.97 | 0.155 |
| 15 | 1 | 1.34 | 1.50 | 0.96 | 0.272 |
| | 2 | 1.67 | 0.28 | 0.98 | 0.130 |
| | 3 | 2.65 | 1.52 | 0.99 | 0.431 |
| 18 | 1 | 2.47 | 1.32 | 0.98 | 0.161 |
| | 2 | 2.24 | 0.40 | 0.94 | 0.116 |
| | 3 | 1.52 | 1.10 | 0.89 | 0.344 |

*Table 6A.5.* Estimation results for the export demand functions in the different commodity groups ($t$-values in parentheses).[a]

| *Country group* | *Commodity* | $a_1$ | $a_2$ | $a_3$ | *DW* | *Durbin's m* | *R2C* | *SEE* |
|---|---|---|---|---|---|---|---|---|
| 11 | 1 | 0.915 | 0.018 | 0.018 | 2.53 | 1.22 | 0.98 | 0.094 |
| | | (10.60) | (1.88) | (1.87) | | | | |
| | 2 | 0.645 | 0.061 | 0.053 | 1.98 | 0.04 | 0.97 | 0.212 |
| | | (3.60) | (4.75) | (4.23) | | | | |
| | 3 | 0.160 | 0.769 | 0.707 | 2.59 | 1.33 | 0.99 | 6.665 |
| | | (0.87) | (6.19) | (4.33) | | | | |
| | 4 | | 0.005 | | 1.22 | 1.57 | 0.95 | 0.622 |
| | | | (18.95) | | | | | |
| 12[b] | 1 | | 0.010 | 0.176 | 1.71 | 0.58 | 0.99 | 0.321 |
| | | | (2.22) | (9.26) | | | | |
| | 2 | 0.701 | 0.084 | 0.137 | 3.32 | 1.98 | 1.00 | 0.150 |
| | | (6.29) | (6.28) | (2.68) | | | | |
| | 3 | 0.534 | 0.011 | 0.153 | 2.13 | 0.25 | 1.00 | 0.204 |
| | | (2.35) | (3.09) | (6.01) | | | | |
| | 4 | 0.846 | 0.074 | 0.197 | 1.98 | 0.04 | 1.00 | 0.996 |
| | | (8.92) | (5.03) | (5.87) | | | | |
| India | 1 | 0.757 | 0.012 | 0.009 | 1.89 | 0.09 | 0.97 | 0.120 |
| | | (4.84) | (1.05) | (0.79) | | | | |
| | 2 | 0.450 | 0.005 | 0.002 | 1.67 | 0.31 | 0.92 | 0.074 |
| | | (3.81) | (4.77) | (1.89) | | | | |
| | 4 | 0.586 | 0.008 | 0.007 | 1.13 | 1.74 | 0.98 | 0.246 |
| | | (3.96) | (2.55) | (2.06) | | | | |
| 13 | 1 | 0.651 | 0.064 | 0.060 | 2.37 | 0.80 | 0.99 | 0.136 |
| | | (9.63) | (5.05) | (4.54) | | | | |
| | 2 | | 0.018 | 0.004 | 1.58 | 0.56 | 0.94 | 0.126 |
| | | | (2.44) | (0.53) | | | | |
| | 3 | 0.426 | 0.007 | 0.003 | 2.03 | 0.02 | 0.99 | 0.109 |
| | | (2.94) | (3.60) | (1.13) | | | | |
| | 4 | 0.821 | 0.011 | 0.011 | 2.67 | 1.34 | 0.85 | 0.325 |
| | | (7.47) | (2.70) | (2.63) | | | | |
| 14 | 1 | 0.458 | 0.038 | 0.024 | 1.64 | 0.38 | 0.97 | 0.274 |
| | | (2.28) | (1.48) | (0.91) | | | | |
| | 2 | 0.854 | 0.016 | 0.013 | 1.57 | 0.94 | 0.95 | 0.143 |
| | | (3.51) | (1.68) | (1.09) | | | | |
| | 3 | 0.558 | 0.014 | 0.008 | 2.81 | 1.67 | 0.99 | 0.203 |
| | | (3.84) | (3.62) | (2.52) | | | | |
| | 4 | 0.579 | 0.012 | 0.010 | 2.18 | 0.43 | 0.95 | 0.367 |
| | | (2.55) | (2.04) | (1.85) | | | | |
| 15 | 1 | | 0.084 | | 2.26 | 0.71 | 0.98 | 0.546 |
| | | | (65.78) | | | | | |
| | 2 | 0.620 | 0.019 | 0.014 | 1.98 | 0.04 | 0.98 | 0.271 |
| | | (5.23) | (3.37) | (2.12) | | | | |
| | 4 | 0.834 | 0.006 | 0.003 | 1.43 | 1.22 | 0.99 | 0.307 |
| | | (5.79) | (1.04) | (0.54) | | | | |

*Table 6A.5.* Continued.

| *Country group* | *Commodity* | $a_1$ | $a_2$ | $a_3$ | $DW$ | *Durbin's* $m$ | $R2C$ | $SEE$ |
|---|---|---|---|---|---|---|---|---|
| 18 | 1 | 0.817 | 0.010 | 0.009 | 1.55 | 1.07 | 0.90 | 0.088 |
| | | (7.71) | (1.19) | (1.07) | | | | |
| | 2 | 0.819 | 0.027 | 0.025 | 2.37 | 0.80 | 0.97 | 0.096 |
| | | (16.35) | (4.71) | (4.10) | | | | |
| | 3 | 0.936 | 0.012 | 0.011 | 2.30 | 0.73 | 0.99 | 0.137 |
| | | (8.21) | (3.94) | (2.89) | | | | |
| | 4 | | 0.002 | | 1.20 | 1.65 | 0.99 | 0.079 |
| | | | (60.07) | | | | | |

[a]For the specification see equation (6.8) in the text. All calculations performed using the MIDIS method for the period 1965–1982, except for Group 11 commodities 1 and 2 (1967–1982) and 3 (1967–1981).

[b]Variable $a_4$ results for Group 12 were obtained as follows: commodity 2 = 0.071 (6.14), commodity 3 = 0.018 (3.23), commodity 4 = 0.098 (5.67).

*Table 6A.6. Ex post* simulations, 1966–1981: mean absolute percentage errors.

| | *Country group* | | | | | | |
|---|---|---|---|---|---|---|---|
| *Variables* | *11* | *12* | *India* | *13* | *14* | *15* | *18* |
| GDP (real) | 4.9 | 1.3 | 1.2 | 2.9 | 1.0 | 3.9 | 2.8 |
| GDP deflator | 3.1 | 3.9 | 6.6 | 6.9 | 4.1 | 3.8 | 5.2 |
| Investment (real) | 4.9 | 1.3 | 1.2 | 2.9 | 1.0 | 3.9 | 2.8 |
| Consumption (real) | 5.9 | 3.0 | 1.8 | 3.1 | 1.6 | 6.4 | 3.8 |
| Imports (real) | 7.9 | 2.5 | 8.6 | 5.7 | 6.6 | 6.9 | 7.7 |
| Imports price index | 1.7 | 2.2 | 7.2 | 4.5 | 2.0 | 3.2 | 3.0 |
| Exports (real) | 9.3 | 3.5 | 10.8 | 7.4 | 5.2 | 9.7 | 11.6 |
| Exports price index | 3.2 | 2.1 | 5.7 | 4.7 | 5.8 | 6.7 | 5.6 |

## References

Chen, E.K.Y. (1979), *Hyper-growth in Asian Economics.* (London: Macmillan).

Driscoll, M.J. and Lahiri, A.K. (1983), Income Velocity of Money in Agricultural Developing Economies, *Review of Economics and Statistics*, **65** (3), pp. 393–401.

Welsch, H. (1986), Habit Persistence and the Structure of International Commodity Trade: A Demand System Analysis for Nine Industrialized Countries. University of Bonn [mimeo].

Welsch, H., (1987), An Aggregate Import Demand Model for Long-Term Projections, *Jahrbücher für Nationalökonomie und Statistik*, **202/203** pp. 372–389.

Wong, C. (1977), Demand for Money in Developing Countries, *Journal of Monetary Economics*, **3**(1), pp. 59-86.

CHAPTER 7

# Growth in an Interdependent World Economy: Linking the National Models through International Trade

*Rumen Dobrinsky and István Székely*

## Summary

This chapter analyzes the effects of different national growth patterns on world trade in real and monetary terms. The world model that underlies our estimations connects the national models via export and import functions in such a way that total world exports equal total world imports in monetary and real terms. The results of different simulation exercises are presented.

## 7.1. Introduction

One specific aspect of the Bonn–IIASA Project is the analysis of the interdependence of the world economy. It is a well recognized fact that national economies are more mutually dependent now than ever before and the world economy can no longer be regarded as a set of autarchic national entities. The interdependence of the world economy is manifested in a variety of forms from the more traditional, such as international trade and financial flows, to the more modern, such as technological transfer. A specific form of economic integration and the exercising of joint economic policies is the international economic alliance, such as the EC, CMEA, etc. Interdependence is a reality that cannot be neglected in the analysis of growth and structural change of the world economy and its components. On the other hand, as already mentioned, this very complex phenomenon can hardly be modeled completely in a study with the limited scope of ours.

In this chapter we analyze some aspects of the interdependence of the world economy in the context of growth and structural change, as reflected by our general approach. We put our attention here on the linking of the national models and the foreign trade models since the description of these models *per se* and the analysis of domestic factors of growth and structural change are presented in the special chapters devoted to those topics. Due to the limitations of our project, we concentrate only on the interdependence of national economic growth, on the one side, and international trade, on the other. Moreover, since the country models are basically supply-driven growth models, we can focus mainly on the effects of different domestic growth patterns on the world trade. An important aspect of our analysis is the interdependence of growth and structural change in real and monetary terms, and especially the effect of real growth on relative foreign trade prices and vice versa.

The tool we are using for these investigations is a world model that integrates the previously described country (or group) models. For manageability, these country models are to a certain extent simplified. The simplifications mainly relate to foreign trade, and they are discussed in Section 7.2. Since the model covers the whole world's trade (with assumptions made about the foreign trade of the countries not involved in the model), it is designed to ensure that world exports equal world imports in both real and nominal terms. The general idea behind and the specification of the subsystem describing the foreign trade flows, and ensuring that these consistency conditions are satisfied, are also given in Section 7.2. Finally, in Section 7.3 the results of the different simulation exercises are presented. Since our model is designed for medium- and long-term forecasts, the results presented in this section can also highlight the different nature of short-, medium- and long-term effects. Better understanding of these aspects is essential for an economic policy aimed at balanced growth.

Knowing the limitations of such models, it is our responsibility to emphasize not only those aspects where our model can be a useful device, but also the reasons for a cautious interpretation of the results presented here. First of all, it should be mentioned that because the world model describes only the aggregate export–import flows, changes in the (commodity) structure of foreign trade are not explicitly reflected. The rather simple form of exchange rate determination used in the model should also be mentioned. All this is to suggest that the simulation results presented here and the consequences drawn from them indicate certain tendencies rather than describe exactly the future development of the world economy.

## 7.2. The Integrated System of International Trade

### 7.2.1. Basic propositions

The basic idea of our approach is to reflect the existing interdependence of the world economy in a simultaneously interdependent world model. Since in our approach the country models are linked by their corresponding trade flows, the appropriate tool for this analysis is an integrated system of international trade which provides a simultaneous link between generated real world trade flows and foreign trade prices and domestic developments (growth and monetary changes) in the countries (or groups of countries). Two basic approaches have emerged from the past experience: the bilateral (on the basis of trade or trade share matrices) approach and the pooling approach.[1]

However, our Integrated System of International Trade (ISIT) was specifically designed to meet some particular requirements of the Bonn–IIASA Project. The first and, perhaps, the most important point is that our system is used for long-term projections. Stemming from that fact, though maybe not apparent at first sight, is the essential requirement for complete internal consistency of the foreign trade subsystem, i.e., total world exports must sum up to total world imports in both nominal and real terms at any time point. Whereas in short-term projections one can afford to neglect some aspects of internal consistency, arguing that the generated discrepancies are comparable in size with the statistical discrepancies arising from the imperfect measurement of the world trade,[2] in long-term projections in the presence of dynamic relations the generated discrepancies may become intolerably high if no special measures are taken to avoid them. (In our work we experienced the validity of this statement.) Besides, the consistency conditions have to be incorporated as a property of the linking system in the design stage, and this is by no means a trivial task.

Stemming from the long-term character of our approach, we focused our attention on the stable or permanent interdependency links, taking the liberty of sometimes neglecting links that we thought would cause mainly short-term fluctuations.

Another basic requirement that had to be fulfilled is the simultaneous interdependence of domestic growth and foreign trade: domestic output (GDP or NMP) of the countries is generated at the same time, exercising its influence on, and being influenced by, international trade (monetary effects being present as well). Whereas the domestic part of the country models is basically supply-driven, we thought it indispensable to incorporate some principles of a supply-and-demand equilibrium in the system of international trade as a market clearing mechanism to determine the world foreign trade prices.

Last, but not least, is the requirement to produce an easily solvable and manageable model. The result, described here, is a compromise between these technical limitations (and available resources!) and the desire for more analytical power. Due to the limitations imposed, some important aspects of international trade are not reflected in the linking system. Some elements of the domestic country models were also excluded from the linked world model. So the system described in this chapter should be regarded as a prototype model.

Following the general lines of the Bonn–IIASA project, we assume that the main driving forces for international trade are the primary domestic factors of growth in the countries or groups of countries. However, we differentiate the directions of influence of these factors in order to realize the principle of supply-and-demand equilibrium of international trade, as indicated earlier. Thus, imports are (mainly) *demand-driven*, the volume and value of any country's imports being determined on the basis of the projected domestic growth and the relative price of imported goods. The generated total world (nominal) imports are then distributed as exports among the countries, the relative export prices being an essential factor for this distribution. Export prices, in their turn, are basically *supply-driven*, the export price functions being in fact inverted export supply functions (see Chapter 8 in this volume). Import prices are then determined on the basis of the average export prices of the main trading partners. [From now on we shall only speak of aggregated imports (exports), meaning the totals of all countries' imports (exports) of goods and services. The same refers to the corresponding price deflators.] So there is a full symmetry in the two sides of the foreign trade system: the generated import demand is later converted into exports, whereas the generated export prices are subsequently converted into import prices.[3] This approach is applied in its pure form to the models of the developed market economies. Some specific features of the export price determination of the CMEA and developing countries (or groups of countries) are discussed later.

The system of international trade is balanced when the total world imports are equal to the total world exports both in nominal and in real terms, with a tolerance level of the order of the observed statistical discrepancy. (One of these conditions may be replaced by the requirement that the general world export price deflator be equal to the general world import price deflator.) Since our system is designed so that the value condition is fulfilled by definition, it remains to guarantee the volume condition. The role of the equilibrating market clearing power in our system is played by import prices, which are scaled in such a way as to provide the required consistency condition.

Another closely related problem is that of currency conversion in the model. Consistency conditions on the world level can be implemented only in the presence of a numeraire unit of valuation of world trade. In our model (as it is usually assumed) this is the US dollar – in current and 1975 prices. The conversion problem connected to the modeling of US dollar exchange rates is treated in the following way in our model. Exchange rates (actually, indexes of the exchange rates) of the developed market economies (and some of the developing countries) are modeled in accordance with a weak purchasing power parity (PPP) assumption. In spite of the controversy surrounding this approach, it does not seem unreasonable for long-term projections. Our analysis of the yearly exchange rates figures from 1971 (the period of free fluctuation) shows that the deviation of the exchange rates indices from the true PPP indices is within the limits of a trend. A similar finding is reported in Klein (1983). However, it must be kept in mind, that the PPP assumption also means that we shall not deal with the short- and medium-term regulating role of the exchange rates. (A more sophisticated approach to the modeling of exchange rates is presented in Chapter 8 in this volume.) As for the exchange rates of the CMEA and some developing countries, they were estimated as some geometric averages of a PPP index and a lagged value (see Section 7.2.2).

Given this type of exchange rate and the character of the pricing subsystem, it follows that the main driving forces of structural changes in the foreign trade prices in our model are: the inflation rate in the USA (following our general approach, in the final run it is dominated by the growth rate of money supply in the USA) and export prices of the developed market economies (especially the characteristics of their long-term performence).

### 7.2.2. Specification of the system

The two pillars of the foreign trade system, embodying the principles of supply and demand, are the import demand and the export price functions.

The *import demand function* is based on a dynamic version of the Linear Expenditure System. The general form of the equation is:

$$\begin{aligned} IM_i = a_1 IM_{i,-1} + a_2 \frac{I_i \cdot P'I_i}{P'IM_i} + a_3 \frac{C_i \cdot P'C_i}{P'IM_i} \\ + a_4 \frac{EX_i \cdot P'EX_i}{P'IM_i} - a_5 \frac{GDP_{i,-1} \cdot P'GDP_i}{P'IM_i} + u \end{aligned} \tag{7.1}$$

where $IM$ denotes real imports; $I$, $C$, and $EX$ are real investment (gross capital formation), real consumption, and real exports, respectively; $GDP$ stands for real GDP; and $P'IM$, $P'I$, $P'C$, $P'EX$ and $P'GDP$ are the corresponding price deflators, all expressed in domestic currency terms. Subscript $i$ refers to the country under consideration. The theoretical derivation of this general form can be found in Welsch (1987). The specific forms used and the estimation results for the the different groups of countries are presented in the previous chapters describing the country models.

The *export price function* for the developed market economies is conceptually an inverted export supply function:

$$P'EX_i = a_1 P'GDP_i^{a_2} P'IM_i^{a_3} P'IM_{i,-1}^{a_4} P'EX_{i,-1}^{a_5} + u \tag{7.2}$$

The theoretical derivation of this specification is based on the profit-maximizing behavior of a representative firm, described in detail in Chapter 8. The estimation results for the developed market economies can also be found there.

For the CMEA and developing countries, price-taking behavior was assumed:

$$P'EX_i^{\$} = a_1 P'EX_{i,-1}^{\$}{}^{a_2} P'EX_{W1}^{\$}{}^{a_3} + u \tag{7.3}$$

where subscript $W1$ refers to the developed market economies and superscript \$ indicates that the price deflators are expressed in US\$ terms. For the oil-exporting developing countries, instead of the export prices of the developed market economies, world market oil prices were taken:

$$P'EX_i^\$ = a_1 P'EX_{i,-1}^\$ {}^{a_2} P_{oil}^\$ {}^{a_3} P_{oil,-1}^\$ {}^{a_4} + u \tag{7.4}$$

The estimation results for the developing countries are reported in Chapter 6.

According to our general approach described above, the subsystem generating the *nominal exports* of the countries and groups is meant to distribute the pooled import demand. The framework applied is the Linear Expenditure System. Assuming habit persistence and using the same reasoning as in the determination of the import demand functions, the following form can be derived:

$$EX'N_i^\$ = a_{1i} P'EX_i^\$ EX_{i,-1}^\$ + a_{2i} IM'N_W^\$$$

$$- \sum_{\substack{k=1 \\ k \neq i}} (a_{k+2,i} P'EX_k^\$ EX_{k,-1}^\$) + u \tag{7.5}$$

$$\sum_{i=1}^{n} a_{2i} = 1 \quad \text{and} \quad \frac{a_{1i}}{1 - a_{2i}} = \frac{a_{i+2,k}}{a_{2k}} \quad i,k = 1,\ldots,n \quad i \neq k$$

where $IM'N_W^\$$ is the sum of the imports of the countries involved in the model plus the net imports of the rest of the world; $EX'N^\$$, $EX^\$$, and $P'EX^\$$ are nominal and real exports and export price index in US\$ terms, respectively; subscript $i$ refers to the country under consideration; and subscript $k$ to those different from the one under consideration. The restrictions on the parameters are to guarantee that the sum of exports equals the sum of imports.

The equations were estimated by the nonlinear FIML method. Due to the limitations of our software package (IAS Bonn System) the system was reformulated in the following way: in the first step world imports were distributed among the USA, the FRG, Japan, the rest of the developed market economies, the CMEA countries as a whole, and the developing countries as a whole; in the second step the latter three were further distributed among the countries and groups involved. The results presented in *Table 7A.1* in the Appendix refer to the first step, and those presented in *Tables 7A.2* to *7A.4* to the second one.

This export determination guarantees the first consistency condition – namely, that total world exports in nominal terms $EX'N_W^\$$ are equal to total world imports in nominal terms $IM'N_W^\$$.

The second consistency condition requires that the total world imports should equal total world exports in real terms as well:

$$IM_W = \sum_i \frac{IM'N_i^\$}{P'IM_i^\$} = \sum_i \frac{EX'N_i^\$}{P'EX_i^\$} = EX_W^\$ \tag{7.6}$$

It is obvious that this condition depends crucially on the relation between the export and import price vectors. If no special measures are undertaken, equation (7.6) will not necessarily hold. One possible way out would be to replace the estimated import price equations by direct calculation of a consistent import price vector. [4] However, this normally requires the endogenous determination of the complete trade share matrix and implies pure price-taking behavior on the import side – two problems that we wanted to avoid. In principle one could try to estimate the system of import price equations jointly imposing cross-equation restrictions which would guarantee that equation (7.6) holds. However, given the present state of the art, this hardly seems possible, taking into account that the system should perform in the structually changing environment of world trade.

We have chosen another approach to deal with the problem. We introduce into the system two vectors of import prices:

(1) A vector of prices $P'M_i^\$$ which are explained by the average export price of the countries of origin [see equations (8.28)–(8.29) in Chapter 8]:[5]

$$P'M_i^\$ = a_1 + a_2 P'IM_{i,-1}^\$ + a_3 PM'EX_i^\$ + a_4 PM'EX_{i,-1}^\$ + u \tag{7.7}$$

For the CMEA and developing countries, equation (7.7) is expressed and estimated in US$ terms which reflects price-taker behavior; for OECD countries, equation (7.7) is expressed and estimated in domestic currency which reflects a mixture of price-maker and price-taker behaviors (i.e., $PM'EX_i^\$$ are adjusted by the exchange rates).

The prices $P'M_i^\$$ reflect the mechanism of formation of the import prices, but they do not necessarily satisfy equation (7.6).

(2) A vector of prices $P'IM_i^\$$ which are scaled in such a way that the condition (7.6) holds. $P'IM_i^\$$ are calculated from $P'M_i^\$$ by using a scaling factor $d_p$ which is the same for all countries:

$$P'IM_i^\$ = d_p \cdot P'M_i^\$ \tag{7.8}$$

$P'IM_i^\$$ are taken as the actual import prices in the model. The scaling factor $d_p$ is defined as:

$$d_p = \frac{1}{EX_W^\$} \cdot \sum_i \frac{IM'N_i^\$}{P'M_i^\$} \tag{7.9}$$

It is easy to check that, if the import prices are determined from (7.8) and (7.9), then equation (7.6) holds:

$$IM_W^{\$} = \sum_i \frac{IM'N_i^{\$}}{P'M_i^{\$} \frac{1}{EX_W^{\$}} \sum_i \frac{IM'N_i^{\$}}{P'M_i^{\$}}}$$

$$= EX_W^{\$} \frac{1}{\sum_i \frac{IM'N_i^{\$}}{P'M_i^{\$}}} \sum_i \frac{IM'N_i^{\$}}{P'M_i^{\$}} = EX_W^{\$} \qquad (7.10)$$

The reasoning behind this approach is as follows: from a technical point of view, we would like to determine the vector of consistent prices (7.8) simultaneously in one solution step. However, due to interdependence in the system, a change in the import prices, such as the scaling, causes repercussions throughout the whole system, and the solution can be obtained only iteratively. During this iterative process, the prices $P'M_i^{\$}$ remain little changed, while $P'IM_i^{\$}$ and the scaling factor $d_p$ (its initial value is 1) iterate until consistency is reached. It can be shown (and this is also borne out by our experience) that, due to the high degree of simultaneity, the attempt to iterate only one vector of import prices will not necessarily guarantee complete consistency.

One attractive feature of this approach is that the iterative process can be simulated within the Gauss–Seidel algorithm, which is used to solve the whole world model.

It remains to be said that although, at first glance, the price scaling may seem to be a purely technical operation, it has a fine theoretical interpretation. Actually, remembering that our import equations reflect import *demand* and our export prices reflect export *supply*, the iterative codetermination of the foreign trade prices and quantities is analogous to the familar cobweb market clearing trajectory, only proceeding in a multi-dimensional space. Its performance can be illustrated graphically with respect to total world imports and exports (real) $IM'^{\$}_W$ and $EX_W^{\$}$ and the corresponding prices $P'IM_W^{\$}$ and $P'EX_W^{\$}$ (see ***Figures 7.1*** and ***7.2***; for simplicity, we omitted the \$ sign from the notation of the figures).

The numbers (1,2,...,$n$) refer to the iterations. Within each iteration we have one value of the total world nominal imports $IM'N_W^{\$}$ which by definition is equal to the nominal world exports $EX'N_W^{\$}$. These values for each iteration are presented by their isoquants. However, when the real parts are not balanced, the import demand and the export supply are represented by different points on the isoquants. For example, for iteration 1 these are the points $I^1$ ($IM_W^1$, $P'IM_W^1$) and $E^1$ ($EX_W^1$, $P'EX_W^1$). This determines the scaling factor $d_p$ [see equation (7.9)] and the price vector for the next iteration [see equation (7.10)]. When the prices change, we get a new value for $IM'N_W = EX'N_W$ and two new points on its isoquant – $I^2$ and $E^2$ and so on. During these iterations the points $I^1$, $I^2$,..., define a hypothetical "world import demand" curve (as a

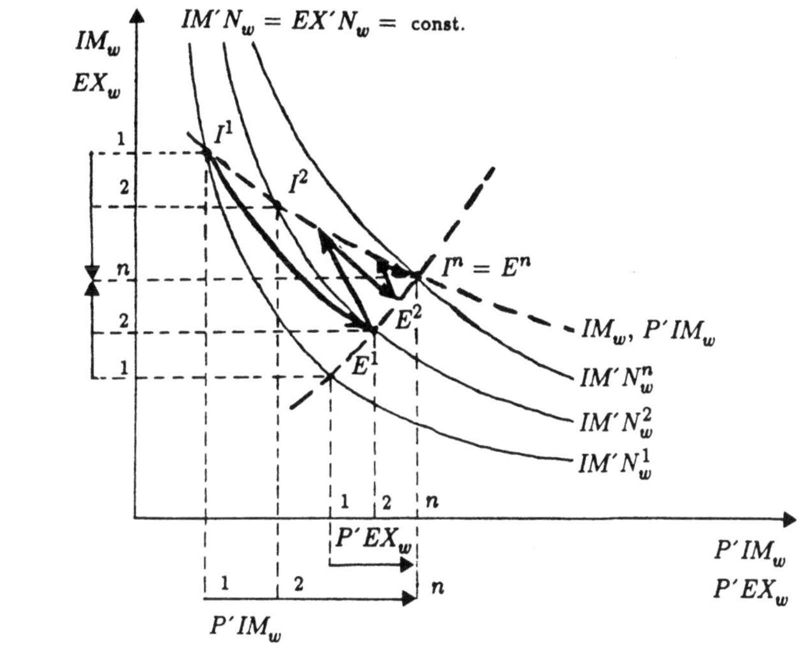


*Figure 7.1.* Iterative determination of equilibrium – first version.

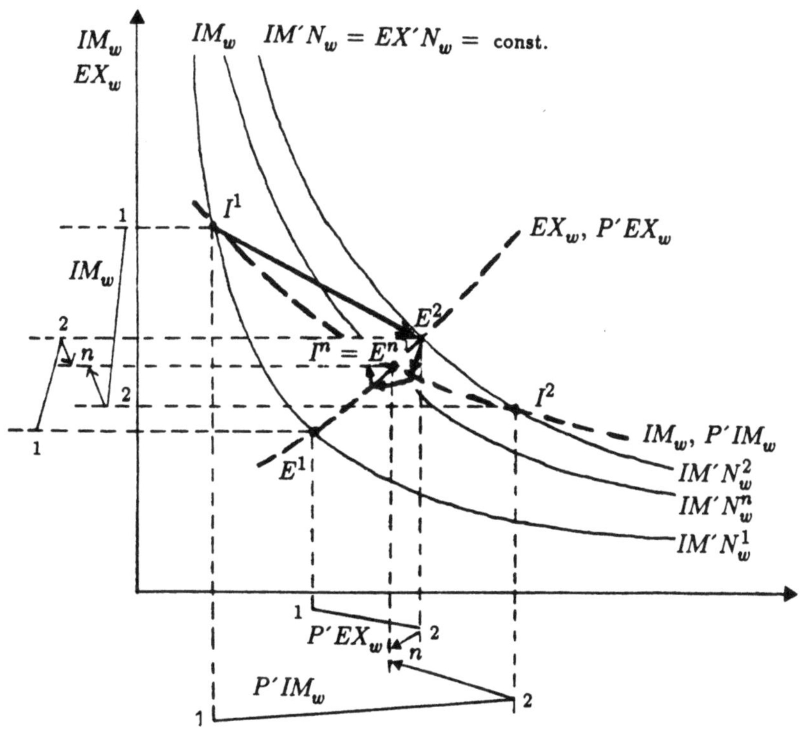


*Figure 7.2.* Iterative determination of equilibrium – second version.

price–quantity relation) while $E^1, E^2, \ldots,$ define in a similar way a "world export supply" curve. These two curves are indicated with the broken lines. Finally in our movement we reach the equilibrating point of intersection $I^n = E^n$, which provides our consistency condition (7.6). The convergence of this procedure can be proved for our system under some natural assumptions.

*Figures 7.1* and *7.2* reflect two possible types of iterative movement with respect to quantities and prices. However, they both refer to one and the same initial condition of $I^1$ and $E^1$. The reverse initial location of $I^1$ and $E^1$ is also possible: it is symmetrical to the one shown. As one can see, *Figure 7.2* is identical to the cobweb model.

Finally, we can also interpret the direction of movement within the iterative process (whether we get the equilibrium point to the left or to the right of the initial isoquant). This depends on the changing structure of the world trade and reflects the changing relative importance of the countries in the formation of world trading prices. As follows from the determination of our scaling factor $d_p$, a movement to the *right* of the initial curve corresponds to a situation when countries with relatively high export (and/or import) prices (i.e., those whose prices grow faster than the world average) *expand* at the same time their share in the world trade. A movement to the left corresponds to the reverse situation. The cobweb type of movement will take place when a combination of the above occurs within subsequent iterations. So the feedback mechanism built into our iterative procedure and illustrated on the figures actually reflects the spillover effect of the structural changes of world trade on average trading prices.

The conversion of national currency units to US$ and vice versa is performed on the basis of the US$ exchange rates which, in our system, are modeled in their index form $I'FX_i$. For OECD (and some developing) countries, a weak PPP form for the indices of the exchange rates is used:

$$I'FX_i = I'FX_{i,-1} \cdot \left( \frac{1 + W'P_i}{1 + W'P_{USA}} \right) \tag{7.11}$$

where $W'P_i$ and $W'P_{USA}$ are the rates of change of the general price level in country $i$ and in the USA, respectively.

For the CMEA (and some developing) countries, the indices of the exchange rates [6] are estimated as some geometric averages of a lagged value and the expected weak PPP value (but formulated with respect to the average foreign trade price):

$$I'FX_i = a_1 I'FX_{i,-1}{}^{a_2} \cdot \left[ I'FX_{i,-1} \cdot \left( \frac{1 + W'P_i}{1 + W'P^{\$}_{FT}} \right) \right]^{a_3} \tag{7.12}$$

where $W'P^{\$}_{FT}$ is the rate of change of the average prices of foreign trade (imports plus exports, in US$ terms). This specification reflects a habit persistence in the formation of the actual level of the exchange rates.

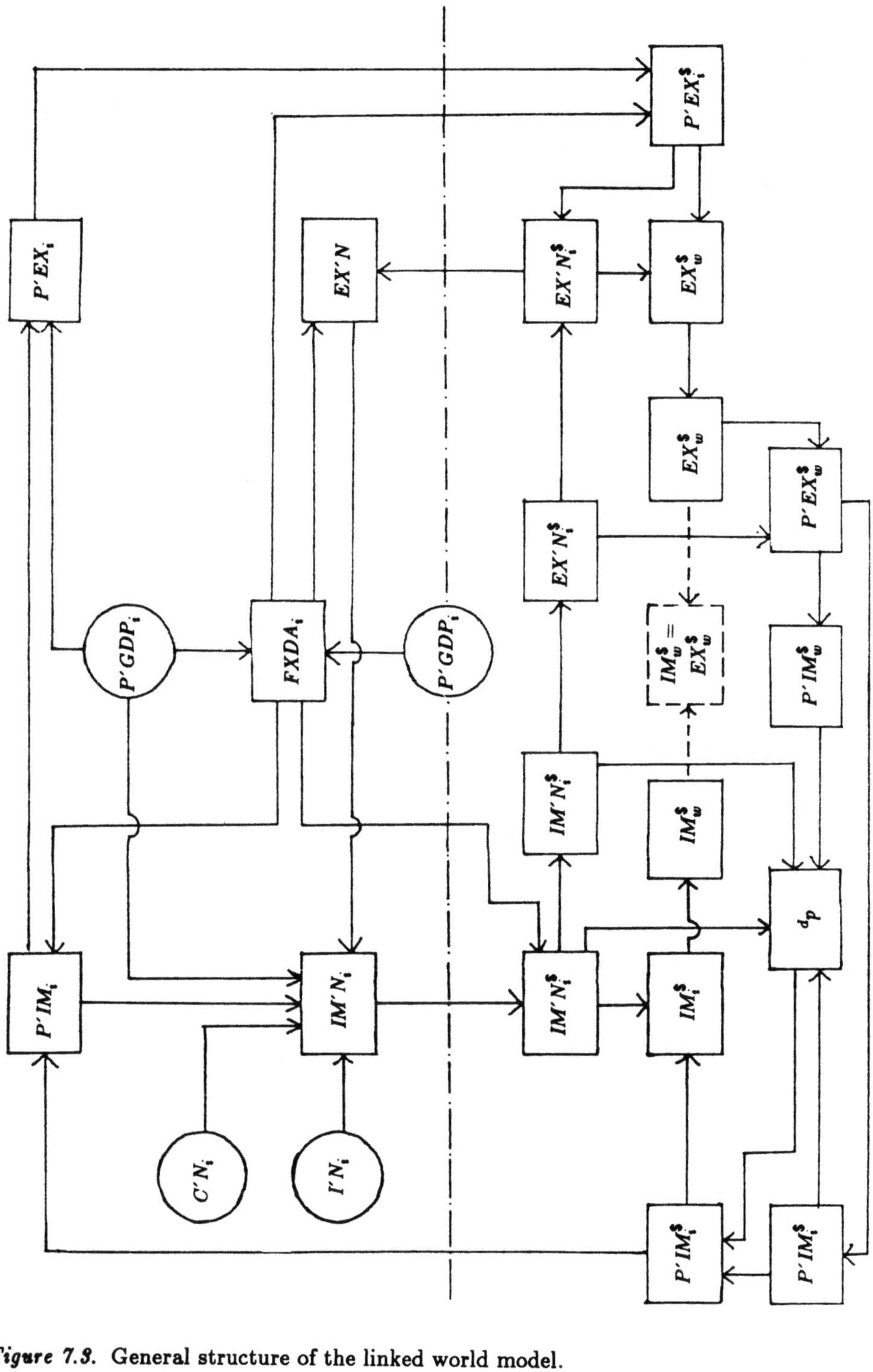


***Figure 7.3.*** General structure of the linked world model.

To illustrate the empirical work connected with the implementation of our approach, we present in Appendix 7A the complete estimation results of the foreign trade subsystem for the CMEA countries which is used in the ISIT. *Table 7A.5* reports on the estimated demand functions, *Table 7A.6* the export price equations, *Table 7A.7* the import price equations and *Table 7A.8* the estimated equations for the indices of the exchange rates (the indices of the conversion factors). The export equations for the CMEA countries are contained in *Table 7A.3*.

The general structure of the linked world model is illustrated on *Figure 7.3*. The upper part (above the axis) corresponds to the domestic part: the models of the countries and groups of countries. The lower part (below the axis) is actually the ISIT. Circles denote variables that are not simultaneously determined within the ISIT.

There are several important loops in the model. Import demand (in domestic currency) $IM'N_i$ is converted into US$ terms $IM'N_i^\$$ on the basis of the exchange rate $FXDA_i$. Then all $IM'N_i^\$$ are summed up to form $IM'N_W^\$$ ($= EX'N_W^\$$). Total world exports are distributed among the countries $EX'N_i^\$$ on the basis of the export prices $P'EX'^\$_i$. Then the countries' exports are converted into domestic currency $EX'N_i$ and the latter is one of the explanatory variables in the import demand equation. The desired export prices $P'EX_i$ (in domestic terms) are generated on the basis of the general price level in the country $P'GDP_i$ and the import prices $P'IM_i$ (in domestic terms). $P'EX_i$ are then converted into US$ terms – $P'EX_i^\$$. The latter, as mentioned, drive the distribution of the total exports and, simultaneously, explain the formation of import prices $P'M_i^\$$ (unscaled). This link in *Figure 7.3* is indicated only schematically through the general world export price level $P'EX_W^\$$. Actually, a more sophisticated link exists in this part of the model in which the introduced "average export prices abroad" $PM'EX_i^\$$ are determined individually for each country. The $P'M_i^\$$ vector is then "scaled" by the scaling factor $d_p$ to form the actual import prices $P'IM_i^\$$ (in US$ terms), and the latter are converted into domestic currency $P'IM_i$. The iterative scaling of the import prices (the scaling factor $d_p$ is also endogenously determined) guarantees that in real terms the total world imports $IM_W^\$$ are equal to the total world exports $EX_W^\$$.

## 7.3. Simulation Experiments with a Linked Model of the World Economy

### 7.3.1. The model and its *ex post* performance

The linked model of the world economy is constructed in accordance with the general principles discussed in the previous section. It comprises the models of the 24 countries and groups of countries of the Bonn–IIASA Project (with some simplifications) which are connected among themselves by the ISIT. The model also computes aggregates of some important variables (e.g., GDP, imports, exports, etc.) for the world total (actually, the part covered by our model) and three subtotals: "OECD total" (the 9 OECD countries treated separately within the project plus Group 10 which aggregates the rest of the rest of the developed market economies); "CMEA total" (the 7 European CMEA countries); "Developing total" or "LDC total" (the 7 groups of developing countries).[7]

All in all, the model contains 1069 endogenous variables (equations) and 332 exogenous variables. Most of the latter are auxiliary variables. Actually, we use three main sets as scenario variables: rate of technical progress, savings ratio and growth rate of money supply (for the CMEA countries we use growth rate of nominal wages).

The performance of the model was tested in a number of *ex post* simulation runs. It should be mentioned that due to the rather simplified modeling of the exchange rates of the OECD countries (which are appropriate for the long-term but not for the short-term simulations), we decided to exclude the exchange rate determination from these tests. So the *ex post* simulation results reported here are based on a version of the model where all exchange rates are taken exogenously as the actual observations.

*Tables 7.1.* and *7.2* give a very brief summary of the *ex post* results from a simulation run for the period 1971–1981. As the performance of some components of the model is analyzed in other chapters, we focus our attention here

*Table 7.1. Ex post* simulation, 1971–1981: mean absolute percentage errors of world totals and subtotals.

| *Variable* | *OECD* | *CMEA* | *LDC* | *World* |
|---|---|---|---|---|
| GDP (real) | 0.5 | 0.8 | 2.7 | 0.3 |
| Imports (real) | 1.9 | 2.4 | 6.5 | 1.4 |
| Exports (real) | 2.1 | 6.1 | 11.6 | 1.4 |
| GDP price index | 1.6 | 1.1 | 2.0 | 1.0 |
| Import price index | 2.7 | 1.8 | 2.0 | 1.9 |
| Export price index | 1.3 | 2.8 | 4.6 | 1.9 |

*Table 7.2. Ex post* simulation, 1971–1981: mean absolute percentage errors of export part, OECD countries.

| | *Variable* | | |
|---|---|---|---|
| *Country* | *Real exports* | *Nominal exports* | *Export price index* |
| USA | 4.0 | 6.9 | 3.0 |
| FRG | 1.8 | 2.9 | 1.8 |
| Japan | 6.8 | 6.3 | 1.7 |
| France | 7.4 | 6.5 | 2.1 |
| UK | 9.2 | 5.8 | 4.1 |
| Italy | 3.7 | 3.4 | 2.6 |
| Netherlands | 7.3 | 8.1 | 1.9 |
| Belgium/ Luxembourg | 7.9 | 8.6 | 2.2 |
| Canada | 4.7 | 6.4 | 3.0 |
| Rest of DMEs | 1.9 | 2.3 | 1.4 |

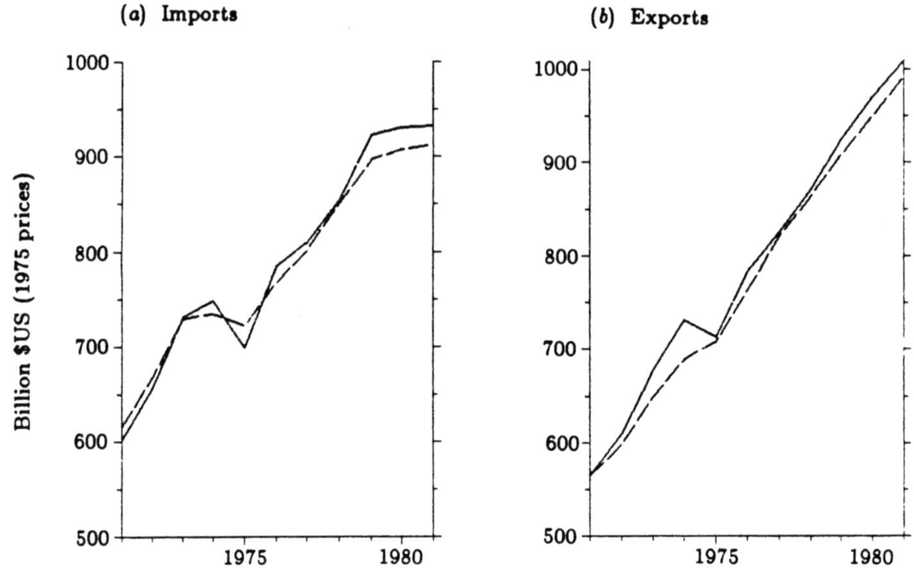


*Figure 7.4.* OECD: real imports and exports (solid lines = observed data; broken lines = simulated).

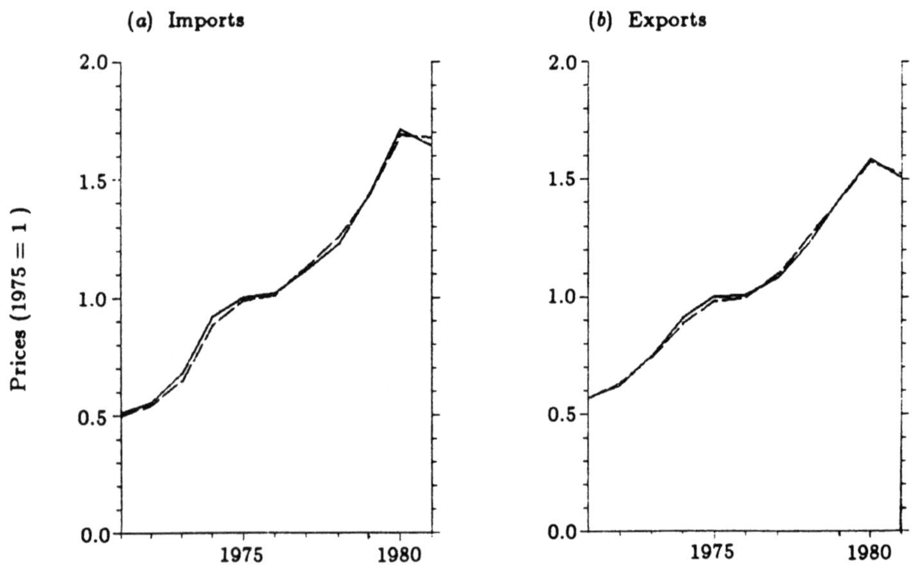


*Figure 7.5.* OECD: indexed import and export prices (solid lines = observed data; broken lines = simulated).

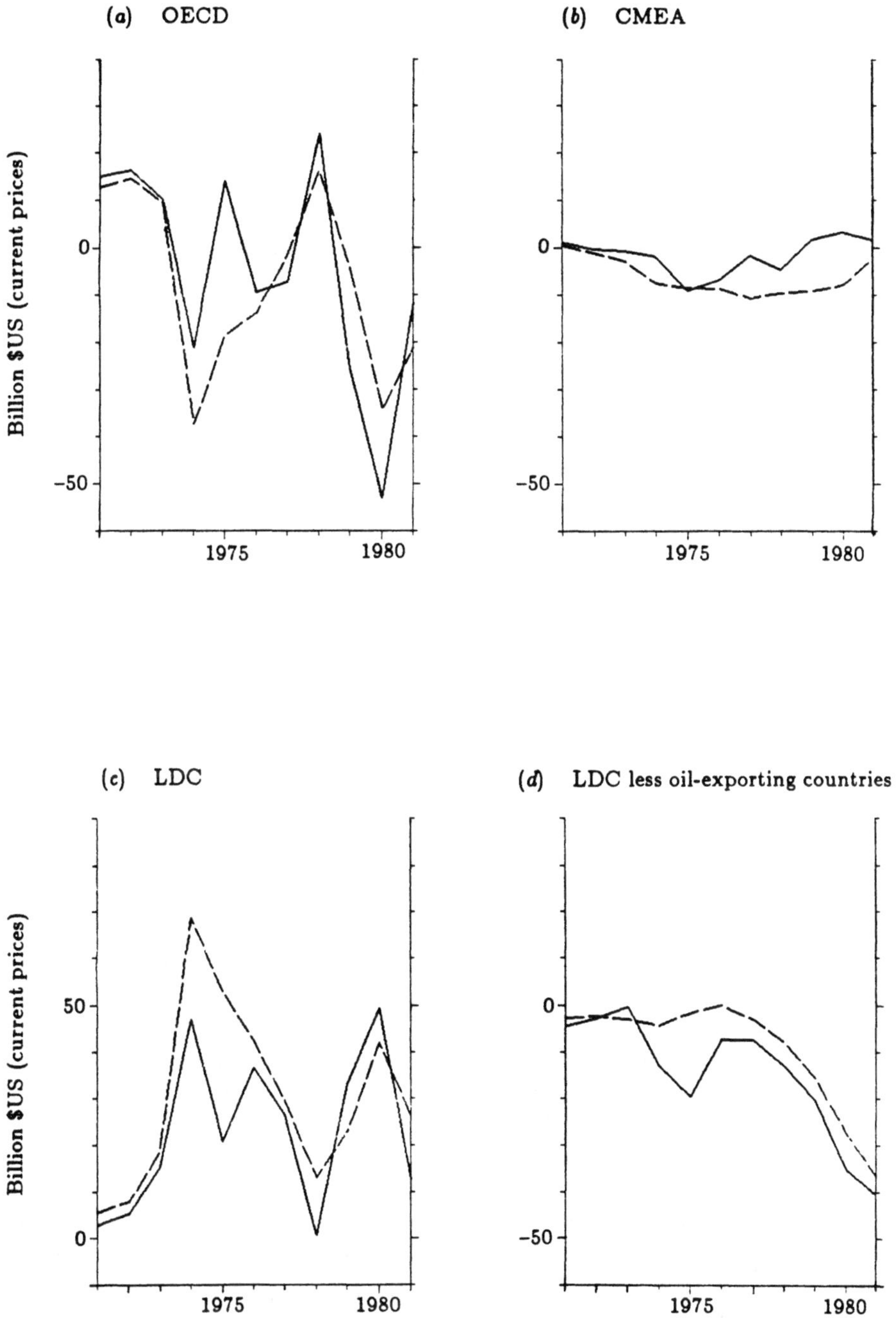


*Figure 7.6.* Foreign trade balances by country group (solid lines = observed data; broken lines = simulated).

mainly on the world totals and subtotals which are calculated in the linked model (*Table 7.1*) and on the performance of the exports subsystem which was designed and estimated especially for this model (this is illustrated in *Table 7.2* by the simulation results for the OECD countries). The level of the MAPEs reported in these tables can be rated as rather satisfactory, taking into consideration the complexity and the large scale of the model as well as the fact that the simulation period covers the two oil shocks.

A closer insight into the *ex post* performance is given by *Figures 7.4–7.6*. *Figures 7.4* and *7.5* illustrate the *ex post* performance with respect to the OECD foreign trade aggregates: total imports and exports (real) and the corresponding price indices. *Figure 7.6* presents the simulation results of a very sensitive indicator – the foreign trade balance, in current prices for (*a*) OECD, (*b*) CMEA (*c*) LDC and (*d*) LDC excluding group 11 (the oil-exporting countries). One can see that the model was capable of capturing the main developements in this important (and difficult to trace!) indicator.

A special simulation run was designed to test the efficiency of the price-scaling mechanism, which contributes to the consistency of the foreign trade system. It was identical to the one discussed above except for the fact that the price-scaling module was "switched off" during its execution. This test run indicated that the balancing procedure alone substantially improves the *ex post* performance of the model. Thus, the MAPE for real world exports in the "unbalanced" run was 2.8% as compared with 1.4% in the "balanced" run; for real world imports these figures are 1.7% and 1.4%, correspondingly; for the world import price index, they are 2.9% and 1.9%, etc. This confirms the importance of the consistency conditions for the adequate long-term simulation of world foreign trade.

### 7.3.2. Interdependence and cross-country effects

As was stated in the very beginning one of our goals is to analyze the international propagation of effects to indicate the interdependence of the world economy. The purest indicators of the cross-country transmission mechanism in a system like ours are the cross-country multipliers. They are analyzed in this section.

Several introductory comments have to be made. In general, the cross-country multiplier measures the induced response of an endogenous variable of a country to a change in the level of an exogeneous variable of another country. The variables where the change is introduced are usually scenario- or policy-related. Following the general approach of the Bonn–IIASA Project, we have selected two exogeneous variables for this purpose: the level of technical progress (one of the most important driving forces of economic growth) and the level of money supply (the exogenous monetary driving force). Further, we analyze the effects of two types of changes in these variables: (1) a single "shock" in the *level* of the variables taking place in one selected year; (2) a sustained shift in the *growth rate* of the variables for the whole simulation period. The latter form of change (which reflects a shift to another growth path for the variable concerned) was specifically chosen taking into consideration the long-term character of our study.

For most of the variables analyzed, we calculated the dynamic multipliers in level form (which we denote as $m_y^t$):

$$m_y^t = \frac{y_s^t - y_c^p}{y_c^t} \cdot 100\% \tag{7.13}$$

where $y_c^t$ is the value of $y$ in some control solution; $y_s^t$ is the value in the shocked or shifted solution. In all cases we used as the control solution the medium scenario of the linked model (see Annex 3); the shock year (or the initial year of the shift) was always 1985. As the system is highly nonlinear and dynamic, the multipliers are calculated from numerical simulations within the system.[8]

From the solution the multipliers $m_y^t$ are calculated in a form of a time series for the whole simulation period. Although the shape of the multiplier curve is also of interest, we concentrate mainly on the *short-term effect* (the value of the multipliers in the initial year of the shock or shift, i.e., 1985) and on the *long-term effect* (the value of $m_y^t$ in the last simulation year – in our case, 1999). All the shock multipliers have been calculated for a 10% shock level in 1985. Accordingly, the values of these multipliers in 1985 divided by 10 will also be exactly the value of the dynamic elasticity multiplier for this year.[9]

To study the transmission effects in the linked model, we simulate a shock or shift in one country and calculate the multipliers for all countries. So, actually, we have two types of multipliers: the "own" multiplier (the induced response of the country's own variables) and the "cross-country" multipliers (the induced response of the other countries' variables). However, the own multiplier accounts for both the strictly own response and also for the feedback effect from the induced response of other countries to the shocked country.

In *Tables 7.3* to *7.7* we present some of the calculated dynamic multipliers. For lack of space, we can show only a selection from these results. We have chosen to show the measured response of the world system to shocks or shifts in countries with a relatively high weight in the world economy, such as the USA, the USSR, Japan, and the FRG. As for the responses, also due to space considerations, we present only the multipliers of our aggregates – OECD, CMEA, LDC and World – and not the actual country multipliers (except for the own multipliers, which are also listed). However, one can consider that these aggregated multipliers represent the typical response of a country belonging to the corresponding area. It should be pointed out that the calculated area multipliers do not account for the own response of the shocked country when it belongs to this area. For example, when we introduce a shock in USA, the corresponding OECD multipliers refer only to the rest of the area: OECD less USA. However, the World multipliers do account for the own response of the shocked country as well.

*Table 7.3.* Dynamic multipliers for GDP (NMP for CMEA): percentage response to changes in the level of technical progress.

(*a*) 10% shock of $\tau$ taking place in 1985.

| Shock occurs in: | Short-term response (1985) | | | | | Long-term response (1999) | | | | |
|---|---|---|---|---|---|---|---|---|---|---|
| | Own | OECD | CMEA | LDC | World | Own | OECD | CMEA | LDC | World |
| USA | 7.27 | 0.04 | 0.01 | 0.47 | 1.86 | 0.06 | −0.01 | 0 | 0.05 | 0.02 |
| FRG | 7.36 | 0.01 | 0 | 0.25 | 0.53 | 0.03 | 0 | 0 | −0.03 | −0.01 |
| Japan | 8.51 | 0.01 | 0 | 0.18 | 0.82 | 0.02 | 0 | 0 | 0.01 | 0 |
| USSR | 12.40 | 0.01 | 0 | 0.04 | 1.96 | 0.22 | 0 | 0 | 0 | 0.03 |

(*b*) Sustained shift of $w_\tau$ to a higher level from 1985 on.

| Shock occurs in: | Short-term response (1985) | | | | | Long-term response (1999) | | | | |
|---|---|---|---|---|---|---|---|---|---|---|
| | Own | OECD | CMEA | LDC | World | Own | OECD | CMEA | LDC | World |
| USA | 0.89 | 0.01 | 0 | 0.06 | 0.23 | 20.8 | 0.3 | 0.1 | 1.1 | 4.7 |
| FRG | 2.42 | 0 | 0 | 0.08 | 0.17 | 53.8 | 0.4 | 0.1 | 3.7 | 4.2 |
| Japan | 2.96 | 0 | 0 | 0.06 | 0.29 | 63.8 | 0.3 | 0.1 | 4.0 | 7.4 |
| USSR | 1.02 | 0 | 0 | 0 | 0.16 | 43.6 | 0 | 0 | 0.1 | 7.3 |

*Table 7.4.* Dynamic multipliers for real imports: percentage response to changes in the level of technical progress.

(*a*) 10% shock of $\tau$ taking place in 1985.

| Shock occurs in: | Short-term response (1985) | | | | | Long-term response (1999) | | | | |
|---|---|---|---|---|---|---|---|---|---|---|
| | Own | OECD | CMEA | LDC | World | Own | OECD | CMEA | LDC | World |
| USA | 10.20 | 1.08 | 0.47 | 1.38 | 2.64 | −0.58 | −0.38 | 0.03 | −0.25 | −0.35 |
| FRG | 9.51 | 0.45 | 0.19 | 0.50 | 1.25 | −0.28 | −0.09 | −0.04 | −0.08 | −0.10 |
| Japan | 10.70 | 0.32 | 0.14 | 0.36 | 0.91 | −0.18 | −0.02 | 0.01 | −0.03 | −0.03 |
| USSR | 3.61 | 0.06 | 0.02 | 0.07 | 0.19 | −0.24 | −0.04 | −0.02 | −0.03 | −0.05 |

(*b*) Sustained shift of $w_\tau$ to a higher level from 1985 on.

| Shock occurs in: | Short-term response (1985) | | | | | Long-term response (1999) | | | | |
|---|---|---|---|---|---|---|---|---|---|---|
| | Own | OECD | CMEA | LDC | World | Own | OECD | CMEA | LDC | World |
| USA | 1.28 | 0.13 | 0.06 | 0.18 | 0.33 | 20.8 | 10.3 | 4.7 | 7.7 | 10.9 |
| FRG | 3.15 | 0.12 | 0.07 | 0.17 | 0.41 | 38.6 | 12.1 | 9.2 | 8.8 | 13.9 |
| Japan | 3.75 | 0.08 | 0.05 | 0.12 | 0.32 | 50.6 | 8.5 | 7.3 | 7.4 | 12.7 |
| USSR | 0.30 | 0.01 | 0 | 0.01 | 0.02 | 3.6 | 0.5 | 0.3 | 0.4 | 0.6 |

*Table 7.5.* Dynamic multipliers for real exports: percentage response to changes in the level of technical progress.

(*a*) 10% shock of $\tau$ taking place in 1985.

| Shock occurs in: | *Short-term response (1985)* | | | | | *Long-term response (1999)* | | | | |
|---|---|---|---|---|---|---|---|---|---|---|
| | *Own* | *OECD* | *CMEA* | *LDC* | *World* | *Own* | *OECD* | *CMEA* | *LDC* | *World* |
| USA | 1.80 | 3.24 | 1.01 | 2.59 | 2.65 | 0.08 | –0.62 | –0.09 | 0.15 | –0.35 |
| FRG | 1.47 | 1.33 | 0.56 | 1.49 | 1.25 | –0.07 | –0.10 | –0.04 | –0.14 | –0.10 |
| Japan | 0.65 | 1.04 | 0.41 | 1.08 | 0.91 | –0.06 | –0.02 | 0.03 | –0.05 | –0.03 |
| USSR | 0.12 | 0.19 | 0.06 | 0.22 | 0.19 | –0.03 | –0.05 | –0.02 | –0.05 | –0.05 |

(*b*) Sustained shift of $w_\tau$ to a higher level from 1985 on.

| Shock occurs in: | *Short-term response (1985)* | | | | | *Long-term response (1999)* | | | | |
|---|---|---|---|---|---|---|---|---|---|---|
| | *Own* | *OECD* | *CMEA* | *LDC* | *World* | *Own* | *OECD* | *CMEA* | *LDC* | *World* |
| USA | 0.22 | 0.40 | 0.13 | 0.33 | 0.33 | 10.3 | 13.5 | 6.1 | 5.9 | 10.9 |
| FRG | 0.42 | 0.41 | 0.19 | 0.50 | 0.41 | 8.4 | 14.5 | 9.4 | 16.5 | 13.9 |
| Japan | 0.23 | 0.34 | 0.14 | 0.38 | 0.32 | –3.0 | 14.1 | 8.7 | 16.5 | 12.7 |
| USSR | 0.01 | 0.02 | 0 | 0.02 | 0.02 | 0.5 | 0.6 | 0.3 | 0.6 | 0.6 |

We would like to mention that our system makes it possible to analyze the international and cross-country transmissions of *real-term shocks* or shifts into *monetary responses.* A synthetic long-term indicator of these transmissions is the behavior of the cumulated foreign trade balance (CFTB) of the countries. Due to the possible change of sign in the value of CFTB, the standard multiplier form (7.13) is not suitable for this indicator. Therefore, we calculate a discrepancy level for it:

$$d^t_{CFTB} = CFTB^t_s - CFTB^t_c \tag{7.14}$$

where $CFTB^t_s$ and $CFTB^t_c$ are the values of CFTB in the shocked and in the control solution, correspondingly.

*Tables 7.3* to *7.6* reflect the induced response of the model to changes in the level of technical progress ($\tau$) in the four selected countries: USA, FRG, Japan and USSR (the ordering is by the computer codes used in the model). The first part of each table shows the short- and long-term value of the multipliers corresponding to a single positive shock of 10% in the level of technical progress in these countries taking place in 1985. The second part of the tables contains the multipliers resulting from a sustained positive shift in the growth rate of technical progress ($w_\tau$) in the countries. The actual values of $w_\tau$ used for this

*Table 7.6.* Absolute response of the cumulated foreign trade balance to changes in the level of technical progress.

(*a*) 10% shock of $\tau$ taking place in 1985.

| *Country where* | *Short-term response (1985)* | | | | | | *Long-term response (1999)* | | | | | |
|---|---|---|---|---|---|---|---|---|---|---|---|---|
| *shock occurs* | *Own* | *OECD* | *CMEA* | *LDC* | *USA* | *LDC*[a] | *Own* | *OECD* | *CMEA* | *LDC* | *USA* | *LDC*[a] |
| USA | –30.7 | 18.2 | 1.5 | 11.0 | –30.7 | 2.1 | 33.5 | –15.6 | –1.4 | –16.5 | 33.5 | 14.0 |
| FRG | –15.1 | 9.5 | 0.8 | 4.8 | 1.9 | 1.3 | 13.9 | –10.6 | –0.9 | –2.4 | –4.2 | –0.5 |
| Japan | –11.6 | 7.5 | 0.6 | 3.5 | 1.2 | 1.0 | 6.8 | –5.4 | –0.3 | –1.1 | –2.0 | –0.5 |
| USSR | –2.8 | 2.1 | 0 | 0.7 | 0.4 | 0.2 | 5.1 | –3.4 | 0 | –1.7 | –1.0 | –0.8 |

[a]A second look at LDC less group 11 (oil-exporting countries).

(*b*) Sustained shift of $w_\tau$ to a higher level from 1985 on.

| *Country where* | *Short-term response (1985)* | | | | | | *Long-term response (1999)* | | | | | |
|---|---|---|---|---|---|---|---|---|---|---|---|---|
| *shock occurs* | *Own* | *OECD* | *CMEA* | *LDC* | *USA* | *LDC*[a] | *Own* | *OECD* | *CMEA* | *LDC* | *USA* | *LDC*[a] |
| USA | –3.9 | 2.3 | 0.2 | 1.4 | –3.9 | 0.2 | –739 | 347 | 30 | 362 | –739 | 211 |
| FRG | –4.8 | 2.9 | 0.3 | 1.6 | 0.5 | 0.4 | –1076 | 609 | 65 | 402 | 219 | 165 |
| Japan | –4.0 | 2.6 | 0.2 | 1.2 | 0.4 | 0.3 | –820 | 426 | 55 | 339 | 133 | 125 |
| USSR | –0.23 | 0.17 | 0 | 0.06 | 0.03 | 0.01 | –50.5 | 32.1 | 0.6 | 17.8 | 9.0 | 8.2 |

[a]A second look at LDC less group 11 (oil-exporting countries).

*Table 7.7.* Percentage responses of real imports/exports and their prices; and absolute level response of CFTB to changes in the level of US money supply.

(*a*) 10% shock in the level of US money supply taking place in 1985.

| | *Short-term response (1985)* | | | | | *Long-term response (1999)* | | | | |
|---|---|---|---|---|---|---|---|---|---|---|
| *Responding variable* | *USA* | *OECD*[a] | *CMEA* | *LDC* | *World* | *USA* | *OECD*[a] | *CMEA* | *LDC* | *World* |
| Real imports | –0.09 | 0.13 | –0.18 | –0.62 | 0.05 | 0.15 | 0.05 | 0.22 | 0.06 | 0.08 |
| Real exports | 0.12 | –0.28 | 0.18 | 0.55 | –0.05 | 0.64 | –0.27 | 0.18 | 0.64 | 0.08 |
| Import prices | 7.17 | 9.53 | 5.94 | 7.14 | 8.26 | 0.90 | 0.98 | 0.84 | 0.98 | 0.96 |
| Export prices | 6.93 | 10.21 | 5.18 | 4.72 | 8.26 | 0.31 | 1.27 | 0.83 | 0.43 | 0.96 |
| CFTB | –4.5 | 8.0 | –0.4 | –3.1 | 0.2 | –19.2 | 20.5 | –0.2 | –1.1 | –4.9 |

[a]Less USA.

(*b*) Sustained shift of the growth rate of US money supply to a higher level from 1985 on.

| | *Short-term response (1985)* | | | | | *Long-term response (1999)* | | | | |
|---|---|---|---|---|---|---|---|---|---|---|
| *Responding variable* | *USA* | *OECD*[a] | *CMEA* | *LDC* | *World* | *USA* | *OECD*[a] | *CMEA* | *LDC* | *World* |
| Real imports | –0.04 | 0.06 | –0.05 | –0.08 | –0.02 | –1.25 | –1.99 | 2.91 | –3.88 | –1.9 |
| Real exports | 0.03 | –0.06 | 0.05 | 0.04 | –0.02 | –0.64 | –3.64 | 2.52 | 1.17 | –1.9 |
| Import prices | 2.23 | 2.64 | 1.69 | 2.11 | 2.30 | 49.8 | 52.2 | 40.0 | 49.6 | 50.5 |
| Export prices | 2.00 | 2.72 | 1.48 | 1.90 | 2.30 | 48.4 | 54.4 | 39.4 | 44.2 | 50.5 |
| CFTB | –1.9 | 2.8 | –0.1 | –0.8 | 0 | –143 | 62 | 23 | 58 | –228 |

[a]Less USA.

purpose were: USA and USSR – twice the medium scenario value of $w_\tau$; FRG and Japan – 2.5 times the medium scenario value of $w_\tau$. (The medium scenario values of $w_\tau$ are available in the relevant country group chapters.) These levels of $w_\tau$ are close to those prevailing in the corresponding countries during the 1960s.

The calculated cross-country multipliers for GDP (*Table 7.3*), as might be expected, are very low. This is a direct consequence of our basically supply-driven growth models. The measured responses result only from the reaction of the production functions to the induced changes in the level of imports of raw materials and intermediate goods in the countries. The only area with a relatively high response is LDC, but this is mainly due to the partially demand-driven determination of GDP in Group 11 (see Chapter 6).

However, the foreign trade part of the model is much more sensitive to changes in the level of technical progress in the countries. The dynamic multipliers for imports (*Table 7.4*) and exports (*Table 7.5*) indicate a variety of cross-country effects caused by shocks or shifts in $\tau$. Besides, owing to the dynamic specification of the foreign trade equations, even the single shocks in $\tau$ induce a long-term response. It is interesting to note that the sustained positive shift of $w_\tau$ in FRG and Japan in the long run induces the largest (among the four countries analysed) increase in the total world trade. If we compare the world export response in the case of FRG and Japan to that for USA, we can see that in the latter case the response is mainly concentrated in OECD whereas in the former it is highest in LDC.

*Table 7.6* shows the generated responses in terms of the discrepancy level of CFTB [10] which, as we mentioned, synthesizes both the real and the nominal effects. The figures in this table show the pure gain (positive value) or loss (negative value) of the countries or areas in terms of CFTB in the shocked or shifted runs *with respect to the medium scenario.* In this table, in addition to the previous ones, we show separately the results for USA and LDC without group 11 since these figures can be considered of special importance with respect to the overall world economic performance.

A general feature is the different sign of the own long-term effect in the shocked and shifted scenarios (and this applies to all analyzed countries!). In all cases the sustained shift to a higher growth rate induces a large negative value of $d_{CFTB}$. The corresponding values for OECD, CMEA, LDC indicate the distribution of this effect (remembering that the own effect is excluded from the area to which the country belongs). Note, too, that the short-term responses of CFTB coincide with the response of FTB itself.

In *Table 7.7* we present the multipliers calculated after a shock or shift in the level of money supply of USA. As we pointed out, the US money supply plays a special role in our model due to the exchange rate determination. The shocked level was again 10% above the medium scenario level, taking place in 1985. The sustained shift of the growth rate of money supply was 3% above the medium scenario value (12% versus 9%). The most important outcome from these simulations is that both the real and the nominal parts of the model responded to changes in the level of US money supply. As for the effect on the

CFTB, it is always negative for USA and positive for OECD (without USA). For the other areas the results are mixed, but we should note the large negative long-term effect of the sustained shift on the CFTB for the area LDC without group 11.

### 7.3.3. Alternative scenarios

The three scenarios presented in the previous chapters (called low, medium, and high) were designed to characterize the different possible growth patterns of the world economy in the future. Assumptions about the driving forces of economic growth (that is, on investment ratios and on rates of technical progress) in the different countries and groups of countries were varied together; to put it in another way, the growth paths of the different regions moved together, leaving the differences among the growth rates at nearly the same level in the different scenarios. The results of these scenarios and the consequences drawn from them might throw some light on the contours of economic growth and structural change in the world economy to the end of this century. With the scenario analysis presented in this section, we attempt to broaden and deepen our understanding of the nature of interdependency in the world economy.

*Effects of changing growth patterns in the developed market economies*

It is widely emphasized that the export performance (and, consequently, the debt situation) of the CMEA and developing countries is very strongly influenced by the domestic development of the developed market economies. Scenarios A and B are designed to indicate the size of these effects and the differences among the relative situations of the different regions. In Scenario A for the developed market economies, the assumptions of the high scenario were taken, whereas for the CMEA and developing countries those of the medium scenario. (For definitions of the different scenarios see Chapter 2). In Scenario B for the developed market economies, the assumptions of the low scenario were taken. The results of these alternative scenarios are reported in *Table 7.8*. For comparison, the same results for the (original) medium scenario are also given.

When comparing the results of Scenario A and the medium scenario in *Table 7.8*, it can be observed that an increase of 0.8–0.9% in the growth rate of real GDP of the developed market economies caused an increase of 1.1–1.2% in the growth rate of world trade in real terms. (Since, due to the specification of the model, world real exports equal world real imports, by definition the term "world trade" refers to both of them.) The changes in the rate of growth of world trade in nominal terms are rather different: while the difference in the first period is 0.8%, at the end of the simulation period it diminishes almost to zero. The explanation for this is to be found largely in the fact that the rate of monetary expansion (rate of change of money supply M2) in the developed

*Table 7.8.* Results of alternative scenarios A and B (average values in percentages).

| | Scenarios | | | | | | | | |
|---|---|---|---|---|---|---|---|---|---|
| | *A* | *M*[a] | *B* | *A* | *M*[a] | *B* | *A* | *M*[a] | *B* |
| *Region* | *1986–1990* | | | *1991–1995* | | | *1996–2000* | | |
| World growth rates | | | | | | | | | |
| real GDP | 4.0 | 3.6 | 3.0 | 4.1 | 3.5 | 3.0 | 4.1 | 3.5 | 3.0 |
| trade in real terms | 5.4 | 4.3 | 3.3 | 5.7 | 4.4 | 3.1 | 5.6 | 4.4 | 2.9 |
| trade in nominal terms | 6.7 | 5.9 | 5.2 | 10.0 | 9.6 | 9.3 | 10.3 | 10.2 | 10.4 |
| Developed market economies' growth rates | | | | | | | | | |
| real GDP | 3.5 | 2.7 | 2.1 | 3.7 | 2.8 | 1.9 | 3.7 | 2.8 | 1.9 |
| real exports | 4.9 | 3.7 | 2.7 | 5.2 | 3.9 | 2.6 | 5.3 | 3.9 | 2.3 |
| real imports | 5.8 | 4.5 | 3.3 | 5.7 | 4.2 | 2.8 | 5.6 | 4.2 | 2.6 |
| USSR | | | | | | | | | |
| growth rate of real GDP | 4.3 | 4.3 | 4.2 | 3.9 | 3.9 | 3.9 | 3.7 | 3.7 | 3.7 |
| growth rate of real exports | 3.8 | 2.9 | 2.1 | 4.7 | 3.6 | 2.4 | 4.9 | 3.6 | 2.1 |
| growth rate of real imports | 3.5 | 3.0 | 2.6 | 4.7 | 3.9 | 3.1 | 5.5 | 4.5 | 3.4 |
| trade balance/GPD | –0.1 | –0.2 | –0.2 | 0.2 | 0.0 | –0.1 | 0.4 | 0.1 | –0.1 |
| trade balance/exports | –1.3 | –2.9 | –4.4 | 2.8 | 0.2 | –2.5 | 4.8 | 1.9 | –1.4 |
| growth rate of real exports | 5.0 | 4.5 | 4.1 | 5.8 | 5.1 | 4.4 | 6.2 | 5.4 | 4.6 |
| growth rate of real imports | 4.2 | 3.8 | 3.4 | 4.7 | 3.9 | 3.2 | 5.6 | 4.6 | 3.6 |
| trade balance/GPD | 0.6 | 0.5 | 0.4 | 0.8 | 0.6 | 0.5 | 1.0 | 0.8 | 0.7 |
| trade balance/exports | 4.2 | 4.2 | 3.7 | 5.9 | 5.1 | 4.2 | 7.1 | 6.2 | 5.3 |
| Oil-exporting LDCs | | | | | | | | | |
| growth rate of real GDP | 7.6 | 6.6 | 5.7 | 7.6 | 6.8 | 6.1 | 6.6 | 6.2 | 5.6 |
| growth rate of real exports | 9.6 | 8.5 | 7.4 | 9.6 | 8.7 | 7.6 | 7.0 | 6.7 | 6.1 |
| growth rate of real imports | 0.2 | –1.0 | –2.2 | 7.3 | 5.8 | 4.2 | 6.0 | 4.7 | 3.0 |
| trade balance/GPD | 3.0 | 1.8 | 0.7 | 6.8 | 4.9 | 3.0 | 7.3 | 5.0 | 2.6 |
| trade balance/exports | 10.5 | 6.6 | 2.6 | 22.2 | 17.5 | 11.9 | 23.9 | 18.3 | 10.8 |
| Other LDCs | | | | | | | | | |
| growth rate of real GDP | 5.3 | 5.3 | 5.2 | 5.8 | 5.8 | 5.7 | 5.7 | 5.6 | 5.5 |
| growth rate of real exports | 3.8 | 3.0 | 2.2 | 5.8 | 4.6 | 3.3 | 6.3 | 4.9 | 3.4 |
| growth rate of real imports | 6.8 | 6.4 | 6.0 | 6.1 | 5.4 | 4.8 | 5.8 | 5.1 | 4.3 |
| trade balance/GPD | –0.4 | –0.7 | –0.9 | –1.7 | –2.2 | –2.8 | –1.7 | –2.6 | –3.4 |
| trade balance/exports | –2.6 | –4.3 | –5.9 | –10.6 | –15.5 | –20.7 | –10.6 | –18.7 | –28.0 |

[a] *M* refers to the (original) medium scenario (see Chapter 2).

market economies was kept at the same level. (Although this is generally true because of the determination of the exchange rates and that of the export prices, it is of special interest in the case of the USA.)

If we look at the growth rates of real GDP in the CMEA and developing countries, it can be seen that the patterns are in general not influenced by the faster growth in the developed market economies. This is a reflection of the supply-driven nature of the country models. The only exception is the group of oil-exporting countries because the production of the oil sector is demand-driven. (For definition of the country groups, see Annex 1.) As to real exports, the picture is rather different. Both the CMEA and the developing countries benefit from the faster growth of the world trade. On the other hand, the faster growth of real exports contributes to the faster growth of real imports. The changes in

*Table 7.9.* Results of alternative scenarios C and D (average values in percentages).

| | *Scenarios* | | | | | | | | |
|---|---|---|---|---|---|---|---|---|---|
| | *C* | *M*[a] | *D* | *C* | *M*[a] | *D* | *C* | *M*[a] | *D* |
| *Region* | *1986–1990* | | | *1991–1995* | | | *1996–2000* | | |
| World growth rates | | | | | | | | | |
| real GDP | 4.0 | 3.6 | 3.0 | 4.1 | 3.5 | 3.0 | 4.2 | 3.5 | 3.0 |
| trade in real terms | 5.6 | 4.3 | 3.1 | 5.7 | 4.4 | 3.3 | 5.7 | 4.4 | 3.1 |
| trade in nominal terms | 7.2 | 5.9 | 4.8 | 10.9 | 9.6 | 8.4 | 11.6 | 10.2 | 8.9 |
| Developed market economies' growth rates | | | | | | | | | |
| real GDP | 2.8 | 2.7 | 2.7 | 2.8 | 2.8 | 2.7 | 2.8 | 2.8 | 2.7 |
| real exports | 5.0 | 3.7 | 2.6 | 5.1 | 3.9 | 2.8 | 5.2 | 3.9 | 2.7 |
| real imports | 5.7 | 4.5 | 3.5 | 5.5 | 4.2 | 3.1 | 5.6 | 4.2 | 3.0 |
| USSR | | | | | | | | | |
| growth rate of real GDP | 5.4 | 4.3 | 3.1 | 5.1 | 3.9 | 2.6 | 5.0 | 3.7 | 2.4 |
| growth rate of real exports | 4.0 | 2.9 | 2.0 | 4.8 | 3.6 | 2.5 | 4.9 | 3.6 | 2.4 |
| growth rate of real imports | 4.0 | 3.0 | 2.1 | 5.0 | 3.9 | 2.9 | 5.7 | 4.5 | 3.4 |
| trade balance/GPD | –0.1 | –0.2 | –0.2 | 0.1 | 0.0 | –0.1 | 0.3 | 0.1 | 0.0 |
| trade balance/exports | –1.1 | –2.9 | –4.2 | 2.1 | 0.2 | –1.3 | 4.0 | 1.9 | –0.2 |
| Other CMEA countries | | | | | | | | | |
| growth rate of real GDP | 5.0 | 3.8 | 2.6 | 5.4 | 4.0 | 2.7 | 5.4 | 4.0 | 2.7 |
| growth rate of real exports | 5.2 | 4.5 | 3.9 | 6.0 | 5.1 | 4.3 | 6.6 | 5.4 | 4.5 |
| growth rate of real imports | 5.5 | 3.8 | 2.2 | 5.0 | 3.9 | 2.9 | 5.7 | 4.6 | 3.7 |
| trade balance/GPD | 0.2 | 0.5 | 0.8 | 0.1 | 0.6 | 1.1 | 0.3 | 0.8 | 1.4 |
| trade balance/exports | 1.8 | 4.2 | 6.6 | 1.0 | 5.1 | 9.1 | 2.2 | 6.2 | 10.3 |
| Oil-exporting LDCs | | | | | | | | | |
| growth rate of real GDP | 9.5 | 6.6 | 4.4 | 8.3 | 6.8 | 5.1 | 7.5 | 6.2 | 4.8 |
| growth rate of real exports | 10.0 | 8.5 | 6.5 | 9.1 | 8.7 | 7.5 | 8.0 | 6.7 | 5.3 |
| growth rate of real imports | 2.6 | –1.0 | –4.4 | 7.2 | 5.8 | 4.9 | 6.0 | 4.7 | 3.6 |
| trade balance/GPD | –1.2 | 1.8 | 4.5 | 1.0 | 4.9 | 8.0 | 1.3 | 5.0 | 7.9 |
| trade balance/exports | –5.0 | 6.6 | 17.2 | 3.9 | 17.5 | 28.3 | 4.9 | 18.3 | 28.0 |
| Other LDCs | | | | | | | | | |
| growth rate of real GDP | 6.4 | 5.3 | 4.5 | 6.7 | 5.8 | 5.0 | 6.5 | 5.6 | 4.8 |
| growth rate of real exports | 4.1 | 3.0 | 2.0 | 5.8 | 4.6 | 3.4 | 6.3 | 4.9 | 3.7 |
| growth rate of real imports | 7.7 | 6.4 | 4.6 | 6.4 | 5.4 | 4.6 | 6.2 | 5.1 | 3.9 |
| trade balance/GPD | –2.5 | –0.7 | 1.0 | –3.8 | –2.2 | –0.5 | –4.0 | –2.6 | –1.0 |
| trade balance/exports | –16.3 | –4.3 | 6.4 | –26.4 | –15.5 | –3.7 | –28.7 | –18.7 | –7.2 |

[a]*M* refers to the (original) medium scenario (see Chapter 2).

the foreign trade balance to exports (or to GDP) ratios are considerable. The strongest influence is on the non-oil-exporting developing countries, where the difference is some 8% at the end of the simulation period. The USSR manifests a similar improvement of its foreign trade balance/export ratio, while the other CMEA countries seem to gain less. A comparison of the real and nominal indicators reveals the importance of the effects of changes in relative export (import) prices.

The effects of slower growth in the developed market economies (Scenario B) can also be assesed from *Table 7.8*. A comparison of the results of Scenarios A and B suggests that slower growth in the developed market economies affects the developing countries more strongly than accelerated growth.

*Internal development and external position in the CMEA and developing countries*

While we have stressed the importance of the external conditions of growth in the CMEA and developing countries, we should not lose sight of the importance of the internal development of these economies. Accelerated growth, in general, leads to an increase in import demand and, if the external conditions do not change, to a worsening of the foreign trade balance. With Scenarios C and D we attempt to assess the extent of this effect. Starting again from the medium scenario, in Scenario C for the CMEA and developing countries the assumptions of the high scenario are taken, whereas in Scenario D those of the low one. The results are summarized in *Table 7.9.*

The results of Scenario C in *Table 7.9* show that the extent of changes in the growth rates of world real GDP and world trade in real terms are very similar to those in Scenario A (and the same is true for Scenarios B and D). However, this does not hold for world trade in nominal terms. Again, this is due to the internal price determination in the models of the developed market economies and to the determination of the export prices and exchange rates. Similarly to Scenarios A and B, the growth of the developed market economies was not much influenced by the faster growth of the CMEA and developing countries.

Regarding the effects on the different regions, it can be observed that, with the exception of the USSR, faster growth in these economies considerably worsened the foreign trade balance to export (or to GDP) ratios. In the case of the non-oil-exporting developing countries, an increase of 0.9–1.1% in the growth rate of real GDP leads to a 10–12% decrease in this ratio. The behavior of the oil-exporting developing countries is similar, indicating that these countries share this common feature of the developing economies. In the case of the USSR, the increase in its own import demand is compensated by an even larger increase in its exports generated in this scenario. This, together with the improvement of the terms of trade, leads to an increase of 1.8–2.1% in the foreign trade balance/export ratio. A comparison of the real and nominal indicators reveals again the importance of effects originating in changes in relative export (import) prices.

The consequences of slower growth in the CMEA and developing countries can also be seen from *Table 7.9* (Scenario D). Not surprisingly, with the exception of the USSR, foreign trade balance/export ratios improved substantially.

Taken as a whole, our scenario analysis tends to suggest that the larger the difference between the growth rates of the developed market economies and the developing countries, the more dramatically the foreign trade balances of the latter deteriorate and their debt burdens increase. Considering this general finding, together with the facts that the external indebtedness in most of the developing ountries has already reached the limits of manageability and that most studies known to us project rather moderate growth in the developed market economies in the next decade, the growth prospects for the developing countries as a whole do not seem to be too promising, if no substantial changes take place in the world economy.

## Notes

[1] Though tentatively, as with any attempt at classification, we could point out that a representative of the first approach is the well known Project LINK (Waelbroeck, 1976; Sawyer, 1979). The same approach is also used in some of the UN studies [e.g., UNCTAD's System for Interlinked Global Modeling and Analysis (SIGMA) – see this volume, Chapter 26]. The pooling approach has been applied in the UN-sponsored study of Leontief *et al.* (1977) and other similar models.

[2] The separate measurement of world exports and imports by aggregating country data leaves a statistical discrepancy of about 1% of the value of world trade.

[3] A similar approach is applied in the foreign trade models of Project LINK, though on the commodity level; see Klein (1983).

[4] This is how this problem is tackled in Project LINK, where the import prices are weighted sums of partners' export prices, the weights being the columns of the trade share matrix. See Sawyer (1979).

[5] A simplified version of this equation is used with three aggregated areas: OECD countries, CMEA countries, developing countries. $PM'EX_i^{\$}$ is the weighted average of the export prices of these areas, the weights being the columns of an aggregated trade share matrix (in *ex ante* simulations they are kept constant at the average level of the last five available observations).

[6] For the CMEA countries, these are actually the conversion factors which "transform" the value of total imports and exports from domestic currency into US$. These conversion factors are calculated artificially by dividing the two time series (in domestic currency and in US$).

[7] For short, further on we shall refer to these totals (subtotals) as: "World", "OECD", "CMEA", "LDC". On some plots and tables they may be indicated by their computer codes which are "W0", "W1", "W2", "W3", correspondingly. The abbreviation "OECD total" (or "OECD") is not very precise because these subtotals include the whole Group 10, which incorporates some non-OECD countries as well.

[8] It should be recalled that in nonlinear systems the values of the multipliers also depend on the initial conditions, so we do not ascribe generality to the calculated values of our multipliers.

[9] Remember that the dynamic elasticity multipliers are defined as:

$$m_{e,y}^t = \frac{y_s^t - y_c^t}{y_c^t} : \frac{x_s^t - x_c^t}{x_c^t}$$

where $x_s^t$ and $x_c^t$ are the values of the shock variable in the shocked and in the control solution, respectively.

[10] The foreign trade balance (FTB) is defined as exports of goods and services less imports of goods and services, in current US$. The medium scenario solution for FTB is given in Annex 3.

# Appendix 7A. Export and Import Functions

*Table 7A.1.* Export functions (first step).[a]

| Variable | USA | FRG | Japan | Rest of DME | CMEA |
|---|---|---|---|---|---|
| $a_1$ | 0.802 | 0.684 | 0.800 | 0.510 | 0.924 |
| | (13.91) | (21.91) | (13.69) | (11.15) | (12.71) |
| $a_2$ | 0.108 | 0.145 | 0.086 | 0.410 | 0.035 |
| | (3.76) | (12.70) | (4.46) | (12.20) | (1.70) |
| $a_3$ | 0.086 | 0.130 | 0.077 | 0.368 | 0.031 |
| | (3.58) | (11.45) | (4.22) | (10.10) | (1.64) |
| $a_4$ | 0.094 | 0.127 | 0.069 | 0.328 | 0.028 |
| | (3.50) | (11.69) | (4.29) | (12.61) | (1.72) |
| $a_5$ | 0.093 | 0.125 | 0.074 | 0.359 | 0.030 |
| | (3.51) | (10.04) | (4.04) | (8.59) | (1.70) |
| $a_6$ | 0.103 | 0.139 | 0.082 | 0.392 | 0.030 |
| | (3.27) | (8.81) | (4.08) | (14.37) | (1.75) |
| $a_7$ | 0.089 | 0.120 | 0.071 | 0.339 | 0.029 |
| | (3.53) | (9.68) | (4.27) | (10.87) | (1.69) |
| DW | 1.96 | 1.79 | 1.22 | 1.49 | 1.99 |
| R2C | 0.99 | 1.00 | 0.99 | 1.00 | 1.00 |
| SEE | 4.517 | 1.657 | 4.454 | 5.449 | 2.216 |

[a]Values in parentheses are $t$-values. The specification of the estimated system is given by equation (7.5) in the text, but note that the equation for the developing countries as a whole is not estimated directly. All estimates use the FIML method for the period 1962–1981.

*Table 7A.2.* Export functions for developed market economies (second step).[a]

| Variable | France | UK | Italy | Netherlands | Belgium/ Luxembourg | Canada |
|---|---|---|---|---|---|---|
| $a_1$ | 0.553 | 0.743 | 0.591 | 0.600 | 0.457 | 0.697 |
| | (18.43) | (30.23) | (10.34) | (20.87) | (15.73) | (15.49) |
| $a_2$ | 0.203 | 0.105 | 0.137 | 0.119 | 0.144 | 0.095 |
| | (15.21) | (10.25) | (8.42) | (14.80) | (17.12) | (7.90) |
| $a_3$ | 0.168 | 0.070 | 0.091 | 0.080 | 0.096 | 0.063 |
| | (12.31) | (7.89) | (9.04) | (12.02) | (14.77) | (7.08) |
| $a_4$ | 0.139 | 0.072 | 0.113 | 0.099 | 0.119 | 0.079 |
| | (9.48) | (7.00) | (8.23) | (13.70) | (15.20) | (8.05) |
| $a_5$ | 0.138 | 0.072 | 0.093 | 0.082 | 0.099 | 0.065 |
| | (12.13) | (8.96) | (8.37) | (8.80) | (8.81) | (5.98) |
| $a_6$ | 0.108 | 0.056 | 0.073 | 0.064 | 0.098 | 0.065 |
| | (12.85) | (8.26) | (8.01) | (13.13) | (14.82) | (7.32) |
| $a_7$ | 0.156 | 0.081 | 0.105 | 0.092 | 0.111 | 0.051 |
| | (9.96) | (8.99) | (7.12) | (10.32) | (10.38) | (6.65) |
| $a_8$ | 0.175 | 0.091 | 0.118 | 0.103 | 0.124 | 0.082 |
| | (11.59) | (9.10) | (7.96) | (11.64) | (12.25) | (7.97) |
| DW | 0.65 | 2.22 | 2.19 | 1.08 | 0.73 | 0.82 |
| R2C | 1.00 | 1.00 | 1.00 | 1.00 | 0.99 | 0.98 |
| SEE | 1.990 | 1.211 | 1.532 | 1.070 | 1.844 | 2.820 |

[a]Values in parentheses are $t$-values. The specification of the estimated system is given by equation (7.5) in the text, but note that the equation for group 10 is not estimated directly. All estimates use the MIDIS method for the period 1962–1984.

*Table 7A.3.* Export functions for CMEA countries (second step).[a]

| *Variable* | *USSR* | *Bulgaria* | *CSSR* | *GDR* | *Hungary* | *Poland* |
|---|---|---|---|---|---|---|
| $a_1$ | 0.292 | 0.899 | 0.874 | 0.884 | 0.941 | 0.894 |
| | (10.50) | (139.24) | (37.50) | (18.75) | (23.97) | (106.89) |
| $a_2$ | 0.643 | 0.054 | 0.069 | 0.080 | 0.031 | 0.059 |
| | (26.53) | (7.98) | (7.85) | (4.01) | (3.07) | (6.67) |
| $a_3$ | 0.611 | 0.044 | 0.056 | 0.066 | 0.025 | 0.048 |
| | (26.53) | (6.27) | (7.08) | (4.18) | (3.00) | (5.57) |
| $a_4$ | 0.603 | 0.051 | 0.065 | 0.076 | 0.030 | 0.056 |
| | (23.07) | (7.25) | (7.85) | (4.01) | (3.07) | (6.67) |
| $a_5$ | 0.618 | 0.052 | 0.066 | 0.075 | 0.029 | 0.055 |
| | (21.74) | (6.70) | (7.20) | (4.03) | (3.06) | (6.23) |
| $a_6$ | 0.624 | 0.053 | 0.067 | 0.078 | 0.030 | 0.056 |
| | (18.32) | (7.28) | (7.52) | (4.01) | (3.03) | (5.86) |
| $a_7$ | 0.611 | 0.051 | 0.065 | 0.076 | 0.030 | 0.057 |
| | (26.53) | (7.98) | (7.85) | (4.01) | (3.07) | (6.25) |
| $a_8$ | 0.553 | 0.047 | 0.059 | 0.069 | 0.027 | 0.051 |
| | (15.54) | (7.02) | (8.30) | (3.68) | (3.03) | (6.08) |
| DW | 1.43 | 1.99 | 1.20 | 1.08 | 1.90 | 1.35 |
| R2C | 1.00 | 1.00 | 1.00 | 0.99 | 0.99 | 0.99 |
| SEE | 0.606 | 0.173 | 1.158 | 0.520 | 0.199 | 0.393 |

[a]Values in parentheses are $t$-values. The specification of the estimated system is given by equation (7.5) in the text, but note that the equation for Romania is not estimated directly. All estimates use the FIML method for the period 1962–1985.

*Table 7A.4.* Export functions for developing countries (second step).[a]

| *Variable* | *11* | *12* | *India* | *13* | *14* | *15* |
|---|---|---|---|---|---|---|
| $a_1$ | 0.150 | 0.851 | 0.529 | 0.910 | 0.833 | 0.900 |
| | (4.50) | (68.31) | (8.71) | (480.27) | (19.27) | (121.06) |
| $a_2$ | 0.731 | 0.105 | 0.045 | 0.011 | 0.028 | 0.052 |
| | (22.99) | (7.98) | (6.29) | (5.31) | (3.94) | (6.67) |
| $a_3$ | 0.695 | 0.058 | 0.025 | 0.006 | 0.016 | 0.029 |
| | (22.99) | (4.36) | (4.03) | (3.72) | (3.13) | (4.11) |
| $a_4$ | 0.405 | 0.058 | 0.043 | 0.010 | 0.027 | 0.050 |
| | (8.00) | (5.74) | (6.29) | (5.31) | (3.94) | (6.67) |
| $a_5$ | 0.673 | 0.096 | 0.042 | 0.006 | 0.016 | 0.029 |
| | (22.99) | (7.98) | (6.29) | (4.47) | (4.10) | (5.19) |
| $a_6$ | 0.627 | 0.090 | 0.039 | 0.009 | 0.026 | 0.048 |
| | (14.22) | (7.33) | (6.44) | (5.10) | (3.94) | (6.67) |
| $a_7$ | 0.695 | 0.099 | 0.043 | 0.010 | 0.027 | 0.045 |
| | (22.99) | (7.98) | (6.29) | (4.47) | (4.10) | (5.19) |
| $a_8$ | 0.589 | 0.084 | 0.037 | 0.009 | 0.023 | 0.042 |
| | (13.62) | (6.28) | (5.56) | (4.70) | (3.93) | (5.58) |
| DW | 1.39 | 1.46 | 0.67 | 2.20 | 1.15 | 1.54 |
| R2C | 1.00 | 1.00 | 0.99 | 0.98 | 0.99 | 0.99 |
| SEE | 2.653 | 1.442 | 0.482 | 0.444 | 0.483 | 1.372 |

[a]Values in parentheses are $t$-values. The specification of the estimated system is given by equation (7.5) in the text, but note that the equation for group 18 is not estimated directly. All estimates use the FIML method for the period 1962–1981.

*Table 7A.5.* Import demand equations, CMEA countries.[a]

| *Country* | $a_1$ | $a_2$ | $a_3$ | $a_4$ | *DW* | *SEE* | *Est.* |
|---|---|---|---|---|---|---|---|
| Bulgaria | 0.188 | 0.514 | 0.857 | 0.556 | 1.51 | 280.318 | 61–85 |
| | (1.47) | (5.61) | (6.40) | (5.13) | | | OLS |
| CSSR | 0.515 | 0.285 | 0.474 | 0.290 | 2.01 | 3.719 | 61–85 |
| | (5.72) | (3.73) | (4.96) | (3.43) | | | OLS |
| GDR | 0.424 | 0.283 | 0.582 | 0.279 | 1.90 | 1.548 | 61–82 |
| | (3.80) | (5.90) | (4.17) | (5.17) | | | FIML |
| Hungary | 0.403 | 0.528 | 0.611 | 0.545 | 2.27 | 7.039 | 61–85 |
| | (3.80) | (7.50) | (4.39) | (6.40) | | | OLS |
| Poland | 0.641 | 0.217 | 0.559 | 0.255 | 1.61 | 26.745 | 61–85 |
| | (10.58) | (9.30) | (4.61) | (11.92) | | | OLS |
| Romania | 0.591 | 0.221 | 0.428 | 0.236 | 1.83 | 4.716 | 61–85 |
| | (7.88) | (6.32) | (4.87) | (6.13) | | | FIML |
| USSR | 0.749 | 0.067 | 0.453 | 0.071 | 2.26 | 2.183 | 61–85 |
| | (4.08) | (0.70) | (1.54) | (0.68) | | | OLS |

[a]Values in parentheses are $t$-values. Throughout, R2C = 0.99. Estimated equation:

$$IM = a_1 IM_{-1+a_2} \frac{A'N + C'N}{P'IM} + a_3 \frac{EX'N}{P'IM} - a_4 \frac{Y_{-1} \cdot P'Y}{P'IM} + u$$

where

| | |
|---|---|
| $IM$ | = total imports, in domestic currency, constant prices |
| $P'IM$ | = price index of total imports, in domestic currency |
| $C'N$ | = total consumption, in current prices |
| $A'N$ | = total gross accumulation, in current prices |
| $EX'N$ | = total exports, in domestic currency, current prices |
| $Y$ | = domestic material output, in constant prices |
| $P'Y$ | = price index of domestic material output |

Note that the equations for GDR and Romania were estimated with additional restrictions on the value of the habit persistence parameters.

*Table 7A.6.* Export price equations, CMEA countries.[a]

| *Country* | $a_1$ | $a_2$ | $a_3$ | $a_4$ | *DW* | *R2C* | *SEE* |
|---|---|---|---|---|---|---|---|
| Bulgaria | 0.330 | 0.473 | – | 1.033 | 1.85 | 0.99 | 0.024 |
| | (4.47) | (8.90) | | (190.18) | | | |
| CSSR | 0.314 | 0.495 | – | 1.042 | 2.48 | 0.98 | 0.037 |
| | (3.48) | (7.44) | | (125.50) | | | |
| GDR | 0.420 | 0.347 | 0.093 | 1.099 | 1.66 | 0.99 | 0.033 |
| | (5.41) | (3.29) | (3.88) | (102.53) | | | |
| Hungary | 0.591 | 0.147 | – | 1.033 | 1.16 | 0.93 | 0.036 |
| | (5.93) | (4.60) | | (113.71) | | | |
| Poland | 0.642 | 0.212 | – | 0.986 | 0.71 | 0.91 | 0.081 |
| | (3.06) | (1.33) | | (39.29) | | | |
| Romania | 0.062 | 0.681 | 0.138 | 1.077 | 1.52 | 0.97 | 0.068 |
| | (0.54) | (4.39) | (2.98) | (48.44) | | | |
| USSR | 0.200 | 0.770 | 0.076 | 1.112 | 1.44 | 0.98 | 0.072 |
| | (1.99) | (6.15) | (1.64) | (40.31) | | | |

[a]Values in parentheses are *t*-values. Estimated equation:

$$P'EX^{\$} = {P'EX^{\$}_{-1}}^{a_1}\, {P'EX^{\$}_{W1}}^{a_2}\, {P'EX^{\$}_{11}}^{a_3} \cdot a_4 + u$$

where

$P'EX^{\$}$ = price index of total exports, in \$ terms

$P'EX^{\$}_{W1}$ = price index for total exports, OECD total, in \$ terms

$P'EX^{\$}_{11}$ = price index for total exports, Group 11, in \$ terms (used as an explanatory variable for some CMEA countries with relatively high exports of oil products)

Note that the equations for GDR and Romania were estimated with additional restrictions on the value of the habit persistence parameters. All estimates use the FIML method for the period 1962–1984, except Hungary, for which the period is 1962–1982.

*Table 7A.7.* Import price equations, CMEA countries.[a]

| *Country* | $a_1$ | $a_2$ | $a_3$ | $a_4$ | *DW* | *R2C* | *SEE* |
|---|---|---|---|---|---|---|---|
| Bulgaria | 0.926 | 0.683 | –0.550 | –0.020 | 2.78 | 0.99 | 0.048 |
| | (6.52) | (6.14) | (2.68) | (0.81) | | | |
| CSSR | 0.897 | 0.476 | –0.304 | –0.024 | 2.63 | 1.00 | 0.027 |
| | (9.94) | (6.86) | (2.29) | (1.56) | | | |
| GDR | 0.992 | 0.639 | –0.577 | –0.013 | 2.26 | 1.00 | 0.034 |
| | (8.49) | (7.80) | (3.54) | (0.53) | | | |
| Hungary | 0.463 | 0.526 | –0.237 | 0.228 | 2.05 | 0.98 | 0.035 |
| | (2.13) | (5.95) | (1.42) | (2.38) | | | |
| Poland | 0.576 | 1.044 | –0.817 | 0.128 | 2.87 | 0.95 | 0.063 |
| | (3.03) | (5.79) | (5.19) | (2.43) | | | |
| Romania | 0.211 | 1.316 | –0.454 | –0.068 | 2.39 | 0.99 | 0.059 |
| | (0.87) | (10.08) | (1.56) | (1.80) | | | |
| USSR | 0.772 | 0.894 | –0.642 | –0.006 | 2.71 | 1.00 | 0.022 |
| | (4.20) | (14.97) | (2.92) | (0.52) | | | |

[a]Values in parentheses are *t*-values. Estimated equation:

$$P'IM^{\$} = a_1 P'IMa^{\$}_{-1} + a_2 PM'EX^{\$} + a_3 PM'EX^{\$}_{-1} + a_4 + u$$

where

$P'M^{\$}$ = price index for total imports, in \$ terms
$PM'EX^{\$}$ = average of export prices abroad, in \$ terms

All estimates use the OLS method for the period 1962–1984.

*Table 7A.8.* Equations for the indexes of exchange rates.[a]

| *Country* | $a_1$ | $a_2$ | $a_3$ | *DW* | *R2C* | *SEE* |
|---|---|---|---|---|---|---|
| Bulgaria | 0.535 | 0.426 | 1.000 | 2.29 | 0.98 | 0.033 |
| | (4.63) | (4.09) | (104.36) | | | |
| CSSR | 0.325 | 0.456 | 1.029 | 1.80 | 0.85 | 0.039 |
| | (2.16) | (4.28) | (93.53) | | | |
| GDR | 0.450 | 0.468 | 1.018 | 2.17 | 0.99 | 0.021 |
| | (6.21) | (7.31) | (155.37) | | | |
| Hungary | 0.522 | 0.488 | 1.004 | 1.41 | 0.77 | 0.045 |
| | (3.01) | (3.58) | (93.82) | | | |
| Poland | 0.300 | 0.705 | 0.994 | 2.27 | 0.98 | 0.061 |
| | (7.03) | (22.57) | (72.67) | | | |
| Romania | 0.640 | 0.321 | 1.003 | 1.75 | 0.88 | 0.059 |
| | (5.08) | (2.71) | (76.54) | | | |
| USSR | 0.710 | 0.244 | 1.006 | 1.76 | 0.90 | 0.033 |
| | (7.79) | (3.73) | (122.56) | | | |

[a]Values in parentheses are $t$-values. Estimated equation:

$$I'FX = I'FX_{-1}{}^{a_1} \cdot \left[ I'FX_{-1} \cdot \left( \frac{1 + W'P}{1 + W'P^{\$}_{FT}} \right) \right]^{a_2} \cdot a_3$$

where

$I'FX$ = index of the exchange rate
$W'P$ = rate of change of the general price level in the country
$W'P^{\$}_{FT}$ = rate of change of the average foreign trade (imports + exports) price, in \$ terms.

All estimates use the MIDIS method for the period 1962–1985, except Poland, for which the period is 1962–1984.

## References

Klein, L. (1983), *The Economics of Supply and Demand.* (Oxford: Basil Blackwell).
Leontief, W. *et al.* (1977), *The Future of the World Economy. A United Nations Study.* (New York and Oxford, UK: Oxford University Press).
Sawyer, J. (1979), *Modelling the International Transmission Mechanism.* (Amsterdam: North-Holland.)
Waelbroeck, J. (ed.) (1976), *The Models of Project Link.* (Amsterdam: North-Holland).
Welsch, H. (1987), An Aggregate Import Demand Model for Long-Term Projections, *Jahrbücher für Nationalökonomie und Statistik*, **203/204**, pp. 372–382.

CHAPTER 8

# Modeling Exchange Rates and Foreign Trade of Developed Market Economies

*Heinz Welsch*

## Summary

This chapter is concerned with that part of the Bonn–IIASA world econometric model that deals with exchange rates and foreign trade of developed market economies. The variables explained are aggregate imports and exports, imports and exports disaggregated according to commodity groups, corresponding prices, and exchange rates. The exogenous explaining variables are GDP, nominal consumption, nominal investment, the GDP deflator, the interest rate and the rate of change of money supply of the countries considered and of the rest-of-the-world category applicable in each case. The model structure as well as the estimation results and the *ex post* tracking performance are presented.

## 8.1. Introduction

Foreign trade and financial flows provide the basic economic linkages between national economies. Therefore their analysis has to play a major part in the analysis of economic growth and structural change in the world economy. However, financial flows (including changes in foreign reserves) are only a mirror image of trade flows, due to the balance of payments identity. For this reason, the scope of this chapter will be restricted to the modeling of foreign trade and one of its major determinants, the exchange rate. In this context, the development of aggregate foreign trade refers to the concept of economic growth whereas variations in the commodity composition of trade flows will be identified with structural change.

According to the general principle of the Bonn–IIASA Research Project the ultimate driving forces behind economic growth and structural change are the rate of growth of the labor force, the (net) savings ratio, and the rate of technical change in the individual countries. They determine the internal development of the economies as characterized by variables such as GDP, consumption, investment, domestic price level, and interest rates. These variables, in turn, influence the trade flows between the countries, where their impact is transmitted by export and import prices and exchange rates. Therefore, the latter variables have to be modeled, too.

The effect of the ultimate driving forces of economic growth and structural change on the internal development of the industrialized countries was the subject of Chapter 3. The present chapter is concerned with aggregate imports and exports and their commodity composition, the corresponding prices, and the exchange rates of these countries. In the following section the general structure of this (partial) model is presented. Section 8.3 deals with the specification and the estimation results of the behavioral equations, and in Section 8.4 the *ex post* tracking performance of the model is demonstrated. The chapter does not contain *ex ante* simulations because, due to the partial character of the model, too many assumptions about the variables exogenous to this part would be required. For *ex ante* simulations of foreign trade and exchange rates of Developed Market Economies (DMEs), the reader is referred to Chapter 7, which deals with solutions of (a simplified version of) the complete, interdependent world model.

## 8.2. General Structure of the Model

One of the basic features of the model of exchange rates and foreign trade of the DMEs is that it consists of a set of interdependent *two-country models*. This means that each country is considered *vis-à-vis* the rest of the world (or some important part thereof). With respect to foreign trade this implies that no bilateral trade relations are considered. For exchange rate determination it means that for each currency only an exchange rate index relative to a basket of foreign currencies is explained. The indexes then serve as a basis for the determination of the dollar rates of the individual currencies.

Another important feature is the *top-down approach* adopted in modeling foreign trade. This means that in a first step aggregate foreign trade is modeled. This is then disaggregated into imports and exports of various commodity categories.

The theoretical basis of the *exchange rate equations* used in the model is the balance of payments approach to exchange rate determination. This approach is designed to explain the exchange rate in a two-country model in which each country produces a specific good and has its specific money and an interest-bearing asset. For each of these items demand functions are specified on the basis of utility or portfolio optimization, where one of the arguments of these functions is the exchange rate. Substituting the appropriate demand functions into the balance of payments equation yields an equation which can be interpreted as an equilibrium condition for the foreign exchange market. The exchange rate equation is then obtained by solving this condition for the exchange rate. This approach is used to determine the exchange rate indices mentioned above.

Regarding *aggregate foreign trade* in goods and services, the general principle used is that volumes are modeled by import and export demand functions whereas the corresponding prices are determined from the supply side. The import demand functions are based on a dynamic version of the Linear Expenditure System which allocates the income of the country under consideration to domestic and imported goods. The same approach is applied to the rest of the world to derive export demand functions. Export prices are modeled by inverted export supply functions, and import prices are explained by weighted averages of the export prices of the countries from which the imports originate, and by exchange rates.

The *structure of foreign trade* is determined by first separating total import and export expenditures into expenditures on commodities and services (as a residual) and then allocating the expenditures on commodities to the various commodity groups, using again the dynamic Linear Expenditure System. The corresponding prices are explained by the prices of aggregate foreign trade and their own lagged values.

In total, the model consists of 24 behavioral equations for each country [1] and for the summary category of other DMEs. The *exogenous explaining variables* are GDP, nominal consumption, nominal investment, the GDP deflator, the interest rate and the rate of change of money supply of the countries considered and of the rest-of-the-world category applicable in each case.

The directions of influence among the *endogenous variables* of the foreign trade and exchange rate model for DMEs are graphically represented in *Figure 8.1.* (For the notations used, the reader is referred to Appendix 8A of this chapter. The behavioral and definitional equations underlying the figure are given in the subsequent sections.)

## 8.3 Specification and Estimation of the Behavioral Equations

### 8.3.1. Exchange rates [2]

For each of the countries considered we define an exchange rate index as weighted average of its exchange rates relative to all other countries:

$$FX_i = \sum_{\substack{j=1 \\ j\neq i}}^{n} \alpha_{ij}\, FX_{ij}\,, \quad i = 1,\ldots,n, \quad \alpha_{ij} \geq 0, \quad \sum_{\substack{j=1 \\ j\neq i}}^{n} \alpha_{ij} = 1 \tag{8.1}$$

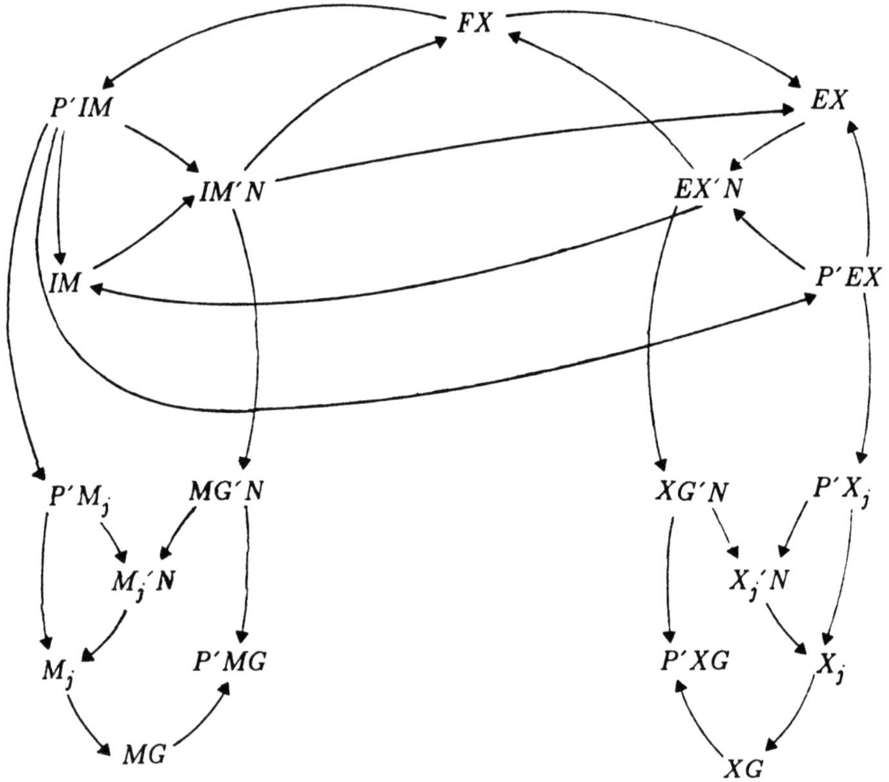


*Figure 8.1.* Graphical representation of the model structure.

where $FX_{ij}$ is the price of the currency unit of country $j$ in terms of the currency of country $i$ and $\alpha_{ij}$ is an indicator of the importance of $j$ as a trading partner of $i$. We assume that the following arbitrage condition is fulfilled (which is in fact approximately true):

$$FX_{ij} = \frac{FX_{ik}}{FX_{jk}}, \quad i, j, k = 1,...,n, \quad FX_{ii} = 1 \tag{8.2}$$

Thus, all exchange rates $FX_{ij}$ and hence all exchange rate indices $FX_i$ can be expressed in terms of the exchange rate *vis-à-vis* one arbitrary currency, say, the currency of country 1. It follows that only the $n-1$ exchange rates of all currencies with respect to currency 1, $FX_{i1}$, $i \neq 1$, have to be determined. All the other exchange rates are then fixed. The exchange rates $FX_{i1}$, $i \neq 1$, in turn, can be recovered from the $n-1$ exchange rate indices $FX_i$, $i \neq 1$. This can be seen by rewriting the last $n-1$ equations of the equation system (8.1) as follows:

$$\begin{aligned} FX_2 &= \alpha_{21}\, FX_{21} + \alpha_{23}\, \frac{FX_{21}}{FX_{31}} + \ldots + \alpha_{2n}\, \frac{FX_{21}}{FX_{n1}} \\ FX_n &= \alpha_{n1}\, FX_{n1} + \alpha_{n2}\, \frac{FX_{n1}}{FX_{21}} + \ldots + \alpha_{n,n-1}\, \frac{FX_{n1}}{FX_{n-1,1}} \end{aligned} \tag{8.3}$$

Dividing the first equation by $FX_{21}, \ldots$, the last equation by $FX_{n1}$, and rearranging, the system (8.3) may be written as

$$A \cdot \frac{1}{FX} = \alpha \quad \text{where } A = \begin{pmatrix} FX_2 & -\alpha_{23} & \ldots & -\alpha_{2n} \\ \cdot & & & \\ \cdot & & & \\ \cdot & & & \\ \alpha_{n2} & -\alpha_{n3} & \cdots & FX_n \end{pmatrix},$$

$$\frac{1}{FX} = \begin{pmatrix} \frac{1}{FX_{21}} \\ \cdot \\ \cdot \\ \cdot \\ \frac{1}{FX_{n1}} \end{pmatrix}, \quad \alpha = \begin{pmatrix} \alpha_{21} \\ \cdot \\ \cdot \\ \cdot \\ \alpha_{n1} \end{pmatrix} \tag{8.4}$$

If $|A| \neq 0$ we may solve (8.4) for the reciprocal values of all $FX_{i1}$:

$$\frac{1}{FX} = A^{-1}\alpha \tag{8.5}$$

In the empirical application, all exchange rates are normalized with respect to the US dollar.

It remains to specify the estimation equations for the *exchange rate indices* as defined by (8.1). As outlined in Section 8.2 they are obtained by substituting demand functions for the various items of the balance of payments equations.

Using the index $i$ for the country under consideration and the index $F$ for all foreign countries, the balance of payments equation of country $i$ (expressed in the foreign currency) can be written as

$$X_{iF} \frac{p_i}{FX_i} - X_{Fi} p_F + \frac{1}{FX_i} \Delta K_{iF}$$

$$- \Delta K_{Fi} - \Delta M_{Fi} = \frac{1}{FX_i} \Delta M_{iF} = 0 \tag{8.6}$$

where $FX_i$ = price of one unit of currency $F$ in terms of currency $i$; $X_{jk}$ = commodities of origin $j$ purchased by country $k$, $M_{jk}$ = money of origin $j$ held by country $k$; $p_j$ = price level of country $j$ (in currency of country $j$), and $j, k \in \{i,F\}$.

The arguments of the demand functions to be substituted for $X_{jk}$, $K_{jk}$ and $M_{jk}$ are available income, the terms of trade $p_F/(p_i/FX_i)$, the inflation rate disparity $\pi = w'p_F - w'p_i$ [3], the interest rate disparity $\rho = r_F - r_i$, the growth rate disparity of real GDP $\Delta w'y = w'y_F - w'y_i$, the disparity in the rate growth of money supply $\Delta w'm$, and the relative creditor or debtor position of country $i$ is $\lambda$, which is measured by the cumulated balance of trade in goods and services, divided by nominal GDP.

After these functions have been substituted into (8.6) the resulting equation is solved for $FX_i$, where use is made of some additional assumptions and some approximations and linearizations (see Krelle and Welsch, 1985).

The result is

$$FX_i = \frac{p_i}{p_F} (a_0 + a_1\pi + a_2\rho + a_3\rho_{-1} + a_4\rho_{-1}\pi + a_5\Delta w'y$$

$$+ a_6\Delta w'm + a_7\lambda). \tag{8.7}$$

This equation states that long-run equilibrium growth (with constant $\pi$, $\rho$, $\Delta w'y$, $\Delta w'm$, $\lambda$) determines the exchange rate by relative purchasing power parity. Deviations from it are due to monetary disequilibria, differences in growth rates and foreign indebtedness.

The equation was estimated for the period 1970–1982, using the method of OLS. The results are depicted in *Table 8.1.*

In recovering the *dollar exchange rates* from the exchange rate indices according to (8.5), it turned out that the matrix $A$ was sometimes badly conditioned, leading to considerably larger errors in the dollar rates compared with the exchange rate indices. Therefore, we chose to replace the system of definitional equations (8.5) by a system of behavioral equations which explain the dollar rate $FX_{i1}$ by the ratio of the exchange rate indices of country $i$ and country 1 (= USA) and the ratio of the weighting factors $\alpha_{1i}/\alpha_{i1}$:

$$\log FX_{i1} = a_0 + a_1 \log(FX_i/FX_1) + a_2 \log(\alpha_{1i}/\alpha_{i1}), \quad i \neq 1 \tag{8.8}$$

These equations provide an approximation to the system of equations (8.5), but they definitely work better. Their key meaning is that the dollar rate of country $i$ should reflect the relative strength of currency $i$ and the dollar as expressed by the ratio of their exchange rate indices, moderated by the strength of the bilateral trade relations $(\alpha_{1i}/\alpha_{i1})$.

The estimation results can be found in *Table 8.2* (1970–1982, OLS).

### 8.3.2. Aggregate foreign trade

Aggregate exports and imports of the DMEs are determined by demand functions.

In choosing among the various specifications of *import demand functions* [4] to be found in the literature, one usually refers to the criteria of compatibility with general demand theory and goodness of fit. In addition to these, the recent literature focuses on the application of statistical specification tests to check the reliability of these functions.

Following Thursby and Thursby (1984), the import demand specification that is (in general) most appropriate, according to these criteria, is a log-linear one where the regressors are the ratio of domestic prices to import prices, a domestic activity variable and, possibly, the lagged dependent variable.

The problem with this specification in the context of long-term forecasting models is that it tends to produce a power coefficient of the activity variable which is larger than one. This is due to the fact that the observed import ratio generally increased substantially over the last decades. Of course, using this specification with a power coefficient larger than unity implies that asymptotically the import ratio grows beyond all limits, a very unreasonable property. Therefore, our task was to find a specification which, besides fulfilling the requirements of compatibility with general demand theory and statistical reliability, gives a good description of the past and shows reasonable long-term behavior.

This is being achieved by taking into account the budget constraint which import demand has to fulfill. Moreover, the effect of predetermined demand components on this budget constraint is explicitly considered.

The basic assumption is that a country allocates its available income to the purchase of imports (*IM*) and domestic goods (*GDP*). Available income of an open economy consists of the value of domestic and foreign sales ($GDP'N$ and $EX'N$, respectively), net transfer and capital income ($B'TRF$, $B'I$) and the net increase in debts ($B'C$ = balance of capital flows). Thus, denoting the import deflator by $P'IM$ and the GDP deflator by $P'GDP$, the budget constraint (which is, of course, equivalent to the balance of payments equation) reads

*Table 8.1.* Parameters of the exchange rate equation (8.7).

| | *USA* | *FRG* | *Japan* | *France* | *UK* | *Italy* | *Netherlands* | *Belgium* | *Canada* | *Other* |
|---|---|---|---|---|---|---|---|---|---|---|
| $a_0$ | 1.11 | 2.32 | 0.872 | 1.07 | 0.863 | 0.925 | 1.40 | 1.40 | 1.31 | .433 |
| | (61.30) | (8.19) | (9.73) | (40.30) | (20.30) | (24.39) | (20.71) | (22.37) | (18.33) | (1.75) |
| $a_1$ | – | – | – | – | – | −0.016 | – | −0.025 | – | – |
| | | | | | | (4.06) | | (4.10) | | |
| $a_2$ | 0.0115 | 0.043 | – | – | – | 0.008 | 0.039[a] | 0.0167 | 0.028 | 0.036 |
| | (1.46) | (2.88) | | | | (1.28) | −0.028[a] | (1.19) | (1.64) | (1.26) |
| $a_3$ | −0.050 | 0.057 | 0.005 | 0.042 | – | – | – | – | – | – |
| | (6.01) | (5.32) | (2.87) | (2.44) | | | | | | |
| $a_4$ | −0.016 | – | – | 0.014 | −0.002 | −0.001 | 0.007 | | | −0.026 |
| | (2.54) | | | (2.08) | (2.00) | (2.33) | (6.06) | | | (2.37) |
| $a_5$ | −0.023 | – | – | – | −0.015 | – | – | – | – | – |
| | (4.68) | | | | (1.55) | | | | | |
| $a_6$ | −0.024 | – | – | – | – | −0.006 | – | −0.008 | −0.004 | −0.009 |
| | (4.68) | | | | | (4.53) | | (2.34) | (1.41) | (1.45) |
| $a_7$ | – | −6.72 | −4.51 | −2.50 | −3.75 | −1.10 | −1.64 | −3.00 | −3.35 | −5.93 |
| | | (5.20) | (3.55) | (3.63) | (5.96) | (2.38) | (7.39) | (8.29) | (4.38) | (2.20) |
| $\bar{R}^2$ | 0.671 | 0.961 | 0.874 | 0.914 | 0.874 | 0.993 | 0.829 | 0.782 | 0.841 | 0.580 |
| $DW$ | 2.80 | 1.72 | 1.45 | 2.23 | 1.97 | 1.93 | 1.57 | 2.93 | 1.67 | 1.70 |

[a] For the Netherlands we substituted $a_{21}\, r_F\, (= 2.69) + a_{22}\, r_i (= 1.84)$ for $a_2 \rho$ .

*Table 8.2.* Parameters of the dollar exchange rate equation (8.8).

| | *FRG* | *Japan* | *France* | *UK* | *Italy* | *Netherlands* | *Belgium* | *Canada* | *Other* |
|---|---|---|---|---|---|---|---|---|---|
| $a_0$ | −0.056 | −0.301 | −0.197 | −0.139 | −0.580 | −0.160 | −0.185 | −0.564 | −0.086 |
| | (2.44) | (2.27) | (4.73) | (2.67) | (7.10) | (1.58) | (1.14) | (5.27) | (2.39) |
| $a_1$ | 0.867 | 0.748 | 1.068 | 0.984 | 0.929 | 1.27 | 1.25 | 0.314 | 0.865 |
| | (21.30) | (17.60) | (13.10) | (17.20) | (40.68) | (10.34) | (7.21) | (2.70) | (8.38) |
| $a_2$ | 0.572 | 0.260 | 0.379 | 0.164 | 0.617 | 0.260 | 0.267 | 0.822 | 0.366 |
| | (4.88) | (2.17) | (5.02) | (3.04) | (7.79) | (1.81) | (1.41) | (5.31) | (2.39) |
| $\bar{R}^2$ | 0.989 | 0.991 | 0.964 | 0.969 | 0.993 | 0.986 | 0.964 | 0.919 | 0.944 |
| $DW$ | 1.72 | 1.07 | 1.67 | 1.00 | 2.00 | 1.68 | 1.22 | 1.39 | 0.980 |

$$P'IM \cdot IM + P'GDP \cdot GDP = GDP'N + EX'N + B'TRF + B' + B'C \tag{8.9}$$

On the other hand, from the national accounts identity we have

$$P'IM \cdot IM + P'GDP \cdot GDP = C'N + I'N + EX'N \tag{8.10}$$

where $C'N$ and $I'N$ denote the value of consumption and investment, respectively. Therefore, instead of the RHS of (8.9) we can use the RHS of (8.10) as income variable.

We now assume that actual demand for imports is the sum of "free demand" $IM^*$ and base demand, which is proportional to lagged demand [5]:

$$IM = IM^* + \lambda IM_{-1}, \quad \lambda \geq 0 \tag{8.11}$$

An analogous assumption holds for the demand for domestic goods.

Free demand for imports and domestic goods is determined by maximizing a Cobb–Douglas utility function subject to the constraint that its value equals available income minus the expenditures on base demand. The resulting import demand equation is

$$IM = aIM_{-1} + b\,\frac{C'N + I'N + EX'N}{P'IM} - c\,\frac{P'GDP}{P'IM}\,GDP_{-1} \tag{8.12}$$

where $IM'N$ denotes the value of imports and $a$, $b$, $c$ are non-negative coefficients (for details see Welsch, 1987).

In empirical applications of this equation we admit the possibility that there is not just one decision making unit allocating its income $C'n + I'N + EX'N$ to imports and domestic goods but that we have consumption goods, investment goods and export goods sectors with different propensities to import. In this case equation (8.12) becomes

$$IM = aIM_{-1} + b_1\frac{C'N}{P'IM} + b_2\frac{I'N}{P'IM} + b_3\frac{EX'N}{P'IM} - c\,\frac{P'GDP}{P'IM}\,GDP_{-1} \tag{8.13}$$

This is the basic import demand equation in our model. The equations actually used are special cases which are obtained by imposing zero or equality restrictions on certain parameters. The estimation results are listed in *Table 8.3* (1962–1982, OLS).

The approach to import demand modeling just outlined is also used to derive *export demand functions.* In order to do this one has to take the view that all countries different from the one under consideration are lumped together. Exports of any one country are then determined by a function describing the demand of all these foreign countries for imports from the country considered. Equation (8.12) can immediately be interpreted as describing this demand if one takes the variables included in this equation as referring to the total of foreign countries.

In order to write down the export demand function explicitly, we introduce the notation that variables referring to the group of countries different from the country considered are indicated by the suffix "F". As the common currency of this group we choose the US dollar. We assume that the import price faced by the group is the export price of the country considered ($P'EX$) divided by the dollar exchange rate of this country ($FX$). Then the export function can be written as

$$\begin{aligned} EX = aEX_{-1} &+ b\frac{C'NF + I'NF + EX'NF}{P'EX/FX} \\ &- c\frac{P'GDPF}{P'EX/FX}GDPF_{-1} \end{aligned} \tag{8.14}$$

It should be noted that the variable $EX'NF$ appearing on the RHS is, of course, given by the term $IM'N/FX$.

In the empirical application of this function it turned out to be useful to split up the income components $C'NF$ and/or $I'NF$ according to subgroups of the group of countries different from the country considered. Moreover, in the equations actually used, zero restrictions for some of the coefficients were imposed. The estimated equation has the form

$$\begin{aligned} EX = aEX_{-1} &+ b_1\frac{C'NF_1 + I'NF_1 + IM'N/FX}{P'EX/FX} + b_2\frac{C'NF_3}{P'EX/FX} \\ &+ b_3\frac{I'NF_3}{P'EX/FX} - c\frac{P'GDPF}{P'EX/FX}GDPF_{-1} \end{aligned} \tag{8.15}$$

where $C'NF_1$ and $I'NF_1$ refer to DMEs (different from the country considered) and $C'NF_3$ and $I'NF_3$ to developing countries. Consumption and investment of CMEA countries were not found to be significant. Estimation results are depicted in *Table 8.4* (1962–1981, OLS).

*Table 8.3.* Parameters of the import equation (8.13).

| | USA | FRG | Japan[a] | France | UK | Italy | Netherlands | Belgium | Canada | Other |
|---|---|---|---|---|---|---|---|---|---|---|
| $a$ | 0.895 | 0.694 | 0.656 | 0.594 | 0.564 | 0.541 | 0.359 | 0.312 | 0.459 | 0.537 |
| | (25.39) | (9.80) | (7.38) | (6.37) | (8.94) | (10.53) | (6.57) | (5.67) | (5.79) | (6.80) |
| $b_1$ | 0.132 | 0.300 | 0.222 | 0.527 | 0.225 | 0.185 | 0.612 | 0.520 | 0.148 | 0.387 |
| | (4.58) | (5.22) | (3.22) | (5.31) | (3.41) | (2.21) | (8.21) | (6.12) | (1.79) | (2.32) |
| $b_2$ | $b_2 = b_1$ | 0.251 | – | 0.472 | 0.384 | 0.361 | 0.373 | 0.248 | 0.510 | 0.445 |
| | | (5.34) | – | (4.82) | (7.10) | (5.52) | (6.57) | (1.78) | (8.42) | (4.94) |
| $b_3$ | $b_3 = b_1$ | 0.316 | 0.629 | 0.570 | 0.443 | 0.533 | 0.563 | 0.686 | 0.518 | 0.646 |
| | | (3.27) | (3.49) | (5.34) | (6.47) | (8.58) | (8.07) | (11.99) | (5.58) | (5.93) |
| $c$ | 0.135 | 0.291 | 0.067 | 0.557 | 0.257 | 0.241 | 0.528 | 0.464 | 0.234 | 0.455 |
| | (4.35) | (6.60) | (1.62) | (5.84) | (4.66) | (4.00) | (7.04) | (5.27) | (3.24) | (3.05) |
| $\bar{R}^2$ | 0.980 | 0.998 | 0.991 | 0.997 | 0.992 | 0.996 | 0.999 | 0.999 | 0.996 | 0.995 |
| $DW$ | 1.96 | 1.80 | 2.66 | 1.67 | 2.51 | 2.04 | 1.97 | 2.18 | 1.81 | 2.50 |

[a]In the case of Japan a slightly more general approach was used. It includes the additional terms $-0.421(3.38) \cdot IM'N_{-1}/P'IM$ and $-0.082(1.64) \cdot GDP'N_{-1}/P'IM$.

*Table 8.4.* Parameters of the export equation (8.15).

| | USA | FRG | Japan | France | UK | Italy | Netherlands | Belgium | Canada | Other |
|---|---|---|---|---|---|---|---|---|---|---|
| $a$ | 0.656 | 0.778 | 0.455 | 0.773 | 0.824 | 0.651 | 0.891 | 0.892 | 0.755 | 0.933 |
| | (10.53) | (8.94) | (2.92) | (11.99) | (19.73) | (6.58) | (34.30) | (23.14) | (14.34) | (45.50) |
| $b_1$ | 0.019 | 0.030 | 0.006 | 0.021 | 0.011 | 0.015 | 0.014 | 0.014 | 0.013 | 0.028 |
| | (6.48) | (2.84) | (1.56) | (5.70) | (2.58) | (3.23) | (4.69) | (4.19) | (4.41) | (4.96) |
| $b_2$ | – | – | 0.051 | – | – | – | – | – | – | – |
| | | | (4.08) | | | | | | | |
| $b_3$ | – | 0.083 | $b_3 = b_2$ | 0.051 | – | 0.049 | – | – | – | – |
| | | (2.27) | | (2.74) | (2.71) | (2.71) | | | | |
| $c$ | – | 0.020 | 21.05 | 0.014 | 0.006 | 0.009 | 0.009 | 0.009 | 0.008 | 0.018 |
| | | (2.68) | (2.60) | (5.29) | (1.78) | (2.90) | (4.31) | (3.92) | (3.57) | (4.52) |
| $\bar{R}^2$ | 0.993 | 0.991 | 0.992 | 0.998 | 0.991 | 0.993 | 0.994 | 0.990 | 0.993 | 0.997 |
| $DW$ | 1.95 | 2.54 | 1.71 | 2.15 | 2.64 | 2.41 | 2.30 | 2.39 | 2.03 | 2.08 |

Since export and import quantities are determined by demand functions, we model the corresponding prices in a way which reflects the supply side of international trade. This means that conceptually *export price functions* are inverted export supply functions. They can be derived as follows (for a similar approach see Stevens *et al.*, 1984). Suppose that each country produces a homogeneous good ($X$), which can either be used domestically or be exported:

$$X = A + EX \tag{8.16}$$

where $A$ = absorption (= consumption + investment).

A representative firm is assumed to maximize its profit $P'A \cdot A + P'EX \cdot EX - C(A + EX)$ with respect to $A$ and $EX$, where $C(A + EX)$ is the cost function and $P'A$, $P'EX$ denote the absorption and export deflator, respectively. The well-known result in the case of perfect competition is that prices are given by

$$P'A = P'EX = MC \tag{8.17}$$

and in the case of discriminating monopolistic behavior by

$$\begin{aligned} P'A &= \left(1 + \frac{1}{\varepsilon_{A,P'A}}\right)^{-1} \cdot MC, \\ P'EX &= \left(1 + \frac{1}{\varepsilon_{EX,P'EX}}\right)^{-1} \cdot MC \end{aligned} \tag{8.18}$$

where $MC$ denotes the marginal costs and $\varepsilon_{A,P'A}$, $\varepsilon_{EX,P'EX}$ are the price elasticities of domestic absorption and exports.

In order to get an estimatable equation for the export price, we have to specify the marginal cost function and, in the monopolistic case, the markup term $(1 + 1/\varepsilon_{EX,P'EX})$ in terms of observable variables.

Regarding marginal costs it can be assumed that the output $X$ is produced according to a Cobb–Douglas production function in GDP and imported goods:

$$X = \alpha GDP^{\beta}\, IM^{1-\beta} \tag{8.19}$$

This specification could be tracked back to a production function for $X$, where the inputs are capital, labor and imports, and a production function for GDP in labor and capital alone, where both of these functions are linear homogeneous and of Cobb–Douglas type and an additional consistency assumption is required.

Minimizing the expenditures

$$C = P'GDP \cdot GDP + P'IM \cdot IM \tag{8.20}$$

with respect to $GDP$ and $IM$ subject to (8.19), substituting the resulting demand functions for $GDP$ and $IM$ into (8.20) and taking the derivative of this with respect to $X$ gives the marginal cost function dual to (8.19):

$$\begin{aligned} MC &= \frac{1}{\alpha} \left[\frac{P'GDP}{\beta}\right]^{\beta} \left[\frac{P'IM}{1-\beta}\right]^{1-\beta} \\ &= \tilde{\alpha}\, P'GDP^{\beta} \cdot P'IM^{1-\beta} \end{aligned} \tag{8.20}$$

where $\tilde{\alpha}$ is a function of $\alpha$ and $\beta$.

Inserting this into the second equation of (8.18) yields

$$P'EX = \tilde{\alpha} \left[1 + \frac{1}{\varepsilon_{EX,P'EX}}\right]^{-1} P'GDP^{\beta} \cdot P'IM^{1-\beta} \tag{8.22}$$

Next, the markup term $(1 + 1/\varepsilon_{EX,P'EX})^{-1}$ may be simplified in a specific way. To do this, it should be noted that the export price elasticity $\varepsilon_{EX,P'EX}$ according to the export demand equation (8.15) has the form

$$\varepsilon_{EX,P'EX} = a\, \frac{EX_{-1}}{EX} - 1\,, \quad a \geq 0 \tag{8.23}$$

From this it can be seen that $\varepsilon_{EX,P'EX}$ decreases for increasing $EX/EX_{-1}$. Consequently, according to (8.22), the markup of export prices over marginal costs also decreases as $EX/EX_{-1}$ increases.

Therefore, we choose to substitute for the markup term in equation (8.22) the expression $(EX/EX_{-1})^{\delta}$, where $\delta$ is expected to be negative [6]:

$$P'EX = \tilde{\alpha} \left[\frac{EX}{EX_{-1}}\right]^{\delta} P'GDP^{\beta} \cdot P'IM^{1-\beta} \tag{8.24}$$

Furthermore, it may be assumed that export prices cannot be adjusted instantaneously to the current marginal costs. This means that lagged marginal costs, which are a function of lagged $P'GDP$ and lagged $P'IM$, affect the export price. Thus, equation (8.24) describes the desired price, whereas the actual export price is given by

$$P'EX = \tilde{\alpha}\left(\frac{EX}{EX_{-1}}\right)^{\delta}\left(P'GDP^{\beta}\cdot P'IM^{1-\beta}\right)^{\gamma_1}\cdot B^{\gamma_2} \tag{8.25}$$

where $B$ is a function of the lagged values of $P'GDP$ and $P'IM$. We specify $B$ as follows:

$$\log B = b\sum_{i=1}^{\infty}\lambda^{i}\log P'GDP_{-i} + c\sum_{i=1}^{\infty}\lambda^{i}\log P'IM_{-1}\,, \quad \lambda\in(0,1) \tag{8.26}$$

which means that the lagged values of log $P'GDP$ and log $P'IM$ influence log $B$ with geometrically decreasing weights. Substituting (8.26) into the logarithmic version of (8.25) and applying the Koyck transformation (i.e., deducting $\lambda$ log $P'EX_{-1}$ and solving for log $P'EX$) gives

$$\log P'EX = a_0 + a_1\log\left(\frac{EX}{EX_{-1}}\right) + a_2\log P'GDP + a_3\log P'IM + a_4\log P'GDP_{-1} + a_5\log P'IM_{-1} + a_6\log P'EX_{-1} \tag{8.27}$$

where all coefficients are functions of the basic structural parameters. It should be noted that the signs of $a_4$ and $a_5$ are undetermined.

This is our basic estimation equation. In the empirical application zero restrictions were used for some of the coefficients. In particular, $a_1$ was set to zero for most countries, which means that in these cases there is no significant monopolistic power. The estimation results are given in *Table 8.5* (1962–1982, OLS).

*Import prices* are also supply-determined. They are explained by the average export price of the countries of origin, adjusted by the exchange rates. Denoting these adjusted average export prices by $PM'EX$ our basic assumption on import price determination can be written as

$$P'IM = \alpha_0 + \alpha_1\, PM'EX + \alpha_2\sum_{i=1}^{\infty}\mu^{i}\, PM'EX_{-i} \tag{8.28}$$

*Table 8.5.* Parameters of the export price equation (8.27).

| | *USA* | *FRG* | *Japan* | *France* | *UK* | *Italy* | *Netherlands* | *Belgium* | *Canada* | *Other* |
|---|---|---|---|---|---|---|---|---|---|---|
| $a_0$ | −0.003 | −0.010 | −0.093 | 0.036 | 0.055 | 0.065 | 0.069 | −0.003 | −0.055 | −0.007 |
| | (0.52) | (3.00) | (10.34) | (6.72) | (10.15) | (10.95) | (5.17) | (1.24) | (2.98) | (0.39) |
| $a_1$ | −0.167 | −0.107 | – | – | – | – | – | – | – | −0.102 |
| | (11.17) | (2.32) | | | | | | | | (1.15) |
| $a_2$ | 0.467 | 0.497 | 1.057 | 0.100 | 0.315 | 0.141 | – | 0.227 | 0.274 | 0.341 |
| | (3.09) | (23.63) | (9.93) | (1.50) | (6.84) | (3.07) | | (7.33) | (2.52) | (5.09) |
| $a_3$ | 0.448 | 0.405 | 0.271 | 0.503 | 0.407 | 0.497 | – | 0.771 | 1.06 | 0.625 |
| | (16.57) | (18.28) | (10.18) | (10.35) | (18.30) | (15.23) | | (22.70) | (6.52) | (6.94) |
| $a_4$ | – | – | −0.949 | – | – | – | 0.278 | – | – | – |
| | | | (8.10) | | | | (3.55) | | | |
| $a_5$ | – | – | – | −0.296 | – | – | 0.646 | – | −0.702 | −0.237 |
| | | | | (2.46) | | | (6.47) | | (3.74) | (1.48) |
| $a_6$ | – | – | 0.227 | 0.698 | 0.269 | 0.307 | – | – | 0.372 | 0.232 |
| | | | (2.93) | (3.59) | (6.38) | (6.86) | | | (1.94) | (0.80) |
| $\bar{R}^2$ | 0.999 | 0.999 | 0.994 | 0.999 | 1.000 | 1.000 | 0.978 | 0.999 | 0.997 | 1.000 |
| $DW$ | 2.02 | 1.64 | 2.12 | 1.50 | 2.14 | 1.59 | 1.76 | 1.81 | 1.85 | 1.82 |

*Table 8.6.* Parameters of the import price equation (8.29).

| | *USA* | *FRG* | *Japan* | *France* | *UK* | *Italy* | *Netherlands* | *Belgium* | *Canada* | *Other* |
|---|---|---|---|---|---|---|---|---|---|---|
| $a_0$ | −0.036 | −0.129 | −0.064 | 0.077 | −0.008 | 0.017 | −0.022 | −0.086 | 0.748 | 0.010 |
| | (2.36) | (3.52) | (1.41) | (1.82) | (0.74) | (1.16) | (0.61) | (6.92) | (11.00) | (0.51) |
| $a_1$ | 0.926 | 0.621 | 1.16 | 0.813 | 0.823 | 0.890 | 0.903 | 0.782 | 0.679 | 1.28 |
| | (14.82) | (4.34) | (28.93) | (10.83) | (9.92) | (11.66) | (9.64) | (11.70) | (13.32) | (1.81) |
| $a_2$ | −0.356 | – | −0.595 | −0.485 | −0.475 | −0.406 | −0.621 | – | 0.282 | −1.16 |
| | (1.64) | | (1.93) | (1.70) | (2.28) | (1.06) | (2.05) | | (4.70) | (6.11) |
| $a_3$ | 0.460 | 0.535 | 0.502 | 0.576 | 0.698 | 0.487 | 0.742 | 0.328 | – | 0.842 |
| | (2.35) | (3.68) | (1.81) | (2.01) | (4.35) | (1.33) | (3.28) | (4.54) | | (5.33) |
| $\bar{R}^2$ | 0.998 | 0.997 | 0.997 | 0.996 | 0.998 | 0.998 | 0.994 | 0.997 | 0.999 | 1.000 |
| $DW$ | 1.94 | 1.44 | 1.62 | 1.48 | 2.30 | 1.77 | 1.60 | 1.71 | 1.94 | 1.32 |

This means that import prices are a linear function of actual and lagged average export prices of the suppliers, where the weights of the lags decrease geometrically. (In the special case $\alpha_1 = \alpha_2$ the geometrical decrease starts in the present period).

The reasoning behind this equation is that due to time lags (transportation and storage time, etc.) it may happen that prices for traded goods enter the statistics of the exporting country before the corresponding prices appear in the statistics of the importing country. Depending on which fraction of trade is delivered within one period and which fraction only later, $P'IM$ is composed of current and lagged $PM'EX$. The nature of this composition is clearly a linear one since no qualitative transformation of goods (production) is involved. (We abstract from transportation as a process which changes the quality of a good.)

Due to the fact that the assumed lag distribution is a geometrical one, equation (8.28) can easily be made estimable via the Koyck transformation. This gives

$$\begin{aligned} P'IM &= \alpha_0\,(1-\mu) + \alpha_1\, PM'EX + \mu(\alpha_2 - \alpha_1) PM'EX_{-1} + \mu P'IM_{-1} \\ &= a_0 + a_1\, PM'EX + a_2\, PM'EX_{-1} + a_3\, P'IM_{-1} \end{aligned} \tag{8.29}$$

where $a_2$ will be negative for $\alpha_1 > \alpha_2$. Equation (8.29) is our basic estimation equation for import prices. The estimation results can be seen in *Table 8.6* (1962–1981, OLS).

### 8.3.3. Structure of foreign trade

A large part of the foreign trade model of DMEs is concerned with the disaggregation of imports and exports according to their commodity composition. Since aggregate foreign trade, as considered in the previous subsection, includes goods as well as services we have to separate, in a first step, these two broad categories. This could be done by using, again, the demand system approach described above. However, to do this, deflators for trade in services would be required, which are not available. Therefore, we modify the dynamic Linear Expenditure System by assuming that for services there are no base quantities, only base expenditures, proportional to lagged expenditures. Then we get the following demand equation for *total commodity exports* (in expenditure form):

$$XG'N = a\; P'XG \cdot XG_{-1} + b\; EX'N - c\,(EX'N_{-1} - XG'N_{-1}) \tag{8.30}$$

where $XG'N$ denotes total expenditures on commodity exports, $P'XG$ and $XG$ are the corresponding deflator and volume, and $EX'N$ is expenditures on exports of commodities and services. A completely analogous equation applies to *total commodity imports.* The estimation results for these equations are very good. We do not reproduce them here due to space considerations.

Our next task is to allocate total commodity export and import expenditures to the various *commodity categories.* The categories considered are 1 = *AGR* = *SITC* 0 + 1, 2 = *CRU* = *SITC* 2 + 4, 3 = *MFL* = *SITC* 3, and 4 = *IND* = *SITC* 5+ ,..., + 9. Again, the dynamic Linear Expenditure System is applied.

The estimation equations (in expenditure form) for export demand for commodity category $j \in \{AGR, CRU, MFL, IND\}$ are [7]:

$$X_j'N = a\ P'X_j \cdot X_{j,-1} + b\ XG'N - \sum_{k \neq j} c_k\ P'X_k \cdot X_{k,-1} \tag{8.31}$$

The equations for imports are analogous. The coefficients $a$, $b$ and $c_k$ are functions of the structural parameters that characterize each commodity group within the demand system: the habit persistence parameters $\lambda_j$ and the marginal expenditure shares $\beta_j$. These functional relationships imply cross-equation restrictions for the estimation coefficients $a$, $b$ and $c_k$ which were imposed as side conditions in the estimation procedure. The systems were estimated using the method of Full Information Maximum Likelihood (FIML), for the period 1963–1983. The $\bar{R}^2$ and $DW$ statistics are given in *Table 8.7*, while *Table 8.8* gives an overview of the implied structural parameters. ($M1$,..., $M4$ and $X1$,..., $X4$ refer to imports and exports, respectively, of the four categories mentioned above).

It remains to specify equations for the *prices of the commodity categories.* We chose the following specification:

$$\log P'X_j = a_0 + a_1 \log P'EX + a_2\ P'EX_{-1} + a_3 \log P'X_{j,-1} \tag{8.32}$$

and the analogue for imports. The estimation results are very good, but we do not reproduce them, due to space limitations.

## 8.4. Performance of the Model

To assess the performance of the complete model of exchange rates and foreign trade of the DMEs, a dynamic *ex post* simulation was run for the period 1970–1981 (which is the overlapping interval of all reference periods underlying the estimation).

The system solved comprises the behavioral equations discussed in section 3 and the following definitional equations:

$$A = A_{-1} + EX'N - IM'N \tag{8.33}$$

*Table 8.7.* Test statistics of the demand system.

| | | *USA* | *FRG* | *Japan* | *France* | *UK* | *Italy* | *Netherlands* | *Belgium* | *Canada* | *Other* |
|---|---|---|---|---|---|---|---|---|---|---|---|
| $M1$ | $\bar{R}^2$ | 0.987 | 0.989 | 0.990 | 0.979 | 0.996 | 0.991 | 0.997 | 0.995 | 0.996 | 0.998 |
| | $DW$ | 2.54 | 2.03 | 2.32 | 1.73 | 2.28 | 1.89 | 1.97 | 1.98 | 2.39 | 1.89 |
| $M2$ | $\bar{R}^2$ | 0.986 | 0.977 | 0.992 | 0.987 | 0.963 | 0.974 | 0.993 | 0.969 | 0.983 | 0.994 |
| | $DW$ | 1.17 | 1.83 | 2.04 | 1.16 | 3.01 | 1.27 | 2.05 | 1.67 | 0.91 | 0.73 |
| $M3$ | $\bar{R}^2$ | 0.983 | 0.993 | 0.997 | 0.998 | 0.992 | 0.994 | 0.978 | 0.986 | 0.968 | 0.994 |
| | $DW$ | 1.56 | 1.65 | 1.64 | 0.65 | 2.15 | 1.55 | 1.93 | 1.67 | 1.16 | 1.11 |
| $M4$ | $\bar{R}^2$ | 0.980 | 0.996 | 0.953 | 0.998 | 0.996 | 0.995 | 0.971 | 0.999 | 0.997 | 0.992 |
| | $DW$ | 1.42 | 1.78 | 1.81 | 1.44 | 2.15 | 1.74 | 2.15 | 1.55 | 1.12 | 2.10 |
| $X1$ | $\bar{R}^2$ | 0.991 | 0.996 | 0.840 | 0.995 | 0.992 | 0.993 | 0.997 | 0.994 | 0.982 | 0.994 |
| | $DW$ | 2.38 | 1.69 | 0.21 | 2.25 | 2.43 | 1.74 | 2.61 | 1.32 | 2.32 | 1.13 |
| $X2$ | $\bar{R}^2$ | 0.983 | 0.986 | 0.934 | 0.987 | 0.988 | 0.973 | 0.983 | 0.992 | 0.974 | 0.980 |
| | $DW$ | 2.02 | 1.01 | 1.34 | 1.51 | 1.75 | 2.09 | 1.11 | 2.33 | 2.05 | 0.39 |
| $X3$ | $\bar{R}^2$ | 0.926 | 0.984 | 0.925 | 0.979 | 0.989 | 0.974 | 0.982 | 0.967 | 0.989 | 0.995 |
| | $DW$ | 1.77 | 1.62 | 1.77 | 1.52 | 2.68 | 2.28 | 0.92 | 1.83 | 1.44 | 1.22 |
| $X4$ | $\bar{R}^2$ | 0.999 | 1.000 | 1.000 | 0.999 | 0.997 | 1.000 | 0.989 | 0.999 | 0.993 | 0.990 |
| | $DW$ | 2.29 | 1.29 | 0.23 | 2.01 | 2.30 | 2.05 | 1.17 | 2.10 | 2.40 | 1.92 |

*Table 8.8.* Structural parameters of the demand system.

| | *USA* | *FRG* | *Japan* | *France* | *UK* | *Italy* | *Netherlands* | *Belgium* | *Canada* | *Other* |
|---|---|---|---|---|---|---|---|---|---|---|
| $\beta_{M1}$ | 0.0675 | 0.0513 | 0.1029 | 0.1951 | 0.0447 | 0.0638 | 0.0328 | 0.0540 | 0.0206 | 0.0363 |
| $\beta_{M2}$ | 0.0539 | 0.0576 | 0.2339 | 0.0183 | 0.0698 | 0.0487 | 0.0455 | 0.0742 | 0.0395 | 0.0561 |
| $\beta_{M3}$ | 0.3203 | 0.2785 | 0.3260 | 0.1631 | 0.1491 | 0.1939 | 0.4464 | 0.0897 | 0.0757 | 0.1529 |
| $\beta_{M4}$ | 0.5583 | 0.6127 | 0.3373 | 0.6236 | 0.7365 | 0.6937 | 0.4753 | 0.7821 | 0.8642 | 0.7547 |
| $\lambda_{M1}$ | 0.9772 | 0.9872 | 1.0262 | 0.0000 | 0.9493 | 0.6992 | 1.0507 | 0.9212 | 0.9884 | 0.8757 |
| $\lambda_{M2}$ | 0.9256 | 0.9499 | 0.9760 | 0.8472 | 0.8680 | 0.6764 | 1.0260 | 0.8226 | 0.9405 | 0.7178 |
| $\lambda_{M3}$ | 0.8486 | 0.8114 | 0.9574 | 0.6022 | 0.7263 | 0.5557 | 1.0026 | 0.8325 | 0.8123 | 0.6769 |
| $\lambda_{M4}$ | 1.0238 | 0.9383 | 1.0390 | 0.4456 | 0.8776 | 0.0000 | 1.0412 | 0.7472 | 0.8998 | 0.6757 |
| $\beta_{X1}$ | 0.1591 | 0.0246 | 0.0272 | 0.1207 | 0.0341 | 0.1337 | 0.1117 | 0.0284 | 0.1031 | 0.1991 |
| $\beta_{X2}$ | 0.1988 | 0.0331 | 0.0071 | 0.0438 | 0.0271 | 0.0441 | 0.0958 | 0.0210 | 0.0624 | 0.2214 |
| $\beta_{X3}$ | 0.0151 | 0.0362 | 0.0031 | 0.0372 | 0.0140 | 0.1297 | 0.1414 | 0.0755 | 0.1428 | 0.0173 |
| $\beta_{X4}$ | 0.6270 | 0.9061 | 0.9627 | 0.7983 | 0.9249 | 0.6925 | 0.6511 | 0.8752 | 0.6917 | 0.5622 |
| $\lambda_{X1}$ | 0.3963 | 0.9183 | 0.0000 | 0.6401 | 0.6280 | 0.1974 | 0.7619 | 0.9947 | 0.6667 | 0.3171 |
| $\lambda_{X2}$ | 0.0000 | 0.6485 | 0.7456 | 0.5202 | 0.2031 | 0.0000 | 0.2102 | 0.9153 | 0.8606 | 0.0000 |
| $\lambda_{X3}$ | 0.8570 | 0.6692 | 0.5437 | 0.4532 | 1.0260 | 0.0790 | 0.6186 | 0.7975 | 0.5323 | 0.9480 |
| $\lambda_{X4}$ | 0.5161 | 0.7480 | 0.6086 | 0.4732 | 0.0000 | 0.6816 | 0.4012 | 0.8391 | 0.5225 | 0.6285 |

*Table 8.9.* Mean absolute percentage errors of the foreign trade and exchange rate model of DMEs. (For definitions of terms, refer to Appendix 8A.)

| *Term* | *USA* | *FRG* | *Japan* | *France* | *UK* | *Italy* | *Neth.* | *Belg.* | *Can.* | *Other* |
|---|---|---|---|---|---|---|---|---|---|---|
| *FX* | 1.97 | 4.71 | 6.14 | 3.04 | 11.73 | 0.79 | 2.60 | 3.87 | 3.59 | 2.87 |
| *FX'dollar* | – | 3.78 | 5.34 | 3.63 | 11.71 | 2.09 | 3.26 | 3.77 | 2.52 | 1.46 |
| *IM* | 3.70 | 2.39 | 5.11 | 2.55 | 3.67 | 3.70 | 3.92 | 1.88 | 1.48 | 3.62 |
| *IM'N* | 3.87 | 5.79 | 6.90 | 3.96 | 7.15 | 2.72 | 4.18 | 5.02 | 2.85 | 2.46 |
| *P'IM* | 3.66 | 6.34 | 6.72 | 4.66 | 10.57 | 4.02 | 3.72 | 4.54 | 2.61 | 1.80 |
| *EX* | 3.33 | 2.38 | 6.18 | 1.53 | 2.72 | 2.62 | 2.66 | 2.71 | 2.38 | 4.91 |
| *EX'N* | 3.26 | 4.61 | 8.18 | 2.96 | 6.65 | 2.68 | 5.20 | 5.64 | 2.73 | 4.07 |
| *P'EX* | 1.82 | 2.55 | 2.96 | 2.46 | 4.60 | 2.07 | 4.24 | 3.40 | 3.06 | 1.62 |
| *MG* | 3.43 | 2.30 | 3.71 | 3.88 | 2.99 | 4.06 | 2.21 | 2.23 | 4.19 | 2.61 |
| *MG'N* | 1.96 | 0.91 | 1.19 | 0.80 | 2.02 | 0.59 | 0.72 | 0.98 | 1.69 | 1.30 |
| *P'MG* | 3.11 | 1.60 | 3.54 | 4.05 | 2.65 | 4.13 | 1.94 | 1.83 | 4.44 | 2.68 |
| *MAGR* | 4.09 | 3.54 | 4.60 | 9.84 | 1.82 | 7.75 | 2.98 | 3.17 | 4.03 | 2.94 |
| *MAGR'N* | 5.38 | 3.66 | 5.66 | 10.21 | 4.79 | 11.24 | 3.64 | 4.38 | 5.72 | 3.97 |
| *P'MAGR* | 4.15 | 2.75 | 4.28 | 4.79 | 5.19 | 3.95 | 3.50 | 3.44 | 3.29 | 4.09 |
| *MCRU* | 1.47 | 4.06 | 6.13 | 7.49 | 6.43 | 12.80 | 2.69 | 5.59 | 7.43 | 4.96 |
| *MCRU'N* | 4.65 | 4.40 | 6.83 | 9.71 | 6.99 | 14.40 | 4.19 | 4.24 | 8.32 | 4.35 |
| *P'MCRU* | 4.47 | 3.29 | 4.28 | 3.93 | 6.21 | 3.18 | 3.34 | 4.86 | 8.46 | 1.74 |
| *MMFL* | 12.77 | 8.94 | 4.88 | 10.48 | 6.71 | 14.79 | 9.03 | 5.96 | 9.63 | 10.87 |
| *MMFL'N* | 12.60 | 6.87 | 8.25 | 11.26 | 10.16 | 14.74 | 8.88 | 7.48 | 18.29 | 12.63 |
| *P'MMFL* | 8.57 | 5.78 | 7.95 | 6.39 | 11.95 | 7.96 | 5.43 | 4.79 | 19.08 | 15.01 |
| *MIND* | 4.92 | 2.04 | 7.59 | 2.53 | 4.76 | 2.97 | 2.84 | 3.04 | 4.76 | 2.91 |
| *MIND'N* | 4.47 | 1.21 | 5.47 | 2.06 | 4.65 | 2.24 | 2.63 | 1.75 | 2.23 | 2.28 |
| *P'MIND* | 4.79 | 1.74 | 3.16 | 2.70 | 3.54 | 2.50 | 2.09 | 2.47 | 5.01 | 1.63 |
| *XG* | 1.95 | 1.49 | 1.50 | 3.71 | 4.21 | 2.26 | 2.42 | 2.79 | 2.44 | 2.53 |
| *XG'N* | 1.31 | 0.81 | 0.97 | 1.33 | 1.97 | 0.63 | 0.89 | 1.21 | 1.03 | 1.96 |
| *P'XG* | 1.24 | 1.28 | 0.83 | 3.59 | 3.74 | 2.51 | 1.85 | 2.53 | 2.79 | 1.51 |
| *XAGR'N* | 5.77 | 3.96 | 46.91 | 4.23 | 6.22 | 10.51 | 1.64 | 3.47 | 7.98 | 4.55 |
| *XAGR'N* | 8.88 | 4.56 | 43.16 | 5.59 | 6.21 | 7.64 | 3.90 | 4.18 | 6.07 | 5.07 |
| *P'XAGR* | 3.61 | 4.02 | 4.22 | 3.09 | 3.38 | 7.24 | 3.47 | 3.73 | 4.92 | 2.58 |
| *XCRU* | 5.04 | 5.73 | 14.54 | 7.82 | 7.27 | 14.26 | 7.53 | 3.65 | 7.63 | 12.50 |
| *XCRU'N* | 4.84 | 4.54 | 12.99 | 7.39 | 7.02 | 14.62 | 5.73 | 5.32 | 9.48 | 11.81 |
| *P'XCRU* | 4.20 | 5.21 | 6.22 | 5.04 | 6.53 | 7.39 | 6.79 | 3.52 | 4.77 | 4.76 |
| *XMFL* | 14.83 | 7.37 | 14.29 | 16.54 | 21.75 | 10.06 | 10.42 | 21.09 | 12.47 | 13.64 |
| *XMFL'N* | 12.03 | 7.62 | 14.93 | 17.01 | 24.06 | 10.35 | 9.83 | 23.84 | 11.45 | 14.61 |
| *P'XMFL* | 4.57 | 6.13 | 4.48 | 8.65 | 12.69 | 8.94 | 4.63 | 9.58 | 8.51 | 12.00 |
| *XIND* | 1.37 | 1.37 | 1.77 | 4.15 | 4.52 | 1.92 | 2.55 | 3.35 | 4.52 | 4.53 |
| *XIND'N* | 1.00 | 0.76 | 1.09 | 2.19 | 2.28 | 1.04 | 1.86 | 1.71 | 1.42 | 4.21 |
| *P'XIND* | 1.24 | 1.20 | 0.85 | 3.84 | 3.95 | 2.54 | 1.19 | 2.93 | 4.22 | 1.80 |

where $A$ is the cumulated balance of trade in goods and services that enters the exchange rate equation (8.7), and

$$IM'N = P'IM \cdot IM \tag{8.34}$$

$$EX'N = P'EX \cdot EX \tag{8.35}$$

$$M_j = M_j'N/P'M_j \tag{8.36}$$

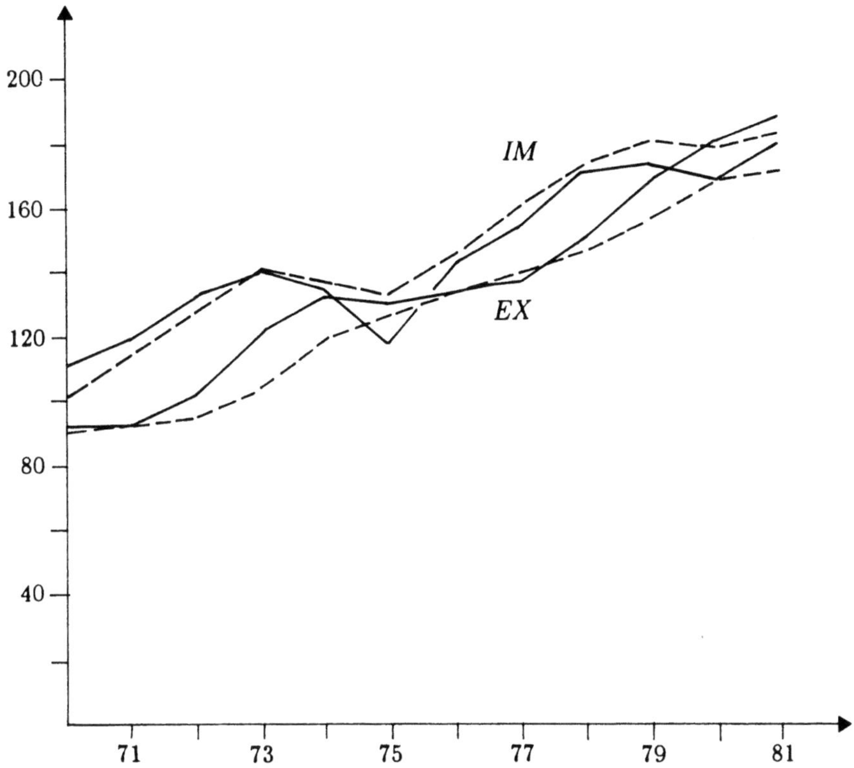


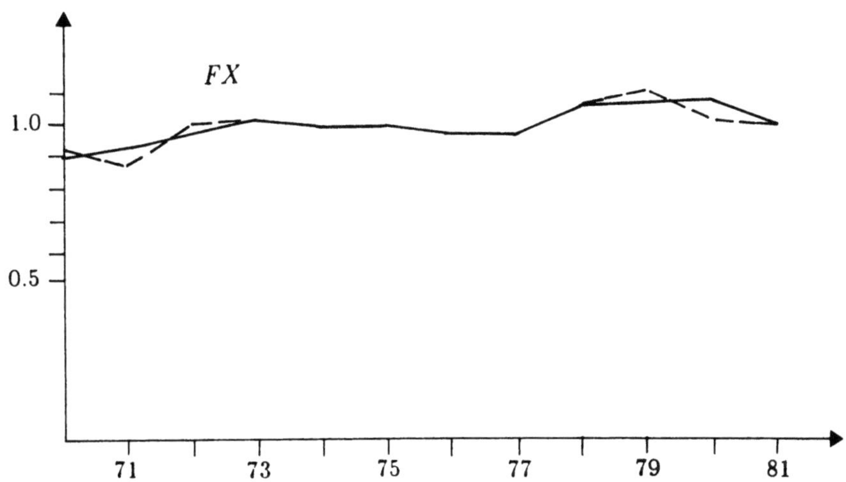


***Figure 8.2. Ex post*** **forecast for the USA: solid lines = observed data; broken lines = computed data.**

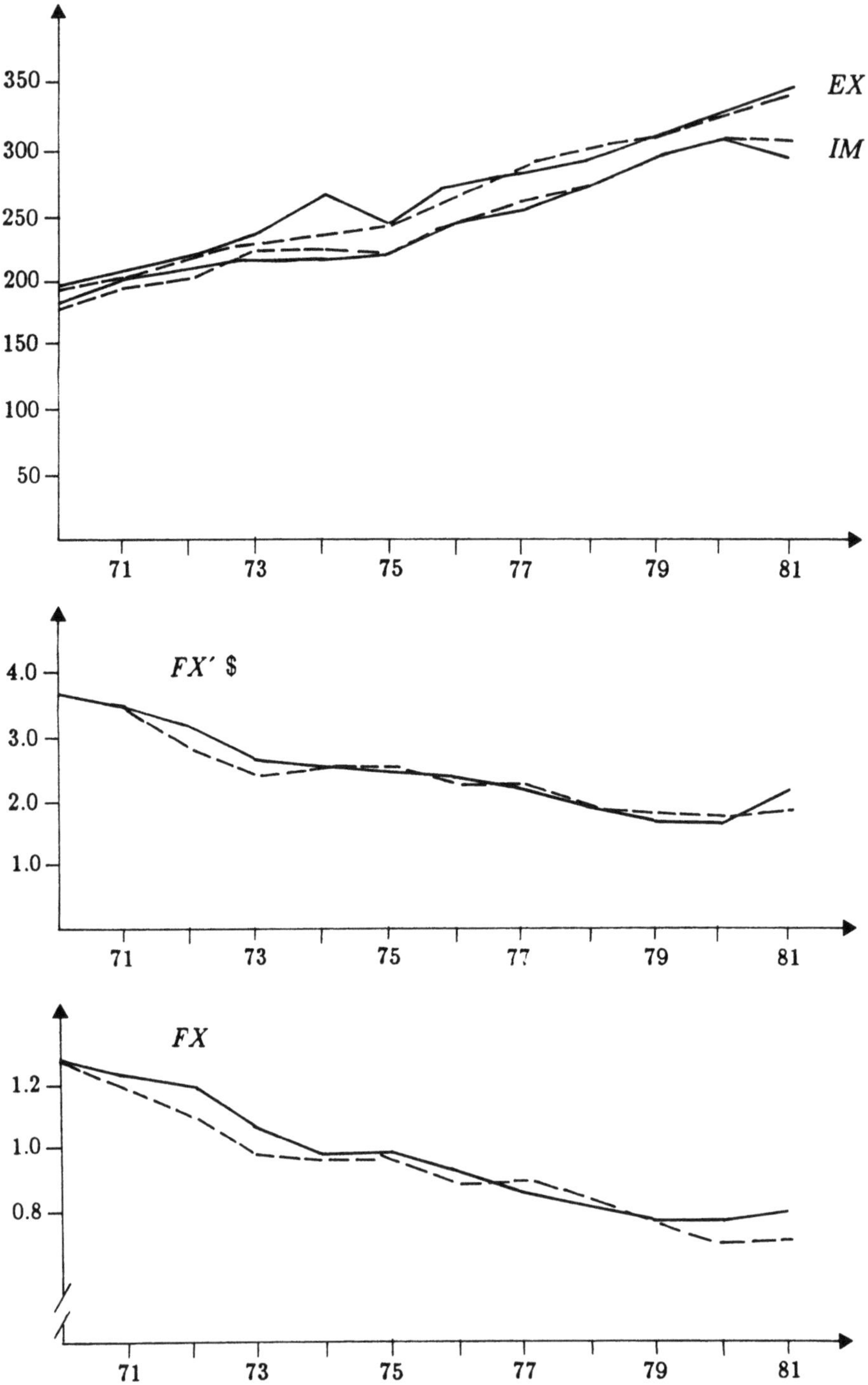


***Figure 8.3.*** ***Ex post*** **forecast for FRG: solid lines = observed data; broken lines = computed data.**

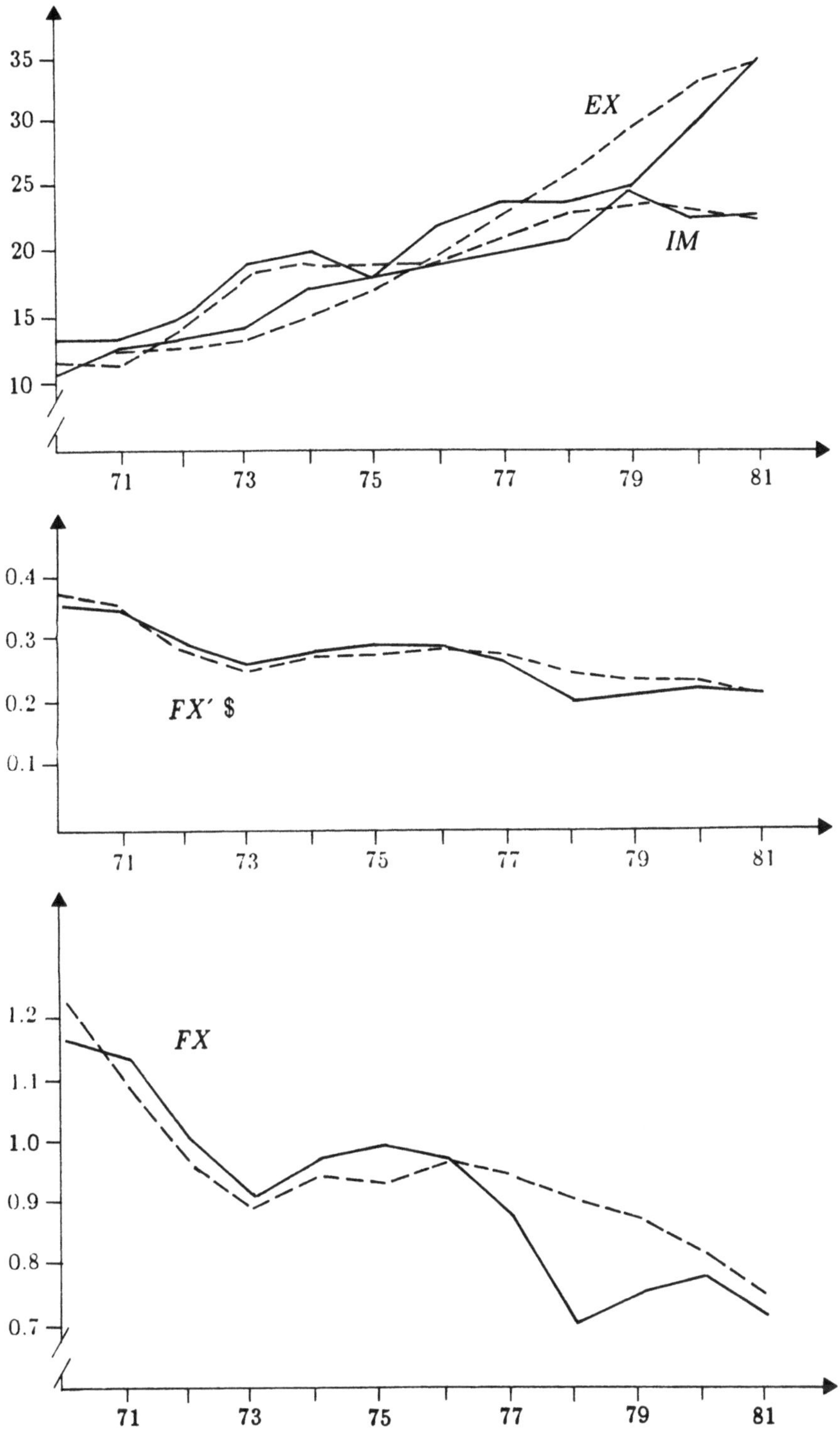


***Figure 8.4.*** ***Ex post*** **forecast for Japan: solid lines = observed data; broken lines = computed data.**

$$X_j = X_j'N/P'X_j \tag{8.37}$$

$$MG = \sum_j M_j \tag{8.38}$$

$$XG = \sum_j X_j \tag{8.39}$$

$$P'MG = MG'N/MG \tag{8.40}$$

$$P'XG = XG'N/XG \tag{8.41}$$

The mean absolute percentage errors of the *ex post* forecast are given in *Table 8.9.* In *Figures 8.2–8.4* the performance of the model is graphically demonstrated by plotting the actual and computed values of real exports and imports as well as the exchange rate indices and dollar rates of the USA, FRG and Japan.

Our conclusion is that the model is able to trace the past behavior of foreign trade and exchange rates of the DMEs in a satisfactory way.

## Notes

[1] For the USA only 23 equations are used.

[2] This subsection is based on Krelle and Welsch (1985).

[3] For each variable $z$, we denote its rate of change by $w'z$.

[4] The following paragraphs on import demand functions are based on Welsch (1987).

[5] This approach is based on the attempt of Pollak (1970) to dynamize the Linear Expenditure System proposed by Klein and Rubin (1947) and extensively used by Stone (see, e.g., Stone, 1954a).

[6] This substitution is also motivated by the fact that (8.18) is only valid for absolute price elasticities larger than unity, an assumption which need not hold empirically. Qualitatively, however, it is evident that the markup over marginal costs depends inversely on the absolute price elasticity. This relationship is captured by (8.24). Moreover, the pressure of competition is important for price setting. This influence is reflected by $P'IM$. Hence, the impact of this variable on $P'EX$ might be higher than is suggested by the above derivation.

[7] The derivation is as described in Subsection 8.3.2 for aggregate foreign trade. For details regarding the application to the structure of commodity trade, see Welsch (1986).

## Appendix 8A: Notation

| | |
|---|---|
| $EX$ | = real exports of goods and services |
| $EX'N$ | = nominal exports of goods and services |
| $FX$ | = exchange rate index relative to currency basket ($\hat{=} FX_i$) |
| $FX'\$$ | = exchange rate relative to US dollar ($\hat{=} FX_{i1}$) |
| $IM$ | = real imports of goods and services |
| $IM'N$ | = nominal imports of goods and services |
| $MG$ | = real imports of commodities |
| $M_j$ | = real imports of commodity group $j$ |
| $MG'N$ | = nominal imports of commodities |
| $M_j'N$ | = nominal imports of commodity group $j$ |
| $P'EX$ | = price index of exports of goods and services |
| $P'IM$ | = price index of imports of goods and services |
| $P'MG$ | = price index of commodity imports |
| $P'M_j$ | = price index of imports of commodity group $j$ |
| $P'XG$ | = price index of commodity exports |
| $P'X_j$ | = price index of exports of commodity group $j$ |
| $XG$ | = real exports of commodities |
| $X_j$ | = real exports of commodity group $j$ |
| $XG'N$ | = nominal exports of commodities |
| $X_j'N$ | = nominal exports of commodity group $j$ |

The index $j$ refers to commodity groups:

$j = AGR = SITC\ 0 + 1$
$j = CRU = SITC\ 2 + 4$
$j = MFL = SITC\ 3$
$j = IND = SITC\ 5 + ,..., + 9$

For instance, $P'MAGR$ denotes the price index of imports of $SITC\ 0 + 1$.

### References


Klein, L.R. and Rubin, H. (1947), A Constant-Utility Index of the Cost-of-Living, *Review of Economic Studies*, **15**, pp. 84–87.

Krelle, W. and Welsch, H. (1985), Exchange Rate Determination for Interdependent Economies, *Zeitschrift für Nationalökonomie/Journal of Economics*, **45**, pp. 373–393.

Pollak, R.A. (1970), Habit Formation and Dynamic Demand Functions, *Journal of Political Economy*, **78**, pp. 745–763.

Stevens, G.V.G, Berner, R.B., Clark, P.B., Hernández-Catá, E., Howe, H.J., and Kwack, S.Y. (1984), *The US Economy in an Interdependent World, A Multicountry Model.* (Washington, DC: Board of Governors of the Federal Reserve System, Publications Department).

Stone, R. (1954), Linear Expenditure Systems and Demand Analysis: An Application to the Pattern of British Demand, *Economic Journal*, **64**, pp. 511–527.

Thursby, J. and Thursby, M. (1984), How Reliable are Simple, Single Equation Specifications of Import Demand? *Review of Economics and Statistics*, **66**, pp. 120–128.

Welsch, H. (1986), Habit Persistence and the Structure of International Commodity Trade. A Demand System Analysis for Nine Industrialized Countries. Discussion Paper B-47, Sonderforschungsbereich 303, University of Bonn.

Welsch, H. (1987), An Aggregate Import Demand Model for Long-Term Projections, *Jahrbücher für Nationalökonomie und Statistik*, 203/204, pp. 372–389.

CHAPTER 9

# Structural Change in Foreign Trade of the CMEA Countries

*Jan B. Gajda*

## Summary

This chapter presents the model of foreign trade for the seven East European members of the CMEA. Some specific problems of the CMEA foreign exchange are discussed: the dual currency (transferable rouble and dollar) system and their exchange rates into "valuta" currencies, insulation of domestic from the world market prices, and foreign debt and its approximation. The valuta exchange rates were found to follow predominantly the behavior of the strength of US $ measured against SDR. Exports are explained mainly from the supply point of view, an important element in a model oriented toward simulation and scenario analysis; imports are considered from the demand side.

## 9.1. Introduction

We start with a short discussion of differences between the centrally planned economy and the free market one as far as the foreign trade subsystem is concerned.

The first important difference is associated with the exchange rates and currencies used to express the foreign trade of the CMEA countries. The rates refer to so-called "valuta" currency, not directly related to the domestic currency. Imported or exported goods having the same dollar price, say, may have different domestic prices and vice versa. This is roughly equivalent to variable subsidy and taxation rates applied to different items in foreign trade. No published data exist which allow us to consistently transfer the value given in the "valuta" prices into the domestic market prices.

The system of "valuta" currency introduces some kind of insulation of domestic prices from the world market prices. Basically the value of exports and imports is published in "valuta" currencies only, being in this way comparable with foreign trade of the rest of the world, but without being directly compatible with other indicators of domestic activity, such as GDP or its counterpart NMP (net material product, which is widely used in centrally planned economies), consumption, etc., expressed in domestic prices of the country.

Smaller economies bothered with foreign debt payments find total insulation from the world prices unacceptable, especially when the world prices change rapidly. The Hungarians (in 1976) and the Poles (in 1982) found it useful as well as necessary to introduce a direct connection between world and domestic prices. They abolished the "valuta" currency system and introduced direct exchange rates for their domestic currency. The exchange rates are determined by the central banks on the basis of several factors, therefore some form of insulation still exists. Nevertheless one may assume that current exchange rates in these countries reflect the differences between domestic and foreign costs and demands more precisely than in the case of countries with the "valuta" exchange rates.

An additional problem arises due to the fact that the intra-CMEA trade used to be expressed in terms of the transferable rouble. There exists a transferable rouble/dollar exchange rate (recently about 0.7 rouble per dollar). Poland and Hungary have set direct exchange rates of their domestic currency for the rouble and the dollar. The rates do not obey the arbitrage condition. As an example one may quote the Polish rates; at the end of 1985 the dollar rate was 147.88 zloty and the rouble rate 88 zloty. In Hungary too the dollar rate happens to be higher than the rouble rate. As a result the dollar value of foreign trade calculated from roubles using the direct rouble/dollar exchange rate differs considerably from the same dollar value obtained through initial conversion from roubles to the domestic currency and then from the domestic currency to dollars.

Actually, the intra-CMEA prices change according to a pattern different from that of the world prices (much slower and with considerable delay). If it were not for the compactness of the world model (and lack of properly disaggregated data) the model of foreign trade for the CMEA countries would be disaggregated into two parts: the first describing intra–CMEA trade and the second describing trade with the Rest of the World.

One might expect that for the Soviet economy the rouble/dollar exchange rate would be free from such biases. However, massive reductions of price in the Soviet domestic market made in the 1960s, supposedly owing to a decrease in the domestic costs, were not reflected in the rouble/dollar exchange rates of this period. The differences in valuations apparently accumulated in the state budget, which interferes with the economic activity through subsidies and taxes, insulating the domestic from the world market prices.

While analyzing the patterns of behavior of the "valuta" exchange rates with respect to the dollar, one can observe some similarities between the CMEA countries. Before 1972 the rates were kept at constant levels. Since 1972 they started to move in quite a similar fashion – increasing until 1979–80 and decreasing afterwards (see *Table 9.1*).

*Table 9.1.* Exchange rates of seven European CMEA countries (US/SDR rate and US GDP deflators given for comparison).

| | | | | | | | | *U S A* | |
|---|---|---|---|---|---|---|---|---|---|
| *Year* | *USSR* | *Bulgaria* | *CSSR* | *GDR* | *Hungary* | *Poland* | *Romania* | *$ per SDR* | *GDP Deflator* |
| 1960 | 0.90 | 1.17 | 7.20 | 4.20 | 56.96 | 95.21 | 6.00 | 1.00 | 0.54 |
| 1961 | 0.90 | 1.17 | 7.20 | 4.20 | 56.96 | 95.21 | 6.00 | 1.00 | 0.55 |
| 1962 | 0.90 | 1.17 | 7.20 | 4.20 | 56.96 | 95.21 | 6.00 | 1.00 | 0.56 |
| 1963 | 0.90 | 1.17 | 7.20 | 4.20 | 56.96 | 95.21 | 6.00 | 1.00 | 0.57 |
| 1964 | 0.90 | 1.17 | 7.20 | 4.20 | 56.96 | 95.21 | 6.00 | 1.00 | 0.58 |
| 1965 | 0.90 | 1.17 | 7.20 | 4.20 | 56.96 | 95.21 | 6.00 | 1.00 | 0.59 |
| 1966 | 0.90 | 1.17 | 7.20 | 4.20 | 56.96 | 95.21 | 6.00 | 1.00 | 0.61 |
| 1967 | 0.90 | 1.17 | 7.20 | 4.20 | 56.96 | 95.21 | 6.00 | 1.00 | 0.63 |
| 1968 | 0.90 | 1.17 | 7.20 | 4.20 | 56.96 | 95.21 | 6.00 | 1.00 | 0.66 |
| 1969 | 0.90 | 1.17 | 7.20 | 4.20 | 56.96 | 95.21 | 6.00 | 1.00 | 0.69 |
| 1970 | 0.90 | 1.17 | 7.20 | 4.20 | 56.96 | 95.21 | 6.00 | 1.00 | 0.73 |
| 1971 | 0.90 | 1.17 | 7.20 | 4.20 | 56.96 | 95.21 | 6.00 | 1.00 | 0.77 |
| 1972 | 0.83 | 1.08 | 6.63 | 3.87 | 52.46 | 87.55 | 5.53 | 1.09 | 0.80 |
| 1973 | 0.74 | 0.97 | 5.83 | 3.34 | 46.19 | 79.52 | 5.02 | 1.19 | 0.84 |
| 1974 | 0.75 | 0.97 | 5.84 | 3.48 | 44.39 | 79.01 | 4.97 | 1.20 | 0.92 |
| 1975 | 0.72 | 0.97 | 5.59 | 3.48 | 41.61 | 79.01 | 4.97 | 1.21 | 1.00 |
| 1976 | 0.75 | 0.97 | 5.77 | 3.48 | 41.62 | 79.01 | 4.97 | 1.16 | 1.06 |
| 1977 | 0.74 | 0.95 | 5.65 | 3.48 | 40.98 | 79.01 | 4.97 | 1.17 | 1.12 |
| 1978 | 0.68 | 0.89 | 5.42 | 3.48 | 38.08 | 75.36 | 4.56 | 1.25 | 1.20 |
| 1979 | 0.66 | 0.87 | 5.31 | 3.48 | 35.61 | 73.53 | 4.47 | 1.29 | 1.31 |
| 1980 | 0.65 | 0.88 | 5.38 | 3.30 | 32.48 | 72.69 | 4.47 | 1.30 | 1.43 |
| 1981 | 0.72 | 0.92 | 5.89 | 5.32 | 34.43 | 80.00 | 4.47 | 1.18 | 1.56 |
| 1982 | 0.73 | 0.95 | 6.12 | 3.46 | 36.85 | 85.11 | 4.47 | 1.10 | 1.67 |

Notes: "Valuta" exchange rates except for those of Hungary and Poland, which were calculated from data published in *UN Yearbook of International Trade Statistics*; for Hungary and Romania, in *IMF International Financial Statistics*; for Poland, in *WS Data Bank* and national yearbook, various issues. Minor discrepancies with other sources, due to rounding errors and existence of difference valuta courses for exports and imports, are possible.

The behavior of the rates exhibits some differences for different countries. One group of countries makes annual updates of their exchange rates. These are: Bulgaria, Czechoslovakia, Hungary and the Soviet Union. Other countries tried to keep the exchange rates stable for longer periods; only at the end of the 1970s did their exchange rates start to show annual adjustments. Romania is somewhat exceptional in this group since its exchange rate has undergone two three-year periods of annual updates interwoven with even longer periods of a stable rate; it is the only country for which the exchange rate did not decrease in 1981 and 1982.

For Romania we have also a commercial exchange rate reported to the IMF. In *Table 9.2* both Romanian rates are shown – the one reported by the IMF and the one reported by the UN. The two series exhibit a tendency to be related in a certain proportion. Since the commercial exchange rate "is in fact a coefficient so as to transform foreign currency values into domestic units of accounts . . . and a noncommercial exchange rate is used for invisibles, mainly tourism, and therefore closer to purchasing power parity criteria" (Colombatto, 1983), we shall use the noncommercial one in our further research.

*Table 9.2.* Comparison of the commercial exchange rate reported by Romania to IMF with the rate shown in *Table 9.1* (in lei per $).

| *Year* | *UN data* | *IMF data* | *Ratio of UN/IMF rates* |
|---|---|---|---|
| 1970 | 6.000 | 20.253 | 3.376 |
| 1971 | 6.000 | 20.253 | 3.376 |
| 1972 | 5.530 | 20.253 | 3.662 |
| 1973 | 5.023 | 20.253 | 4.032 |
| 1974 | 4.970 | 20.000 | 4.024 |
| 1975 | 4.970 | 20.000 | 4.024 |
| 1976 | 4.970 | 20.000 | 4.024 |
| 1977 | 4.970 | 20.000 | 4.024 |
| 1978 | 4.559 | 18.355 | 4.026 |
| 1979 | 4.470 | 18.000 | 4.027 |
| 1980 | 4.470 | 18.000 | 4.027 |
| 1981 | 4.470 | 15.000 | 3.356 |
| 1982 | 4.470 | 15.000 | 3.356 |

Due to the different role played by the "valuta" currencies in the centrally planned economies, and (at first glance) arbitrary decisions setting the "valuta" exchange rates, it was not at all obvious that one can describe the behavior of these rates in a pattern similar to that of the exchange rates of market economies. Below we show that some consistent pattern, although different from that of market economies, has been found for the exchange rates of the CMEA countries.

The second problem is associated with exports. Except for the Soviet Union and Bulgaria, all the other members of the CMEA have to cope with considerable foreign debt. This suggests that the governments of these countries should exhibit a kind of pro-export activity, resulting in an increase in the country's marginal propensity to export. The change may be assumed to be roughly proportional to the burden associated with the debt, as measured, say, by the debt per unit of exports.

Many models of the CMEA countries assume that exports are demand determined [see, e.g., Welfe (1982)]. Our explanation of the behavior of exports takes into account the fact that in the CMEA countries goods supplied for exports are in many cases of better quality than the same goods produced for the domestic market. One can assume that the fraction of goods of the "first" quality (usually supplied for exports) is quite stable, at least in the short and medium range, the total supply of them increasing as the relevant economy expands. As the world demand increases, goods of the "second" (lower) quality can be sold as well, thus shifting the export supply curve. The above statement leads us to supply-driven equations describing exports, with the world demand variable responsible for the shifts in the supply curve.

The third problem is related to imports, especially in the case of a country bothered by a heavy foreign debt. The earning capacities of exports, diminished by the debt interest and capital payments, impose strong restrictions on the value of imports. The behavior of imports in our model is demand driven with a budget constraint (see Welsch, 1986).

The fourth problem is connected with the foreign debt (net foreign assets) and the balance of payments. The information on the balance of payments is available from IMF sources for Romania and Hungary only. For Romania the data cover the years 1971–81, and for Hungary the years 1970–82. In both cases the series are expressed in US dollars. There exists some information on the balance of payments for Poland (for 1970 and 1978–81 in "valuta" zloty and for 1981–83 in both domestic zloty and dollars). For the other countries no data are available and some approximations have to be used. Following the research plan (Krelle, 1985) let us recall the basic equation for the balance of payments:

$$EX^* \; - \; IM^* \; + \; B.TR^* \; + \; B.K^* \; - \; B.FK^* \; = \; 0$$

It seems that the most important item here is the balance of capital flows $B.K^*$. One can determine it from the above equation as follows:

$$\begin{aligned} B.K^* &= \text{-}EX^* + IM^* \; \text{-}B.TR^* + B.FX^* \\ &\simeq \text{-}EX^* + IM^* \end{aligned}$$

where the balance of transfers $B.TR^*$ and the change of the stock of gold and foreign exchange $B.FX^*$ are assumed to be equal to zero. In a more elaborate version of the approximation one may assume that the two items are proportional to, say, $GDP$, $EX$ or $IM$.

Analyzing the published data, one can see that these items are absent from the Romanian balance of payments. In the case of Hungary they vary from \$36 million to \$99 million i.e., from at best 1.5% of exports $EX^*$, usually less than 1%. For Poland the balance of transfers in some exceptional years reaches 10% of exports, being dominated by the transfers of money from Polish emigrants to their relatives in the country. Other CMEA countries either do not have so many emigrants in western countries, or contacts between relatives are looser than in the case of Polish and Hungarian emigration. In some countries (Hungary and Poland, for example) the citizens are allowed to possess foreign currency; they can even get some interest on their foreign currency deposits in banks. In other countries regulations require that the citizens exchange any foreign currency for the domestic one. Sometimes they obtain special coupons to be cashed in "valuta" shops. The above-mentioned differences may have quite strong impacts on the size of inflow of the currency from abroad.

The inhabitants of the CMEA countries basically do not have income-producing property abroad, and other items entering the balance of the transfer payments are of negligible importance. Hence, the assumption of zero balance of transfers seems quite reasonable; in the case of Hungary or Poland one may replace it with the assumption of some proportionality between this inflow and exports (at, say, the 1%–5% level).

Not much is known about operations on gold and foreign currency reserves in the CMEA. These countries do not produce significant amount of gold. The only exception is the Soviet Union; however, we lack any information on this topic. The assumption of zero balance of gold and foreign currency reserves has thus a rather pragmatic character.

The suggested reformulation of the equation for the net foreign assets $A^{E,\text{net}}$, understood as a proxy for the foreign debt, has the following form:

$$A_t^{E,\text{net}} = A_{t-1}^{E,\text{net}} + (EX_t^* - IM_t^*) + zA_{t-1}^{E,\text{net}}$$

where $z$ denotes the interest rate charged on international credits. As a first approximation one may set $z \simeq 0.1$ (not much is known about the actual interest rates charged). A more sophisticated approach would introduce $z = z^E$ where $z^E$ is an "external" interest rate, calculated on the basis of the interest rates charged by the main creditors, and/or, say, the LIBOR interest rate. Such approximations are, nevertheless, of limited precision since the interest rate charged depends on the time horizon.

In some other models for the CMEA countries (see Gajda, 1983) a similar approach was tested in the late 1970s, when the data on foreign debt of the CMEA countries were neither published nor collected in one source. The accumulated foreign trade balance was used as a proxy for the foreign debt, even without the corrective term $zA_{t-1}^{E,\text{net}}$. In the early 1980s, when the estimated size of foreign debt was published, it turned out to be quite close to the series of proxies. Moreover, the proxies correlated with other variables of the IES models better than the debt series evaluated by some western sources. It may be of importance that the debt is defined in many foreign currencies like dollar, pound sterling, French and Swiss francs, German mark, etc. In most cases it is to be repayed in the same currency. When expressed in dollars, its value is sensitive to variation in exchange rates between these currencies.

## 9.2. The Initial Specification

The initial assumptions specified in the research plan (see Krelle, 1985) are briefly reviewed below for convenience of further exposition.

It was assumed that the imports satisfying the needs of domestic production, consumption and investment are proportional to the real national income $Y$ (in the case of the CMEA countries NMP – net material product – rather than GDP was taken as the income):

$$IM = \beta^{**}Y \tag{9.1}$$

Exports are function of world demand, represented by real income abroad $Y^E$:

$$\begin{aligned} EX &= \beta^* Y^E \\ &= \beta^* \sum_{i \subset N} \alpha_i^* \, Y_i \end{aligned} \tag{9.2}$$

where $N =$ a set of indices of countries trading with the given country; $\alpha_i^* =$ a measure of importance of country $i$ in the trade with the country under consideration; and $Y_i =$ the real income of country $i$.

The "variable parameter" $\alpha_i$ is to be explained as follows:

$$\alpha_i = a_i + a_2 \frac{(p/e)^E}{(p/e)} + a_3(w_{K/L} - w_{K/L}^E) + a_4(w_Y - w_Y^E) \tag{9.3}$$

where $w_l$ denotes the rate of growth of $l$; $K/L =$ the capital/labor ratio; $p =$ price deflator (in this case, the domestic one); $e =$ exchange rate; the superscript $E$ denotes the variables "abroad", i.e., calculated from the data of the other countries, similar to the real-income-abroad variable $Y^E$ shown above. The "variable parameter" $\beta_i$ of (9.2) is to be explained in a similar way.

The export price deflator $p_{EX}$ depends on the price level $p$ of the country under consideration:

$$p_{EX} = \gamma \, p \tag{9.4}$$

the "variable parameter" $\gamma$ is to be explained similarly to (9.3).

Prices of imports depend on the export prices abroad (i.e., charged by other countries) $p_{EX}^E$:

$$p_{IM} = a_i + a_2 \, p_{EX}^E + a_3 \, p_{EX,t-1}^E + \cdots + u_p \tag{9.5}$$

Thus, the exports of the country under consideration depend on the demand of other countries, while the imports depend on domestic demand, characterized by the level of domestic activity, both modified accordingly to measures of the relative dynamics and prices in the given country and in the world. Prices of exports are in turn determined by the level of domestic prices; prices of imports, by prices abroad.

The exchange rates stem from the hypothesis of purchasing power parity of the domestic currency and US dollar:

$$e_i = \gamma^* \frac{p_i}{p_{01}} \tag{9.6}$$

where the "varying parameter" $\gamma^*$ depends on the measures of purchasing power disparity, interest rate disparity, interest rate abroad, etc., and $p_{01}$ is the US economy price deflator.

The balance of payments equation

$$EX^* - IM^* + B.TR^* + B.K^* - B.FX^* = 0 \tag{9.7}$$

contains (accordingly to SNA approach) the following variables: $EX^*$ = exports, $IM^*$ = imports, defining the current account balance $EX^* - IM^*$, and $B.TR^*$ = balance of transfers, $B.K^*$ = balance of capital flows, $B.FX^*$ = change in stock of gold and foreign reserves, amounting to the foreign trade balance. The endogenous variable $B.K^*$ is to be calculated as a residual.

## 9.3. Model Equations

### 9.3.1. Exports

Estimations of equations explaining the behavior of exports began with the initial specification, according to the research plan (see Krelle, 1985). The results were of nonuniform quality; the values predicted by the model equations had higher variability than that of the empirical values of modeled variables. The final specification of the export equations was thus chosen as follows:

$$EX = a_i + a_2\, NMP + a_3 \left( \frac{NFA\$N_{t-1}}{EX\$N_{t-1}} \right) + a_4\, NMP^*dum + a_5 dum + u_{EX} \tag{9.8}$$

where $EX$ = real exports, domestic (or valuta) currency; $NMP$ = real net material product, domestic currency; $NFA\$N$ = accumulated foreign trade balance, US \$; $EX\$N$ = nominal exports in US \$; $dum$ = dummy variable (described in the text).

We argued that, in the behavior of the CMEA exports, supply plays an important role that should not be overlooked. The well-known problem of these countries is the lack of goods of quality high enough for foreign customers, accompanied by the lack of a network of services assuring repair. Products supplied for domestic markets frequently happen to have lower quality than the same products exported (see Mejstrik, 1983). If one assumes that the shares of goods of the "first", "second" and "third" quality levels in the total production stay relatively constant, then obviously the supply of these goods increases as the domestic output increases. On the other hand, the share of these countries in the world's trade is relatively small. Thus, one can expect that in most cases the increased supply of goods of the highest quality can be sold on the world market without throwing the market out of balance (such special goods as copper, oil or sugar seem to be exceptions). Hence, in general we will be inclined to explain exports as being driven by the domestic output, measured by NMP.

It is assumed that the higher the burden associated with the foreign debt, the harder the administration tries to increase the marginal propensity to export. The debt pressure is characterized by the ratio of the last year's debt (for the reasons explained earlier, this is measured as the accumulated foreign trade balance *NFA$N*) to the last year's exports. One can interpret the absolute value of this ratio as the number of years the country would need to repay its debt using all its earnings from exports (assumed to be equal to the last year's earnings). Observe that an increase of debt is associated with negative balance of payments (or its approximation – foreign trade balance); hence, the positive debt is reflected in negative values of the accumulated foreign trade balance *NFA$N* (and thus *NFASN/EX$N*). The Soviet Union, having a systematic positive foreign trade balance and simultaneously being in debt to western banks, is an exception. The probable reason is that the Soviet Union's trade with the developing countries more than offsets the negative balance with the developed West. This supports the opinion that disaggregation of the CMEA foreign trade into two or perhaps three subgroups (intra-CMEA, Developed Countries, Rest of the World) would be of importance if one wants to analyze the causal relationships in foreign trade more precisely.

The third variable is the increase (first difference) of the world trade. This variable represents the influence of changes in world demand, which "shifts" the export supply curve, as we argued above.

An inspection of the data indicates that some countries (the Soviet Union, Bulgaria and Romania in particular) have changed their propensity to export in the late 1970s and early 1980s. The changes were caused by some special policy measures which the explanatory variables do not convey any information about. We captured this phenomenon by using interactive variables of the following form: *NMP** (*dummy*), where the dummy variable had value 1 for the relevant years, and zero otherwise. The estimation results presented in *Table 9.3* show that the propensity to export changed from 0.046 to $(0.046 + 0.118) = 0.164$ for the Soviet Union, in the years 1980–83, from 0.397 to $(0.397 + 0.450) = 0.847$ for Bulgaria in the years 1977–83, from 0.123 to 0.105 (decreased) for Romania in the years 1977–83. The changes were significant, with the $t$–ratios varying from 19.67 for Bulgaria, through 6.39 for the Soviet Union, to 3.96 for Romania.

The interpretation of the absolute values of these changes is not straightforward since the explained and explanatory variables were measured in different units. The case of Bulgaria is supported by the data in yearbooks, showing that its exports (measured in US$ per capita) increased by more than 60% since 1977. Even if one takes into account that the dominant fraction of the Bulgarian export trade is directed to other CMEA countries, with all the measurement

*Table 9.3.* Estimates of equations for exports ($t$-ratio shown in parentheses below parameter estimate).

$$EX = a_1 + a_2\, NMP + a_3 \left[NFA\$N_{t-1}/EX\$N_{t-1}\right] + a_4\, NMP^{*}dum + a_5\, dum + u_{EX}$$

| *Country* | $a_1$ | $a_2$ | $a_3$ | $a_4$ | $a_5$ | $D-W$ | $\bar{R}^2$ |
|---|---|---|---|---|---|---|---|
| USSR | 0.249 | 0.046 | | 0.118 | –55.413 | 1.55 | 0.99 |
| | (0.76) | (43.31) | | (6.39) | (6.47) | | |
| | | | | *dum80–83*NMP*† | *dum80–83*† | | |
| Bulgaria | –1086.30 | 0.397 | | 0.450 | –7308.8 | 1.7 | 1.00 |
| | (13.74) | (53.29) | | (19.67) | (16.64) | | |
| | | | | *dum77–83*NMP*† | *dum77–83*† | | |
| CSSR | –0.552 | 0.142 | –9.598 | | | 0.77 | 0.98 |
| | (0.13) | (11.73) | (1.89) | | | | |
| GDR | –13.615 | 0.431 | 8.612 | –5.730 | –7.876 | 0.87 | 0.99 |
| | (7.86) | (34.37) | (5.71) | (3.90) | (5.28) | | |
| | | | lagged† | *dum80*† | *dum81*† | | |
| Hungary | –79.447 | 0.663 | –23.291 | | 29.389 | 1.40 | 0.99 |
| | (10.33) | (25.22) | (1.60) | | (3.38) | | |
| | | | | | *dum82*† | | |
| Poland | –163.268 | 0.190 | –121.441 | | –70.10 | 0.99 | 0.98 |
| | (5.61) | (18.99) | (3.28) | | (1.39) | | |
| Romania | –1.266 | 0.123 | –1.864 | –0.018 | | 1.56 | 0.98 |
| | (0.53) | (17.31) | (1.40) | (3.96) | | | |
| | | | | *dum77–83*NMP*† | | | |

problems associated with recalculation of the transferable rouble into dollars, the dynamics are impressive.

For two countries, the Soviet Union and Bulgaria, the estimates of parameters associated with the variable $NFA\$N/EX\$N$ were nonsignificant; hence, the variables were dropped from the final version of the equations. On the other hand, the estimates of coefficients associated with NMP show a consistently high $t$-ratio, confirming high precision of estimation.

### 9.3.2. Imports

Three basic specifications of import equations explaining the total imports of the CMEA countries have been tested.

The first one follows propositions of the research plan (see Krelle, 1985). Following the discussion contained in a report of Bonn–IIASA Group (Gajda, 1985) the estimated equation was specified with additive disturbances:

$$IM = \left[a_1 + a_2 \frac{(p/e)^E}{(p/e)} + a_3 \left(w_{K/L} - w^E_{K/L}\right) + a_4 \left(w_Y - w^E_Y\right)\right] Y^E + u_{IM} \qquad (9.9)$$

rather than:

$$IM = \beta^* \; Y^E \qquad (9.10)$$

where:

$$\beta^* = a_i + a_2 \frac{(p/e)^E}{(p/e)} + a_3 (w_{K/L} - w^E_{K/L}) + a_4 (w_Y - w^E_Y) + u_\beta \qquad (9.11)$$

The estimation results were not of uniform quality. The equation fit was reasonable, even when some parameters were "zeroed" due to incorrect signs of their estimates. There exist strong indications of autocorrelation of residuals, as judged by the Durbin–Watson statistics. The inspection of plots of empirical versus fitted values of endogenous variables shows that the empirical values behave rather smoothly, while the fitted values show high period–to–period variability.

The second approach, based on the Stone–Geary demand function [see Welsch (1986) for an exposition], will be reviewed here briefly to indicate some changes due to the specifics of the economies of Eastern Europe. We assume the two-commodity model, with the first commodity being imports ($x_1$) while the second commodity consists of goods of domestic origin ($x_2$). The demand for these commodities is described by the Stone linear expenditure system:

$$x_i = \hat{x} + a_i \,/\, p_i \left[Y p_Y - \hat{x}_1 p_1 - \hat{x}_2 p_2\right] \qquad i = 1, 2 \qquad (9.12)$$

where $\hat{x}_i$ denotes some autonomous part of the demand for the $i$th commodity and $p$ are the price deflators of the commodities and income $Y$. The demand for the good $i$ is assumed to be a sum of an autonomous demand $\hat{x}_i$ and a component $x_i^*$ which depends on prices of both goods $p_1$, $p_2$ and the income $(Yp_Y - \hat{x}_1 p_1 - \hat{x}_2 p_2)$ available after the autonomous demands have been satisfied. It can be derived from the maximization of the Cobb–Douglas-type utility function with constant demand parts subject to the budget constraint

$$x_1 p_1 + x_2 p_2 \;=\; Y p_Y \qquad (9.13)$$

(see Welsch, 1986). Assuming the autonomous demand to be a fraction of the past demand, one may write

$$x_{it} = x_{it}^* + \lambda_i \; x_{i,t-1} \qquad \lambda_i \geq 0 \qquad (9.14)$$

where $\lambda_i$ denotes a habit-forming coefficient. A fraction, $\alpha_i$ say, of the predetermined demand is paid according to the current prices, while for the rest the lagged prices are valid (due to preexisting contracts, etc.). The resulting equations have the following form:

$$x_{it} = \lambda_i x_{i,t-1} + \beta_i \frac{p_{Yt}}{P_{it}} Y - \beta_i \lambda_i x_{i,t-1} - \beta_i (1 - \alpha_i) \lambda_i \frac{p_{i,t-1}}{p_{it}} x_{i,t-1}$$

$$- \beta_i \alpha_j \lambda_j \frac{p_{jt}}{p_{it}} x_{j,t-1} - \beta_i (1 - \alpha_j) \lambda_j \frac{p_{j,t-1}}{p_{jt}} x_{j,t-1} \tag{9.15}$$

where $j = 2$ when $i = 1$ and vice versa.

The last two elements of the equation (9.15) are strongly correlated. First estimations made it clear that, unless one of the variables were omitted, there exists no chance of obtaining correct signs of the estimates. Since the dominant part of foreign exchange of the CMEA consists of intra-CMEA trade, with prices negotiated well in advance on the basis of past world and current CMEA prices, we chose the elimination of the influence of the term with current prices in favor of the lagged ones as a plausible simplification. Further modification stems from the observation that the income variable $Y$ consists of items as different as consumption and investment goods. Henceforeward, in the final version this variable was split into three elements: consumption, investment and exports (within the above-mentioned framework, the income consists of the values of domestic sales and exports, while demand refers to the domestic production and imports). This allows different propensities to import to be associated with these different groups of goods. The splitting has an additional positive effect in the case of countries having imports and exports measured in "valuta" currency, since the sum of *NMP* (in domestic currency) and *IM* (in "valuta" currency) has no meaning. The version suitable for estimation is as follows:

$$IM = a_1 IM_{-1} + a_2 \frac{p_{INV}}{p_{IM}} INV + a_3 \frac{p_{EX}}{p_{IM}} EX + a_4 \frac{p_{Y,-1}}{p_{IM}} Y_{-1} + u_{IM} \tag{9.16}$$

This approach produces a good fit, with signs consistent with the theory and significant estimates (see *Table 9.4*). On the other hand, the adding up property of the original system is now fulfilled only approximately.

In some cases a restricted estimation, assuming some coefficients *a priori* equal to zero, was performed. In this way consumption was removed from all these equations; in the case of the Soviet Union the estimates of effects of investment and lagged *NMP* had an unsatisfactory quality and were thus set equal to zero. In all other equations the effects of exports and investment have been captured separately.

*Table 9.4.* Estimates of equations for imports ($t$-ratio shown in parentheses below parameter estimate).

$$IM = a_1 IM_{t-1} + a_2 \frac{p_{INV}}{p_{IM}} INV + a_3 \frac{p_{EX}}{p_{IM}} EX + a_4 \frac{p_{NMP,t-1}}{p_{IM}} NMP_{t-1} + u_{IM}$$

| *Country* | $a_1$ | $a_2$ | $a_3$ | $a_4$ | $D\text{-}W$ | $\bar{R}^2$ |
|---|---|---|---|---|---|---|
| USSR | 0.526<br>(2.17) | | 0.373<br>(2.23) | | 2.09 | 0.98 |
| Bulgaria | 0.601<br>(2.89) | 0.277<br>(1.68) | 0.594<br>(1.91) | –0.082<br>(1.14) | 1.79 | 0.99 |
| CSSR | 0.344<br>(1.91) | 0.149<br>(3.54) | 0.765<br>(3.28) | –0.034<br>(2.78) | 1.63 | 0.99 |
| GDR | 0.638<br>(5.95) | 0.494<br>(2.91) | 0.627<br>(3.12) | –0.122<br>(2.56) | 1.59 | 0.99 |
| Hungary | 0.463<br>(3.13) | 0.619<br>(5.19) | 0.463<br>(2.82) | –0.129<br>(3.34) | 1.59 | 0.99 |
| Poland | 0.737<br>(9.25) | 0.178<br>(8.34) | 0.478<br>(2.27) | –0.050<br>(5.12) | 1.80 | 0.99 |
| Romania | 0.499<br>(2.85) | 0.190<br>(4.14) | 0.616<br>(1.80) | –0.041<br>(2.77) | 1.64 | 0.97 |

The import equations rank among the best in the model of foreign trade of the CMEA countries. Given the usually high correlation between *NMP* and *IM*, the simultaneous presence of both explanatory variables increases the risk of harmful multicollinearity effects. The estimates obtained do not seem to show many signs of them – we can interpret this fact as an argument in favor of plausibility of the underlying theory.

### 9.3.3. Export price deflator

Equations for export prices have been estimated in the form proposed in the research plan, with the substitution of the hyperbolic $(1/T)$ trend variable in the place of the capital/labor ratio:

$$p_{EX} = \left[a_1 + a_2 \frac{(p/e)^E}{(p/e)} + a_3 (1/T)\right] p_Y + u_p \tag{9.17}$$

The results are shown in *Table 9.5.* Except for Bulgaria and Hungary, the equations fit the observations quite well. What troubles us is the indication of the presence of autocorrelation for all countries but the Soviet Union and Romania. One should stress, however, that the domestic deflators $p_Y$ in these countries are very stable, slightly lower than unity before 1975 and slightly higher afterward;

thus the variability of the deflator is to be explained predominantly by the "variable" parameter. The problem of the internal consistency of these series, mentioned earlier (mostly due to the valuation of almost half of the foreign trade in intra-CMEA prices in transferable roubles while foreign trade with the other countries is valued in dollars, basically, at world market prices), diminishes the precision of the estimations. This applies to the import price deflator as well.

*Table 9.5.* Estimates of equations for export price deflator ($t$-ratio shown in parentheses below parameter estimate).

$$p_{EX} = \left[a_1 + a_2 \frac{(p/e)^E}{(p/e)} + a_3\,(1/T)\right] p_{NMP} + u_p$$

| *Country* | $a_1$ | $a_2$ | $a_3$ | $a_4$ | $D\text{-}W$ | $\bar{R}^2$ |
|---|---|---|---|---|---|---|
| USSR | -3.282 | 3.419 | 129.709 | | 1.88 | 0.98 |
| | (5.07) | (14.63) | (3.66) | | | |
| Bulgaria | -1.155 | 1.616 | 43.906 | | 0.77 | 0.77 |
| | (1.28) | (3.63) | (1.19) | | | |
| CSSR | 1.013 | 0.176 | -78.230 | | 0.91 | 0.97 |
| | (4.91) | (13.58) | (7.11) | | | |
| GDR | 0.552 | 0.419 | -71.555 | | 0.74 | 0.90 |
| | (0.52) | (3.61) | (1.41) | | | |
| Hungary | -0.195 | 0.010 | 54.649 | -0.089 | 0.91 | 0.70 |
| | (0.69) | (2.84) | (2.05) | (2.00) | | |
| | | | | $dum76\text{–}82^{\dagger}$ | | |
| Poland | 1.260 | 0.007 | -69.642 | | 0.72 | 0.94 |
| | (5.98) | (2.47) | (4.69) | | | |
| Romania | -0.706 | 0.335 | 2.930 | | 1.67 | 0.98 |
| | (1.84) | (15.35) | (0.14) | | | |

The structure of our model requires us to link the export prices to the behavior of the domestic prices (and therefore to the costs of production). Hence, the problems just mentioned remain. On the other hand, the economic reforms pursued currently in the CMEA countries make this specification much more realistic in simulations of the future.

### 9.3.4. Import price deflator

The equations for the import price deflator were estimated in accordance with the initial specification (9.5) with a minor modification: instead of lagged world prices, lagged own import prices were introduced.

$$p_{IM} = a_1 + a_2\, p_{EX}^{E} + a_3\, p_{IM,-1} + u_p \qquad (9.18)$$

The reason was the fact, mentioned earlier, that a considerable proportion of the CMEA foreign trade is intra-CMEA trade where the prices of the previous period persist, due to the pricing system of the intra-CMEA exchange. *Table 9.6* shows our results. In some cases further improvements are necessary, especially when the Durbin–Watson statistics, already biased toward two by the presence of the lagged explained variable, suggests closeness to the inconclusive region. In the equation for Poland, the estimate associated with the lagged explained variable has low precision and the current export prices abroad seem to explain the variable sufficiently. This is in strong opposition to the other equations, where these coefficients are estimated with high $t$–statistics.

*Table 9.6.* Estimates of equations for import price deflator ($t$-ratio shown in parentheses below parameter estimate).

$$p_{IM} = a_1 + a_2 p^E_{EX} + a_3 p_{IM,-1} + u_p$$

| *Country* | $a_1$ | $a_2$ | $a_3$ | $a_4$ | $D2\ W$ | $\bar{R}^2$ |
|---|---|---|---|---|---|---|
| USSR | –0.084<br>(4.20) | 0.269<br>(5.48) | 0.880<br>(19.74) | | 1.82 | 0.99 |
| Bulgaria | –0.047<br>(1.49) | 0.112<br>(3.75) | 0.996<br>(19.40) | –09.096<br>(3.47)<br>*dum81*† | 1.35 | 0.99 |
| CSSR | –0.078<br>(4.62) | 0.127<br>(4.84) | 1.018<br>(29.51) | | 1.23 | 0.99 |
| GDR | 0.002<br>(0.11) | 0.224<br>(6.03) | 0.837<br>(18.14) | | 1.93 | 0.99 |
| Hungary | –0.207<br>(1.41) | 0.250<br>(3.23) | 0.939<br>(5.43) | 0.102<br>(2.11)<br>*dum76–81*† | 2.04 | 0.95 |
| Poland | 0.251<br>(3.46) | 0.489<br>(5.37) | 0.203<br>(1.13) | 1.029<br>(11.11)<br>*dum82*† | 1.32 | 0.99 |
| Romania | –0.045<br>(1.36) | 0.360<br>(4.25) | 0.768<br>(9.64) | | 2.50 | 0.98 |

## 9.3.5. Exchange rates

The statistical analysis suggested that the item dominating the behavior of the variable coefficient $\gamma$ in equation (9.6) of the initial specification is the value of dollar in the international market, as measured by US$ per unit of SDR (denoted *US$/SDR*). The estimated version has the following form:

$$e = (a_1 + a_2\ US\$/SDR)\frac{p}{p_{01}} + a_3 + u_e \tag{9.19}$$

Further analysis has shown that there are some incompatibilities between the mean values of the left- and the right-hand side variables; the introduction of an additive term $a_3$ improved the fit of equations enormously. As one can see from *Table 9.7*, the $t$–statistic for $a_3$ exceeds 8, except for Hungary.

*Table 9.7.* Estimates of equations for exchange rates ($t$-ratio shown in parentheses below parameter estimate).

$$e = (a_1 + a_2\ US\$/SDR)\frac{p}{p_{01}} + a_3 + u_e$$

| *Country* | $a_1$ | $a_2$ | $a_3$ | $D-W$ | $\bar{R}^2$ |
|---|---|---|---|---|---|
| USSR | 0.701 | –0.482 | 0.615 | 2.70 | 0.90 |
| | (7.41) | (4.89) | (16.79) | | |
| Bulgaria | 0.840 | –0.557 | 0.791 | 2.50 | 0.91 |
| | (7.44) | (4.62) | (17.28) | | |
| CSSR | 6.477 | -5.068 | 5.430 | 2.09 | 0.87 |
| | (7.56) | (5.78) | (17.44) | | |
| GDR | 2.328 | -1.439 | 2.947 | 1.35 | 0.79 |
| | (4.67) | (2.73) | (13.30) | | |
| Hungary | 60.591 | -15.851 | –0.669 | 2.06 | 0.96 |
| | (11.02) | (2.54) | (1.17) | | |
| Poland | 43.438 | –47.868 | 56.627 | 1.30 | 0.97 |
| | (10.11) | (5.97) | (8.29) | | |
| Romania | 3.679 | -1.753 | 3.387 | 1.86 | 0.93 |
| | (6.82) | (3.08) | (14.89) | | |

The other test statistics of our equations are astonishingly good. We write "astonishingly" since the rates used must be understood as figures fixed in a somewhat arbitrary way, as conversion coefficients of dollars and/or transferable roubles and the "valuta" currency. The equations in *Table 9.7* prove that, although the decision to change the exchange rates as well as the decision to fix their initial values on *its absolute value* might have been a political one, the changes, once introduced, were economically justified and not as *ad hoc* as popular opinion claimed.

## 9.4. Final Remarks

The successful estimation of equations explaining the functioning of foreign exchange of the CMEA countries in a uniform manner for all regions of the world model suggests that the differences in the mechanisms of the foreign trade between countries with different economic systems are dominated by the general rules of economic thinking, at least at this highly aggregated level. The successful estimation of the "valuta" exchange rates suggest that it may be possible to estimate approximate exchange rates for the domestic currency of the CMEA countries. The solution seems to be associated with the solution of the problem of the demand pressure index (see also Dobrinsky, 1986).

A set of auxiliary identities was used to incorporate the model of the CMEA foreign exchange into the larger world model. More detailed data for these identities have been prepared (see Gajda and Welsch, 1985a, b, c). As they obey the general rules set in the research plan and are reported elsewhere in this volume we do not reproduce them in this place.

## Acknowledgment

The author was not able to complete the full term with the Central Group affiliated with Bonn University. Thanks are due to the colleagues who continued the work: Dr. H. Welsch, H. Ross, Dr. J.J. Sztaudynger (who continued the work on the commodity breakdown of the CMEA foreign trade), and Dr. R. Dobrinsky (who linked the model of foreign exchange with the models of the CMEA economies and did the final simulations and analyzed the structural change). The description of these results is obviously left in the hands of these authors.

## References

Colombatto, E. (1988), CMEA, Money and Ruble Convertibility, *Applied Economics*, **15**, pp. 479–506.

Dobrinsky, R. (1986), *The Macroeconomic Data Base for the CMEA Countries*, Working Paper WP-86-54. (Laxenburg, Austria: International Institute for Applied Systems Analysis).

Gajda, J.B. (1983). The System IES2 of Minimodels of European Centrally Planned Economies. A New Version, paper prepared for the Seventh International Conference, "Models and Forecasts '83", organized by the Polish Academy of Sciences and the Institute of Econometrics and Statistics, University of Lodz, Poland. [Reprinted in: *Macromodels and Forecasts of National Economy* (1986), Warsaw and Lodz: Polish Scientific Publishers.]

Gajda, J.B. (1985), A Note on Linear vs. Nonlinear Aspects of Specification when Forecasting is the Main Target of the Model, Bonn–IIASA Research Group Report.

Gajda, J.B. and Welsch, H. (1985a), Time Series of Bilateral Commodity Export Shares for 16 Countries and 8 Groups of Countries 1960–1982, Bonn–IIASA Research Group Report.

Gajda, J.B. and Welsch, H. (1985b), Time Series of Bilateral Commodity Import Shares for 16 Countries and 8 Groups of Countries 1960–1982, Bonn–IIASA Research Group Report.

Gajda, J.B. and Welsch, H. (1985c), Time Series of Bilateral Commodity Trade Flows for 16 Countries and 8 Groups of Countries 1960–1982, Bonn–IIASA Research Group Report.

Krelle, W. (1985), The World Model in Detail (second version), Bonn–IIASA Research Group Report.

Mejstrik, M. (1983), Quality and Economic Effects of Export in Simple Model, paper prepared for the Seventh International Conference, "Models and Forecasts '83", organized by the Polish Academy of Sciences and the Institute of Econometrics and Statistics, University of Lodz, Poland.

Welfe, W. (1983), Models of the Socialist Economy, in: L.R. Klein (ed.). *Lectures in Econometrics.* (Amsterdam: North Holland).

Welsch, H. (1986), An Aggregate Import Demand Model for Long-Term Projections. Bonn-IIASA Research Group Report. [Reprinted in: *Jahrbücher für Nationalökonomie und Statistik* (1987), **203** (4), pp. 372–389.]

# Part II

# Analysis and Forecast of Structural Change on the Basis of Individual Country Models: Results of Collaborating Research Groups in Different Countries

CHAPTER 10

# US Economic Growth in Retrospect and Prospect

*Bert G. Hickman*


## Summary

This chapter presents another view of the major factors affecting US economic growth both historically and prospectively. The Hickman–Coen (HC) annual growth model is used to investigate the determinants of economic growth and productivity change during 1956–1982 and to forecast the principal macroeconomic variables for 1985–2000. Our approach stresses the interaction of aggregate supply and demand in the growth process instead of concentrating exclusively on reduced-form analysis of the production function. It leads to a different view of the driving forces of economic growth, as summarized in the concluding section.


## 10.1. An Empirical Growth Model

The HC annual growth model for the US economy (Hickman and Coen, 1976) is a dynamic nonlinear simultaneous equation system combining Keynesian and neoclassical elements and allowing for departures from the full-employment growth path owing to gradual price adjustments. The key assumptions underlying this property are as follows.

(1) Firms are imperfectly competitive and set prices as a markup over normal unit labor costs, with allowance for the prices of imported inputs and the degree of capacity utilization.

(2) Given prices, output is determined by effective demand, which is disaggregated into three categories of investment, six of consumption, federal and state and local purchases, exports, and imports.
(3) Firms choose capital and labor inputs to minimize cost, conditional on expected output and factor prices. The desired long-run or equilibrium inputs are derived by minimizing the cost of producing the expected output. Only part of the gap between actual and desired inputs is closed each year, however, owing to adjustment costs, so that the short-term factor demand functions contain lags which may prevent the attainment of full long-term equilibrium inputs for the given level of production.
(4) Labor force participation is a function of the real after-tax consumption wage and the ratio of employment to population (Coen and Hickman, 1980a). The latter variable is included to capture the "discouraged worker effect", a nonprice signal which induces potential workers to withdraw from the labor market when unemployment rises. The labor force participation model is disaggregated into 16 age–sex groups, so that the aggregate labor force depends on the composition of population as well as its level.
(5) The model includes an "expectations-augmented" Phillips curve for nominal wage inflation as a function of the gap between the actual and natural unemployment rates and the lagged rate of consumer price inflation.
(6) The model can be solved for potential as well as actual output. Potential GNP is defined as that output which would be realized each year if the markets for labor and capital were continuously cleared at the natural rate of unemployment (Coen and Hickman, 1980b). A key characteristic of this concept is that potential output is unaffected by deviations of actual output, factor inputs, and real factor prices from their full-employment values. It is truly a measure of productive potential in which output is constrained only by available technology and factor supplies, and labor and capital are assumed to be fully employed *each period* along the growth path. Departures from the natural path imply disequilibrium in the factor markets, as the quantities of capital and labor deviate from their full-employment levels, but these temporary deviations do not affect potential output in subsequent periods, since they can be offset by future changes in investment and employment.

## 10.2. Factor Demands

The demands for labor and capital are interrelated in the model, since they are jointly derived on the assumption that firms minimize production costs subject to a long-run or planning Cobb–Douglas production function with constant returns to scale:

$$XNR_t^* = Ae^{\beta t}\ (K_t^*)^{\alpha}\ \ (MH_t^*)^{1-\alpha}, \quad A,\alpha,\beta > 0 \tag{10.1}$$

where $XNR^*$ is expected output (gross private nonresidential product), $K^*$ is desired business fixed capital stock, $MH^*$ is desired manhours, and $\beta$ is the rate of Hicks-neutral technical progress.

Minimizing production cost subject to (10.1) gives the long-run factor demand functions:

$$MH_t^* = [\alpha/(1-\alpha)]^{-\alpha} A^{-1} [(W^*/Q^*)_t]^{-\alpha} XNR_t^* e^{-\beta t} \tag{10.2}$$

$$K_t^* = [\alpha/(1-\alpha)]^{1-\alpha} A^{-1} [(W^*/Q^*)_t]^{1-\alpha} XNR_t^* e^{-\beta t} \tag{10.3}$$

where $Q^*$ is the expected implicit rental price of capital and $W^*$ is the expected nominal before-tax wage rate. The implicit rental price is defined by $Q \equiv PI(r+d)\,T$, where $PI$ is the investment price deflator, $r$ is the after-tax rate of return, $d$ is the depreciation rate, and $T$ symbolizes the tax treatment of investment expenditure.

Adjustment costs prevent firms from accommodating immediately to variations in the desired inputs. These adjustment costs include external purchase costs and internal installation costs for capital goods and hiring, training, and layoff costs for labor. They are represented implicitly by exponential partial adjustment processes:

$$MH_t/MH_{t-1} = (MH_t^*/MH_{t-1})^f, \quad 0 < f \leq 1 \tag{10.4}$$

$$K_t/K_{t-1} = (K_t^*/K_{t-1})^g, \quad 0 < g \leq 1 \tag{10.5}$$

where $f$ and $g$ are the adjustment speeds for labor and capital.

Combining the desired input and adjustment hypotheses yields the short-term or disequilibrium demand functions:

$$MH_t = \{[\alpha/(1-\alpha)]^{-\alpha} A^{-1} [(W^*/Q^*)_t]^{-\alpha} XNR_t^* e^{-\beta t}\}^f MH_{t-1}^{1-f} \tag{10.6}$$

$$K_t = \{[\alpha/(1-\alpha)]^{1-\alpha} A^{-1} [(W^*/Q^*)_t]^{1-\alpha} XNR_t^* e^{-\beta t}\}^g K_{t-1}^{1-g} \tag{10.7}$$

Joint estimation of the short-run demand functions (10.6) and (10.7) yields estimated values of the adjustment speeds $f$ and $g$ and of the structural parameters of the production function (10.1) and long-run factor demand functions (10.2) and (10.3).

## 10.3. Determination of Potential Output

The first step is to specify the natural unemployment rate in order to estimate the full-employment labor supply. We follow Wachter (1976) in determining the natural rate as a weighted average of the full-employment unemployment rates in our 16 age–sex groups. The latter are determined by a set of regressions on the prime-age male group (to remove cyclical variations) and the population proportion of the age–sex group (to capture structural shifts from demographic factors). It is important to emphasize that the natural unemployment rate (*UF*) in the HC model is not a nonaccelerating inflation rate of unemployment (*NAIRU*), since it is not calculated from a Phillips curve by imposing the nonacceleration constraint. It does take account of changes over time in factors affecting structural and frictional unemployment, and the related changes in labor market tightness, but it does not impose the assumption of a vertical long-run Phillips curve.

The complete labor supply model includes the 16 labor force participation equations and an equation for average hours of work. With the addition of the natural rate equation, the system can be solved simultaneously for the natural unemployment rate *UF* and the corresponding full-employment supply of manhours *MHF*, conditional on given values of the real wage, the population and its age–sex distribution, and the size of the armed forces and government employment.

### 10.3.1. The natural growth path

Along the natural growth path of potential output both labor and capital are fully employed. Making use first of the labor market condition, potential output is defined as the level of output that would equate labor supply and demand at the natural rate of unemployment and full-employment wage rate. Since the labor demand function (10.6) relates manhours to output and relative factor prices, an expression may be derived for potential output, conditional on the wage/rental ratio, by substituting *MHF* for *MH* and solving the equation for output:

$$XNRP_t = A[\alpha/(1-\alpha)]^{\alpha} \, [(W^*/Q^*)_t]^{\alpha} \, e^{\beta t} \, MHF_t^{1/f} \, MHF_{t-1}^{-(1-f)/f} \qquad (10.8)$$

where *XNRP* is potential output. Since full employment would prevail each period along the natural growth path, note that *XNRP* depends on *MHF* in the current and preceding year irrespective of whether the economy actually operated at full employment in the preceding year.

It remains to determine $W^*/Q^*$. We assume that real wage expectations would be realized along the natural growth path, so that the full-employment wage/rental ratio is

$$(WF/QF)_t = (WF/PF)_t \, [(d+r)\, T]_t^{-1} \tag{10.9}$$

where $WF/PF$ is the real wage on the equilibrium path. Note that the last term is an autonomous component, since $d$, $r$ and $T$ are all exogenous. Finally, the potential real wage is assumed to grow at the same rate as potential labor productivity:

$$(WF/PF)_t/(WF/PF)_{t-1} = (XNRP/MHF)_t/(XNRP/MHF)_{t-1} \tag{10.10}$$

This last assumption is both observationally realistic and consistent with the theoretical structure of the model.

Rewriting (10.8) in productivity form as

$$(XNRP/MHF)_t = Be^{\beta t} \, [(WF/QF)_t]^{\alpha} \, (MHF_t/MHF_{t-1})^{(1-f)/f} \tag{10.11}$$

where $B = A[(1-\alpha)/\alpha]^{-\alpha}$, and using (10.9) and (10.10) to determine $WF/QF$ and the labor supply system to determine $MHF$, one can solve simultaneously for the full-employment values of labor productivity, output, labor force, employment, unemployment, hours of work, and the real wage and rental/wage ratio along the natural growth path, for exogenous values of the demographic and policy variables (population, armed forces, government employment, and tax parameters).

Capital as well as labor must be fully employed along the natural growth path. In the standard neoclassical model, a constant fraction of output is saved and automatically invested. In contrast, the HC model includes an explicit investment demand function, equation (10.7), which determines actual business fixed investment, and hence saving need not equal investment *ex ante* in the model solution for current output. The natural growth path of potential business fixed capital stock is calculated by setting current and lagged capital stock, the wage/rental ratio, and aggregate output equal to their full-employment values on the right-hand side of the investment function. Gross investment is then obtained from the capital stock identity

$$IF_t \equiv KF_t - (1-d)\, KF_{t-1} \tag{10.12}$$

where $IF$ is potential gross business fixed investment or gross nonresidential fixed investment, and $KF$ is the potential net stock of business fixed capital. Thus the quantities of net and gross business fixed investment required to sustain the natural growth path are fully determined. A greater flow of saving could not be profitably absorbed in business fixed capital formation under the given investment conditions, and a smaller flow would be inadequate to attain the required rate of capital deepening to equilibrate the capital/labor and rental/wage ratios. Inventory investment and residential construction are determined elsewhere in the model and do not affect the path of potential (non-residential) output, although they are added to the latter to determine potential GNP.

### 10.3.2. The production function and potential output

In the derivation of the desired input functions (10.2) and (10.3), the production function (10.1) is viewed as a planning relation between equilibrium input and output levels. The corresponding disequilibrium production function for actual current output is

$$XNR_t = Ae^{\beta t} (k_t K_{t-1})^{\alpha} (m_t MH_t)^{1-\alpha} \tag{10.13}$$

where $k_t$ and $m_t$ are indexes of the intensity of utilization, respectively, of the measured inputs $K_{t-1}$ and $MH_t$. The intensity of use of capital stock, for example, can be increased by operating equipment at a faster rate, increasing the number of machine-hours per day or week, or diminishing downtime by postponing maintenance. Similarly, a manhour may represent a greater amount of effective labor input as workers are induced to work at a faster pace and with fewer or shorter breaks, although the scope for variations in intensity of use of labor input is smaller than for capital.

Variations in the intensity of factor utilization occur in the process of adjusting manhours and capital stock to changes in the desired or equilibrium quantities. Because adjustment costs render uneconomic the instantaneous adaptation to changes in desired quantities, the observed inputs of $K$ and $MH$ must be used at intensities that differ from the equilibrium rates of factor utilization. Since the principal source of variation in factor utilization is a discrepancy between the actual and desired inputs, our premise is that the intensities of factor use are revealed by firms' investment and employment decisions. Specifically, we assume that the intensity of use of each factor is proportional to the extent of its disequilibrium:

$$m_t = MH_t^*/MH_t \tag{10.14}$$

$$k_t = K_t^*/K_{t-1} \tag{10.15}$$

Moreover, since the observed changes in measured inputs are proportional to the desired changes by hypotheses (10.4) and (10.5), the intensity indexes may be measured in terms of observable variables:

$$m_t = (M_t/M_{t-1})^{(1-f)/f} \tag{10.16}$$

$$k_t = (K_t/K_{t-1})^{1/g} \tag{10.17}$$

Potential output can now be determined from the production function (10.13) by substituting full-employment manhours and capital stock and measuring the effective capital and labor inputs at their natural intensities $m^n$ and $k^n$:

$$XNRP_t = Ae^{\beta t}\,(k_t^n KF_{t-1})^{\alpha}\,(m_t^n MHF_t)^{1-\alpha} \tag{10.18}$$

It might be thought that natural utilization intensities would be unity, but this is the case only under stationary conditions. Since adjustment costs lead firms to adjust measured inputs to desired levels with some lag, we would expect to observe firms using factor inputs which are below desired levels even if the economy were experiencing steady growth. Thus the natural intensities associated with *growth* equilibrium will exceed unity by amounts which depend on the natural growth rates of output and the wage/rental ratio and on the speeds of adjustment of capital and labor. The natural utilization intensities are determined endogenously in the natural path solution as:

$$m_t^n = (MHF_t/MHF_{t-1})^{(1-f)/f} \tag{10.19}$$

$$k_t^n = (KF_t/KF_{t-1})^{(1-g)} \tag{10.20}$$

Cyclical fluctuations of output relative to the natural growth path will be accompanied, of course, by corresponding fluctuations of the actual utilization intensities relative to their natural levels.

Although algebraically equivalent to (10.8), the production function expression (10.18) for potential output is more than a redundant curiosity. It serves as a reminder that capital as well as labor requirements must be satisfied along the natural growth path, even though capital stock does not appear explicitly in equation (10.8).

## 10.4. Estimation

The model was estimated over the sample period 1949–1982. The primary data are from official US agencies. A distinction is made between output produced in the housing and government sectors, on the one hand, and the nonresidential sector on the other, with corresponding differences for capital and labor inputs. Thus output in the aggregate production and factor demand functions is measured by gross private nonresidential product, the capital stock is net of housing and inventories, and labor input excludes the services of government employees. The level of potential GNP is then determined by adding real housing services and real income originating in government to the model's prediction of potential gross private nonresidential product. A separate housing model determines residential construction for inclusion in the GNP identity and the saving–investment process to determine actual GNP.

Most equations were estimated by OLS, with correction for first-order autocorrelation in the residuals where indicated. The key factor demand equations (10.6) and (10.7) were estimated jointly by the seemingly unrelated regression method, however, since they share common parameters from the production function.

### 10.4.1. Expectations

The variables entering the factor demand functions are not actual but expected output and factor prices. The expected price of capital goods is proxied by an autoregression on the investment goods deflator. The expected wage is determined from a labor market Phillips curve, assuming that agents know the parameters and estimate the expected wage on the basis of the unemployment gap in the previous period. Expected output is equal to actual output in all the models, since experiments with autoregressive output expectations failed to improve the estimates of the factor demand systems

### 10.4.2. Technical progress

It is well established that a decline in the rate of growth of technical progress or total factor productivity is partly responsible for the slow growth of labor productivity in the OECD countries in recent years (Kendrick, 1983). In the HC model this influence is captured in the estimates of the factor demand equation, by allowing for piecewise breaks in the exponential trend term. A significant reduction in the rate of technical progress is estimated for the USA after 1968 and a further deceleration occurred beginning in 1974.

Both factor demand functions for the USA are adjusted for serially correlated errors. The estimated capital elasticity in the production function is 0.25, and the estimated adjustment speeds of labor and capital are 0.65 and 0.19, respectively. The estimated technical progress rate slows from 1.86% per annum in 1950–1968 to 1.39% in 1969–1973 and 0.57% thereafter.

## 10.5. Productivity and Growth: Potential and Realized

Equation (10.11) provides an expression for potential labor productivity in terms of the cumulative level of technical progress, the weighted potential wage/rental ratio, and the rate of labor utilization. It may be rewritten as

$$PRODL_t = C1_t \times AWQL_t \times UTL_t \qquad (10.21)$$

where

$$PRODL_t = (XNRP/MHF)_t,\ \ C1_t = Be^{\beta t},\ \ AWQL_t = [(WF/QF)_t]^{\alpha}$$

and

$$UTL_t = (MHF_t/MHF_{t-1})^{(1-f)/f}$$

Working with the short-run labor demand function (10.6), one may derive a similar expression for realized or observed productivity. Because the short-term labor demand function is stochastic, however, the productivity equation contains an error term:

$$PROD = C1_t \times AWQE_t \times UTMH_t \times ERMH_t \tag{10.22}$$

where the constant term is the same as before, $AWQE$ depends on $WE/QE$, $UTMH$ on actual labor utilization, and $ERMH$ on the stochastic disturbance in the partial adjustment process for labor input [equation (10.4)].

A final set of analytical productivity measures is obtained by dividing (10.22) by (10.21) to yield:

$$PRODR_t = AWQR_t \times UTLR_t \times ERMH_t \tag{10.23}$$

where

$$PRODR = PROD/PRODL,\ AWQR = AWQE/AWQL,$$

and

$$UTLR = UTMH/UTL$$

These measures facilitate a term-by-term comparison of the sources of the annual deviations of realized productivity from potential productivity.

Another useful interpretive expression may be obtained by rearranging (10.22) and (10.21) and taking their ratio:

$$\begin{aligned} UTPR_t &= AWQR_t \times UTLR_t \times ERMH_t \times MHR_t \\ &= PRODR_t \times MHR_t \end{aligned} \tag{10.24}$$

where $UTPR = XNR/XNRP$ is the ratio of actual to potential output (potential utilization rate) and $MHR = MH/MHF$ is the ratio of actual and full-employment manhours. Thus the output gap is related quantitatively to the corresponding productivity and manhour gaps. The latter gaps are clearly interdependent, however, so that care must be exercised in interpreting the numerical measures based on (10.24).

## 10.6. Productivity and the Growth Slowdown, 1956–1982

As shown in the second half of *Table 10.1*, the average annual rates of increase of the total population, the working-age population, and the labor force were about equal during 1956–1968. Pronounced rises occurred thereafter both in the proportion of the population which is of working age and in the labor force participation rate, so that the labor force increased much faster after 1968 than before, despite the slowdown which occurred in overall population growth. A corresponding acceleration occurred in employment, although the gain was moderated after 1973 by a rising unemployment rate.

Meanwhile, the growth rate of labor productivity, as measured by real GNP per worker, was about halved in 1968–1973 and dropped almost to zero in 1973–1982. During 1968–1973 the acceleration in employment largely offset the deceleration in productivity, so that real GNP increased nearly as rapidly as before, but real growth decelerated sharply in 1973–1982 as both components declined.

Corresponding calculations are shown in the top half of the table for the estimates of annual growth rates along the natural growth path. The patterns of potential and realized growth are much the same, but it will be seen that the potential gains in employment and productivity were larger than those actually realized after 1968, and especially after 1973, so that after 1973 unemployment substantially exceeded the natural level and real GNP per capita increased at only half its potential rate.

The investment requirements to sustain potential output along the natural growth path are also shown in *Table 10.1.* The average share of net investment decreased after 1973, owing to the deceleration of real growth, but the required gross share was stable because of an offsetting rise in potential replacement investment. A similar pattern is observed for the realized investment shares. All of these figures refer to business fixed investment and exclude housing construction, inventory accumulation, and net foreign investment.

The proximate sources of the productivity slowdowns during 1968–1973 and 1973–1982 are quantified in *Table 10.2*, which is based on equations (10.21) and (10.22). Since the aggregate production function excludes inputs and outputs from government employment and the housing stock, these new breakdowns refer to a universe which is moderately smaller than in the previous table. Another difference is that labor productivity is measured per manhour in *Table 10.2* and per worker in *Table 10.1.*

Potential manhour productivity ($PRODL$) increased at an annual rate of 2.76% during 1956–1968, 2.05% in 1968–1973, and 0.65% in 1973–1982. Technical progress ($C1$) accounted for the largest portion of the gain during each period and its deceleration explains about two-thirds of the decline in the growth rate of

*Table 10.1.* USA: Natural and realized growth rates (percent per annum).

| Period | Population Total | Population Work age | Labor force | Employment | Real GNP | GNP per worker | GNP per capita | Unemployment rate | Investment net | Share gross |
|---|---|---|---|---|---|---|---|---|---|---|
| A. Natural | | | | | | | | | | |
| 1956–68 | 1.46 | 1.42 | 1.61 | 1.51 | 3.51 | 2.00 | 2.05 | 4.72 | 2.40 | 9.50 |
| 1968–73 | 1.09 | 1.95 | 2.68 | 2.55 | 3.79 | 1.24 | 2.70 | 5.77 | 2.37 | 10.01 |
| 1973–82 | 1.01 | 1.72 | 2.25 | 2.30 | 2.70 | 0.40 | 1.69 | 5.91 | 1.79 | 10.06 |
| B. Realized | | | | | | | | | | |
| 1956–68 | 1.46 | 1.42 | 1.47 | 1.54 | 3.66 | 2.12 | 2.20 | 5.00 | 2.42 | 9.58 |
| 1968–73 | 1.09 | 1.95 | 2.55 | 2.27 | 3.40 | 1.13 | 2.32 | 4.98 | 2.60 | 10.49 |
| 1973–82 | 1.01 | 1.72 | 2.32 | 1.74 | 1.88 | 0.14 | 0.87 | 7.24 | 2.02 | 10.77 |

*Table 10.2.* USA: Productivity growth and determinants (percent per annum).

| | A. Natural | | | | B. Realized | | | | |
|---|---|---|---|---|---|---|---|---|---|
| Period | PRODL | C1 | AWQL | UTL | PROD | C1 | AWQE | UTMH | ERMH |
| 1956–68 | 2.76 | 1.86 | 0.91 | –0.01 | 2.85 | 1.86 | 0.93 | –0.11 | 0.17 |
| 1968–73 | 2.05 | 1.39 | 0.52 | 0.14 | 1.91 | 1.39 | 0.52 | 0.24 | –0.24 |
| 1973–82 | 0.65 | 0.57 | 0.13 | 0.05 | 0.43 | 0.57 | 0.26 | 0.32 | –0.08 |

*Table 10.3.* USA: Growth rates of output, productivity and employment (percent per annum).[a]

| | A. Natural | | | | | B. Realized | | | | |
|---|---|---|---|---|---|---|---|---|---|---|
| Period | XNRP | PRODL | MHF | AHF | EPF | XNR | PROD | MH | AH | EP |
| 1956–68 | 3.36 | 2.76 | 0.60 | –0.50 | 1.12 | 3.56 | 2.85 | 0.71 | –0.43 | 1.14 |
| 1968–73 | 4.10 | 2.05 | 2.05 | –0.37 | 2.42 | 3.59 | 1.91 | 1.68 | –0.47 | 2.15 |
| 1973–82 | 2.65 | 0.65 | 2.00 | –0.34 | 2.34 | 1.60 | 0.43 | 1.17 | –0.61 | 1.78 |

[a] Exponential growth rates between endpoints of intervals, except for unemployment rate and investment shares, which are period averages. See text for definitions of variables and relations.

potential labor productivity between 1956–1968 and 1968–1973 and 60% of the reduction from 1968–1973 to 1973–1982. The wage/rental component ($AWQL$) was also important, however, especially before 1973. Reductions in the growth rate of the potential wage/rental ratio account for roughly half of the deceleration in potential manhour productivity in 1968–1973 and one-fourth in 1973–1982. Moderate swings in potential labor utilization ($UTL$) also contributed to the calculated changes in potential productivity.

Realized productivity growth ($PROD$) exceeded its potential during 1956–1968, but fell short thereafter. The contribution of the wage/rental component was virtually the same for realized and potential productivity growth until 1973. After 1973, however, the expected actual wage/rental ratio ($AWQE$) increased faster than potential, so that realized productivity was augmented on that account. It nonetheless decreased more rapidly than did potential productivity, owing to a substantial reduction in the intensity of labor utilization ($UTMH$).

The growth rate of potential gross private nonresidential output ($XNRP$) is reported in the first column of *Table 10.3* and split between its potential labor productivity ($PRODL$) and potential manhour ($MHF$) components in columns 2 and 3. The breakdown of potential manhour growth between its average hours ($AHF$) and employment ($EPF$) components is presented in columns 4 and 5. The figures demonstrate strikingly how the sustained growth of the potential labor force and employment cushioned the deceleration of potential output in the face of the drastic decline of potential productivity growth after 1973.

The corresponding figures for realized inputs and output are also shown in *Table 10.3*, part B. The output ($XNR$) shortfall from potential after 1968 is reflected primarily in corresponding shortfalls in the growth rates of manhours ($MH$) and employment ($EP$), although realized productivity growth ($PROD$) was also affected adversely.

Realized and potential labor productivity are equally affected by the underlying rate of technical progress. As shown in equation (10.23), the ratio of actual to potential labor productivity ($PRODR$) therefore depends only on the ratio of the actual and potential wage/rental terms ($AWQR$), the intensity of labor utilization ($UTLR$), and the stochastic disturbance in the labor demand function ($ERMH$). The multiplicative contributions of each of these factors to the annual productivity ratios are shown in columns 1–4 of *Table 10.4*. The table also includes the breakdown from equation (10.24), giving the ratio of actual to potential output ($UTPR$) in column 6 as the product of the productivity ratio in column 4 and the ratio ($MHR$) of actual to full-employment manhours in column 5. The procyclical movement of realized productivity, which in the HC model is largely a reflection of the lagged adjustment of labor input to changes in output, is readily apparent from a comparison of columns 4 and 6.

Until the mid-1960s, the deviations of actual from potential productivity were generally moderate and the contributions of the three components were of roughly equal importance. In later years, the variations in $PRODR$ were dominated by the relative price term $AWQR$. Since the expressions for $QE$ and $QF$ share the same tax parameters and discount and depreciation rates, the variation in $AWQR$ is due solely to deviations of the expected and potential real wage

*Table 10.4.* USA: Realized and potential productivity, manhours and output (relatives).[a]

| *Year* | *AWQR* (1) | *UTLR* (2) | *ERMH* (3) | *PRODR* (4) | *MHR* (5) | *UTPR* (6) |
|---|---|---|---|---|---|---|
| 1956 | 1.0065 | 1.0040 | 0.9901 | 1.0004 | 1.0140 | 1.0144 |
| 1957 | 0.9961 | 0.9900 | 1.0157 | 1.0016 | 0.9949 | 0.9965 |
| 1958 | 0.9934 | 0.9813 | 1.0179 | 0.9923 | 0.9600 | 0.9526 |
| 1959 | 0.9937 | 1.0114 | 0.9917 | 0.9967 | 0.9808 | 0.9775 |
| 1960 | 0.9946 | 1.0004 | 0.9760 | 0.9711 | 0.9815 | 0.9531 |
| 1961 | 0.9969 | 0.9906 | 0.9937 | 0.9812 | 0.9640 | 0.9459 |
| 1962 | 0.9968 | 1.0068 | 0.9899 | 0.9934 | 0.9765 | 0.9701 |
| 1963 | 1.0016 | 1.0021 | 0.9875 | 0.9911 | 0.9804 | 0.9717 |
| 1964 | 0.9979 | 1.0050 | 0.9902 | 0.9930 | 0.9896 | 0.9827 |
| 1965 | 1.0019 | 1.0087 | 0.9947 | 1.0052 | 1.0059 | 1.0112 |
| 1966 | 1.0115 | 1.0071 | 1.0202 | 1.0392 | 1.0195 | 1.0594 |
| 1967 | 1.0162 | 0.9998 | 1.0081 | 1.0243 | 1.0190 | 1.0438 |
| 1968 | 1.0116 | 1.0008 | 1.0145 | 1.0271 | 1.0206 | 1.0482 |
| 1969 | 1.0198 | 1.0033 | 0.9911 | 1.0141 | 1.0270 | 1.0415 |
| 1970 | 1.0239 | 0.9876 | 0.9939 | 1.0050 | 1.0030 | 1.0080 |
| 1971 | 1.0171 | 0.9899 | 1.0058 | 1.0126 | 0.9840 | 0.9964 |
| 1972 | 1.0080 | 1.0035 | 0.9945 | 1.0059 | 0.9904 | 0.9963 |
| 1973 | 1.0117 | 1.0061 | 1.0026 | 1.0204 | 1.0018 | 1.0223 |
| 1974 | 1.0302 | 0.9928 | 0.9744 | 0.9966 | 0.9881 | 0.9847 |
| 1975 | 1.0357 | 0.9777 | 0.9930 | 1.0055 | 0.9469 | 0.9520 |
| 1976 | 1.0065 | 1.0067 | 1.0003 | 1.0136 | 0.9590 | 0.9720 |
| 1977 | 1.0185 | 1.0073 | 1.0090 | 1.0352 | 0.9723 | 1.0065 |
| 1978 | 1.0249 | 1.0107 | 1.0078 | 1.0440 | 0.9920 | 1.0357 |
| 1979 | 1.0290 | 1.0024 | 1.0063 | 1.0380 | 0.9966 | 1.0344 |
| 1980 | 1.0319 | 0.9879 | 1.0088 | 1.0284 | 0.9740 | 1.0016 |
| 1981 | 1.0284 | 0.9938 | 0.9977 | 1.0198 | 0.9626 | 0.9816 |
| 1982 | 1.0235 | 0.9820 | 0.9954 | 1.0005 | 0.9302 | 0.9307 |

[a]See text equations (10.23) and (10.24) for variable definitions and relations.

rates. When positive wage gaps became the rule after 1965, they were usually accompanied by corresponding positive gaps between realized and potential productivity. When the expected real wage is above the potential wage, additional capital deepening is induced, raising measured productivity above the potential level. With the demand for labor conditional on output, in other words, a higher real wage reduces manhour demand and induces an equivalent increase in realized labor productivity. Variations in *UTLR* or *ERMH* may mitigate or augment this effect, but in most years it is the magnitude of *AWQR* which accounts for that of *PRODR*. Hence the principal lesson from *Table 10.4* is that a positive productivity gap stemming from a positive wage gap is a symptom of labor market disequilibrium rather than a lasting productivity gain.

In summary, the productivity slowdown appears along the potential growth path as well as in the realized performance of the economy. Moreover, the shortfall of realized productivity growth was not pronounced in any of the periods. In particular, slack demand had only a minor effect on realized productivity between 1973 and 1982. Positive wage gaps did frequently elevate realized productivity levels, but this is symptomatic of adverse employment effects rather than fundamental improvements in output per worker or per manhour. What slack demand could and did affect adversely after the early 1970s was full utilization of labor and capital resources, as actual output lagged behind potential and unemployment increased. The development of sizable real wage gaps during the 1970s also aggravated unemployment.

## 10.7. A Forecast of US Growth, 1985–2000

The forecast shown in *Table 10.5* was made in early 1985 for a study of the financial viability of the social security system (Coen and Hickman, 1985). The complete model was used to generate endogenous forecasts of nominal as well as real variables and of the actual as well as the potential path of the economy.

Realized growth is expected to average 2.8% in 1986–1990 and 2.4% in 1991–1995, as compared with potential rates of 2.5% and 2.2%. Unemployment falls correspondingly, with some narrowing of the gap between the actual and natural rates. In the last five years of the forecast, however, actual growth declines a little, potential growth rises somewhat, and unemployment increases modestly. Inflation averages 3.4% during the remainder of this decade before rising to 4% or more in the 1990s. Long-term interest rates, as represented by Moody's Corporate Average, decline substantially over the forecast horizon. The federal deficit also falls over time, from a 1985 ratio to GNP of 4.9% to an average of 0.9% during 1996–2000.

The gradual diminution of the growth rate over the forecast horizon reflects both the demographic and economic constraints on aggregate supply and the assumption of an anti-inflationary policy stance inhibiting the growth of aggregate demand. A capsule summary of these forces is presented in *Table 10.6.*

To begin with labor supply, the growth rate of the noninstitutional population aged 16 or over is shown in column 10. The noninstitutional population provides the pool of potential entrants into the labor force, and it is an exogenous variable in our model, based on official demographic projections. The decision to participate in the labor force is endogenous, however, and the aggregate participation rate *LPC* is forecast to rise until the year 2000, so that labor force ($LC1$) growth remains high despite the fall in population growth.

Were the demand for labor to increase at the same rate as the labor force, the unemployment rate would remain constant. In the forecast, however, civilian employment $E$ is seen in column 7 to rise faster than the labor force in the last half of the 1980s and the first half of the 1990s, resulting in the unemployment reductions previously observed in *Table 10.5.* Similarly, an excess of labor force over employment growth results in some increase of unemployment in 1996–2000.

The growth of labor demand depends on the growth rate of output and manhour productivity. The relevant aggregate demand variable is gross private nonresidential product *XNR*, shown in column 2 of *Table 10.6.* The wedge between *GNP* and *XNR* is provided by real income originating in government employment and in housing rent. A comparison of columns 1 and 2 shows that

*Table 10.5.* Key macroeconomic variables.[a]

| Year or Period | GNP growth rates: annual averages (%) | | | Unemployment rate: annual average (%) | | Interest rate: annual average (%) | Federal surplus[b] | |
|---|---|---|---|---|---|---|---|---|
| | Real | Potential | Implicit deflator | Actual | Natural | | Billions $ | % of GNP |
| 1984 | 6.8 | 1.9 | 3.8 | 7.5 | 5.2 | 13.5 | –176 | –4.8 |
| 1985 | 2.9 | 2.9 | 3.5 | 7.4 | 5.0 | 12.7 | –189 | –4.9 |
| 1986 | 2.3 | 2.5 | 3.1 | 7.4 | 4.9 | 11.3 | –180 | –4.4 |
| 1987 | 3.3 | 2.4 | 3.0 | 7.1 | 4.7 | 10.1 | –171 | –3.9 |
| 1988 | 2.9 | 2.5 | 3.6 | 6.7 | 4.6 | 9.2 | –160 | –3.4 |
| 1989 | 2.8 | 2.4 | 3.7 | 6.5 | 4.5 | 8.7 | –164 | –3.3 |
| 1990 | 2.9 | 2.7 | 3.9 | 6.3 | 4.4 | 8.3 | –155 | –2.9 |
| 1985–1990 | 2.8 | 2.5 | 3.4 | 6.8 | 4.6 | 9.5 | –166 | –3.5 |
| 1990–1995 | 2.4 | 2.2 | 4.3 | 5.4 | 4.2 | 7.8 | –134 | –2.1 |
| 1995–2000 | 2.2 | 2.3 | 4.0 | 5.6 | 4.0 | 7.2 | –82 | –0.9 |

[a] Annual rates of growth calculated between endpoints of intervals. Source: Coen and Hickman (1985).
[b] N.I.A. concept.

*Table 10.6.* Factors affecting the growth path of real GNP (annual growth rates in percent).[a]

| Year or Period | GNP (1) | XNR (2) | PROD (3) | MH (4) | AH (5) | EP (6) | E (7) | LC1 (8) | LPC (9) | NNI (10) |
|---|---|---|---|---|---|---|---|---|---|---|
| 1985 | 2.9 | 2.9 | 1.6 | 1.3 | –0.2 | 1.6 | 1.5 | 1.3 | 0.2 | 1.1 |
| 1985–1990 | 2.8 | 2.9 | 1.4 | 1.5 | –0.2 | 1.7 | 1.5 | 1.3 | 0.4 | 0.9 |
| 1990–1995 | 2.4 | 2.4 | 1.2 | 1.2 | –0.2 | 1.3 | 1.3 | 1.1 | 0.4 | 0.7 |
| 1995–2000 | 2.2 | 2.2 | 1.2 | 1.0 | –0.3 | 1.2 | 1.2 | 1.3 | 0.4 | 0.9 |

[a] Column headings are defined in the text. See *Table 10.5* for units. Source: Coen and Hickman (1985).

the growth rates of *XNR* and *GNP* are virtually the same, in view of the modest uptrends of government employment and housing rent in the forecast.

Given the growth rate of *XNR* as determined by aggregate demand, one may subtract the rate of increase of labor productivity in column 3 to obtain the growth rate of private manhours in column 4. Since productivity growth is virtually constant after 1990, manhours decelerate with output. The growth rate of private employment *EP* falls accordingly, although it continues to exceed that of private manhours by about 0.2 percentage points per year, owing to our endogenous forecast of a downtrend in average hours per worker *AH* at that rate. Finally, we note that total civilian employment *E* increases a little less rapidly than private employment during the late 1980s, owing to our exogenous assumptions about a slowdown of government employment.

Labor productivity, which is a key linking variable between the aggregate demand for output and the derived demand for manhours and employment in *Table 10.6*, is itself endogenous in the HC model. The growth of labor productivity is basically governed by the rate of capital deepening (rate of increase of the capital/labor ratio) and the rate of Hicks-neutral technical change (growth of total factor productivity in the production function). The rate of capital deepening in turn depends on the income share of capital and the rate of change of the real wage. (It also depends on tax policies affecting the rental price of capital, but these policies are held constant in the forecast under existing law as specified in the Economic Recovery Tax Act of 1981 and the Tax Equity and Fiscal Responsibility Act of 1982.) In this projection, the rate of Hicks-neutral technical change has been fixed at 0.9% per year, or 0.3% above the rate for 1973–1982 as estimated from our factor demand functions. Had we simply extrapolated the earlier estimate, the output growth rates in *Table 10.6* would have been 0.3% lower.

The growth of real aggregate demand in the forecast is conditioned by our assumptions on macroeconomic policies and foreign economic developments. *Table 10.7* summarizes the key assumptions on fiscal and monetary policies.

The growth rate of nominal federal purchases of goods and services is set exogenously and is assumed to decrease sharply after 1984–1985, in order to decrease the federal deficit. After allowing for inflation, the result is an immediate sharp reduction in the rate of increase of real federal purchases, from a peak of 6.7% in 1985 to an average annual rate of 0.9% in 1986–1990 and negative growth thereafter.

With regard to monetary policy, the growth rate of M1 is assumed to be reduced gradually over the forecast horizon, at a pace consistent with falling interest rates and a relatively stable income velocity of money.

State and local purchases of goods and services are determined in the model as a function of federal grants-in-aid, real personal income per capita, and population growth. A declining pattern in nominal grants-in-aid is assumed, again because of the desire to reduce the federal deficit. The negative effect of reduced grants-in-aid on state and local spending was partly offset by adjusting real purchases upward in the 1980s and 1990s, to forestall an incipient growth in the state and local surplus. The net result of these policy assumptions is a deceleration of growth in real state and local spending after 1995.

*Table 10.7.* Forecasts of fiscal and monetary variables (annual growth rate in percent).[a]

| Year or Period | Nominal federal expenditures | | | Real government purchases | | Money stock | |
|---|---|---|---|---|---|---|---|
| | *Defense* | *Other G and S* | *GAIDs* | *Federal* | *State and local* | *M1* | *Velocity* |
| 1984 | 10.2 | 9.4 | 8.0 | 5.8 | 2.2 | 6.4 | 4.3 |
| 1985 | 13.4 | 10.7 | 8.4 | 6.7 | 2.5 | 7.0 | –0.5 |
| 1986 | 6.3 | 3.2 | –7.0 | 1.2 | 2.3 | 7.0 | –1.6 |
| 1987 | 6.7 | 4.1 | 2.0 | 1.4 | 0.8 | 6.5 | –0.2 |
| 1988 | 6.3 | 4.1 | 2.0 | 1.1 | 0.9 | 6.5 | –0.1 |
| 1989 | 6.4 | 4.3 | 2.0 | 0.7 | 1.2 | 6.0 | 0.4 |
| 1990 | 5.1 | 4.2 | 1.0 | 0.2 | 1.4 | 6.0 | 0.7 |
| 1985–1990 | 6.1 | 4.0 | 0.0 | 0.9 | 1.3 | 6.4 | –0.2 |
| 1990–1995 | 4.6 | 4.6 | 0.0 | –0.9 | 1.6 | 6.0 | 0.6 |
| 1995–2000 | 4.6 | 4.5 | 0.0 | –0.6 | 0.6 | 6.0 | 0.0 |

*Table 10.8.* Forecasts of foreign sector variables (annual percentage changes unless otherwise noted).[a]

| Year or Period | Exports | | | Imports | | | Nominal net exports | | Import prices | |
|---|---|---|---|---|---|---|---|---|---|---|
| | *Value* (1) | *Volume* (2) | *Price* (3) | *Value* (4) | *Volume* (5) | *Price* (6) | *Level ($ bil.)* (7) | *% of GNP* (8) | *Oil* (9) | *Other* (10) |
| 1984 | 8.5 | 5.1 | 3.3 | 24.7 | 27.0 | –1.8 | –64 | –1.7 | –2.5 | –1.0 |
| 1985 | 5.5 | 2.4 | 3.0 | 12.0 | 9.7 | 2.1 | –95 | –2.4 | –1.0 | 2.6 |
| 1986 | 10.3 | 6.8 | 3.3 | 13.1 | 6.5 | 6.2 | –118 | –2.9 | 0.0 | 7.0 |
| 1987 | 10.2 | 6.7 | 3.3 | 9.1 | 2.9 | 6.0 | –123 | –2.8 | 5.0 | 6.2 |
| 1988 | 11.0 | 7.2 | 3.5 | 6.5 | 1.6 | 4.9 | –110 | -2.4 | 6.0 | 4.8 |
| 1989 | 11.0 | 7.2 | 3.6 | 5.6 | 1.2 | 4.4 | –89 | -1.8 | 6.5 | 4.2 |
| 1990 | 11.0 | 7.2 | 3.6 | 5.7 | 1.3 | 4.4 | –63 | –1.2 | 6.5 | 4.2 |
| 1985–1990 | 10.1 | 7.0 | 3.5 | 8.0 | 2.7 | 5.2 | –101 | –2.2 | 4.8 | 5.3 |
| 1990–1995 | 10.0 | 6.1 | 3.7 | 6.9 | 2.6 | 4.1 | 29 | 0.1 | 4.4 | 4.2 |
| 1996–2000 | 9.8 | 6.5 | 3.1 | 8.3 | 4.7 | 3.4 | 125 | 1.4 | 3.6 | 3.6 |

[a] Annual rates of growth calculated between endpoints of intervals. Source: Coen and Hickman (1985).

The key exogenous variables in the foreign sector are nominal export expenditures and the price of oil and other imports (*Table 10.8*). We assume an average growth rate of nominal export demand of about 10% between 1986 and 2000. After allowing for export price inflation, as determined endogenously by domestic factors, real export growth is found to increase from 2.4% in 1985 to an average rate of 7% in 1986–1990, before declining to about 6% in the 1990s.

The dollar price of imports is forecast exogenously to increase sharply during the next few years, owing to an assumed depreciation of the exchange rate. After peaking in 1986–1987, the rate of import price inflation is assumed to fall gradually over the duration of the forecast (column 6). The path of demand for real imports, as determined endogenously by changes in real income and the relative price of imported and domestic goods, is shown in column 5 of the table, and the nominal value of imports appears in column 4. Finally, the current deficit on net exports (column 7) is forecast to be eliminated in the early 1990s and to move into surplus thereafter.

In summary, the growth path of real output in this long-term forecast is primarily determined by supply factors, including population growth, labor force participation decisions, technical progress, and the economic incentives for capital deepening. Fiscal and monetary policies and international developments are also influential, however, since they affect the adjustment of aggregate demand to aggregate supply, with important consequences for the inflation and unemployment forecasts. No attempt is made in this forecast to alter tax rates to increase supply incentives or to augment aggregate demand by influencing consumption or investment decisions, but discretionary changes in federal expenditures do importantly affect economic performance and the federal deficit.

## 10.8. Driving Forces of Economic Growth

As in neoclassical theory, the major exogenous variables in the Bonn multicountry growth model are the labor force, the aggregate saving rate, and the rate of technical progress. Neoclassical theory is also the starting point for the HC annual growth model, but our treatment varies in significant ways from that of the Bonn group.

### 10.8.1. The production function

In the HC model, the aggregate production function is defined for gross private nonresidential product and includes only capital and labor inputs. The Bonn models utilize a three-factor function including imported goods as an additional input and referring to total production minus domestic secondary inputs.

### 10.8.2. Labor input

Population growth is exogenous in our model, but labor force participation rates and the natural rate of unemployment are both endogenous variables. Changes in the aggregate participation rate drive a wedge between population and labor force, whereas those in the natural rate affect the amount of potential employment for a given potential labor force. As shown in *Table 10.1*, the potential labor force increased about half of a percentage point faster than the noninstitutional population during 1973–1982, and about the same margin is predicted in future years (*Table 10.6*). However, the rate of population growth itself is projected to fall in the official sources. Simple trend extrapolation would overpredict labor force growth, although extrapolation at the same rate as population growth would not.

The growth of potential labor input also depends on the trend of average working hours. The HC average hours equation implies a continuing downtrend at about the historical rate (*Tables 10.3* and *10.6*).

The net result of these various forces is a predicted decline in the annual growth rate of potential manhours from 1.5% per year in 1985–1990 to 1.2% in 1990–1995 and 1.0% in 1995–2000. The corresponding annual growth rate of labor input in the Bonn model of the US averages 1.4% and shows a rising pattern during 1985–2000.

### 10.8.3. Capital

The distinction in our model between output origination in housing and general government, on the one hand, and gross private nonresidential product, on the other, leads to a corresponding distinction on the side of capital and labor inputs. As discussed above, business fixed investment is endogenous in the model, and its equilibrium share of potential output is part of the solution for the potential growth path. The equilibrium share for gross fixed business investment was 10% in 1968–1982. The corresponding ratio in the Bonn model is much higher because it includes residential construction with business fixed capital.

### 10.8.4. Technical progress

The principal differences between the two models concern the treatment of technical progress or total factor productivity. In the Bonn model, Harrod neutrality is assumed, and the equilibrium rate of progress, which varies annually, is estimated by correcting the realized change in manhour productivity each year for the associated changes in the capital/output and import/output ratios, using factor shares as weights. When smoothed with a five-year moving average, the resulting annual estimates of technical progress follow a pronounced long swing during 1956–1980, which is identified with a Kondratieff long wave or cycle (Krelle, 1986b, 1987). In the HC model, the rate of technical change is estimated econometrically from the interrelated system of factor demands and the production function. It is assumed to be Hicks-neutral and to follow a piecewise exponential trend rather than a long swing.

The following observations are offered in support of the specification used in the HC model.

The growth-accounting approach to estimation of an annual rate of technical progress relies on the counterfactual assumption that firms are always on the production function for observed inputs of capital and labor. In contrast, our model rests squarely on the well-established fact that factor utilization and, especially, capital utilization fluctuate cyclically and may often depart from full utilization for years at a time. Business firms are normally off the production function for measured inputs of labor and capital, which must therefore be adjusted for varying rates of utilization in order to derive the correct effective inputs either for growth accounting or direct econometric estimation of the production function.

As a corollary, realized labor productivity varies directly with changes in capacity utilization and deviates from the path of potential productivity. Fluctuations of aggregate demand induce procyclical changes in realized productivity as firms adjust labor inputs with a lag to changes in output and relative factor prices.

It was shown in column 6 of *Table 10.4* that the US economy operated at only 95% of potential GNP in the recessions of 1958 and 1961. A strong and lengthy recovery began in 1962 under the stimulus of the Kennedy tax cuts and was greatly accelerated by the Vietnam War beginning in 1965. The potential utilization rate averaged 105% during 1965–1969 before declining to an average of 101% in 1970–1973. These demand-induced output fluctuations in turn induced parallel movements in the ratio of actual to potential manhours (column 5) and in the ratio of actual to potential labor productivity (column 4). The latter ratio also responded to induced changes in the expected wage/rental ratio (column 1), but was unaffected by the underlying trend in technical progress since the latter impinges equally on realized and potential labor productivity.

Against this background, the doldrums of the 1970s are seen to result from a reduction in the secular rate of technical progress and sporadic shocks to aggregate supply and aggregate demand, rather than as an endogenous sequel to a long upswing of technical progress in the 1960s. *Figure 10.1*, based on a similar chart by Krelle (1987), does show a pronounced swing in the smoothed growth rate of actual labor productivity during the 1960s. No such swing is apparent in the similarly smoothed growth rate of potential labor productivity, however. Potential productivity varies moderately from year to year because of changes in the potential wage/rental ratio, and from period to period because of the trend breaks in the rate of technical progress in 1969 and 1974, but it does not follow a long cycle. The swing in realized productivity growth is an artifact of the major demand shocks of the 1960s rather than symptomatic of an underlying long cycle in technical progress.

In the HC model, the trend rate of Hicks-neutral progress during 1973–1982 is estimated at 0.6% per year. The corresponding Harrod-neutral rate is found by dividing the Hicks-neutral rate by the estimated labor elasticity of production – a procedure which is justified by the multiplicative form of the Cobb–Douglas function – and equals 0.8% per year. As noted earlier, however, the Hicks-neutral rate was set judgmentally to 0.9% in our forecast for 1985–2000, which

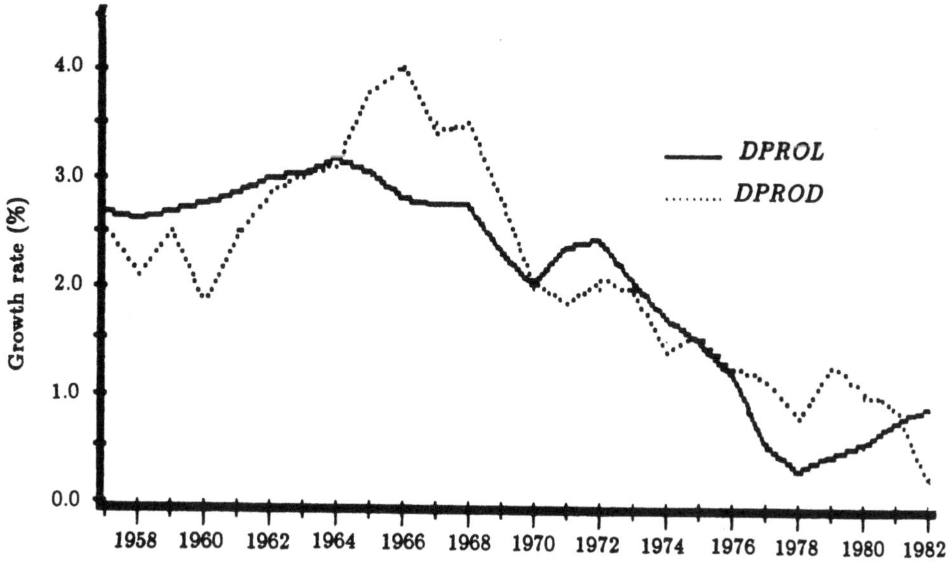


*Figure 10.1.* Actual ($DPROD$) and potential ($DPROL$) growth rates of labor productivity (five-year moving averages).

implies a Harrod-neutral rate of 1.2%. As shown in *Table 10.6*, the growth rate of labor productivity converges to the exogenous Harrod-neutral rate of progress in the 1990s. When combined with the falling pattern projected for labor input, the result is a decline in the growth rate of potential GNP from 2.5% in 1985–1990 to 2.2% in 1990–1995 and 2.3% in 1995–2000, and corresponding projections for actual GNP of 2.8%, 2.4% and 2.2%.

With regard to the Bonn model forecast, it was decided not to impose the postulated Kondratieff upswing in technical progress in the growth projections. Rather, Harrod-neutral technical progress at a constant trend rate of 0.8% per year during 1985–1985 was assumed in the forecasts for the US economy. Together with the projected average growth rate of labor input of 1.4%, this results in an average growth rate of real GNP which is nearly constant at 2.2% per year, although drifting upward slightly over the period owing to the gradual rise in labor input.

## References


Coen, R.M. and Hickman, B.G. (1980a), *A Disaggregated Model of Labor Supply and Unemployment, 1951–2000*, Working Paper WP-80-15. (Laxenburg, Austria: International Institute for Applied Systems Analysis).

Coen, R.M. and Hickman, B.G. (1980b), *The Natural Growth Path of Potential Output*, Working Paper WP-80-132. (Laxenburg, Austria: International Institute for Applied Systems Analysis).

Coen, R.M. and Hickman, B.G. (1980c), *Testing Factor Demands for Monetary Influences and Technical Change in the Postwar Economy*, Research Memorandum No. 241. (Palo Alto, CA: Center for Research in Economic Growth, Stanford University).

Coen, R.M. and Hickman, B.G. (1983), *Energy Shocks and Macroeconomic Activity: Simulation Results from the Hickman-Coen Model*, EMF 7.11. (Palo Alto, CA: Energy Modeling Forum, Stanford University).

Coen, R.M. and Hickman, B.G. (1984), Tax Policy, Federal Deficits, and US Growth in the 1980s, *National Tax Journal*, **3**, pp. 89–104.

Coen, R.M. and Hickman, B.G. (1985), *Social Security and Macroeconomic Activity in 1985–2010: A Forecast with the Hickman-Coen Annual Growth Model*, CEPR Publication No. 50. (Palo Alto, CA: Center for Economic Policy Research, Stanford University).

Easterlin, R.A. (1978), What Will 1984 Be Like? Socioeconomic Implications of Recent Twists in Age Structure, *Demography* **15**(4), pp. 397–432.

Hickman, B.G. and Coen, R.M. (1976), *An Annual Growth Model of the US Economy*. (Amsterdam: North-Holland).

Kendrick, J.W. (1983), International Comparisons of Recent Productivity Trends, in: S.H. Schurr, S. Sonenblum and D.O. Wood (eds.), *Energy, Productivity, and Economic Growth*. (Cambridge, MA: Oelgeschlager, Gunn, and Hain).

Krelle, W. (1986a), The Bonn–IIASA Research Project on Economic Growth and Structural Change. Distributed at task force meeting, Sofia, June 23–25 (mimeographed).

Krelle, W. (1986b), Growth, Decay, and Structural Change. Paper read at the IIASA Conference on Economic Growth and Structural Change, November 24/25, International Institute for Applied Systems Analysis, Laxenburg, Austria.

Krelle, W. (1987), Long-term Fluctuations of Technical Progress and Growth, *Journal of Institutional Economics* [*Zeitschrift für die gesamte Staatswissenschaft*], **143**, pp. 379–401.

Wachter, M.L. (1976), The Changing Cyclical Responsiveness of Wage Inflation, *Brookings Papers on Economic Activity* **1**, pp. 115–159.

CHAPTER 11

# Sectoral Output, Employment, and Real Wages: Long-Run Trends for the US Economy

*Peter Pauly*

## Summary

This chapter reports on the results of an empirical examination of sectoral employment dynamics for the US economy. We present a new model of labor demand which includes the effects of real wage dispersion, approximated by the conditional variance of real wages across industrial sectors. The model is based on new econometric techniques to model conditional heteroskedasticity. Pooled time-series cross-section results for 33 industrial sectors in the period 1964 to 1984 form the basis for a projection of sectoral employment levels for the remainder of the decade.

## 11.1. Introduction

Structural changes continuously affect all sectors and markets in the economy. Few consequences of these changes are, however, felt by economic agents as directly as adjustments in the sectoral demand for labor. Shifts in final demand, technologies, and relative prices lead to effects on output, employment, and real wages that reflect the differences in sectoral sensitivities to the forces of structural change. The purpose of this chapter is, therefore, to focus on one particular aspect of this process for the US economy: the sectoral employment dynamics. In particular, we examine the dependence of employment on real wages and real wage differentials across sectors.

The effects of real wages on employment dynamics have been examined in numerous studies [e.g., Bils (1985), Grubb *et al.* (1983), Sachs (1983)]. In addition to providing additional evidence regarding these effects at a sectoral level, we also focus the discussion on the role of real wage dispersion. It has been argued that real wage differentials across sectors are an indication of the degree of labor market flexibility. According to this view, a large variance of real wages *per se* contributes to higher employment levels since it indicates the availability of low-wage job opportunities to absorb laid-off workers from other sectors [see Bell and Freeman (1985), Lawrence and Lawrence (1985), or Wachter and Wascher (1983)]. We shall therefore examine whether, empirically, such a positive effect of dispersion can be found at the sectoral level, in addition to standard real wage effects.

The chapter is organized as follows. In Section 11.2 we examine the causal structure between employment, output, and real wages. In Section 11.3, we present estimation results from labor demand equations for a cross-section of 33 industries (which are listed in detail in *Table 11.2*). The novel aspect here is the use of the ARCH-M model [Engle (1982), Engle *et al.* (1987), Diebold and Pauly (1988)], to model wage dispersion endogenously as the conditional variance of a time-series process representation of real wages. Finally, in Section 11.4 we report some predictions of sectoral employment dynamics based on this model.

## 11.2. Causality Tests

In principle, any sensible analysis of employment dynamics on an aggregate level must be concerned with the feedbacks from employment to output and real wages. On a sectoral level, however, in particular for an economy characterized by centralized wage bargaining, it is not at all clear whether the real wage–productivity nexus requires a simultaneous treatment.

To address that issue, we set up a simple Granger causality test in a panel framework [Chamberlain (1982)] to explore the causal structure between manhours (*MH*), real value added ($Q$), and real wages ($W/p$). This involves the estimation of a vector-autogressive system of the form

$$\begin{pmatrix} (W/p)_t \\ MH_t \\ Q_t \end{pmatrix} = A(L) \begin{pmatrix} (W/p)_{t-1} \\ MH_{t-1} \\ Q_{t-1} \end{pmatrix} + \varepsilon_t \tag{11.1}$$

where $\varepsilon_t = (\varepsilon_{t1}, \varepsilon_{t2}, \varepsilon_{t3})'$ and $A(L)$ is a matrix polynominal in the lag operator, which can be partitioned into typical elements $A_{ij}(L)$ $(i,j = 1,2,3)$. The null hypothesis of weak exogeneity can be tested as a zero constraint on the parameter set $A_{ij}(L)$ $\forall_{i,j}$ $(i \neq j)$. The test statistic is distributed as $\chi^2(q)$, where $q$ denotes the number of constraints.

The results of the relevant likelihood ratio tests can be summarized as follows:

| | *Hypothesis* | *Test statistic* |
|---|---|---|
| (i) | wages and output endogenous vs. both exogenous | 4.23 |
| (ii) | wages exogenous, output endogenous vs. both exogenous | 7.31 |
| (iii) | wages endogenous, output endogenous vs. both exogenous | 12.18 |

The test statistics are well below the 95% critical values. We therefore conclude that, based on this parameterization, real value added and real wages are strictly causally prior to manhours. On a sectoral level, it is thus appropriate to estimate labor demand equations in a single equation framework.

## 11.3. A Model of Sectoral Demand for Labor

Traditionally, estimates of labor demand are based on a model that explains the rate of change of manhours ($M\hat{H}$) as a function of real value added growth ($\hat{Q}$) and the rate of change of real product wages ($\hat{W}/e = \hat{w}$), i.e.,

$$M\hat{H} = \alpha + \beta\hat{Q} + \beta\hat{Q} + \gamma\hat{w} + u \tag{11.2}$$

A preliminary pooled cross-section time-series estimate of (11.2) for 33 sectors and a sample of 21 years (1964–1984) yielded the following result ($t$-values in parentheses):

$\alpha = 0.74$ (3.5)

$\beta = 0.80$ (45.7)

$\gamma = -0.62$ (35.7)

The variance decomposition yields a cross-section component of 0.182, and time-series and error components of 0.0004 and 0.0041, respectively.

While the model provides a rather satisfactory representation of the data set at hand, such an aggregate estimate conceals the extent to which firms' behavior differs on a sectoral level. Individual time-series for these sectors yield production elasticity estimates which range from 0.05 to 1.45, and real wage elasticities in the range between 0 and –1.52. It is thus obvious that detailed sectoral projections must be based on sector-specific estimates. In addition, there appears to be a fair amount of cyclical variability in these estimates. A sequence of cross-section estimates, while showing little variation in the output elasticity (0.93 to 0.99) over 21 years, exhibits significant fluctuations in the real wage

*Table 11.1.* Model.

$$M\hat{H}_{i,t} = \alpha_i + \beta_i \hat{Q}_{i,t} + \gamma_i \hat{w}_{i,t} + \delta \sigma^2_{w,i,t} + u_{i,t} \tag{11.3}$$

with

$$\sigma^2_{w,t} = \begin{pmatrix} \sigma^2_{w,1,t} \\ \sigma^2_{w,2,t} \\ \cdot \\ \cdot \\ \cdot \\ \sigma^2_{w,k,t} \end{pmatrix} = \begin{pmatrix} a_{01} \\ a_{02} \\ \cdot \\ \cdot \\ \cdot \\ a_{0k} \end{pmatrix} + \begin{pmatrix} a_{11} & & & 0 \\ & a_{22} & & \\ & & \ddots & \\ 0 & & & a_{kk} \end{pmatrix} \begin{pmatrix} \hat{\varepsilon}_{1,t-1} \\ \hat{\varepsilon}_{2,t-1} \\ \cdot \\ \cdot \\ \cdot \\ \hat{\varepsilon}_{k,t-1} \end{pmatrix}$$

and

$$\hat{W}_t = R(L)\,\hat{W}_{t-1} + \varepsilon_t \tag{11.4}$$

where

$$\hat{W}_t = (w_{1,t}, w_{2,t}, \ldots, w_{k,t})'$$

and

$$\varepsilon_t = (\varepsilon_{1,t}, \varepsilon_{2,t}, \ldots, \varepsilon_{k,t})'$$

Notation:

$M\hat{H}_{i,t}$ = Rate of change of manhours in sector $i$
$\hat{Q}_{i,t}$ = Rate of change of real value added in sector $i$
$\hat{w}_{i,t}$ = Rate of change of real wages in sector $i$
$\sigma^2_{w,i,t}$ = Conditional variance of real wage increases in sector $i$

elasticity (–0.47 to –1.22). An interesting aspect of these preliminary results is that the importance of real wage changes, after a high in the early 1970s, seems to have been on a continuous decline since 1974, the year of the first oil shock. [Detailed results are available upon request to the author.]

The model to be used in the remainder of this chapter seeks to combine the standard labor demand model with a proper representation of the effects of real wage variability across sectors (dispersion). Econometrically, the important point here is to provide a satisfactory measure of dispersion. Earlier studies used unconditional variances of real wages, generally in the form of moving averages. Such a measure is, however, quite inappropriate since it does not reflect agents' information sets. Rather, what is required here is a representation of the conditional variance. Unfortunately, in general, little information is available about the appropriate conditioning sets. A convenient approximation can under those conditions be provided by a version of Engle's (1982) ARCH model. Here the conditional variance of a variable is approximated as an autoregressive process of the forecast error variance of the structural representation of the variable in question. The extension of the concept that applies in the present context is that the conditional variance also enters as a regressor, thus affecting the mean of the dependent variable. This is the ARCH-M model discussed in Engle, Lilien and Robins (1987) and Diebold and Pauly (1988).

The model is summarized in *Table 11.1*. The labor demand equation (11.3) includes as a regressor the conditional variance of a real wage model (11.4), which in this case is a simple $AR(q)$ model on a sectoral basis. The ARCH part models dispersion as a first-order autoregressive process. The model could obviously be generalized by modeling sectoral real wages as a vector-autoregressive process. Lack of degrees of freedom prevents us, however, from estimating such a fully specified model.

The estimation results are quite encouraging. In about 60% of the sectors, real wage dispersion enters significantly with the correct (positive) sign. The fit, on the sectoral level, improves marginally, compared with the standard model. *Table 11.2* summarizes results for the two crucial elasticities in the model [detailed results are available upon request]. Again, the most striking result is the enormous variability of elasticities across sectors. The range of output elasticities is very similar to the one obtained in the standard model, while the variance of real wage elasticities is somewhat reduced. An important result is that, almost uniformly across sectors, the introduction of a significant effect of dispersion tends to reduce the real wage elasticity. Note that we obtain virtually no wage effect at all for some of the services sectors.

## 11.4. Sectoral Employment Projections

The model estimated in the previous section will now be the basis for projections of sectoral employment paths for the period 1985–1995. Note that the estimation provides us with three sets of sectoral equations: an autoregressive representation of sectoral real wage dynamics, an ARCH model to describe the conditional variance, and a manhours equation to determine labor input as a function of output, real wages, and the dispersion measure.

For the purposes of these projections, sectoral output forecasts are taken from Wharton Econometrics' annual model. A summary of historical growth rates and the relevant projections is given in *Table 11.2*. In general the growth rates reflect a further decline of average rates of expansion for a large number of sectors. Examples of growth acceleration compared with the 1970s include the construction industry, metal industries, motor vehicles, and chemical production.

The results of these projections are summarized in *Figures 11.1–11.8*. In these graphs, we report the shares of output and employment for various sectoral groupings. Employment is obtained by applying trend rates of decline of average working hours to the projections of total manhours.

*Table 11.2.* Estimation results. (Sources: ***Wharton Econometrics Long-Term Outlook***; own calculations.)

| *Sector code* | *Sector title* | *Output elasticity* | *Real wage elasticity* | *Real value added: annual growth rates (%)* | | |
|---|---|---|---|---|---|---|
| | | | | *1961–1971* | *1971–1981* | *1985–1995* |
| 0211 | Metal mining | 0.89 | –0.47 | 1.6 | 0.1 | –1.0 |
| 0221 | Coal mining | 0.88 | –0.78 | 2.6 | 3.6 | 2.7 |
| 0231 | Oil and gas extraction | 0.29 | –0.41 | 3.4 | 1.8 | –0.7 |
| 0241 | Nonmetal mining, except fuels | 1.09 | –0.88 | 3.4 | 1.3 | 1.9 |
| 0300 | Construction | 0.47 | –0.61 | 4.2 | –3.8 | 1.0 |
| 0503 | Lumber and wood products | 0.71 | –0.81 | 6.3 | 2.9 | 2.3 |
| 0507 | Furniture and fixtures | 0.74 | –0.69 | 3.6 | 4.0 | 3.1 |
| 0515 | Stone, clay and glass products | 0.93 | –0.77 | 3.0 | 1.4 | 2.9 |
| 0531 | Primary metal industries | 0.67 | –0.98 | 12.4 | 1.0 | 2.0 |
| 0535 | Fabricated metal products | 0.95 | –1.05 | 3.9 | 2.2 | 3.1 |
| 0541 | Machinery, except electrical | 0.71 | –0.71 | 4.5 | 5.4 | 5.9 |
| 0545 | Electrical and electronic equipment | 0.65 | –0.34 | 7.2 | 5.5 | 3.9 |
| 0555 | Motor vehicles and equipment | 1.06 | –0.46 | 8.1 | 1.5 | 3.4 |
| 0557 | Other transportation equipment | 0.76 | –0.48 | 5.7 | 3.3 | 2.2 |
| 0561 | Instruments and related products | 0.28 | –0.50 | 5.5 | 6.0 | 4.2 |
| 0571 | Misc. manufacturing industries | 0.11 | –0.13 | 3.7 | 3.1 | 2.4 |
| 0603 | Food and kindred products | 0.64 | –0.48 | 3.2 | 2.7 | 2.1 |
| 0605 | Tobacco manufactures | 1.07 | –1.13 | 1.5 | 1.4 | 0.96 |
| 0607 | Textile mill products | 0.95 | –1.21 | 7.0 | 1.6 | 1.4 |
| 0609 | Apparel and related products | 0.53 | –0.59 | 2.7 | 3.3 | 2.6 |
| 0611 | Paper and allied products | 0.64 | –0.36 | 4.3 | 2.5 | 2.3 |
| 0613 | Printing and publishing | 0.36 | –0.28 | 3.4 | 2.4 | 2.5 |
| 0615 | Chemical and allied products | 0.35 | 0.05 | 6.7 | 1.8 | 2.6 |
| 0617 | Petroleum and coal products | 0.87 | –0.47 | 3.8 | 0.3 | 1.6 |
| 0619 | Rubber and misc. plastic products | 1.58 | –1.33 | 7.0 | 3.6 | 3.3 |
| 0621 | Leather and leather products | 0.27 | –0.01 | 0.3 | –0.3 | –0.5 |
| 0700 | Transport and public utilities, total | 0.60 | –0.41 | 5.8 | 2.7 | 3.0 |
| 0800 | Communication, total | 0.64 | –0.40 | 8.0 | 7.4 | 5.5 |
| 0900 | Electric, gas and sanitary services | 0.63 | –0.26 | 6.0 | 4.1 | 2.2 |
| 1030 | Wholesale trade | 0.66 | 0.01 | 4.4 | 3.0 | 2.8 |
| 1060 | Retail trade | 1.02 | –0.01 | 4.8 | 3.2 | 2.8 |
| 1105 | Banking | 1.00 | 0.01 | 4.2 | 4.1 | 3.0 |

*Figures 11.1* and *11.2* represent two of the declining sectors. For both mining and transportation, output and employment shares decline continuously. The productivity trends are, however, quite different. While average productivity in mining continues to go down significantly during the forecast period, the output share of transportation stabilizes with declining employment shares.

The aspect of structural change that has received most of the attention in the past is the apparent decline in manufacturing (industrial) activity and the shift toward services. That is, naturally, also reflected in our results. *Figures 11.3* and *11.4* illustrate the steep reduction in the number of manufacturing jobs, from a share of more than 35% in the early 1950s to just about 15% projected for the end of the century; the present share is about 20%. In contrast, the output share remained remarkably stable over the past 30 years, and is expected to continue to be in the range of approximately 20%. These results reinforce the important effects of productivity gains on employment structures. While the qualitative results are greatly similar for both durable and nondurable manufacturing, there appears to be much more cyclical variability in the former than in the latter sector.

*Figures 11.5* to *11.7* cover various service sectors. The situation here is entirely different. Rather than being characterized by constant output shares and declining employment shares, these sectors are generally on a steady growth path leading to increasing shares for both employment and output. For the entire service sector, the employment share is expected to reach about 25% by the end of the projection period, growing at roughly the same rate as during the entire postwar period. Output shares grow much more slowly, indicating that in these sectors job creation is aided by declining average productivities.

A slightly more complex picture emerges for the government sector. Contrary to popular perceptions, the output share of government has been on an almost continuous decline since the early 1950s. At the same time the government employment share had been rising until 1975, when a reversal occurred. The present projection is for a continuation of this downward trend in the share of government employment, though at a somewhat reduced rate.

Overall, our results suggest that there is little indication of change in the pace of structural adjustment of employment shares over the next decade or so. The decline of manufacturing shares will continue, while the service share will increase at about the same rate experienced during the 1970s. Government's share in employment will continue to fall, thus extending the trend established since the mid-1970s.

## 11.5. Summary

In this chapter, we have examined long-run trends in employment shares for 33 sectors of the US economy. The particular focus was on sectoral differences in the effects of real wages and on the importance of real wage dispersion for intersectoral adjustment flows. A model of sectoral labor demand was estimated, which incorporates an endogenous variability measure into a traditional employment function. Pooled estimates confirm the importance of dispersion effects,

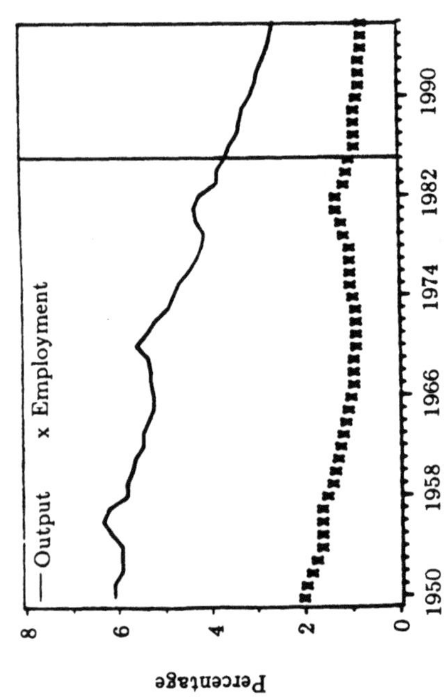


*Figure 11.1.* Mining: employment and output shares.

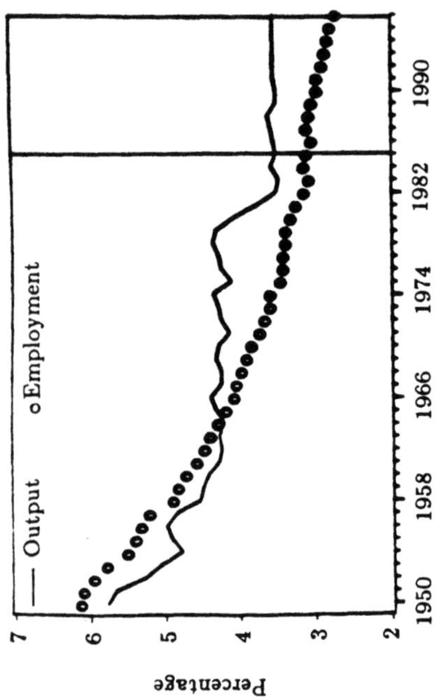


*Figure 11.2.* Transportation: employment and output shares.

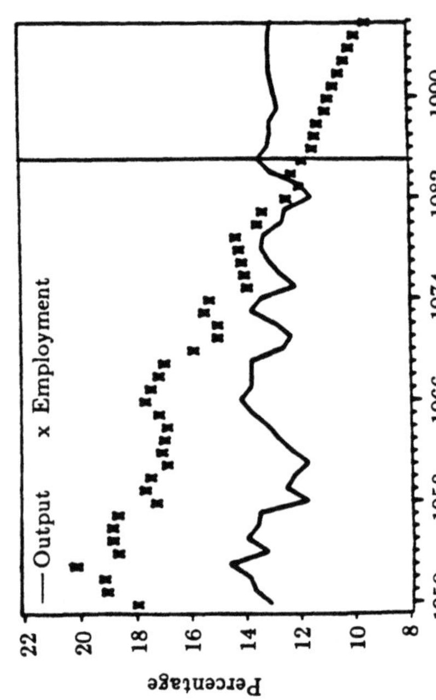


*Figure 11.3.* Durable manufacturing: employment and output shares.

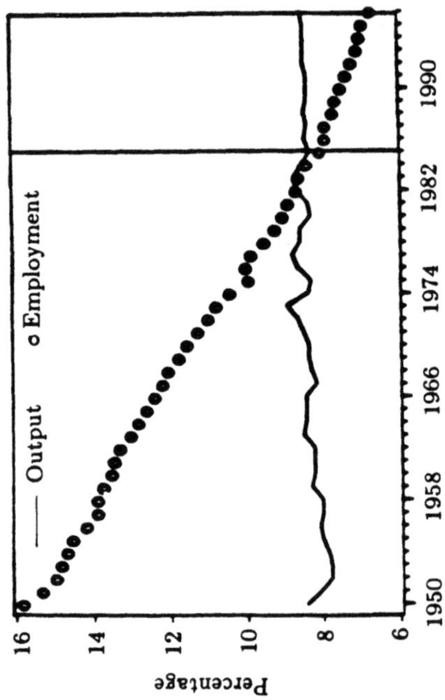


*Figure 11.4.* Nondurable manufacturing: employment and output shares.

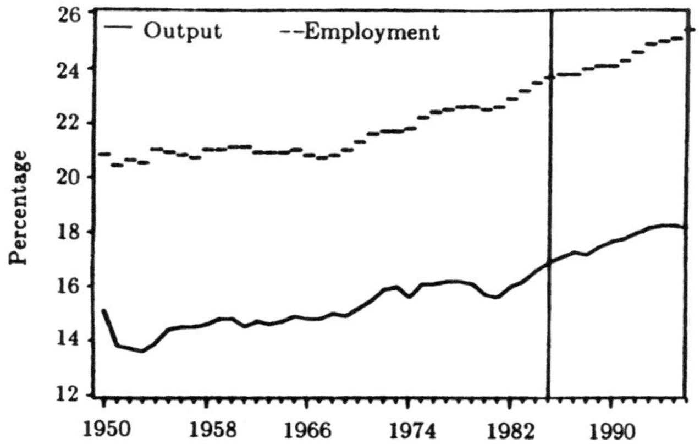


*Figure 11.5.* Wholesale and retail trade: employment and output shares.

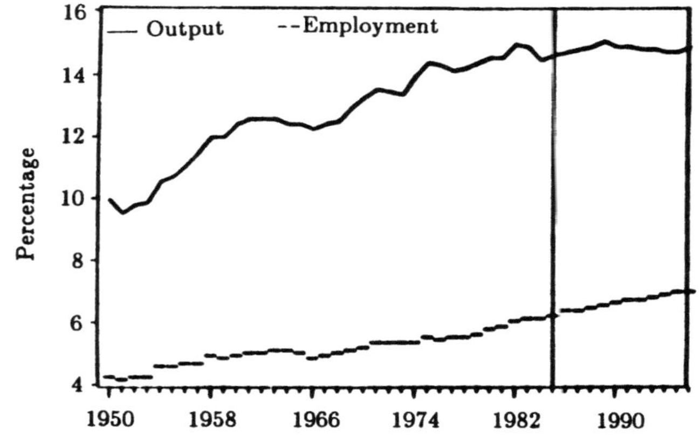


*Figure 11.6.* Finance, insurance and real estate: employment and output shares.

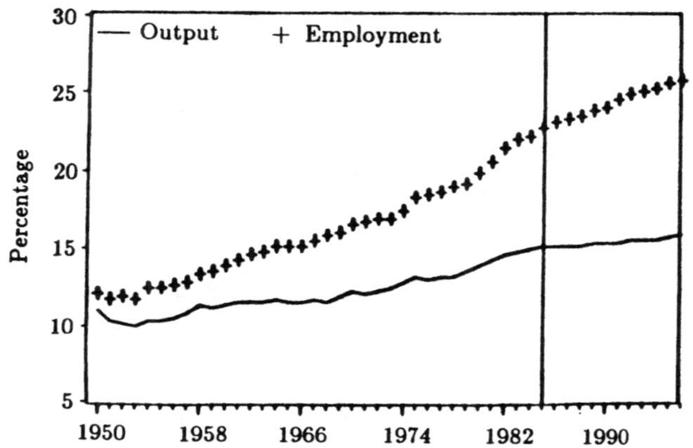


*Figure 11.7.* Services (total): employment and output shares.

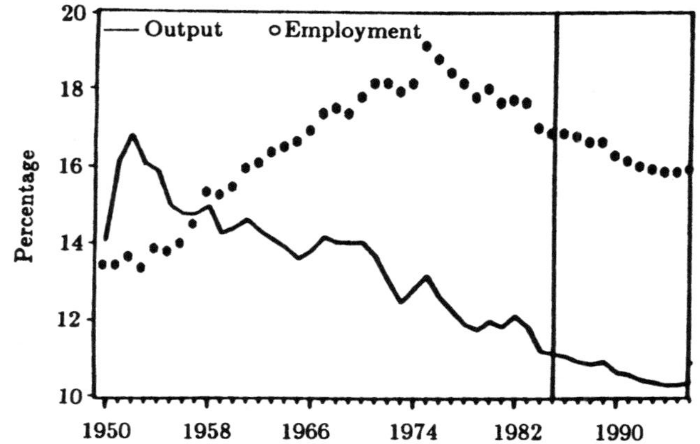


*Figure 11.8.* Government (total): employment and output shares.

and the variance of real wage elasticities in this model is somewhat smaller than in previous sectoral employment studies.

On the basis of this model, sectoral employment trends are then projected for the period 1985–1995. The results suggest a continued increase in employment shares for services, at the expense of manufacturing industries. Despite a moderate decline in average growth rates for almost all sectors, it appears that no significant shift in the speed of intersectoral reallocations can be expected. Employment patterns during the decade ahead will basically follow those trends established in the 1970s and early 1980s.

## References

Bell, L. and Freeman, R. (1985), Does a Flexible Industry Wage Structure Increase Employment? The US Experience, Discussion Paper, Harvard University, January.

Bils, M. (1985), Real Wages over the Business Cycle: Evidence from Panel Data, *Journal of Political Economy*, **93**, pp. 666–689.

Chamberlain, G. (1982), Panel Data, in: Z. Griliches and M.D. Intriligator (eds.), *Handbook of Econometrics*, Vol. 2. (Amsterdam: North Holland).

Grubb, D., Jackman, R., and Layard, R. (1983), Wage Rigidity and Unemployment in OECD Countries, *European Economic Review*, **21**, pp. 11–49.

Diebold, F.X. and Pauly, P. (1988), Endogenous Risk in a Portfolio-Balance Rational-Expectations Model of the DM/$ Rate, *European Economic Review*, Vol. 32, No. **1**, pp. 27–53.

Engle, R.F. (1982), Autoregressive Conditional Heteroskedasticity with Estimates of the Variance of UK Inflation, *Econometrica*, **50**, pp. 394–419.

Engle, R.F., Lilien, D.M., and Robins, R. (1987), Estimating Time-Varying Risk Premia in the Term Structure: The ARCH-M Model, *Econometrica* **55**, pp. 391–407.

Lawrence, C. and Lawrence, R.Z. (1985), Manufacturing Wage Dispersion: An End Game Interpretation, *Brookings Papers on Economic Activity*, **1**, pp. 47–106.

Lilien, D. (1982), Sectoral Shifts and Cyclical Unemployment, *Journal of Political Economy*, **90**, pp. 777–793.

Ross, S.M. and Wachter, M.L. (1973), Wage Determination, Inflation, and the Industrial Structure, *American Economic Review*, **63**, pp. 675–692.

Sachs, J.R. (1983), Real Wages and Unemployment in the OECD Countries, *Brookings Papers on Economic Activity*, **1**, pp. 255–304.

Wachter, M.L. (1970a), Cyclical Variation in the Interindustry Wage Structure, *American Economic Review*, **60**, pp. 75–84.

Wachter, M.L. (1970b), Relative Wage Equations for U.S. Manufacturing Industries 1947–1967, *Review of Economics and Statistics*, **52**, pp. 405–410.

Wachter, M.L. and Wascher, W.L. (1983), Labor Market Policies in Response to Structural Changes in Labor Demand, pp. 177–216 in: *Industrial Change and Public Policy*. (Kansas City: Federal Reserve Bank of Kansas City).

CHAPTER 12

# Long-Range Modeling of the USSR Economy

*S.V. Dubovsky and O.A. Eismont*


## Summary

In this chapter the economic development of the USSR in the latest decades is analyzed. Causes of the slowdown in economic growth are shown to be natural resources depletion and relative decrease in the efficiency of new technologies. A simulation model of economic growth, taking into account natural resources depletion and endogenous technical progress, is described. Economic growth forecasts for the USSR are presented for two scenarios: the "inertial" one with the last decade's trends holding true for the future and the "accelerated" one with increased labor activity and intensified technical progress.


## 12.1. Introduction

The statistics of the USSR's economic development up to the present time can hardly be satisfactorily described and explained by means of traditional production functions which use only three factors of production – labor, capital and constant exogenous technical progress.

As is shown in *Table 12.1*, in the last 20 years the growth rates of the USSR's economic development have slowed down. One of the major causes of the slowdown is the decreasing rate of technical progress. The amounts of newly developed machines and equipment have dropped from 4636 in 1961–1965 to 3134 in 1981–1985 (*Statistical Yearbook*, 1981 and 1986), which is an indicator of a decrease in the replacement rate of old technologies by new ones. Simultaneously, capital renewal rates have decreased due to the drop of average growth rates in machinery and equipment production industries from 12.4% in 1961–1965 to 6% in 1981–1985. The depreciation rate has dropped almost by half. Thus the assumption of a constant technical progress rate in the production function is not valid for the present study.

*Table 12.1.* USSR national income growth rates.

| | *1966–1970* | *1971–1975* | *1976–1980* | *1981–1985* |
|---|---|---|---|---|
| *Average annual growth rates* (%) | 6.9 | 5.0 | 4.2 | 3.1 |

During the last 10–15 years, as more accessible and less costly natural resources in the central regions of the country have become considerably exhausted, it has become necessary to move to less accessible and more costly resources in the eastern parts of the country, much more distant from major consumption centers, which is another major cause of the slowdown of economic growth in the USSR.

Among all natural resources, energy takes a special place since the economy as a whole is strongly dependent on it. On the other hand, the energy production sector consumes a large fraction of material, labor and financial resources and thus is dependent on the economy: the energy production sector accounts for about 20% of investment (35% of industrial investment), 25% of industrial capital and 2.5 million workers.

In the last 15 years (1971–1985) in all energy subsectors the output/capital ratio has been decreasing: in electrical power generation by 22% (from 0.38 to 0.30), in the fuel industry by 52% (from 1.19 to 0.57), and in the oil industry by 64%. Besides that, extraction costs (excluding capital costs) have also been rising: by 2.8 times in oil production, 3.8 times in natural gas production, 1.6 times in coal production, 1.3 times in electricity production. In addition, the quality of energy resources is deteriorating. For example, average ash contents in coal produced in the USSR in these years has increased from 22.6% to 26.7% (Lalayanz, 1986).

As a result, while the energy/output ratio has been decreasing by 0.8%/yr the costs of energy resources have been increasing by 1.8%/yr, which means that technical progress could not compensate for the deterioration of conditions of energy production.

Keeping all the above factors in mind we must conclude that, when modeling the economic growth of the USSR, which is the world's main producer and one of the world's main exporters of energy resources, it is necessary to take into account processes of energy resource depletion. Introduction of endogenous technical progress and natural resources depletion into the model might better explain the existing statistics.

This chapter deals with economic growth modeling which takes into account the aforementioned factors and the results of simulations.

## 12.2. Technical Progress and Economic Growth

Technical progress is considered as a process of permanent renewal of technologies and equipment. Each technology is associated with the capital of a specific age. Thus in order to describe formally the process of technology renewal it is necessary to know the age distribution of capital.

This distribution is described as follows [Dubovsky (1984); see also Dubovsky (1982), Dubovsky *et al.* (1983)].

$$\frac{\partial \rho}{\partial t} + \frac{\partial \rho}{\partial \tau} = -\mu(t, \tau)\rho$$

$$\rho(t_0, \tau) = \psi(\tau), \quad \rho(t, 0) = \dot{F}(t) \tag{12.1}$$

where $t$ = current time, $\tau$ = capital age, $\rho(t, \tau)$ = age distribution of capital, $\psi(\tau)$ = initial age distribution, $\dot{F}(t)$ = new capital introduction rate and $\mu(t, \tau)$ = capital depreciation rate.

In the majority of economic growth models capital age distribution $\rho(t, \tau)$ is replaced by capital described in an aggregate form:

$$K(t) = \int_0^\infty \rho(t, \tau) d\tau$$

If $\mu(t, \tau) = \mu(t)$, then while $\rho(t, \infty) = 0$ from (12.1) the well-known equation for $K(t)$ follows

$$\dot{K}(t) = \dot{F}(t) - \mu(t)K(t) \tag{12.2}$$

If $\mu(t, \tau) = \mu(\tau)$, then while $0 \leq \tau \leq \infty$,

$$\rho(t_0, \tau) = \dot{F}(t_0 - \tau) \exp\left[- \int_0^\tau \mu(x) dx\right]$$

it follows from (12.1) that:

$$\dot{K}(t) = \dot{F}(t) - \int_0^\infty \mu(\tau) \dot{F}(t - \tau) \exp\left[- \int_0^\tau \mu(x) dx\right] d\tau \tag{12.3}$$

Let the total capital be characterized by the average technical level $U(t)$. This level is usually dependent on labor efficiency, capital/labor ratio, energy/output ratio, etc.

Next, the relationship between the average technical level $U(t)$ and the technical level of newly introduced capital will be derived. It is assumed that the technical level of capital of age $\tau$ depends on the time of capital formation $t - \tau$ and is defined as $u(t - \tau)$.

If the age distribution of capital is given by the function $\rho(t, \tau)$, then the average technical level of capital is given by the following expression:

$$U(t) = \int_0^\infty \frac{\rho(t, \tau)u(t - \tau)}{K(t)}\, d\tau \tag{12.4}$$

Taking the equality $\partial u/\partial t = -\partial u/\partial \tau$ into account together with relationship (12.1), we get from (12.4) the equation

$$\dot{U}(t) = -\frac{\dot{K}(t)}{K(t)}\, U(t) + u(t - 0)\,\frac{\dot{F}(t)}{K(t)} - \frac{1}{K(t)} \int_0^\infty \mu(t, \tau)\rho(t, \tau)u(t - \tau)\, d\tau$$

from which two different types of equations are derived, depending on how the depreciation rate is defined.

In the first case $\mu(t, \tau) = \mu(t)$:

$$\dot{U}(t) = \frac{\dot{F}(t)}{K(t)}\,[u(t - 0) - U(t)] \tag{12.5}$$

In the second case $\mu(t, \tau) = \delta(\tau - \tau_1)$ ($\delta =$ Dirac function, $\tau_1 =$ capital lifetime):

$$\dot{U}(t) = \frac{\dot{F}(t)}{K(t)}\,[u(t - 0) - U(t)] - \frac{\dot{F}(t - \tau_1)}{K(t)}\,[(u(t - \tau_1) - U(t)] \tag{12.6}$$

Equation (12.6) can be rewritten in the following form:

$$\frac{\dot{U}(t)}{U(t)} = \frac{\dot{F}(t)}{K(t)}\, x_0 - \frac{\dot{F}(t - \tau_1)}{K(t)}\, x_1, \tag{12.7}$$

$$x_0 = [u(t - 0) - U(t)] \,/\, U(t), \quad x_1 = [u(t - \tau_1) - U(t)] \,/\, U(t)$$

where parameters $x_0$ and $x_1$ can be readily interpreted in the following way: $x_0$ = relative increment of the newest technical level of capital compared to the average one, $x_1$ = relative decrement of the obsolete technical level compared to the averge one.

Thus, according to (12.7) the growth rate of the technical level of capital as a whole depends on the capital renewal rate and the relative economic efficiency of the newest technologies compared to the obsolete ones.

The assumption that the relative economic efficiency of newly introduced technologies and the depreciation rate remain constant [$x_0$, $\mu$ = const] leads to the traditional interpretation of technical progress in production functions as an integral of equations (12.2) and (12.7)

$$\frac{U(t)}{U(t_0)} = \left[\frac{K(t)}{K(t_0)}\right]^{x_0} \exp[\mu x_0(t - t_0)] \tag{12.8}$$

However, this assumption is not always true. Using known statistics of average indicators characterizing introduction of new technologies and discharge of obsolete ones, it is possible to use equation (12.7) for estimating values of $x_0$, $x_1$ as functions of time.

According to the capital growth, labor productivity and capital/labor ratio statistics for the USSR industrial sector given in the *Statistical Yearbook* (1981 and 1986), the values of $x_0$ for four five-year periods were estimated using a simplified formula:

$$\frac{\Delta U}{U} = x_0 \left(\frac{\Delta K}{K} + \mu\right)$$

where $\Delta U/U$ = relative increment of technical level in five years, $\Delta K/K$ = relative capital growth in five years, $\mu$ = relative capital depreciation in five years. The results are given in *Table 12.2.*

*Table 12.2.* Relative increment in cost of a newly created job and its productivity for the industrial sector of the USSR compared to the preceding average level (in percent).

| | *1966–1970* | *1971–1975* | *1976–1980* | *1981–1985* |
|---|---|---|---|---|
| Increment in cost of a new job | 60 | 68 | 70 | 76 |
| Increment in productivity of a new job | 56 | 55 | 31 | 37 |

When analyzing these estimates, it should be noted that the increase of job costs in 1976–1985 was not followed by the equivalent increase of labor productivity. The hypothesis $x_0$ = const did not correspond to reality, either.

Forecasting the dynamics of the technical level of new technologies presents a special problem. Dubovsky (1984) analyzed different dynamic patterns of the technical level. In the simplest case $x_0$ is either a constant or is an exogenous function of time. This is equivalent to the introduction into the production function of "free of charge" but embodied technical progress.

A more sophisticated approach takes into account the R&D expenditures (technological investment). In this case equation (12.5) can be reduced to the following form:

$$\dot{U} = C_u(U_k U_u)^{\alpha} / YU/K \tag{12.9}$$

where $Y =$ GDP (FP), $C_u$, $\alpha$ = parameters ($0.5 \leq \alpha \leq 1$), $U_k$, $U_u$ = relative shares in GDP (FP) of productive and technological investment.

## 12.3. Natural Resources Depletion and Economic Growth

Now let us consider the problem of modeling natural resources depletion and its influence on economic growth. It is assumed that the society extracts and consumes first the most easily accessible and less costly natural resources and, as these are exhausted, turns to less accessible and more costly ones. (It is ignored that, in real practice, several reserves with different extraction costs are exploited simultaneously.)

To illustrate this assumption, the specific extraction cost distribution density for natural resources $q(C_R, t)$ is introduced, where $C_R$ = specific extraction cost. Natural resource extraction costs include labor and capital costs and exclude all kinds of rent. In *Figure 12.1* the qualitative graph of function $q(C_R, t)$ is given, $q(C_R, t_0)$ being the initial extraction cost distribution. Solid lines correspond to the actual process of natural resources extraction, when resources of different costs are extracted simultaneously; broken lines correspond to the model process, when at each moment only the least costly type of natural resource is extracted.

It is assumed that natural resources extraction costs depend on the cumulative extraction $Q$ of natural resources and on the technical level $U_R$ of the resource sector in the following form:

$$C_R = C_{R_0} \left[\frac{Q}{Q_0}\right]^{\beta} \left[\frac{U_R}{U_{R_0}}\right]^{-\gamma} \tag{12.10}$$

where

$$Q = \int_0^t R_{pr}\, \mathrm{d}t + Q_0,\; Q_0 = \int_{-\infty}^{0} R_{pr}\, \mathrm{d}t$$

where $R_{pr}$ = natural resource production rate.

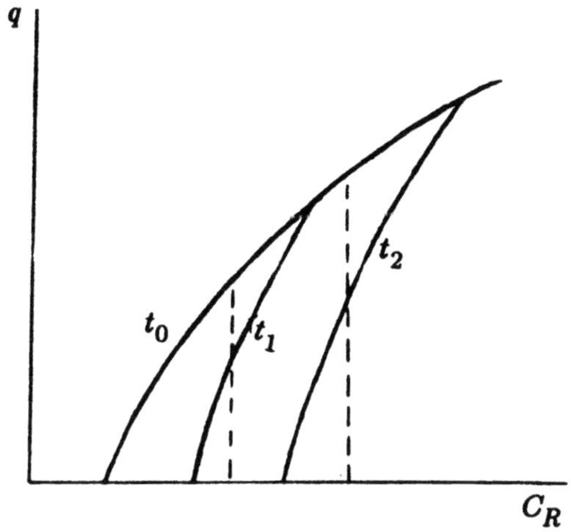


*Figure 12.1.* Extraction cost distribution density for natural resources: solid line = actual process, broken line = model process, $t_2 > t_1 > t_0$.

It is assumed that the natural resource is not imported and all domestic demand as well as all export needs are met by domestic production.

In *Figure 12.2* the results of estimation of equation (12.10) for aggregate energy resource in the USSR are presented ($U_R = U$, $\beta \sim 2.2$, $\gamma \sim 1.2$). Equation (12.10) refers to the case when the natural resource is unsubstitutable. In this case the extraction costs may grow indefinitely. In real life, however, another case is more realistic, i.e., where the natural resource is substitutable. In this case the aggregate energy resource can be associated with the final energy (electricity, heat, motor fuel, etc.) which is used for GDP (FP) production. Final energy extraction costs can be defined accordingly. In this case due to interfuel substitutability and introduction of practically inexhaustible energy resources (for example, solar and fusion energy), the extraction costs will be limited. The corresponding qualitative behavior of function $C_R(Q)$ is presented in *Figure 12.3.*

## 12.4. A Macromodel of Economic Growth

Let $Y$ be the production function associated with GDP (FP). $Y$ is assumed to depend on production factors of capital $K$, labor $L$, natural resource consumption rate $R_c$ which is used for GDP (FP) production, technical level $U$, and accumulated amount of produced natural resource $Q$:

$$Y = Y(K, L, R_c, U, Q)$$

The production function is assumed to be linear homogeneous in $K$, $L$, $R_c$:

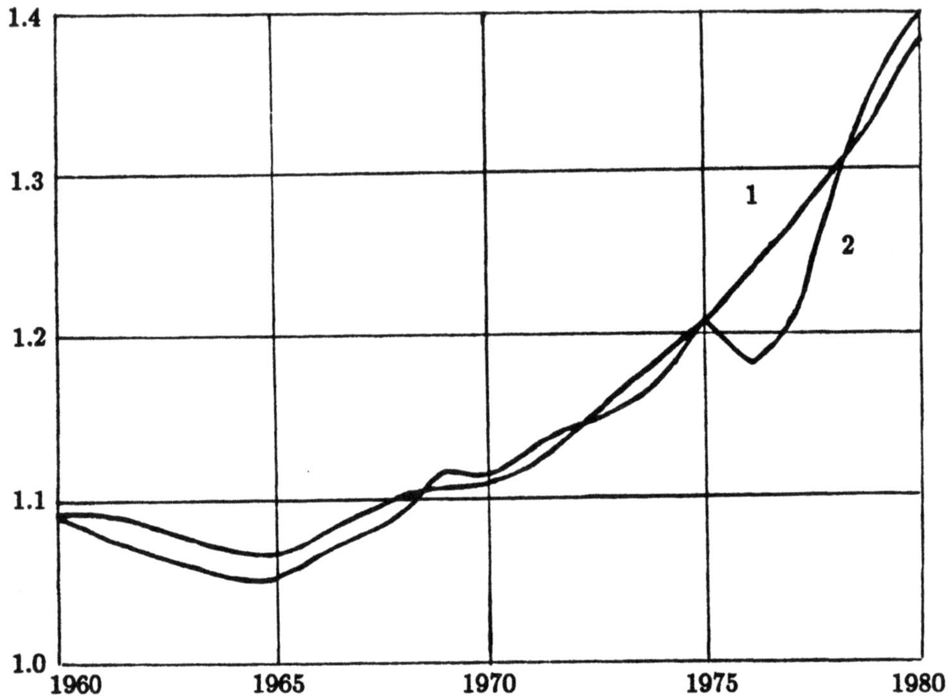


*Figure 12.2.* Primary energy production cost in the USSR: 1 = model, 2 = statistics.

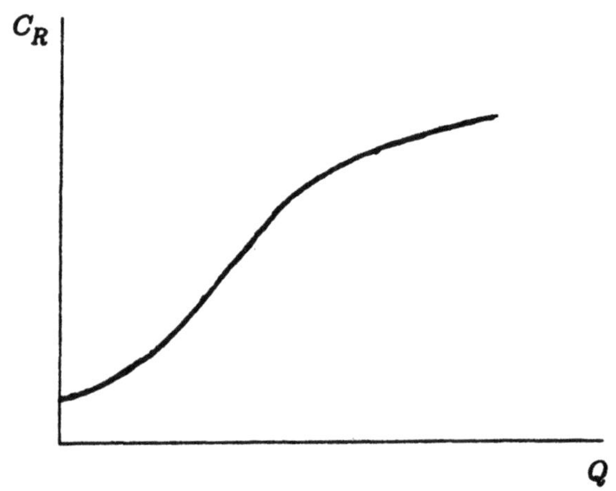


*Figure 12.3.* Final energy cost as a function of its cumulative production.

$$Y = \frac{\partial Y}{\partial K}K + \frac{\partial Y}{\partial L}L + \frac{\partial Y}{\partial R_c}R_c \tag{12.11}$$

The following hypothesis concerning the influence of resource depletion on GDP (FP) is introduced:

$$\frac{\partial Y}{\partial Q} =- R_{pr} \frac{\partial C_R}{\partial Q} \tag{12.12}$$

The production function is assumed to be linear homogenous in capital and the technical level of technology:

$$Y = \frac{\partial Y}{\partial K} K + \frac{\partial Y}{\partial U} U \tag{12.13}$$

Profit maximization condition is used

$$Y - p_L L - p_K K - p_R R_c \rightarrow \max_{K,L,R_c} \tag{12.14}$$

where $p_L$ = average wage, $p_K$ = capital cost, $p_R$ = cost of the *in situ* natural resource. The market price of natural resource will be

$$p = p_R + C_R \tag{12.15}$$

From (12.14) follows

$$\frac{\partial Y}{\partial L} = p_L, \quad \frac{\partial Y}{\partial K} = p_k, \quad \frac{\partial Y}{\partial R_c} = p_R \tag{12.16}$$

If the natural resource is used only for GDP (FP) production, then from (12.12) follows

$$Y = F(K, L, R_c, U) - R_c C_R \tag{12.17}$$

where function $F$ can be interpreted as gross output of the economy composed of two sectors: the resource-producing one and the rest of the economy.

On the other hand, if the natural resource is not a production factor and is used only as a consumer product [its consumption being proportional to GDP (FP)]:

$$R_c = a_R Y \tag{12.18}$$

where $a_R$ = resource to GDP(FP) ratio, then from (12.12) it follows:

$$Y = \varphi(K, L, U) \exp(-a_R\, C_R) \tag{12.19}$$

Using relationships (12.11)–(12.13), and (12.16) we can get the following equation:

$$\begin{aligned} \frac{\dot{Y}}{Y} &= \left[1 - \frac{p_L L}{Y} - \frac{p_R R_c}{Y}\right] \frac{\dot{K}}{K} + \frac{p_L L}{Y}\frac{\dot{L}}{L} + \frac{p_R R_c}{Y}\frac{\dot{R}_c}{R_c} \\ &+ \left[\frac{p_L L}{Y} + \frac{p_R R_c}{Y}\right] \frac{\dot{U}}{U} - \frac{R_{pr}^2}{Y}\frac{\partial C_R}{\partial Q} \end{aligned} \tag{12.20}$$

If we assume that natural resource production cost does not depend on the technical level [i.e., $C_R = C_R(Q)$], then we can get

$$\begin{aligned} \frac{\dot{Y}}{Y} &= (1 - U_L - U_R)\frac{\dot{K}}{K} + U_L\frac{\dot{L}}{L} + U_R\frac{\dot{R}_c}{R_c} \\ &+ (U_L + U_R)\frac{\dot{U}}{U} - U_c\frac{\dot{C}_R}{C_R} \end{aligned} \tag{12.21}$$

where

$$U_L = \frac{p_L L}{Y}, \quad U_R = \frac{p_R\, R_c}{Y}, \quad U_c = \frac{C_R R_{pr}}{Y}$$

When $U_L$ = const, $U_R$ = const, $U_c$ = const, equation (12.21) has the first integral, which is a Cobb–Douglas production function with labor- and resource-saving technical progress:

$$\frac{Y}{Y_0} = \left[\frac{K}{K_0}\right]^{1 - U_L - U_R} \left[\frac{L}{L_0}\right]^{U_L} \left[\frac{R_c}{R_{c_0}}\right]^{U_R} \left[\frac{U}{U_0}\right]^{U_L + U_R} \left[\frac{C_R}{C_{R_0}}\right]^{-U_c} \tag{12.22}$$

Using (12.8) we get a production function which is not linearly homogeneous in capital, labor and the natural resource rate:

$$\frac{Y}{Y_0} = \left(\frac{K}{K_0}\right)^{1 - (U_L + U_R)(1 - x_0)} \left(\frac{L}{L_0}\right)^{U_L} \left(\frac{R_c}{R_{c_0}}\right)^{U_R} ,$$

$$\left(\frac{C_R}{C_{R_0}}\right)^{-U_c} \exp\left[\mu x(t - t_0)\right] \tag{12.23}$$

If instead of (12.13) the following hypothesis is used:

$$Y = \frac{\partial Y}{\partial U} U \tag{12.24}$$

then under the condition $U_L$, $U_R$, $U_c$ = const we get the following first integral:

$$\frac{Y}{Y_0} = \left(\frac{K}{K_0}\right)^{1 - U_R - U_L} \left(\frac{L}{L_0}\right)^{U_L} \left(\frac{R_c}{R_{c_0}}\right)^{U_R} \left(\frac{C_R}{C_{R_0}}\right)^{-U_c} \frac{U}{U_0} \tag{12.25}$$

In this case $U$ can be interpreted as labor-, resource- and capital-saving technical progress.

Now let us consider the problem of optimal extraction of natural resources. If the natural resource is a private property (in a market economy), then the resource owner tends to maximize his discounted profits, i.e.,

$$\int_0^\infty \left[\rho - C_R\,(Q, U)\right] R_{pr} \exp\,(-rt)dt \rightarrow \max \tag{12.26}$$

where $r$ is a discount factor. Taking into account that

$$\dot{Q} = R_{pr} \tag{12.27}$$

from (12.25), using (12.27) and the corresponding Euler equation we can get:

$$\dot{p} = rp - \dot{C}_R \tag{12.28}$$

Equation (12.28) coincides with the classical Hotelling equation [see Hotelling (1931)] when there are no extraction costs.

In case that the natural resource is a state-owned property (in a centrally planned economy), the goal is to maximize discounted FP. Then the maximization problem will be as follows:

$$\int_0^\infty Y(K, L, R_c, U, Q) \exp(-rt)\mathrm{d}t \rightarrow \max \tag{12.29}$$

It should be noted that the natural resource is used for GDP(FP) production $R_c$, for consumer needs $R_f$ and for exports $R_{ex}$

$$R_{pr} = R_c + R_f + R_{ex} \tag{12.30}$$

Then using (12.27), (12.30) and the Euler equation corresponding to problem (12.29), we get:

$$\frac{\mathrm{d}}{\mathrm{d}t}\left[\frac{\partial Y}{\partial R_c} \exp(-rt)\right] - \frac{\partial Y}{\partial Q} \exp(-rt) = 0 \tag{12.31}$$

From this equation, using (12.12) and (12.16), we get equation (12.28). Thus, irrespective of the type of natural resource ownership, the equation describing the behavior of its price is the same – namely, (12.28).

## 12.5. Computational Experiments

The model which has been described was combined with the corresponding demographic and foreign trade models [Dubovsky (1980), Grechuha *et al.* (1985)] and used in computational experiments performed for a whole range of scenarios starting with the so-called "inertial" growth to the "accelerated" one. (The programming, estimation and simulation were executed by V.M. Vasiliev and A.F. Mironichev.)

The "inertial" scenario is characterized by the slowdown of economic growth rates, which is caused not so much by the slowdown of labor growth rates and natural resources depletion but mainly by a relative decrease of new technologies' efficiencies.

The new situation in the USSR and the goals set up by the new Soviet leadership presume acceleration of socioeconomic development of the USSR based on speeding up technical progress and thus raising economic efficiency. The "accelerated" scenario, which is formed on the basis of fundamental changes in the economic control mechanism, was analyzed. In this scenario the relative increment of new jobs efficiency compared to the past average is assumed to

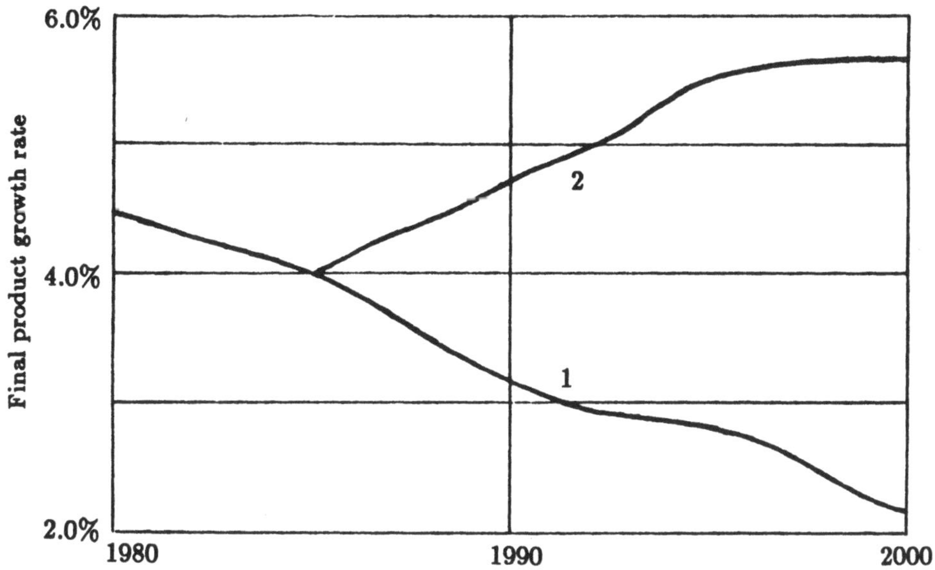


*Figure 12.4.* "Inertial" (1) and "accelerated" (2) scenario of the USSR economic growth.

grow threefold in the years 1986–2000. At the same time, technical progress should compensate for the rising costs of natural resources extraction and, as a result, the energy/output ratio (expressed in cost terms) should not rise. Labor activity is also assumed to grow: in 1986–1990 by 5%, in 1991–1995 by 10%, in 1995–2000 by 7.5% (the overall growth in 15 years will be 22.5%).

The simulation results for the two scenarios are presented in *Figure 12.4*. The comparison of these scenarios shows the considerable potential possibilities of technical progress in speeding up long-range economic growth and justifies optimistic views on economic development of the USSR in the period up to the year 2000, provided the measures aimed at perfecting the economic control mechanism are fulfilled.

## References

Dubovsky, S.V. (1980), Simulation Model of International Multiproduct Trade within Global Modeling System, in: *Systems Modeling of Global Development Processes*, Vol. 14. (Moscow: VNIISI). [In Russian].

Dubovsky, S.V. (1982), Dynamic Simulation Input–Output Model, in: *Man-Computer System for Modeling Global Development Processes*, Vol. 9. (Moscow: VNIISI). [In Russian].

Dubovsky, S.V., Vasiliev, V.M., and Eismont, O.A. (1983), Modeling the USA Economic Growth: Taking into Account New Tendencies, in: *Elements of Man–Computer System for Modeling Global Development Processes*, Vol. 3. (Moscow: VNIISI). [In Russian].

Dubovsky, S.V. (1984), Dynamics of Technological Structure and Economic Growth, in: *Global Development Processes: Modeling and Analysis*, Vol. 3. (Moscow: VNIISI). [In Russian].

Grechuha, V.A., Scherbov, S.Y., and Visotskaya, N.V. (1985), Main Tendencies in the World Demographic Development, in: *Systems Analysis of Global Development Processes*, Vol. 3. (Moscow: VNIISI). [In Russian].

Hotelling, H. (1931), The Economics of Exhaustible Resources. *Journal of Political Economy*, **39**, pp. 137–175.

Lalayanz, A. (1986), Energy-fuel Complex of the USSR in 1986–1990, *Planovoye chozyaistvo*, **7**, pp. 12–21. [In Russian].

*Statistical Yearbook* (1981), USSR National Economy in 1980. (Financy i Statistica Publishers). [In Russian].

*Statistical Yearbook* (1986), USSR National Economy in 1985. (Financy i Statistica Publishers). [In Russian].

CHAPTER 13

# Driving Forces of Economic Structural Change: The Case of Japan in the Last Decade

*Chikashi Moriguchi*

## Summary

Major economic structural changes in the near future are not likely to take place either in robotization or in bio-related industries. On the other hand, changes in Japanese agriculture, mining and some of the domestic service sectors are going to be major or massive ones. The prolonged process of "internal" adjustment will bring about economic frictions between Japan and its trading partners.

## 13.1. Introduction

In order to discuss economic structural change, we had better define what we mean by a "structure". Here I should like to confine the concept to a narrow and simple meaning, i.e., the relative size of industrial sectors that form the industrial base of an economy. I am not going to deal with the so-called structural parameter changes of economic models, which are dealt with by many authors elsewhere in this work.

I should like to stress the impact of relative price changes that can be said to be a necessary condition for economic structural change to take place. Then I am going to deal with the degree of competition among firms as the real driving force behind structural change.

Finally, I shall touch upon technical progress as an endogenous factor that is affected by relative price changes and market competition. Some implications of the recent robot revolution will be discussed as a case in point.

## 13.2. Relative Price Changes as a Source of Structural Change

Let us make a quick survey of the trend of relative prices among major factors of production.

### 13.2.1. Labor and capital

The relative price of labor and capital is defined by per capita wage earnings divided by price of capital goods. The latter is given by a fixed investment deflator. Here the modification by interest rate, corporate tax rate and the rate of capital consumption is neglected for the sake of simplified international comparison.

As *Figure 13.1* indicates, there has been a marked upward tendency of the relative price of labor against capital goods in Japan, whereas in the USA the same relative price change stays within a much smaller range. This difference is generally attributed to a difference in labor productivity growth between the two countries.

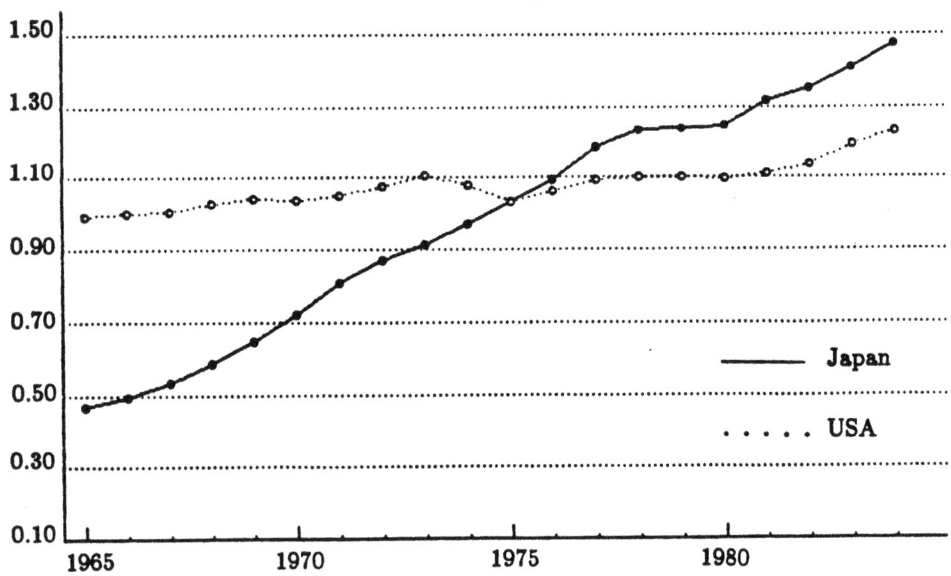


*Figure 13.1.* Average wage earnings divided by the deflator of fixed investment. (Sources: for Japan – *National Income Account, Annual Report* (Economic Planning Agency); for the USA – *Economic Report of the President* (Council of Economic Advisors.)

If there were a need for labor-saving investment or investment for capital/labor substitution, it would certainly be Japan, not the USA, that faced a stronger investment pressure. And this process is self-sustaining: high investment raises both labor productivity and the wage rate, which increases the wage/capital goods price ratio.

### 13.2.2. Capital and energy

It goes without saying that the relative price of energy plays a crucial role in determining the direction of technological development and industrial structure. During the rapid growth of the 1960s, we saw the relative price of energy steadily decline, stimulating a massive shift to energy-intensive heavy industrialization. This was a worldwide phenomenon, although the magnitude of change varied among different countries. The relative price of energy and capital, for instance, declined much faster in Japan than in any other country.

This can be seen from *Figure 13.2*, which compares Japan with the USA. Japan made a fast shift to oil as its major source of energy. Other countries with domestic coal mining industries had to pay substantial social costs for their transition to oil. Japan's pace of adjustment was the fastest simply because it had no major domestic primary energy industry.

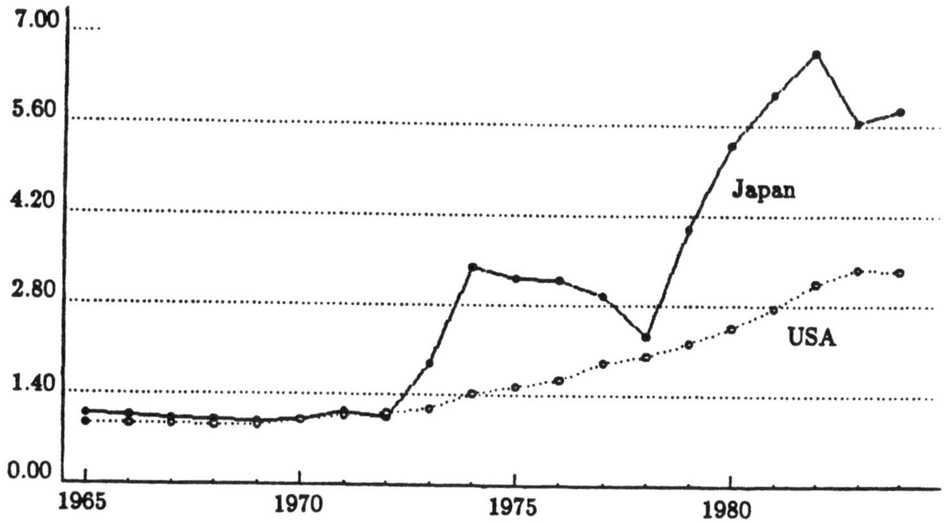


*Figure 13.2.* Oil price divided by investment deflator in Japan and average fuel price divided by investment deflator in the USA.

During the period of rapid increases of oil prices, Japan once again had to adjust quickly because its dependency on overseas oil had reached more than 70% of total energy consumption. That made Japan's ratio of oil price to capital goods prices the highest in the world, which compelled the country to pursue a new wave of energy-saving capital formation. In all industries energy-saving efforts were made, and within ten years energy input per economic unit of gross output was reduced by more than 30%.

One thing should be noted: the period of nationwide energy-saving efforts coincided with a marked change in the import structure. Part of the energy-intensive sector yielded to imports. Aluminum was a typical example. Japan started to import energy that is embodied in aluminum and other energy-intensive intermediate products, which made a significant contribution to the reduction of energy consumption in Japan.

We have discussed capital goods prices and energy prices, but it would be insufficient to look at these two inputs alone. Some empirical studies have revealed that energy and capital (service) are not gross substitutes but complementary, while energy and labor as well as labor and capital are in fact substitutes for one another [see Ban (1985), Kuroda (1985), and Murota (1983)]. This finding might seem to conflict with the observation that a relative increase in energy cost stimulated energy-saving investment. However, the observed complementarity between energy and capital service use is based on *ex post* relations with respect to a given stock of capital. It does not say everything about *ex ante* capital investment plans.

### 13.2.3. Capital and other manufactured goods

The above observations seem to suggest the hypothesis that the larger the rate of change in relative prices, the higher is the rate of capital formation.

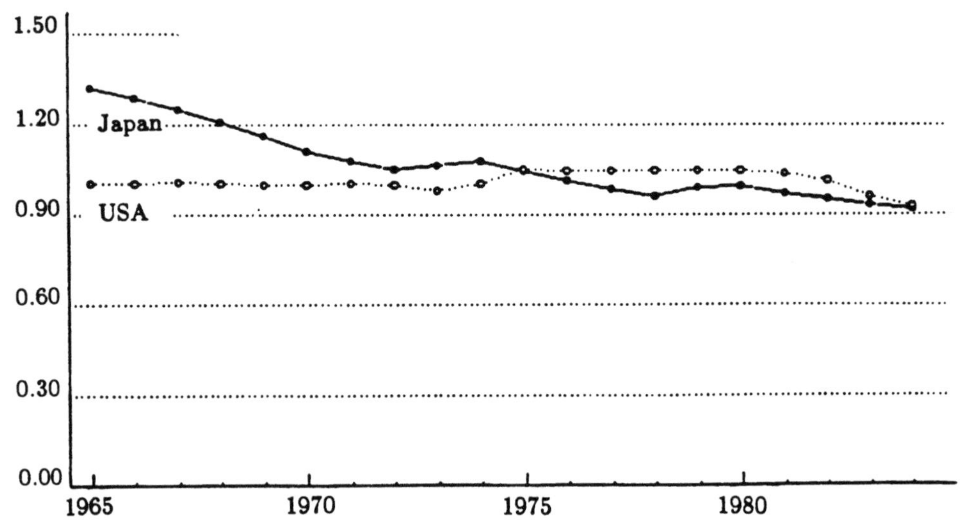


*Figure 13.3.* Investment deflator/GNP deflator.

When we look further into the behavior of relative prices of capital goods, we find that in Japan they have kept a downward trend for more than twenty years (see ***Figure 13.3***). The trend seems to have been strengthened by the microelectronics revolution of these years. On the other hand, we do not see any similar trend in the United States, where the relative price of capital goods has stayed almost invariant, except for the recent few years when overvaluation of the dollar brought down the import price of machinery and equipment.

What explains this large difference between the two countries? One underlying element seems to be the following fact: Japan is a resource-poor country – natural resources imports do not compete with domestic resources. Thus, import prices are exogenous variables, being determined in the world commodity market. In the USA, on the other hand, most resource imports compete with domestically produced goods. All prices are more or less determined endogenously; even the domestic oil price tends to be determined by domestic factors, at least in part. Thus, prices tend to fluctuate more proportionately in the USA than in Japan. To put it in a different way, the scope of endogeneity of the economy is wider in the USA than in Japan.

## 13.3. Competition as the Driving Force behind Structural Change

So far I have emphasized the importance of relative price changes as a major driving force behind economic structural change. However, there must be an economic entity that implements and pursues it, and this entity is the firm that competes.

There are several alternative ways to define the degree of competition among firms in a market. The number of engaged firms, the degree of variations of market price in a certain time period, and the degree of concentration are some of them. Whichever measure is applied, it seems clear that Japanese industries are the most competitive among major industrialized countries. Steel, chemical and petrochemical industries are highly competitive with six to ten corporations engaged in active competition. Automobile, electrical and electronic machinery and precision instruments industries are growing with many competing firms. In these industries, Japan exceeds any other industrialized country in terms of number of engaged firms or degree of concentration. Japan's semiconductor industry is a new example of fierce oligopolistic competition.

The question is: what are the instruments of competition? Competition by developing new products is a most successful strategy. But when one firm succeeds, the competition ends. The Japanese strategy seems to have been based more heavily on strong investment, inching up one's own relative share of the market through the expansion of production capacity. This occurs particularly when economies of scale prevail in the market.

The Japanese style of strong competition by means of capacity expansion has resulted in the high speed of adjustment when sudden change in relative prices occurs. A vacuum is quickly filled by keen competition, which expands to the point of excess capacity. We can cite a long list of industries that have followed this pattern: steel, chemical fertilizers, petrochemicals, and synthetic fibers, with semiconductors becoming a newer example.

The role of exports for Japanese economic growth has been a target of discussion and criticism. It seems an undeniable fact that sometimes excess productive capacity expansion caused the so-called "export drive". Active investment among competitive firms is a basic condition for high growth. But this dilemma of investment is that it creates demand, on one hand, and additional supply capacity, on the other. Particularly when there are economies of scale, competitive investment creates a larger supply capacity than might otherwise exist. Under these circumstances, expansion in the world markets rescues the competing firms from the consequences of overproduction. This explains, at least in part, why Japanese corporate firms engaged in export markets with so much aggressiveness.

### 13.3.1. Fast deployment abroad

The export-orientedness of Japanese corporations should be studied more carefully. Certainly, MITI's strong policy encouraged the growth of new export industries, but more truth seems to lie in the fact that competition among firms highlighted the ample opportunities for profit in overseas markets.

The characteristics of competition in domestic markets are distinctly different. While Japanese competitive firms are engaged in full-fledged free competition in the world market, at home they seem to be subject to more cooperative discipline, paying careful attention to domestic "order". In the world market each competing firm comes much closer to behaving as an atomistic, free-wheeling entity than in the home market. This would apply to any country, but in the case of Japan the contrast seems to be remarkable.

### 13.3.2. Did domestic competition come to an end?

Thanks to the fairly large home market of more than 100 million people, domestic competition under protection against imports did not have to end up with a situation of monopoly or of a high degree of oligopoly. The government could have employed trade liberalization policies earlier to maintain market competition, to the delight of potential foreign competitors. In fact, when the government or MITI announced a "deadline" for trade liberalization, Japan's home manufacturers stepped up their investment programs so as to improve their competitiveness against the potential "invasion forces" from abroad!

However, many observers overestimate MITI's ability to predict the direction of economic structural change and draw up a new plan for Japanese industries to adjust to it. Looking back at 30 years of economic development, one can easily pick up a certain number of counterexamples in which business firms did not act in accordance with MITI's view and still achieved great success at their own risk.

## 13.4. Technical Progress and Structural Change

Relative price changes are frequently brought about by technical progress that occurs in particular industrial sectors. During the 1960s Japan imported various technologies from overseas that centered around metal, chemical and petrochemical industries. Heavy industrialization was accelerated by building large plants, which yielded economies of scale. It was apparent that technical progress took place exogenously in the form of imported or "borrowed" technology.

In the 1970s, the situation changed when the sources of new technologies had been exhausted and new challenges of environmental control and energy price shock arose. By that time Japan had accumulated its own expertise and know-how, and in terms of the balance of technology trade Japan's deficit was shrinking.

In the latter half of the 1970s there came a wave of revolutionary microelectronics and robotics in which Japan displayed its comparative advantage in R&D activity. Some of the technical progress at this time might have been the result of the energy shock and other exogenous factors. The direction of technical development was toward the so-called "kei shou tan haku" (Japanese words for "lighter, smaller, shorter and thinner"). The characteristics of the Japanese attitude toward craftsmanship were most suitable; the manpower policy for science and engineering that had started in the 1960s (as in the USA and major European countries) returned unexpected fruits toward the end of the 1970s. [See Peck and Tamura (1976)].

As exemplified by the semiconductor industry, Japanese economic growth and technical progress in the last decade were typically borne by major corporations in Japan. Electronics giants, such as Hitachi, NEC and Toshiba, have made and stepped up their investment plans competitively; the computer industry involved a certain number of competing firms of almost similar size and market share. Robots were developed by hundreds of manufacturing firms from large shipbuilders to small machine tool makers.

The structural change that took place in this decade was characterized by the growth of the machinery industry and the decline of primary metal and other "smokestack" industries. Total manufacturing employment did not show any sizable increase or decrease, but within industries only the machinery sector showed increased employment.

The new structure was most dramatically observed in international markets, where Japan's relative share of machinery exports jumped. This was partly because the US economy grew fast under the Reagan administration, and income-elastic imports of home electronics goods and industrial machinery exploded. On the other hand, the world market for steel had reached a stage of stagnation: the US domestic steel market was being protected, and newly industrialized countries started to export steel.

All of the industrialized countries increased their exports to the USA. Their dependency on exports jumped, with the exception of the UK. However, Japan (and Asian NICs) topped the group because of their faster adjustment – that is to say, their fast investment in high-technology industries.

### 13.4.1. Robotics

Let me take as an example Japan's investment in robots. Starting from the early 1970s this investment has grown exponentially (see *Figure 13.4*). A simple log-linear function is specified with three explanatory variables: (1) total business fixed investment; (2) relative price of robots over wage earnings, and (3) the number of different kinds of robots as a measure of the rate of technical progress in robotics (see *Table 13.1*). Regression estimates show that investment in robots is taking place four times faster than general business fixed investment. The relative price of robots seems to play a significant role, although the elasticity is around unity.

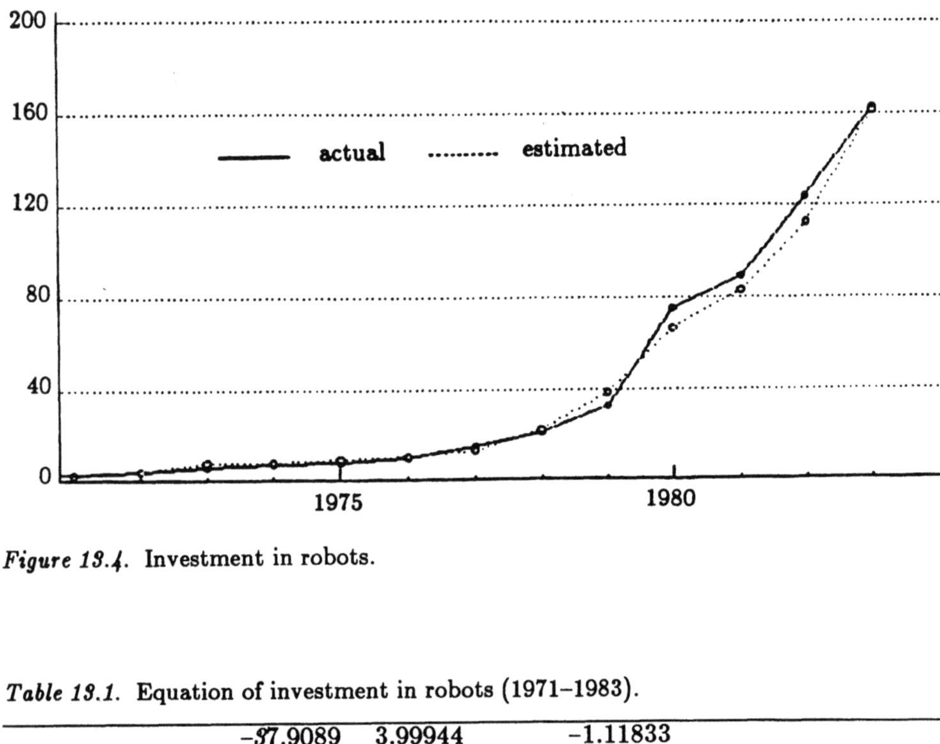


*Figure 13.4.* Investment in robots.

*Table 13.1.* Equation of investment in robots (1971–1983).

| | | |
|---|---|---|
| LOG IFROB080 | = | -37.9089 (-5.42) + 3.99944 (5.68) LOG IP -1.11833 (-3.07) LOG ROBOECON<br>+ 0.20402 above (0.84) LOG NKINDROB<br>(0.84) |
| $R^{**2}$ | = | 0.9939 $se = 0.18$, $DW = 1.510$ |
| IFROB080 | = | Purchase of robots as deflated by the price index of playback robot |
| IP | = | Total business fixed investment in 1980 prices |
| ROBOECON | = | Price of playback robot divided by average wage earnings |
| NKINDROB | = | Number of different kind of industrial robots |

As a matter of fact, various kinds of robots are being introduced for various purposes. Labor-saving accounts only for a small portion of all changes. Improving the quality of products by means of consistent, flawless work of robots seems to be a major reason for introducing them. The number of direct production workers who are being replaced by robots is not large. Many robots have to be "taught" by skilled workers so that they can simulate human work; and while robots play back the work they are programmed to do, humans work to develop a more efficient production system or handle peripheral equipment which is necessary to make robots work more efficiently.

In this way robots are not creating any major danger for employment in general. To the contrary, an interesting phenomenon is the revival of small firms which otherwise were dying due to shortage of young workers. (In Japan there is an element of dual structure in the labor market: young workers have a strong preference for employment offered by large corporations, which leaves only a residual class of workers for small companies where the turnover rate is quite high. However, this problem seems to have been at least partly solved by the introduction of robots.) MITI set up a financial cooperative, named JAROL, which aims at supplying low-interest-rate loans to those small firms that intend to install industrial robots.

In the sphere of export markets, robotics seem to be changing the picture of the international division of labor. Some of the labor-intensive export industries of Japan that have suffered from rising wage costs and upward valuation of the yen are making a revival by installing industrial robots to ease their high payrolls. For example, by using robots, the cutlery and silverware industry is now specializing in producing low-priced exports that can match keen price competition from Korea and Taiwan.

### 13.4.2. The role of government in R&D activity

During the period of importing technology, MITI intervened in the private sector with respect to royalties or other conditions of contract so that a competitive bid for buying overseas technology would not end up with an unfavorable outcome for Japan. At the time of environmental control and the energy shock, MITI tried to accelerate technological progress through investment tax credits. The auto industry had to meet low air pollution standards, and competition for government approval spurred the industry to enter the world market for automobiles. Research cooperatives for developing VLSI (very dense types of large-scale integrated circuits) is a widely publicized case, though one may argue as to how much it contributed to the development of highly condensed microchips.

It should be emphasized that MITI could take advantage of the highly competitive situation of Japanese industries. Cooperation and competition do not seem compatible; however, in the circumstances of oligopolistic competition, MITI's tactics worked well. But when MITI must deal not only with Japanese firms but also foreign firms competing in Japan, MITI's hitherto-useful tactics may not work any more. At the final stage of Japan's internationalization, the driving forces of economic structural change might assume quite new features.

## 13.5. Concluding Remarks

Major economic structural changes expected in the near future are not likely to take place either in robotization or in bio-related industries. R&D will be quite robust in these developing fields; but from the viewpoint of visible structure, changes in Japanese agriculture, mining, and some of the domestic service sectors are likely to be major or massive ones. These industrial sectors have been under heavy pressure as a result of wide price differentials between domestic and overseas markets. For agricultural products, worldwide excess supply situations coupled with the rising yen present a huge price gap. The situation is similar for domestic coal mining.

Structural change will certainly be accompanied by the so-called negative or backward adjustment policy, which will not enhance Japan's comparative advantage in the public policy sphere. It will be a time-consuming process, and the prolonged process of "internal" adjustment will bring about economic frictions between Japan and its trading partners. The period of Japan's swift structural change on the basis of expanding export industries seems to have come to an end.

## References

Ban, K. (1985), *Energy Price and Corporate Factor Demand – Case of Japanese Manufacturing Industry*. (Tokyo: Report of Special Energy Study, Ministry of Education, Government of Japan).

Kuroda, M. (1985), *Economic Growth and Total Factor Productivity – A Japan–US Comparison*. (Tokyo: Special Energy Study, Government of Japan).

Murota, Y. with Itoh, H. (1983), *A Macroeconomic Model with Trans-log Cost Functions*. (Japan Economic Research).

Peck, M.J. with Tamura, S. (1976), Technology, pp. 525–585 in: H. Patrick and H. Rosovsky (eds.), *Asia's New Giant: How the Japanese Economy Works*. (Washington, DC: Brookings Institution).

Rosenberg, N. (1976), *Perspectives on Technology*. (London and New York: Cambridge University Press).

CHAPTER 14

# A Long-Term Simulation of the French Economy

*Jean-Louis Brillet*


## Summary

A long-term scenario for the French economy is based on the Mini-DMS model, a reduced version of the Dynamic Multi-Sectoral model used by the of the National Institute for Statistics and Economic Studies (INSEE) for its medium-term forecasts. This scenario, covering 1987–2000, starts from assumptions derived by aggregation of recent hypotheses used by DMS for its own forecasts (for 1986–1991); the consecutive values follow constant trends. Thus, the final evolution should be regular, as close as possible to a stable growth path. The most important features and main properties of the Mini-DMS model are described, as well as the assumptions used in the simulation. The results show that the French inflation and growth rates, in the medium term, converge to the foreign rates.


## 14.1. Introduction

The Mini-DMS model, a smaller version of the Dynamic Multi-Sectoral model [Fouquet *et al.* (1978)], has been used for some time as a forecasting tool by INSEE's medium-term forecasting unit, the Service des Programmes, in quantitative studies made by the French Commissariat Général au Plan. This model has proved to be particularly convenient in the case of long-term forecasts, because its limited size simplifies the stipulation of assumptions, and its stabilizing properties produce coherent results over a long period.

The model has recently been reestimated over the period 1961–1985 [see Brillet (1986)]. The econometric coefficients as well as some equations have been changed. Moreover, the general outlook on the long-term evolution of some crucial variables, such as the franc/dollar ratio and the price of oil, changed in 1986 by a wide margin. Thus, we first used the modified model to produce a new simulation; afterward, we checked its quality and used the favorable case as an actual scenario of the long-term evolution of the French economy.

## 14.2. The Mini-DMS Model

### 14.2.1. Goal

As stated before, the Mini-DMS model is a reduced version of the Dynamic Multi-Sectoral model, built in 1976–1978 at the National Institute for Statistics and Economic Studies, as a medium term macroeconomic forecasting tool. DMS is still used today (of course, in a much-improved version) to produce scenarios of two types: more controlled ones for planning purposes (for the Commissariat Général au Plan), and more spontaneous ones for an annual publication in association with the Bureau International de Provisions Economiques (BIPE) or for particular customers (such as the French Senate).

The first version of Mini-DMS was designed in 1978 to study the mathematical properties of DMS, using a set of mathematical methods, mostly based on eigenvalue analysis, which were too expensive in terms of computer time (and also in terms of human time needed to study the results) to be applied to the full model. [For a description of the original model, see Fouquet *et al.* (1978).] Then the range of applications grew steadily: first the model was used for teaching quantitative macroeconomics, then to explore the implications of potential changes in the formulations of the larger model.

This increased sophistication (and number of equations) improved the quantitative properties of the model, until it was considered to be reliable enough to be used on its own for operational forecasts. Of course, these forecasts do not have the same detail, both in the assumptions and the results, as DMS or any other large model; but the costs are also lower, due to the simplified assumptions, the shorter time needed to reach a satisfactory solution, and the reasonable amount of results to study. For instance, a study on social policy might not need the sectoral decomposition of large models.

### 14.2.2. Main features

We first give some general information. Mini-DMS is

- Annual.
- Dynamic: many of its formulations involve lagged influences (growth rates, inertia factors).

- Medium-long term: medium-term mechanisms are privileged, although short-term and structural ones are present.
- Medium-sized: it has 190 equations, 65 of which are estimated, and 108 exogenous variables.
- It presents two activity sectors (or products): industrial/nonindustrial (plus nonmarket).
- Its philosophy is mostly Keynesian, but supply-side influences prove to be very strong in the medium term; it has almost no financial (or monetary) sector.

As to particular features, the most important ones are:

(1) The production function uses a Cobb–Douglas Putty–Clay formulation, defining desired values of employment and investment.
(2) Actual employment and investment adjust partially to these desired values.
(3) Unemployment is estimated by using a constant activity flexion ratio.
(4) The savings ratio is influenced positively by revenue growth, inflation (savings are considered in real terms), and unemployment (security savings).
(5) The partition of household consumption into products follows a Houthakker–Taylor formulation.
(6) Imports and exports depend on demand (domestic or foreign), on price competitiveness, and on the rate of use of productive capacity.
(7) A Phillips-type equation determines the wage rate.
(8) The price of value added is determined by a reduced form equation which takes into account the unit wage cost, the profit ratio of the previous period (a low ratio induces firms to recover their margins through a price increase), and the rate of capacity utilization (if it is low, firms will try to sell more by decreasing prices).
(9) The exchange rate can be exogenous, or follow the PPP (purchasing power parity) hypothesis, or be determined by an econometric equation.

Thus two instantaneous loops appear in the model: a Keynesian one allowing the equilibrium between real demand and supply, and a second one relating wages to prices.

We will not present the quantitative properties of the model in detail. The most important ones appear in the presentation of the forecast. Suffice it to say that a dynamic simulation over the estimation period gives good results (at least comparable with the ones obtained by other similar models), and that the value of the multipliers (both demand and supply ones) are acceptable from a theoretical point of view and coherent with the results obtained with models of the same class (in particular, with DMS).

## 14.3. The Assumptions

### 14.3.1. Medium-term

The medium-term assumptions used by the model on the 1986–1991 period are inspired by the assumptions established for a forecast of the DMS model itself.

This forecast, produced in June 1986 for the French Senate, was designed to give a general picture of the evolution of the French economy and to point out the problems which might arise in the future.

### 14.3.1.1. International environment

The following assumptions are used:

- Price of oil: growth from $18/barrel in 1986 to $22/barrel in 1991.
- GDP of our main partners: growth of 2.7% until 1989, 2.8% thereafter.
- Exchange rate: appreciation of the franc compared to the currency of our main partners:

| *1986* | *1987* | *1988* | *1989* | *1990* | *1991* |
|---|---|---|---|---|---|
| –1.0 | 1.2 | –0.1 | –0.7 | –0.5 | –0.3 |

- Foreign inflation: for GDP, 2.6% in 1987, 3.2% in 1988, 3.35% thereafter; for world trade, 1.1% in 1987, 2.9% thereafter.

### 14.3.1.2. Policy assumptions

The main policy assumptions are:

- For investment, prices and wages, the equations of the model are used as they are estimated. Considering the uncertainty of the political orientation of France, we suppose that the influence of recent policies (deregulation, freedom of prices, denationalizations) will not produce behavior different from that estimated over the sample period.
- The number of hours worked per week does not decrease from the current level of 38.5 hours.
- In the industrial sector, the productivity of capital decreases gradually, by about 2% per year. This stems from the hypothesis that investment is mostly (70%) designed not to increase productive capacity, but to replace old equipment.

As to actual policy decisions, we assume that:

- The number of civil servants is kept constant over the period (this covers a decrease in State employees and an increase in local administrations).
- Government investment grows slower (around 1% per year on the average) than consumption (about 2% on the average).

- The different VAT rates stay constant over the period.
- The income tax rate decreases somewhat, especially at the end of the period (from 6.1% to 5.9% with respect to the current rate).
- In order to help balance the social security account, the benefits (especially unemployment ones) will be lowered a bit and the contributions increased. This is postulated to be realized by a slight decrease of the employer rate (0.5% on average) and a significant increase of the worker rate (by 3%, from 19 to 22%).

### 14.3.2. Long-term

The assumptions used over the 1992–2000 period mostly represent a continuation of the evolution observed at the end of the medium-term period, except when this evolution was seen as being too strong to be sustained in the long run.

#### 14.3.2.1. International environment

- Price of oil: $22/barrel for the whole period.
- GDP of our main partners: growth of 2.75% per year.
- Exchange rate: depreciation of the franc compared to the currency of our main partners, of 0.3% per year (this covers a yearly appreciation of the mark at 1.2% and a depreciation of the dollar at –0.8%).
- Inflation: for GDP, 3.35%; for world trade, 2.95% over the whole period.

#### 14.3.2.2. Policy assumptions

- The number of hours worked per week stays at a level of 38.5 hours.
- In the industrial sector, the productivity of capital keeps decreasing, but at a lower rate – about 1.2% per year – leading to a 25% decrease between 1986 and 2000.

As to actual policy decisions:

- The number of civil servants is still kept constant over the period.
- Government investment is restored to a 2% growth rate.
- The different VAT rates remain stable.
- The income tax rate decreases somewhat, especially at the end of the period (from 6.1% to 5.9% as compared to the actual rate).
- As to social security, the amount covered by contributions keeps decreasing slightly (by something like 10% over the period), and the worker rate keeps increasing, while the employer rate is stable.

## 14.4. The Simulation

### 14.4.1. Results

The results of the simulations are presented in *Tables 14.1 to 14.4*. They show, respectively, the evolution of:

(1) Growth rates of main financial indicators: consumption, investment, imports, exports, GDP (compared with foreign GDP).
(2) Production factors: capital, labor (for each of the two products), tensions in the labor market and productive capacity.
(3) Prices and wages: different measures of prices (of consumption, production, exports, imports, global demand and its foreign equivalent); real and current wage rates.
(4) Other elements of the accounts of households, firms and government plus indicators for the rest of the world.

Each of the tables shows first the complete annual evolution over the 1987–2000 period; then a summary presents the average growth rates for specified subperiods.

In addition, *Figures 14.1* through *14.8* give a visual interpretation of the most important evolutions.

*Table 14.1.* Growth rates of selected indicators, 1987–2000.

| *Indicator* | *1987* | *1988* | *1989* | *1990* | *1991* | *1992* | *1993* |
|---|---|---|---|---|---|---|---|
| GDP: | | | | | | | |
| Foreign | – | 2.63 | 2.73 | 2.77 | 2.77 | 2.74 | 2.74 |
| Growth differential | – | -0.23 | -0.17 | 0.15 | 0.08 | 0.07 | 0.03 |
| Total | – | 2.40 | 2.56 | 2.92 | 2.85 | 2.81 | 2.77 |
| Market | – | 2.49 | 2.67 | 3.08 | 2.99 | 3.03 | 2.97 |
| Imports | – | 5.30 | 5.55 | 6.02 | 4.97 | 5.61 | 5.74 |
| Household consumption | – | 1.99 | 2.45 | 2.98 | 3.01 | 2.82 | 2.80 |
| Investments: | | | | | | | |
| Lodgings | – | 2.42 | 2.61 | 1.97 | 2.14 | 3.81 | 1.99 |
| Productive | – | 6.27 | 6.74 | 6.70 | 3.93 | 4.70 | 5.35 |
| Industrial | – | 7.83 | 8.81 | 9.54 | 6.25 | 6.61 | 7.56 |
| Government | – | 2.53 | 2.53 | 2.53 | 2.53 | 2.53 | 2.53 |
| Financial | – | 3.50 | 3.69 | 3.83 | 3.94 | 4.24 | 4.40 |
| Total | – | 4.69 | 5.04 | 4.93 | 3.38 | 4.21 | 4.24 |
| Exports | – | 4.71 | 4.94 | 5.05 | 5.33 | 5.63 | 5.61 |
| Stock variations[a] | 17.07 | 18.87 | 18.88 | 20.82 | 20.94 | 22.11 | 23.35 |

[a]In constant 1970 francs.

### 14.4.2. Interpretation

Let us first consider the development of GDP and its components, shown in *Table 14.1* and *Figures 14.1* and *14.2*. We first observe that the French GDP develops quite similarly to the foreign one. The difference in the medium term (after 1990) is almost constant and lower than 0.1% (but always positive). We must admit that this close relation would not have been obtained in a straightforward solution of the equations, and that some corrections had to be made with respect to the assumptions. But the limited nature of these corrections (concerning only the purchasing power of the wage rate and the level of productive investment) shows that the model is stable.

The development of the components of GDP is also rather stable:

(1) Household consumption growth, in the medium term, increases at a constant level of 2.8%, with a slight deceleration toward the end (to 2.55% in 2000). This decrease is partially offset, concerning household expenditures, by a symmetrical evolution of housing investment.
(2) Productive investment growth, more sensitive to short-term fluctuations, nonetheless keeps close enough to an average value of 5% per year.
(3) Imports and exports converge in the medium term to an almost identical evolution, the common deceleration being of course due to the slower growth of world trade.

*Table 14.2* describes the evolution of the production process, by sector. We observe in particular that industrial value added grows faster, as was indeed the

*Table 14.1.* Continued.

| *1994* | *1995* | *1996* | *1997* | *1998* | *1999* | *2000* | *1987–1990* | *1990–1995* | *1995–2000* |
|---|---|---|---|---|---|---|---|---|---|
| 2.74 | 2.74 | 2.74 | 2.74 | 2.74 | 2.74 | 2.74 | 2.68 | 2.74 | 2.74 |
| 0.06 | 0.06 | 0.08 | 0.09 | 0.05 | 0.04 | 0.01 | –0.08 | 0.06 | 0.05 |
| 2.80 | 2.80 | 2.81 | 2.83 | 2.78 | 2.78 | 2.74 | 2.60 | 2.81 | 2.79 |
| 2.99 | 2.99 | 3.00 | 3.01 | 2.96 | 2.96 | 2.92 | 2.72 | 3.00 | 2.97 |
| 5.52 | 5.48 | 5.51 | 5.56 | 5.33 | 5.42 | 5.26 | 5.56 | 5.47 | 5.42 |
| 2.84 | 2.87 | 2.87 | 2.82 | 2.72 | 2.65 | 2.55 | 2.45 | 2.87 | 2.72 |
| | | | | | | | | | |
| 3.02 | 3.11 | 3.20 | 3.26 | 3.30 | 3.37 | 3.41 | 2.31 | 2.81 | 3.31 |
| 4.41 | 4.35 | 4.61 | 5.14 | 4.74 | 5.38 | 5.15 | 6.50 | 4.55 | 5.00 |
| 6.37 | 6.28 | 6.30 | 6.64 | 5.97 | 6.68 | 6.35 | 8.63 | 6.61 | 6.39 |
| 2.53 | 2.53 | 2.53 | 2.53 | 2.53 | 2.53 | 2.53 | 2.51 | 2.53 | 2.53 |
| 4.55 | 4.67 | 4.79 | 4.89 | 4.98 | 5.06 | 5.13 | 3.63 | 4.36 | 4.97 |
| 3.91 | 3.91 | 4.10 | 4.45 | 4.23 | 4.65 | 4.53 | 4.83 | 3.93 | 4.39 |
| 5.66 | 5.59 | 5.51 | 5.43 | 5.41 | 5.30 | 5.28 | 4.85 | 5.57 | 5.39 |
| 24.22 | 24.92 | 25.59 | 26.29 | 26.83 | 27.42 | 28.01 | | | |

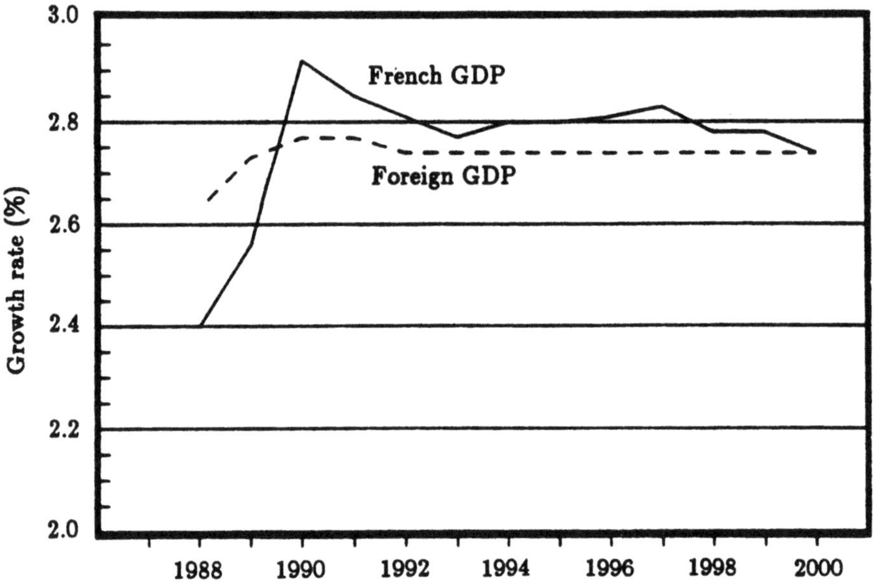


*Figure 14.1.* French and foreign GDP evolutions.

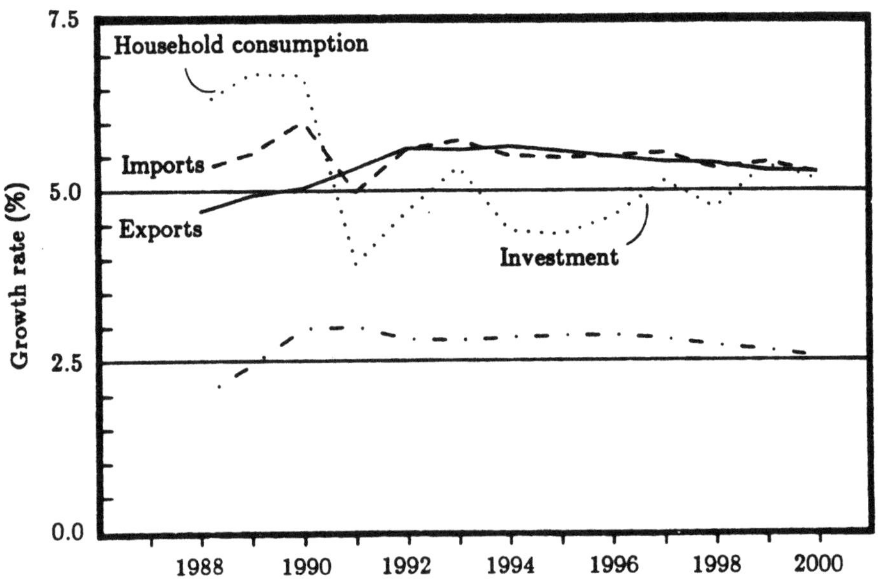


*Figure 14.2.* Elements of GDP

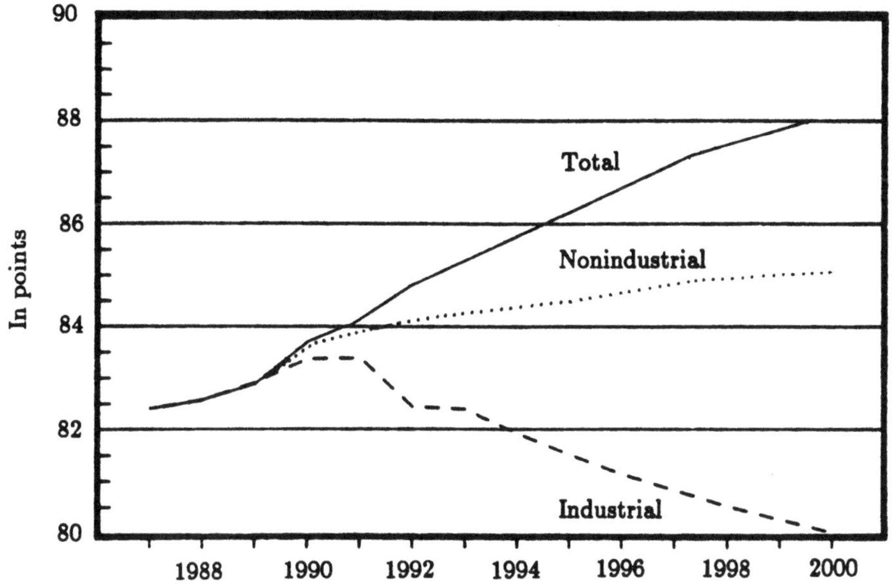


*Figure 14.3.* Rate of use of productive capacity.

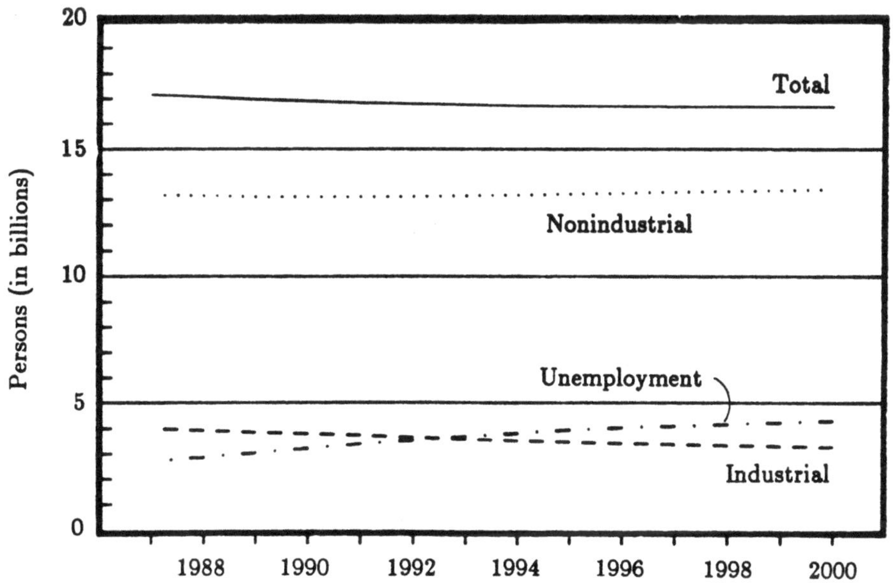


*Figure 14.4.* Employment.

*Table 14.2.* Growth rates of productive factors, 1987–2000, by sector.

| *Factor* | *1987* | *1988* | *1989* | *1990* | *1991* | *1992* | *1993* |
|---|---|---|---|---|---|---|---|
| Industry | | | | | | | |
| Value added | – | 2.84 | 2.88 | 3.22 | 2.96 | 3.35 | 3.48 |
| Production capacity | – | 2.48 | 2.63 | 2.95 | 4.03 | 4.05 | 4.08 |
| Rate of use | 82.410 | 82.584 | 82.905 | 83.384 | 83.395 | 82.846 | 82.395 |
| Capital | – | 4.82 | 5.12 | 5.22 | 5.54 | 5.60 | 5.68 |
| Apparent productivity | | –1.88 | –2.13 | –1.90 | –2.44 | –2.13 | –2.03 |
| Hourly labor productivity | – | 4.93 | 4.81 | 4.92 | 4.73 | 5.43 | 5.51 |
| Activity | – | –1.99 | –1.84 | –1.62 | –1.69 | –1.98 | –1.92 |
| Work duration | – | –0.00 | –0.00 | –0.00 | –0.00 | –0.00 | –0.00 |
| Employment | – | –1.99 | –1.84 | –1.62 | –1.69 | –1.98 | –1.92 |
| In millions | 4.01 | 3.93 | 3.85 | 3.79 | 3.73 | 3.65 | 3.58 |
| Nonindustrial | | | | | | | |
| Value added | – | 2.35 | 2.53 | 2.94 | 2.95 | 2.88 | 2.76 |
| Production capacity | – | 2.19 | 1.95 | 2.50 | 2.09 | 2.26 | 2.31 |
| Rate of use | 82.411 | 82.584 | 82.887 | 83.712 | 84.132 | 84.852 | 85.321 |
| Capital | – | 2.06 | 2.06 | 2.07 | 2.15 | 2.08 | 2.08 |
| Apparent productivity | – | 0.28 | 0.47 | 0.85 | 0.78 | 0.79 | 0.67 |
| Hourly labor productivity | – | 2.57 | 2.80 | 2.99 | 2.98 | 2.66 | 2.52 |
| Activity | – | –0.22 | –0.26 | –0.05 | –0.03 | 0.21 | 0.23 |
| Work duration | – | –0.01 | –0.01 | –0.01 | –0.01 | –0.00 | 0.00 |
| Employment | – | –0.21 | –0.26 | –0.04 | –0.02 | 0.21 | 0.23 |
| In millions | 13.17 | 13.15 | 13.11 | 13.11 | 13.10 | 13.13 | 13.16 |
| Total | | | | | | | |
| Value added | – | 2.49 | 2.63 | 3.02 | 2.95 | 3.01 | 2.96 |
| Production capacity | – | 2.29 | 2.18 | 2.65 | 2.75 | 2.88 | 2.93 |
| Rate of use | 82.411 | 82.584 | 82.893 | 83.600 | 83.879 | 84.154 | 84.292 |
| Capital | – | 2.94 | 3.04 | 3.11 | 3.29 | 3.28 | 3.34 |
| Apparent productivity | – | –0.43 | –0.40 | –0.09 | –0.33 | –0.26 | –0.36 |
| Hourly labor productivity | – | 3.14 | 3.28 | 3.44 | 3.37 | 3.29 | 3.20 |
| Activity | – | –0.63 | –0.62 | –0.41 | –0.40 | –0.27 | –0.23 |
| Work duration | – | –0.00 | –0.00 | –0.01 | –0.01 | 0.00 | 0.00 |
| Employment | – | –0.62 | –0.62 | –0.40 | –0.40 | –0.27 | –0.23 |
| In millions | 17.18 | 17.07 | 16.97 | 16.90 | 16.83 | 16.79 | 16.75 |
| Tensions | | | | | | | |
| Total employment | – | 0.508 | 0.505 | 0.327 | 0.321 | 0.219 | 0.190 |
| Offers of jobs | 80.0 | 80.0 | 80.7 | 81.9 | 78.6 | 68.5 | 60.3 |
| Unemployment in millions | 2.673 | 2.853 | 3.033 | 3.210 | 3.386 | 3.537 | 3.680 |
| Unemployment rate | 11.23 | 11.95 | 12.67 | 13.35 | 14.02 | 14.58 | 15.10 |

case in past years. But the main difference concerns the evolution of factors: in industry, a sustained investment level, combined with a constant decrease of employment, leads to a fast growth of the share of capital. The productive capacity increases a little faster than production. Thus, we reach rather low values of the utilization rate of capacity in the later periods. For the nonindustrial sector, the opposite is true. With a continuous growth of the two factors (a rather slower one for employment), the resulting productive capacity increases now more slowly than production, in spite of a higher level (0.7%) of the autonomous technical progress.

*Table 14.2.* Continued.

| *1994* | *1995* | *1996* | *1997* | *1998* | *1999* | *2000* | *1987–1990* | *1990–1995* | *1995–2000* |
|---|---|---|---|---|---|---|---|---|---|
| 3.50 | 3.49 | 3.50 | 3.51 | 3.46 | 3.46 | 3.44 | 2.95 | 3.36 | 3.47 |
| 4.01 | 3.97 | 3.91 | 3.87 | 3.79 | 3.76 | 3.71 | 2.66 | 4.03 | 3.81 |
| 81.932 | 81.525 | 81.154 | 80.842 | 80.525 | 80.269 | 80.017 | | | |
| 5.85 | 5.89 | 5.93 | 5.96 | 6.02 | 6.02 | 6.07 | 5.00 | 5.71 | 6.00 |
| –2.22 | –2.27 | –2.29 | –2.31 | –2.41 | –2.41 | –2.49 | –1.95 | –2.23 | –2.38 |
| 5.41 | 5.21 | 5.03 | 4.84 | 4.63 | 4.45 | 4.28 | 4.84 | 5.26 | 4.64 |
| –1.81 | –1.63 | –1.46 | –1.27 | –1.12 | –0.95 | –0.80 | –1.80 | –1.81 | –1.12 |
| 0.00 | 0.00 | 0.00 | 0.00 | 0.00 | 0.00 | 0.00 | –0.00 | –0.00 | 0.00 |
| –1.81 | –1.63 | –1.46 | –1.27 | –1.12 | –0.95 | –0.80 | –1.80 | –1.81 | –1.12 |
| 3.52 | 3.46 | 3.41 | 3.37 | 3.33 | 3.30 | 3.27 | | | |
| | | | | | | | | | |
| 2.78 | 2.78 | 2.78 | 2.78 | 2.74 | 2.72 | 2.67 | 2.58 | 2.83 | 2.74 |
| 2.26 | 2.26 | 2.30 | 2.38 | 2.40 | 2.46 | 2.48 | 2.19 | 2.24 | 2.40 |
| 85.768 | 86.253 | 86.743 | 87.202 | 87.555 | 87.869 | 88.085 | | | |
| 2.11 | 2.08 | 2.04 | 2.04 | 2.09 | 2.12 | 2.20 | 2.04 | 2.10 | 2.10 |
| 0.66 | 0.68 | 0.72 | 0.73 | 0.63 | 0.58 | 0.47 | 0.52 | 0.72 | 0.63 |
| 2.54 | 2.51 | 2.49 | 2.47 | 2.45 | 2.45 | 2.45 | 2.76 | 2.64 | 2.46 |
| 0.24 | 0.26 | 0.29 | 0.30 | 0.28 | 0.26 | 0.22 | –0.18 | 0.18 | 0.27 |
| 0.00 | 0.00 | 0.00 | 0.00 | 0.00 | 0.00 | 0.00 | –0.01 | –0.00 | 0.00 |
| 0.24 | 0.26 | 0.29 | 0.30 | 0.28 | 0.26 | 0.22 | –0.17 | 0.18 | 0.27 |
| 13.19 | 13.23 | 13.26 | 13.31 | 13.34 | 13.38 | 13.41 | | | |
| | | | | | | | | | |
| 2.98 | 2.98 | 2.99 | 2.99 | 2.95 | 2.94 | 2.90 | 2.68 | 2.98 | 2.95 |
| 2.88 | 2.87 | 2.88 | 2.92 | 2.91 | 2.94 | 2.94 | 2.35 | 2.86 | 2.92 |
| 84.404 | 84.554 | 84.714 | 84.872 | 84.957 | 85.038 | 85.058 | | | |
| 3.45 | 3.48 | 3.50 | 3.55 | 3.64 | 3.69 | 3.79 | 3.00 | 3.37 | 3.63 |
| –0.45 | –0.48 | –0.50 | –0.53 | –0.66 | –0.73 | –0.86 | –0.31 | –0.38 | –0.66 |
| 3.19 | 3.12 | 3.06 | 3.01 | 2.95 | 2.91 | 2.88 | 3.25 | 3.23 | 2.96 |
| –0.20 | –0.14 | –0.07 | –0.02 | –0.00 | 0.02 | 0.02 | –0.55 | –0.25 | –0.01 |
| 0.00 | 0.00 | 0.00 | 0.00 | 0.00 | 0.00 | 0.00 | –0.00 | 0.00 | 0.00 |
| –0.20 | –0.14 | –0.08 | –0.02 | –0.00 | 0.02 | 0.02 | –0.54 | –0.25 | –0.01 |
| 16.71 | 16.69 | 16.68 | 16.67 | 16.67 | 16.68 | 16.68 | | | |
| | | | | | | | | | |
| 0.164 | 0.113 | 0.062 | 0.015 | 0.002 | 0.017 | 0.015 | 0.442 | 0.201 | 0.009 |
| 52.0 | 44.7 | 38.0 | 32.4 | 26.8 | 22.2 | 17.8 | | | |
| 3.817 | 3.946 | 4.028 | 4.103 | 4.174 | 4.241 | 4.306 | | | |
| 15.60 | 16.05 | 16.34 | 16.60 | 16.84 | 17.06 | 17.27 | | | |

Aggregating the two sectors, we get a small rise in the utilization rate of capacity (see *Figure 14.3*), and a constant decrease of employment (see *Figure 14.4*) which leads in the long term to a quasi-stabilized loss of 500,000 jobs (–740,000 in industry, +240,000 in the complement). Simultaneously, unemployment rises by 1,600,000 persons. This must be seen against the background of a growth of active population (which decelerates late in the period). It is due to the fact that the loss of a job in the industrial sector is not compensated by an equal increase in the nonindustrial sector, where a new job can not easily be occupied by a previously unemployed (sometimes unqualified) person.

*Table 14.3.* Growth rates of prices and wages, 1987–2000.

| *Indicator* | *1987* | *1988* | *1989* | *1990* | *1991* | *1992* | *1993* |
|---|---|---|---|---|---|---|---|
| Price index | | | | | | | |
| Of foreign GDP | – | 3.18 | 3.35 | 3.35 | 3.35 | 3.36 | 3.36 |
| Of GDP | – | 3.46 | 3.84 | 3.31 | 3.27 | 3.38 | 3.34 |
| Differential | – | 0.28 | 0.49 | –0.04 | –0.08 | 0.03 | –0.01 |
| Of value added | – | 3.43 | 3.68 | 3.20 | 3.15 | 3.24 | 3.20 |
| Of production | – | 3.18 | 3.54 | 3.19 | 3.16 | 3.29 | 3.25 |
| Of global use (excluding VAT) | – | 3.09 | 3.45 | 3.09 | 3.08 | 3.32 | 3.27 |
| Of consumption | – | 3.30 | 3.57 | 3.22 | 3.21 | 3.17 | 3.15 |
| Of exports | – | 2.77 | 3.29 | 3.04 | 2.94 | 2.91 | 2.82 |
| Of imports | – | 1.99 | 2.71 | 2.56 | 2.53 | 2.85 | 2.77 |
| Domestic price | – | 2.94 | 3.48 | 3.21 | 3.17 | 3.03 | 2.89 |
| Foreign price | – | 2.88 | 2.96 | 2.93 | 2.90 | 2.95 | 2.95 |
| Price differential | – | 0.06 | 0.51 | 0.27 | 0.26 | 0.08 | –0.06 |
| Exchange rate | – | 0.12 | 0.67 | 0.46 | 0.29 | 0.00 | 0.00 |
| Wages | | | | | | | |
| Nominal hourly rate | – | 5.35 | 6.64 | 6.01 | 6.02 | 6.16 | 6.27 |
| Real hourly rate | – | 1.99 | 2.97 | 2.71 | 2.72 | 2.90 | 3.03 |
| Nominal individual wages | – | 5.35 | 6.64 | 6.00 | 6.01 | 6.16 | 6.27 |
| Real individual wages | – | 1.99 | 2.96 | 2.70 | 2.72 | 2.90 | 3.03 |
| Total wages | – | 4.93 | 6.20 | 5.77 | 5.79 | 5.94 | 6.09 |
| Nominal available income | – | 5.72 | 6.51 | 6.75 | 6.23 | 6.00 | 6.03 |
| Real available income | – | 2.34 | 2.84 | 3.43 | 2.93 | 2.75 | 2.79 |

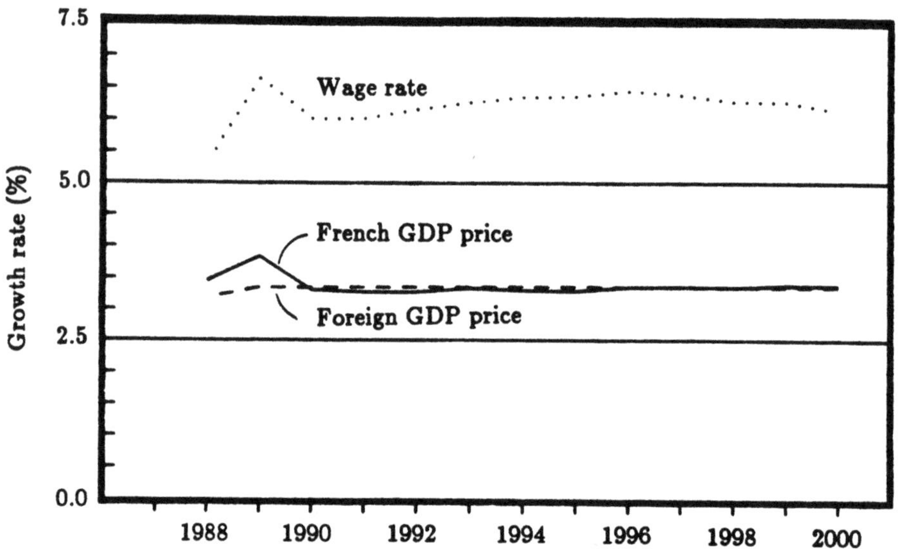


*Figure 14.5.* Prices and wages.

*Table 14.3.* Continued.

| *1994* | *1995* | *1996* | *1997* | *1998* | *1999* | *2000* | *1987–1990* | *1990–1995* | *1995–2000* |
|---|---|---|---|---|---|---|---|---|---|
| 3.36 | 3.36 | 3.35 | 3.36 | 3.36 | 3.36 | 3.36 | 3.26 | 3.36 | 3.36 |
| 3.50 | 3.29 | 3.36 | 3.36 | 3.35 | 3.39 | 3.38 | 3.50 | 3.32 | 3.37 |
| –0.06 | –0.06 | 0.01 | 0.00 | –0.01 | 0.04 | 0.03 | | | |
| 3.15 | 3.15 | 3.23 | 3.25 | 3.25 | 3.32 | 3.35 | 3.40 | 3.18 | 3.28 |
| 3.22 | 3.22 | 3.29 | 3.30 | 3.32 | 3.37 | 3.38 | 3.27 | 3.23 | 3.33 |
| 3.24 | 3.25 | 3.31 | 3.30 | 3.31 | 3.34 | 3.34 | 3.18 | 3.23 | 3.32 |
| 3.12 | 3.12 | 3.19 | 3.20 | 3.22 | 3.27 | 3.29 | 3.33 | 3.15 | 3.23 |
| 2.76 | 2.75 | 2.78 | 2.79 | 2.79 | 2.83 | 2.84 | 3.00 | 2.83 | 2.81 |
| 2.78 | 2.78 | 2.79 | 2.78 | 2.81 | 2.80 | 2.82 | 2.39 | 2.74 | 2.80 |
| 2.81 | 2.79 | 2.84 | 2.84 | 2.85 | 2.90 | 2.91 | 3.18 | 2.94 | 2.87 |
| 2.95 | 2.95 | 2.95 | 2.96 | 2.96 | 2.96 | 2.96 | 2.89 | 2.94 | 2.96 |
| –0.14 | –0.16 | –0.11 | –0.12 | –0.11 | –0.05 | –0.05 | 0.28 | –0.00 | –0.09 |
| 0.00 | 0.00 | 0.00 | 0.00 | 0.00 | 0.00 | 0.00 | 0.41 | 0.06 | 0.00 |
| | | | | | | | | | |
| 6.35 | 6.36 | 6.45 | 6.38 | 6.28 | 6.27 | 6.15 | 5.94 | 6.23 | 6.30 |
| 3.13 | 3.15 | 3.16 | 3.08 | 2.97 | 2.90 | 2.77 | 2.53 | 2.98 | 2.98 |
| 6.35 | 6.37 | 6.45 | 6.39 | 6.29 | 6.27 | 6.15 | 5.94 | 6.23 | 6.31 |
| 3.14 | 3.15 | 3.16 | 3.09 | 2.97 | 2.90 | 2.77 | 2.53 | 2.99 | 2.98 |
| 6.19 | 6.27 | 6.41 | 6.40 | 6.32 | 6.32 | 6.20 | 5.57 | 6.05 | 6.33 |
| 6.07 | 6.08 | 6.15 | 6.11 | 6.02 | 6.02 | 5.91 | 6.26 | 6.08 | 6.04 |
| 2.87 | 2.87 | 2.87 | 2.82 | 2.71 | 2.66 | 2.54 | 2.84 | 2.84 | 2.72 |

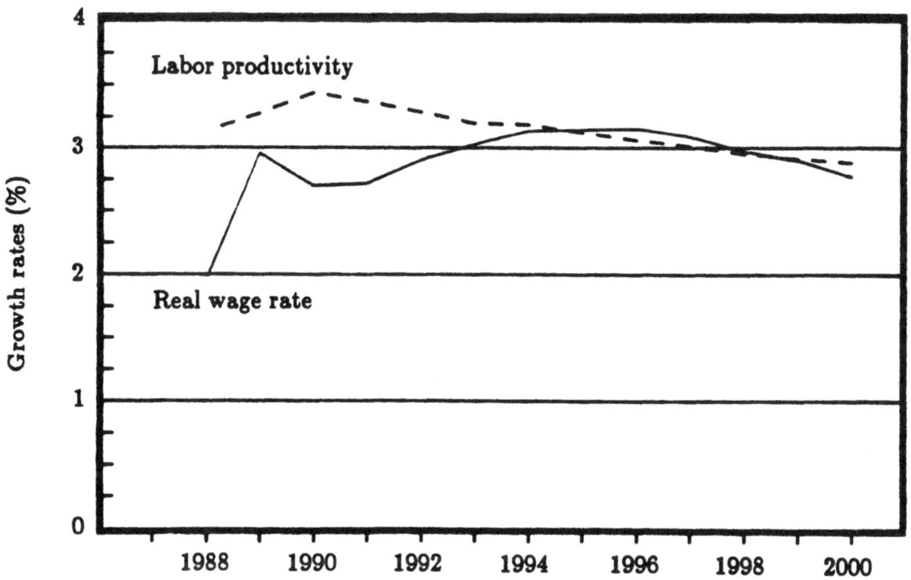


*Figure 14.6.* Real wage rate and labor productivity.

*Table 14.4.* Growth indicators for households, firms, and governments, 1987–2000.

| *Indicator* | *1987* | *1988* | *1989* | *1990* | *1991* | *1992* | *1993* |
|---|---|---|---|---|---|---|---|
| Households | | | | | | | |
| Social security payments | – | 8.19 | 8.87 | 8.27 | 8.06 | 8.30 | 8.40 |
| Benefits | – | 6.41 | 7.31 | 6.65 | 6.64 | 6.67 | 6.70 |
| Savings ratio (in points) | 14.84 | 15.23 | 15.66 | 16.03 | 16.02 | 16.05 | 16.05 |
| Financing capacity | 136.2 | 150.2 | 168.3 | 195.8 | 202.7 | 204.9 | 216.3 |
| in points of GDP | 2.62 | 2.73 | 2.87 | 3.14 | 3.06 | 2.91 | 2.89 |
| Firms | | | | | | | |
| Total wages | – | 4.94 | 6.20 | 5.84 | 5.87 | 6.17 | 6.33 |
| Social security payments | – | 5.62 | 6.62 | 5.48 | 4.59 | 6.12 | 6.26 |
| Autofinancing | – | 6.16 | 8.55 | 8.89 | 9.33 | 7.35 | 6.61 |
| Profits ratio (in points) | 7.39 | 7.43 | 7.60 | 7.86 | 8.16 | 8.22 | 8.24 |
| Industrial profits ratio | 4.93 | 5.11 | 5.40 | 5.87 | 6.22 | 6.36 | 6.43 |
| Financing capacity | –48.54 | –66.53 | –71.45 | –76.24 | –50.38 | –56.35 | –73.64 |
| in points of GDP | –0.93 | –1.21 | –1.22 | –1.22 | –0.76 | –0.80 | –0.98 |
| Government | | | | | | | |
| VAT earnings | – | 5.57 | 6.35 | 6.34 | 6.05 | 6.83 | 6.73 |
| Other indirect taxes | – | 6.08 | 6.64 | 6.62 | 6.57 | 6.40 | 6.35 |
| Income tax | – | 4.84 | 9.61 | 3.97 | 2.83 | 5.20 | 5.02 |
| Tax on firm profits | – | 17.46 | 0.92 | 8.16 | 9.12 | 9.45 | 6.98 |
| Wages paid | – | 5.01 | 6.31 | 5.68 | 5.69 | 5.43 | 5.54 |
| Consumption | – | 5.07 | 5.56 | 5.37 | 5.35 | 5.59 | 5.55 |
| Investment | – | 6.29 | 6.81 | 5.73 | 5.74 | 6.18 | 6.20 |
| Interest paid (in billions) | –104.2 | –110.3 | –117.9 | –158.4 | –166.8 | –174.8 | –182.4 |
| Balance (in billions) | –113.8 | –101.5 | –102.8 | –134.9 | –147.7 | –139.9 | –134.2 |
| Balance in points of GDP | –2.19 | –1.84 | –1.75 | –2.16 | –2.23 | –1.99 | –1.80 |
| Earnings as part of GDP | 44.84 | 45.14 | 45.32 | 45.30 | 45.14 | 45.36 | 45.57 |
| Rest of the world | | | | | | | |
| Interest balance (in billions) | –17.8 | –19.6 | –20.7 | –21.4 | –22.3 | –23.2 | –23.8 |
| Real export/import ratio | 91.86 | 91.34 | 90.82 | 89.99 | 90.30 | 90.32 | 90.21 |
| Terms of trade | – | 0.77 | 0.57 | 0.46 | 0.39 | 0.05 | 0.05 |
| Commercial balance (in billions) | 31.7 | 36.6 | 39.6 | 36.5 | 50.9 | 56.5 | 60.0 |
| National financing capacity | 14.9 | 19.8 | 23.2 | 21.7 | 37.3 | 44.3 | 49.8 |
| in points of GDP | 0.29 | 0.36 | 0.40 | 0.35 | 0.56 | 0.63 | 0.67 |

The unemployment rate in the year 2000 reaches a level of more than 17%, which, although it is technically feasible, could prove to be unacceptable from a social point of view. It might call for a previously unobserved structural change in the behavior of the labor market, favoring, for instance, part-time work or work in teams (this would mean in fact an acceleration of an observed trend). Indeed, the assumption of a constant weekly work time, though it is confirmed by recent data and in accordance with medium-term anticipations, could hardly be sustained in the context of this long-term study.

*Table 14.4.* Continued.

| *1994* | *1995* | *1996* | *1997* | *1998* | *1999* | *2000* | *1987–1990* | *1990–1995* | *1995–2000* |
|---|---|---|---|---|---|---|---|---|---|
| 8.46 | 8.49 | 8.59 | 8.54 | 8.41 | 8.37 | 8.21 | 8.35 | 8.34 | 8.42 |
| 6.72 | 6.66 | 6.62 | 6.51 | 6.37 | 6.32 | 6.17 | 6.72 | 6.68 | 6.40 |
| 16.11 | 16.16 | 16.21 | 16.26 | 16.30 | 16.37 | 16.43 | | | |
| 225.8 | 234.5 | 242.6 | 250.3 | 256.1 | 263.1 | 267.2 | | | |
| 2.84 | 2.78 | 2.71 | 2.63 | 2.53 | 2.45 | 2.34 | | | |
| | | | | | | | | | |
| 6.45 | 6.55 | 6.71 | 6.72 | 6.64 | 6.64 | 6.51 | 5.60 | 6.27 | 6.64 |
| 6.37 | 6.44 | 6.57 | 6.56 | 6.47 | 6.47 | 6.34 | 5.84 | 5.95 | 6.68 |
| 5.95 | 5.39 | 4.94 | 4.69 | 4.53 | 4.39 | 4.27 | 7.78 | 6.92 | 4.56 |
| 8.20 | 8.12 | 8.00 | 7.86 | 7.70 | 7.54 | 7.36 | | | |
| 6.42 | 6.34 | 6.23 | 6.08 | 5.92 | 5.75 | 5.57 | | | |
| −88.96 | −110.00 | −141.09 | −184.43 | −229.36 | −289.86 | −355.70 | | | |
| −1.12 | −1.30 | −1.57 | −1.94 | −2.27 | −2.70 | −3.12 | | | |
| | | | | | | | | | |
| 6.80 | 6.85 | 6.96 | 6.99 | 6.91 | 6.97 | 6.91 | 6.02 | 6.65 | 6.95 |
| 6.35 | 6.36 | 6.45 | 6.48 | 6.44 | 6.49 | 6.46 | 6.38 | 6.40 | 6.47 |
| 5.02 | 5.05 | 5.05 | 5.11 | 5.06 | 4.97 | 4.96 | 6.05 | 4.62 | 5.03 |
| 6.09 | 5.69 | 5.41 | 5.27 | 4.91 | 4.81 | 4.90 | 8.55 | 7.46 | 5.06 |
| | | | | | | | | | |
| 5.62 | 5.63 | 5.71 | 5.64 | 5.54 | 5.52 | 5.40 | 5.61 | 5.58 | 5.56 |
| 5.53 | 5.53 | 5.59 | 5.59 | 5.61 | 5.65 | 5.66 | 5.28 | 5.51 | 5.62 |
| 6.18 | 6.19 | 6.26 | 6.25 | 6.26 | 6.29 | 6.29 | 6.21 | 6.10 | 6.27 |
| −189.5 | −195.9 | −201.3 | −205.0 | −206.8 | −206.0 | −202.2 | | | |
| −126.4 | −113.8 | −94.1 | −66.3 | 32.0 | 10.9 | 63.8 | | | |
| −1.59 | −1.35 | −1.05 | −0.70 | −0.32 | 0.10 | 0.56 | | | |
| 45.78 | 46.02 | 46.25 | 46.49 | 46.71 | 46.91 | 47.11 | | | |
| | | | | | | | | | |
| −24.1 | −24.0 | −23.4 | −22.3 | −20.4 | −17.8 | −14.4 | | | |
| 90.33 | 90.42 | 90.42 | 90.31 | 90.38 | 90.28 | 90.30 | | | |
| −0.02 | −0.03 | −0.01 | 0.00 | −0.02 | 0.03 | 0.02 | 0.59 | 0.09 | 0.00 |
| 67.4 | 74.7 | 80.7 | 84.5 | 93.1 | 98.5 | 107.7 | | | |
| 59.6 | 70.2 | 80.2 | 89.0 | 103.7 | 116.2 | 134.0 | | | |
| 0.75 | 0.83 | 0.90 | 0.93 | 1.02 | 1.08 | 1.17 | | | |

Considering prices (*Table 14.3, Figures 14.5* and *14.6*), we observe a very stable evolution of each individual index, including the price of French GDP, which grows at the same rate as its foreign counterpart. As to the wage rate, the similarity of its development in real terms with that of labor productivity is another factor of stability in the present situation.

*Table 14.4* completes the study by showing, as expected, that social benefits grow at a significantly lower rate than the contributions of workers (by something like 2% per year). If we go into the details of the government account, we see that while the global expenditures grow at about 6%, almost the same as GDP, indirect taxes (VAT and others) get in the long term more and more

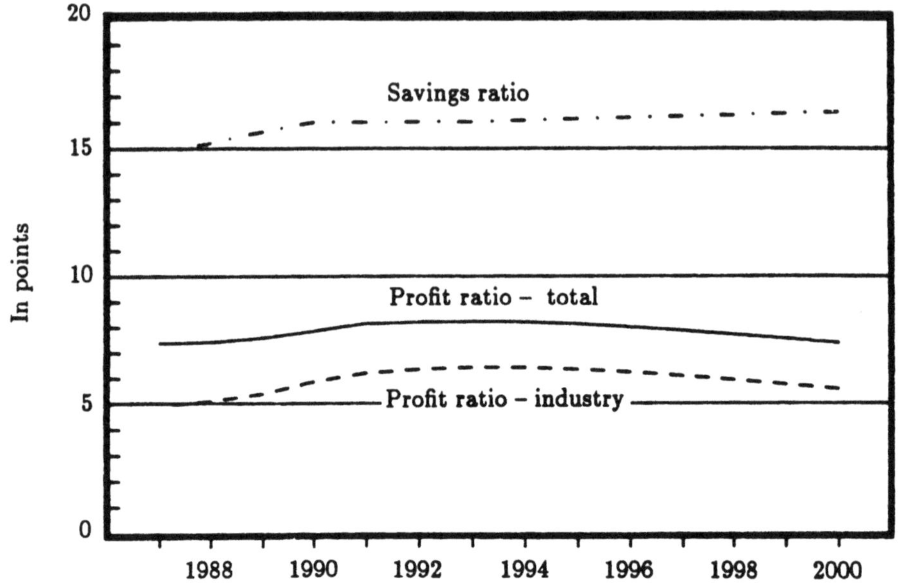


*Figure 14.7.* Profits and savings ratios.

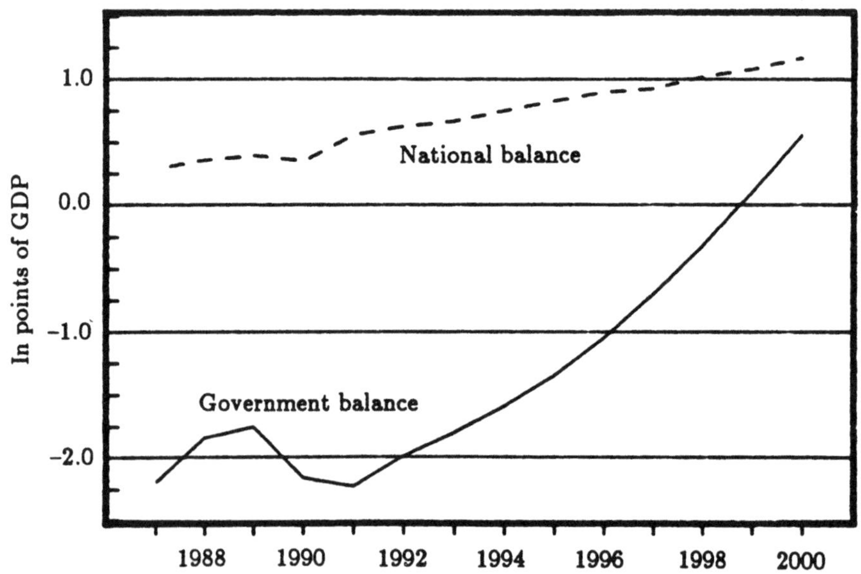


*Figure 14.8.* National account balances.

weight, particularly compared with the income tax. But as VAT is mostly paid by households, the substitution between the two kinds of taxes should not much affect purchasing power.

***Figure 14.7*** suggests that profits and savings ratios remain virtually on the same level over the forecasting period.

Finally, the balances (***Figure 14.8***) show a constant improvement over the period. The current account balance is always positive. In the long term it reaches a value of over 1% of GDP (the trade balance shows the same development). The government balance improves even more, from an initial deficit of more than two points to a small positive value in the last periods. This evolution, while actually representing instability of an important ratio, is consistent with the goals set by government itself. If the results of this forecast can be considered reliable, they mean that government budget equilibrium can be reached in the long term without forcing the French economy to suffer a lower growth rate than its foreign trading partners.

## 14.5. Conclusion

With this chapter, we hoped to achieve two goals: (1) present a coherent forecast, conditional on rather moderate external assumptions, of the structural evolution of French economy; and (2) prove that the Mini-DMS model is reliable enough to be used in long-term studies, in spite of its reduced size and the fact that it was not built for that purpose.

### References

Brillet, J.-L. (1986), *Le Modèle Mini-DMS en 1985*, Service des Programmes, Working Paper 320/16, 31 January.

Fouquet, D., Charpin, J.M., Guillaume, H., Muet, P.A., and Vallet, D. (1978), *DMS, Modèle Dynamique Multisectoriel.* (Paris: Collections de l'INSEE, Serie C, No. 64–65).

CHAPTER 15

# Implications of a Modernization Strategy for the United Kingdom

*Michael Landesmann and Andrew Snell*


## Summary

The economic consequences of modernizing the UK's tradeables sector are analyzed by computer simulations using Cambridge University's Multisectoral Dynamic Model. These simulations assume that a combination of economy-wide and industry-specific policy measures gradually improve the competitiveness and quality of British goods and that productivity growth is brought into line with that of the major West European competitors such that the UK's share of total OECD exports no longer declines. Attention is focused on whether expansionary fiscal policies can reduce the level of unemployment while, for a number of reasons, the pressures on the balance of payments get released only very slowly. Success in reducing unemployment levels then crucially hinges upon whether the benefits from increased productivity growth can be diverted from real earnings increases toward job creation (almost entirely in the services sector). A detailed breakdown of employment generation possibilities in the non-manufacturing sector is presented and analyzed as to the policy measures required to sustain productivity growth.


## 15.1. Introduction

The secular decline of the British economy and in particular of its manufacturing sector in relation to its main industrial competitors is now well documented. Since World War II the economy has been characterized by relatively low productivity and output growth. Between 1979 and 1982 this gradual decline turned into a nosedive, as manufacturing output fell by nearly 20% relative to its

1979 peak to stand at its lowest level for 15 years. The gravity of the situation was exposed further when in 1983, for the first time in its history, the UK suffered a trade deficit in manufactured goods.

There is much debate as to the exact cause of this economic crash. Treasury officials view it as an inevitable consequence of oil exports displacing manufactured goods on the current account while others disagree, pointing to high interest rates and an overvalued currency as the prime culprits. However, on the causes of the secular decline there is more of a consensus. Levels of labor productivity substantially lower than those of our competitors and a steady relative deterioration of the non-price competitiveness of British goods have been identified in many studies [see, for example, Blackaby (1979), and the House of Lords Select Committee on Overseas Trade (1985)] as the key causes of a steadily worsening (non-oil) visible trade balance and low GDP growth.

In this chapter we examine the economic consequences of a modernization strategy. It is envisaged that a combination of macro (economy-wide) and micro (industry-specific) policy measures initiated by a government strongly committed to an active industrial policy in 1988 gradually improves the competitiveness and quality of British goods until, by the mid-1990s, the rate of productivity growth is brought into line with that of the major Western European countries, and that the share of UK exports in the OECD total no longer declines as these markets grow.

Such a strategy is envisaged as taking place after the events of 1979 to 1981, when the sharp drop in manufacturing output and employment levels (and an even sharper drop in investment activity) was followed by a period of unusual productivity growth over the years 1982 to 1986. These developments are discussed in some detail in Appendix 15A [see also the discussion in Muellbauer (1986)], and their interpretation is of great relevance for an evaluation of whether a catching-up process of the UK tradeables sector with its competitors can be or already has been set into motion.

It must be stressed, however, that the presentation of a detailed package of policy measures together with an econometric assessment of their effects is not attempted here. [For an attempt to do this, see, e.g., Landesmann and Pierse (1986).] Evidence from past experience in this respect is likely to be scant and, even if it is available, may lead to ambiguous conclusions. We consider a number of interventionist scenarios that envisage substantial increases in the growth of government spending both as an important component of an industrial policy package (infrastructural investment, other investment, training and R&D support) as well as initiating a fiscal expansion of the economy as a whole.

The scenarios are simulated using the Cambridge Growth Project's Multisectoral Dynamic Model [MDM; see Barker and Peterson (1987) for a full description of the version used in this chapter]. The model is solved from 1986 to the year 2000 and the results are compared to a base run which closely corresponds to Cambridge Econometrics' 1986 forecast.

Broadly speaking, the scenarios have been designed to address the following question: If a modernization strategy were to be "successful" in the manner described in the preceding paragraphs, how much scope would a government then have for reducing unemployment and to what extent would their actions in this direction be constrained by the level of wage settlements?

Section 15.2 presents the details of the strategy and introduces the scenarios and their assumptions. In Section 15.3 the latter are examined in more detail, and then an overview of the results for the interventionist scenarios is provided. The focus here is primarily on the macro variables and in particular the relationships between unemployment, the balance of payments, the PSBR and inflation.

Section 15.4 presents details of a market-oriented strategy and its associated scenario, the "Lawson Scenario". Here the benefit of the doubt is given to the current government's claim that in the correct environment (one of weakened trade unions, deregulated markets, etc.) industry will modernize itself and market forces will drive the economy back to competitiveness. Section 15.5 deals with compositional aspects of the results and compares the interventionist scenarios with those of the market-orientated strategy. There are two subsections. The first analyzes the components of expenditure, helping to throw light on the dynamics behind the macroeconomic results of the previous two sections. The second provides details of structural change over the period to 2000. The focus here is on the relative growth of the manufacturing and service sectors, an issue which is currently the center of much debate. [See, e.g., *Bank of England Quarterly Bulletin* (1986).]

Section 15.6 asseses the degree to which the halving of the price of crude oil at the start of 1986 affects the results. Finally, Section 15.7 provides a summary and conclusion.

## 15.2. Details of a Modernization Strategy

The underlying assumption of the modernization strategy (MS) simulations is that the average productivity growth (PG) rate for manufacturing industry is about 1% to 1.2% p.a. higher than its historical trend, bringing it into line with the average Western European performance. The results of these simulations will be compared with a scenario which does not envisage a reversal in the trend decline in the competitive position of the British economy. This is the position taken in the Cambridge Econometrics Forecast of the British economy, which is very similar to our base run. [See Cambridge Econometrics (1986) for a full discussion of this scenario. See Appendix 15A for productivity growth comparisons with a number of OECD countries.]

The following questions must be posed when MS is envisaged:

(1) How could it be achieved?
(2) What other effects would accompany it?
(3) What is the likely course of developments in the rest of the world and how sensitive are the results to this?

We answer these separately in the following subsections.

### 15.2.1. The road to modernization

*An interventionist strategy*

For this strategy major industrial policy programs would be introduced by an interventionist government by the year 1988. These would include massive public investment to overhaul the infrastructure (roads, rail, inner city facilities, etc.), important initiatives in the area of training and retraining, an extension and streamlining of public support for research and development (R&D) facilities, various investment support schemes, incentives to introduce new technologies, measures to support export activity (in the area of product development and marketing), etc.

*A market-orientated strategy*

Here we allowed for a longer-term effect of the present government's market-orientated strategy upon sustained higher productivity growth. Relatively high productivity growth rates have been achieved since 1981, but there is considerable controversy as to whether or not these rates are the short-run responses to once-for-all scrapping and reorganization processes which took place in the extremely deep recession of 1979 to 1981. This controversy is expanded in some detail in Appendix 15A; see also the recent discussion in Muellbauer (1986). In this scenario we give the benefit of the doubt to such a strategy and assume that sustained improvements in the competitive performance of the market sector can be achieved through continued industrial relations legislation leading to a further weakening of the trade union bargaining position in plants and industries and a change in the tactics of a number of trade unions along the lines of the engineering unions; through the incentive effects of reduced taxation and deregulation; and through the continued shrinking of public sector services and the privatization of publicly owned enterprises. The reason for exploring this scenario is to analyze the implications for employment, income distribution and industrial structure of this type of strategy in the longer run.

### 15.2.2. The consequences of modernization for wage behavior and export performance

When considering a fundamental shift in the productivity performance of the UK tradeable sectors two issues arise: one is the impact upon wages, and the other concerns an accompanying change in the non-price attributes of British goods.

*The effect on wages*

Wage behavior in MDM is built around what is known as "real wage resistance" behavior [see Lawson (1987)]. The idea behind this hypothesis is that underlying wage bargaining there is an inherent expectation of rising living standards of around 2% p.a.; this trend has been shown to be persistent in the UK for almost a century [see Tarling and Wilkinson (1982)]. Real (post-tax) wage demands deviate from this trend only as a result of sharp changes in the rate of unemployment, and – as long as these are successful – as a result of incomes policies. However, in both these cases reductions below the trend can be shown historically to be temporary, leading to pent-up wage demands which exert renewed pressure as soon as the level of unemployment stabilizes and the incomes policy is either lifted or breaks down. A productivity performance of the UK economy below or near the 2% mark leads to distributional conflicts and can be used to explain the underlying inflationary processes in the UK in the past.

There are broadly two reasons to assume that a change in the long-term productivity performance would also affect the trend demand for real wage growth. First, the modernization processes themselves (the introduction of new techniques, new work practices, etc.) and also the accompanying job losses will only get the tacit support or agreement of the remaining or new work forces if they are accompanied by prospects for higher real wage growth. Furthermore, if productivity growth outstrips real wage growth for some time, this will show up in higher profitability, which would provide a strong incentive for unions in the affected industries to claim a share in these improvements.

Second, if one were to compare the trend growth rates of real wages in advanced economies, one would find that they are linked to their respective productivity growth performance. [See, e.g., Kendrick (1984, pp. 177, 181) where the inflationary implications are also pointed out in situations in which the trend growth rates in productivity and real earnings diverge, as in the UK after 1973.] Thus, if the UK economy moves toward a productivity growth path more in line with its Western European competitors, it is quite likely that real wage growth will adjust similarly. As we shall see, the relationship between additional real wage growth and improved productivity growth (PG) is a crucial one for the different MS scenarios.

### *The effect on non-price competitiveness*

The increase in the rate of PG will affect wage demands (in the manner discussed above) and prices and the exchange rate (through the responses of the model). These in turn will determine the price competitiveness of the different UK industries in domestic and world markets. However, studies of the reasons for the deteriorating UK trade performance in the past have pointed out the importance of non-price factors [see, e.g., NEDO (1977)].

In MDM the effects of these non-price factors show up in the estimated export equations in the form of low elasticities of UK exports with respect to world demand growth. In our MS simulations we have taken account of an accompanying improvement in non-price competitiveness by a gradual adjustment of the demand elasticities in the export equations, so that by 1997 these elasticities are broadly in line with those of our European competitors. By 1997, then, the share of UK exports in OECD trade is assumed to have stopped its historical trend decline, and to be in a transition phase where there is some scope for regaining lost shares. (See Appendix 15B for a discussion as to whether there are grounds to believe that such improvements in export performance have already occurred since 1982.)

Studies have shown that UK income elasticities of demand for imports (except for particular branches of industry) are on average not particularly high in relation to our main competitors. Also, a modernization process might entail a widening of UK consumers' purchasing patterns, which would make a fall in demand elasticities for imports unrealistic. Hence, no adjustments have been made with respect to the estimated set of import equations.

### 15.2.3. Developments in the rest of the world

Any forecasts of developments in the world economy at the present time, and in particular long-run forecasts, have to be approached with great caution. The world economy is presently experiencing dramatic changes in the level, structure and technological character of industrial production across the globe; the role of the public sector is being fundamentally reviewed in many countries; and there is little unanimity in the economics profession about the basis for fiscal and monetary policy. Trade policy scenarios and, of course, developments in international financial relationships, are far from certain and seem to exhibit features of increasing instability. These developments greatly complicate the design of policy simulation exercises for national economies.

In simulation exercises using MDM the views about developments in the world economy enter fundamentally in three ways: first, through the forecasts of world demand (composed of demand in the major export markets of the UK economy) and of world inflation; second, through exchange rate movements, relative financial policies (in particular, interest rate policies) and international capital flows insofar as they are relevant for the financial position of the UK economy; and third, through movements in the oil price, with its implications for North Sea oil production and the UK balance of payments.

In our MS scenarios we made use of international forecasts concerning development in world activity (see, e.g., OECD's *Economic Outlook*), and we show in Section 15.6 the sensitivity of our results with respect to an upward revision of world activity forecasts following the lower oil price scenario. We also made sure that implicit consideration is given to the fact that developments since the mid-1970s have also had strong effects upon industrial restructuring and upon inflationary tendencies in competitor countries, and it is the *relative* speed and strength of modernization processes which matter in evaluating its effects upon the UK's competitive stance. Implicit in our simulations, therefore, is a particular view about productivity growth rates and wage inflation in competitor countries.

The analysis of exchange rate movements in the present international environment is extremely difficult. MDM uses an exchange rate equation which determines the value of sterling relative to a trade-weighted basket of other currencies as a function of the balance of trade, the difference between US and UK interest rates, inflation rates in the UK and those in its export markets and the value of UK oil reserves. For details on the exchange rate equation used in MDM, see Snell (1987).

Interest rates in this model are seen as a tool of financial policy to relieve some part of the pressure on sterling when it is subject to strong upward or downward pressures due to a deterioration or improvement in the UK's balance-of-payments (BOP) position. These movements are a direct consequence of persistent borrowing to finance BOP deficits (and the reverse in the case of sustained BOP surpluses).

## 15.3. Results for the Modernization Strategies

### 15.3.1. The different scenarios

*Table 15.1* summarizes the different assumptions which underlie the different MS scenarios.

The modernization scenarios A to E can thus be grouped in the following way:

| *Real earnings growth* | *Scenario* | *Fiscal expansion* |
|---|---|---|
| high | A | strong |
| | B | moderate |
| medium | C | strong |
| | D | weak |
| low | E | strong |

*Figure 15.1* gives an idea of how the productivity performance over the simulation period differs in the MS scenarios A and C from that in base run F and how they compare with previous historical periods.

It should be noted, with reference to *Figure 15.1*, that adjustments to trend productivity growth rates in MDM only partially determine actual productivity growth rates. They are also influenced by changes in the rates of growth of production levels and (in some cases) by the amount of investment undertaken in the different industries. The outcome for actual productivity growth rates will thus differ in scenarios A to E owing to different assumptions about real earnings growth and government expenditure.

Against the background of the adjustments in the different scenarios of real earnings targets, we analyzed the scope for fiscal expansion to relieve the severe unemployment problem in the UK. The mix of real earnings growth and fiscal expansion in scenarios A to E explores the range of possible outcomes which could be derived from an improved competitive performance of the UK economy in the long run.

*Table 15.1.* Assumptions in modernization strategy (MS) scenarios.

| | Model adjustments[a] | | Fiscal expansion[b] | | |
|---|---|---|---|---|---|
| *Scenario* | *Productivity and export competitiveness* | *Real earnings growth* | *1988–1992* | *1993–1996* | *1997–2000* |
| A | + 1–1.5% increase | | +5 | +1.5 | +1.5 |
| B | in trend productivity growth[c] | 0.5 | +3 | +3.0 | +1.5 |
| C | *plus* adjustments of demand elasticities | +0 | +5 | 3.0 | +1.5 |
| D | cities of export equations (see | | +5 | +1.5 | +1.5 |
| | | 1.0 | | | |
| E | section 15.2.2) | | +3 | +1.5 | +1.5 |
| F | – | 0 | Base run assumptions[d] | | |

[a]Expressed in percentage differences from base run F.
[b]Annual percentage growth rates in real government current and capital expenditures.
[c]The adjustments in productivity trends are differentiated by manufacturing and non-manufacturing sectors. For details, see Section 15.5.
[d]For details, see Cambridge Econometrics (1986)

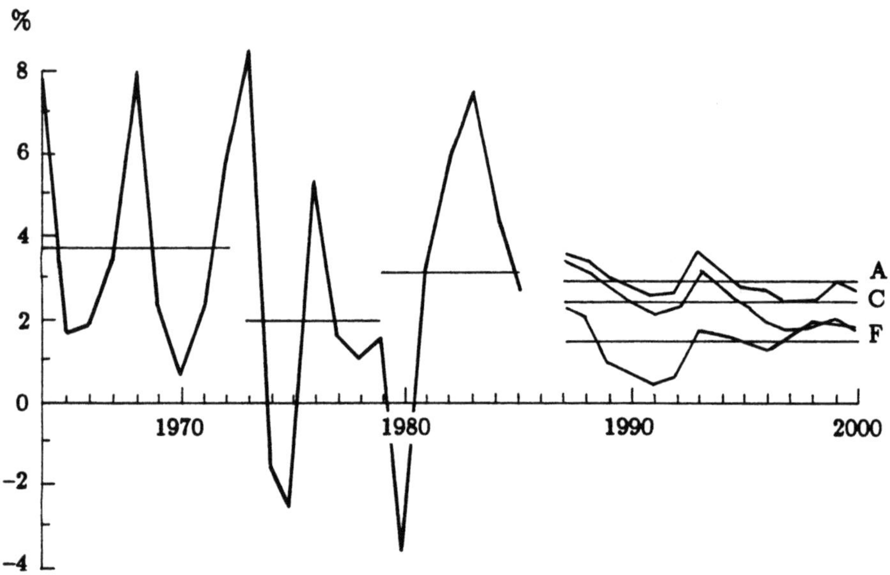


*Figure 15.1.* Productivity growth: output per person employed in UK manufacturing – past experience and projections for base run F and "modernization strategy" runs A and C.

*Table 15.2.* Unemployment and the balance of payments constraint.

| *Scenario result* | *1988* | *1992* | *1997* | *2000* |
|---|---|---|---|---|
| Unemployment (in thousands) | | | | |
| A | 3,390 | 2,667 | 2,430 | 2,203 |
| B | 3,484 | 3,384 | 3,069 | 2,890 |
| C | 3,357 | 2,568 | 1,717 | 1,272 |
| D | 3,421 | 2,762 | 2,461 | [a] |
| E | 3,516 | 3,473 | 3,628 | [a] |
| F | 3,283 | 3,383 | 3,857 | 3,852 |
| BOP/GDP (in %) | | | | |
| A | −0.55 | −2.06 | −2.33 | −2.69 |
| B | −0.35 | −1.29 | −2.51 | −2.53 |
| C | −0.41 | −1.77 | −2.48 | −2.36 |
| D | −0.69 | −2.29 | −3.04 | [a] |
| E | −0.49 | −1.58 | −2.55 | [a] |
| F | +0.39 | −0.94 | −2.55 | −2.38 |

| GDP average annual growth rate | *1987–92* | *1992–97* | *1987–97* |
|---|---|---|---|
| A | 2.90 | 2.99 | 2.90 |
| B | 2.23 | 3.02 | 2.62 |
| C | 2.78 | 3.15 | 2.97 |
| D | 3.02 | 3.29 | 3.16 |
| E | 2.35 | 2.82 | 2.59 |
| F | 1.07 | 1.36 | 1.21 |

[a]The values for scenarios D and E over the last period have been omitted because of the unsustainability of the developments associated with these runs (see discussion in text).

### 15.3.2. Fiscal expansion and the unemployment problem

*Table 15.2* presents some of the macroeconomic results of scenarios A to F. The basic dilemma that will confront the UK economy over the period 1988–1997 clearly emerges from this table.

It turns out that *a major fiscal expansion set against the background of real wage rate targets being revised in the light of the improved productivity performance* (by 0.5% p.a. in simulation A and 1% p.a. in simulation D) *will be unsustainable.* Over the period 1988–1992 the balance of payments deteriorates to such an extent (accentuated by the fact that over this period the UK will become a net importer of oil again) that a drastic cutback in the government's expenditure program would have to occur. Even then, the BOP continues to worsen, with the deficit/GDP ratio rising even after 1997. Certainly in scenario D this would lead to another crisis point before the year 2000.

On the other hand, if a more cautious stance is taken with respect to government expenditure in the early period against the background of increased target rates of real earnings, as in simulations B and E, unemployment remains above the 3 million mark by 1998 in scenario B and actually worsens in scenario E. Only in scenario C, in which the target rate of real wage growth remains on its historical trend path in spite of the increase in the trend growth rate of productivity, is there enough leeway for the government to mount a major attack on the problem. In this scenario unemployment drops to 1.3 m by the year 2000 and the BOP/GDP ratio begins to fall toward the end of the simulation period.

### 15.3.3. The balance-of-payments constraint

In *Table 15.3* we present the average annual rates of change of exports and imports (both at constant prices), and changes in terms of trade over the periods 1987–92, 1993–97, and 1997–2000. We can see that, with regard to the longer-term improvement in the UK's trading performance, there is indeed evidence that the catching-up processes introduced into our modernization strategy scenarios A to E yield the desired effects. From about 1993 onward, growth rates of exports in real terms generally outstrip growth rates in imports.

However, these trends are not always dominant as far as the development of the balance of payments is concerned. As a result of strong devaluations of sterling after 1988/89, the negative terms-of-trade effect outstrips the positive real movements of exports and imports for a number of years. This "*J*-curve" effect, which describes an initial deterioration in the balance-of-payments position following a devaluation, should be followed by an (overall) positive effect on the BOP, as exports respond favorably to the devaluation.

On the other hand, our simulations, which start in the late 1980s with a rapidly deteriorating BOP situation, show a continuing pressure toward further devaluations, so that the negative initial effects of a sequence of devaluations get superimposed upon each other. The situation worsens further as the international borrowing required to finance the deficits leads to increasing flows of interest payments abroad over the 1990s and, hence, to further pressures on sterling.

This sequence of events reaches crisis point in those simulations in which the BOP/GDP ratio (see *Table 15.2*) continues to rise in the second half of the 1990s (scenarios A, D and E). In scenario B that ratio stablizes, and in scenario C it begins to fall, so that continued high growth is possbile without a worsening in the BOP position.

The simulations show that if, during the important transition phase a BOP crisis can be avoided, the fruits of a successful modernization strategy can be reaped as a result of a reversal in the historically deteriorating trade performance. The precise effects upon employment depend upon a number of trade-offs to be discussed below. The pressures on the BOP in the longer term will be reduced if the visible signs of success in the British tradeables sector attract international investors to the UK. In the shorter and medium term, however, during the critical transition phase, policies to avoid a crisis situation will be important, since, otherwise, the entire strategy might have to be abandoned.

*Table 15.3.* Trade balance, terms of trade and exchange rates (average annual growth rates).

| Scenario | (1) Export growth 1987–1992 | (1) 1992–1997 | (1) 1997–2000 | (2) Import growth 1987–1992 | (2) 1992–1997 | (2) 1997–2000 | (3) Change in exchange rate[a] 1987–1992 | (3) 1992–1997 | (3) 1997–2000 | (4) Change in balance of trade[b] 1987–1992 | (4) 1992–1997 | (4) 1997–2000 |
|---|---|---|---|---|---|---|---|---|---|---|---|---|
| A | 5.06 | 5.86 | 7.20 | 5.66 | 4.69 | 5.17 | –2.78 | –5.85 | –9.28 | –1.60 | –0.38 | 0.23 |
| B | 4.50 | 5.48 | 7.00 | 4.95 | 4.87 | 4.87 | –1.39 | –4.96 | –8.71 | –1.12 | –0.51 | |
| C | 4.90 | 5.91 | 7.06 | 5.46 | 4.86 | 4.85 | –2.04 | –5.60 | –8.56 | –1.48 | –0.24 | 0.37 |
| D | 5.22 | 6.64 | – | 5.86 | 5.21 | – | –3.52 | –7.88 | – | –1.73 | –0.26 | – |
| E | 4.66 | 5.49 | 6.73 | 5.15 | 4.77 | 5.78 | –2.15 | –5.50 | –11.53 | –1.25 | –0.38 | –0.27 |
| F | 2.94 | 2.70 | 4.66 | 3.48 | 3.15 | 3.21 | +0.78 | –1.86 | –6.27 | –0.57 | –0.88 | 0.21 |

[a] Calculated as £/$.
[b] Calculated as (1) – (2) + change in terms of trade. The change in the balance of trade results from differential growth rates in exports and imports at constant prices *plus* changes in the terms of trade (changes in export price minus changes in import prices, both expressed in the same exchange rate).

*Table 15.4.* Public sector borrowing requirement (PSBR) and inflation.

| Scenario | PSBR/GDP (in %) 1987 | 1988 | 1989 | 1992 | 1997 | 2000 | Inflation rate (annual averages) 1987–1992 | 1992–1997 | 1997–2000 |
|---|---|---|---|---|---|---|---|---|---|
| A | 2.55 | 3.81 | 3.85 | 4.38 | 4.23 | 4.44 | 6.4 | 7.7 | 5.7 |
| B | 2.55 | 3.66 | 3.56 | 3.87 | 4.60 | 4.75 | 5.5 | 6.9 | 5.4 |
| C | 2.53 | 3.79 | 3.80 | 4.30 | 4.38 | 4.27 | 5.5 | 7.0 | 5.4 |
| D | – | – | – | – | – | – | 5.5 | 9.7 | – |
| E | 2.57 | 3.68 | 4.08 | 3.95 | 4.45 | – | 6.5 | 7.8 | – |
| F | 2.26 | 2.90 | 2.47 | 1.93 | 2.20 | 2.04 | 4.7 | 5.5 | 5.0 |

### 15.3.4. The PSBR and inflation constraints

Apart from the balance-of-payments problem, there are two other types of constraint which could restrict expansionary policy options in the UK – namely, the size of the PSBR and inflation.

*Table 15.4* gives the figures for the development of the PSBR/GDP ratio under the assumption that the standard rate of income tax is maintained at 28% from 1988 onward after having fallen to that level in 1986–1987. These projections reflect the intentions of the opposition parties (Labour and Alliance) to (fully or partially) reverse any further reductions in the standard rate of income tax introduced in the March 1987 Budget so as to concentrate on measures which reduce the unemployment level.

The absolute increases in government current and capital expenditures reflect the imposed rate of growth in real terms, already reported in *Table 15.1*. The increases in the PSBR/GDP ratio in the early period 1988–1992 are quite rapid, but they then stabilize at around the 4–4.5% mark from 1992 onward. This increase reverses the decline under the Conservative administration; but in historical terms and also in comparison with other industrial economies, the levels reached by 1992 are not extraordinarily high. Although it is important to evaluate how financial markets would react to such an increase, such an evaluation is beyond the scope of this chapter.

As regards inflation, we can see from *Table 15.4* that there is an upward movement in inflation rates over the first part of the period, when the economy absorbs the imported price increases associated with the devaluation of sterling. However, in the late period (1997 to 2000), and in spite of continued devaluation, inflation rates come down again as the cumulative increases in productivity levels make sufficient room for high real income growth. Only the very high real wage growth scenarios (D and E) experience strong inflationary pressures leading to an inflation–devaluation spiral which the economy could not sustain without a drastic change in the government's fiscal stance.

## 15.4. The Market-oriented Approach

We now turn to the market-oriented strategy referred to above, which we could call the "Lawson Scenario" (LS), after the present Chancellor of the Exchequer. This scenario envisages a continuation of the Conservative government's policies aimed at secularly reducing the share of public spending in total expenditure. The emphasis is on a tight control of inflationary pressures and on preventing sharp devaluations of the currency.

We have already mentioned that there is much debate as to the permanence of the recent jump in PG. Although the effects of the present strategy upon long-run PG are far from clear (see Appendix 15B), we assume here that the market-oriented approach is successful on this account and that therefore the economy is subjected to the same adjustments in PG rates and in non-price competitiveness as are the MS scenarios. It remains a "low wage growth" scenario (it shares this feature with scenario C) in that, in spite of higher PG, no adjustments to the "target rate of real wages" have been made, leaving room for an increase in profitability. The results for the LS are presented in *Table 15.5*.

As we can see from *Table 15.5*, and in comparison with the results for the other MS scenarios given in *Table 15.2*, the Lawson scenario will not exploit fully the room that exists for additional output growth which could be obtained from higher productivity growth. Given the structure of MDM, economic growth in this scenario is "demand constrained" since increased investment and export demand will not make up for the loss of immediate consumption demand and the relatively tight control of public expenditure.

*Table 15.5.* Results from a "Lawson scenario".

| *Development* | *1988* | *1992* | *1988–1992* |
|---|---|---|---|
| GDP growth (average annual rate) | | | 1.28 |
| Unemployment (in thousands) | 3,587 | 4,202 | |
| Price inflation (average annual rate) | | | 3.75 |
| BOP/GDP | +0.4 | –.39 | |
| PSBR/GDP | 3.52 | 3.35 | |
| TGE/GDP (in %):[a] | | | |
| in current prices | 23.8 | 24.1 | |
| in constant prices | 23.2 | 23.2 | |
| \$/£ [a] | 1.29 | 1.34 | |

[a]For comparison, we present the figures for (total government expenditure) TGE/GDP and exchange rate movements for one of our MS scenarios – scenario C:

| | *1988* | *1992* |
|---|---|---|
| TGE/GDP (in %): | | |
| in current prices | 24.7 | 28.2 |
| in constant prices | 24.0 | 27.0 |
| \$/£ | 1.27 | 1.15 |

Why are the other components of demand not compensating for low consumption and low public expenditure? The profitability effects on investment are small and insufficient to compensate for the lack of sales expectations, and there are two conflicting effects on exports compared with the other MS scenarios. Because of low real wage growth (comparable to scenarios C and F), the effects on unit costs and low inflation rates should improve export prospects relative to the other MS scenarios. However, low domestic demand improves the balance-of-payments position, especially in the first five years of the simulation period, and thus leads to much less pressure for a depreciation of sterling. The consequence is that the beneficial effects of a series of strong devaluations of sterling upon export growth are lost in comparison with the other MS runs. (The Chancellor of the Exchequer might, of course, decide to induce further devaluations in sterling through additional interest rate reductions, thus causing a deterioration in the BOP in the short run, as in the MS scenarios. Developments in this direction depend upon the government's strategy for exchange rate stability, worry about inflationary pressures which devaluations could induce, etc.)

We must emphasize that MDM in its present form is unable to track developments in the UK economy well enough for an evaluation of a Lawson strategy upon employment levels. The behavioral relationships embodied in the structure of MDM (in particular, those in the wage and employment equations) have been estimated from historical data [see W. Peterson in: Barker and Peterson (1987)] and they might be bad indicators for the responses, particularly those in the labor market, which such a scenario would evoke.

Measures such as the abolition of the Wages Councils, tough industrial relations legislation, and developments similar to the increase in the number of people who are prepared to work part-time, the availability of a large pool of unemployed, deregulation and various measures to support small firms, etc., might all lead to the emergence of a "low wage–low tech" sector of the economy which might absorb a much larger number of the unemployed than a simulation of MDM would suggest. In addition, given low employment prospects, the participation rate would fall dramatically and with it the recorded unemployment figure. However, even if all this were true, the results for the scenario indicate the enormity of the task to be faced in overcoming the unemployment problem.

## 15.5. Where Will Jobs Come From? An Industrial Analysis

The most important issues here are the relative employment losses and gains in the different sectors of the economy, the sensitivity of the results to PG assumptions in some crucial non-manufacturing industries, and the features of the sectoral pattern of output growth and employment creation against the background of the different expenditure patterns.

*Table 15.6* presents the employment losses and gains in different sectors of the economy over the period 1987 to 2000. The first part of the table presents industrial employment; the second part, government employment figures. It also gives a breakdown of public sector employment for education and health services which together are responsible for the creation of 1.5 million additional jobs by the year 2000 in scenario C.

We can see that over the period 1987 to the year 2000 there will be, in almost all scenarios, employment losses in all the sectors except for distribution and the different service industries. In fact, the employment-creating potential of the distributive and service activities is impressive, especially considering that in 1987 half the labor force (50.2%) is employed in these activities. In the highest employment scenario (C), the creation of over 2 million new jobs in this area is projected (58% of the work force will be employed in it). The productivity growth projections underlying these trends in the different service activities will be discussed below. Losses of employment in the manufacturing sector are projected to level off over the period. In the high employment scenario (C), the manufacturing sector gains 200,000 jobs by the year 2000 as import penetration is kept under control, a direct result of the particular expenditure pattern associated with the scenario (i.e., a high ratio of public versus private expenditure). Further, UK industries become increasingly more competitive through PG, quality improvements and the continuous devaluation of sterling.

*Table 15.6.* Employment creation and job losses by sector.

| | *Employment levels (in thousands) for scenario C* | | *Employment increases/decreases (in thousands) 1987 to 2000 for scenario:* | | | | |
|---|---|---|---|---|---|---|---|
| *Sector* | *1987* | *2000* | *C* | *A* | *E* | *F* | *LS* |
| Agriculture | 606 | 275 | –331 | –309 | –284 | –131 | –280 |
| Mining | 293 | 164 | –129 | –137 | –146 | –102 | –122 |
| Manufacturing | 5,530 | 5,723 | +193 | –174 | –658 | –53 | –811 |
| Construction | 1,448 | 1,217 | –231 | –444 | –652 | –385 | –702 |
| Utilities | 326 | 155 | –171 | –178 | –141 | –61 | –181 |
| Rail, road, other transport | 983 | 758 | –225 | –187 | –182 | –108 | –298 |
| Communications | 430 | 382 | –48 | –62 | –86 | –9 | –133 |
| Distribution | 3,583 | 4,563 | +980 | +854 | +648 | +307 | +467 |
| Business services | 1,631 | 2,151 | +520 | +471 | +382 | +316 | +162 |
| Professional services | 1,274 | 1,382 | +109 | +19 | –94 | +34 | –225 |
| Misc. services | 3,385 | 4,001 | +616 | +718 | +853 | +912 | –749 |
| Total (adjusted) | 19,648 | 20,931 | +1,283 | +571 | –402 | +719 | –1,374 |

| *Total government employment under scenario:* | *1987* | *1992* | *1997* | *2000* |
|---|---|---|---|---|
| A | 4,529 | 5,651 | 6,104 | 6,416 |
| B | 4,529 | 5,135 | 5,879 | 6,180 |
| E | 4,529 | 5,135 | 5,547 | 5,831 |
| F and LS | 4,529 | 4,433 | 4,448 | 4,474 |
| C | 4,529 | 5,651 | 6,469 | 6,800 |
| of which (for Scenario C): | | | | |
| National Health Service | 1,306 | 1,630 | 1,859 | 1,849 |
| Education | 1,446 | 1,814 | 2,139 | 2,293 |

Job creation in the commercial sector as a whole is very dependent upon assumptions made about PG, as is borne out by the difference between scenarios F and LS (the two low-government-expenditure scenarios, with and without PG increases, respectively). However, while technological change and associated PG is well understood in the manufacturing sector and some other sectors, such as agriculture and mining, the situation is quite different with respect to distribution and service activities.

Because of the high levels of employment in these industries, the lack of information and the uncertainty concerning the future course of PG is very serious for economy-wide forecasts of employment. Slight differences in PG projections in these activities generate large differences for overall employment, especially if projected over a long enough time span.

The PG assumptions underlying the MS scenarios (and also the Lawson scenario) were that the increases assumed for two of the service industries (business and professional services) would be the same as those for the manufacturing sector. The reason for this assumption was that these two service industries are experiencing the rapid introduction of new technologies, particularly of information processing and communication technologies. On the other hand, distributive trades and miscellaneous services (the latter composed of such activities as hotel and catering, private health care, recreational and cultural services, dry cleaning, hairdressing, specialized research, etc.) were exempted from any productivity increases beyond their historically estimated trend growth rates. The overall output and the PG figures in manufacturing and the service sector which emerge in the different scenarios based upon these assumptions are presented in *Table 15.7.*

*Table 15.7.* NOG – net output growth (average annual rates at constant prices) and PG – productivity growth – in manufacturing and services.

| | | *Manufacturing* | | | *Services* | | |
|---|---|---|---|---|---|---|---|
| *Growth item* | *Scenario* | *87–92* | *92–97* | *87–97* | *87–92* | *92–97* | *87–97* |
| NOG | A | 1.82 | 2.87 | 2.34 | 2.83 | 3.33 | 3.08 |
| | B | 1.60 | 2.68 | 2.14 | 2.17 | 3.25 | 2.71 |
| | C | 1.75 | 2.91 | 2.33 | 2.64 | 3.47 | 3.06 |
| | D | 1.88 | 3.19 | 2.54 | 3.03 | 4.01 | 3.52 |
| | E | 1.67 | 2.67 | 2.17 | 2.36 | 3.20 | 2.78 |
| | F | 0.70 | 1.03 | 0.86 | 1.66 | 1.42 | 1.29 |
| | LS | 1.23 | 1.82 | 1.52 | 1.20 | 1.80 | 1.50 |
| PG | A | 2.90 | 2.98 | 2.94 | 1.42 | 2.16 | 1.79 |
| | B | 2.96 | 3.31 | 3.13 | 1.03 | 2.18 | 1.61 |
| | C | 2.58 | 2.46 | 2.52 | 1.24 | 2.15 | 1.69 |
| | D | 3.21 | 3.38 | 3.30 | 1.61 | 2.66 | 2.14 |
| | E | 3.27 | 3.69 | 3.48 | 1.21 | 2.23 | 1.72 |
| | F | 1.02 | 1.61 | 1.31 | -0.04 | 0.86 | 0.41 |
| | LS | 2.69 | 3.38 | 3.03 | 0.37 | 1.37 | 0.87 |

We can see that higher real income growth and the disproportionate increase in expenditure on services leads to a rapid growth in output in the service sector, particularly over the period 1992–1997. In general, this growth outstrips that of manufacturing (except in the Lawson scenario). This high overall growth, as well as compositional changes within the service sector, also lead to a speeding-up of productivity growth, narrowing the gap to the traditionally higher productivity gains made in manufacturing.

The employment implications of these projections have already been presented in *Table 15.6.* However, in order to demonstrate how sensitive the overall employment projections are to slight changes in the PG assumptions in these sectors, we carried out a simulation in which these two industries, distribution and miscellaneous services, experienced an additional 0.5% p.a. increase in trend PG (about half the amount assumed for manufacturing and business and professional services in the MS runs). This change resulted in a difference of *minus 827,000 jobs* in a scenario similar to A, so that we would obtain a negative net increase in overall employment in this scenario!

## 15.6. Changes in Energy Prices and World Activity

In this section we examine how sensitive our results are to changes in assumptions concerning world output growth, world inflation and oil prices in the medium term. Since this chapter was started, oil prices have halved and, for the purpose of our sensitivity exercise, it is assumed that they maintain this low level for the next five or six years. World GDP is increased by about 0.7% p.a. and world inflation reduced by about 1.3% for the exercise. These adjustments are consistent with one another; lower oil prices may stimulate growth and would almost certainly lead to lower inflation in the economy (the latter is already occurring). We have chosen scenario B as our base run. Not only is it representative of the MS scenarios, but it is also the most useful run for analyzing the sensitivity of the external balance to our proposed changes. The simulation is medium-term and runs to 1992, but long-run implications may be drawn from it. A summary of the results (and assumptions) of the simulation is presented in *Table 15.8.*

*Table 15.8.* Assumptions and results for a low oil price – high world demand scenario.[a]

| | *1986–1992*[b] | *1986* | *1987* | *1988–1992*[b] |
|---|---|---|---|---|
| ***Assumptions:*** | | | | |
| World inflation* | –1.3 | | | |
| World demand* | +0.7 | | | |
| Real oil price | –30.0 | | | |
| ***Results:*** | | | | |
| Real GDP (level) | | +0.2 | +0.7 | +0.6 |
| BOP (£1980 m)* | | –1927 | +1014 | +1113 |
| BOP/GDP (%)* | | –0.8 | +0.4 | +0.4 |
| Unemployment (1000s)* | | –2.3 | –3 | +73 |
| Inflation | | +1.1 | +1.7 | –0.4 |
| Real exchange rate | | –1.8 | +1.1 | +4.6 |
| Exports (volume) | | +2.2 | +3.4 | +4.0 |

[a]Figures indicate percentage above (+) or below (–) base, except those items marked (*), which indicates the difference in the respective quantities.
[b]Yearly average.

In what follows we use the words "above", "surplus", etc., to indicate "above base", "surplus over base", etc. The impact of lower oil prices on the BOP is unfavorable in the first two years. The UK is currently a net oil exporter so that the real value of our oil exports is halved in the first year. This feeds into the exchange rate equation and depresses both the nominal and real value of the currency. The boost to exports from higher world activity is insufficient to avoid a large deficit in the first year; but as world demand continues to grow, surpluses eventually appear and the balance grows to over £1 bn (1980 prices) in the latter periods. However, the growth in exports is stifled by the appreciating currency (a direct result of the surpluses), and they fail to keep pace with world trade.

The other components of demand are boosted little, with both consumption and investment ending the simulation a mere 1% above base. The absence of supply-side effects in MDM are much to blame here. (Many economists would argue that supply-side effects would be most important in the context of lower oil prices.)

The net result of minor improvements in investment and consumption and only modest gains in exports is that GDP stands less than 1% higher at the end of the simulation (where world output is 5% higher) and unemployment is reduced by a paltry 100,000.

How is it that the UK economy is left sinking in a world of buoyant demand? The answer lies in the fact that, despite large BOP surpluses, government expenditure remains passive. In reality, we would expect some fiscal response. This would stimulate GDP directly through multiplier effects on investments and consumption, and indirectly by removing the BOP surpluses (stemming the rise in the currency) and so stimulating exports.

Finally, following large positive jumps in the first two years (the result of a currency depreciation), the domestic rate of inflation eventually becomes equal to that of the rest of the world.

## 15.7. Conclusions

Simulations with the Cambridge Growth Project's model have shed light on a number of important issues which would emerge were one seriously to consider embarking on a "modernization strategy" for the United Kingdom.

*Productivity and growth*

An improvement in the long-term productivity performance would succeed to some extent in overcoming the historical constraints which have characterized the UK economy in the past. In particular, there would be a good deal more space for increased real income growth without renewed strong inflationary processes being generated; and the adversary trends in export and import growth in real terms, which in the past led to the accelerated erosion of the positive balance in manufacturing trade, could be reversed. The implications for unemployment, income distribution and the BOP would, however depend to a large extent on the precise nature of the MS and accompanying developments.

*Unemployment*

The question whether greater productivity and a better competitive performance by UK industries would help to solve the unemployment problem is a complicated issue. The simulations have shown that the creation of more employment depends upon the degree to which the additional resources available from higher productivity growth can be used for increased real income growth for those already employed.

The trade-off between real wage compensation from increased productivity performance and higher public expenditure to create more jobs was the one trade-off which could be shown explicitly by using the Growth Project's model. These are, of course, not the only redistributive measures which can be envisaged to free additional resources for the creation of jobs for the unemployed. Unfortunately, a quantitative evaluation of the extent to which other redistributive measures (such as increased taxation of the well-to-do) could contribute toward job creation is not within the scope of this chapter. It must be stressed, however, that any measures which have a negative impact upon investment activity, an integral and essential part of any modernization and development process of UK industry, would be counterproductive.

*Composition of employment*

As regards the pattern of job creation, our scenarios have shown that most new jobs which could absorb the unemployed will have to be generated in the non-manufacturing sector. The determinants of job creation in that sector are complex and different views contradict one another.

One view is that a "low wage" scenario with additional measures which remove "rigidities" in the labor market (such as removing inhibitions for relative wage rate movements; freedom to use labor in the production process and to introduce new technologies as management sees fit; reducing restrictions in the dismissal, hiring, and retraining of workers; etc.) will automatically lead to employment expansion in both "high tech" and "low tech" sections of industry.

The other view is more skeptical about whether the above scenario would in fact lead to more employment creation, and if so, whether these would be the types of jobs that would safeguard the welfare aspirations of the British people. An orientation toward a "low wage – low tech" sector would be counterproductive for any attempt to move "upstream" in terms of attracting skill-intensive, high value-added industries to the UK within the international division of labor. Furthermore, the emergence of such a sector would lead to a larger segment of the population depending upon insecure, unskilled and low-paid jobs and a widening income gap between them and the rest of the community.

The different scenarios showed that employment and output growth patterns, particularly in the non-manufacturing sector, differed depending upon which strategy was adopted. Since employment growth showed strong sensitivity toward productivity growth projections in the different non-manufacturing industries (especially in services) and the particular expenditure patterns linked to the different scenarios, the forms of organization of these industries and the backing of particular activities through public expenditure will be crucial for employment growth in these areas. Although it is difficult to make exact numerical calculations at the moment, there is no doubt that the employment implications will be very different depending on whether this sector will in the future be largely organized along private market service lines or around a state-supported welfare sector directed, in the first instance, toward the provision of social needs.

*Balance of payments*

Although the remarks made above about sectoral composition and particular expenditure patterns associated with the different scenarios have implications for the balance of payments, it is worth reemphasizing that the balance-of-payments problem will reemerge as the binding constraint upon UK economic policy options in the 1990s.

The predicament of the UK economy by the late 1980s, when its North Sea oil production starts to decline and it reaches a position with a negative balance in manufacturing trade, has been widely debated recently [see, e.g., House of Lords Select Committee on Overseas Trade (1985), Coutts *et al.* (1986), and the thorough discussion of balance-of-payments issues confronting the UK economy in Rowthorn and Wells (forthcoming)]. Here we limit ourselves to pointing out that, in spite of the fact that our modernization strategy was designed in such a way as to improve greatly the trade performance in the longer run (from 1993 onward the growth of exports exceeds the growth of imports in real terms at historically high levels of domestic activity), *the balance-of-payments problem remains critical* on the transition path until the late 1990s. As we have shown, the gradual character of the catching-up process, and the large amounts of resources (including imports) needed to embark on such a strategy, suggest a delicate path will have to be followed between an expansionist scenario (which facilitates the modernization process) and the extent of the balance-of-payments disequilibrium (which could conceivably be financed and tolerated by international financial markets over the transition path).

The simulations discussed in this chapter show one thing very clearly. A movement of the UK economy toward international competitiveness is not in itself, under the present circumstances, sufficient to cope with the severe economic and social problems. In particular, resolution of the unemployment problem *would not be automatic*. It depends crucially upon the readiness of those in work to share the gains of a successful modernization process with those presently unemployed, and upon the determination of a government to pursue two essential targets at the same time: employment creation *and* modernization.

# Appendix 15A. Productivity Growth under Thatcher and Prospects under the Different Scenarios

The following graphs and tables give some information about the productivity performance (output per person-hour and output per person employed) in UK manufacturing industry.

*Table 15A.1*, reproduced from OECD's *Economic Outlook* (1986), shows the pattern of productivity growth (output per person-hour) in manufacturing industry for a number of OECD economies including the UK. The well-known picture of the slow growth in labor productivity in the USA and the UK relative to the other OECD countries over the 1960s and 1970s emerges clearly. However, over the period 1979 to 1985 (and also over the subperiods 1979–82 and 1982–85) UK productivity growth (measured as output per person-hour) outstrips that of its main industrial competitors with the exception of Japan (see also *Figure 15A.5*).

*Table 15A.1.* Labor productivity growth in manufacturing (gross value added/total hours worked).

| *Country* | *1960–69* | *1969–79* | *1979–82* | *1982–85* | *1979–85* |
|---|---|---|---|---|---|
| UK | 5.8[a] | 2.2 | 3.5 | 4.8 | 4.1 |
| USA | 4.3 | 2.5 | 1.7 | 4.2 | 2.9 |
| Germany | 6.9 | 4.5 | 1.7 | 4.1 | 3.5 |
| France | 7.7 | 5.4 | 2.9 | 4.1 | 3.5 |
| Italy | 7.4[b] | 5.1 | 3.6 | 3.9 | 3.5 |
| Sweden | 7.6 | 3.6 | 2.0 | 5.7 | 3.8 |
| Japan | 11.7[a] | 8.3 | 7.0 | 4.8 | 5.9 |

[a]1966–69.
[b]1965–69.
Source: OECD, *Economic Outlook* (May 1986).

*Figure 15A.1* and *Figure 15A.2* present, on a log-scale, time series on output, employment, output per person employed and output per person-hour in UK manufacturing over the period 1963 to 1985. The following features emerge from the graphs:

(1) Over the period 1963 to 1973, growth in output and growth in output per person employed move, insofar as trends are concerned, more or less in parallel. Cyclically, the relationship is more complicated, but this does not concern us here. The average figures for productivity growth (growth in output per person employed) over the period 1964 to 1972 is 3.1% p.a. From 1973 to 1979, overall productivity growth rates in manufacturing declined markedly and the level of output in manufacturing was no higher in 1979 than in 1973. The average productivity growth rates (output per person employed) per annum are 2.09% for the period 1973 to 1979 (2.0% p.a. for output per person-hour) and 0.9% p.a. for 1974 to 1979 (1.08% p.a. for output per person-hour).

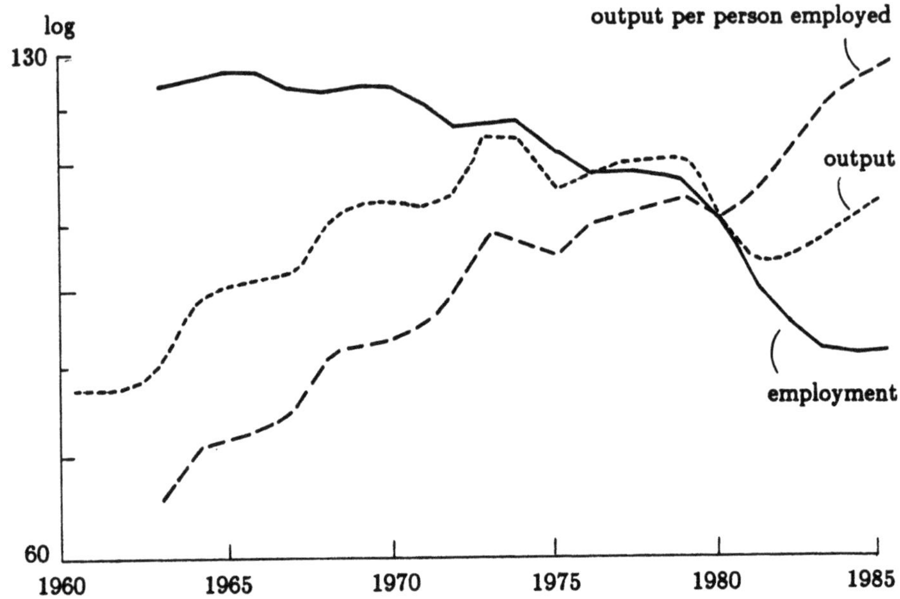


*Figure 15A.1.* Output, employment, and output per person employed in manufacturing (1980 = 100).

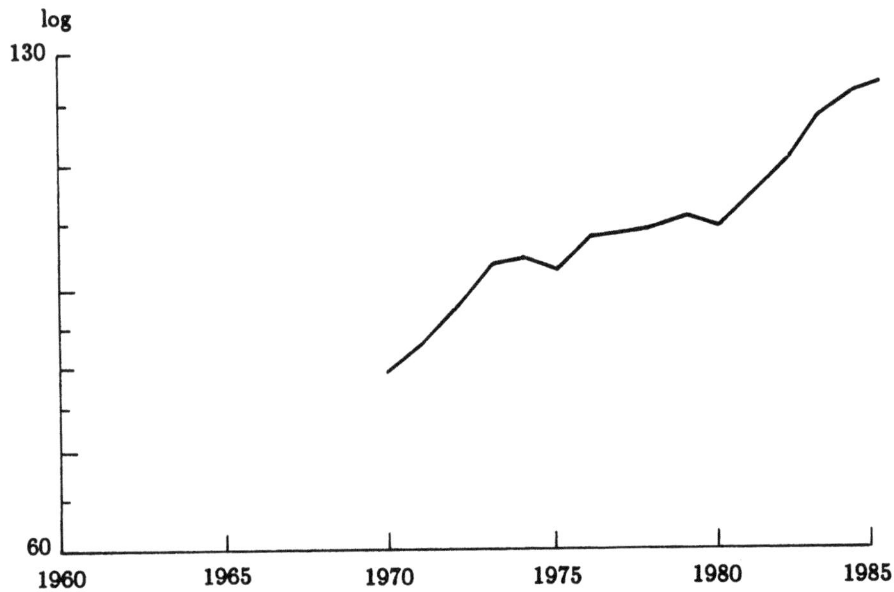


*Figure 15A.2.* Output per person-hour in manufacturing (1980 = 100).

(2) Productivity performance under Thatcher is marked by annual rates of output growth per person-hour and per person employed, over the years 1979 to 1986, as shown in *Table 15A.2*. The year-to-year changes in output and output per person employed have been plotted in *Figure 15A.3*. We can clearly see the much sharper decline in output than in output per person employed in the years 1979 to 1981 (the years in which a major "shake-out" of labor occurred), after which the growth rates in output and labor productivity started to move in parallel. The productivity growth rates achieved in the period 1981 to 1984 are indeed historically very high, similar to those achieved in the peak periods of the 1960s and early 1970s, and far beyond anything achieved over the period 1973 to 1979. As output growth slowed down in 1985 and 1986, productivity growth declined even more (3.06% in 1985 and 1.80% in 1986 for output per person-hour), with evidence, however, of output and productivity growth recovering toward the end of 1986 and the beginning of 1987.

*Table 15A.2.* Productivity growth in UK manufacturing, 1979–1985.

| *Output growth (%)* | *1979* | *1980* | *1981* | *1982* | *1983* | *1984* | *1985* | *1979–1986* | *1982–1986* | *1986* |
|---|---|---|---|---|---|---|---|---|---|---|
| Per person-hour | 1.68 | –1.48 | 4.80 | 5.25 | 7.80 | 4.45 | 3.06 | 1.80 | 3.42 | 4.47 |
| Per person employed | 1.57 | –3.94 | 3.50 | 6.57 | 8.52 | 5.51 | 3.48 | 1.91 | 3.39 | 5.70 |

Source: *Employment Gazette* (March 1987).

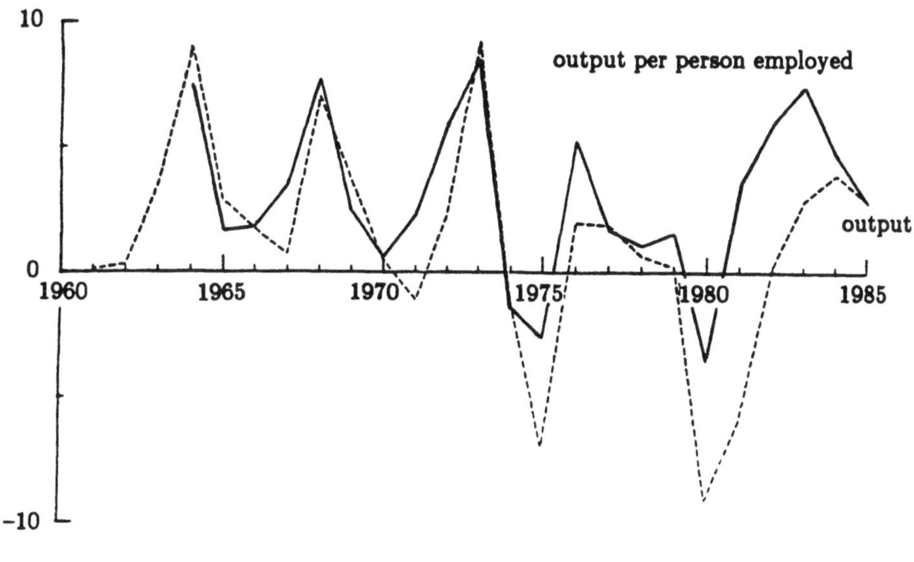


*Figure 15A.3.* Yearly growth rates of output and output per person employed in UK manufacturing.

There are basically two interpretations of the productivity growth experience since 1979. [On the productivity performance under the Conservative government since 1979, see Jones (1983), Mendis and Muellbauer (1983), Smith-Gavine and Benett (1985), Coutts *et al.* (1986) and Muellbauer (1986).]

One view is that the recovery of productivity growth since 1981 is the result of a major shake-out of labor and of a once-for-all additional scrapping process of old capacity which has changed the plant and firm composition of a number of manufacturing industries. However, according to this view, this change amounts only to a once-for-all level change and will leave long-term rates of productivity growth unaffected. The critics of the Conservative government point furthermore to the abysmal performance of investment in manufacturing – see *Figures 15A.4* and *15A.5* – and believe that an investment slump of that magnitude will be followed by very low rates of productivity growth. Only high rates of growth of output and of net investment can ensure high productivity growth in the future.

The other view is that something fundamental has changed in British industry over the 1979 to 1986 period which will – independent of any fiscal expansion by the government – have long-term implications for the growth and productivity performance of the UK economy. Proponents of this view point toward the change in the industrial relations climate partly induced by changes in industrial relations legislation, in work practices, and attempts made by the government to deregulate industry and privatize publicly owned enterprises. They also point out that compositional changes (mentioned in the previous paragraph) can have once-for-all effects not only on industrial performance but also upon longer-term growth.

In our view, the information so far available is still insufficient to judge the long-term effects of the Thatcher experiment upon productivity growth in manufacturing. The recent closing of the gap between output and productivity growth, however, indicates that, after the major labor shake-out period, any further sustained improvement in productivity performance will again be closely linked to output growth and a sustained recovery in investment activity.

In our simulations we have taken both the above interpretations of recent productivity performance on board, and they are reflected in the projections for productivity growth in the different scenarios as shown in *Figures 15A.6* and *15A.7*.

As discussed in the main text, scenarios F and LS represent two versions of the effects of restrictive government policies upon the long-term growth of the UK economy. Scenario F represents a pessimistic view of the longer-term productivity growth outlook in such conditions (the productivity growth rates are similar to those representative for the period 1973 and 1979), while scenario LS (the "Lawson scenario") allows for continued higher productivity growth into the 1990s much more in line with the pre-1973 experience.

The latter is also true for scenarios A to E which represent MS scenarios with expansionary government expenditure programs and different real wage growth rates (see main text for discussion).

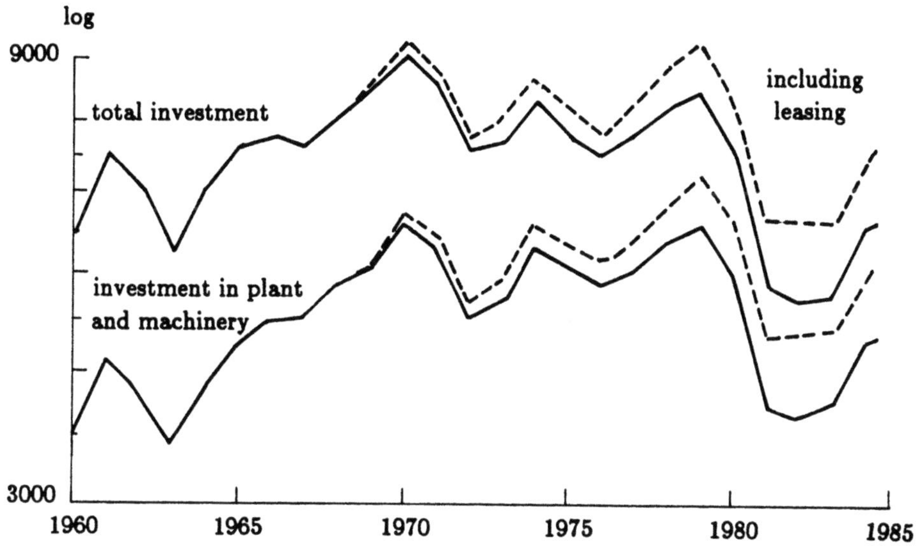


*Figure 15A.4.* Investment in manufacturing.

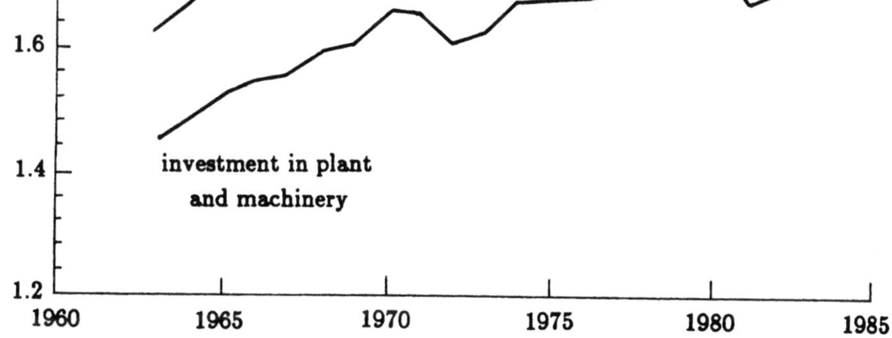


*Figure 15A.5.* Investment per person employed in manufacturing.

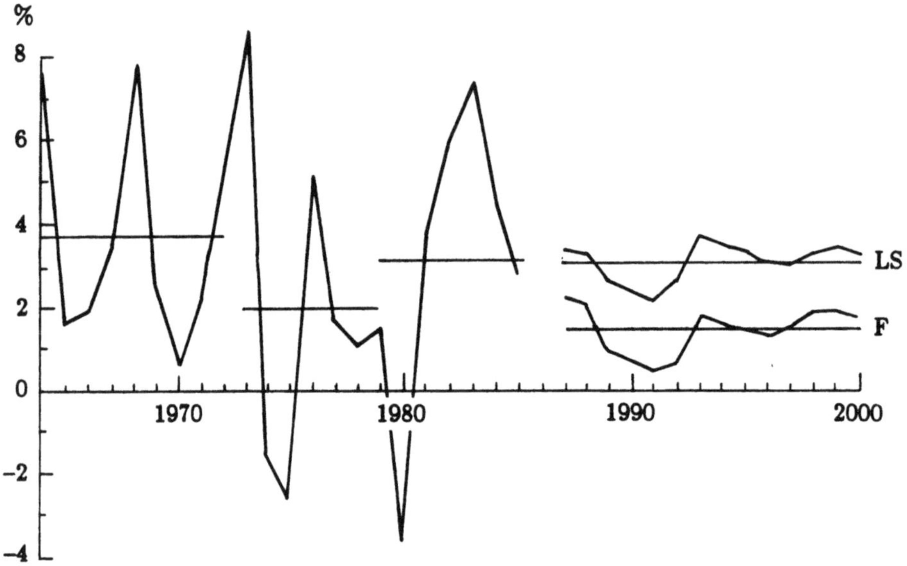


*Figure 15A.6.* Productivity growth (output per person employed) in UK manufacturing – past experiences and projections in simulations LS and F.

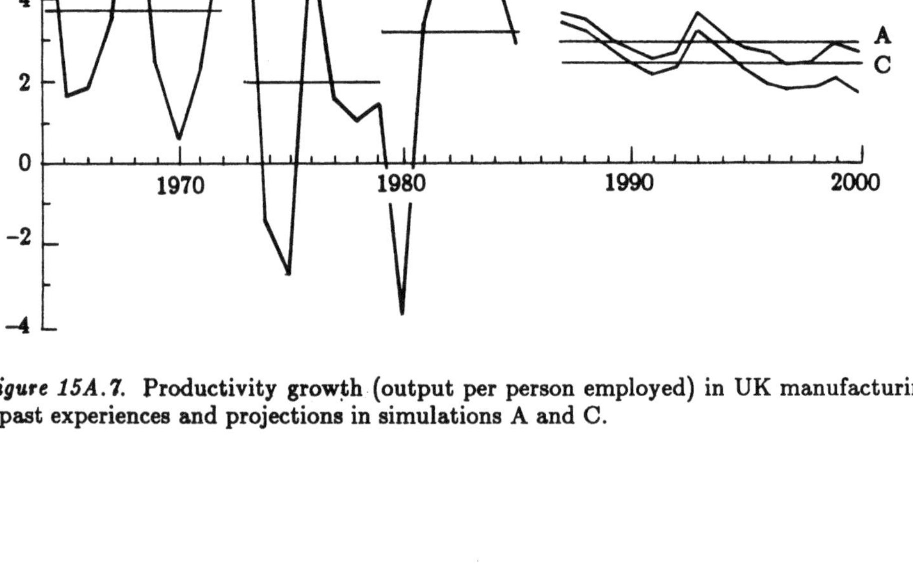


*Figure 15A.7.* Productivity growth (output per person employed) in UK manufacturing – past experiences and projections in simulations A and C.

# Appendix 15B. Is There Evidence of an Improvement in UK Export Performance?

In this appendix we give some empirical substance to the hypothesis that the recent turnround in the UK's productivity performance in manufacturing has gone hand in hand with a transformation of the demand for its exports of manufactured goods. We hope that this goes some way toward justifying the adjustments we made to MDM's export equations for the MS scenarios. We investigated the hypothesis here, by analyzing the behavior of UK manufacturing exports over the past 14 years through the "eyes" of a simple but adequate demand equation.

Explicitly, we used the following specification:

$$DX_t = \alpha + \beta\ WX_t + \sum_{i=0}^{5} C_{t-i} \tag{15B.1}$$

where

$C_t$ = an index of competitiveness
$DX_t$ = the volume of UK manufacturing exports
$WX_t$ = the volume of world manufacturing exports
$t$ = time measured in quarters

All variables are natural logs. The $C_t$ index (taken from CSO's *Economic Trends*) measures UK to world export prices in dollars and reflects the relative importance of each foreign country in the world market place. $WX_t$ was constructed by taking the exports of the nine largest OECD economies (in dollars) and dividing by an index of their prices. The former series (taken from OECD's *International Trade Statistics*) accounts for over 90% of OECD exports while the latter (taken from IMF's *International Financial Statistics*) pertains to all OECD countries. $DX_t$ was taken directly from the CSO's *Monthly Digest of Statistics.* Although $DX_t$ is a seasonally adjusted series, $C_t$ and $WX_t$ show no seasonal pattern so seasonality should not raise any problems for the estimation of (15B.1).

Henceforth, "manufacturing" is suppressed, and we refer to just "exports". $C_t$ and $WX_t$ were found to be nearly orthogonal to a set of seasonal dummies.

Specification (15B.1) has a dual interpretation: (a) it is an *ad hoc* explanation of the share of UK exports in the OECD total (corrected for competitiveness); or (b) it is an export demand function. For (b) it is necessary to envisage a direct link between exports and world GDP. No such complications arise in (a). In (a)'s view (15B.1) provides a direct assessment of how the UK has fared against its competitors in the battle for a share of the world market, and it is this interpretation that we prefer.

An assumption common to both views, however, is that market reactions to competitiveness (relative prices) are spread across many periods, whereas reactions to the size of the market are instantaneous. This is plausible for two reasons. First, price changes involve substitution between goods, and this may be costly or difficult to invoke immediately. No such cross-substitution is involved when the market expands or contracts. Secondly (and perhaps less convincingly), relative price changes may take time to detect.

As a starting point, we estimated (15B.1) for the whole period of 58 quarters starting in 1972(1) and ending in 1986(2). Omitting wholly insignificant terms we get

$$\underset{}{DX_t} = \underset{(19.1)}{6.45} + \underset{(14.7)}{0.54}\ WX_t - \underset{(4.9)}{0.36}\ C_{t-4} + \underset{(1.6)}{0.40}\ C_{t-5} \tag{15B.2}$$

$$R^2 = 0.80,\quad \bar{R}^2 = 0.80,\quad DW = 1.3\quad (t\text{-ratios in parentheses})$$

If we look only at the fit and $t$-ratios, we might be quite satisfied with (15B.2). However, the $DW$ statistic indicates substantial error autocorrelation; the long-run effect of competitiveness is perverse; and an inspection of the residuals (not presented here for lack of space) shows that the model overpredicts in 1979 and 1980, but underpredicts in the last few years of the sample. This is not inconsistent with our hypothesis that underlying changes in trend PG fundamentally affect export demand; our data straddle the productivity divide associated with the Thatcher administration (documented in Appendix 15A).

In the light of all this it seems natural to split the data into a pre-Thatcher and a Thatcher era and estimate (15B.1) for both subperiods. Using data from 1972(1) to 1979(2), and dropping insignificant competitiveness terms, gives

$$DX_t = \underset{(8.2)}{5.82} + \underset{(10.0)}{0.59}\ WX_t - \underset{(2.6)}{0.62}\ C_{t-4} + \underset{(1.7)}{0.40}\ C_{t-5} \tag{15B.3}$$

$$R^2 = 0.82,\quad \bar{R}^2 = 0.79,\quad DW = 2.2$$

Statistic (15B.3) does not exhibit error autocorrelation, has a reasonable fit and passes structural stability tests. Dropping five observations from the beginning and end of the sample for a Chow test gives $F$ (5,21) values of 1.3 and 2.2, which are satisfactory. Further, the "income elasticity" (henceforth referred to as $\beta$) is significantly less than unity ($t_{26} = 6.4$), and this reflects the secular decline in export shares of the pre-Thatcher era documented in the text and elsewhere.

Estimating (15B.1) for the 1979(3) to 1986(2) subperiod (again dropping insignificant terms) gives

$$DX_t = \underset{(7.6)}{5.79} + \underset{(5.6)}{0.58}\ WX_t - \underset{(1.3)}{0.22}\ C_{t-2} \qquad (15B.4)$$

$$R^2 = 0.68,\ \bar{R}^2 = 0.66,\ DW = 0.57$$

The estimate of $\beta$ is identical to that of the earlier period and, if we knew nothing about econometrics, the story would end there and our hypothesis would fail. However, even the most casual statistical inspection shows that (15B.4) is grossly inadequate. The competitiveness term is not very significant, the $R^2$ is unsatisfactory (considering that $DX_t$ and $WX_t$ are trended), the model is very unstable and there is substantial error autocorrelation. Dropping the first five observations for a Chow test gave an $F$ (5,22) value of 4.1.

To shed light on this subperiod, we allowed the parameters to vary over the sample by the use of a simple technique. We looked at (15B.4) through a "window" of 17 observations. Moving the window one quarter at a time across the sample gives 12 regressions from the 28 observations and, hence, 12 estimates of $\beta(\hat{\beta})$. This gives an idea of the nature of any parameter instability in the subperiod. Choice of window size is never a trivial matter. A smaller window gives a better idea of the parameter regime at a particular point in time, but only at the expense of precision from having too few degrees of freedom. Thus, while we should let the data "breathe" and allow the parameters to vary, we should not let them "hyperventilate" and suffer imprecise parameter estimates. The choice of 17 quarters was a compromise between these two extremes. Shorter windows of 12 and 14 observations were tried, but made little qualitative difference to the results.

Because our hypothesis primarily concerns the elasticity $\beta$, we present the results for this coefficient only. The right-hand graph of ***Figure 15B.1*** plots the progress of the estimate $\hat{\beta}$ as we move the window of 17 quarters from 1979(3) startpoint [1983(3) endpoint] to a 1982(2) startpoint [1986(2) endpoint]. For completeness, the left-hand graph presents the corresponding result for the pre-Thatcher sample. Care must be taken when interpreting the varying estimate $\hat{\beta}$: it is *not* an estimate of $\beta$ at *one* point in time but an estimate of its "average" value (in some sense) over 17 quarters.

The results are very informative. While $\hat{\beta}$ is reasonably stable in the pre-Thatcher subsamples (averaging just below 0.6), the equation breaks down completely in subsamples that use any data from the 1979(3) to 1980(2) period. A stable relationship reemerges in the post-1980(4) subsamples, but this time the values for $\hat{\beta}$ vary around unity rather than 0.6.

The 1979(3) to 1980(4) period was a tortuous time for UK manufacturing. In the face of a very severe recession, output fell by nearly 20% and manufacturing exports by about 10%. This may explain the breakdown of our relationship

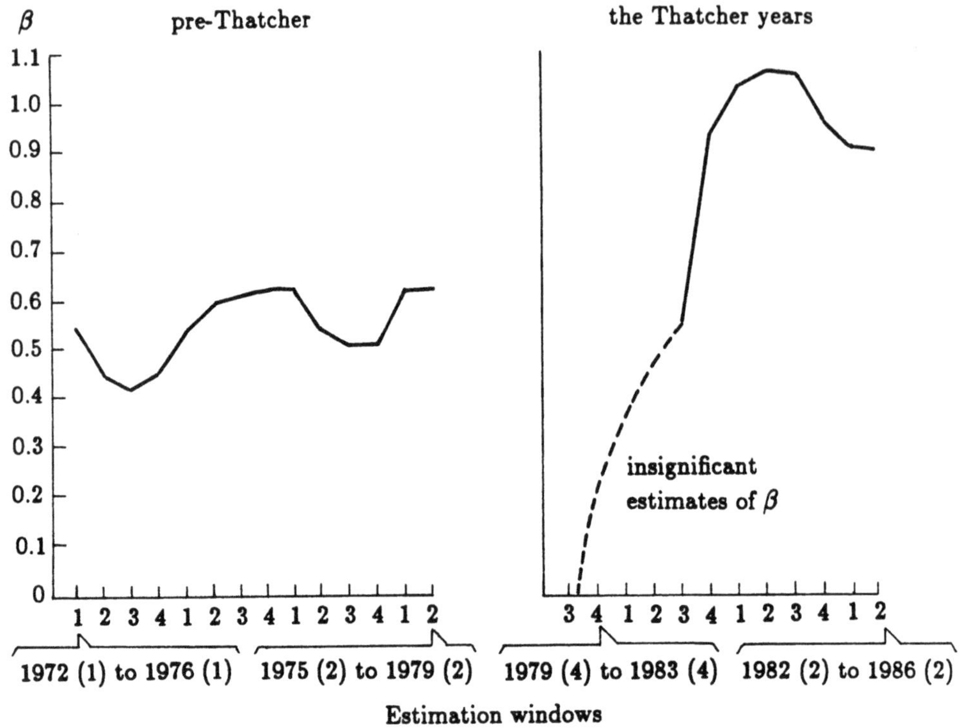


*Figure 15B.1.* The time path of the estimate of $\beta$ from 17-quarter subsamples.

in subsamples that include these quarters. However, it was also a time of restructuring and consolidation, a process that continued into the 1980s and which may have generated the high PG rates experienced from 1981(1) onward. If our empirical work is to be believed, this productivity "miracle" has been accompanied by a transformation in the nature of the demand for UK exports. Taking interpretation (a) or (b), above, of our export specification, it would imply that UK goods are no longer inferior in the world markets (their income elasticity of demand having risen from 0.6 to unity). As a result, the secular decline in market share suffered by UK goods prior to 1979 has been arrested (our exports now grow in line with world trade).

Finally, we can model the "recovery" depicted in the graph by means of a time trend on $\beta$ starting in 1979(3) and ending in 1984(4) leaving $\beta$ constant from 1985(1) to 1986 (2). This choice of trend was suggested by an inspection of *Figure 15B.1.* Incorporating this scheme for $\beta$ in (15B.4) gives

$$DX_t = \underset{(11.9)}{6.5} + (\underset{(8.7)}{0.64} + \underset{(5.2)}{0.022}T)\ WX_t - \underset{(2.9)}{0.35}\ C_{t-2} \qquad (15B.5)$$

$$R^2 = 0.83, \ \bar{R}^2 = 0.82, \ DW = 1.82$$

where $T = 1$ in 1973(3); $T = 2$ in 1979(4); . . .; $T = 22$ in 1984(4); but then stays constant – $T = 22$ in 1985(1); . . .; $T = 22$ in 1986(2).

(15B.5) has no error autocorrelation, is remarkably stable [dropping five observations from the beginning and end of the sample gives $F$ (5,19) values of 0.8 and 1.4, respectively], has a correctly signed and significant role for competitiveness and fits quite well. It gives an initial estimate for $\beta$ [for the *point* in time 1979(3)] of 0.66 and a final value [for the *point* in time 1986(2)] of just over unity.

Although the results here were *qualitatively* robust to changes in measurement (various measures of export volume, world export volume and competitiveness were tried), they are indicative rather than conclusive. A more detailed study would use disaggregated data and incorporate qualitative information. The current results are nonetheless very encouraging for future work.

## References

*Bank of England Quarterly Bulletin* (1986), Services in the UK economy: September.

Barker, T.S. and Peterson, A.W.A. (forthcoming), *The Cambridge Multisectoral Dynamic Model.* (London and New York: Cambridge University Press).

Blackaby, F. (ed.) (1979), *De-industrialisation.* (London: Heinemann).

Cambridge Econometrics (1986a), *Forecast* 86/1. (Cambridge, UK).

Cambridge Econometrics (1986b), *The British Economy 1985–2000: Alternative Scenarios for the World Growth and Labour Productivity.* (Cambridge: UK).

Coutts, K., Godley, W., Rowthorn, B., and Ward, T. (1986), *The British Economy: Recent History and Medium-term Prospects. A Report on the Thatcher Experiment.* (Cambridge, UK: Faculty of Economics and Politics, Cambridge University).

Jones, D.I.H. (1983), Productivity and the Thatcher Experiment, *Socialist Economic Review*, pp. 27–45.

House of Lords Select Committee on Overseas Trade (1985), *Report*, CMND 9697. (London: HMSO).

Kendrick, J.W. (1984). *International Comparisons of Productivity and Causes of the Slowdown.* (Cambridge, MA: Ballinger).

Landesmann, M. and Pierse, R. (1986), The Effects of Industrial Policies in the United Kingdom, 1973 to 1981, in A. del Monte and S. Vinci (eds.), *Industrial Policies in Europe and in the United States.* (Naples: Liguori Editore).

Lawson, A. (forthcoming), Incomes Policy and Earnings, Ch. 15 in T.S. Barker and A.W.A. Peterson (eds.), *The Cambridge Multisectoral Dynamic Model.* (London and New York: Cambridge University Press).

Mendis, L. and Muellbauer, J. (1983), *Has There Been a British Productivity Breakthrough? Evidence from an Aggregate Production Function for Manufacturing*, Discussion Paper No. 170. (Centre for Labour Economics, London School of Economics).

Muellbauer, J. (1986), Productivity and Competitiveness in British Manufacturing, *Oxford Review of Economic Policy*, **2**, No. 3.

NEDO (1977), *International Price Competitiveness, Non-price Factors and Export Performance*. (London: D. Stout).

Rowthorn, B. and Wells, J. (forthcoming), *Deindustrialization and Britain's Changing Role in the World Economy*. (London and New York: Cambridge University Press).

Smith-Gavine, A.N. and Benett, A. (1985), "Technological" Productivity Growth: The Latest Reckoning, *P.U.L. Bulletin*, **49**.

Snell, A. (1987), Sterling Exchange Rate, Ch. 19 in T.S. Barker and A.W.A. Peterson (eds.), *The Cambridge Multisectoral Dynamic Model*. (London and New York: Cambridge University Press).

Tarling, R. and Wilkinson, F. (1982), The Movement of Real Wages and the Development of Collective Bargaining in the Period 1855 to 1920, in *Contributions to Political Economy* (annual supplement of the *Cambridge Journal of Economics*).

CHAPTER 16

# Prospective Structural Changes in the West German Economy

*Jürgen B. Donges, Henning Klodt, and Klaus-Dieter Schmidt*

## Summary

Projections up to the year 2000 show that the West German economy will undergo deep transformations as per capita income rises. Service activities will expand at a rate above average, while manufacturing and building industries are going to shrink in relative terms. We expect labor displacement in industry to continue, whereas the labor absorption capacity of the services sector might grow steadily. As industrial employment in West Germany has been found to be still higher than one would derive from "normal patterns", the projected decline of industry's share in labor input implies that, by the year 2000, the overall employment structure should be closer to the country's level of development than it is nowadays.

## 16.1. Introduction

The West German economy, once praised for its exceptionally successful postwar development, has been struggling with deep-seated problems over the past 15 years. Since the early 1970s, the growth of potential output slowed down markedly (from about 4–5% a year during the 1960s to less than 2% in the early 1980s); an extended period of full employment came to an end (unemployment increased sharply, affecting more than 2 million workers or 9% of the labor force); and industry lost ground in certain high-technology areas (such as the information technologies). The problems are structural rather than cyclical; their main roots are on the supply side and not on the demand side. As has happened in most Western European countries, the West German economy has also unduly delayed needed adjustment to rapidly changing global economic conditions, energy saving being the notable exception to this rule [Fels and Schmidt (1980), Schmidt *et al.* (1984), and Wolter (1984)].

The pressure to adjust is likely to remain strong in the years to come, if only for such reasons as the shrinking and aging population, sudden increases and decreases in energy prices, continued fluctuations of real interest rates and exchange rates, rising costs of environmental control, new waves of revolutionary technologies, and further growing intensity in international competition. The West German economy has to pass, with success, the test of its capacity to adjust, should there be a chance for achieving sustained growth and for improving employment opportunities over the long term.

It is against this background that the present chapter proposes to explore how the structure of production and employment of the West German economy will have changed by the year 2000, under alternative assumptions about the rate of long-term growth. The analysis, an extended version of which has been carried out by the authors elsewhere [Donges *et al.* (1986)], follows the familiar Clark–Fourastié–Kuznets–Chenery "three-sector hypothesis" and constitutes a reappraisal of comprehensive projections of structural patterns for West Germany up to 1980, which were undertaken at the Kiel Institute in the early 1970s [Fels *et al.* (1971); hereafter called the FSW study].

## 16.2. Structural Change in Retrospect

One message of the FSW study was, as it may be recalled, that West Germany became over-industrialized during the 1950s and 1960s. The share of industry in gross domestic product (more than 40% on average) was permanently higher than what would be expected of a "representative" country with similar per capita income and population; moreover, this share increased over time, whereas it declined in other advanced European countries, such as France and the UK. In anticipation of the profound exchange rate adjustments which were to occur during the 1970s after the collapse of the Bretton–Woods system, and by expecting major shifts in the international division of labor, which were to be induced by the emergence of new competitors on the world market, the FSW study predicted that West Germany's industry would begin to decrease its relative weight in the sectoral composition of production from the mid-1970s onward.

As *Table 16.1* shows, this prediction has been confirmed by actual developments, at least as far as the reversal of industry's trend is concerned, to which, however, several factors additional to those mentioned in the FSW study (e.g., wage explosion, profit squeeze, sluggish investment) contributed to a significant extent [Schmidt *et al.* (1984)]. Since 1970 the share of industry in both the gross value added and the labor force fell sharply. The pattern of resource allocation shifted from industry (and manufacturing) to services (whereas the primary sector continued the trend previously observed). But the shift was not accomplished at full employment. On the contrary, while the industrial sector lost 2.3

*Table 16.1.* Structure (%) of production and employment, 1960–1985.

| *Indicator* | *1960* | *1970* | *1980* | *1985* |
|---|---|---|---|---|
| Gross value added[a] | | | | |
| Primary sector[b] | 8.6 | 4.6 | 3.1 | 2.8 |
| Secondary sector[c] | 50.4 | 50.4 | 43.5 | 41.1 |
| of which: Manufacturing | 40.3 | 40.2 | 33.9 | 33.2 |
| Tertiary sector[d] | 49.0 | 45.0 | 53.4 | 56.1 |
| | | | | |
| Active labor force | | | | |
| Primary sector[b] | 15.9 | 9.7 | 6.4 | 6.3 |
| Secondary sector[c] | 45.8 | 47.7 | 43.2 | 40.1 |
| of which: Manufacturing | 36.9 | 38.1 | 34.2 | 32.0 |
| Tertiary sector[d] | 38.3 | 42.6 | 50.4 | 53.6 |

[a] At current market prices.
[b] Agriculture, forestry and fishing; mining and quarrying.
[c] Manufacturing; electricity, gas and water; construction.
[d] Trade, transport and communications; banking, insurance and real estate; public administration; other services.

Sources: *Statistisches Bundesamt*, Volkswirtschaftliche Gesamtrechnungen, Fachserie 18, Reihe S.8: Revidierte Ergebnisse 1960–84 (Stuttgart, 1985); *Idem*, Reihe 1: Konten und Standardtabellen 1985 (Stuttgart, 1986); own calculations.

million jobs in the period 1970–1985 (manufacturing: 1.9 million), the tertiary sector absorbed only a fraction of the displaced workers – and this in spite of the extraordinary expansion of employment in public administration by more than 1 million [Schatz and Wolter (1982); Schmidt (1984)].

### 16.2.1. Changes in final demand

In the long run a country's capability to achieve sustained economic growth and to bring about the necessary structural changes is determined by its propensity to invest. *Table 16.2* provides information about the trends in aggregate expenditures during the period 1960–1985. While the shares of private and public consumption increased and that of foreign trade (exports minus imports) remained almost constant, the share of gross fixed capital formation decreased. This shift from investment to consumption is a most remarkable aspect indeed. It helps to explain why the West German economy has been lagging in modernizing productive capacities and in creating new jobs until recently [Schmidt *et al.* (1984)].

Shifts in the demand structure have not been confined to aggregate expenditures. It is well known, for instance, that there are long-term changes in the pattern of private consumption on account of Engel's law: the steady improvement in standards of living have altered the style of life of the West Germans and shifted preferences of private households away from food and other basic needs toward durable consumer goods and services. This unquestionably has clouded the development prospects of the producers facing an income elasticity of demand of less than unity. *Table 16.3* shows how markedly the pattern of

*Table 16.2.* Distribution (%) of gross domestic product at current market prices, 1960–1985.

| *Demand component* | *1960* | *1970* | *1980* | *1985* |
|---|---|---|---|---|
| Private consumption | 52.0 | 54.4 | 56.3 | 56.4 |
| Government consumption | 18.6 | 18.4 | 19.6 | 19.8 |
| Gross fixed capital formation | 24.2 | 24.2 | 21.5 | 19.5 |
| Change in stocks | 2.2 | 2.1 | 1.2 | 0.4 |
| Exports of goods and non-factor services | 17.0 | 22.6 | 29.5 | 35.0 |
| Less: imports of goods and non-factor services | 14.0 | 21.6 | 28.1 | 31.1 |
| Gross domestic product | 100.0 | 100.0 | 100.0 | 100.0 |

Sources: As for *Table 16.1.*

*Table 16.3.* Structure (% of total expenditure) of private household consumption, 1960–1985.

| *Expenditure item* | *1960* | *1970* | *1980* | *1985* |
|---|---|---|---|---|
| Agricultural products | 33.3 | 27.3 | 22.3 | 20.9 |
| Manufactured goods | 36.9 | 40.7 | 43.1 | 41.9 |
| Services | 29.8 | 32.0 | 34.6 | 37.2 |

Sources: As for *Table 16.1.*

private household demand changed. In a sense, these shifts are consistent with the changing structure of production and employment mentioned above: the increase in purchases of foodstuffs has been slower and that of services has been sharper than that of consumption in general, in particular since 1970.

Having said this, it should be noted that the shifts on the demand side cannot completely explain the gains and losses of sectors in terms of production and employment. Above all, the relative decline of manufacturing since the early 1970s is not in line with the stable shares of manufactured goods in total consumption during the past 15 years. An explanation for this apparent contradiction is that German manufacturers lost shares in world trade, mainly with regard to investment goods rather than consumer goods and particularly in the domestic markets rather than in third markets.

### 16.2.2. Changes in the international division of labor

The difficulties encountered by West German producers of manufactured goods during the 1970s reflect, broadly, a weakening of the competitive strength and the emergence of new competitors such as Japan and a number of newly industrializing countries [Donges (1984)]. The share in world exports, which had risen from 3.6% in 1950 to 9.7% in 1960 and to 11.9% in 1970, has slightly decreased over the past 15 years. A loss of share in international trade is a normal phenomenon for a mature industrial country. In the case of the West German economy, one can even argue that the previously excessive weight, by international standards, of the tradeables sector in comparison to the non-tradeables sector was corrected to some extent [Fels and Schmidt (1980)].

However, this process of adjustment becomes evident more on the import side than in exports. As *Table 16.4* shows, the share of final manufactured goods in exports has remained, more or less constantly, on a high level (predominantly investment goods); by contrast, final manufactures have increased sharply their weight in imports, which is a consequence of the rising market penetration by developing countries, in particular the newly industrializing ones (textiles, clothing, electronic consumer goods and the like). These shifts in the composition of trade in manufactures are ignored in a recent paper by Lawrence (1986), who has stressed that "domestic use, rather than international trade performance, accounts for the weakness of manufacturing output" (p. 56).

*Table 16.4.* Structure (%) of foreign trade by major product categories, 1960–1985.

| *Product categories* | *1960* | *1970* | *1980* | *1985* |
|---|---|---|---|---|
| Exports | | | | |
| Foodstuffs | 3.5 | 3.5 | 5.3 | 5.2 |
| Raw materials | 4.5 | 2.6 | 1.9 | 1.6 |
| Semi-manufactures | 10.3 | 7.7 | 8.8 | 7.6 |
| Final products: | 81.7 | 86.2 | 84.0 | 85.6 |
| consumer goods | 9.3 | 10.6 | 11.5 | 11.7 |
| investment goods | 52.1 | 54.7 | 51.4 | 54.0 |
| Imports | | | | |
| Foodstuffs | 26.6 | 19.4 | 12.8 | 12.5 |
| Raw materials | 21.9 | 13.7 | 17.5 | 12.4 |
| Semi-manufactures | 19.0 | 16.3 | 17.8 | 18.7 |
| Final products: | 32.5 | 50.6 | 51.9 | 56.4 |
| consumer goods | 10.2 | 22.5 | 22.9 | 14.1 |
| investment goods | 11.0 | 13.2 | 15.4 | 27.1 |

Sources: *Statistisches Bundesamt*, Aussenhandel, Fachserie 7, Reihe 1, various issues; *Idem*, Jahrbuch für die Bundesrepublik Deutschland, various issues; own calculations.

The ongoing changes in the international division of labor have three fundamental implications for the West German economy:

(1) As Germany has achieved a prominent position in technological development, it cannot benefit as much as before from importing and applying technology developed in other countries. Having come closely to the technology frontier, competitive strength can only be maintained, or restored, if the country's own technological base is deepened adequately.

(2) In the course of their catching up, the newly industrializing countries have intensified competition in high-tech markets. Hence, pressure to accelerate the pace of scientific innovations and of product and process innovations as well is likely to increase further.

(3) The developing countries embarking upon industrialization will continue to exploit, to the largest extent possible, their comparative advantage deriving from the almost unlimited supply of low-skilled labor. Thus, traditional labor-intensive industries in West Germany will be subjected to marked import competition in the future, too.

### 16.2.3. Changes in the global technological environment

Typically, a country's ability to invent and innovate, and to effectively use technologies developed abroad, is a key factor in the patterns of specialization in foreign trade. (West) Germany has a long tradition as a technologically leading country. Indicative of this strength is the robust competitive position which human capital- and research-intensive branches, such as chemicals, pharmaceuticals, medical technology, engineering goods and motor cars, have been displaying even during the 1970s, when the world economy was struck with many turbulences.

And yet, the West German economy (as in other European countries) has found it hard to keep abreast with the microelectronics and other revolutionary technologies, even allowing for the fact that a notable competitive strength could have been achieved early in some important segments (such as videotex or sensors and systems for robots) and that there was considerable success in nuclear power, aircraft industries, military equipment, and rocket launching (backed by massive government support). As can be seen in *Table 16.5*, West Germany lagged behind the United States and Japan regarding the robotization of the economy or the extent of use of microelectronics. Even in numerically controlled machine tools, in which German companies have long been strong leaders, Japan has caught up. Moreover, new technologies were often applied with the main purpose of reducing labor costs, i.e., process innovations were given priority over product innovations by many German firms. The consequences have been both losses in foreign sales and increased import competition in several high-tech fields [Härtel and Langer (1984); Klodt (1984)].

In recent years much of the technological gap seems to have been made up. Intensified R&D efforts by firms, in conjunction with more aggressive marketing strategies, were indispensable for this catching-up process to happen, but it should also be noted that the West German high-tech industry benefited a great deal from the strong US dollar (until early 1985). However it happened, the technology challenge will not vanish in years to come. On the contrary, the new technologies carry an enormous potential for developing new markets for entirely new goods and services facing a high income elasticity of demand, for locational decentralization of economic activities, for inducing fresh investment, and for improving job opportunities in branches which are not themselves high-tech [Giersch (1982); Perez (1985)].

To put it another way: new technologies such as robotics, computers, and telecommunications will rapidly alter the production and employment structures as well as the organization of work life both in factories and offices of the West German economy. If one is to characterize future structural change in terms of

*Table 16.5.* Indicators of technological change in West Germany, the USA and Japan.

| *Technology* | *Year* | *West Germany* | *USA* | *Japan* |
|---|---|---|---|---|
| Integrated circuits: | | | | |
| Consumption per capita | 1977 | 5 | 13 | 8 |
| (value in US $) | 1980 | 13 | 21 | 19 |
| | 1985 | 19 | 53 | 39 |
| Share in world | 1982 | 6.5 | 48.9 | 25.9 |
| consumption (%) | 1985 | 5.0 | 56.0 | 21.0 |
| Share in world | 1982 | 2.0 | 69.5 | 23.4 |
| production (%) | 1985 | 1.9 | 66.8 | 26.7 |
| Electronic goods: | | | | |
| Share in world | 1980 | 9.1 | 40.0 | 15.7 |
| production (%) | 1986 | 7.2 | 41.8 | 16.3 |
| Industrial robots[a] | | | | |
| Installed units | 1974 | 0.1 | 1.2 | 1.5 |
| (thousand) | 1980 | 1.2 | 4.5 | 6.0 |
| | 1985 | 7.5 | 20.0 | 25.0 |
| Units per million | 1974 | 5 | 15 | 30 |
| employees | 1980 | 50 | 50 | 100 |
| | 1985 | 300 | 200 | 400 |
| NC machine tools: | | | | |
| Installed units | 1975 | 8 | 40 | 14 |
| (thousand) | 1980 | 25 | 80 | 28 |
| | 1985 | 40 | 100 | 120 |
| Units per million | 1975 | 300 | 400 | 250 |
| employees | 1980 | 950 | 800 | 500 |
| | 1985 | 1,600 | 1,000 | 1,000 |

[a]Here we define industrial robots as handling or working machines which are freely programmable with three or more motion axes.
Sources: OECD (1985) *Verband Deutscher Maschinen- und Anlagenbau*, Fachgemeinschaft Montage, Handhabung, Industrieroboter im VDMA (unpublished data); *Idem*, Fachgemeinschaft Werkzeugmaschinen im VDMA (unpublished data); Franzmeyer *et al.* (1986); own calculations and estimates.

new technologies, this might be sketched as follows: more small-scale and less large-scale production; more automation and less mechanization; more engineering skills (brainpower) and less routine (manpower); more product-orientation and less processing-orientation; more services and less manufacturing.

## 16.3. Prospects for Structural Change

### 16.3.1. Model and data

Structural change and economic growth are interrelated: the redeployment of production factors away from shrinking activities into expanding sectors promotes the growth of the whole economy, and a rapid pace of economic development facilitates structural adjustments. Since this chapter is concerned with future structural trends in West Germany, it states one specific direction of causality – namely, that the level of per capita income determines those trends.

Thus, the sectoral structure of the West German economy will be regarded as an endogenous variable, whereas the explanatory variable is assumed to incorporate the relevant, and often interacting, basic supply and demand factors underlying economic growth.

Beyond all doubt, there is a variety of factors determining shifts of the sectoral structure of an economy. Economic growth is only one among others. In West Germany, e.g., reconstruction after World War II, the integration into the European Common Market, the realignment of exchange rates in the early 1970s, the two oil price shocks in 1973–1974 and 1979–1980, and the rise and decline of the US dollar in the 1980s all might have had a significant impact on the performance of different industries. Analyses of past trends of structural change should obviously take account of those effects, as some indeed have done [Fels and Schmidt (1980); Schmidt *et al.* (1984)]. However, as such singular events are difficult to predict (if they are predictable at all), an assessment of future trends must concentrate on more permanent factors. A long list of econometric studies of structural change, which were pioneered by Chenery (1960) and include also the Kiel FSW study [Fels *et al.* (1971)], has revealed that the level of per capita income satisfied this requirement for long-term projections.

As is usual in this type of analysis, we derive the projections from "normal patterns" of structural change that occur in the process of economic growth. These patterns are measured by OLS regression analyses, using the following equation [Donges *et al.* (1986)]:

$$\ln\left(\frac{A_{ij}}{100 - A_{ij}}\right) = a + b\ Y_j + c\ \ln\ Y_j + D$$

where $A_{ij}$ = percentage share of industry $i$ in total output (or total employment) of country $j$; $Y_j$ = per capita income of country $j$; $D$ = raw material dummy, in order to take account of the impact on the industrial structure of a rich natural resource endowment in some sample countries. The "dummy" takes the value of 1.0 for Canada, Australia, the Netherlands and Norway, in addition to the UK since 1979; for all other countries the dummy variable is 0.

The data on sectoral structure were taken from OECD statistics on national accounts. We used the information regarding 24 different industries in 20 countries. Luxembourg has been excluded due to insufficient sectoral disaggregation of data. For an analysis of the production structure, the sectoral contributions to gross domestic product seem to be appropriate. The structure of labor input is measured by the sectoral shares in total employment. The annual data have been pooled for the period 1970–1982. As compared to simple cross-section analyses, this combination of cross-section and time-series data has two main advantages: (1) distortions of estimates due to singular events in particular years are reduced; (2) an overestimation of sectoral shifts is avoided, which could occur if pure cross-section parameters were used for projections over time because cross-section elasticities are higher than time-series elasticities [Kuznetz (1966)].

Per capita incomes as indicator of the sample countries' level of economic development are converted to US dollars at 1975 prices. In order to avoid distortions from short-term fluctuations of exchange rates, the international comparisons are calculated at purchasing power parities, which are taken from Summers and Heston (1984). This source provides information on the period up to 1980. For more recent years, the data on per capita income were derived from IMF statistics (current issues).

### 16.3.2. International "normal patterns" at a sectoral level

For the three major sectors of the economy, the estimated coefficients are given in *Table 16.6.* The regression results for 20 branches are presented in Fels *et al.* (1971).

*Table 16.6.* Estimated coefficients for the relationship between the sectoral structure and the level of per capita income (OECD sample)[a]

| *Share of:* | $a$ | $b \times 10^{-3}\ Y_j$ | $c \ln Y_j$ | $D$ | $\bar{R}^2$ | $F$ | $n$ |
|---|---|---|---|---|---|---|---|
| GDP | | | | | | | |
| Primary | 15.48 | 0.18 | −2.22 | 0.55 | 0.65 | 138.4 | 225 |
| sector | (6.72) | (2.35) | (−7.05) | (8.32) | | | |
| Secondary | −12.36 | −0.34 | 1.59 | −0.25 | 0.41 | 56.7 | 243 |
| sector | (−11.11) | (−9.47) | (10.49) | (−7.60) | | | |
| Tertiary | −0.37 | −0.10 | 0.01 | 0.07 | 0.55 | 93.3 | 232 |
| sector | (−0.43) | (3.49) | (0.06) | (2.84) | | | |
| Total employment | | | | | | | |
| Primary | 69.67 | 1.02 | −8.98 | −0.14 | 0.65 | 88.8 | 142 |
| sector | (6.03) | (3.75) | (−5.92) | (−1.56) | | | |
| Secondary | −27.22 | −0.74 | 3.57 | −0.15 | 0.43 | 36.8 | 141 |
| sector | (−8.85) | (−9.47) | (8.76) | (−3.75) | | | |
| Tertiary | 7.58 | 0.44 | −1.15 | 0.23 | 0.67 | 92.7 | 137 |
| sector | (2.29) | (5.20) | (−2.62) | (6.66) | | | |

[a]Coefficients for the equation

$$\ln \left( \frac{A_{ij}}{100 - A_{ij}} \right) = a + b \times 10^{-3}\ Y_j + c \ln Y_j + D$$

The coefficient of determination $\bar{R}^2$ has been adjusted for the degrees of freedom; $F$ gives the $F$-value for the whole regression; the figures in brackets refer to $t$-values; $n$ is the number of observations.

Each of the regressions for the three major sectors is statistically significant, as indicated by the $F$-values; according to the $t$-statistics most of the coefficients are significant, too. This lends support to the fundamental hypothesis that there is a close statistical relationship between structural change and economic development. The estimated equations, therefore, could be regarded as the description of a kind of "normal" changes in the economic structure of the OECD countries, as per capita income rises.

A graphic illustration of this pattern could help in interpreting the coefficients. The lines in *Figure 16.1* represent the regression results for the production structure. According to these results an economy faces a permanent decline of its primary sector during the process of economic growth (Engel's law effects), whereas the secondary sector expands its share as per capita income rises. Beyond a \$4,500 US income level, the secondary sector also shrinks. Such a level of income was achieved in West Germany in the late 1960s; nowadays the per capita income in Italy has reached this point. The tertiary sector shows a continuously increasing share in gross domestic product. These results obviously fit the three-sector hypothesis described above.

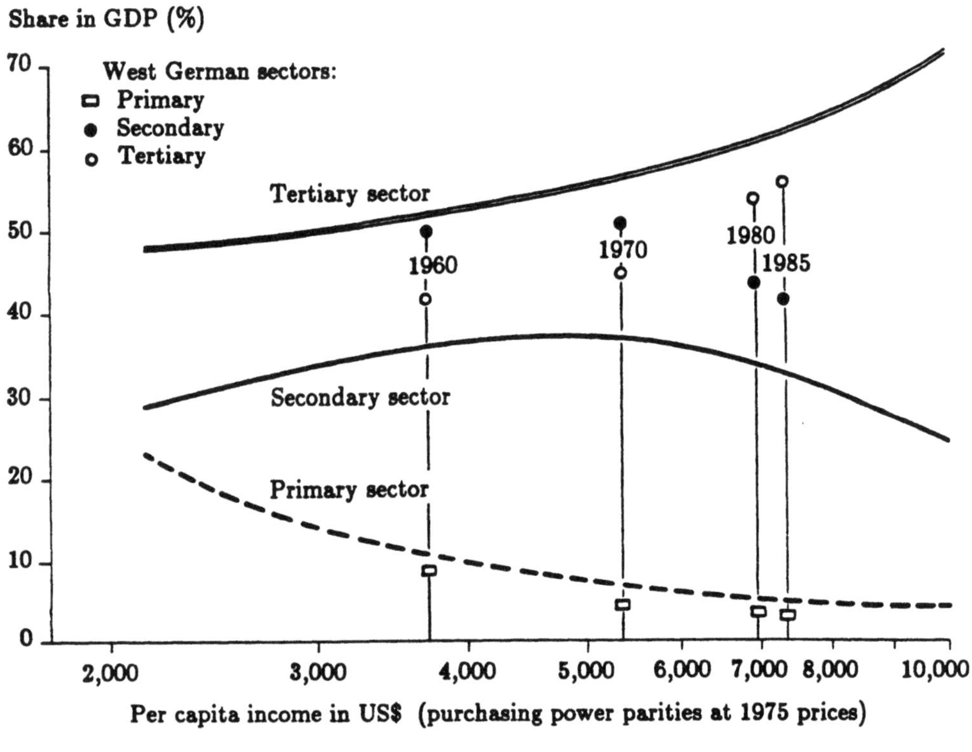


*Figure 16.1.* Per capita income and sector shares in gross domestic product for an OECD sample. (Sources as for *Table 16.1.*)

Furthermore, *Figure 16.1* shows the deviations of the West German economy from "normal patterns" for selected years. As in other OECD countries, the relative importance of the primary sector has diminished, the tertiary sector was constantly growing and the secondary sector has passed a maximum. It should be noted, however, that the level of the secondary sector's share is still substantially above the "normal" levels, whereas the primary and tertiary sectors are of a comparatively small size. This deviation from the international average, already emphasized in the FSW study, was reduced during the 1970s, when structural adjustment in West German industry consisted, to a large extent, in scaling down the level of activity. In the 1980s, the pace of shrinkage slackened, and the gap between the sectoral structure of West Germany and the international average has not been narrowed.

### 16.3.3. Projected structural changes in West Germany

The production and employment structures that one can conceive for the West German economy in the year 2000 are sensitive, according to the model used, to the rate of growth of per capita income. Three alternative scenarios have been chosen:

(1) In a "base scenario" we assume that gross domestic product per capita, in real terms, will increase at an annual rate of 3%. This assumption is, by and large, consistent with the growth trend recorded during the past 25 years and implies that the policy reforms initiated in recent years will be maintained.

(2) In a more "pessimistic scenario" the annual rate of growth is assumed to be 1% lower than in the "base scenario". The underlying premise is that inappropriate domestic macro- and microeconomic policies generate distortions, which hold back economic growth.

(3) In a more "optimistic scenario" we expect significant policy improvements domestically, great headway in completing the internal common market in the EC and a substantial trade liberalization resulting from the new GATT round of multilateral negotiations (the "Uruguay Round"), which would stimulate a faster growth rate than in the "base scenario". The projected rate of growth is 4%.

The main tools for projecting structural trends are the regressions for "normal patterns" discussed above. Nevertheless, it would surely be inappropriate to base the forecasts on those regression results alone. This would imply, for example, the prediction that West Germany's primary sector is going to increase its weight in the economy, since the estimated share of this sector in gross domestic product at high income levels is about 4%, whereas the actual share in West Germany is below 3% (*Table 16.1*). There is nothing in the foreseeable coal-mining and agricultural policies which could lead us to expect a reversal in the historical structural trend of this sector.

Therefore, the results of the regression analyses were adjusted according to information available from other Kiel research projects on structural change [Fels and Schmidt (1980); Schmidt *et al.* (1984 and 1986)]. ***Table 16.7*** shows the predicted sectoral shares in gross domestic product and in labor input for the three scenarios in terms of overall economic growth as well as the actual shares in selected previous years.

*Table 16.7.* West Germany: Sectoral and branches shares (%) of GDP and employment.

| | Gross domestic product | | | | | Employment | | | | |
|---|---|---|---|---|---|---|---|---|---|---|
| | Actual | | Predicted for 2000 [a] | | | Actual | | Predicted for 2000 [a] | | |
| *Sectors/branches* | *1970* | *1985* | *A* | *B* | *C* | *1970* | *1985* | *A* | *B* | *C* |
| Primary sector[b] | 4.6 | 2.8 | 2.5 | 2.3 | 2.3 | 9.7 | 6.3 | 4.3 | 3.7 | 3.6 |
| Secondary sector[b] | 50.4 | 41.1 | 33.0 | 30.2 | 28.8 | 47.7 | 40.1 | 32.1 | 29.8 | 28.5 |
| Tertiary sector[b] | 45.0 | 56.1 | 64.5 | 67.5 | 68.9 | 42.6 | 53.6 | 63.6 | 66.5 | 67.9 |
| Agriculture, forestry and fishing | 3.4 | 1.7 | 1.5 | 1.4 | 1.4 | 8.5 | 5.4 | 3.5 | 3.0 | 2.9 |
| Mining and quarrying | 1.3 | 1.1 | 1.0 | 0.9 | 0.9 | 1.2 | 0.9 | 0.8 | 0.7 | 0.7 |
| Manufacturing: | 40.2 | 33.2 | 27.5 | 25.2 | 24.0 | 38.1 | 32.0 | 27.2 | 25.1 | 23.9 |
| Intermediate goods | 11.8 | 8.8 | 6.2 | 5.5 | 5.0 | 8.4 | 6.5 | 4.7 | 4.2 | 3.8 |
| Investment goods | 15.9 | 15.7 | 14.8 | 14.0 | 13.7 | 16.3 | 15.9 | 15.1 | 14.3 | 14.0 |
| Consumer goods | 12.5 | 8.7 | 6.5 | 5.7 | 5.3 | 13.3 | 9.6 | 7.4 | 6.6 | 6.1 |
| Electricity, gas and water | 2.2 | 2.8 | 2.5 | 2.3 | 2.2 | 0.9 | 1.1 | 0.9 | 0.9 | 0.9 |
| Construction | 8.0 | 5.1 | 3.0 | 2.7 | 2.6 | 8.7 | 7.0 | 4.0 | 3.8 | 3.7 |
| Wholesale and retail trade, restaurants and hotels | 11.4 | 10.9 | 9.7 | 9.4 | 9.2 | 15.1 | 16.2 | 15.9 | 15.7 | 15.5 |
| Transport, storage and communications: | 5.9 | 5.9 | 6.6 | 7.4 | 7.7 | 5.3 | 5.6 | 6.4 | 7.1 | 7.4 |
| Transport and storage | 4.1 | 3.4 | 3.1 | 2.9 | 2.7 | 3.6 | 3.6 | 3.3 | 3.0 | 2.8 |
| Communications | 1.8 | 2.5 | 3.5 | 4.6 | 5.0 | 1.7 | 2.0 | 3.1 | 4.1 | 4.6 |
| Finance, insurance, real estate, and business services: | 10.7 | 15.9 | 21.6 | 22.4 | 23.4 | 4.3 | 6.2 | 12.2 | 13.2 | 14.1 |
| Financial institutions | 2.6 | 4.5 | 5.2 | 5.5 | 5.7 | 1.5 | 2.3 | 3.0 | 3.3 | 3.5 |
| Insurance | 0.7 | 1.1 | 1.4 | 1.4 | 1.5 | 0.7 | 0.8 | 1.2 | 1.2 | 1.3 |
| Real estate and business services | 7.4 | 10.3 | 15.0 | 15.5 | 16.2 | 2.1 | 3.1 | 8.0 | 8.7 | 9.3 |
| Personal services, other private producers | 7.2 | 11.8 | 15.2 | 17.1 | 17.7 | 6.7 | 9.6 | 13.3 | 14.8 | 15.4 |
| Government services | 9.7 | 11.6 | 11.4 | 11.2 | 10.9 | 11.2 | 16.0 | 15.8 | 15.7 | 15.5 |

[a] The predicted percentages of the year 2000 refer to three different assumptions on the expected annual growth rate of per capita income: A = 2%; B = 3%; C = 4%. For method of projection, see text.
[b] See *Table 16.1* for definition. Source: As for *Table 16.1*.

At the three-sector level, the primary sector will continue to lose relative weight in the West German economy, both in terms of value added and employment. Moreover, there will be a fundamental shift in production and employment from the secondary to the tertiary sector, and this the more so the faster the economy grows. This is consistent with the ongoing move toward an information- and communication-based economy.

At a more disaggregated level, manufacturing and construction emerged as the main losers. In the case of construction, the decline of population in combination with a comparatively high degree of saturation in housing and road building provide a reasonable explanation for the decrease. In the case of manufacturing, a major impact might come from the unrelenting industrialization effort in the Third World and the concomitant changes in the international division of labor; hence, consumer goods industries will face a faster relative shrinkage than investment goods industries.

The expanding sectors of the future are to be found among service industries. But not all of the service industries are expected to increase their share in gross domestic product or in employment. The predicted shares of wholesale and retail trade, restaurants and hotels and of transport and storage are below the level of 1985. The major winners are communications, business services and personal services. It should be noted that the share of these activities in employment increases, which should help to dissipate apprehensions (quite widespread in this country) about the job-killing effects of the new revolutionary technologies

## References

Chenery, H.B. (1960), Patterns of Industrial Growth, *American Review*, **50**, pp. 624–654.

Donges, J.B. (1984), Die Exportorientierung der deutschen Wirtschaft: Erfahrungen, Probleme, Perspektiven, pp. 11–41 in: E. Dichtl and O. Issing (eds.), *Exporte als Herausforderung für die deutsche Wirtschaft.* (Köln: Deutscher Instituts-Verlag).

Donges, J.B., Klodt, H., and Schmidt, K.-D. (1986), *The West German Economy Toward the Year 2000. An Analysis of Structural Change*, Kiel Working Papers No. 268 (May).

Fels, G., Schatz, K.-W., and Wolter, F. (1971), Der Zusammenhang zwischen Produktionsstruktur und Entwicklungsniveau. Versuch einer Strukturprognose für die westdeutsche Wirtschaft, *Weltwirtschaftliches Archiv*, **106**, pp. 240–278.

Fels, G. and Schmidt, K.-D. (1980), *Die deutsche Wirtschaft im Strukturwandel.* (Tübingen: J.C.B. Mohr, for the Kiel Institute of World Economics).

Franzmeyer, F. *et al.* (1986), *Strukturpolitische Konzeptionen für die Industrie im internationalen Vergleich.* (Berlin: Deutsches Institut für Wirtschaftsforschung).

Giersch, H. (ed.) (1982), *Emerging Technologies: Consequences for Economic Growth, Structural Change, and Employment.* (Tübingen: J.C.B. Mohr, for the Kiel Institute of World Economics).

Härtel, H.-H. and Langer, C. (1984), *Internationale Wettbewerbsfähigkeit und strukturelle Anpassungserfordernisse.* (Hamburg: Verlag Weltarchiv).

Klodt, H. (1984), Deutsche Investitionsgüter auf dem Weltmarkt – Ist die Wettbewerbsfähigkeit bedroht? *Die Weltwirtschaft*, No. 1, pp. 64–78.

Kuznets, S. (1966), *Modern Economic Growth: Rate, Structure, and Spread.* (New Haven, CT: Yale University Press).

Lawrence, R.Z. (1986), International Competitiveness as a Constraint on European Growth. Paper presented at the Brookings Institution Conference on Impediments to European Economic Growth, 9–10 October, Washington, DC.

OECD (1985), *The Semiconductor Industry. Trade Related Issues.* (Paris).

Perez, C. (1985), Microelectronics, Long Waves and World Structural Change: New Perspectives for Developing Countries, *World Development*, **13**, pp. 441–463.

Schatz, K.-W. and Wolter, F. (1982), *International Trade, Employment and Structural Adjustment: The Case Study of the Federal Republic of Germany.* International Labor Organization, World Employment Programme Research Working Paper No. 19, October.

Schmidt, K.-D. (1984), *Arbeitsmarkt und Bildungspolitik.* (Tübingen: J.C.B. Mohr, for the Kiel Institute of World Economics).

Schmidt, K.-D., *et al.* (1984), *Im Anpassungsprozess zurückgeworfen – Die deutsche Wirtschaft vor neuen Herausforderungen.* (Tübingen: J.C.B. Mohr, for the Kiel Institute of World Economics).

Schmidt, K.-D., Gundlach, E., and Klodt, H. (1986), *Im Strukturwandel vorangekommen?* Kiel Discussion Papers No. 122 (May).

Summers, R. and Heston, A. (1984), Improved International Comparisons of Real Product and Its Composition: 1950–1980, *Review of Income and Wealth*, **30**, pp. 207–262.

Wolter, F. (1984), From Economic Miracle to Stagnation: On the German Disease, pp. 95–121 in: A.C. Harberger (ed.), *World Economic Growth.* (San Francisco, CA.: Institute for Contemporary Studies).

CHAPTER 17

# Growth of the West German Economy: Forecast by the Bonn Model 11

*Wilhelm Krelle and Hermann Sarrazin*


## Summary

The Bonn Model 11 solutions for the FRG are compared with those of the Bonn–IIASA world model. The models were constructed for different purposes and according to different principles. In general, the forecasts of the Bonn Model 11 are more pessimistic than those of the world model.


## 17.1. The Bonn Model 11

The model consists of 480 equations, 151 of which are behavior equations. It is a yearly model constructed to deliver medium-term forecasts. *Table 17.1* indicates the submodels and the number of definitional and stochastic equations within these submodels. The table shows that the model emphasizes the government sector and the social insurance system. The administrative regulations are modeled in detail, following the administrative laws and rules of behavior.

The model as it stands now is a pure economic one in the sense that political decisions as well as the variables describing the world economy are taken as exogenous. There are four important ***domestic exogenous variables***:

(1) The nominal wage rate agreed upon by employers and trade unions, which is the result of a bargaining process that has not been modeled. The actual wage rate differs from it by the wage drift, which is endogenous. For the forecasting period we assumed a yearly rate of growth of 4.5% of the nominal standard wage rate.

(2) Public investment at current prices, which depends on political decisions that cannot be explained only by the economic situation. For the forecasting period we assumed a yearly rate of growth of 5.4% of the value of government investment.
(3) The central bank's supply of high powered money. For the forecasting period we assumed a yearly rate of growth of 5%.
(4) The discount rate, which depends on the monetary policy of the Bundesbank. For the forecasting period it is kept constant at the 3.5% level.

*Table 17.1.* Number of behavioral and definitional equations in the submodels of the Bonn Model 11.

| *Submodel* | *Behavioral equations* | *Definitional equations* | *Total* |
|---|---|---|---|
| Domestic product and national income | 13 | 42 | 55 |
| Private final demand | 5 | 20 | 25 |
| Prices | 17 | 39 | 56 |
| Capital stocks, capacity and utilization ratio | 2 | 11 | 13 |
| Labor market | 12 | 44 | 56 |
| Foreign trade and balance of payments | 12 | 35 | 47 |
| Monetary sector | 29 | 8 | 37 |
| Public sector including | 61 | 130 | 191 |
| Total public sector | (1) | (46) | (47) |
| Federal, state, local government | (34) | (40) | (74) |
| Social insurance system | (26) | (44) | (70) |
| Total | 151 | 329 | 480 |

Tax rates and the rates of social security contributions are also held at the actual level.

There are four important *foreign exogenous variables*:

(1) The real world market demand. We chose total world imports (in billions of 1975 dollars) as an indicator and used the forecasts of the world model; see *Table 20* in Annex 3.
(2) The price level on the world market. We chose as an indicator the price level of total world imports (in $) as forecast by the world model. This series is not reproduced in Annex 3. We present the growth rates in *Table 17.2.*
(3) The foreign interest rates. We kept the long-term foreign interest rates constant at 8%, the short-term and the Eurodollar rate at 6%.
(4) The exchange rate index of the DM with respect to all other OECD currencies. We assumed a 1% appreciation of the DM per year. This, of course, is a very crude approximation which is only used for the purpose of comparison. Otherwise, we use a special subsystem to determine the exchange rates; see Krelle and Sarrazin (1985).

*Table 17.2.* Rate of inflation on the world market = rate of growth of the price index of total world imports.[a]

| *Year* | *Rate of inflation* (%) | *Year* | *Rate of inflation* (%) |
|---|---|---|---|
| 1980 | 18.18 | 1990 | 3.86 |
| 1981 | –2.97 | 1991 | 4.36 |
| 1982 | –3.58 | 1992 | 4.74 |
| 1983 | –0.51 | 1993 | 5.04 |
| 1984 | –1.02 | 1994 | 5.25 |
| 1985 | –6.37 | 1995 | 5.42 |
| 1986 | –2.89 | 1996 | 5.53 |
| 1987 | 1.11 | 1997 | 5.60 |
| 1988 | 2.51 | 1998 | 5.64 |
| 1989 | 3.29 | 1999 | 5.66 |

[a]Estimated by the Bonn–IIASA world model, see Annex 2. Up to 1984: observation; from 1985–1999: forecasts.

Thus, the Bonn Model 11 is only connected with the world model insofar as it takes over the world market forecasts from this model. In all other respects it works independently.

It is not possible to describe the real part of the model in detail, due to lack of space. We can only briefly sketch the most important behavior equations. The consumption function follows the permanent income hypothesis; the investment functions (as far as equipment and nonresidential structures are concerned) are of the neoclassical Coen-type. The model explains investment in residential structures, in nonresidential structures, in equipment and in inventories. The perpetual inventory method is used for calculating the corresponding capital stocks. They form an important link between the demand and supply sides of the economy, because the maximal production capacity of the private sector depends on the gross real capital stock and on the corresponding productivity.

Prices are determined by cost–push and demand–pull elements, the latter being approximated by a measure of the capacity utilization. The value-added tax burden has been considered separately for the various components.

Gainfully occupied persons are subdivided into persons employed in the private and in the public sector, the first group again into employed and self-employed persons. Employment (in manhours) in the private sector follows from real production (which is determined by real demand) and from a labor coefficient. The trend of the labor coefficient depends on the trend of the factor price ratio and on the state of technology which in turn is explained by the average capital/labor ratio and a time trend. Short-term fluctuations of the labor coefficient during the business cycle are taken care of by introducing the growth factor of private GNP as an additional explanatory variable. Employment (in number of employees) follows from employment in manhours and from the average yearly working hours per employee which varies over the cycle and is explained endogenously. The actual labor supply results from the product of the potential labor force (exogenous) and of the corresponding labor participation rate. This rate is determined by a behavioral equation. It takes into account hidden unemployment by using the "discouraged" and "additional" worker hypothesis. The number of unemployed follows from the labor market identity as a residual.

Imports and exports are subdivided into goods, capital services, labor services, and other services. They are explained by domestic and world market prices, the exchange rate, appropriate activity variables and other items.

The monetary part of the model explains the allocation of financial wealth and various domestic interest rates. The holdings of money, call deposits, saving and time deposits of private households, of enterprises and of the government sector are estimated by behavioral equations which follow from portfolio theory. The ratio of each of these assets to the total wealth of the sector in question depends on interest and inflation rates, activity variables and other items. The monetary part of the model explains different domestic interest rates on the basis of the discount rate, the free liquid reserves of the commercial banks, foreign interest rates, the inflation rate and other items. For details, see Krelle and Sarrazin (1985) and Sarrazin (1986).

## 17.2. Solution of the Bonn Model 11

We solved the model for three world trade scenarios – optimistic, medium and pessimistic – which coincide with the solutions of the world model for the optimistic, medium and pessimistic scenarios. Of course, this does not mean that solutions of the model for the FRG, which is part of the world model, and those of the Bonn Model 11 are consistent. The underlying assumptions are different for the two models in other respects. If they arrived at similar solutions nevertheless, we would feel encouraged to accept the results.

*Table 17.3* shows some comparisons. We see from *Table 17.3(a)* that the Bonn Model 11 forecasts more modest growth rates. The reason is that the rate of technical progress is endogenous in the Bonn Model 11 and declines slowly in the future (as it did in the past), whereas the rate of technical progress is exogenous in the world model and is supposed to recover from the decline in the past (see *Figure 2.1* in Chapter 2). This may be seen from *Table 17.3(d)*: the growth rate of labor productivity decreases in the Bonn Model 11 whereas it increases in the world model, with the exception of the pessimistic case. The results of the pessimistic scenario are rather similar in both models as far as the growth rates of GDP and of labor productivity are concerned. The assumptions of the optimistic scenario in the world model seem to be too optimistic. Thus, we should concentrate on the medium and pessimistic scenarios.

In *Table 17.3(c)*, the consequences of the different approaches with respect to the exchange rates used in the models become visible. In the Bonn Model 11 the exchange rate is an exogenous variable. Thus we are in a regime of fixed exchange rates. A higher world market demand (as in the medium and optimistic scenarios) yields higher prices [see *Table 17.3(b)*] but cannot influence the exchange rate. Therefore the current account surplus is larger in the optimistic case, and the deficit is larger in the pessimistic case, than in the world model where the exchange rate is endogenous and shields the country from external influences to a certain degree. Given the actual exchange rate system now, the

*Table 17.3.* Comparison of forecasts: the Bonn Model 11 versus the Bonn–IIASA World model.

| | *Bonn Model 11 scenario* | | | *Bonn–IIASA World Model scenario* | | |
|---|---|---|---|---|---|---|
| | *Optimistic* | *Medium* | *Pessimistic* | *Optimistic* | *Medium* | *Pessimistic* |
| (*a*) Real GDP growth rates (%) | | | | | | |
| 1985 | 2.6 | 2.6 | 2.6 | 3.5 | 2.9 | 2.3 |
| 1990 | 2.7 | 2.2 | 1.5 | 3.9 | 3.0 | 2.1 |
| 1995 | 2.3 | 2.0 | 1.4 | 4.0 | 3.0 | 1.9 |
| 1999 | 2.4 | 2.0 | 1.4 | 4.0 | 3.0 | 1.9 |
| (*b*) Rate of inflation = rate of growth of GDP deflator (%) | | | | | | |
| 1985 | 2.2 | 2.2 | 2.2 | 3.5 | 3.9 | 4.3 |
| 1990 | 3.2 | 2.6 | 2.1 | 2.2 | 3.4 | 4.5 |
| 1995 | 4.0 | 3.1 | 2.2 | 1.1 | 2.7 | 4.2 |
| 1999 | 3.6 | 3.2 | 2.3 | 1.0 | 2.5 | 4.1 |
| (*c*) Foreign trade balances (billion DM)[a] | | | | | | |
| 1985 | 44.6 | 44.6 | 44.6 | 48.6 | 25.1 | 5.7 |
| 1990 | 75.5 | 38.0 | 3.9 | 99.8 | 48.7 | 6.5 |
| 1995 | 169.2 | 44.7 | –58.4 | 106.9 | 42.3 | –6.6 |
| 1999 | 319.9 | 78.5 | –148.4 | 116.8 | 39.1 | –15.9 |
| (*d*) Growth rate of GDP per working hour = growth rate of labor productivity (%) | | | | | | |
| 1985 | 2.8 | 2.8 | 2.8 | 4.0 | 3.3 | 2.7 |
| 1990 | 2.1 | 2.5 | 2.8 | 4.3 | 3.4 | 2.5 |
| 1995 | 1.3 | 1.8 | 2.3 | 4.4 | 3.4 | 2.4 |
| 1999 | 1.3 | 1.5 | 2.0 | 4.4 | 3.4 | 2.3 |

[a] For the Bonn Model 11: current account balance [Leistungsbilanzsaldo]. For the Bonn–IIASA World Model: trade balance, including services, estimated by multiplying the current $ trade balances (including services) from *Table 25* in Annex 3 with the $/DM exchange rate forecast in *Table 29*.

results of the world model are more realistic than those of the Bonn Model 11. But even here corrective actions would be taken if the current account balance increased as in the optimistic case or decreased as in the pessimistic case. As already stated in Chapter 2 and as is apparent from *Table 29* in Annex 3, the preliminary exchange rate equations used in the world model should be replaced by more sophisticated ones along with lines of the exchange rate system which we originally wanted to use and which is presented in Chapter 8.

The effects of the different approaches to explain the price level are visible in *Table 17.3(b)*. We assume in both models that the monetary authorities follow the monetarist line and fix the rate of money supply in advance. In the world model the price level is explained by Fisher's equation. Therefore, an increase of GDP will decrease the price level, if the velocity of money does not change too much. This comes true in the forecasts of the world model: the more optimistic scenarios yield lower rates of inflation. In the Bonn Model 11 the general price level follows from prices which are determined by costs and by a mark-up which varies in relation to the degree of capacity utilization. Costs are

*Table 17.4.* Unemployment forecasts from the Bonn Model 11.

| | *Optimistic scenario* | | *Medium scenario* | | *Pessimistic scenario* | |
|---|---|---|---|---|---|---|
| *Year* | *Unemployed (millions)* | *Rate* (%)[a] | *Unemployment (millions)* | *Rate* (%)[a] | *Unemployment (millions)* | *Rate* (%) |
| 1985 | 2.30 | 8.27 | 2.30 | 8.27 | 2.30 | 8.27 |
| 1986 | 2.25 | 8.05 | 2.35 | 8.41 | 2.43 | 8.72 |
| 1987 | 2.23 | 7.97 | 2.46 | 8.82 | 2.67 | 9.58 |
| 1988 | 2.10 | 7.46 | 2.45 | 8.76 | 2.77 | 9.96 |
| 1989 | 1.93 | 6.99 | 2.45 | 8.75 | 2.88 | 10.36 |
| 1990 | 1.81 | 6.38 | 2.41 | 8.59 | 2.96 | 10.68 |
| 1991 | 1.59 | 5.59 | 2.31 | 8.22 | 2.98 | 10.77 |
| 1992 | 1.39 | 4.88 | 2.23 | 7.94 | 3.02 | 10.93 |
| 1993 | 1.17 | 4.10 | 2.12 | 7.56 | 3.03 | 10.99 |
| 1994 | 0.90 | 3.15 | 1.95 | 6.98 | 2.98 | 10.88 |
| 1995 | 0.63 | 2.20 | 1.78 | 6.37 | 2.93 | 10.74 |
| 1996 | 0.35 | 1.24 | 1.59 | 5.70 | 2.97 | 10.55 |
| 1997 | 0.09 | 0.32 | 1.40 | 5.04 | 2.81 | 10.37 |
| 1998 | 0.00 | 0.00 | 1.20 | 4.30 | 2.73 | 10.13 |
| 1999 | 0.00 | 0.00 | 0.92 | 3.50 | 2.64 | 9.83 |

[a]With respect to the working population (Erwerbspersonen).

predominantly labor costs. They also increase if demand goes up. Thus, we have the opposite effect: in the optimistic case the rate of inflation is higher than in the pessimistic one. The Bonn Model 11 seems to be more realistic than the world model in this respect. The world model does not have a submodel to explain the nominal wage rate for the OECD countries. This implies that this rate is supposed to have no influence on the price level, and this means that money wages follow the value of the marginal product of labor.

The world model does not provide unemployment figures directly. They are interesting, of course. Thus, we present these figures as forecast by the Bonn Model 11 in *Table 17.4*.

We see that in the optimistic case unemployment will disappear only in the middle of the 1990s. In the medium scenario it will decrease very slowly, but still be present at the end of this century. In the pessimistic scenario the unemployment rate will still increase and only decrease from the middle of the 1990s, but not much. Thus, in each case we have to live with unemployment for a long time if the regulations and institutions of the labor market are not changed substantially, which is not assumed in our model.

## 17.3. Conclusions

The comparison of the solutions of both models shows that the models forecast qualitatively similar developments of the German economy, with the exception of the rate of inflation. The Bonn Model 11, in which the rate of technical progress is an endogenous variable, comes up with more pessimistic forecasts since the rates of technical progress assumed in the world model are higher. It is apparent that the exchange rate equations should be improved in both models and that the nominal wage rate should be introduced into the world model for OECD countries. The ranges of possible future development of the West German economy which are spanned by the world model include, by and large, those which are covered by the Bonn Model 11, with the notable exception of the current account balance for which the world model yields much tighter bounds.

## References

Krelle, W. (ed.) (1986), *Ökonomische Prognose-, Entscheidungs- und Gleichgewichtsmodelle.* Ergebnisse aus dem gleichnamigen Sonderforschungsbericht der Universität Bonn, Weinheim, VCH.

Krelle, W. and Sarrazin, H.T. (1985), Simultaneous Determination of Capital Flows, the Exchange Rate and Interest Rates in the Bonn-Model 11, pp. 142–161 in: H. Schneeweiss and H. Strecker (eds.), *Contributions to Econometrics and Statistics Today.* (Berlin and Heidelberg: Springer).

Sarrazin, H.T. (1986), Das Bonner Modell 11, pp. 76–88 in: W. Krelle (ed.), *Ökonomische Prognose-, Entscheidungs- und Gleichgewichtsmodelle.* Ergebnisse aus dem gleichnamigen Sonderforschungsbericht der Universität Bonn, Weinheim, VCH.

CHAPTER 18

# Strategic Planning and Economic Growth in Bulgaria

*O. Panov and J. Djarova*

## Summary

The main characteristics and factors in the economic development of Bulgaria, as well as their manifestation over a period of nearly 30 years, are considered. The results of a macroeconomic model can serve the goals of strategic planning, enabling an analysis and forecasting of various national policies that show promise of leading to economic growth.

## 18.1. Introduction

The problem of maintaining high and steady rates of growth is a problem for every economy at each stage of its development. Economic growth is one of the most important criteria of an economic system's vigor and of its capacity for progressive development. Basically, economic growth relates to three factorss of production, namely: labor resources, production funds, and natural conditions and wealth. While in the beginning of its socialist development, mainly in order to speed up its industrialization, Bulgaria paid special attention to the management of production volume and to the coordination of the factors of economic growth, during the past 15 years economic policy has been reoriented to reveal and activate qualitative impacts. For the increase of the final social product, priority has been given to labor productivity, to the effective use and development of productive resources and to the fulfillment of the individual and collective needs in society.

Combining the qualitative and quantitative aspects to study economic growth is a necessary condition for reaching the right results and conclusions and for making meaningful management decisions. The goal set to achieve economic growth for our country can be defined in the widest terms as a striving after fuller satisfaction of the material, intellectual and cultural needs of society. This will be fulfilled by growth of production and productive resources and, particularly, by the efficiency of their utilization and the distribution of the social product in line with the social economic tasks at the economy's stage of development and the goals of society as a whole.

When the goal is to plan and manage the process of economic growth, the mere analysis and management of the main production factors is not sufficient. The factors of production show up in a specific way; in a particular environment they create the driving forces of economic growth. Organizational and management factors play an important role in this environment. The organizational factors are mainly related to economic structures, while the management factors are related to the strategic planning of the socioeconomic development of the country. Strict differentiation is not possible, just as it is not possible to separate the economic factors of production from one another. The type of structures, their dynamics and structural changes are an outcome of strategic decisions. At each stage the strategy has to conform to the actual state of the structures. The two types of factors of economic growth – the productive factors and organizational–management factors – do not exist independently of each other, either. The productive factors create an environment of structural changes and strategic planning. At the same time, we cannot speak of strategic planning or structural changes apart from the concrete state of production. Connections between the two types of factors turn economic growth into a process, organized and managed in a planned way.

## 18.2. Strategic Decisions – An Integrating Factor in Economic Growth

The management of economic growth as a planned dynamic and continuous process is mainly related to strategic decisions made toward the social and economic development of a country. Strategic planning reveals the main problems of each stage of economic development that are connected with the weakening of the impact of some factors or with the predominant influence of some factors of economic growth; one also seeks to discover the structures in the national economy that are most appropriate for reaching the goals of the country's economic development. The strategic decisions are made in response to the problems revealed, and form the nucleus of the country's social and economic policy.

As far as Bulgaria is concerned, some important strategic decisions can be shown to have influenced, and still to be positively influencing, economic growth. These decisions activated the impact of one or another factor at a certain period of time according to the concrete need.

The transition from extensive to intensive development emphasizes the importance of the qualitative aspects of the factors of economic growth, mainly: the efficient use of labor resources and fixed assets. The main goal of the strategy of introducing scientific–technical progress on a wide basis is to increase production efficiency. Maintaining high and steady rates is mainly related to the faster development and adoption of modern technologies in the country. A complete restructuring of the economy and of production is planned on the basis of computerization, of biotechnology and of new materials. The renovation of the present technical equipment of the country is the main tool to implement the goals of economic development in order to ensure a drastic increase of labor productivity, an improvement of product quality, a reduction of the material costs, environmental protection and, as a final result, the rapid increase of total productive efficiency.

At present, the strategy of scientific–technical development has a complex impact on the factors of economic growth. Qualitative changes in technology and in the organization of production increase labor productivity, productive efficiency and product quality. Changes in production technologies and new product development aim at solving the problems of reducing the use of raw materials, of other materials and of energy. Technology improvement is related to labor productivity growth to compensate for diminishing labor resources and to permit the restructuring of the labor force.

The development of the Bulgarian economy during the last 10 years, based mainly on the newest achievements of science and technology, can be characterized as an intensive development. The productivity of the new technology is nearly 40–50% higher than the average productivity in the country. Improved skills of labor contributed significantly to the increase in national income. Thus, at a constant assets/persons employed ratio, the growth of the national income as a result of scientific–technical progress is 65% due to the improvement of skills and 35% due to the higher productivity of the fixed assets [Angelov (1985)]. The accelerated renovation of production in recent years has led to the increase of the relative share of improved and new products in the volume of net material product from 5.5% in 1980, to 13.3% in 1985, and 8.9% in 1986 [*Statistical Reference Book* (1987)].

An important strategic document directly related to economic growth is the "Program for Raising the Living Standard" of the country. The relationship between social development (i.e., of man and society) and economic growth is bilateral. Accelerated social development is a condition for the full reproduction of labor resources and for increasing the efficiency of the persons employed in production. On the other hand, the increased efficiency of the labor force as a result of the influence of the social program accelerates economic growth. Studies show that the personal consumption of people employed in material production as an indicator of the living standard has a positive effect on economic growth. On the other hand, economic growth determines the dynamics of change of the two basic funds, "accumulation" and "consumption", and is a precondition for increasing personal consumption. Thus, during the recent 10 years at an average annual rate of growth of the national income of nearly 6.3%, the consumption fund has increased by 5.1% and the real income per capita has increased by 4.1% [*Factors of Efficiency...* (1985)].

The program of increasing living standards influences the labor force and helps to overcome the highly diminished potential for an extensive economic growth. It creates conditions for turning labor force efficiency into an active factor.

The regional allocation of productive factors – to develop the country's regions – is also of strategic importance for the economic development of the country. The regional policy, formed on the basis of a complete accounting for natural wealth and historical heritage and aims, has disclosed the possibilities for a most effective development, in spite of its complexity. The regional allocation of the productive factors restructures them in such a way as to bring production closer to labor resources and to sources of raw materials and energy. This increases production efficiency by coordinating the productive factors in a favorable way.

The regional policy's social orientation strengthens the impact of each region's concrete conditions for the reproduction of labor resources and improves the distribution of the financial resources for the socioeconomic development of the region. The complex influence of the regional development strategy is a premise for bringing all regional rates of growth closer together. The increased economic independence of the regions is a premise for the search of specific forms of economic and social vigor in the specific territory. Thus, the greater part of the small and medium-sized advanced technological enterprises in the country have been established in the regions.

It can be said, in summary, that the economic development policy of Bulgaria has provided dynamic balance of the social and economic goals of growth. According to the specific historical period, economic growth was the objective of many strategic decisions. These decisions created conditions for the mobilization of some factors of production and, above all, for increasing their efficiency.

## 18.3. Characteristics of Economic Growth in Bulgaria

Since its founding, the People's Republic of Bulgaria has achieved great successes in the development of its economy. The gross national product has increased more than 18 times, and the national income more than 14 times. Compared to 1952, in 1985 the fixed assets had increased more than 10 times, the foreign trade exchange more than 29 times, and exports alone over 30 times. The social productivity of labor has increased 15 times. Nearly all the annual growth of the national income is due to increased labor productivity. Industry now produces 69.9% of the GNP and 60.6% of the national income. As a result of the growth of the industries most closely connected with scientific and technical progress, machine-building and the chemical industry became the basic industries in the structural composition of industry. About 15% of the total industrial production is attributed to ferrous/nonferrous metallurgy, to the energy industry, and to the petrochemical industry [*Statistical Reference Book* (1986)].

The volume of production of the traditional Bulgarian sector – agriculture – has increased nearly 3 times in the period 1952–1985.

In the same period, social care increased substantially. Social funds per capita increased nearly 12 times, and real income per capita 4.3 times.

The analysis of the real economic growth in Bulgaria is very important for revealing the possibilities to control the factors of future growth. (All figures in the text relate to the People's Republic of Bulgaria, which was founded in 1944. Data and comparisons in the chapter refer to the period 1950–1985.) In our analysis, we use the dynamics of the national income as a main indicator of the economic growth. The factors to be considered in quantitative and qualitative terms include (1) size of labor resources and labor productivity, and (2) size of means of production and efficiency of use.

The dynamics of the country's national income are characterized by the following growth rates (based on data from the statistical annals of CMEA member-countries):

| | | | | | |
|---|---|---|---|---|---|
| 1951 | 0.357 | 1963 | 0.117 | 1974 | 0.070 |
| 1952/54 | 0 | 1964 | 0.089 | 1975 | 0.085 |
| 1953 | 0.182 | 1965 | 0.067 | 1976 | 0.063 |
| 1955 | 0.049 | 1966 | 0.110 | 1977 | 0.062 |
| 1956 | 0.016 | 1967 | 0.087 | 1978 | 0.053 |
| 1957 | 0.118 | 1968 | 0.058 | 1979 | 0.064 |
| 1958 | 0.067 | 1969 | 0.069 | 1980 | 0.055 |
| 1959 | 0.189 | 1970 | 0.073 | 1981 | 0.050 |
| 1960 | 0.073 | 1971 | 0.069 | 1982 | 0.043 |
| 1961 | 0.030 | 1972 | 0.073 | 1983 | 0.028 |
| 1962 | 0.019 | 1973 | 0.079 | 1984 | 0.044 |

These data suggest:

(1) By 1955, an increase of the national income at a very high rate.
(2) By 1960, a tendency toward a decrease in the rate of growth but still at a relatively high level.
(3) In 1961–1965, the first fall in the rate of growth.
(4) By 1970, an increase in the rate of growth of the national income.
(5) After 1970, a constant decrease of the rates of growth – they reached their lowest value at the end of 1970 and the beginning of 1980s.

The rates of growth of the national income are an indicator of the speed of growth, while its absolute growth determines the possibilities to enlarge production and to introduce qualitative changes in economic development. With the exception of 1952 and 1954, the absolute growth of the national income has been positive throughout the years of the country's development.

## 18.4. Economic Growth and Labor Resources

Labor resources as a factor of economic growth have their concrete features, connected with the biological laws of labor force reproduction and its cycle. The analysis of the demographic processes in Bulgaria in the past shows that economic growth has been only modestly influenced by the availability of labor (defined as persons employed in material production). There is a tendency for the number of those employed in material production to decrease in the period considered, excluding the years 1965–1975. The exhaustion of the quantitative impact of labor resources in the second half of the 1970s and the beginning of the 1980s influenced the growth of the national income and had its impact on the absolute decrease of growth. At the same time, in the largest material branch – industry – more than 40% of the national income growth produced in it can be attributed to the growth of the labor force. The positive impact of the extensive growth of those employed in industry, building, and transport is mainly a result of the restructuring of labor resources at the industry level.

The reasons for the smaller impact of the labor resources factor with respect to its size are the following:

(1) Decline of both the birth rate and of the death rate.
(2) Fall in the growth rate of the active population as a result of the slowing down in the rate of growth of the youngest age group (0–15) and its share in the total population, and as a result of a growth in the share of the age group over 60. (In Bulgaria the active age is 16–60 for men and 16–55 for women).
(3) An increase in the number of employees in the unproductive sphere at the expense of the total engaged in material production.

The trend toward a decreased impact of extensive labor force growth on economic growth can be observed in all developed countries, more or less. In this situation the role of labor efficiency becomes increasingly important. Productivity as a main measure of labor force efficiency depends directly on the level of skills, on the perfection of production and labor organization, on the technological level of production, and the right balance of those employed in industry compared to those employed in other sectors. In the period considered, average annual growth rates (%) of Bulgarian labor productivity were:

| | | | | | |
|---|---|---|---|---|---|
| 1951–1955 | 13.0 | 1961–1965 | 7.2 | 1971–1975 | 7.7 |
| 1956–1960 | 11.2 | 1966–1970 | 8.2 | 1976–1980 | 6.2 |

which had the following percentage effects on the growth of the country's national income (in %) [source: Angelov (1985)]:

| | | | | | |
|---|---|---|---|---|---|
| 1951–1955 | 105.7 | 1961–1965 | 105.1 | 1971–1975 | 96.2 |
| 1956–1960 | 114.4 | 1966–1970 | 83.3 | 1976–1980 | 96.2 |

Studies show that the ratio assets/persons employed in material production had a considerable effect on labor productivity; the first grew faster than the latter. The relationship of the two depends directly on the use of the production funds.

## 18.5. Economic Growth and Fixed Assets

Fixed assets are the basis for the growth of the national income (*Table 18.1*). Observations show that the fixed assets have increased in volume at relatively high rates according to the country's stage of development. The quantitative effect of that factor on the growth of the national income was, and still is, positive.

*Table 18.1.* Dynamics of the fixed assets and growth of the national income as a result of basic assets growth.

| *Indicator* | *1961–1965* | *1966–1970* | *1971–1975* | *1976–1980* |
|---|---|---|---|---|
| Growth of basic assets | 162.2 | 168.1 | 153.4 | 147.0 |
| Growth of national income based on growth of basic assets[a] | 162.0 | 131.0 | 117.0 | 135.0 |

[a]Estimated according to the formula: growth of fixed assets (in %) × 100 divided by the growth of national income (in %).
Source: Angelov (1985).

In terms of quantity, fixed assets have a direct impact on the volume of the final product; and in terms of quality, they influence the national income through the growth of labor productivity. An important characteristic of the efficiency of growth is the output/capital ratio in the material sphere measured as the ratio of the net material product to the value of the productive fixed assets. In spite of the fact that the qualitative factors of production are now the main source of the country's economic growth, the output/capital ratio is decreasing in the present stage. This tendency is mainly related to the following circumstances:

(1) The cost of the modern means of production is steadily growing. This creates a problem for the efficiency of the basic assets.
(2) The renovation of assets is not in correspondence with their age structure at each stage of economic development.

As a result, the age structure of the fixed assets deteriorates. Thus, in 1970 the renovation coefficient was 12.7%, in 1975 it was 6.57%, and in 1982 it reached 7.57%. To improve the quality of fixed assets, a strategy was pursued in the 1970s which gave priority to the use of investment for reconstruction and modernization. Some 70–75% of the total volume of investment in material production has been used for that purpose. But adoption of the new technologies is not always the rule, and sometimes it is not implemented on a high technological basis. Moreover, fixed assets are not always used to their capacities.

These factors, which influence the qualitative level of the fixed assets in a negative way, also have a negative effect on labor productivity, which is the main source of economic growth. Therefore, the policy of the country with respect to science and technology is directed toward overcoming the restriction of extensive growth of the fixed assets by influencing their quality level positively.

## 18.6. Forecasting the Economic Growth of Bulgaria

The main devices to support the strategic planning of economic growth are models based on production functions or functionals. In Bulgaria the IMPRESS model has been used as a tool for strategic planning. The name of the model is an acronym for Interactive Model for Forecasting the Economic Development of the Socialist countries (in Russian). The model is a result of joint work at the International Research Institute of Management Problems (in Moscow), the International Institute of Economic Problems of the Socialist Countries (in Moscow), and the Institute of Social Management (in Sofia.)

### 18.6.1. General model characteristics

IMPRESS differs from traditional models of economic growth based on the use of production functions and two main factors of growth – the number of persons employed in material production and the size of fixed assets – by the following features:

(1) The model incorporates as main factors the skilled labor force and the personal consumption of persons employed in material production. It allows for a quantitative expression of the social factors of labor productivity growth and identifies the effect of the complex factor of scientific and technical progress expressed in live labor. The nature of the impact of these factors on the issues of economic development is determined by the use of elasticity coefficients. They are nonlinear functions depending on many other factors.

(2) The growth path is described by a growth functional (18.1), in which each factor is defined by its own trajectory. The functional is a nonlinear relationship describing the connection between fixed assets growth rates and national income growth rates.

As a differential expression the functional can be presented as follows:

$$\frac{\mathrm{d}Y}{Y} + S_L \frac{\mathrm{d}L}{L} + S_K \frac{\mathrm{d}K}{K} + S_{L_1} \frac{\mathrm{d}L_1}{L_1} + S_c \frac{\mathrm{d}C}{C} \tag{18.1}$$

where $Y$ is the produced national income; $L$ is the number of persons employed in material production; $L_1$ is the number of specialists in higher education, employed in material production and research; $C$ is the size of the personal consumption of those employed in material production; $S_L$, $S_K$, $S_C$, $S_{L_1}$ are nonlinear functions of $L$, $K$, $C$, $L$, which express the elasticity of national income with respect to the corresponding variables.

The elasticity coefficients present nonlinear functions of various other factors (fixed assets/persons employed, personal consumption, etc.). They are mostly exponential functions of the following form:

$$S_L = 1 - S_K - S_{L_1} - S_C \tag{18.2}$$

$$S_K = 1 - e^{q_1 \cdot F_1} \tag{18.3}$$

$$S_{L_1} = q_2 \cdot F_1 \cdot e^{q_4 \cdot F_1 + q_5 \cdot F_3} \tag{18.4}$$

$$S_c = q_3 \cdot F_1 \cdot e^{q_4 \cdot F_1 + q_5 \cdot F_3} \tag{18.5}$$

where

$$F_i = q(u_1 \cdot f_i), \quad i = 1, 2, 3$$

$$q(x) = \begin{matrix} \ln x, & x \geq 1 \\ 0\ , & 0 \leq x < 1 \end{matrix}$$

$$f_1 = \frac{K}{L}, \quad f_2 = \frac{L_1}{L}, \quad f_3 = \frac{C}{L}$$

The decision (scenario) variables which determine the concrete solution of the model are:

(1) The share of personal consumption in national income.
(2) The share of productive investment (investment in the material sphere) in national income.

Apart from the hypotheses on the dynamics of these control parameters for the specific forecast, exogenous estimates of the demographic situation and of the skills of the workers in material production are also used.

Insofar as the model considers the change of the consumer's value expressed by the physical value of the corresponding indicator, constant prices are used. The parameters of the functional result from an estimation on the basis of the economic development of Bulgaria in 1950–1984 in real terms.

We tried different versions (scenarios). As a result some tentative forecasts have been obtained for the years 1985–2004. They relate to the following group of indicators:

(1) Absolute values, indexes and average annual rates of growth of national income, fixed assets, and personal consumption of the population.
(2) Absolute values and average annual rates of growth of the efficiency as return of funds (national income/fixed assets ratio, labor productivity, assets/persons employed in the material production ratio).
(3) The contribution of each growth factor to the growth of labor productivity.

### 18.6.2. Use of the model

The forecasts are obtained by determining different scenarios for the decision and exogenous variables. In designing these scenarios, the socioeconomic development of Bulgaria by the year 2000 has been considered, and a strategy of high intensification of production has been supposed as a basis for the increased role of the intensive factors of growth. Scientific–technical progress should become the leading factor of development. The strategy implies a continuous improvement of the living standard. Moreover, data from demographic forecasts to the year 2000 are used for the country.

Many versions of the model have been used. They form groups of characteristic versions. At a constant share of the national income used for personal consumption and for fixed assets in material production, different groups of scenarios have been experimented with:

(1) An increase every five years in the relative share of national income used for personal consumption.
(2) An increase every five years in the relative share of the national income used for fixed assets.
(3) Mixed scenarios where the strategies described in the first two postulates take turns every five years.

The model is solved for the period 1984–2000. The results presented below are given as a value index with respect to 1984.

The basic version follows the tendencies of the economic development that took place in 1950–1984. In 1990 it ensures an increase of the national income of about 20% of the social productivity of labor as well as a growth of investment of 33% and personal consumption per capita of 21%. It shows an increase of 1.62 times in the national income in the year 2000 with an average annual rate of growth of 2.5%. Labor productivity increases 1.57 times. According to this version, the tendency toward a fall in the average annual rates of growth of the national income continues, the return of funds (national income/fixed assets ratio) decreases but at a slower rate, and labor productivity grows more slowly than both personal consumption per capita and national income.

The forecasts are best if the relative share of national income used for personal consumption is maintained in the first five years, and then grows by 0.05% every year during the second five years, and by 0.1% every year until the end of the forecast period. By the year 2000 national income and labor productivity more than double as compared to 1985. Labor productivity grows faster than personal consumption per capita. The return of funds decreases, but at a slower rate, and stabilizes at the end of the period. All growth rates stabilize. The rate of growth of national income in the year 2000 is 4.2%. This coincides with the rate of growth observed in the year 1985.

Practical experience shows that strategic planning should be oriented to manage the key factors at each stage of the country's economic development. Forecasting models can serve as a basis for decision making.

## Acknowledgment

The authors express their warm gratitude for their successful collaboration to the Soviet scientists Professor Dr. Vladimir Samovol and Dr. Vitalii Molostvov.

## References

Angelov, T. (1985), Problems of Economic Growth in Bulgaria and in European Socialist Countries, *Nauka i Izkustvo* [in Bulgarian].

*Factors of Efficiency and Growth: Problems, Use and Reserves* (1985), [in Bulgarian].

*Intensification of Economic Growth: Problems, Tendencies, Solutions* (1984), [in Bulgarian].

Shimon, D. and Sanoval, B. (1981), On the Functional of Economic Growth, *Economica i matematicheskie metodi*, **17**, No. 2 [in Russian].

*Statistical Reference Book* (1986), Sofia.

*Statistical Reference Book* (1987), Sofia.

CHAPTER 19

# The Growth of the Polish Economy and Prospects for Structural Change

*Adam B. Czyżewski and Wladyslaw Welfe*

## Summary

Sources of economic growth and slump in the Polish economy in the 1970s and the early 1980s are analyzed, with major emphasis placed on the role of foreign trade. The model W-5 and the foreign debt submodel, briefly presented, are used to perform simulations and construct forecasts of economic growth and structural change during the next 10–15 years. Their dependence on foreign debt servicing is analyzed in some detail. Then the assumptions and results of the forecasts are discussed, with stress laid on the role played by imports of technology, under alternative assumptions of foreign credit availability. Increasing rates of growth can be expected by the end of the present century, assuming structural changes toward development of export-oriented, high-tech industries.

## 19.1. Introduction

The long-term prospects of the Polish economy are naturally determined by the expected development of the main factors of growth, including capital equipment and labor, their productivity being dependent on technical and organizational progress, implementation of industrial research and development, and effective management motivation systems. There are, however, barriers to growth of these factors – the social and technical constraints in expanding investments, labor force problems, declining resources of nonrenewable raw materials (coal) as well as environmental constraints. We know from past experience that the utilization of potential capacity may undergo serious fluctuations accompanied by deep disequilibria. More specifically, we must acknowledge that in the middle of

*Table 19.1.* Main indicators of economic activity, 1970–1985. (For each variable, the first row = absolute values expressed in billions of 1982 zlotys; the second row = growth rates in percentages.)

| *Variable* | *1970* | *1971* | *1972* | *1973* | *1974* | *1975* | *1976* | *1977* | *1978* | *1979* | *1980* | *1981* | *1982* | *1983* | *1984* | *1985* |
|---|---|---|---|---|---|---|---|---|---|---|---|---|---|---|---|---|
| Real NMP | 3309 | 3577 | 3955 | 4383 | 4841 | 5189 | 5945 | 5822 | 5997 | 5860 | 5508 | 4848 | 4581 | 4856 | 5128 | 5292 |
| | 5.2 | 8.1 | 10.6 | 10.8 | 10.4 | 7.2 | 6.9 | 5.0 | 3.0 | –2.3 | –6.0 | –12.0 | –5.5 | 6.0 | 5.6 | 3.2 |
| Real NMP originating | 1401 | 1523 | 1682 | 1881 | 2125 | 2371 | 2599 | 2808 | 2888 | 2844 | 2763 | 2361 | 2254 | 2384 | 2511 | 2609 |
| in industry | 6.9 | 8.7 | 10.4 | 11.8 | 13.0 | 11.6 | 9.6 | 8.0 | 2.8 | –1.5 | –2.8 | –14.5 | –4.2 | 5.8 | 5.3 | 3.9 |
| National income | 3231 | 3501 | 3902 | 4426 | 4441 | 5293 | 5620 | 5882 | 6015 | 5868 | 5611 | 5023 | 4498 | 4748 | 4988 | 5152 |
| distributed | 4.9 | 8.4 | 11.5 | 13.4 | 11.6 | 7.1 | 6.2 | 4.7 | 2.3 | –2.4 | –4.4 | –10.5 | –10.5 | 5.6 | 5.1 | 3.3 |
| Net investments | 752 | 784 | 968 | 1195 | 1451 | 1519 | 1487 | 1671 | 1692 | 1548 | 1326 | 1005 | 805 | 881 | 989 | 1032 |
| | 0.5 | 4.3 | 23.5 | 23.5 | 21.4 | 4.7 | –2.1 | 12.4 | 1.1 | –8.5 | –12.0 | –24.2 | –19.9 | 9.4 | 12.3 | 4.3 |
| Personal consumption | 1952 | 2089 | 2273 | 2467 | 2634 | 2938 | 3193 | 3402 | 3435 | 3547 | 3651 | 3504 | 2989 | 3176 | 3295 | 3361 |
| | 3.9 | 7.0 | 8.0 | 8.5 | 6.8 | 11.5 | 8.7 | 6.5 | 0.1 | 3.3 | 2.2 | –4.0 | –14.7 | 6.3 | 3.7 | 2.0 |
| Exports | 546 | 583 | 673 | 752 | 844 | 913 | 927 | 1001 | 1059 | 1130 | 1083 | 877 | 951 | 1048 | 1148 | 1166 |
| | 8.5 | 6.8 | 15.4 | 11.7 | 12.2 | 8.2 | 1.5 | 7.8 | 5.8 | 6.7 | –4.2 | –19.0 | 8.4 | 10.2 | 9.5 | 1.2 |
| Imports | 558 | 634 | 773 | 947 | 1083 | 1133 | 1242 | 1241 | 1257 | 1242 | 1217 | 1006 | 903 | 963 | 1045 | 1120 |
| | 12.3 | 13.6 | 21.9 | 22.5 | 14.4 | 4.6 | 9.6 | 0.0 | 1.3 | –1.2 | –2.0 | –17.3 | –10.2 | 6.6 | 8.5 | 7.2 |
| Balance of current accounts (in millions of US$): | | | | | | | | | | | | | | | | |
| Hard currency area | 135 | 99 | –194 | –1309 | –2292 | –2747 | –2768 | –2830 | –2692 | –3396 | –2612 | –2095 | –1016 | 62 | 787 | 322 |
| Rouble area | –15 | –123 | –124 | –203 | 14 | 351 | 0 | –236 | 34 | –105 | –685 | –1455 | –513 | –701 | –927 | –1097 |

Source: Data base of the W-5 model.

the 1980s – the starting point for future projections – the underutilization of industrial capacities and shortages in commodity supplies were two of the main characteristics of the Polish economy.

The integrated annual model W-5 of the Polish economy serves as the main tool for the quantitative analysis and projections. The framework of the model, including 13 groups of industries, determines the level of disaggregation of the analysis [1]. The chapter begins by discussing the background of the development of the Polish economy. Then the most likely structural changes on the macro-level are outlined and the prospects for growth of the Polish economy until the year 2000 are estimated. Special attention is given to the trade balance deficit, which is analyzed by using a special submodel [Czyżewski (1986)].

## 19.2. Economic Background of the Polish Crisis

The 1970s were dominated by the policy of "accelerated growth", adopted in the first half of the decade, yielding 10% rates of growth of the net material product (NMP). By 1976, this was replaced by the policy of "economic maneuver" in an effort to slow down the investment process and the foreign debt increase. Its failure was one of the main sources of the decline of all economic activities in the late 1970s and the early 1980s, a decline that was especially deep in the investment sector. The program of recovery, realized during the years 1983–1985, brought a steady increase of economic activities, with the investment rates of growth exceeding the planned levels. However, by 1985 the general activity levels had not reached the previous (1978) values (see *Table 19.1* and *Figures 19.1–19.6*).

One of the main reasons for the economic crisis was the failure of the policy to expand exports in the 1970s, which were necessary to cover the foreign debt servicing and imports. First, foreign demand proved to be insufficient to expand the output of new plants to their full capacity levels (deep depressions being due to oil crises and to trade and political restrictions, which were especially severe in the early 1980s). Second, the pressure of domestic markets weakened export propensities – this could be clearly observed in the meat market. The motivation system did not favor production of exports. As a result, the growth rates of exports were far below those of imports, especially during the second half of the 1970s, which led to rapidly growing hard currency indebtedness.

The balance of payments crisis with regard to the hard currency area, which occurred in 1981/1982, and the lack of further foreign financing for either current trade transactions or debt servicing, especially after the introduction of political restrictions, caused a sharp decline in imports – exercizing a strongly negative impact on the utilization of capacities and, at the same time, producing a positive balance of trade. Since that time, despite the increased supplies of raw materials and intermediate commodities from the USSR, imports remained the main constraint for both higher utilization of growing capacities and investments and implementation of new technology as well (see *Table 19.2*).

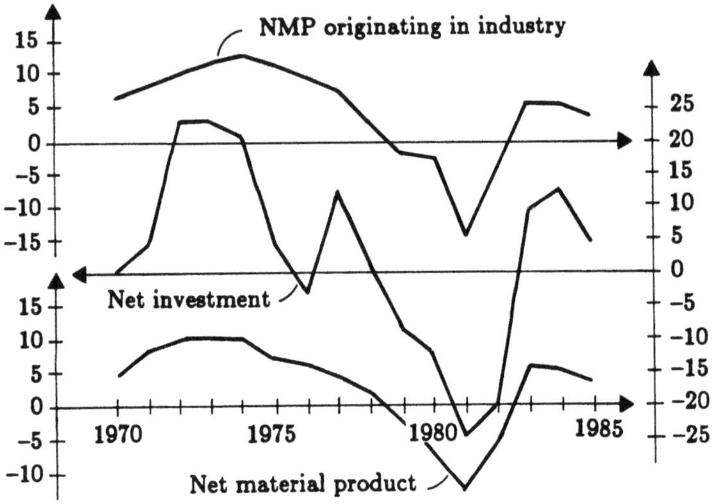


*Figure 19.1.* Growth rates of net material product (NMP), net investments, and NMP originating in industry, 1970–1985 (constant 1982 prices).

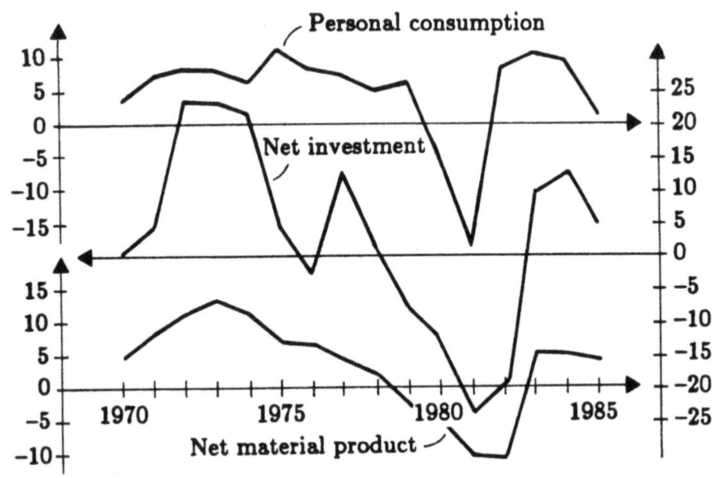


*Figure 19.2.* Growth rates of distributed national income, net investment, and personal consumption, 1970–1985 (constant 1982 prices).

Reduced possibilities of imports from developed countries in the late 1970s affected, first of all, imports of consumer goods. Absence of efficient policies aimed at slowing down the rate of growth of domestic demand led to the overall domestic disequilibria, beginning with the consumer markets. Because of price rigidities, excess demand easily spread from separate markets (durables,

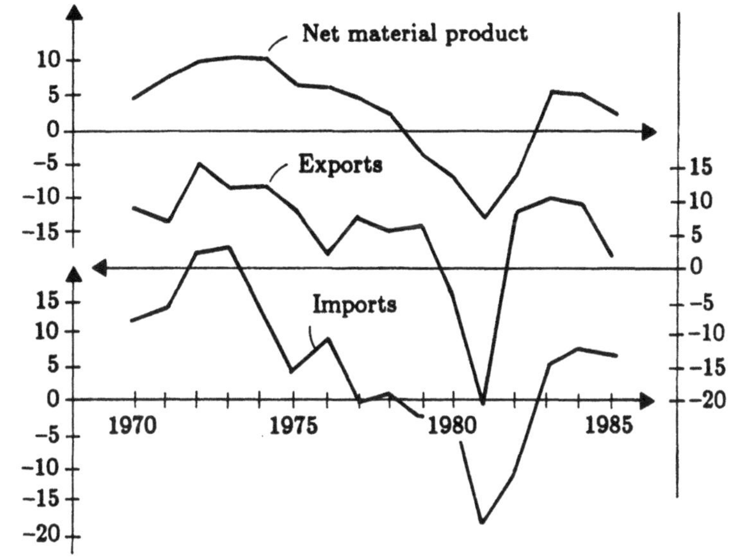


*Figure 19.3.* Growth rates of imports, exports, and net material product, 1970–1985 (constant 1982 prices).

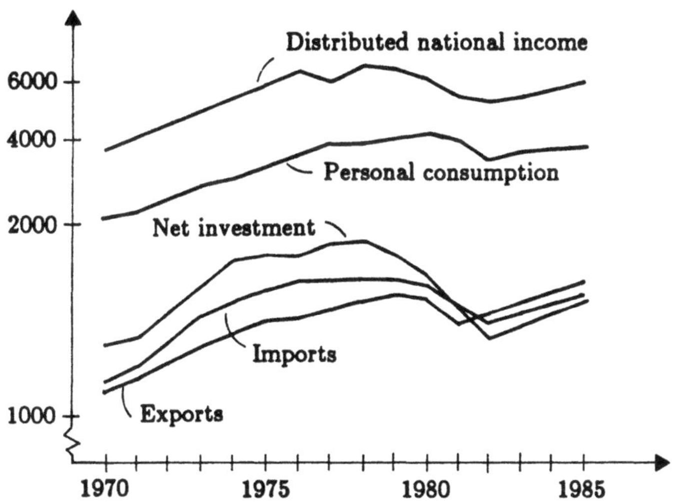


*Figure 19.4.* Logarithmic growth of imports, exports, net investment, personal consumption, and distributed national income, 1970–1985 (constant 1982 prices).

foodstuffs – see *Table 19.3*) to the overall market for consumer goods, leading to rationing or quasi-rationing associated with an enormous increase in forced savings. Retail price increases, started in the late 1970s, were not sufficient to clear the markets.

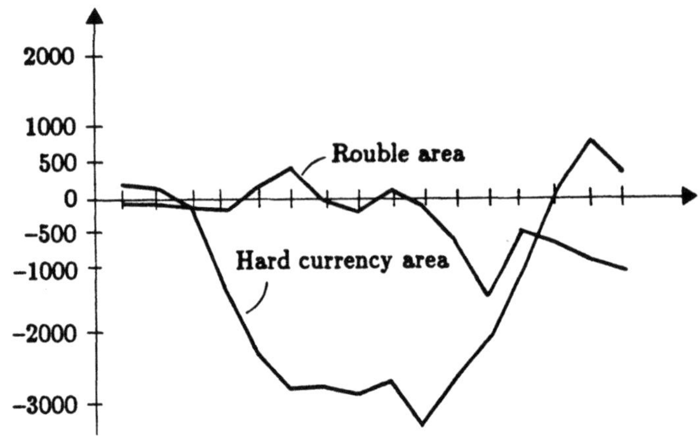


*Figure 19.5.* Balance of current accounts, 1970–1985.

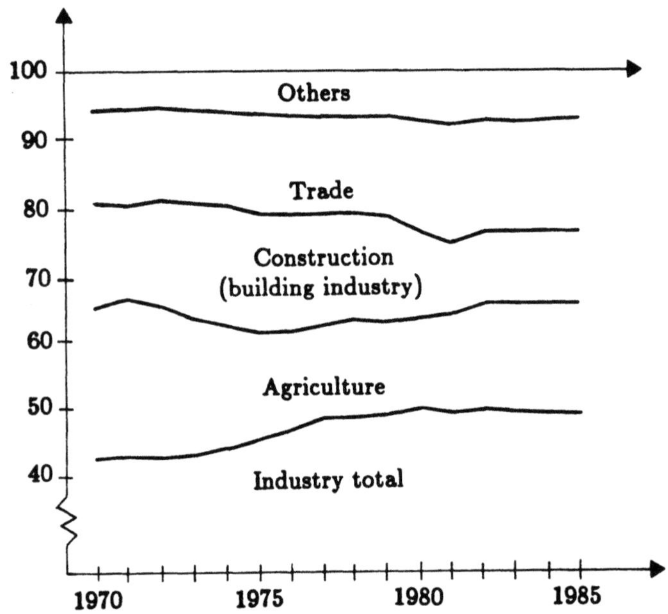


*Figure 19.6.* Sectoral shares of net material product, 1970–1985 (constant 1982 prices).

These inflationary phenomena were accompanied by compensatory increases in wages (29% annually in the years 1980–1985) and incomes, contributing to the inflationary spiral. More specifically, a serious decline (9.8%) of supplies of consumer goods in 1981–1982, reflecting substantial import cuts (by 10.2%) and decline of domestic output with rising nominal wages (51%), brought about deep market disequilibria. For social reasons, the price increases,

*Table 19.2.* Utilization rates (in %) of capacity (capacity defined as the amount of machinery).

| *Industry* | *1965* | *1970* | *1975* | *1977* | *1978* | *1979* | *1980* | *1981* | *1982* | *1983* | *1984* |
|---|---|---|---|---|---|---|---|---|---|---|---|
| Fuel and power | 100.2 | 100.0 | 96.9 | 104.1 | 102.9 | 93.1 | 96.8 | 44.8[a] | 40.7[a] | 50.9 | 48.3 |
| Metallurgical, chemical | 98.5 | 89.8 | 99.2 | 98.2 | 96.3 | 88.4 | 84.2 | 79.4 | 81.1 | 108.8 | 113.8 |
| Electro-engineering | 96.6 | 97.6 | 98.7 | 92.6 | 91.7 | 96.5 | 94.8 | 83.4 | 87.4 | 105.8 | 110.1 |
| Light | 95.2 | 97.7 | 98.0 | 92.4 | 91.1 | 87.7 | 85.2 | 81.8 | 76.0 | 114.9 | 120.5 |
| Food processing | 97.0 | 97.1 | 96.4 | 98.4 | 92.0 | 78.3 | 75.2 | 77.4 | 67.9 | 52.7 | 53.5 |
| Agriculture | 9.3 | 97.1 | 98.1 | 93.4 | 94.3 | 92.1 | 87.9 | 91.0 | 92.5 | 168.5 | 176.1 |
| Building | 98.9 | 101.4 | 104.4 | 103.7 | 93.8 | 82.9 | 88.0 | 70.7 | 47.6 | 54.2 | 58.1 |
| Transportation | 102.7 | 96.5 | 102.5 | 100.7 | 100.4 | 100.0 | 95.0 | 85.3 | 77.6 | 140.3 | 150.8 |

[a]Underestimated, owing to the change of constant prices
Source: calculations based on the W-5 model.

*Table 19.3.* Effective demand as percentage of real household expenditures.

| *Commodity group* | *1965* | *1970* | *1977* | *1978* | *1979* | *1980* | *1981* | *1982* | *1983* | *1984* | *1985* |
|---|---|---|---|---|---|---|---|---|---|---|---|
| Foodstuffs | 102.7 | 100.0 | 105.2 | 103.0 | 101.2 | 102.2 | 110.3 | 111.4 | 108.2 | 107.6 | 110.6 |
| Alcoholic beverages | 100.0 | 100.0 | 100.0 | 100.0 | 100.0 | 100.0 | 146.7 | 117.9 | 112.7 | 110.4 | 100.0 |
| Textiles, apparel, shoes | 119.4 | 106.2 | 107.8 | 114.5 | 115.8 | 113.0 | 112.7 | 141.3 | 113.7 | 104.7 | 112.2 |
| Durables | 104.3 | 116.3 | 100.0 | 100.0 | 104.2 | 103.9 | 102.7 | 130.2 | 101.6 | 102.7 | 108.1 |
| Other nondurables | 103.2 | 100.0 | 107.8 | 102.5 | 101.1 | 100.0 | 101.0 | 104.2 | 100.0 | 100.5 | 113.8 |
| Services | 100.0 | 102.7 | 125.6 | 120.7 | 119.2 | 117.4 | 122.9 | 130.4 | 103.2 | 100.6 | 100.7 |

Source: calculations based on the W-5 model.

although quite high (in 1982 they doubled, and in the next years they increased by 25% and 15%, respectively) were not sufficient to restore equilibrium.

The policy launched in the early 1980s, aimed at introducing important changes in the management and planning system, is known as the economic reform. It was expected to expand substantially the decision making at the plant or enterprise level (including wages and, partly, price formation), leaving for the central planner less direct commands and more financial instruments. The policy was intended to ensure more efficient use of production factors, and promote a more rational investment process.

This policy was implemented in a rather unfavorable environment, as the disequilibria prevailed over the early 1980s, with diminishing strength, related to the general anticrisis policy. The latter, started in 1982, was aimed at increasing exports (loosely import-dependent, such as coal) and extending the working hours; it brought a partial recovery. High rates of growth of the NMP by 6.0% in 1983, and by 5.6% in 1984 followed. However, a decline in the growth rate to 3.2% in 1985 showed that the new economic mechanisms were not fully at work and the economy suffered from several weaknesses, the most important being:

(1) Notorious underutilization of capacities.
(2) Aging capital stock.
(3) Persisting inflationary spiral.
(4) Continuing increase of foreign indebtedness.

The above features were to be taken into account by the official programs for the years 1986–1990 and those up to the year 2000.

## 19.3. The Integrated Model W-5

Economic development of Poland in the late 1970s and at the beginning of the 1980s brought a general decline in all activities, i.e., it amounted to an economic crisis. As this was not the consequence of the lack of effective demand, no demand-determined models were used. Nor was it a result of underdevelopment of production capacities. Therefore, the traditional approach to the construction of econometric macro-models of the socialist countries, which assume full utilization of production capacities, proved to be unsuitable to explain and forecast the setback of economic activity and its expected revival.

The characteristics of the Polish crisis thus necessitated a search for new solutions in construction of macro-models. They should be capable of integrating long-term mechanisms of economic growth of the socialist economy and adjustment behavior patterns of more and more autonomous economic entities to short-term constraints producing shortages and disequilibria. This motivated the construction of the model W-5. Its special version, which makes an intensive use of input–output relationships, called the integrated model W-5 of the Polish economy, is briefly described below.

### 19.3.1. The structure and main characteristics of the model

The W-5 model is a simulation-type forecasting model with more than 1,000 equations; among them, more than 400 are stochastic. There are no more than 60 exogenous variables, excluding the dummies. They embrace strictly exogenous variables (characteristics of international trade, weather conditions and demographic phenomena) as well as variables which are instruments of the economic policy. The data for the model cover annual time series for the years 1960–1982, expressed, for the most part, in constant prices of 1982. Regression equation parameters were estimated by OLS on the basis of annual data for 1963–1982.

The model covers not only real processes of production and distribution of the NMP, but also – its distinctive feature – financial processes, together with a subsystem generating the movement of prices. It was divided into 15 blocks of equations which encompass:

(1) Employment, time worked, shifts worked.
(2) Investment outlays, investments put into operation.
(3) Fixed assets, scrappings, depreciation.
(4) Capacity output in terms of potential net output, degree of utilization.
(5) Net output, gross output and sales, deflators.
(6) Distributed national income.
(7) Market deliveries, retail sales, household purchases, deflators.
(8) Stocks of inventories, stock increase, deflators.
(9) Exports, imports, deflators.
(10) Money incomes, expenditures of households and the nonsocialized sector.
(11) Wages, wage bill, pensions and other incomes; personal incomes, savings, deflators.
(12) Financial accumulation of enterprises, other nonmaterial costs.
(13) State budget: income, expenditures.
(14) Balance of payments.
(15) Prices.

The degree of disaggregation is not particularly deep (it is lower than in the standard I–O tables). It embraces the following sectors and branches in the sphere of material production: construction; agriculture divided into plant, animal and other production; forestry; industry divided into groups of branches encompassing mining and power, raw material industries, i.e., metallurgy, chemical and mineral industry, electro-engineering, light industry, food processing and other branches; transport and communications; and trade. In the sphere of nonmaterial services, housing and communal services were distinguished. Specific classifications of product flows directed to the groups of final users were adopted. They refer to households (7 groups of expenditures), investments (3 kinds), stocks of inventories and foreign sector (5 groups according to the SITC classification). The model generates information concerning the above blocks in both constant (1982) and current prices and in US dollars for the foreign trade.

The feature that distinguishes the W-5 model from other models is that its basic blocks can generate information about not only the resources and flows realized in economic processes, but also about potential values of variables that are unobservable. This refers to capacity output and its utilization and, especially, demand and supply and, additionally, excess demand. As for consumer goods, it allows estimation of the excess demand for particular groups of these goods as well as their global level (forced savings).

We should underline here that the W-5 model is the first large model for the socialist economy which explicitly poses and tries to solve the above problems [2]. In particular, it was proved that the traditional use of a two-factor production function is not sufficient to explain the decline of output in a period of economic crisis. The role of material inputs must be analyzed and treated as the crucial constraining factor which determines the utilization of capacities. Thus, a clear distinction was drawn between the concept of the production function suitable for long-term analysis – used to generate capacity output – and short- and medium-term adjustments in the output levels reflected in the degree of capacity output utilization. These adjustments, which are mainly due to the constraints in supply of production factors, have been mostly interpreted in the spirit of Kalecki's (1963) theory of growth barriers. As a result, we were able to estimate jointly the adjusted production function parameters, including those showing the impact of the capacity output utilization, and to generate not only the expected output (offer), but also (unobservable) capacity and the degree of its utilization (see *Table 19.2*).

The second important characteristic of the W-5 model stems from the recognition that the traditional understanding of a demand function (for consumer goods as well as for investment goods and materials – both domestic and imported) is not sufficient to explain the real phenomena under disequilibrium conditions. Thus, we suggest generalizing the concept of a demand function. The effective demand – besides its notional value (being in accordance with equilibrium conditions) – must contain additional elements showing spillovers resulting from:

(1) Forced substitution – intergroup transfers.
(2) Postponement in its fulfillment – intertemporal transfers.
(3) Anticipation of changes in intensity of disequilibria.

The budget constraint must be redefined to include additional sources of demand financing (mainly, forced savings). Moreover, a specific concept of disequilibrium indicators for estimating the excess demand was formulated. The difference between the effective demand function and that of excess demand (both variables are unobservable) yields the equation explaining the expenditures (realizations) which are observable. This enables the estimation of the parameters of this equation and also the separation of the effective and excess demand [3] (see *Table 19.3*).

## 19.3.2. Mechanisms of economic growth and balancing the economy

The description of the mechanisms of growth follows M. Kalecki's analysis (1963). The model identifies the basic relationships that explain investment and production decisions. It defines the feedbacks between investment outlays and output, changes in the efficiency of the investment process and changes in the use of the capital stock by constrained capacities.

There is a "supply-type acceleration" rule defined in the model, which consists of the following feedback. Investment outlays materialize in the form of an increase of the capital stock after the time needed for accomplishment of the investment process. Given the degree of capacity utilization, this induces an expansion in the sector producing investment goods, too, which is further accelerated due to technical progress. The national income growth achieved in this way involves – given the investment/NMP ratio (especially "productive" investment/NMP ratio) – a further increase of investment outlays (also in the investment goods sector, see *Figure 19.7*). This autonomous growth process may be cyclical if, for example, outlays for new start-up investment projects are concentrated in the first years of a five-year plan (which could not be confirmed for Poland). It may be accelerated (delayed) by regulatory measures aimed at:

(1) Change of the investment share in national income and also in the share of productive investment.
(2) Change in the length of the investment cycle.
(3) Change in the degree of utilization of capital stock depending upon, among other things, barriers encountered in the production process.

To properly describe the phenomena of the period of the economic crisis, we included broadly in the model – in the production functions – the growth barriers concerning both fixed assets (equipment) and labor force supply, and also foreign trade barriers (mainly against imports of fuel and raw materials). The model offers the possibility to analyze the expected results of measures aimed at enlarging (or limiting) labor force supply, mainly by means of social policy decisions. Foreign trade barriers are presented in the model by making total imports and their components dependent on exports and an assumed (exogenous) balance of trade (by groups of countries), which obviously depends on external net financing (debt servicing in the last few years). The high intensity of the above barriers explains the decline in capacity output utilization and production itself in the years of the economic crisis and their relaxation – the process of recovery, especially in its initial stage (see *Figure 19.8*). The possibilities of relaxation (or tightening) of supply constraints are also incorporated by introducing material inputs (cost shares by industries) to the model; these depend on the foreign trade constraints and institutional settings.

The modeling of the distribution and utilization of the NMP and national income is closely related to the requirements for analyzing imbalances occurring in the national economy and its particular sectors as well as recovery policies. The model allows direct generation of the supply of consumer goods (deliveries for the market, supply of domestic and imported commodities, retail sales, etc.),

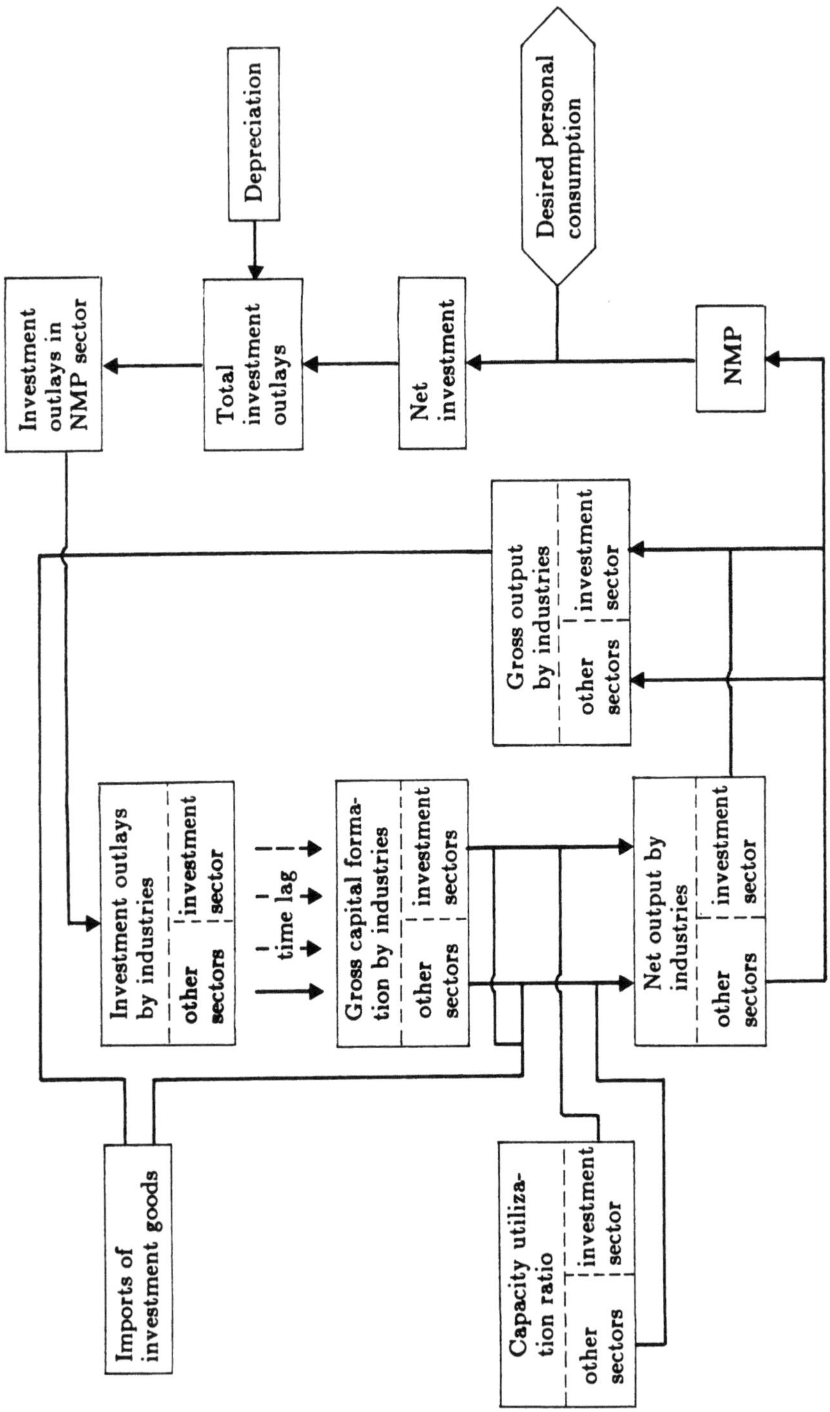


*Figure 19.7.* Net material product: growth and investment loop.

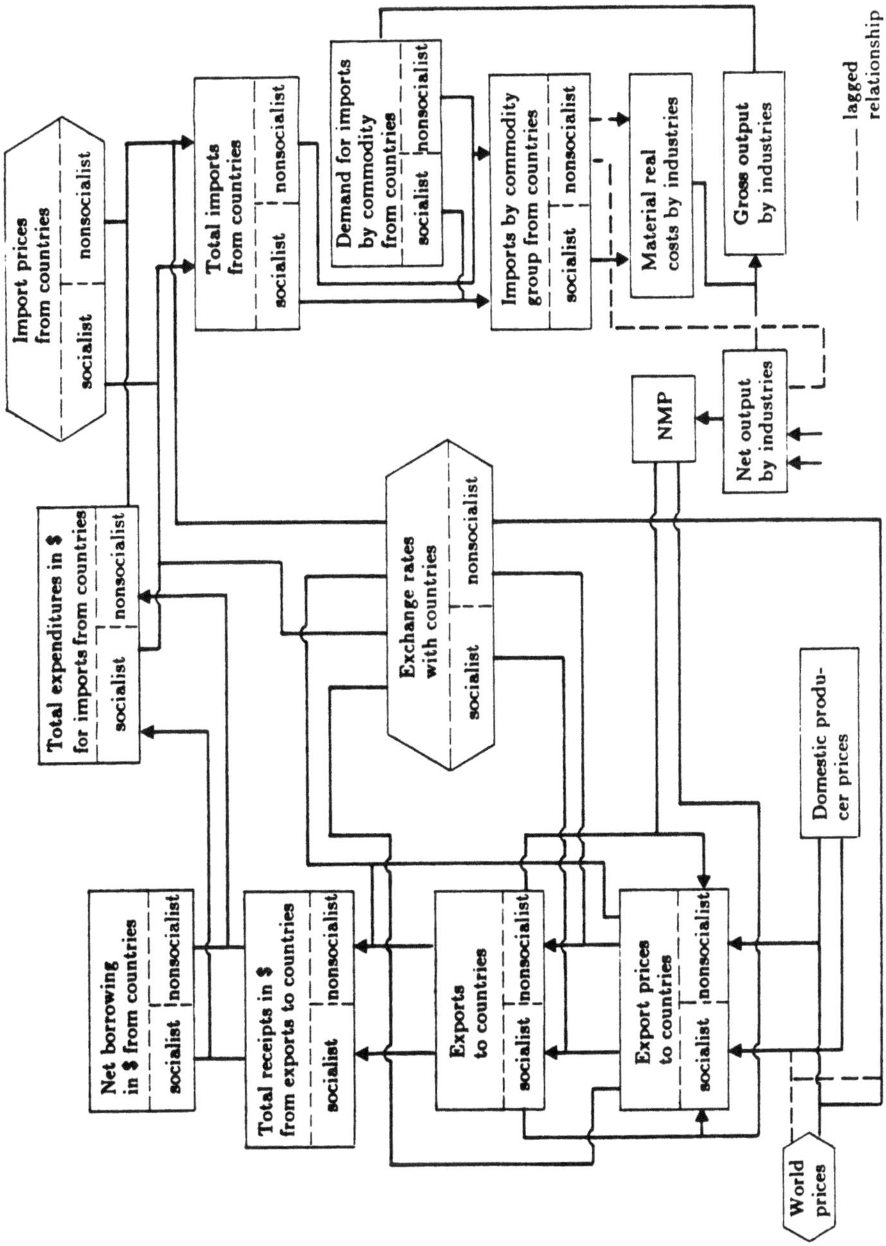


*Figure 19.8.* Foreign trade and production sector.

as well as investment goods. At the same time, the model allows us to generate directly final demand, especially for consumer goods (mostly that of households), investment demand, foreign demand and demand for imported commodities. Total demand for products by industries is obtained by using either the input–output approach (including the use of bridge matrices) or its stochastic approximations. The demand for labor is generated by using equations derived from the production function. The demand for fixed assets is not directly specified.

The possibilities of balancing demand and supply appear in the model in almost all phases of the production process while the number of autonomous, especially market-clearing, price mechanisms of adjustment is limited. It is assumed that price sensitivity is shown mainly by households (demand responses) and production units in connection with changes in prices of imported commodities relative to domestic ones. As a result, the model generates the values of variables, showing directly the degree of balance tensions (mainly, excess demand), (a) by the direct comparison of demand and supply on the consumer goods market, forced savings, and total demand and supply by industries at the producers level; or (b) indirectly by means of specific disequilibrium indicators in exports and material supplies.

It should be underlined that the central planner's measures aimed at restoring equilibrium have been partly endogenized. This takes place in the case of investment goods when it is assumed that initial demand of economic organizations for investment outlays is modified according to the policy targets (e.g., protecting consumption levels during an economic crisis). Similarly, the demand for imported goods by commodity groups is adjusted respectively to the financial possibilities defined by the central planner. In the model, there are also many other options available, including central planner interventions.

As for financial processes, we paid attention to the fact that wage increases, being the main source of personal income increase, were related mostly to the growth of living costs and to the growth of labor productivity but also to the increase in tensions in the labor market. The possibility of the central planner's interference is also considered in the model. The wage increase is one of the main factors of cost and price increases. Thus, the inflationary loop was introduced explicitly. It reflects the fact that growing living costs of household cause a slightly less than proportional growth of nominal wages, which results in an increase in labor costs (partly compensated by the increase of labor productivity) and, therefore, an increase in production costs (if not compensated by saving of energy and materials). These cost pressures finally lead, after some time lag, to an increase in producer prices, wholesale and retail prices, and household living costs (see *Figure 19.9*). The interference of the central planner, by limiting price growth in an administrative way, is of short-term significance only. It is associated with the increase of subsidies, etc. After some time, it is followed by "compensatory" price increases.

It is worth indicating that the model creates modest, yet interesting possibilities for the analysis of the consequences of financial, monetary and fiscal policies, including loans offered by the banks, taxes by the budgetary authorities, and government spending.

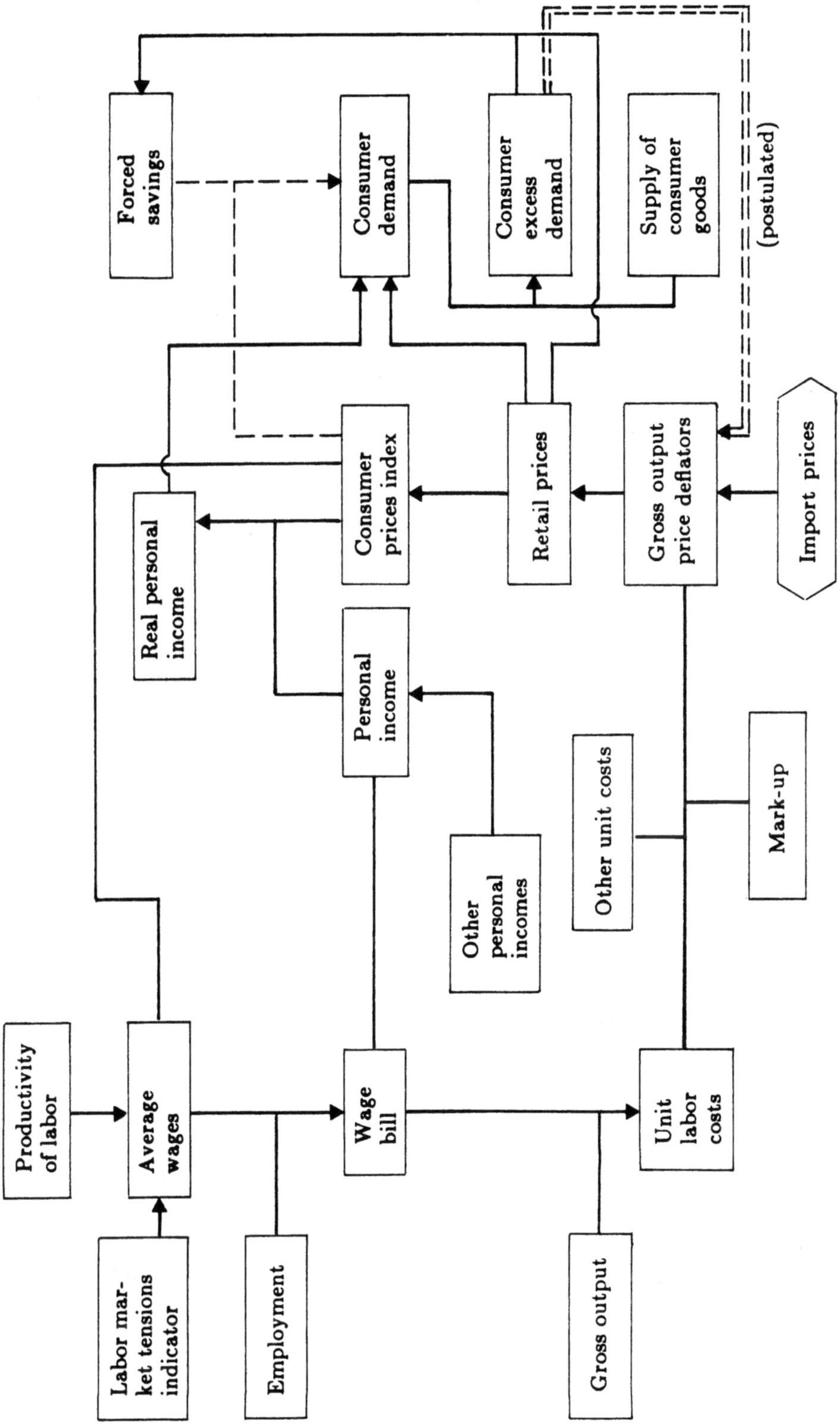


*Figure 19.9.* Consumer goods market: inflationary loop.

## 19.4. External Debt and the Submodel of Foreign Debt

### 19.4.1. Foreign debt model

Observing the dependence of the economy on external conditions was the starting point for elaboration of the long-term development scenario to be simulated on the W-5 model. More specifically, the impact of alternative paths of growth of foreign trade on convertible currencies external debt was one of the main objects of our study. For this purpose, a special foreign debt model was constructed [see Czyżewski (1986)].

This model calculates the expected levels of foreign debt in response to the assumed rates of growth of national income and the elasticities of total exports and imports in convertible currencies with respect to the national income. These three quantities are key characteristics of general economic efficiency in the foreign trade area. Other assumptions, which can be altered, include: foreign trade prices in convertible currencies, interest rates and overhead charges on the assumed interest rates for delayed debt repayments and outstanding interest (delayed principal and outstanding interest payments are explicitly defined in the debt model). Assumed rates of growth of exports and imports in nonconvertible currencies help to determine the rate of growth of domestic expenditures (distributed national income).

Poland's debt model, used for convertible currency debt simulations over the ten years' perspective, is built around the balance of payment equation and contains the following variables.

(1) On the current account side: exports, imports of goods and the balance of trade; balance of services and transfers, and interest on credits received (long- and medium-term only, including refinancing credits).

(2) On the capital account side: long- and medium-term credits received (drawings, repayments) and refinancing credits (drawings, repayments). Unlike many other models of Poland's debt, this model explicitly utilizes information about time distribution of current credit repayments and refinancing credit repayments since different types of arrears are charged different interest rates.

The results provided by the debt model are helpful in understanding what are the most desired general directions of structural changes from the point of view of Poland's external financial commitments. After choosing the most favorable solution from the examined set of various admissible assumptions, an attempt was made to determine the possible mechanisms for turning the solution into economic reality, given the existing structure of the economy and historical development trends.

### 19.4.2. External debt simulations

The analysis of scenarios based on the debt model was preceded by an examination of the sensitivity of the results to isolated changes of individual assumptions which compose a scenario [see Czyżewski (1987)]. The solution which served as a basis for comparisons was defined by the following assumptions:

(1) Annual rate of growth of the national income: 3%; elasticities of convertible currencies exports with respect to the national income: 1.5%; and imports: 1.2%.
(2) For convertible currencies (annual rates of growth): import prices increase by 4.5%; decline of terms of trade by 0.5%; increase of balance of services and transfers by 6%.
(3) Interest rates: for long- and medium-term credits = 7%; for refinancing credits = 9%; and charge for delayed principal and outstanding payments = 10%.

These assumptions were combined with 5.5% and 4% rates of growth of nonconvertible currencies exports and imports, respectively, in order to get implied rates of growth of the national income distributed. *Table 19.4* shows the results of this solution.

Sensitivity analysis led to the conclusion that the debt path is very sensitive to changes in elasticities of exports and imports with respect to national income. High sensitivity was observed especially with respect to changes in the elasticity of exports. An increase in this elasticity from 1.5 to 1.75 caused a decrease in the debt level in 1995 of $5.6 billion (18%), whereas a decrease in imports elasticity from 1.2 to 1.0 produced a decline in the debt level in 1995 of $3.4 billion (11%). It follows that structural changes should aim at increasing the propensity to export. A change in the rate of growth of national income (3.5% as against 3%) resulted in a 1995 debt level which is $2.1 billion lower (7%) than in the base solution, whereas lowering the rate to 2.5% yields a debt level of $2.0 billion (6%) above the baseline solution. Finally, equal rates of growth of export and import prices in convertible currencies lower the debt level in 1995 by $3.9 billion (12%) compared with the base solution.

The above results were used to combine the assumptions into various alternative scenarios. The most interesting scenarios assumed the availability of new credits of $0.5 billion p.a. in the years 1987–1989. Their analysis led to the conclusion that new loans will substantially contribute to setting back the debt increase if the acceleration of economic growth is accompanied by significantly higher rates of growth of convertible currencies exports than imports.

*Table 19.5* shows the results of three scenarios of debt development in the years 1987–1995 which assume an increase in foreign trade efficiency due to foreign credits. These scenarios are denoted by letters A, B and C. The assumptions composing scenario A are as in the base solution (*Table 19.4*), except for interest rates for refinanced credits, delayed principal and outstanding interests repayment (here 8.75%, 9% and 11%, respectively) and for the rate of growth of nonconvertible currencies imports (here 3.6% p.a.). Scenarios B and C

*Table 19.4.* Results of the baseline solution of the debt model (in billions of US$).

| *Variable* | *1987* | *1988* | *1989* | *1990* | *1991* | *1992* | *1993* | *1994* | *1995* |
|---|---|---|---|---|---|---|---|---|---|
| Trade balance (1985 constant prices) | 1.4 | 1.5 | 1.6 | 1.7 | 1.8 | 2.0 | 2.1 | 2.3 | 2.4 |
| Current account balance | −0.6 | −0.6 | −0.5 | −0.4 | −0.4 | −0.3 | −0.1 | −0.0 | 0.3 |
| Debt | 29.7 | 30.3 | 30.8 | 31.2 | 31.6 | 31.9 | 32.0 | 32.0 | 31.7 |

Source: simulation based on debt model POLDEBT.

*Table 19.5.* Scenarios of development of convertible currencies debt.

| *Variable* | *Scenario* | *1987* | *1988* | *1989* | *1990* | *1991* | *1992* | *1993* | *1994* | *1995* |
|---|---|---|---|---|---|---|---|---|---|---|
| *Assumptions:* | | | | | | | | | | |
| National income (% p.a.) | A | 3.0 | | | | | | | | |
| | B & C | 3.0 | 3.2 | 3.4 | 3.6 | | | | | |
| Exports elasticity | A | 1.5 | | | | | | | | |
| | B & C | 1.5 | 1.5 | 1.55 | 1.6 | 1.65 | 1.7 | | | |
| Imports elasticity | A | 1.2 | | | | | | | | |
| | B | 1.2 | 1.3 | 1.4 | 1.5 | | | | | |
| | C | 1.2 | 1.25 | 1.3 | 1.35 | | | | | |
| Utilization of foreign credits[a] | A | none | | | | | | | | |
| | B & C | 0 | 0.5 | 0.5 | 0.5 | 0 | 0 | 0 | 0 | 0 |
| *Results:* | | | | | | | | | | |
| Trade balance (1985 constant prices)[a] | A | 1.4 | 1.5 | 1.6 | 1.7 | 1.8 | 2.0 | 2.1 | 2.2 | 2.4 |
| | B | 1.4 | 1.5 | 1.6 | 1.7 | 1.8 | 2.0 | 2.1 | 2.3 | 2.5 |
| | C | 1.4 | 1.5 | 1.6 | 1.7 | 1.9 | 2.1 | 2.3 | 2.5 | 2.8 |
| Trade balance (current prices)[a] | A | 1.4 | 1.6 | 1.7 | 1.9 | 2.1 | 2.3 | 2.5 | 2.8 | 3.1 |
| | B | 1.4 | 1.6 | 1.7 | 1.9 | 2.0 | 2.3 | 2.5 | 2.8 | 3.1 |
| | C | 1.4 | 1.6 | 1.7 | 1.9 | 2.1 | 2.5 | 2.8 | 3.1 | 3.5 |
| Interest payments due[a] | A | 2.5 | 2.6 | 2.8 | 2.8 | 2.9 | 3.1 | 3.1 | 3.2 | 3.2 |
| | B | 2.5 | 2.6 | 2.7 | 2.8 | 2.9 | 3.0 | 3.0 | 3.1 | 3.1 |
| | C | 2.5 | 2.6 | 2.7 | 2.8 | 2.9 | 3.0 | 3.0 | 3.0 | 2.9 |
| Current account balance[a] | A | −0.6 | −0.5 | −0.4 | −0.3 | −0.1 | −0.0 | 0.2 | 0.5 | 0.8 |
| | B | −0.6 | −0.5 | −0.4 | −0.2 | −0.1 | +0.0 | −0.3 | 0.6 | 0.9 |
| | C | −0.6 | −0.5 | −0.3 | −0.2 | +0.0 | 0.2 | 0.6 | 1.0 | 1.5 |
| Total debt[a] | A | 29.6 | 30.0 | 30.4 | 30.7 | 30.8 | 30.8 | 30.6 | 30.2 | 29.4 |
| | B | 29.6 | 30.0 | 30.4 | 30.6 | 30.7 | 30.7 | 30.4 | 29.9 | 28.9 |
| | C | 29.6 | 30.0 | 30.4 | 30.5 | 30.5 | 30.3 | 29.7 | 28.7 | 27.3 |
| Domestic expenditures (% growth rates) | A | 2.7 | 2.7 | 2.7 | 2.7 | 2.7 | 2.7 | 2.7 | 2.6 | 2.6 |
| | B | 2.7 | 2.9 | 3.1 | 3.3 | 3.3 | 3.3 | 3.3 | 3.3 | 3.2 |
| | C | 2.7 | 2.9 | 3.1 | 3.3 | 3.2 | 3.2 | 3.2 | 3.2 | 3.2 |

[a]In billions of US$.

Source: simulations based on debt model POLDEBT.

anticipate the availability of new credits ($0.5 billion p.a. in 1987–1989) which, since 1988, are associated with increases both in rates of growth of the national income and in the size of export and import elasticities.

## 19.5. Growth and Structural Changes of the Polish Economy to the Year 2000

The simulation experiments performed by using the debt model and forecasting exercises based on econometric macromodels have clearly supported the long-term policy goals, which anticipate deep structural changes of the Polish economy. They should be accompanied by parallel changes in the planning and management system toward increased economic efficiency. The external conditions, especially the enormous foreign debt and the necessity for its repayment, call for directing structural changes toward increased export capability and efficiency. Thus, the criteria of structural changes should be similar to those of the international specialization of production. The additional criterion, or rather constraint, comes from the existence of domestic disequilibria and policies aimed at restoration of the internal equilibrium.

### 19.5.1. Factors of economic growth

The natural point of departure in considering possible future changes in economic structures is, of course, the existing structure of the national economy, which can be analyzed from various aspects. Its most important characteristic is the material structure of output determined both by composition of final demand and availability of production factors. The next step is the identification of structures expected to accomplish the economic and social goals, taking into account potential resources and expected external constraints.

Analysis of the likely paths of growth linking present and future structures reveals that the impact of extensive factors will be limited. Acceleration may be expected solely from intensive factors leading to improved economic efficiency. Employment in the material product (MP) sector, which grew in 1960–1983 by 1.4% annually while the population increased by 0.9%, will remain approximately constant in the period 1986–2000, except for the last five years, as the population is expected to increase by only 0.7% and unfavorable changes in its age structure will occur. The growth of fixed assets will continue, but its momentum will be much lower than in the 1960s and the 1970s. In 1960–1980, fixed assets in the MP sector grew by 7%, whereas the expectations for the coming 15-year period do not exceed 3% in the 1980s and 4% by the end of the century.

On the other hand, the growth of capacity output in the past was characterized by a low utilization rate. In the same period of 1960–1980, despite an increase in employment and adoption of advanced technologies in the 1970s, national income grew only by 5.6%. As for the future, we can assume an improved use of the production potential, due to some extent to restructuring. Thus, taking into account the estimated effects of embodied technical progress

(1% of the increase of fixed assets is associated with an additional 0.6–0.8% increase of output due to technical progress) and the average elasticity of output with respect to fixed assets (= 0.5), we may expect rising efficiency of investment efforts and capital stock use. As a result, the rate of growth of national income should exceed the rate of growth of fixed assets.

The stagnation of employment and the pressure of the balance of payments, which will significantly constrain material and energy inputs, do not allow us to presume that any major increase in the utilization of capacities will occur in the next five years. The pressures to increase investment outlays will thus arise either from new industries (structural changes) or from economic units interested in modernizing their capital stock in order to increase their efficiency (decrease labor and material inputs).

It must not be forgotten that the policy of protecting consumption, pursued in the late 1970s and the early 1980s, led to a continuous drop in investment outlays. it was stopped only in 1983 as its continuation threatened decapitalization and, in retrospect, a likely new stage of the economic crisis. Thus, despite the low level of capacity utilization in the economy, there was a need to increase net investment outlays (by an annual rate of 7%). After 1985, growth rates were squeezed again to 3–4%, which postpones the restructuring process in industry to the late 1980s. Faster growth of investment outlays would presumably lead to deterioration of the economic situation in the short run. Its acceleration seems feasible only in the 1990s, provided new imported technology is available on credit terms. The level of fixed assets, however, should not be considered a significant constraint of economic growth in the next years as the utilization of capacities will probably remain on the low side. Moreover, under such conditions, it might be profitable to slow down the increase of the capital stock by an accelerated scrapping of old equipment and a simultaneous increase in the level of utilization of the modern one (increasing the number of shifts worked in industry, etc.)

Let us briefly examine the requirements with respect to growth rates from the demand side. The minimum rate of growth should ensure maintenance of the present level of consumption per capita. This means a 0.7% growth rate of consumption, equal to the growth rate of the population. Together with the required level of accumulation sufficient for a 7% increase of net investment outlays, this calls for an approximately 3.3% annual rate of growth of distributed national income. The rate of growth of net material product should be equal to at least 3.5% because of the necessity to maintain a positive balance of trade. A comparison of the likely impact of the extensive factors of production on the growth rate of the NMP (estimated at the rate of 1–2%) with the lowest required growth rate shows that at least 1.5–2.5% of the growth rate has to be brought about by improved economic effectiveness, which according to our previous calculations might be provided by embodied technical progress; in fact, it might be considerably higher if we take into account the other sources of efficiency increase.

One of the most important factors determining economic effectiveness is participation in the international division of labor (driven by export expansion), conditioned in the case of the Polish economy by appropriate structural changes toward increasing export capacities. The gains from exports are linked, first of all, with dynamics of the NMP which is, to a large degree, determined by import possibilities.

The required rate of growth of intermediate imports is closely connected with the dynamics of output and of national income. In 1960–1983, intermediate imports grew by 6.2% while national income grew by only 4.4%. Thus, the observed average income elasticity of these imports was equal to 1.4. It is expected that an increase of the economic efficiency accompanied by appropriate structural changes may lower this elasticity to the average level of 1.2 over the period 1986–2000. The necessity to restore and modernize the capital stock creates heavy demand for investment imports, which are the main carriers of embodied technical progress. Thus, the share of investment imports should attain the average level of 15–20%, depending on the rate of growth of national income (in 1983 its share was equal to 7%, while in the 1970s it reached 22–24% and never went below 13%). In practice, the level of investment imports will depend mostly on preferred directions of product specialization. Imports of consumer goods, whose share was comparatively low, should grow slightly to reach the level of 15%, irrespective of the growth rate of the economy. Simultaneously, a significant change in the structure of these imports should occur (now its main component is food). It follows from these assumptions that the share of intermediate imports should go down to 65–70% (depending on the desired rate of growth).

The target dynamics of the Polish imports are conditioned by the availability of foreign currency inflows, with exports being the main source. To estimate the required dynamics of exports, additional circumstances, such as change in capital accounts determined by the net increase of foreign credits and costs of debt servicing, should be considered. Then, terms of trade changes and world inflation rates must be taken into account to link the balance of trade expressed in constant prices with the balance of capital accounts expressed in current prices.

For the 1986–1995 forecast, it might be assumed that Poland's positive balance of trade with nonsocialist countries will be maintained (a slight growth from 1.0 in 1986 to 2.4–2.8 in 1995 of billions of constant US dollars), but its levels will be lower than the interest payments falling due in the forecasting period (see *Table 19.5*). For trade with socialist countries, a slow improvement in the balance was assumed, yielding approximately US$1.0 billion in 1995.

Under these assumptions, a 3.6% growth rate of national income requires a 5–5.5% growth rate of total imports and 6.3–6.5% growth rate of total exports. The results envisage structural changes which lead to an increase of the marginal propensity to export under the assumption of a constant or slightly increasing income elasticity of intermediate imports. However, these results reflect the assumption that imports of technology and consumer goods will be further constrained due to balance-of-payments barriers.

Relaxing the above constraints and allowing for higher rates of growth (as in *Table 19.6*) would lead to an increase in growth rates of the NMP in the 1990s from 4% to 4.8%, accompanied by an increase in exports from 6.5% to 7.5%. The growth rate of total imports would stay at the level of ca. 7%, assuming that

*Table 19.6.* Forecasts of main economic indicators, 1986–1999 (percentage rate of growth in billions of 1982 zlotys, unless otherwise indicated).

| *Variable* | *1983* | *1984* | *1985* | *1986* | *1987* | *1988* | *1989* | *1990* | *1991* | *1992* | *1993* | *1994* | *1995* | *1996* | *1997* | *1998* | *1999* |
|---|---|---|---|---|---|---|---|---|---|---|---|---|---|---|---|---|---|
| Real NMP: | | | | | | | | | | | | | | | | | |
| – in billions of 1982 zlotys | 4856 | 5128 | 5301 | 5566 | 5787 | 6202 | 6450 | 6663 | 6929 | 7207 | 7516 | 7839 | 8193 | 8569 | 8981 | 9412 | 9863 |
| – growth rate | 6.0 | 5.6 | 3.4 | 5.0 | 4.0 | 4.0 | 4.0 | 3.3 | 4.0 | 4.0 | 4.3 | 4.3 | 4.5 | 4.6 | 4.8 | 4.8 | 4.8 |
| NMP deflator | 17.5 | 14.8 | 15.6 | 18.0 | 11.6 | 10.4 | 9.7 | 11.0 | 9.5 | 9.0 | 8.5 | 8.0 | 7.0 | 6.0 | 6.0 | 5.0 | 5.0 |
| Net investment/national income rate (real) | 18.6 | 19.8 | 19.8 | 19.4 | 21.5 | 22.2 | 22.8 | 23.6 | 24.0 | 25.0 | 26.0 | 26.0 | 25.0 | 24.0 | 24.0 | 22.0 | 22.0 |
| Rate of depreciation in MP sector (real): | | | | | | | | | | | | | | | | | |
| – % of fixed assets | 1.7 | 1.7 | 1.7 | 1.8 | 1.8 | 1.8 | 1.9 | 1.9 | 2.0 | 2.0 | 2.0 | 2.1 | 2.1 | 2.2 | 2.2 | 2.2 | 2.3 |
| – % of national income | 7.2 | 7.0 | 7.0 | 6.8 | 6.8 | 6.7 | 6.6 | 6.6 | 6.6 | 6.6 | 6.5 | 6.5 | 6.4 | 6.3 | 6.2 | 6.1 | 6.0 |
| Labor productivity per employee in MP sector | 6.7 | 5.6 | 4.1 | 4.3 | 4.0 | 3.3 | 3.3 | 3.3 | 3.3 | 3.3 | 2.9 | 2.9 | 3.2 | 3.3 | 3.5 | 3.5 | 4.2 |
| NMP originating in industry: | | | | | | | | | | | | | | | | | |
| – in billions of 1982 zlotys | 2364 | 2510 | 2603 | 2718 | 2835 | 2991 | 3148 | 3278 | 3442 | 3614 | 3806 | 4007 | 4219 | 4447 | 4692 | 4945 | 5207 |
| – growth rate | 5.7 | 5.3 | 3.7 | 4.4 | 4.3 | 5.5 | 5.3 | 4.1 | 5.0 | 5.0 | 5.3 | 5.3 | 5.3 | 5.4 | 5.5 | 5.4 | 5.3 |
| Labor productivity in industry per working hour | 6.0 | 6.4 | 3.8 | 4.5 | 4.8 | 6.1 | 6.5 | 5.3 | 6.2 | 5.6 | 6.5 | 6.1 | 6.0 | 6.3 | 6.5 | 6.5 | 6.5 |
| Employment | | | | | | | | | | | | | | | | | |
| – MP sector | 14.6 | 14.6 | 14.5 | 14.6 | 14.6 | 14.7 | 14.8 | 14.8 | 14.9 | 15.0 | 15.2 | 15.4 | 15.6 | 15.8 | 16.0 | 16.2 | 16.3 |
| – industry | 5.0 | 5.0 | 5.0 | 5.0 | 5.0 | 5.0 | 5.0 | 5.0 | 5.0 | 5.0 | 4.9 | 4.9 | 4.9 | 4.8 | 4.8 | 4.8 | 4.7 |
| Average working hours in socialized industry[a] | 1841 | 1840 | 1840 | 1839 | 1830 | 1820 | 1780 | 1760 | 1750 | 1750 | 1750 | 1750 | 1740 | 1740 | 1730 | 1730 | 1720 |
| Exports (real) | 10.3 | 9.5 | 1.3 | 4.6 | 4.6 | 6.0 | 6.3 | 5.9 | 6.4 | 6.5 | 7.0 | 7.0 | 7.2 | 7.4 | 7.5 | 7.6 | 7.6 |
| Imports (real) | 5.2 | 8.6 | 7.9 | 3.6 | 3.1 | 6.6 | 7.6 | 7.0 | 7.0 | 6.9 | 6.9 | 6.9 | 7.0 | 7.2 | 7.2 | 7.2 | 7.2 |

[a]Per employee per year.

accelerated imports of technology will be one of the main factors of increased NMP growth rate.

### 19.5.2. The forecasts

The basic forecast of the main economic indicators is summarized in *Table 19.6* and illustrated in *Figures 19.10–19.12*. The assumptions of the forecast were discussed at length in the last section. It may be seen that employment is constant until the middle of the 1990s and increases thereafter. Fast growth of investment outlays and accelerated scrapping, which allow for the restructuring of capital stock, will induce changes in the sectoral structure of employment. These changes reflect a decrease in employment in industry and simultaneous increase in employment in other branches of the MP sector (see *Table 19.6* and *Figures 19.10* and *19.12*).

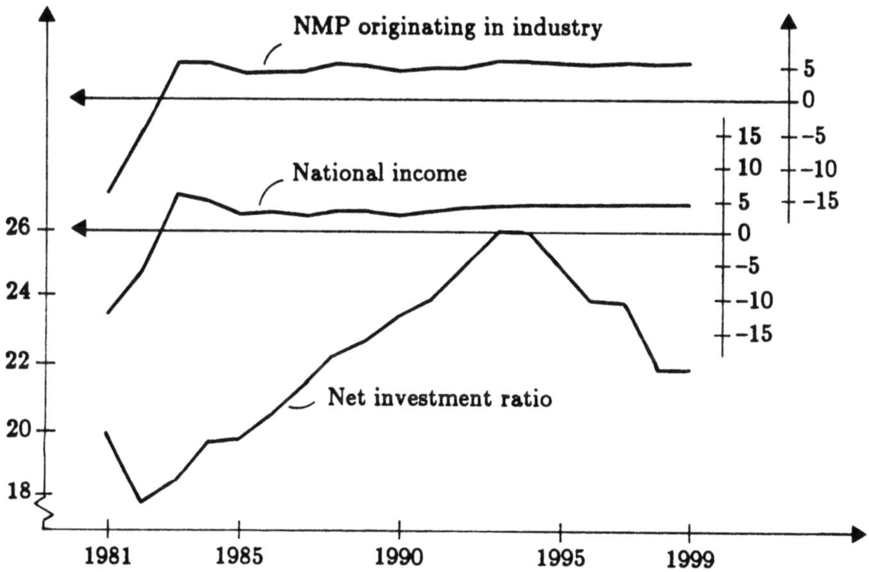


*Figure 19.10.* Net investment ratio, national income, and NMP originating in industry, 1981–1985 (observed) and 1986–1999 (estimated), in constant 1982 prices.

The impact of declining employment on the level of output in industry is compensated for by higher labor productivity in the 1990s in this sector, resulting from intensive factors. Labor productivity in the MP sector is assumed to grow in the last five years of the forecasting period (see *Figure 19.11*). Therefore the growth rate of NMP is expected to rise, from 4% in 1991–1992 to 4.8% in 1999 (*Figure 19.12*). Exports are expected to grow 1.5–1.6 times faster than NMP over the whole forecasting period. Imports will grow at a higher rate than exports until 1993, and then their growth rate will stabilize. Faster growth of imports in the 1980s and at the beginning of the 1990s is connected with the growing share of investment outlays in national income, which in turn requires

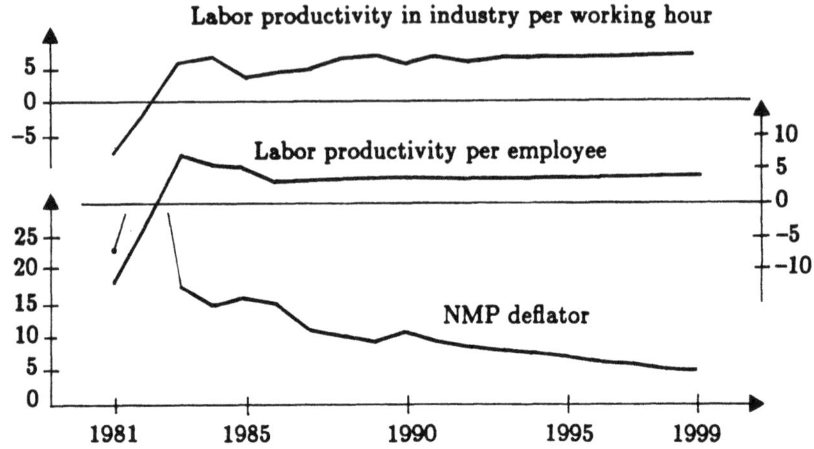


*Figure 19.11.* Growth rates of NMP deflator and labor productivity, 1981–1985 (observed) and 1986–1999 (estimated), in constant 1982 prices.

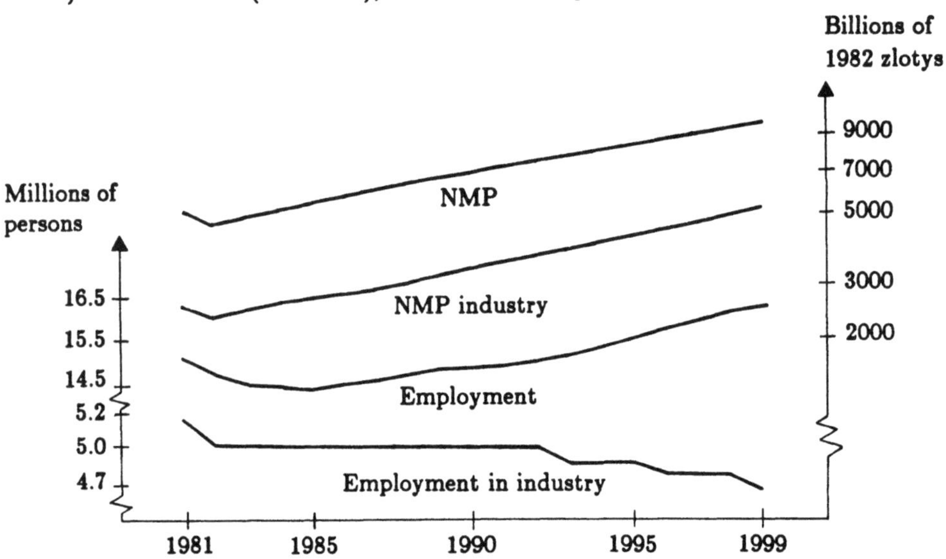


*Figure 19.12.* Logarithmic growth of employment and net material product, 1981–1985 (observed) and 1986–1999 (estimated), in constant 1982 prices.

additional imports of investment goods and also compensating imports of consumer goods. The forecast anticipates the use of new credits in years 1988–1990 to finance the increase of imports.

The picture of the development of Polish economy which we get from the forecast shows that the effects of structural changes which are currently being introduced will manifest themselves no sooner than in the late 1990s.

## Notes

[1] A comprehensive description of the W-5 model is provided in the papers of W. Welfe (1985a, b, c), prepared for IIASA within the project Economic Growth and Structural Change, ECO-I-1 and ECO-II-1.

[2] There are interesting attempts to apply the disequilibrium econometric framework to small macro-models, accentuating the short-term adjustments mostly of the central planner, but neglecting the production process. See Portes *et al.* (1983).

[3] As for demand for consumer goods, this problem has found its widest presentation in the series of studies by A. Welfe (1984, 1985, 1986). We also used here some suggestions of Kornai (1982).

## References

Czyżewski, A. B. (1986), *Free Currency Foreign Debt Model POLDEBT,* Technical Description, IIASA Research Project ECO-I-1 and ECO-II-1 (mimeo). Łódź.

Czyżewski, A.B. (1987), *Simulational Model of Poland's Free Currencies Debt* [in Polish], to be published in *Wiadomości Statystyczene,* **4**.

Kalecki, M. (1963), *Zarys teorii funkcjonowania gospodarki socjalistycznej.* Warsaw: PWN.

Kornai, J. (1982), *Growth, Shortage and Efficiency: A Macrodynamic Model of the Socialist Economy.* Oxford: Basil Blackwell.

Portes, R., Quandt, R.E., Winter, D., and Yeo, S. (1983), *Macroeconomic Planning and Disequilibrium: Estimates for Poland and 1955–1980.* Birkbeck College Discussion Paper 139 (revised).

Welfe, A. (1984), Demand Analysis in Conditions of Disequilibrium [in Polish], *Ekonomista,* **5**.

Welfe, A. (1985), *The System of Demand Equations in a State of Disequilibrium,* Collaborative Paper CP–85–35. (Laxenburg, Austria: International Institute for Applied Systems Analysis).

Welfe, A. (1986), *Intensity of Disequilibrium and Changes of Inventories,* Collaborative Paper CP–86–7. (Laxenburg, Austria: International Institute for Applied Systems Analysis).

Welfe, W. (1985a), Econometric Macromodels of Unbalanced Growth, *Prace IEiS* UL, **52** (Łódź).

Welfe, W. (1985b), *The Concept of an Integrated W-5 Model of the Polish Economy,* IIASA Research Project ECO-I-1 and ECO-II-1, (mimeo). Łódź and Laxenburg, Austria.

Welfe, W. (1985c), *The W-5 Integrated Model of the Polish Economy,* IIASA Research Project ECO-I-1 and ECO-II-1 (mimeo). Łódź.

CHAPTER 20

# Structural Change in the Czechoslovak Economy to the Year 2000

*Karl Zeman*

## Summary

This chapter summarizes the basic findings from analytical and forecasting work on feasible methods of restructuring of the Czechoslovak economy in the period until 2000. The chapter concentrates on changes in the basic proportions of the reproduction process under conditions of intensive economic growth and basic trends in structural development. These trends are tested by a model based on the assumption of linkage between the achieved level of economic development and the structure of net material product and employment.

## 20.1. Introduction

The implementation of the long-term economic strategy for the comprehensive intensification in the reproduction process requires structural changes in the whole Czechslovak economy.

The structure of the national economy represents a socioeconomic system that connects the division of labor and the efficiency of productive resources with economic interests and objectives. Its optimum development is the outcome of necessary changes in the combination and efficiency of utilized internal productive resources, the technical and economic level of the country, its participation in the international division of labor, and its priorities and objectives of socioeconomic development [Komárek (1985)].

The scope of structural changes connected with the intensification of the Czechoslovak economy in the period until the year 2000 can be compared with the scope of structural changes which occurred in the 1950s. They become more demanding when we assume that the implementation of these necessary structural changes will take place essentially on the basis of utilizing internal resources while providing at the same time an increase in living standards and strengthening the social confidence of the people [Potáč (1986)]. When foreign resources are called for, it will be necessary to limit them to specific purposes and to assure their rapid economic repayment.

## 20.2. Changes in the Proportionality of the Reproduction Process under Transition to Intensive Economic Growth

The transition to economic growth with a predominance of intensive factors will change the proportions of the reproduction process, which will have direct consequences for the structure of Czechslovak economy.

In the period 1986–2000 the following changes in the proportionality of the Czechslovak reproduction process will occur [see *Hlavní směry...* (1986), Komárek (1985), Potáč (1986), Šedivý (1985), Věrtelář (1986), and Vintrová 1985)]:

(1) There will be faster rates of growth of produced net material product (NMP) in comparison to the rate of growth of the gross output, which will be reflected in a decrease of productive consumption and an increase in the final product obtained from each unit of raw material and capacity. This trend was already apparent in the period 1981–1985 and is a component of plans for the period 1986–1990 (*Table 20.1*).

A change in the cited proportions of the reproduction process will translate into slower rates of growth in the primary sectors of the economy (agriculture and the extraction of raw materials). Other consequences will include lower requirements for investments in the primary sectors of the economy, which are usually more fixed asset-intensive. A decrease in fixed asset intensity will in turn exert an influence toward decreasing the rate of productive investments and hence slowing down the growth rate of production of the means of production (machines and installations for investments).

(2) There will be more rapid growth of NMP than of industrial production. For the first time in the history of a planned economy in Czechoslovakia, the relation between rates of growth of both these indicators will change in an essential manner. In the period 1986–1990 it is assumed that average annual rates of growth of NMP will be 3.5% and of gross industrial output 2.9%, while in the period 1981–1985 the average annual growth rate of NMP was 2.1% and gross industrial output 2.4%. This development in the relative growth rates of both indicators should also predominate in the 1990s [Věrtelář (1986)].

(3) There will be a slowing down of the growth rate of production of the means of production (a) in relation to growth rates in the production of consumer

goods, and (b) as a consequence of decreasing the share of material costs in the gross output (gross social product) and decreasing the rate of productive investments.

(4) There will be a slowing down of the decrease of the fixed assets/output ratio, which until now has been decreasing in most branches of the economy. It is assumed that in the second half of the 1980s a slowing down of the decrease of effectiveness will occur. Average annual rates of decline of the fixed asset/output ratio (characterized by the volume of NMP per unit of fixed assets in the material sphere), should achieve -1.1% during 1986–1990 as compared to -3.8% during 1981–1985 [Věrtelář (1986)].

(5) There will be the requirement that a decisive part of NMP increments be covered by the productivity of labor: in the period 1986–1990 in the range of 92–95%, as compared to over 80% in the period 1981–1985. According to the findings of the long-term forecast for the Czechoslovak economy for the period 1986–2000, it is assumed that the average annual rate of growth of the social productivity of labor will be approximately 3.4%, while the average annual rate of growth of NMP will reach a minimum of approximately 3.5%, which would essentially be covered by an increase in the social productivity of labor (i.e., by 96%) [*Hlavní směry...* (1986)].

*Table 20.1.* Growth of basic indicators[a] in the Czechslovak reproduction process (average annual percentage change).

| *Indicator* | *1981–1985*[b] | *1986–1990*[c] |
|---|---|---|
| Gross output (gross social product) | 1.7 | 2.2 |
| Productive consumption (without depreciation) | 1.3 | 1.2 |
| Gross NMP produced (including depreciation) | 2.1 | 3.5 |

[a]In comparative prices.
[b]Figures for 1985 represent expectation.
[c]Preliminary.
Source: Věrtelář (1986).

The work which has already been done on the long-term forecast for the Czechoslovak economy, as well as findings of research forecasting analyses, make it possible to identify basic trends and directions which structural changes will take in the period up to the year 2000.

The largest structural change will be the strengthening of the share of the tertiary sector (which includes productive and nonproductive services) and within its framework nonproductive services and an optimization of the share of industry in the structure of the material sphere [*Hlavní směry...* (1986), Komárek (1985), Šedivý (1985), and Vintrová (1985)].

A markedly lower share of the tertiary sector represents the greatest deviation in the structure of the Czechoslovak national economy in the middle of the 1980s. A strengthening of the share of this sector is evidently connected with an essential and long-term trend in the intensification process, which should also be implemented in the Czechoslovak economy.

The growth of the share of the tertiary sector should also be projected onto the allocation of resources, especially labor. Some forecasting hypotheses are based on the view that the absolutely predominant part of increments in employment during the next 15 years (if not their whole increment) should be oriented toward the tertiary sphere [Summers and Heston (1984)]. The implementation of this requirement is in harmony with the assumed objective, according to which the predominant part of the increment of NMP will be covered by an increase in the social productivity of labor [*Hlavní směry...* (1986)]. The growth rate of productivity of labor in the material sphere in this period would set free labor and/or would make it possible to direct increments in the labor force during the next 15 years to the tertiary sector.

It should be assumed, at the same time, that even when the tertiary sector receives priority in the allocation of resources, its share in the Czechoslovak economy in the year 2000 will still not have achieved levels which exist in comparable, developed market economies.

The basic macrostructural change within the material sphere will be the optimization of the share of industry. Changes in the relation between the rate of growth of industrial output and NMP in the period 1986–1990 will reflect this trend, which will continue throughout the 1990s, also as a consequence of assumed changes in the internal structure of Czechoslovak industry [Věrtelář (1986) and Zeman (1984, 1985)].

## 20.3. A Model for Testing Development Trends in the Structure of the Czechoslovak Economy

Our model for testing development trends in the structure of the Czechoslovak economy is based on an understanding of economic growth as a transformation process. In its course, changes in the structure of the Czechoslovak economy depend upon changes in economic (and other) conditions. In this context the linkage between the achieved level of economic development and the structure of NMP produced (and used) is considered to be of basic importance. This transformation process contains elements of both continuity with past developments and of breaks that guide the development of the economy to new trajectories of economic growth.

The strategy of a transition to the intensification of economic growth adopted by the European member countries of the Council for Mutual Economic Assistance (CMEA) has many common features, which are reflected in the structural adaptation of their economies. For this reason we have made use of comparisons with such trends in other selected European CMEA member countries in testing the model of development trends in the structure of the Czechoslovak economy. Our basic critierion for selecting these countries was the level of economic development they have achieved (a higher level as far as the GDR is concerned, and a lower one for Hungary and Poland) and the fact that they are found in the same economic region of Europe as Czechoslovakia.

### 20.3.1. Hypothesis

Our hypothesis for the testing (verification) of development trends in the macroeconomic structure of the Czechoslovak and other selected European CMEA economies was based on the assumption of a linkage between the achieved level of economic development, the structure of NMP, and employment. The existence of this linkage is also reflected in the changes which – on the basis of input–output relations – exist in the structure of factors of economic growth. With the increased development of productive forces, opportunities for combining productive factors change. Opportunities for substitution exist between both factors of economic growth (labor force, fixed assets, natural resources) and the factors which influence the efficiency of the reproduction process as a whole (i.e., the qualification of the labor force, attitudes and values of people in the reproduction process, production technologies and structure of the total economy).

The assumption that each stage of economic development in a specific country is related (ceteris paribus) to a specific structure of the national economic complex [UN (1963, 1977) and UNIDO (1980)] has been strengthened by the theory of economic complexes in the world economy, for the purpose of analyzing the economic development of the socialist countries [Tauchman (1971, 1981].

On the basis of this theory, we can formulate a working hypothesis according to which changes in the structure of national economic complexes in time and space (in relation to other national economic complexes) are the outcome of changes in economic conditions, i.e., the achieved level of development of productive forces (characterized in a summary manner by the indicator of economic level), natural conditions and the status of the given complex relative to the national economic complexes of other countries.

### 20.3.2. Specification of a model

The working hypothesis cited above has been applied with the aid of a model which makes it possible to distinguish two basic types of development of mutual links between the structure of national economic complexes (vertical–input and horizontal–output) and the level of economic development:

(1) A monotonic type of development which takes place when the share of a given part in the whole grows or declines with the growth of the absolute level of economic development.

(2) A nonmonotonic type of development, which takes place when the share of a given part in the whole grows with an increase in the level of economic development and in further stages of development declines and/or is stabilized or vice versa.

On the basis of existing experience, the following regression function has proved to be best suited for expressing a nonmonotonic type of relationship since it makes it possible to project the structure of a given aggregate indicator directly (without any corrections and external intervention, i.e., fulfilling the requirement that the sum of elements or shares in the analyzed aggregate = 1.00) [Scheper and Reichenbach (1973)]:

$$\log \frac{s}{1-s} = \log a + b \log U + n \tag{20.1}$$

where:

log = base-10 logarithms

$s$ = share of partial component in whole

$a$ = const.

$b$ = regression coefficient

$U$ = absolute magnitude of economic level

$n$ = random variable

Regression function (20.1) in a modified form was used for the purpose of making a projection of the structure of the NMP and employment (natural logarithms were used):

$$\ln \frac{S_1}{S_2} = b + a \ln U \tag{20.2}$$

in which $S_1/S_2$ = the relation of complementary groups.

To obtain the dependent variables $s$ and/or $S_1/S_2$, it is always necessary to divide the unit under analysis (NMP, employment) into two complementary groups, because the adding-up constraint is preserved only in that way. Thus, a whole sequence of divisions into two parts was applied [1].

By using function (20.1) and/or (20.2) we can estimate the ratio of shares of complementary subgroups in the higher groups as a function of the absolute magnitude of economic level ($U$). Estimation functions (20.1) and (20.2) also make it possible to identify the nonmonotonic links that "forecast" the development of the individual shares in a logistic trend, which corresponds to the real development of national economic structure and productive forces.

In spite of the fact that this estimation function "forecasts" a slower development of structural change than, for instance, exponential types of functions, if we accept the hypothetical assumption of unlimited economic growth (i.e., a situation where $U \to \infty$), its application would lead to a situation where the share of one industry would approach one and all others would approach zero. This theoretical finding could also lead us to the conclusion that we are, to a certain extent, over-valuing the speed of structural changes, even when we are using this type of function.

The specification of the model makes it clear that the forecasting analysis has been developed on the basis of the genetic method. A model of logistic trends based on analyses of the development of time series has been used. This introduces the assumption that a similar type of development of links between economic level (an explanatory variable) and the shares (relations) of complementary groups (dependent variables), which occurred in the interpolated period (analysis), will also occur in the projected period (extrapolation) of the model.

### 20.3.3. Statistical information

Analysis of development trends in the macroeconomic structure of the Czechoslovak economy (and comparison countries) was developed on the basis of statistical data obtained from IIASA [Dobrinsky (1986)] and CMEA [Statističeskij... (n.d.)] statistical sources, which characterized:

(1) The development of the structure of NMP in national currencies and constant prices during 1960–1982, extrapolated to the year 1985 with the help of a net material product (NMP) index.
(2) The development of the structure of population employed in the material sphere and in the whole national economy.
(3) The level of achieved economic development ($U$) on the basis of an internationally comparable per capita indicator of gross domestic product (GDP) (in comparable currency equivalents – US dollars). Data from international UNO projects have been used for this purpose [Summers and Heston (1984)].

The analysis has been carried out by considering individual branches of the material sphere, i.e., agriculture (including forestry), industry, construction, transportation and communications, trade and others (their sum = 100.0%) and branches of the material and nonmaterial spheres (in terms of employment), i.e., agriculture (including forestry), industry, construction, transportation and communications, trade and the nonmaterial sphere (NMS), according to a CMEA classification of branches.

Interpolation of the model was carried out for two periods, to which two alternative extrapolations of the model (1986–2000) correspond: $A$ = 1960–1985; $B$ = 1970–1985. The choice of these periods was determined by the differentiation (change) of targets and means found in economic growth trajectories and in adaptation of the structure of the national economy to changes of foreign and domestic conditions in Czechoslovakia and the comparison countries.

## 20.4. *Ex ante* Simulation

The purpose of extrapolating the model's parameters was not to compare alternative macroeconomic projections of the Czechoslovakian national economic complex (characterized by the structure of NMP and by employment) with those of the selected European CMEA countries, but rather to identify the consequences of adaptation to changes in domestic and foreign economic conditions. For this reason the cited alternative projections should not be understood as forecasts of the actual development of macroeconomic structures in the countries being compared, but as the verification of changes in their structure and validation of the links between variables in the model for the period of extrapolation (1986–2000) [2].

For interpolation (and extrapolation) requirements, NMP and employment have been divided into complementary groups ($s$) of branches according to criteria derived from our working hypothesis, i.e., according to the development of their share, depending on the achieved level of economic growth, which is characterized by the indicator of economic level ($U$).

The following changes have taken place in the structural proportions of the sphere of material production (i.e., net material product and employment), together with the growth of absolute magnitudes of economic level ($U$ = explanatory variable) during the period 1960–1985:

(1) The share of agriculture has been declining.
(2) A deceleration of increases in the share of industry, including construction, in the structure of material production (without agriculture) has occurred.
(3) Unit increments of the share of industry have been accompanied by decreasing increments in the share of construction.
(4) The structure of productive services (transport, communication and trade) has changed in favor of trade.
(5) As far as the structure of employment within the whole national economy is concerned, an increase in the absolute magnitude has been accompanied by the following changes during the period 1960–1984:
    - The share of material sphere branches has declined.
    - The share of agriculture in the structure of material sphere productive branches has declined.
    - The share of the nonmaterial sphere (services) in the structure of the tertiary sector has increased (material and nonmaterial services).
    - Changes in the share of industry have been accompanied by smaller increments in the share of construction.
    - The structure of productive services (transport, communication, trade) has changed in favor of trade.

Our interpretation of the findings obtained by extrapolating the parameters used in the model for the period 1985–2000 are based on the assumption that these structural trends will continue and will be influenced by the assumed acceleration of economic growth rates in all the countries compared. This acceleration should also be reflected in accelerated annual rates of growth of GDP per capita ($U$) in the period 1985–2000 (CSSR 3.0%, GDR 4.6%, Hungary

3.1%, Poland 3.2%), when compared with the period 1961–1984 (CSSR 2.0%, GDR 3.3%, Hungary 2.6%, Poland 1.9%) [UN (1986); Summers and Heston (1984)].

### 20.4.1. Structural shifts within the material sphere

A characteristic development in the branch structure of the material sphere, both in Czechoslovakia and the comparison countries during the past 25 years (1960–1985), was the creation of a structure that corresponds to the achieved high level of economic and industrial development. Both in Czechoslovakia and the countries compared with it, as the absolute magnitude of the economic level increases, the share of agriculture decreases and the share of industry, in the structure of both net material product (*Table 20.2*) and employment in the material sphere (*Table 20.3*), increases.

The differences in the shares of agriculture and industry that exist among the countries under comparison, to a great extent, reflect differences in their achieved level of economic development ($U$). Interpolation of our model indicates a generally very close linkage between economic growth and declining share of agriculture (including forestry) in the structure of both NMP and employment in the material sphere (details on the econometric estimations are available upon request).

During the period 1970–1985 the tight nature of these links decreased, especially in Hungary and Poland. This was due to the influence of economic policies oriented in a more marked manner toward achieving agricultural self-sufficiency in Czechoslovakia and the GDR, the increased importance of this branch in creating the specialization profile of the economy in Hungary, and dealing with the specific conditions which existed in the national economy of Poland (especially in the period 1980–1984). These influences have been projected into the extrapolated parameters of the model according to alternative $B$ (with the exception of the structure of the NMP produced in Czechoslovakia, where the decline in the share of agriculture has been accelerating).

The real development of the share of agriculture in the structure of the NMP, both in Czechoslovakia and the countries under comparison, in the period up to the year 2000 will probably tend toward a slower decline, depending upon the increased role of agricultural output in creating a specialization profile in these countries (for instance, in Hungary and Poland). Decreases of the share of agriculture in employment in the material sphere will depend on opportunities for increases in the productivity of labor in this branch.

The strategy for economic development in Czechoslovakia and in the countries under comparison for the next 15 years is based on the assumption of maintaining the dominant role of the industrial complex. On the other hand, we can assume that during this period growth rates of the share of this branch in the NMP structure will slow down.

*Table 20.2.* Net material product structure (percentage shares at constant prices).

| *Country/ sector* | *1960* | *1970* | *1985* | *Projection* *1990* *A* | *B* | *2000* *A* | *B* |
|---|---|---|---|---|---|---|---|
| CSSR | | | | | | | |
| Agriculture and forestry | 16.9 | 12.0 | 7.6 | 6.2 | 5.4 | 4.0 | 3.0 |
| Industry | 51.4 | 57.6 | 58.6 | 59.5 | 58.8 | 59.6 | 56.7 |
| Construction | 10.9 | 11.4 | 10.9 | 12.0 | 10.1 | 12.2 | 8.3 |
| Transport and communication | 4.9 | 4.3 | 4.7 | 4.2 | 5.2 | 3.9 | 6.0 |
| Trade | 15.2 | 13.8 | 17.9 | 17.6 | 20.4 | 20.1 | 26.0 |
| GDR | | | | | | | |
| Agriculture and forestry | 17.6 | 12.7 | 8.3 | 6.0 | 6.2 | 3.8 | 4.1 |
| Industry | 58.2 | 62.2 | 69.1 | 70.3 | 71.8 | 73.2 | 76.0 |
| Construction | 5.9 | 6.9 | 6.3 | 6.5 | 5.7 | 6.4 | 4.9 |
| Transport and communication | 5.1 | 4.8 | 4.2 | 4.1 | 4.0 | 3.7 | 3.6 |
| Trade | 10.6 | 10.4 | 9.5 | 9.9 | 9.2 | 9.6 | 8.4 |
| Hungary | | | | | | | |
| Agriculture and forestry | 35.3 | 21.7 | 16.2 | 10.9 | 12.3 | 5.7 | 7.9 |
| Industry | 36.7 | 44.1 | 49.5 | 50.8 | 53.4 | 51.1 | 58.7 |
| Construction | 10.5 | 11.5 | 9.6 | 10.2 | 9.4 | 9.1 | 7.7 |
| Transport and communication | 4.8 | 6.0 | 5.2 | 5.6 | 5.0 | 5.5 | 4.3 |
| Trade | 12.3 | 15.0 | 17.0 | 19.8 | 18.0 | 23.9 | 19.4 |
| Poland | | | | | | | |
| Agriculture and forestry | 38.2 | 25.1 | 17.3 | 14.8 | 15.4 | 7.8 | 9.7 |
| Industry | 31.9 | 42.3 | 49.2 | 49.3 | 47.8 | 53.9 | 47.6 |
| Construction | 13.2 | 15.1 | 11.3 | 15.1 | 16.7 | 15.0 | 22.4 |
| Transport and communication | 3.0 | 3.5 | 4.9 | 4.4 | 4.2 | 5.4 | 4.3 |
| Trade | 13.1 | 13.2 | 15.2 | 14.9 | 14.7 | 16.0 | 14.9 |

Source: Dobrinsky (1986).

*Table 20.3.* Employment in the material sphere structure (percentage shares).

| | | | | Projection | | | |
|---|---|---|---|---|---|---|---|
| | | | | 1990 | | 2000 | |
| *Country/ sector* | *1960* | *1970* | *1985* | *A* | *B* | *A* | *B* |
| CSSR | | | | | | | |
| Agriculture and forestry | 30.2 | 22.7 | 17.5 | 13.9 | 13.6 | 9.8 | 11.0 |
| Industry | 43.7 | 47.0 | 48.1 | 48.3 | 48.4 | 47.7 | 48.2 |
| Construction | 9.8 | 10.9 | 12.2 | 13.6 | 13.3 | 15.2 | 14.0 |
| Transport and communication | 7.1 | 8.4 | 8.4 | 8.7 | 8.3 | 8.9 | 8.2 |
| Trade | 9.2 | 11.0 | 13.8 | 15.5 | 16.4 | 18.4 | 18.6 |
| GDR | | | | | | | |
| Agriculture and forestry | 20.4 | 16.1 | 13.3 | 9.9 | 9.5 | 7.2 | 7.7 |
| Industry | 49.8 | 51.8 | 54.6 | 56.4 | 57.7 | 57.7 | 59.4 |
| Construction | 7.5 | 9.8 | 9.4 | 11.2 | 9.7 | 12.9 | 9.7 |
| Transport and communication | 8.5 | 8.9 | 9.3 | 9.7 | 9.6 | 10.0 | 9.7 |
| Trade | 13.8 | 13.4 | 13.4 | 12.8 | 13.5 | 12.2 | 13.5 |
| Hungary | | | | | | | |
| Agriculture and forestry | 45.4 | 30.8 | 29.0 | 21.8 | 25.3 | 15.6 | 23.5 |
| Industry | 33.2 | 43.1 | 39.0 | 42.8 | 37.2 | 42.8 | 34.0 |
| Construction | 6.5 | 8.6 | 9.3 | 11.3 | 10.0 | 13.4 | 10.1 |
| Transport and communication | 7.2 | 8.0 | 10.0 | 10.6 | 11.7 | 11.8 | 13.2 |
| Trade | 7.7 | 9.5 | 12.7 | 13.5 | 15.8 | 16.4 | 19.2 |
| Poland | | | | | | | |
| Agriculture and forestry | 50.4 | 44.5 | 37.0 | 32.7 | 33.6 | 24.2 | 31.4 |
| Industry | 29.2 | 33.4 | 36.4 | 39.1 | 37.9 | 43.3 | 38.3 |
| Construction | 7.5 | 8.5 | 9.2 | 10.5 | 9.6 | 12.2 | 9.4 |
| Transport and communication | 6.0 | 6.7 | 7.7 | 8.1 | 8.1 | 9.0 | 8.4 |
| Trade | 6.9 | 6.9 | 9.7 | 9.6 | 10.8 | 11.3 | 12.4 |

Source: Scheper and Reichenbach (1973).

Construction's share in the structure of net material product will, to a greater extent, be influenced by that branch's structural adaptations to new requirements for the reconstruction of housing and the modernization of the productive base. Pressures to increase efficiency in this branch in all countries will probably lead to a stabilization of its share in the structure of the NMP and employment in the material sphere.

The continuation of development trends from the past period (1960–1984) into the period for which the model has been extrapolated (*Tables 20.2* and *20.3*) would lead to a reproduction of existing characteristics of structural proportions in the material sphere, i.e., a fairly high share for industry (including construction) and insufficient development of productive services (transportation, communications and trade). Analyses and forecasting studies agree that one of the decisive prerequisites for the intensification of the material sphere, especially in the economically highly developed European CMEA member countries (i.e., Czechoslovakia and the GDR), is the adequate (proportional) development of productive (and nonproductive) services [Komárek (1985), Šedivý (1985), and Vencovský (1986)].

Extrapolations of the model indicate that a continuation of existing production trends would mean a general continuation of the decline in the share of transportation and communications (which is very marked in Czechoslovakia) and of the trade sector in the GDR and Poland. Only in Czechoslovakia and Hungary is a trend toward a continuing increase in the share of trade in the structure of the net material product in evidence. These trends are also apparent in the development of employment structures in the sphere of material production.

### 20.4.2. Structural shifts within the national economy

The available statistical data make it possible to characterize the development of the structure of the whole national economy in Czechoslovakia on the basis of the structural proportions of total employment.

Employment development in Czechoslovakia (and the countries under comparison) is characterized by a significantly close link between economic growth and decline in the share of agriculture (including forestry), a slight decline or no change in the share of industry (including construction) and an increase in the share of the nonmaterial sphere.

The branch structure of employment (*Table 20.4*) in the analyzed countries allows us to identify the influence of differences in the achieved level of economic development in these countries: a lower share of agriculture and a higher share of industry (including construction) in Czechoslovakia and the GDR in comparison with Hungary and Poland. Smaller differences exist among Czechoslovakia, the GDR and Hungary as far as productive and nonproductive services are concerned.

A continuation of existing development trends in the structure of total employment will strengthen an increase in the share of the tertiary sector (i.e., transport and communication, trade and nonproductive services), as well as

*Table 20.4.* Employment structure (percentage shares of total employment). Source: Scheper and Reichenbach (1973).

| | | | | Projection | | | |
|---|---|---|---|---|---|---|---|
| | | | | 1990 | | 2000 | |
| *Country/ sector* | *1960* | *1970* | *1984* | *A* | *B* | *A* | *B* |
| CSSR | | | | | | | |
| Agriculture and forestry | 26.2 | 18.6 | 13.8 | 10.6 | 10.2 | 7.0 | 7.9 |
| Industry | 37.7 | 38.6 | 38.0 | 37.2 | 36.5 | 34.5 | 34.6 |
| Construction | 8.5 | 9.0 | 9.7 | 10.4 | 10.0 | 11.0 | 10.0 |
| Transport and communication | 6.1 | 6.8 | 6.3 | 6.9 | 6.3 | 6.9 | 6.1 |
| Trade | 7.9 | 9.1 | 10.9 | 12.1 | 12.5 | 14.2 | 13.7 |
| Nonmaterial sphere | 13.6 | 17.9 | 21.3 | 22.8 | 24.5 | 26.4 | 27.7 |
| GDR | | | | | | | |
| Agriculture and forestry | 17.3 | 13.0 | 10.6 | 7.6 | 7.5 | 5.2 | 6.0 |
| Industry | 42.2 | 42.0 | 43.5 | 43.5 | 45.4 | 42.8 | 46.0 |
| Construction | 6.3 | 8.0 | 7.5 | 8.8 | 7.6 | 9.6 | 7.5 |
| Transport and communication | 7.2 | 7.2 | 7.4 | 7.6 | 7.6 | 7.7 | 7.7 |
| Trade | 11.6 | 10.9 | 10.6 | 10.0 | 10.8 | 9.4 | 10.8 |
| Nonmaterial sphere | 15.4 | 18.9 | 20.4 | 22.5 | 21.1 | 25.3 | 22.0 |
| Hungary | | | | | | | |
| Agriculture and forestry | 38.9 | 26.1 | 23.5 | 17.2 | 18.9 | 11.7 | 16.2 |
| Industry | 28.4 | 36.6 | 31.6 | 34.0 | 28.4 | 32.0 | 24.4 |
| Construction | 5.6 | 7.3 | 7.4 | 9.0 | 7.6 | 10.1 | 3.3 |
| Transport and communication | 6.2 | 6.8 | 8.1 | 8.7 | 8.9 | 9.7 | 9.4 |
| Trade | 6.6 | 8.0 | 10.3 | 11.1 | 11.9 | 13.5 | 13.7 |
| Nonmaterial sphere | 14.3 | 15.2 | 19.1 | 20.0 | 24.3 | 23.0 | 29.0 |
| Poland | | | | | | | |
| Agriculture and forestry | 45.2 | 39.0 | 30.3 | 27.0 | 27.1 | 18.6 | 24.4 |
| Industry | 26.1 | 29.4 | 29.8 | 32.3 | 30.7 | 33.2 | 29.9 |
| Construction | 6.7 | 7.3 | 7.4 | 8.6 | 7.7 | 9.3 | 7.3 |
| Transport and communication | 5.4 | 5.9 | 6.3 | 6.7 | 6.6 | 7.3 | 6.6 |
| Trade | 6.2 | 6.0 | 8.0 | 8.0 | 8.7 | 9.2 | 9.7 |
| Nonmaterial sphere | 10.4 | 12.4 | 18.2 | 17.4 | 19.2 | 22.4 | 22.1 |

more marked increases in the share of nonproductive services (*Table 20.4*). The share of nonproductive services within the structure of the tertiary sector is increasing in all the countries analyzed (only in Hungary is this share stable). We can assume a continuation of this trend for the period until the year 2000.

The development of the employment structure within the national economy of Czechoslovakia (and the countries compared) indicates that objective trends exist which are characteristic links between degrees of economic development and structure of national economic complexes, just as in other developed countries. But they are modified by the influence of preceding historical developments and the different target function for the economies of the socialist countries.

In the context of the achieved level of economic development in Czechoslovakia (and the countries compared), there will be a more marked tendency toward optimization in the share of agriculture, stabilization in the share of industry and an increase in the share of productive and nonproductive services in the structure of the national economy. The trend toward an increase in the share of the tertiary sector is in harmony with the requirements and demands for improved living standards. The efficient development of productive and nonproductive services also creates the prerequisites for an intensification of production in the primary and secondary sectors. The growth of nonproductive services is of particular significance for the quantitative and qualitative reproduction of the labor force. On the other hand, the development of the tertiary sector depends upon increases in the productivity of labor in the secondary sector (and in the primary sector), especially from the point of view of freeing and transferring labor from this sector to the tertiary sector.

The cited development trends in the structure of employment in the period for which the model was extrapolated will be corrected by the specific economic conditions under which the planned development of the Czechoslovak national economy will be implemented. Under these conditions, transfers of labor force among the three sectors will be influenced by opportunities for increases in the productivity of labor in those sectors which will free labor (especially in various branches of manufacturing and partially also agriculture). We can thus assume that the above-cited alternatives (*A* and *B*) for the development of the employment structure express conditions for the division of labor in the national economy which accompany the growth of the economic level. It seems that a clear relationship exists between the efficient development of the secondary sector (and the whole national economy) and the development of the tertiary sector.

## Notes

[1] The subdivision of NMP is as follows:

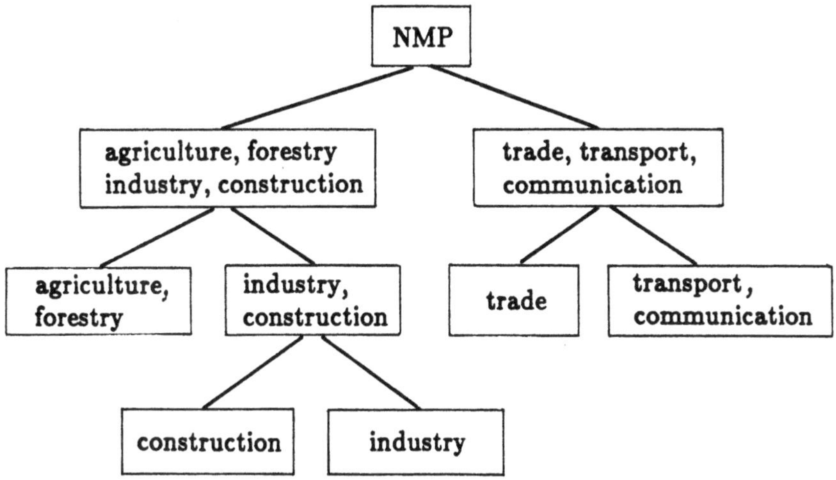


The subdivision of employment follows this scheme.

[2] Regression analysis was performed by J. Hutař, research member of the Central Institute of Economic Research, Prague.

## References

Dobrinsky, R. (1986), *The Macroeconomic Data Base for the CMEA Countries*, Working Paper WP-86-054. (Laxenburg, Austria: International Institute for Applied Systems Analysis.

*Hlavní směry hospodářského a sociálního rozvoje ČSSR na léta 1986–2000 a výhled do r. 2000 (schváleno XVII. sjezdem KSČ)* [Main trends in the economic and social development of Czechoslovakia in the period 1986–2000 and forecasts till the year 2000 (approved by the 17th Congress of the Communist Party of Czechoslovakia)] (1986), *Hospodářské noviny*, No. 15, Appendix.

Komárek, V. (1985), Strukturální změny – ano, ale jaké? [Structural changes – yes, but of what kind?], *Hospodářské noviny*, No. 8.

Potáč, S. (1986), Vstupujeme do nové kvality hospodářského a sociálního rozvoje [We are entering a period of new quality in our economic and social development], *Hospodářské noviny*, No. 14.

Scheper, W. and Reichenbach, H. (1973), Die Entwicklung der Anteile der Wirtschaftsbereiche am Bruttoinlandsprodukt. Eine Strukturprognose, *Weltwirtschaf. Archiv*, **109**, 291–320.

Šedivý, Z. (1985), Základní předpoklady a podmínky intenzifikace čs. ekonomiky v 90. letech [Basic prerequisites and conditions for intensifying the Czechoslovak economy in the nineties], *Nová mysl*, **29**, 23–32.

*Statističeskij ježegodnik stran – členov SEV* (n.d.) (Moskva: SEV).

Summers, R. and Heston, A. (1984), Improved international comparisons of real product and its composition: 1950–1980. *Review of Income and Wealth*, **30**, 215–216, 261.

Tauchman, J. (1971), *Nástin teorie ekonomického komplexu ve světové ekonomice* [Theory of economic complex in the world economy]. (Praha: Academia).

Tauchman, J. (1981), Znovu k teorii ekonomického komplexu ve světové ekonomice [Once again on the theory of economic complex in the world economy]. *Polit. Ekon.*, **20**, 915–924.

UN (1963), *A Study of Industrial Growth.* (New York: United Nations).

UN (1977), *Structural Change in European Industry.* (New York: United Nations).

UN (1986), *Economic Survey of Europe in 1985–1986* (1986), (New York: United Nations).

UNIDO (1980), *World Industry since 1960: Progress and Prospects.* (Vienna: United Nations Industrial Development Organization).

Vencovský, F. (1986), Život prokáže naši připravenost [Life will show how well prepared we are], *Hospodářské noviny*, No. 18.

Věrtelář, V. (1986), Osmá pětiletka [The eighth five-year plan], *Hospodářské noviny*, No. 3.

Vintrová, R. (1985), Reprodukční proporce intenzívního rozvoje (Kaktuálním otázkám dlouhodobé prognózy) [The reproduction process with intensive development (On current problems of long-term forecasts)], *Politická ekonomie*, **33**, 909–921.

Zeman, K. (1984), Prognostická úvaha o vývoji základních strukturálních parametru čs. průmyslu do r. 1955 [Forecasts of the development of basic structural parameters in the Czechoslovak industry till the year 1955], *Trend*, **16**, No. 5.

Zeman, K. (1985), Projekce odvětvové struktury průmyslu evropských členských zemí RVHP do r. 2000 [Projections of the branch structure of industry in the European member countries of the CMEA till the year 2000], *Trend*, **17**, No. 5.

CHAPTER 21

# Economic Development of the CSSR: Analysis and Conditional Forecasts

*Ivan Sujan*

## Summary

This chapter contains the results of analysis and forecasts of the Czechoslovak economy covering the period 1970–2000. We have used an econometric macromodel which explains the relationships between basic indicators of the Czechoslovak economy, as well as the foreign trade flows between European CMEA countries. The *ex post* analysis indicates the sources of slowdown in economic growth: deterioration of some external economic conditions and relative exhaustion of extensive growth resources are identified as the main negative factors. Conditional forecasts show that the planned long-term growth targets would be feasible under certain changes in economic structure and management, e.g., more investment to reduce the material and energy intensity of production.

## 21.1. Introduction

During the past 15 years the Czechoslovak economy has faced serious problems in external and internal economic conditions. Similar to other Eastern European countries, these problems resulted in a slowdown of economic growth [UN (1986)]. The average annual growth rate of gross national income (GNI, i.e., national income originating in the material sphere extended by depreciation charges – the main synthetic planning indicator in Czechoslovakia) declined from 5.85% in 1970–1975 to 3.83% in 1975–1980 and 2.16% in 1980–1985.

In published official documents a certain acceleration of economic growth at an annual rate of at least 3.5% is envisaged for the period 1986–2000 [*Hlavna*

*Table 21.1.* Aggregate factor contributions to the growth rates of selected indicators (in constant prices).

| | *Gross national income (GNI)* | | | *Imports* | | | *Exports* | | | *Domestic use of GNI* | | | *Personal consumption* | | | *Investment* | | |
|---|---|---|---|---|---|---|---|---|---|---|---|---|---|---|---|---|---|---|
| *Group of factors (F)* | *70–80* | *80–85* | Δ | *70–80* | *80–85* | Δ | *70–80* | *80 85* | Δ | *70–80* | *80–85* | Δ | *70–80* | *80–85* | Δ | *70–80* | *80–85* | Δ |
| Total average annual rate of growth (%) | 4.84 | 2.16 | −2.68 | 4.52 | 1.01 | −3.51 | 6.28 | 5.40 | −0.88 | 4.30 | 0.62 | −3.68 | 3.22 | 0.98 | −2.24 | 5.77 | −1.10 | −6.87 |
| Contributions of: | | | | | | | | | | | | | | | | | | |
| External $F$ | 0.39 | −0.64 | −1.03 | 0.72 | −2.62 | −3.34 | 2.04 | 1.48 | −0.56 | 0.02 | −1.99 | −2.01 | 0.42 | −0.68 | −1.10 | −0.61 | −2.18 | −1.57 |
| Intensive growth $F$ | 1.48 | 1.56 | 0.08 | 1.16 | 1.70 | 0.54 | 2.46 | 2.54 | 0.08 | 1.05 | 1.22 | 0.17 | 0.94 | 0.98 | 0.04 | 1.28 | 2.15 | 0.87 |
| Extensive growth $F$ | 1.96 | 0.77 | −1.19 | 1.60 | 0.56 | −1.04 | 0.34 | 0.07 | −0.27 | 2.37 | 0.97 | −1.40 | 2.04 | 0.83 | −1.21 | 2.49 | 0.51 | −1.98 |
| Direct influences of central management | −0.09 | −0.19 | −0.10 | −0.62 | −0.57 | 0.05 | −0.06 | −0.03 | 0.03 | −0.26 | −0.37 | −0.11 | −1.52 | −0.81 | 0.71 | −0.02 | −1.64 | −1.62 |
| Other internal $F$ | 1.10 | 0.66 | −0.44 | 1.66 | 1.94 | 0.28 | 1.50 | 1.34 | −0.16 | 1.12 | 0.79 | −0.33 | 1.34 | 0.66 | −0.68 | 2.63 | −0.06 | −2.57 |

*smery...* (1986)]. This development strategy is mainly based on the assumed transition to an intensive (resource efficient) path of economic growth.

In this chapter some results of an *ex post* analysis and conditional forecasts of the most important indicators of the Czechoslovak economy are presented. The main purpose of the *ex post* analysis (Section 21.3) is to identify the sources of slowdown in the growth rates in 1980–1985 as compared with the period 1970–1980. Conditional forecasts (Section 21.4) show the feasibility of long-term growth rates until the year 2000 in several variants under different scenarios. Both the *ex post* analysis and conditional forecasts are based on computer simulations using our econometric macromodel

## 21.2. The Econometric Macromodel

In our simulation analysis and forecasts, we used an econometric macromodel which consists of two partial models:

(1) A central econometric model of the Czechoslovak economy (version CEM-4 consisting of 218 equations and 349 variables).
(2) A matrix model of foreign trade between European CMEA countries (version MMZO-3.2 consisting of 43 equations and 129 variables).

Both partial models are parts of the System of Models for Macroeconomic Analyses (SMMA), which has been developed in the Institute of Socio-Economic Information and Automation in Management (VUSEI-AR, formerly Computing Research Center UNDP) in Bratislava [Klas (1985)]. The models are stochastic, dynamic and nonlinear.

A description of the models used may be obtained from the author. Information on the earlier (slightly different) versions of these models is given in Sujan (1984) and Sujan and Strauch (1986).

## 21.3. Economic Growth in the CSSR, 1970–1985

### 21.3.1. The simulation procedure

We identified 17 significant factors (see *Table 21.2* below), classified into five groups (see *Table 21.1*), which contributed to the changes in growth rates of the main macroeconomic indicators. Each factor is represented by a group of exogenous variables of the joint econometric macromodel. A set of dynamic simulations was used to separate the contributions of each factor to the growth rates of endogenous variables in the period under investigation.

This simulation procedure was used separately for the periods 1970–1980 and 1980–1985. Positive or negative changes in the factor contributions to the average annual growth rates of endogenous variables in the two periods indicated the sources of acceleration or deceleration in economic growth.

*Table 21.2.* Factor contributions to the growth rates of GMP produced and used (in constant prices).[a]

| | | *Gross national income (GNI)* | | | *Domestic use of GNI* | | |
|---|---|---|---|---|---|---|---|
| *Factors* | | *70–80* | *80–85* | Δ | *70–80* | *80–85* | Δ |
| Total average annual rate of growth (%) | | 4.84 | 2.16 | -2.68 | 4.30 | 0.62 | -3.68 |
| Contributions of: | | | | | | | |
| $A_1$ | World economic activity | 0.15 | -0.05 | -0.20 | -0.36 | -0.20 | -0.16 |
| $A_2$ | Position of the CSSR economy in CMEA | -0.01 | 0.26 | 0.27 | 0.07 | 0.23 | 0.16 |
| $A_3$ | Availability of raw materials at foreign markets | 0.97 | -0.03 | -1.00 | 1.43 | -0.03 | -1.46 |
| $A_4$ | Changes in foreign prices of raw materials | -0.66 | -0.86 | -0.20 | -1.16 | -2.10 | -0.94 |
| $A_5$ | Other external factors | -0.06 | 0.04 | 0.10 | 0.04 | 0.11 | 0.07 |
| $B_1$ | Efficiency of CSSR foreign trade | 0.01 | 0.11 | 0.10 | -0.56 | -0.34 | 0.22 |
| $B_2$ | Changes in CSSR export performance | 0.01 | 0.00 | -0.01 | -0.15 | -0.17 | -0.02 |
| $B_3$ | Changes in material intensity of production | 0.36 | 0.50 | 0.14 | 0.43 | 0.58 | 0.15 |
| $B_4$ | Contribution of technical progress | 1.10 | 0.95 | -0.15 | 1.33 | 1.15 | -0.18 |
| $C_1$ | Changes in employment | -0.07 | 0.01 | 0.08 | -0.02 | 0.04 | 0.06 |
| $C_2$ | Energy consumption in material sphere | 2.03 | 0.76 | -1.27 | 2.39 | 0.93 | -1.46 |
| $D_1$ | Regulation of investment, structure of production and sales | -0.00 | 0.12 | 0.12 | 0.08 | 0.06 | -0.02 |
| $D_2$ | Regulation of wages | -0.12 | -0.06 | 0.06 | -0.24 | -0.02 | 0.22 |
| $D_3$ | Increase in retail prices | 0.03 | -0.25 | -0.28 | -0.10 | -0.41 | -0.31 |
| $E_1$ | Inertial factors from preceding periods | 1.21 | 0.45 | -0.76 | 1.28 | 0.64 | -0.64 |
| $E_2$ | Other internal factors | 0.02 | -0.09 | -0.11 | -0.01 | -0.31 | -0.30 |
| $E_3$ | Random factors | -0.13 | 0.30 | 0.43 | -0.15 | 0.46 | 0.61 |

[a]The method used to estimate these figures is indicated in Appendix 21A.

The aggregated main results of the simulation analysis with respect to the six most important macroeconomic variables are presented in *Table 21.1*. *Table 21.2* contains more detailed results for gross national income produced and used. Each group of factor contributions in *Table 21.1* is an aggregate of individual factor contributions in *Table 21.2*, e.g., group $A = A_1 + A_2 + \cdots + A_5$, etc. The main results of the *ex post* analysis are interpreted and evaluated in the next subsections.

### 21.3.2. Growth of production and contributions of the most significant factors

According to the results of our simulation analysis, one of the main sources of slowdown in the growth of gross national income (GNI) produced was the deterioration of external economic conditions (*Table 21.1*). In the total slowdown of –2.68 percentage points, the contribution of this group of factors was –1.03 points, i.e., 38%. Most damaging were the negative impacts of decreasing availability and increasing prices of imported raw materials (mainly crude oil), and the slowdown in world economic activity (*Table 21.2*).

Another significant source of the slowdown was the unfavorable development of extensive growth resources (44% of total slowdown). Other internal factors accounted for –0.44 points of total slowdown in GNI. Within this group of factors, the negative influence of inertial factors from preceding periods was most significant.

In order to reduce the foreign debt of Czechoslovakia, the direct influences of central management were concentrated on the planned deceleration in the growth of production in 1981 and 1982. In 1983–1985, however, the efforts of central management were oriented to the acceleration of economic growth. Therefore, the total contribution of this factor over the whole period 1980–1985 was almost negligible.

The positive contribution of "intensive" growth factors to the change in the growth rate of production in 1980–1985 versus 1970–1980 was relatively small. Some favorable results were achieved in decreasing the material intensity of production, but the results were still slow and insufficient as compared with the most developed economies. However, due to a considerable decrease in the contributions of extensive growth factors, the share of "intensive" growth factors in the total growth rate of GNI increased from 31% in 1970–1980 to 71% in 1980–1985.

### 21.3.3. Foreign trade, terms of trade and external equilibrium

The rate of growth of Czechoslovak exports (in constant domestic prices) in 1980–1985, 5.40%, was 2.5 times higher than the rate of growth of GNI. On the other hand, Czechoslovak imports grew substantially more slowly, at the rate of 1.01%. The slowdown in the growth rate of imports during the period 1970–1980 was more than –3.5 percentage points.

According to our simulation analysis, external factors accounted for –0.56 percentage points of total slowdown (–0.88 points) in the growth rate of exports in 1980–1985 as compared with the period 1970–1980 (*Table 21.1*). Within this group of factors, the negative contribution of the slowdown in world economic activity was extremely high. It lowered the demand for Czechoslovak exports, the average annual growth rate dropping by -3.5 percentage points. This negative contribution was partly compensated by improving the relative position of the Czechoslovak economy in CMEA (contribution of +1.8 points to the growth rate of exports). Besides, the volume of Czechoslovak exports had to be increased due to the sharp rise in prices of crude oil and other raw materials imported from the USSR.

As expected, the simulation analysis showed that the main part of the slowdown in growth of Czechoslovak imports was caused by external factors (*Table 21.1*). Among them, the negative contributions of both the slowdown in world economic activity and the unfavorable changes in availability and prices of imported raw materials (including fuels) were most significant. On the other hand, the contribution of the position of the CSSR in CMEA was positive.

As a result of maintaining the high growth rate of exports and imposing strong restrictions on imports, the surplus in the Czechoslovak trade balance in constant domestic 1977 prices increased from 9.2 milliard (= $10^9$) crowns in 1980 to 50.7 milliard crowns in 1985, which was about 24% of real exports and 9% of GNI produced. However, our analysis showed that the major part of this real trade surplus (about 32 milliard crowns in 1985) was absorbed by the continuing decrease in the Czechoslovak terms of trade, which was caused mainly by the sharp rise in prices of imported raw materials from the CMEA countries. Therefore, the growth of the trade surplus in current prices was much slower and varied over territories. The trade surplus with the market economies was used to reduce the Czechoslovak foreign debt, but at the same time a growing trade deficit with the USSR was created.

### 21.3.4. Domestic utilization of GNI, internal equilibrium and economic efficiency

As a consequence of the unfavorable development of external economic conditions and the necessity to reduce foreign debt, the growth rate of domestic use of GNI in 1980–1985 was considerably lower than that of GNI produced. Moreover, the sharp slowdown in the growth of imports influenced negatively the growth of production and consequently also its domestic use. The growth rate of the domestic use of GNI in 1980–1985 was only 0.62%, i.e., -3.68 points below the growth rate of the period 1970–1980. During 1981 and 1982 the volume of domestic use of GNI declined by -1.5% a year, which was followed by a slow growth at the rate of about 2% during 1983–1985.

According to our analysis, external factors accounted for 55% of the slowdown in domestic use of GNI. The contributions of other factors were similar to the case of GNI produced.

The slowdown in domestic use of GNI varied over end-use categories. The most dramatic change occurred in investment: a turn from growth at the annual rate of 5.77% in 1970–1980 to decrease at the rate of -1.10% in 1980–1985. At the same time the growth rate of real personal consumption declined from 3.22% to 0.98%. It should be noted that this growth rate of personal consumption was the lowest among all the European CMEA countries (except for Poland) in 1980–1985 [UN (1986)].

Our analysis showed (*Table 21.1*), that the direct influences of central management partly improved the growth rate of personal consumption (by 0.71 percentage points) at the expense of a reduction in investment. The downturn in investment was also considerably influenced by the negative contribution of inertial factors from the preceding periods.

The positive influence of central management on personal consumption in 1980–1985 was achieved by regulation of both the supply of consumer goods and of wages (allowing for faster growth of average wage rates than of labor productivity). At the same time, certain increases in retail prices were necessary with a negative impact on consumer demand. However, these central interventions were not fully balanced and, due to evoked shifts in consumer demand, the share of excess demand in total demand for industrial consumer goods increased (according to our estimates) from 4.6% in 1980 to 6.9% in 1985. Our analysis showed that about 25% of this excess demand was substituted by an additional supply of food and the rest was allocated to forced savings of the population.

The slowdown in the growth of GNI produced and used, as well as the deterioration of terms of trade, resulted in the unfavorable development of economic efficiency. Our analysis showed a considerable downturn in the indicator of total economic efficiency, from the annual growth of 2.64% in 1970–1980 to the decline of –1.57% a year in 1980–1985. It should be noted that one of the main sources of this downturn was the negative contribution of inertial factors from the preceding periods. The positive contributions of intensive growth factors (mainly the decrease in material intensity of production) were not sufficient to overcome those factors by 1985.

## 21.4. Conditional Forecasts of Economic Growth in the CSSR, 1986–2000

### 21.4.1. Scenarios of the conditional forecasts

Scenarios and input data for conditional forecasts have been prepared using information from the published Czechoslovak planning documents [*Hlavne smery...* (1986)], United Nations studies (1986), and many other relevant sources published in Czechoslovakia and in the other CMEA countries. We have taken into consideration also some general assumptions contained in the Bonn–IIASA Research Project papers [Dobrinsky (1985), Krelle (1985), and Zeman (1986)].

Common assumptions in all scenarios for 1986–2000 are as follows:

(1) Acceleration in the growth of world economic activity, especially in the CMEA region.
(2) Stabilization of the position of the Czechslovak economy in CMEA (e.g., the Czechoslovak share in the total CMEA machinery production, etc.).

(3) Continued deterioration of the availability of raw materials in foreign markets.
(4) Slowdown in the growth of the import prices of raw materials, especially in the CMEA market where certain lagged impacts of the recent world price developments are expected.
(5) Stabilization of the efficiency of Czechoslovak foreign trade corresponding to the expected more favorable development of the Czechoslovak terms of trade.
(6) Slight improvement of the Czechoslovak export performance as related to the relevant external and internal conditions.
(7) Continued slowdown in the growth of primary energy resources and energy consumption in the material sphere.
(8) Continued growth of the share of machinery industry (especially electronic industry) in the structure of industrial production.
(9) Structural changes in fixed capital investment: rising share of machinery investment and rising allocation of investment to the nonmaterial sphere.
(10) Accelerated growth of social consumption.
(11) Slower growth of average wage rates in relation to labor productivity.
(12) Slowdown in the growth of the aggregate retail price index.
(13) Relative improvement in the supply of nonfood consumer goods, including imported goods.

Other important exogenous assumptions differ for individual variants.

*Variant 0* (basic scenario)

The basic scenario of our conditional forecasts is supposed to simulate the development of the Czechoslovak economy according to the assumptions of the 8th Five-Year Plan (for the period 1986–1990) and the long-term prospects up to the year 2000 [*Hlavne smery...* (1986)]. The scenario of this variant contains certain important assumptions which are very ambitious regarding the strategy of intensification of the Czechoslovak economy:

(1) Considerable decrease in the energy and material intensity of production: Energy intensity of GNI produced is supposed to decrease in 1985–1990 by the annual rate –2.9%, which is almost twice as fast as in 1980–1985 (–1.5). The overall material intensity of production is supposed to decrease in 1985–1990 by –0.97% a year, i.e., three times faster than in 1980–1985. Almost the same decreasing rates are supposed also for the long-term period up to the year 2000 (see *Table 21.3*). At the same time, the investment ratio is also supposed to decrease by –1.4% a year.
(2) Relatively high Czechoslovak credit participation abroad, especially in the other CMEA countries: Consequently, the transition from negative to positive trade balance (in current foreign prices) with socialist countries is expected. Current positive trade balance with nonsocialist countries is expected to fall gradually, reaching a slightly negative position in the year 2000.

(3) Strictly limited growth of employment in the material sphere: Expected higher growth of total employment in the period 1990–1995 (following from the demographic waves) would be allocated to the nonmaterial sphere. Supposed annual growth rates of employment in 1985–2000 are 0.3% in the material sphere, and 1.8% in the nonmaterial sphere, respectively.

The scenarios of additional forecasting variants contain certain modifications of the above three assumptions. These variants are supposed to simulate impacts of less favorable developments of external and internal economic conditions and to evaluate the eventual necessary changes in the development strategy to cope with the problem.

*Variant 1* (higher energy and material intensity of production)

This variant is subdivided into two subvariants. Taking into account past trends and real possibilities of their future changes, in subvariant 1.1 a slower decrease in the energy intensity (–2.13% a year) and material intensity (–0.52% a year) of production is expected in 1985–2000, with negative impacts on the growth rates of all macroeconomic indicators.

Subvariant 1.2 simulates a possible partial solution of this problem. A certain additional volume of the fixed capital investment is supposed to ensure a faster decrease in the energy and material intensity of production. Of course, the decrease in the investment ratio would be much slower in this case.

*Variant 2* (higher credit participation abroad)

In this variant a higher Czechoslovak credit participation abroad than in scenario 0 is supposed. This assumption follows mainly from the expected higher costs of production and transport of some raw materials imported from the USSR (especially natural gas and iron ore). Also, growing machinery exports to nonsocialist countries would require higher Czechoslovak credit participation in the related investment projects. As a consequence, higher positive trade balances and imports restrictions are expected up to 1995, reversing in the period 1995–2000 as a consequence of higher credit installments.

*Variant 3* (higher employment in the material sphere)

In this variant it is supposed that the absorbing capacity of the nonmaterial sphere with respect to the growth of employment is rather limited. Thus, the assumed growth rate of employment in 1985–2000 in this variant for the nonmaterial sphere is lowered to 1.33% and for the material sphere is raised to 0.49% a year. In this case a slightly faster decrease in working time would be possible. On the other hand, annual labor productivity in the material sphere would be lower than in scenario 0.

*Table 21.3.* Selected results of the conditional forecasts, 1985–2000: average annual percentage growth rates (except where noted) under scenario variants.

| | Variants[a] | | | | | |
|---|---|---|---|---|---|---|
| | *0* | *1.1* | *1.2* | *2* | *3* | *4* |
| *Variables* | *Basic variant* | *Higher energy intensity of production* | *Variant 1.1 with more investment* | *Higher credit participation abroad* | *Higher employment in material sphere* | *Aggregate variant 1.2 + 2 + 3* |
| Energy intensity of GNI | –2.83 | –2.13 | –2.59 | –2.71 | –2.90 | –2.54 |
| Material intensity of GMP | –0.82 | –0.52 | –0.72 | –0.82 | –0.82 | –0.72 |
| Investment/GNI ratio | –1.41 | –1.78 | –0.48 | –1.66 | –1.28 | –0.57 |
| Trade balance, soc. area[b,c] | 9.6 | 10.7 | –0.7 | 13.5 | 8.9 | 2.5 |
| Trade balance, nonsoc. area[b,c] | –1.0 | 1.4 | –2.5 | 2.0 | –0.9 | 0.8 |
| Employment in material sphere | 0.30 | 0.30 | 0.30 | 0.30 | 0.49 | 0.49 |
| Employment in nonmaterial sphere | 1.82 | 1.82 | 1.82 | 1.82 | 1.33 | 1.33 |
| Worked hours per worker a year | –0.49 | –0.49 | –0.49 | –0.49 | –0.69 | –0.69 |
| Gross material product (GMP) | 2.56 | 2.16 | 2.42 | 2.43 | 2.64 | 2.36 |
| Intermediate consumption | 1.72 | 1.63 | 1.67 | 1.59 | 1.80 | 1.62 |
| Gross national income (GNI) | 3.52 | 2.79 | 3.27 | 3.39 | 3.59 | 3.22 |
| Exports | 4.41 | 4.04 | 4.04 | 4.39 | 4.42 | 4.03 |
| Imports | 3.98 | 3.33 | 3.85 | 3.80 | 4.01 | 3.70 |
| Trade balance (real net exports)[c] | 115.8 | 120.6 | 99.6 | 122.3 | 115.0 | 105.6 |
| Domestic use of GNI | 3.27 | 2.38 | 3.12 | 3.07 | 3.36 | 3.01 |
| Personal consumption | 2.57 | 1.91 | 2.35 | 2.43 | 2.55 | 2.19 |
| Social consumption | 6.05 | 5.09 | 5.14 | 6.13 | 5.67 | 5.14 |
| Investment in fixed assets | 2.06 | 0.96 | 2.77 | 1.68 | 2.27 | 2.63 |
| Synthetic efficiency indicator | 2.11 | 1.39 | 1.81 | 2.09 | 1.92 | 1.68 |
| Labor productivity in mat. sphere | 3.21 | 2.48 | 2.96 | 3.08 | 3.09 | 2.72 |
| Average nominal wage rate | 2.00 | 1.30 | 1.77 | 1.87 | 1.93 | 1.57 |
| Gross domestic product (approx.) | 3.85 | 3.08 | 3.49 | 3.76 | 3.85 | 3.45 |
| GDP per employee (approx.) | 3.12 | 2.36 | 2.77 | 3.03 | 0.12 | 2.73 |

[a]In domestic constant 1977 prices, except where noted. [b]Current foreign prices. [c]Absolute values in milliard Czechoslovak crowns in 2000.

*Variant 4* (combined variants 1.2 + 2 + 3)

This combined variant simulates a growth path of the Czechoslovak economy for the case if all three problems outlined in the scenarios of the variants 1.2, 2 and 3 were to occur simultaneously.

### 21.4.2. Main results of the conditional forecasts

The main results of our conditional forecasts of the Czechoslovak economic development up to the year 2000 are presented in *Tables 21.3* and *21.4*.

*Scenario 0* (baseline)

The growth rates of the most important macroeconomic variables simulated in this variant are very close to those in the 8th Five-Year Plan for the period 1986–1990 and the official long-term prospects up to the year 2000 [*Hlvane smery* ... (1986)]. The growth rate of GNI produced (about 3.5%) would be higher by almost one percentage point than that of GMP (*Table 21.3*). That would mean a considerably slower growth of intermediate consumption as a result of intensive (resource efficient) economic growth.

In the period up to 1990, the forecasted growth of exports is considerably higher than that of imports; but after 1990 the forecasted growth rates of exports and imports are very close (*Table 21.4*). Consequently, the growth rate of GNI used in 1985–1990, about 3%, would be about 0.5 percentage points below that of GNI produced, but after 1990 this difference would be only 0.1 points.

Among the end-use categories, social consumption would have the highest growth rate – more than 6% a year. The growth rate of personal consumption (2.57%) would be higher than that of investment (2.06%). The growth rates of these two indicators would be considerably higher than in the last five years, but not as high as the growth rate of GNI. On the other hand, due to fast development of the nonmaterial sphere, the growth rate of the approximation of GDP would be higher (3.85%).

As the accelerated growth of GNI would be achieved at the expense of very slow growth of employment in the material sphere, the growth of labor productivity would accelerate to 3.2%. This would be considerably higher than the forecasted growth of the average nominal wage rate (2%). Consequently, even at the relatively slow growth of consumer prices (0.7% a year), the estimated share of excess demand in the total demand for nonfood consumer goods would fall from 6.8% in 1985 to 5.3% in 2000.

Faster growth of labor productivity, a considerable decrease in the material intensity of production, as well as the stabilization of Czechoslovak terms of trade and fixed capital productivity would result in a transition of the synthetic efficiency indicator from previous decline to growth at the average rate of 2.11% up to the year 2000.

*Table 21.4.* Selected results of the conditional forecasts for five-year periods, 1970–2000: average annual percentage growth rates under scenario variants.

| Variable[a] | Variant | Actual 70–80 | Actual 80–85 | Forecasts 85–90 | Forecasts 90–95 | Forecasts 95–2000 |
|---|---|---|---|---|---|---|
| GMP | 0 | 4.70 | 1.74 | 2.30 | 2.79 | 2.58 |
| | 1.1 | | | 2.13 | 2.34 | 2.01 |
| | 1.2 | | | 2.13 | 2.63 | 2.49 |
| | 2 | | | 2.12 | 2.67 | 2.61 |
| | 3 | | | 2.30 | 3.00 | 2.61 |
| | 4 | | | 1.95 | 2.61 | 2.54 |
| GNI produced | 0 | 4.84 | 2.16 | 3.53 | 3.63 | 3.40 |
| | 1.1 | | | 3.13 | 2.78 | 2.45 |
| | 1.2 | | | 3.13 | 3.42 | 3.26 |
| | 2 | | | 3.35 | 3.40 | 3.42 |
| | 3 | | | 3.53 | 3.84 | 3.41 |
| | 4 | | | 2.96 | 3.40 | 3.29 |
| Exports | 0 | 6.28 | 5.40 | 5.07 | 4.54 | 3.63 |
| | 2 | | | 5.05 | 4.50 | 3.64 |
| | 4 | | | 5.03 | 3.97 | 3.09 |
| Imports | 0 | 4.52 | 1.01 | 3.94 | 4.57 | 3.43 |
| | 2 | | | 2.66 | 4.81 | 3.93 |
| | 4 | | | 2.69 | 4.65 | 3.76 |
| Domestic use of GNI | 0 | 4.30 | 0.62 | 2.99 | 3.52 | 3.30 |
| | 1.1 | | | 2.48 | 2.55 | 2.12 |
| | 1.2 | | | 2.56 | 3.47 | 3.34 |
| | 2 | | | 2.38 | 3.44 | 3.49 |
| | 3 | | | 2.99 | 3.76 | 3.33 |
| | 4 | | | 1.94 | 3.54 | 3.56 |
| Personal consumption | 0 | 3.22 | 0.98 | 2.24 | 2.98 | 2.51 |
| | 1.1 | | | 1.89 | 2.21 | 1.64 |
| | 1.2 | | | 1.88 | 2.77 | 2.41 |
| | 2 | | | 2.00 | 2.74 | 2.53 |
| | 4 | | | 1.65 | 2.50 | 2.42 |
| Investment in fixed assets | 0 | 5.77 | -1.10 | 2.04 | 2.12 | 2.02 |
| | 1.1 | | | 1.53 | 1.08 | 0.28 |
| | 1.2 | | | 2.38 | 2.98 | 2.96 |
| | 2 | | | 1.34 | 1.72 | 1.98 |
| | 4 | | | 1.71 | 3.05 | 3.14 |

[a] In domestic constant 1977 prices.

*Variant 1* (higher energy and material intensity of production)

A supposed slower decrease in the energy and material intensity of production according to the scenario of the subvariant 1.1 would lead to slower growth rates (in comparison with the basic variant) of GNI produced (–0.7 percentage points) and GNI used (–0.9 points). This slowdown would accelerate in time, reaching, e.g., for GNI used, almost –1.2 percentage points in 1995–2000 (see *Table 21.4*).

A simulation according to the scenario of the subvariant 1.2 has shown a possible increase of investment, which would reduce the differences in energy and material intensities of production between variants 0 and 1.1 by about 65%. Similar effects would be reached for the growth rates of GNI and personal consumption in this case. The growth rate of GNI produced from the basic scenario (3.5%) would be lowered in variant 1.1 to 2.8%, but in variant 1.2 to only 3.3%. However, improvements over variant 1.1 would only be reached after 1990.

*Variant 2* (higer credit participation abroad)

In this variant higher positive trade balances connected with slower growth of imports (especially before 1990) would lead to slower growth of GNI produced and used. On the other hand, in the period 1990–2000, imports would grow more rapidly (see *Table 21.4*) due to credit installments with positive impacts on GNI produced and used. However, over the period 1985–2000, the average rates of growth would be slightly worse than in the basic variants (*Table 21.3*).

*Variant 3* (higher employment in the material sphere)

A simulation according to this scenario has shown slightly faster growth of GNI produced and used after 1990 as a consequence of higher employment in the material sphere. On the other hand, labor productivity in the material sphere would be lower, with negative impacts on wages and personal consumption. The growth rate of social consumption would be lower, too, due to less employment in the nonmaterial sphere. However, the partial shift in employment from nonmaterial to material sphere would not affect GDP (*Table 21.3*).

*Variant 4* (combined variants 1.2 + 2 + 3)

The results of this combined forecasting variant represent a certain aggregation of the deviations of the growth rates simulated in variants 1.2, 2 and 3 from those simulated in the basic scenario. As a result, the long-term growth rates of almost all macroeconomic indicators in variant 4 are 0.3–0.5 percentage points lower. On the other hand, the simulated growth rate of investment is higher (2.63% in variant 4 versus 2.06% in variant 0; see *Table 21.3*).

The deviations from the basic variant in the gowth rates are not equal over time (*Table 21.4*). In the period 1985–1990 the growth rates in variant 4 differ from those in variant 0 more significantly, being lower for GNI produced by 0.57%, for GNI used by 1.05%, for personal consumption by 0.59%, etc.

## 21.5. Conclusion

The results of our *ex post* analysis of the Czechoslovak economy have revealed that the main sources of the slowdown in economic growth in 1980–1985 versus 1970–1980 were unfavorable development of external factors (mainly the availability of raw materials from foreign markets) and relative exhaustion of extensive growth resources (mainly the supply of energy to the material sphere). Negative effects of these factors were not sufficiently compensated by positive contributions of intensive growth factors (technical progress, decrease in material intensity of production, etc.). The current planning and management system was not efficient enough to create favorable conditions for intensive growth.

Our results may also be compared with some general assumptions on the driving forces of structural changes and economic growth formulated in the framework of the Bonn–IIASA Research Project [Dobrinsky (1985) and Krelle (1985)]. It is true that capital accumulation, measured by the investment ratio, decreased considerably in the period 1980–1985. However, according to our findings, this is not the primary cause of the slowdown in economic growth. It results from the deterioration of external economic conditions and exhaustion of extensive factors. In such a situation, the efforts to restore the external equilibrium by increasing exports and putting restrictions on imports, as well as the efforts to maintain at least a minimum growth of personal consumption and sufficient growth of social consumption, led inevitably to the considerable decrease in investment.

With respect to labor supply, no substantial changes have been observed. However, changes in its sectoral structure may be significant [sectoral changes in employment and production in the CSSR are analyzed separately in Zeman (1986)].

For the contributions of technical progress we have found an only slight decrease. According to our results, only a small part of the observed decreasing difference between growth rates of production and the weighted average rates of direct factors may be interpreted as a decreasing rate of technical progress. The greater part of this difference reflects the negative contributions of the slowdown in energy resources and imported raw materials – factors that can hardly be replaced by fixed capital or labor.

The results of our conditional forecasts have indicated that the official growth targets for the Czechoslovak economy up to the year 2000 might be feasible, but they depend crucially on substantial structural changes, which may be reached only under efficient changes in the planning and management system. The transition to the intensive path of economic growth requires more time, more investment, and more creative activity of enterprises and individuals. The crucial problem is the substantial decrease in the energy and material intensities of production, which require additional investment. The nature of this problem, as well as the problem of necessary higher Czechoslovak credit participation in certain foreign investment projects, indicates that major positive effects may be expected only after 1990. Hence, a considerable tension follows from the planned targets for 1986–1990 because of the necessity to look for additional growth resources as well as to speed up the efficiency changes in the planning and management system.

# Appendix 21A. Estimation of the Contributions of Individual Factors to the Growth Rates of Macroeconomic Indicators

In the model used for *ex post* analysis, each structural equation contains 3–8 explanatory variables. For parameter estimation, annual time series of 25 observations (1961–1985) have been used.

In a typical structural equation some explanatory variables are exogenous and some are endogenous, explained in other equations which also may contain exogenous and endogenous explanatory variables. Substituting step-by-step exogenous variables for endogenous ones, we can express any endogenous variable as a function of only exogenous variables and lagged endogenous variables (with higher-order lags). This reduced (or final) form of the model explains each of the endogenous variables with a high number of primary factors represented by exogenous variables. Inertia from preceding periods is represented by lagged endogenous variables.

In large dynamic nonlinear models, the analytical expression of the reduced (or final) form of the model may be extremely complicated. Therefore, in our *ex post* analysis we have used a simulation procedure to estimate the contributions of primary factors to the growth rates of macroeconomic indicators. Each primary factor under consideration was represented by a group of relevant exogenous variables. The contributions of each factor were estimated as differences between corresponding growth rates of values of selected endogenous variables generated by two subsequent dynamic simulations. In the first of these simulations the values of exogenous variables representing a factor under consideration were kept constant over the whole analyzed period at the values of the starting year. In the next simulation the actual values of a given factor were used. The values of all other exogenous variables were equal in both simulations.

Using this simulation procedure, we decomposed the growth rates of selected macroeconomic indicators for a given period to the contributions of individual primary factors, inertia factor and random disturbances. The results for periods 1970–1980 and 1980–1985 are given in *Table 21.2*.

## References

Dobrinsky, R. (1985), *Aggregated Macroeconomic Models of the European CMEA Countries*. Paper prepared for the IIASA Workshop on Economic Growth and Structural Change, Lodz, Poland, December 9–10.

Green, D.W. (1977), The Use of Econometric Methods in the Analysis of Centrally Planned Economies: The SRI-WEFA Experience, *Journal of Comparative Economics*, **1**, pp. 259–275.

*Hlavne smery hospodarskeho a socialneho rozvoja CSSR na roky 1986 az 1990 a vyhlad do roku 2000* [Main guidelines for economic and social development of the CSSR for the period 1986–1990 and prospects up to the year 2000], (1986). (Bratislava: Pravda, April 3).

Klas, A. (1985), System modelov pre makroekonomicke analyzy a jeho pouzitie pri simulaciach vyvoja ekonomiky [A system of models for macroeconomic analysis and its use in the simulation of economic development], *Ekonomicko-matematicky obzor*, **21**, pp. 241–265.

Klein, L.R. (1976), Five-Year Experience of Linking National Econometric Models and of Forecasting International Trade, in: H. Glejser (ed.), *Quantitative Studies of International Economic Relations*. (Amsterdam: North-Holland).

Krelle, W. (1985), *The State of the Bonn-IIASA Research Project*. Paper prepared for the IIASA Workshop on Economic Growth and Structural Change, Lodz, Poland, December 9–10.

Lovell, C.A.K. (1968), Capacity Utilization and Production Function Estimation in Postwar American Manufacturing, *Quarterly Journal of Economics*, **83**, pp. 219–239.

Sujan, I. (1982a), Analyza podielov jednotlivych faktorov na spomaleni tempa rastu ceskoslovenskej ekonomiky v rokoch 1975–1980 [An analysis of the factor contributions to the deceleration in the growth rate of the Czechoslovak economy in the period 1975–1980], *Politicka ekonomie*, **30**, pp. 595–610.

Sujan, I. (1982b), *Simulations of Restoring the External Economic Equilibrium of Czechoslovakia*. Paper presented to the 6th International Conference on the Econometric Modeling of the Socialist Economies, Budapest, Hungary, March.

Sujan, I. (1984), *Modeling the Foreign Trade Flows in the System of Models of the European CMEA Countries*. An invited paper for the 8th International Conference on Econometric Modeling of the Socialist Economies, Lodz, Poland, October.

Sujan, I. (1985), *The Use of Econometric Macromodels under Changing Conditions in Centrally Planned Economies*. An invited paper for the 9th International Conference on Models and Forecasts '85, Bratislava, Czechoslovakia, October.

Sujan, I. and Strauch, D. (1986), *Ekonometricky makromodel reprodukcneho procesu CSSR* [Econometric macromodels of the Czechoslovak economy], *Ekonomicko-matematicky obzor*, **22**.

UN (1986), *Economic Survey of Europe in 1985-1986*. (Geneva: United Nations).

Vanous, J. (1980), *An Econometric Model of Soviet and Eastern European Foreign Trade*, Discussion paper 88-30. (Vancouver: University of British Columbia).

Zeman, K. (1986), *The Driving Forces of Structural Change in the Czechoslovak Economy till the Year 2000*. Paper presented at the Task Force Meeting on Economic Growth and Structural Change, Sofia, Bulgaria, June 24–25.

CHAPTER 22

# Analysis and Forecasts of Growth of the Hungarian Economy

*Andras Simon*

## Summary

Hungary has a small open economy, depending heavily on foreign markets. This feature underlies the specification of the model that generates forecasts to the year 2000. Though output is determined by demand, as in market economies, other features distinguish the economy from a typical market economy. Personal income and investments can only be explained by policy behavior. The model of the economy explains domestic variables in constant prices. Only a few price effects are included in the model. Employment is determined by demographic factors independent of the rate of utilization of capacity. Simulations show a 0.2–0.3 elasticity of GDP with respect to world trade and an expected growth rate of 2.0–2.4% for the next 15 years.

## 22.1. The Driving Forces of Structural Change

The model underlying our analysis of the Hungarian economy may seem rather heretical in comparison to the neoclassical models usually applied when modeling growth. As a model of a planned economy, it is even more out of line by exhibiting features of a demand-driven economy. Let us try to justify this specification, not claiming that our approach is the only one that would produce reasonable simulation of actual developments.

For market economies it is usually accepted that short-run changes in output may be better explained by demand, while neoclassical growth models may be adequate tools for capturing more long-run developments. For planned economies output is often considered to be capacity-determined even in the short run; and capacity is modeled by a production function, with capital and labor as the main explanatory variables. Let us review the arguments against this approach by considering the economy in both long-run and short-run perspectives.

### 22.1.1. Demand pull in the short run

It is common to assume that one main difference between a centrally planned and a market economy from a modeling point of view is the following: In a market economy there is always enough free productive capacity to allow the level of output to adjust to an increase in demand. Thus, in a market economy, modeling output usually means modeling demand. A centrally planned economy, on the other hand, operates at full capacity because at given prices, which are set administratively, demand always tends to exceed supply. Hence, when designing an econometric model for a centrally planned economy, it is usually thought to be enough to set up production functions to model capacity, and by the identity of output and capacity we arrive at production. This conclusion, however, is too simple to be fruitfully applicable in modeling. Without going too deeply into theoretical arguments, let us make some refinements regarding the relations of output and capacity.

Excess demand in these countries means shortages in supply. Shortages, however, are stochastic phenomena. They may appear in a market economy, too, in the market for manufactured goods. Here prices are fixed by the firm and kept constant for a longer period. Demand in the model of this market is a random variable which occasionally may exceed supply. The firm, however, has good reasons to set prices sufficiently high that the probability of failing to meet demand is fairly low. This means that it will keep reserve capacities and a sufficient stock of inventory.

In a planned economy, prices are set administratively with less attention to the relation of expected demand to capacity. Two cases of shortages may arise in this economy. In the first case, aggregate demand is higher than aggregate productive capacity, but for each firm the expected value of demand is rather close to capacity. This entails a high probability that the firm cannot meet demand. This results in a situation of frequent shortages, bottlenecks and forced substitution, but it does not mean that output cannot adjust to demand. Several countries with planned economies show this feature; nevertheless, macroeconomic modeling in these countries can use the same approach to explain output as models of market economies. In the second case of shortages, the disequilibrium on the market is more severe. Here aggregate demand is so high, compared with capacity, that demand cannot be met by forced substitution. This is the case where forced saving appears. The production function approach to modeling output here is even less adequate than in the former case. Namely, output is constrained not by the optimal combination of labor and capital, but rather by scarcities in some key intermediate inputs. Labor and capital are not fully utilized, and the level of output may be directly connected to a scarce input, such as imports.

Which kind of shortages are typical for a planned economy? Past experience has shown that industrial output in Hungary does fluctuate over time. These fluctuations are too large to be explained by changes in the level of capital and labor inputs in a production function. This phenomenon raises the following question:

Are output fluctuations the result of variations in the supply of scarce intermediate inputs, or are they results of changes in demand? There is no *a priori* argument for either of these cases. In fact, there is much reason to think that, depending on the actual conditions, sometimes the first explanation, and sometimes the second, is valid. Before analyzing the two cases, let us summarize again some ideas which are relevant in addressing this question.

Centrally planned economies are sometimes characterized as economies in which excess demand constantly prevails. Excess demand in a macroeconomic sense can, however, be defined in several ways. According to our definition, an excess demand situation is one in which prices are kept administratively lower than they would be in a free market. This definition implies a higher rate of capacity utilization in the excess demand case than in the "equilibrium" situation. It does not imply, however, that output necessarily has to be determined by the level of capacity. When looking at the actual experience of most of the centrally planned economies, we see that in normal circumstances excess demand appears in the form that conditions for purchasing a good are difficult: the product to be bought is, for example, out of stock, deliveries are not in accordance with orders, the selection of goods is not the desired one, some of the goods are not available for shorter or longer periods, etc. These conditions are the result of a lack of capacity reserves in the economy, due to higher than equilibrium capacity utilization. In the aggregate, however, there is no forced saving in the economy because aggregate demand is met by aggregate output. There is even some flexibility in supply since higher demand results in higher output. During peaks as well as troughs of demand, shortages exist, although in peaks it is probably more difficult to procure goods than during troughs, but not necessarily. This means that even in a "shortage economy" output can be modeled correctly from the demand side. However attractive the symmetry between the demand-driven market economy and the capacity-driven planned economy may be, it cannot be accepted as a true description of the normal functioning of centrally planned economies.

The fact that the level of output in the short run is determined by demand in these economies does not mean that some disequilibrium situations may not appear. Usually, changes in demand policy are accompanied by changes in import policy. The aim is to keep a normal level of shortage of imports – not lower than that of the domestic goods, but at the same time not so high as to cause bottlenecks in production. This behavior is the basis of imports in the model used for our forecast. These equations explain the level of import demand as a function of output. The balance between demand policy and import policy, however, may be disturbed. Severe import restrictions may result in an unintended disruption of production or in distortions in the structure, quality or quantity of output. These adverse conditions may show up in several areas: in its milder form, the domestic market may adjust to them without forced saving, and only the export performance may be affected; in a more severe case, there may be an unintended fall in output such that forced substitution is not possible, and as a consequence forced saving occurs.

When modeling an individual economy, the specific characteristics in each country have to be considered. There are some countries in which a supply constraint seems to occur over shorter or longer periods. This has been the case in Poland and, more or less, in Romania in recent years. In Poland, constraints on imports forced whole factories to stay idle and put manpower and capital out of work for an extended period.

What is the main difference between the macroeconomic relationships of a demand-pulled economy and those of an import-constrained one? In both cases, output and imports are strongly related to each other. Whether it is output or imports in this relationship that determines the other variable will show up in export performance or in the consumption function. In demand-controlled situations a cutback in domestic use eases tensions on the domestic market and spurs exports. In the acute shortage case a decrease in domestic output may be coupled with a decrease or a leveling off in exports. This behavior of exports can be seen, for example, in the Polish case, while the behavior of Hungarian exports indicates a demand-controlled situation.

### 22.1.2. Demand pull in the long run

What kind of specification is most suitable for a model to explain growth of the Hungarian economy up to the year 2000? Calculations have shown that the current level of industrial output in Hungary is about 10% lower than the level estimated on the basis of a production function fitted on data from the last 25 years. There were several options to handle this discrepancy in a model that is expected to make forecasts for 15 years into the future:

(1) To consider this 10% as a short-term vagary of demand that is going to be corrected either by automatic forces in the economy or by wise demand policy, and to take the values of the production function as forecasts. This option seemed to be too optimistic to be chosen.

(2) To interpret the falling of the level of output below its estimated value as a result of an historical decrease in the rate of technical progress and use a lower value for the parameters of the technical progress in simulations. This option seems to be feasible, and forecasts for output would presumably be low enough to be in line with our expectations; but underlying parameters of the forecast would have to be plucked from the air and not from the model.

(3) To analyze the economic structure and the exogenous variables by a model that creates the 10% discrepancy between capacity and output, to assess whether these exogenous and endogenous factors would continue to prevail, and to make the forecast conditional on these factors. This is the option we strived to take.

What was the reason that output has fallen below its long-term potential in the past 5–10 years? The answer may be found in an increased import intensity, owing to a deterioration in the terms of trade and a slackening export potential. In the present structure of exports there is no hope that exports would pull the economy out of its slump. More than 60% of exports to the hard currency area consists of food and basic materials. The markets of both commodity groups are stagnating, and they are going to continue to stagnate even if the rest of the world economy is expected to boom. If the economy had enough flexibility to substitute for imports and increase competitiveness in more dynamic segments of world economy, it could have compensated for the slow growth of its traditional exports. At present, there are no signs of such an increased flexibility. This causes the growth of the economy to be trapped on a path determined by the elasticity of demand for its exports and by the import elasticity of output. This trap has been holding the economy for a long time and seems to determine its long-run growth. For shorter periods, say, 3–5 years, economic policy can separate growth from this long-run path, but only at the expense of balance-of-payments deficits. The ambitious policy of trying to separate domestic growth from recessions in the rest of the world succeeded temporarily. But the final result was only a delay in the effects of adverse developments in the world market.

All this indicates that the slackening growth of output is not a result of a cyclical or other transitory lack of demand, but is caused by the weak capability of the economy to adapt to new situations.

Is there a reason to expect this deeply rooted feature of the economy to change in a longer perspective of 10–15 years? We do not feel that we have sufficient information for a definite answer. We decided to use the parameters of the model as they are, running "no-change-in-the-system" forecasts.

In this demand-pulled model, export elasticities are manifestations of technical progress in the home country compared with that in the rest of the world. Obviously, this approach has some unattractive features: growth is determined independently of employment. This is a weak point of the model, even if we take into account the peculiar nature of planned economies that guarantee full employment, regardless of the actual utilization of labor.

## 22.2. Selected Features of the Economy

Let us summarize the features of the model that are used to produce forecasts for the Hungarian economy. This description at the same time may serve as a short description of the working of the economy.

### 22.2.1. General structure of the model

As mentioned before, the economy may be characterized as being export-pulled. Thus, the economic growth possibilities of the country are limited by its export potential. Output is highly import-intensive, and imports for the most part are basic materials or high-technology goods for which domestic production is not substitutable. As imports in the longer run are paid for by exports, the rate of growth of the economy depends on how much exports can be sold on the world market. This, in turn, depends on the rate of expansion of the world market and the demand and price elasticities for Hungarian exports. For a shorter period, say, three to five years, the trend of growth can be strongly affected by a policy that has a direct effect on the balance of payments by deliberately separating the trend in exports from that of imports through borrowing, or by external factors such as terms of trade changes, changes in conditions of borrowing, etc. When specifying the model these factors were taken into account.

In a more refined picture, the export pull is less direct. Exports affect the economy through two channels: in the short run, they affect output as a component of demand. In the longer run, as an item in the balance of payments, export earnings help determine the external debt situation of the country, and through this, the main policy decisions of personal income, government consumption and investment. These decisions, however, are also determined by a series of other factors mentioned above.

In the model, these factors are all pooled into one variable: the external debt of the economy. There is always a certain level of indebtedness, which is considered to be tolerable both by the creditors and by the user of credits. Without trying to define the exact limits of the level of acceptable indebtedness by prior argument, an analysis of the actual behavior of policy makers as they responded to various debt levels in the past has been undertaken. The estimated equations show that during the observation period there was a response on the part of policy with a one- or two-year lag. The interaction of policies and the balance-of-payments constraint brought about a cyclical movement in the course of Hungarian economic development.

### 22.2.2. Prices and incomes

Prices and incomes in a planned economy are usually under the direct and strict control of the government. This does not mean that prices are constant or that there are no economic forces behind price changes that do take place. Both relative prices and the general price level change in response to changes in underlying economic conditions. These price changes, however, usually take place within the context of general price reforms, which do not necessarily follow changes in domestic costs or in import prices. However large the economic costs of, for example, keeping domestic fuel prices considerably below their world market prices may be, it is nonetheless a feasible policy. Hungary pursued this policy for five years, and some other countries of the region still have not changed oil prices since 1973. Modeling this policy behavior seems to be beyond the capabilities of econometric methods.

In recent years the Hungarian price system underwent major changes. Decision making became more decentralized and dependent on the behavior of enterprises. Our model, however, is not yet able to include this new feature of the economy.

### 22.2.3. Employment and the labor force

The labor market has very similar features in every centrally planned economy. The demand for labor absorbs all the working force available, independent of the rate of capacity utilization. Major unemployment did not appear even during the Polish crisis, when output dropped by two-digit percentage points. The reasons, for full employment in these countries are not simple. Any explanation would refer to the following factors:

(1) Institutional arrangements: state-owned enterprises comply with the public view that laying off workers is an anti-social and immoral act.
(2) Enterprise behavior: there is a motivation on the part of firms to maximize output rather than profits.
(3) Price policy: owing to subsidized consumer goods and services, the marginal product of labor is much higher than the wage level.

There is no definite answer from economic theory on the question as to which of these factors is the most important. For our limited purposes, fortunately, we do not need an answer. What we have to assume when building a model is that employment is equal to labor supply. With regard to the participation rate of the labor force, any economic explanation is very dubious. Even if economic variables played a role in determining labor force participation, they would be far less important than the role played by demographic and institutional factors. For the purpose of the model, it was assumed that total employment was equal to the total labor force determined by demographic factors.

## 22.3. Expected Economic Development until 2000

The growth rate of the Hungarian economy up to the year 2000 depends largely on the performance of its exports (see *Tables 22.1, 22.2* and *22.3*). In our forecast, we had to make some assumptions beyond the assumed rate of world trade. Both agricultural exports and exports of base materials (mostly oil products) were forecasted independently of aggregate world market demand. For food it was assumed that a rate of growth of 2% can be sustained. This entails a rate of penetration in – mostly West European – markets that does not invoke protectionist steps and is feasible from the supply side. Similar to the case of food, constraints are present for exports of base materials on both the supply and demand side. Here exports are determined in the model by the difference of imports from the rouble area and domestic use, but the resulting growth rate of exports to the hard currency area of 1.5–2.0% is probably just the rate that can be absorbed by a stable world market for these commodities.

About 60% of total exports to the hard currency area are now determined independent of the growth rate in the world economy. In the case of a 2% yearly expansion of total world demand, the aggregation of the assumed and calculated 1.5–2.0% rates in this portion of exports and the higher rates of exports in the rest of Hungarian trade results in a 2.4% rate of growth of total exports.

*Table 22.1.* Forecast for selected variables: assumed growth of world trade = 2%.

| *Variable* | *1983* | *1984* | *1985* | *1986* | *1987* | *1988* | *1989* |
|---|---|---|---|---|---|---|---|
| Growth rates (%): | | | | | | | |
| gross output, industry | 1.25 | 1.83 | 0.80 | 0.30 | 3.92 | 0.91 | 0.94 |
| gross output, agriculture | 0.68 | 2.60 | -6.00 | -0.60 | 2.00 | 1.20 | 1.20 |
| gross output, construction | 0.64 | 1.55 | -7.70 | 0.20 | 0.08 | -2.89 | 2.34 |
| gross output, services | 3.73 | 2.17 | 1.29 | 3.35 | 0.99 | 0.96 | 1.60 |
| exports to non-CPEs, real | 16.35 | 4.33 | -10.37 | -3.51 | 5.75 | 3.28 | 3.16 |
| imports from non-CPEs, real | 5.30 | -1.56 | -0.62 | -1.27 | -2.30 | -0.24 | 3.17 |
| GDP | 1.39 | 1.78 | -0.01 | 0.72 | -0.48 | 0.57 | 1.12 |
| employment | -0.64 | -0.61 | -0.60 | -0.50 | -0.50 | -0.50 | -0.50 |
| productivity | 2.02 | 2.38 | 0.59 | 1.22 | 0.02 | 1.07 | 1.62 |
| personal consumption | 0.51 | 0.88 | 1.26 | 2.90 | -1.66 | -0.75 | -0.17 |
| investment, fixed assets | -2.11 | -0.6 | -0.3 | 0.64 | 1.52 | -1.74 | 3.11 |
| Investment ratio (%) | 22.90 | 22.90 | 22.22 | 22.20 | 22.65 | 22.12 | 22.56 |
| Current acct. balance ($) | 0.30 | 0.33 | -0.02 | -0.41 | -0.17 | -0.07 | -0.09 |

*Table 22.2.* Forecast for selected variables: assumed growth of world trade = 4%.

| *Variable* | *1983* | *1984* | *1985* | *1986* | *1987* | *1988* | *1989* |
|---|---|---|---|---|---|---|---|
| Growth rates (%): | | | | | | | |
| gross output, industry | 1.25 | 1.83 | 0.80 | 0.30 | 4.72 | 1.26 | 1.19 |
| gross output, agriculture | 0.68 | 2.60 | -6.00 | -0.60 | 2.00 | 1.20 | 1.20 |
| gross output, construction | 0.64 | 1.55 | -7.70 | 0.20 | 0.19 | -1.46 | 4.25 |
| gross output, services | 3.73 | 2.17 | 1.29 | 3.35 | 1.22 | 1.24 | 1.87 |
| exports to non-CPEs, real | 16.35 | 4.33 | -10.37 | -3.51 | 8.06 | 3.97 | 3.96 |
| imports from non-CPEs, real | 5.30 | -1.56 | -0.62 | -1.27 | -2.28 | 0.29 | 4.31 |
| GDP | 1.39 | 1.78 | -0.01 | 0.72 | -0.24 | 0.87 | 1.54 |
| employment | -0.64 | -0.61 | -0.60 | -0.50 | .50 | -0.50 | -0.50 |
| productivity | 2.02 | 2.38 | 0.59 | 1.22 | 0.26 | 1.37 | 2.04 |
| personal consumption | 0.51 | 0.88 | 1.26 | 2.90 | -1.62 | -0.58 | 0.13 |
| investment, fixed assets | -2.11 | -0.6 | -0.3 | 0.64 | 1.52 | -0.47 | 4.83 |
| Investment ratio (%) | 24.80 | 22.90 | 22.22 | 22.20 | 22.59 | 22.29 | 23.01 |
| Current acct. balance ($) | 0.30 | 0.33 | -0.02 | -0.41 | -0.08 | 0.02 | -0.01 |

*Table 22.3.* Forecast for selected variables: assumed growth of world trade = 6%.

| *Variable* | *1983* | *1984* | *1985* | *1986* | *1987* | *1988* | *1989* |
|---|---|---|---|---|---|---|---|
| Growth rates (%): | | | | | | | |
| gross output, industry | 1.25 | 1.83 | 0.80 | 0.30 | 5.43 | 1.63 | 1.53 |
| gross output, agriculture | 0.68 | 2.60 | -6.00 | -0.60 | 2.00 | 1.20 | 1.20 |
| gross output, construction | 0.64 | 1.55 | -7.70 | 0.20 | 0.27 | 0.61 | 6.50 |
| gross output, services | 3.73 | 2.17 | 1.29 | 3.35 | 1.37 | 1.54 | 2.24 |
| exports to non-CPEs, real | 16.35 | 4.33 | -10.37 | -3.51 | 10.64 | 4.75 | 4.77 |
| imports from non-CPEs, real | 5.30 | -1.56 | -0.62 | -1.27 | -2.25 | 1.09 | 5.56 |
| GDP | 1.39 | 1.78 | -0.01 | 0.72 | -0.12 | 1.33 | 2.03 |
| employment | -0.64 | -0.61 | -0.60 | -0.50 | -0.50 | -0.50 | -0.50 |
| productivity | 2.02 | 2.38 | 0.59 | 1.22 | 0.62 | 1.83 | 2.53 |
| personal consumption | 0.51 | 0.88 | 1.26 | 2.90 | -1.56 | -0.33 | 0.50 |
| investment, fixed assets | -2.11 | -0.6 | -0.3 | 0.64 | 1.52 | 1.42 | 6.85 |
| Investment ratio (%) | 24.80 | 22.90 | 22.22 | 22.20 | 22.51 | 22.53 | 23.59 |
| Current acct. balance($) | 0.30 | 0.33 | -0.02 | -0.41 | 0.02 | 0.12 | 0.08 |

*Table 22.1.* Continued

| *1990* | *1991* | *1991* | *1993* | *1994* | *1995* | *1996* | *1997* | *1998* | *1999* | *2000* |
|---|---|---|---|---|---|---|---|---|---|---|
| 1.31 | 1.10 | 1.12 | 1.19 | 1.22 | 1.23 | 1.24 | 1.25 | 1.25 | 1.26 | 1.26 |
| 1.20 | 1.20 | 1.20 | 1.20 | 1.20 | 1.20 | 1.20 | 1.20 | 1.20 | 1.20 | 1.20 |
| 1.35 | 2.20 | 2.70 | 3.00 | 3.14 | 3.16 | 3.10 | 3.03 | 2.95 | 2.90 | 2.85 |
| 1.86 | 0.79 | 0.96 | 1.08 | 1.16 | 1.22 | 1.25 | 1.27 | 1.28 | 1.28 | 1.29 |
| 2.84 | 2.59 | 2.54 | 2.48 | 2.45 | 2.43 | 2.42 | 2.41 | 2.40 | 2.39 | 2.38 |
| 1.53 | 1.81 | 2.24 | 2.30 | 2.39 | 2.46 | 2.47 | 2.46 | 2.44 | 2.43 | 2.41 |
| 1.28 | 1.40 | 1.50 | 1.60 | 1.65 | 1.67 | 1.68 | 1.69 | 1.68 | 1.68 | 1.68 |
| –0.50 | –0.50 | –0.50 | –0.50 | –0.50 | –0.50 | –0.50 | –0.50 | –0.50 | –0.50 | –0.50 |
| 1.78 | 1.90 | 2.00 | 2.10 | 2.15 | 2.17 | 2.18 | 2.19 | 2.18 | 2.18 | 2.18 |
| 0.28 | 0.61 | 0.84 | 1.01 | 1.13 | 1.21 | 1.26 | 1.29 | 1.31 | 1.32 | 1.33 |
| 2.04 | 2.37 | 2.80 | 3.04 | 3.15 | 3.16 | 3.11 | 3.04 | 2.97 | 2.92 | 2.89 |
| 22.73 | 22.95 | 23.24 | 23.57 | 23.92 | 24.27 | 24.60 | 24.93 | 25.25 | 25.55 | 25.86 |
| –0.07 | –0.05 | –0.06 | –0.06 | –0.06 | –0.07 | –0.08 | –0.08 | –0.08 | –0.08 | –0.08 |

*Table 22.2.* Continued

| *1990* | *1991* | *1992* | *1993* | *1994* | *1995* | *1996* | *1997* | *1998* | *1999* | *2000* |
|---|---|---|---|---|---|---|---|---|---|---|
| 1.65 | 1.49 | 1.53 | 1.61 | 1.65 | 1.68 | 1.70 | 1.72 | 1.74 | 1.76 | 1.79 |
| 1.20 | 1.20 | 1.20 | 1.20 | 1.20 | 1.20 | 1.20 | 1.20 | 1.20 | 1.20 | 1.20 |
| 3.50 | 4.33 | 4.71 | 4.84 | 4.82 | 4.73 | 4.61 | 4.51 | 4.43 | 4.39 | 4.36 |
| 2.22 | 1.21 | 1.40 | 1.53 | 1.62 | 1.68 | 1.71 | 1.74 | 1.76 | 1.78 | 1.80 |
| 3.60 | 3.36 | 3.35 | 3.32 | 3.32 | 3.32 | 3.33 | 3.34 | 3.34 | 3.34 | 3.34 |
| 2.56 | 2.81 | 3.28 | 3.31 | 3.37 | 3.40 | 3.39 | 3.37 | 3.36 | 3.36 | 3.36 |
| 1.78 | 1.94 | 2.05 | 2.14 | 2.19 | 2.21 | 2.22 | 2.23 | 2.24 | 2.26 | 2.27 |
| –0.50 | –0.50 | –0.50 | –0.50 | –0.50 | –0.50 | –0.50 | –0.50 | –0.50 | –0.50 | –0.50 |
| 2.28 | 2.44 | 2.55 | 2.64 | 2.69 | 2.71 | 2.72 | 2.73 | 2.74 | 2.76 | 2.77 |
| 0.67 | 1.07 | 1.35 | 1.53 | 1.66 | 1.75 | 1.80 | 1.83 | 1.86 | 1.89 | 1.91 |
| 3.93 | 4.24 | 4.55 | 4.65 | 4.62 | 4.53 | 4.42 | 4.33 | 4.27 | 4.23 | 4.21 |
| 23.50 | 24.03 | 24.62 | 25.23 | 25.83 | 26.41 | 26.98 | 27.53 | 28.08 | 28.62 | 29.16 |
| 0.01 | 0.02 | 0.01 | –0.00 | –0.01 | –0.01 | –0.01 | –0.01 | –0.01 | 0.00 | 0.01 |

*Table 22.3.* Continued

| *1990* | *1991* | *1991* | *1993* | *1994* | *1995* | *1996* | *1997* | *1998* | *1999* | *2000* |
|---|---|---|---|---|---|---|---|---|---|---|
| 2.07 | 1.96 | 2.01 | 2.12 | 2.19 | 2.24 | 2.30 | 2.35 | 2.41 | 2.47 | 2.52 |
| 1.20 | 1.20 | 1.20 | 1.20 | 1.20 | 1.20 | 1.20 | 1.20 | 1.20 | 1.20 | 1.20 |
| 5.68 | 6.34 | 6.53 | 6.52 | 6.42 | 6.28 | 6.17 | 6.08 | 6.03 | 6.01 | 6.01 |
| 2.65 | 1.68 | 1.89 | 2.04 | 2.14 | 2.22 | 2.28 | 2.33 | 2.39 | 2.44 | 2.50 |
| 4.44 | 4.25 | 4.29 | 4.30 | 4.34 | 4.38 | 4.42 | 4.46 | 4.49 | 4.52 | 4.55 |
| 3.66 | 3.87 | 4.33 | 4.35 | 4.39 | 4.42 | 4.43 | 4.44 | 4.46 | 4.49 | 4.53 |
| 2.33 | 2.51 | 2.63 | 2.73 | 2.79 | 2.83 | 2.87 | 2.91 | 2.95 | 3.00 | 3.05 |
| –0.5 | –0.50 | –0.50 | –0.50 | –0.50 | –0.50 | –0.50 | –0.50 | –0.50 | –0.50 | –0.50 |
| 2.83 | 3.01 | 3.13 | 3.23 | 3.29 | 3.33 | 3.37 | 3.41 | 3.45 | 3.50 | 3.55 |
| 1.13 | 1.58 | 1.89 | 2.10 | 2.25 | 2.35 | 2.42 | 2.49 | 2.55 | 2.61 | 2.67 |
| 5.86 | 6.02 | 6.17 | 6.15 | 6.05 | 5.93 | 5.83 | 5.76 | 5.72 | 5.72 | 5.72 |
| 24.41 | 25.25 | 26.12 | 26.99 | 27.84 | 28.68 | 29.51 | 30.32 | 31.14 | 31.96 | 32.79 |
| 0.09 | 0.09 | 0.08 | 0.08 | 0.08 | 0.08 | 0.09 | 0.11 | 0.12 | 0.14 | 0.16 |

Although the elasticity of exports with respect to world demand in the rest of Hungarian trade is much higher than 1.0, as a result of aggregating the two groups the *marginal* elasticity of the growth rate of total Hungarian exports to the growth rate of total world trade is not more than around 0.5–0.6.

Elasticity for total imports seem to be around 1.5. Considering the rate of growth of 2.4% in exports, this means that in the baseline forecast an equilibrium path of trade volumes allows a GDP growth of about 1.7% yearly. Increasing the assumed rate of world trade, each percentage point brings about a 0.5–0.6% increase in the rate of exports, and a similar increase of imports makes room for an additional 0.3–0.4% in the growth rate of GDP.

These growth rates have magnitudes that we have become used to in the last five years, or even higher by a fraction of a percentage point, but they are definitely lower than the rates of the late 1970s. Considering the fact, however, that growth in the 1970s was largely based on borrowed foreign resources, and that the forecast includes an estimated debt position that does not deteriorate in the long run, the figures for the period 1987–2000 seem to be in line with developments of the past 10–15 years.

CHAPTER 23

# Structural Adjustment, Productivity and Capital Accumulation in Austria

*Stefan Schleicher*

## Summary

The declining trend of GDP growth, accompanied by a similar tendency in labor productivity and increasing evidence of structural deficiencies in the sectoral composition of GDP, has become a matter of increasing concern for economic policy in Austria. We present empirical evidence and some causal interpretations about structural adjustment, changes in productivity and capital accumulation that have occurred over the past 20 years in the Austrian economy. Extending previous work on Austrian labor productivity (Bayer, 1983; Mitter and Skolka, 1984), we apply a growth accounting framework (Baily, 1981; Nadiri, 1980) to obtain empirical evidence about sectoral changes in productivity.

## 23.1. Analytical Tools and Data Base

In order to evaluate the relationship between factors of production and output, we make use of the following aggregate or sectoral production functions:

$$Q = F * K^b * N^{1-b} \tag{23.1}$$

where: $Q$ = output; $F$ = total (factor) productivity; $K$ = capital input; and $N$ = labor input.

In this production function framework, labor productivity $Q/N$ follows as

$$Q/N = F * (K/N)^b \tag{23.2}$$

where $K/N$ denotes the capital/labor ratio or the capital intensity.

Denoting by $q$, $f$, $k$ and $n$ the relative rates of change of the above defined variables for output, total factor productivity, capital and labor and substituting for the output elasticity of capital $b$ the corresponding factor share, we end up with the usual framework for decomposing the factor contributions to output growth

$$q = f + b * k + (1 - b) * n \tag{23.3}$$

and growth of labor productivity

$$q - n = f + b * (k - n) \tag{23.4}$$

Equation (23.3) is used to evaluate the relative changes of total factor productivity $f$. In this calculation all rates of change of actual data $x$ were smoothed by the following exponential filter to eliminate short-term cyclical fluctuations:

$$y_t = y_{t-1} + 0.3 * (x_t - y_{t-1}) \tag{23.5}$$

These smoothed variables $y$ will be interpreted as potential values in contrast to the actual unsmoothed figures $x$.

Annual data for the Austrian economy disaggregated into 19 sectors from 1964 to 1985 were used; these stem from the WIFO Databank, the data base of the Austrian Institute for Economic Research. Output is measured in value added, labor input comprises employed persons, and gross capital stocks as calculated by Hahn and Schmoranz (1983) are used for capital input.

## 23.2. Aggregate Economic Growth and Productivity

According to *Table 23.1*, the potential value of aggregate economic growth has slowed down from 5% in the early 1970s to 2% in the most recent years.

Using the production function framework presented in the previous section, this potential growth rate can be decomposed into growth of both capital and labor and an unexplained residual, which is usually termed total (factor) productivity. *Table 23.1* reveals that the contribution of capital to economic growth has dropped only slightly from 1.0% in the 1960s to 0.8% in the most recent years. Labor scarcely affected output growth in the 1960s, peaked at 1.4% in

*Table 23.1.* Aggregate economic growth and factor productivity (% change).

| Year | *Actual value added* | *Potential value added* | | | | *Potential labor productivity* | | |
|---|---|---|---|---|---|---|---|---|
| | | *Value added* | *Total product.* | *Capital* | *Labor* | *Labor product.* | *Total product.* | *Capital intensity* |
| 1965 | 3.1 | 4.3 | 3.4 | 1.0 | -0.0 | 4.3 | 3.4 | 1.0 |
| 1966 | 5.2 | 4.6 | 3.6 | 1.0 | 0.0 | 4.5 | 3.6 | 1.0 |
| 1967 | 3.1 | 4.1 | 3.4 | 1.0 | -0.2 | 4.5 | 3.4 | 1.1 |
| 1968 | 4.3 | 4.2 | 3.6 | 1.0 | -0.3 | 4.7 | 3.6 | 1.1 |
| 1969 | 5.7 | 4.7 | 3.8 | 1.0 | -0.1 | 4.8 | 3.8 | 1.0 |
| 1970 | 6.5 | 5.2 | 4.0 | 1.0 | 0.2 | 4.9 | 4.0 | 0.9 |
| 1971 | 4.9 | 5.1 | 3.6 | 1.0 | 0.5 | 4.4 | 3.6 | 0.8 |
| 1972 | 5.9 | 5.4 | 3.4 | 1.1 | 0.9 | 4.1 | 3.4 | 0.8 |
| 1973 | 5.1 | 5.3 | 2.7 | 1.2 | 1.4 | 3.3 | 2.7 | 0.6 |
| 1974 | 4.3 | 5.0 | 2.4 | 1.2 | 1.4 | 3.0 | 2.4 | 0.6 |
| 1975 | -0.4 | 3.4 | 1.3 | 1.2 | 1.0 | 2.0 | 1.3 | 0.8 |
| 1976 | 4.8 | 3.8 | 1.8 | 1.1 | 0.9 | 2.5 | 1.8 | 0.8 |
| 1977 | 4.5 | 4.0 | 1.9 | 1.1 | 1.0 | 2.6 | 1.9 | 0.7 |
| 1978 | 1.1 | 3.2 | 1.2 | 1.1 | 0.9 | 1.9 | 1.2 | 0.7 |
| 1979 | 5.1 | 3.8 | 1.9 | 1.1 | 0.8 | 2.5 | 1.9 | 0.7 |
| 1980 | 3.2 | 3.6 | 1.8 | 1.0 | 0.7 | 2.5 | 1.8 | 0.7 |
| 1981 | 0.2 | 2.6 | 1.0 | 1.0 | 0.6 | 1.7 | 1.0 | 0.7 |
| 1982 | 1.4 | 2.2 | 1.1 | 0.9 | 0.2 | 2.0 | 1.1 | 0.9 |
| 1983 | 2.0 | 2.1 | 1.3 | 0.9 | -0.1 | 2.2 | 1.3 | 0.9 |
| 1984 | 2.3 | 2.2 | 1.3 | 0.8 | 0.0 | 2.1 | 1.3 | 0.8 |
| 1985 | 3.0 | 2.4 | 1.5 | 0.8 | 0.1 | 2.2 | 1.5 | 0.7 |

Source for all tables in this chapter: WIFO Databank.

1974 and slowed down to zero again in the 1980s. By far the most important source for economic growth has been total factor productivity which, however, shows a pronounced downward trend from 4.0% in 1970 to 1.5% in 1985.

*Table 23.1* also contains a similar analysis for potential labor productivity. The decline from almost 5% labor productivity growth in 1970 to approximately 2% in the 1980s is mainly due to a corresponding decline in total factor productivity. The contribution of capital intensity growth to labor productivity growth is fairly stable around 0.8%

## 23.3. Structural Adjustments at the Sectoral Level

The aggregate figures for growth and productivity conceal the considerable adjustments that took place at the sectoral level.

*Table 23.2* indicates that between 1965 and 1985 the share of the primary sector in total value added declined from 7.3% to 5.3%, but the shares of both the secondary sector (from 40.6% to 41.1%) and of the tertiary sector (from 52.1% to 53.6%) expanded.

The corresponding shifts in employment are more pronounced and different, as can be seen from *Table 23.3*. Employment in the primary sector dropped between 1965 and 1985 from a share of 4.9% to 1.6%. The labor share

*Table 23.2.* Structural changes in value added.

| Sector | *Sector shares (%)* 1965 | 1970 | 1975 | 1980 | 1985 | *Changes in sector shares (%)* 1970 | 1975 | 1980 | 1985 |
|---|---|---|---|---|---|---|---|---|---|
| Primary: | | | | | | | | | |
| Agriculture | 6.32 | 6.04 | 5.37 | 5.03 | 4.80 | -0.29 | -0.67 | -0.34 | -0.23 |
| Mining | 0.95 | 0.74 | 0.61 | 0.52 | 0.49 | -0.21 | -0.12 | -0.09 | -0.04 |
| Secondary: | | | | | | | | | |
| Food | 4.76 | 4.91 | 4.82 | 4.90 | 4.72 | 0.15 | -0.09 | 0.09 | -0.18 |
| Textiles and apparel | 3.68 | 3.50 | 2.91 | 2.57 | 2.31 | -0.17 | -0.59 | -0.34 | -0.26 |
| Wood products | 2.13 | 2.19 | 2.12 | 2.24 | 2.22 | 0.06 | -0.07 | 0.13 | -0.02 |
| Paper products | 2.39 | 2.16 | 2.03 | 2.03 | 2.19 | -0.23 | -0.13 | 0.00 | 0.15 |
| Chemical products | 1.65 | 2.15 | 2.52 | 2.85 | 3.00 | 0.50 | 0.37 | 0.33 | 0.15 |
| Petroleum industries | 2.06 | 2.05 | 1.69 | 1.38 | 0.95 | -0.01 | -0.37 | -0.31 | -0.43 |
| Nonmetallic minerals | 1.77 | 1.82 | 1.85 | 1.82 | 1.86 | 0.06 | 0.03 | -0.03 | 0.04 |
| Basic metals | 2.55 | 2.77 | 2.17 | 2.37 | 2.27 | 0.22 | -0.60 | 0.20 | -0.10 |
| Metal products | 8.37 | 9.64 | 9.83 | 10.80 | 11.93 | 1.26 | 0.19 | 0.97 | 1.13 |
| Energy and water | 2.74 | 2.97 | 3.19 | 3.35 | 3.34 | 0.23 | 0.23 | 0.16 | -0.01 |
| Construction | 8.56 | 8.53 | 9.00 | 7.75 | 6.34 | -0.02 | 0.46 | -1.25 | -1.41 |
| Tertiary: | | | | | | | | | |
| Wholesale and retail trade | 13.91 | 13.62 | 14.50 | 13.96 | 14.16 | -0.29 | 0.87 | -0.54 | 0.20 |
| Restaurants and hotels | 3.85 | 3.54 | 3.31 | 2.98 | 2.79 | -0.31 | -0.23 | -0.33 | -0.18 |
| Transport and communication | 4.85 | 5.16 | 5.69 | 6.34 | 6.62 | 0.31 | 0.53 | 0.65 | 0.28 |
| Financial services | 9.16 | 9.84 | 10.79 | 11.80 | 12.83 | 0.69 | 0.95 | 1.01 | 1.03 |
| Other services | 5.36 | 4.41 | 4.20 | 3.94 | 3.81 | -0.95 | -0.21 | -0.26 | -0.13 |
| Public services | 14.95 | 13.96 | 13.41 | 13.37 | 13.38 | -1.00 | -0.55 | -0.04 | 0.01 |

employed in the secondary sector fell from 49.5% to 39.6%. The tertiary sector absorbed the working force no longer needed in the other sectors by expanding its share from 45.6% to 58.8%.

A comparison of the changes in the sector shares in *Tables 23.2* and *23.3* reveals some interesting details. Whereas all sectors in the primary and secondary sector have lost labor shares over the last five years (except for energy and water), a number of industries in the secondary sector have expanded their shares in total value added: metal products, nonmetallic minerals, chemicals and paper products. The latter three industries belong to the raw material- and energy-intensive industries and hint at the fundamental structural problems of Austrian industry.

## 23.4. The Labor Productivity Slowdown at the Sectoral Level

*Table 23.4* contains a concise summary of the labor productivity slowdown at the sectoral level which occurred over the last 15 years. For all 19 sectors relative changes in potential labor productivity are partitioned according to equation (23.4) into components due to relative changes in total factor productivity and relative changes in capital intensity.

*Table 23.3.* Structural changes in employed persons.

| Sector | Sector shares (%) 1965 | 1970 | 1975 | 1980 | 1985 | Changes in sector shares (%) 1970 | 1975 | 1980 | 1985 |
|---|---|---|---|---|---|---|---|---|---|
| Primary: | | | | | | | | | |
| Agriculture | 3.69 | 2.45 | 1.67 | 1.34 | 1.13 | −1.24 | −0.78 | −0.33 | −0.21 |
| Mining | 1.26 | 0.90 | 0.67 | 0.53 | 0.46 | −0.36 | −0.23 | −0.15 | −0.06 |
| Secondary: | | | | | | | | | |
| Food | 4.68 | 4.47 | 4.07 | 3.84 | 3.70 | −0.21 | −0.40 | −0.23 | −0.14 |
| Textiles and apparel | 7.43 | 6.71 | 5.38 | 4.59 | 3.93 | −0.71 | −1.34 | −0.78 | −0.66 |
| Wood products | 3.38 | 3.11 | 3.03 | 3.22 | 3.12 | −0.26 | −0.08 | 0.19 | −0.10 |
| Paper products | 2.80 | 2.68 | 2.41 | 2.17 | 2.02 | −0.12 | −0.27 | −0.24 | −0.15 |
| Chemical products | 2.38 | 2.75 | 2.66 | 2.54 | 2.36 | 0.37 | −0.09 | −0.12 | −0.18 |
| Petroleum industries | 0.38 | 0.33 | 0.36 | 0.35 | 0.32 | −0.05 | 0.02 | −0.01 | −0.03 |
| Nonmetallic minerals | 2.25 | 2.00 | 1.86 | 1.67 | 1.49 | −0.24 | −0.14 | −0.20 | −0.17 |
| Basic metals | 2.76 | 2.61 | 2.60 | 2.44 | 2.13 | −0.15 | −0.01 | −0.16 | −0.31 |
| Metal products | 11.68 | 12.30 | 12.53 | 12.35 | 11.46 | 0.62 | 0.23 | −0.18 | −0.89 |
| Energy and water | 1.17 | 1.23 | 1.21 | 1.13 | 1.23 | 0.06 | −0.03 | −0.08 | 0.10 |
| Construction | 10.55 | 9.88 | 9.70 | 9.21 | 7.89 | −0.67 | −0.18 | −0.48 | −1.33 |
| Tertiary: | | | | | | | | | |
| Wholesale and retail trade | 11.14 | 11.58 | 12.78 | 13.31 | 13.67 | 0.45 | 1.20 | 0.53 | 0.36 |
| Restaurants and hotels | 3.25 | 3.53 | 3.38 | 3.80 | 4.24 | 0.28 | −0.15 | 0.42 | 0.43 |
| Transport and communication | 7.84 | 7.75 | 7.61 | 7.31 | 7.56 | −0.09 | −0.15 | −0.30 | 0.25 |
| Financial services | 4.70 | 5.29 | 2.86 | 3.26 | 3.58 | 0.59 | −2.43 | 0.40 | 0.32 |
| Other services | 3.38 | 3.00 | 11.82 | 12.85 | 14.09 | −0.38 | 8.81 | 1.03 | 1.24 |
| Public services | 15.28 | 17.40 | 13.39 | 14.09 | 15.63 | 2.11 | −4.01 | 0.70 | 1.54 |

The current labor productivity growth rate of 2.2% for the whole economy can be attributed to an increase of 0.7% in capital intensity and an 1.5% rise in total factor productivity. These numbers, however, do not reflect the wide span of sectoral differences.

Highest labor productivity growth can be observed in the secondary sector: metal products (5.7%), paper products (5.4%), chemical products (4.9%) and nonmetallic minerals (4.7%). These branches also show the highest growth rates of capital intensity and of total factor productivity.

On the lower end of current potential labor productivity growth, we find the service sector branches of restaurants and hotels (−1.3%) and other services (−0.5%) together with the petroleum industries (−0.7%). The latter two branches exhibit the disturbing fact of a deteriorating total factor productivity despite an above-average growth of capital intensity.

A look at the dynamics of the relative labor productivity changes and their relationship to the relative movements of total factor productivity and capital intensity suggests that in the early 1970s the capital intensities grew at a slightly higher rate than in the 1980s, but were accompanied by a remarkably higher contribution of total factor productivity. In addition branches with declining output shares seem to exhibit lower growth both in capital intensities and total factor productivity.

*Table 23.4.* Potential labor productivity (% change) by sector: LP = labor productivity, TP = total productivity, and CI = capital intensity.

| | 1970 | | | 1975 | | | 1980 | | | 1985 | | |
|---|---|---|---|---|---|---|---|---|---|---|---|---|
| *Sector* | *LP* | *TP* | *CI* | *LP* | *TP* | *CI* | *LP* | *TP* | *CI* | *LP* | *TP* | *CI* |
| Primary: | | | | | | | | | | | | |
| Agriculture | 10.77 | 7.92 | 2.85 | 8.43 | 6.59 | 1.85 | 6.79 | 5.45 | 1.34 | 4.01 | 2.87 | 1.13 |
| Mining | 7.26 | 6.21 | 1.04 | 3.25 | 2.67 | 0.58 | 5.42 | 4.58 | 0.84 | 3.20 | 2.93 | 0.27 |
| Secondary: | | | | | | | | | | | | |
| Food | 6.23 | 4.68 | 1.55 | 2.88 | 1.58 | 1.29 | 4.24 | 3.13 | 1.11 | 2.36 | 1.53 | 0.83 |
| Textiles and apparel | 6.22 | 5.14 | 1.08 | 2.76 | 1.20 | 1.56 | 3.36 | 2.57 | 0.80 | 3.16 | 2.24 | 0.92 |
| Wood products | 7.13 | 3.72 | 3.41 | 1.98 | –0.39 | 2.38 | 1.95 | 1.04 | 0.90 | 3.88 | 3.13 | 0.75 |
| Paper products | 3.70 | 2.58 | 1.12 | 1.40 | 0.12 | 1.28 | 3.91 | 2.73 | 1.19 | 5.44 | 4.12 | 1.32 |
| Chemical products | 7.46 | 6.41 | 1.05 | 5.24 | 3.59 | 1.65 | 5.46 | 4.45 | 1.01 | 4.93 | 3.95 | 0.97 |
| Petroleum industries | 7.94 | 6.30 | 1.63 | –1.46 | –2.44 | 0.98 | –2.28 | –3.83 | 1.55 | –0.68 | –1.62 | 0.94 |
| Nonmetallic minerals | 7.80 | 5.07 | 2.72 | 3.66 | 1.76 | 1.90 | 4.74 | 3.66 | 1.08 | 4.73 | 3.74 | 0.99 |
| Basic metals | 8.08 | 7.16 | 0.93 | –3.79 | –5.08 | 1.29 | 3.35 | 2.53 | 0.82 | 4.33 | 3.38 | 0.95 |
| Metal products | 7.06 | 6.17 | 0.90 | 2.45 | 1.02 | 1.43 | 4.35 | 3.15 | 1.20 | 5.67 | 4.37 | 1.31 |
| Energy and water | 6.34 | 5.86 | 0.48 | 4.46 | 3.71 | 0.75 | 5.21 | 4.31 | 0.90 | 2.16 | 1.71 | 0.45 |
| Construction | 5.65 | 3.93 | 1.71 | 3.63 | 1.92 | 1.71 | 1.17 | 0.51 | 0.66 | 1.59 | 1.16 | 0.43 |
| Tertiary: | | | | | | | | | | | | |
| Wholesale and retail trade | 4.09 | 2.81 | 1.29 | 1.07 | 0.65 | 0.42 | 0.99 | 0.20 | 0.80 | 1.68 | 0.74 | 0.94 |
| Restaurants and hotels | 1.66 | 0.50 | 1.16 | 0.01 | –1.15 | 1.17 | –0.16 | –1.40 | 1.24 | –1.26 | –2.51 | 1.24 |
| Transport and communication | 7.64 | 6.16 | 1.48 | 3.94 | 2.75 | 1.19 | 5.07 | 3.85 | 1.22 | 3.05 | 2.17 | 0.88 |
| Financial services | 4.34 | 4.37 | –0.03 | –0.04 | 0.69 | –0.73 | 1.11 | 1.45 | –0.34 | 1.90 | 1.66 | 0.24 |
| Other services | 3.54 | 1.96 | 1.58 | –0.53 | –0.38 | –0.15 | 0.09 | 0.02 | 0.07 | –0.53 | –0.79 | 0.27 |
| Public services | 0.62 | 0.54 | 0.08 | –0.84 | –0.89 | 0.05 | 1.30 | 0.90 | 0.39 | 0.09 | 0.02 | 0.07 |
| All sectors | 4.89 | 4.01 | 0.88 | 2.02 | 1.26 | 0.76 | 2.54 | 1.82 | 0.71 | 2.22 | 1.49 | 0.73 |

## 23.5. Econometric Evidence for the Total Factor Productivity Slowdown

The crucial role of total factor productivity growth, which emerged in the decomposition of labor productivity growth, calls for causal interpretations. We will focus on the hypotheses that cyclical behavior of the economy and capital intensity may have an impact on the productivity of a factor.

We tested these propositions with the following econometric specification:

$$f = a_0 + a_1^* f_{-1} + a_2^* (q^{\text{act}} - q) + a_3^* (k - n) + \text{res.} \tag{23.6}$$

Thus, growth of total factor productivity $f$ is to be explained by a constant, lagged factor productivity growth $f_{-1}$ for dynamic adjustments, the difference between actual and smoothed output growth $(q^{\text{act}} - q)$ as an indicator for capacity utilization, growth of capital intensity $(k - n)$ and a residual. *Table 23.5* summarizes the estimation results for the sample period 1966 to 1985.

*Table 23.5.* Regressions for potential total factor productivity 1966–1985.

| | Estimated coefficients | | | | |
|---|---|---|---|---|---|
| *Sector* | *Constant* | *Total production lagged* | *Capacity utilization* | *Capital intensity* | $R^2$ *adjusted* |
| Primary: | | | | | |
| Agriculture | –0.185 | 1.009 | 0.429 | | 0.979 |
| Mining | 0.505 | 0.832 | 0.390 | | 0.921 |
| Secondary: | | | | | |
| Food | 0.197 | 0.951 | 0.401 | | 0.981 |
| Textiles and apparel | –0.627 | 0.830 | 0.280 | 1.017 | 0.931 |
| Wood products | 0.523 | 0.758 | 0.371 | | 0.938 |
| Paper products | 0.634 | 0.752 | 0.329 | | 0.928 |
| Chemical products | 0.496 | 0.749 | 0.324 | 0.717 | 0.904 |
| Petroleum industries | –0.067 | 0.891 | 0.383 | | 0.962 |
| Nonmetallic minerals | 1.408 | 0.623 | 0.306 | | 0.859 |
| Basic metals | 0.526 | 0.831 | 0.396 | | 0.925 |
| Metal products | –0.052 | 0.750 | 0.313 | 0.768 | 0.943 |
| Energy and water | –0.364 | 0.911 | 0.420 | 0.959 | 0.979 |
| Construction | 0.222 | 0.890 | 0.180 | | 0.926 |
| Tertiary: | | | | | |
| Wholesale and retail trade | 0.033 | 0.724 | 0.352 | 0.395 | 0.943 |
| Restaurants and hotels | –3.128 | 0.538 | 0.448 | 2.117 | 0.927 |
| Transport and communication | –0.696 | 0.761 | 0.380 | 1.271 | 0.987 |
| Financial services | 1.099 | 0.640 | 0.402 | 0.938 | 0.882 |
| Other services | –0.934 | 0.150 | 1.678 | | 0.809 |
| Public services | –0.109 | 0.695 | 0.309 | 2.034 | 0.956 |
| All sectors | –0.455 | 0.807 | 0.308 | 1.112 | 0.988 |

For the whole economy the effect of changes in both capacity utilization and capital intensity turned out to be highly significant. The lagged dependent variable indicates the adjustment process which is caused by changes in the other dependent variables.

The sectoral results stress the importance of capacity utilization for explaining total factor productivity. The economic rationale behind this phenomenon is the observation that it is costly for firms to make adjustments to factor inputs. In response to demand shocks, firms may vary factor utilization levels in the short run, leading to changes in observed factor productivity.

To what extent capital formation fosters productivity growth is still rather controversial (Olson, 1982; Lindbeck, 1983). In the data base used for all sectors a significant influence of relative changes in capital intensity on growth of total factor productivity could not be found. This effect is pronounced, however, in textiles, chemicals, metal products, energy and in almost all services sectors.

Equation (23.3) may be interpreted as a labor demand function for each of the 19 sectors considered, the growth rate of total factor productivity being explained by equation (23.6). With actual sectoral output and capital formation determined by effective demand, this yields a neo-Ricardian perspective of a multisectoral economy, which seems to be very adequate for Austria since both capital formation and composition of output have been strongly influenced in the past by the political decision process, which has often overruled economic factors such as changes in structural composition of demand and relative prices.

## 23.6. Prospects and Conclusions

There is no evidence that growth of the Austrian economy was seriously hampered by supply conditions. Over the last 20 years the contribution of capital to economic growth has been very stable. When domestic labor turned scarce at the end of the 1960s foreign workers expanded the stock of the available labor force.

Like almost all other industrial nations, Austria experienced a substantial slowdown of total factor productivity which affected labor and capital productivity. Econometric evidence suggests that this slowdown mainly reflects underutilization of existing production capacities caused both by a drop of overall demand and shifts in the sectoral composition of demand. In a number of manufacturing branches and in almost all services, overall factor productivity has been significantly influenced by capital intensity.

Extrapolating past trends of demand and capital formation, Austria should have no problems in maintaining an average annual GDP growth rate of 2 to 3% over the next decade. More crucial, however, seems to be the decision about the composition of output and the decision about the corresponding production techniques. The econometric results obtained indicate a substantial impact of the decision about size and composition of capital on growth of output, factor productivity and labor demand.

As in other studies (Aiginger, 1987), we want to emphasize that the supply structure of the Austrian economy only sluggishly follows the considerable shifts in the composition of demand. In some resource-intensive industries, such as paper and chemical products, sectoral shares in total value added have even expanded over the last years, which leads us to question the functioning of the structural adjustment process toward a technologically advanced industrialized economy.

## References

Aiginger, K. (ed.) (1987), *Österreichische Strukturberichterstattung*. (Vienna: Österreichisches Institut für Wirtschaftsforschung).

Baily, M.N. (1981), The Productivity Growth Slowdown and Capital Accumulation, *American Economic Review*, **71**(2), pp. 326–331.

Bayer, K. (1983), Produktivitätswachstum österreichischer Industriebranchen im internationalen Vergleich, *Monatsberichte des Österreichischen Instituts für Wirtschaftsforschung*, pp. 630–638.

Hahn, F. and Schmoranz, I. (1983), Schätzung des österreichischen Kapitalstocks nach Wirtschaftsbereichen. *Monatsberichte des Österreichischen Instituts für Wirtschaftsforschung*, pp. 40–52.

Hickman, B.G. and Coen, R.M. (1976), *An Annual Growth Model of the U.S. Economy*. (Amsterdam: North-Holland).

Lindbeck, A. (1983), The Recent Slowdown of Productivity Growth, *The Economic Journal*, **83**, pp. 13–34.

Mitter, P. and Skolka, J. (1984), Labour Productivity in Austria between 1964 and 1980, *Empirical Economics*, **9**, pp. 27–49.

Nadiri, M.I. (1980), Sectoral Productivity Slowdown, *American Economic Review*, **70**(2), pp. 349–352.

Nadiri, M.I. and Schankerman, M.A. (1981), Technical Change, Returns to Scale and the Productivity Slowdown. *American Economic Review*, **71**(2), pp. 314–319.

Olson, M. (1982), *The Rise and Decline of Nations*. (New Haven, CT: Yale University Press).

# PART III

# Analysis and Forecast of World Trade, the Debt Situation, and North–South Economic Relations

CHAPTER 24

# Structural Change in World Trade: Analyses and Forecasts of the FIND Model

*Georg Erber*

## Summary

This chapter presents six scenarios run with the FIND submodel of North–South trade for different commodity groups. The results show that different assumptions about real growth of world trade and world inflation lead to considerable structural shifts in the composition of the commodity structure of intra- and interregional trade relations of both country groups until the year 2000. A key role in balancing the North–South trade relationship is played by the development of world energy prices and the possibility of the developing countries entering the markets of the developed economies for manufactured products.

## 24.1. Introduction

The Forecasting Interindustrial Development Model (FIND) is a large-scale econometric model for the Federal Republic of Germany, developed by a research group of the German Institute for Economic Research (Deutsches Institut für Wirtschaftsforschung) in Berlin. Its development started in 1983 in cooperation with the Special Research Department 21 of Bonn University [cf. Krelle *et al.* (1982)]. A number of publications summarizes the current state of the research project: Erber (1986a, 1986b), Erber *et al.* (1983, 1986), Horn (1986), Meier (1986), and Nakamura (1986).

FIND is an annual input–output model of 51 industry sectors. Its modeling concept rests on a bottom-up approach by modeling each single sectoral activity on the demand and supply side. An overview of the input–output structure of the model is given by Erber (1986a). As far as possible, microeconomic foundations for deriving factor demand systems, household demand systems, or other simpler sectoral single-equation demand functions are used. The model explains real and price variables. It includes a detailed investment and capital accounting model based on a capital vintage approach. The capital and capacity utilization submodel distinguishes between equipment and structure. For the explanation of foreign trade, German export and import activities are linked with a small world trade model by two country groups – developed and developing market economies, and their inter- and intraregional trade matrix by seven one-digit SITC classifications [cf. Erber 1986b)]). Its fairly detailed government and social insurance submodel consists of an integration of the already available parts of the aggregated long-run forecasting model of the German Institute for Economic Research [cf. Blazejczak 1986)]. The complete model consists of about 7,000 behavioral and definitional equations.

The FIND model is applied to analyze the past development of structural change [cf. DIW (1983), Erber and Haas (1983)]. Additionally, it is now used for current research projects considering questions of consumer tax harmonization in the EEC, and the analysis of the impacts of the technological change on employment (META–Study II of the Ministry of Research and Technology). The world trade model was used in the following study to run different scenarios for the future development of North–South trade relations until the year 2000.

## 24.2. Methodological Aspects of Long-Range Forecasting with Econometric Models

Since the publication of the book *Limits to Growth* by Meadows *et al.* (1972) and the *The Global 2000 Report to the President* by Barney (1980), there has developed a growing awareness on the long-run impacts of changes in global socioeconomic trends by a broader public. Many people now understand that these developments should be regarded in current decision making to avoid long-run structural disparities and negative consequences, on the one hand. On the other hand, forecasts over a long time horizon are always conditional on assumptions on future trends that are difficult to assess – owing to instability of international political conditions caused by lack of appropriate international policy coordination [cf. Artis and Ostry (1986)], unexpected basic innovations such as the microelectronic revolution, sudden booms and depressions triggered by speculative bubbles in the international finance markets with long-lasting effects or general changes in cultural preferences shifting demand patterns in an unexpected way. This shattered the belief in the usefulness of quantitative forecasts. The critical evaluation of long-range forecasts [e.g., Hugger (1974)] strengthened the view that results of such investigations should not be taken as unassailable truths.

Despite uneasiness about reliability of forecasts and how much trust should be placed in quantitative forecasts, there is still the necessity to use all the best available information in current decision making. That this IIASA–Bonn Research Project on Economic Growth and Structural Change is subject to doubts and questions is obvious to the participating members [cf. Ross (1986)], but we hope that this model will nonetheless contribute to the understanding of future possible developments and give insights that cannot be obtained by other approaches.

## 24.3. Forecasts of Structural Shifts in North–South Trade by Commodity Groups

Builders of national forecasting models always face the problem of modeling foreign trade relationships of a sectoral (disaggregated) nature. Since the development of world trade has a major impact on the performance of the domestic economy, a forecast can be severely biased by bad predictions of the competing foreign demand and supply conditions, exchange rate movements, and transfer of capital or technology. Therefore, it is necessary to forecast disaggregated variables of world market activities, which give a closer link to the single markets of the industry of a country. Aggregate variables of world trade activities in many cases give unsatisfactory results if used as explanatory variables in sectoral import and demand equations. To solve this forecasting problem more satisfactorily and in a consistent manner without having to predict a great number of world trade variables in an *ad hoc* fashion, a small world market model was constructed in the context of the FIND model.

### 24.3.1. Specification of the world trade model

The world trade model of the FIND project explains the inter- and intraregional trade shares of two country groups, developed and developing market economies, by seven different commodity groups of the SITC classification. Since it needs only four aggregate world trade variables – the total volume of imports of the developed and developing market economies together with the corresponding price indices – it can be run independently from the rest of the FIND model, which uses the results as inputs for its sectoral export and import equations.

The SITC submodel has 224 equations. There are 56 behavioral equations explaining trade shares in current prices and prices by commodity groups. The remaining 168 equations are definitional equations partly necessary to guarantee consistency of the accounting framework and partly adding up the different aggregates to get the corresponding aggregate world trade variables for all market economies or all commodities. *Table 24.1* shows the classification used by the model.

For this classification two import demand systems were estimated. As Armington (1969) has proposed, demand is derived by a two-stage decision process. One country or country group decides at the first stage how to allocate its total import demand between different goods, and thereafter it decides from which country it orders what amount for each commodity or commodity group. *Figure 24.1* illustrates this two-stage allocation process. The behavioral equations of the demand system are derived from the almost ideal demand framework of Deaton and Muellbauer (1980).

*Table 24.1.* Classification used by the world trade model.

| | | |
|---|---|---|
| *I.* | *Country Groups:* | *Developed market economies*<br>*Developing market economies* |
| *II.* | *SITC goods classification:* | |
| | 0 + 1 | Food and live animals (0), and<br>Beverages and tobacco (1) |
| | 2 + 4 | Crude materials, inedible, except fuels (2), and<br>Animal and vegetable oils and fats (4) |
| | 3 | Mineral fuels, lubricants and related materials |
| | 5 | Chemicals |
| | 6 + 8 | Manufactures classified chiefly by material (6), and<br>Miscellaneous manufactured articles (8) |
| | 7 | Machinery and transport equipment |
| | 9 | Commodities and transactions not classified |

Source: UN (various years), covering the period 1960–1980. Base year for price and volume indices: 1976 = 100.

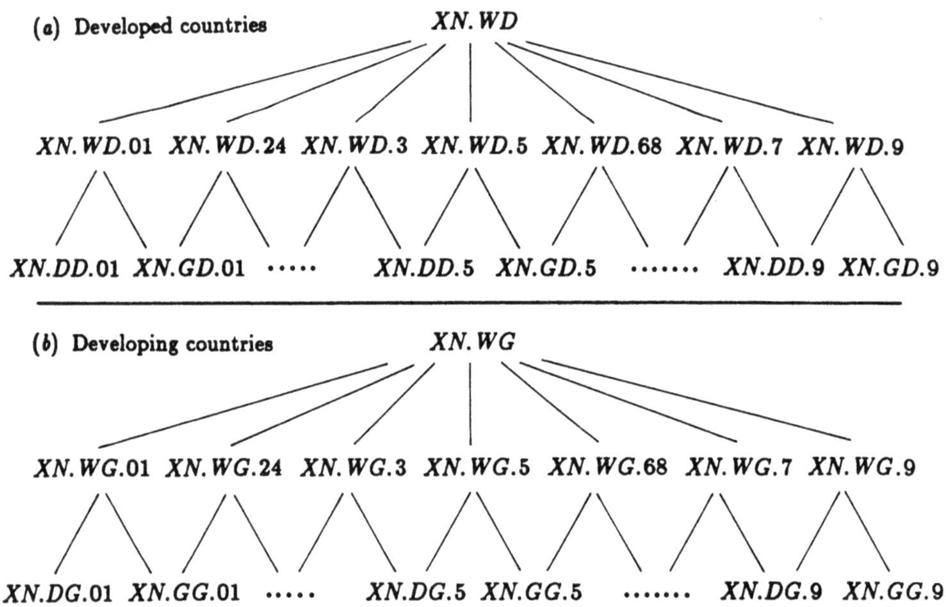


*Figure 24.1.* Simplified scheme of the FIND–SITC submodel, for (*a*) developed countries, and (*b*) developing countries.

The symbols of *Figure 24.1*, also used in the equations below, are:

XN = exports in current prices
W = all markets economies
D = developed market economies
G = developing market economies

The import demand systems were estimated by the full information maximum likelihood (FIML) method. *Tables 24.2* and *24.3* summarize the results of the coefficient estimates. For a discussion of the results, see Erber (1986b).

The second stage of the import demand systems was chosen from a static and dynamic, almost ideal demand system with respect to the results obtained after the estimation.

$$\frac{XN.GD_i}{XN.WD_i} = a_i + b_i \log \left[\frac{P.XGD_i}{P.XDD_i}\right] + c_i \log \left[\frac{XN.WD_i}{P_{\text{Stone}}}\right] \tag{24.1}$$

$$\frac{XN.GD_i}{XN.WD_i} = a_i + b_i \log \left[\frac{P.XGD_i}{P.XDD_i}\right] + c_i \log \left[\frac{XN.WD_i}{P_{\text{Stone}}}\right] + (1 - d_i)\left[\frac{XN.GD_i}{XN.WD_i t-1}\right] \tag{24.2}$$

## 24.3.2. Results of the different Bonn–IIASA scenarios

A number of alternative scenarios proposed by the Bonn–IIASA project for the future development of the world market were run with this model. Since currently the model has data available for all variables only from 1960 to 1980, we used the already available data for the aggregates of the two country groups published by the International Monetary Fund (1985, Table 19 of the Statistical Appendix, which also contains estimates for 1986) for the time period 1981 to 1986. The dynamic simulation of the model begins in 1975 in order to let the model adjust to its dynamics. From 1981 to 1986, the IMF data were used to forecast the endogenous variables. From 1987 to 2000, the assumptions of *Table 24.4* were applied to the volume and price indices of the total imports of each country group.

Because it is impossible to present all results in this chapter we selected some outcomes to indicate that the model generates reasonable results.

First, the results for the structural development of world trade by the seven commodity groups are discussed. The following tables present the results for the years 1980, 1990, and 2000 for the scenarios II, IV, and V. To compare the forecasted development with the historical structure, the values for the year 1960,

*Table 24.2.* Estimation results of the first stage of the AIDS models determining the import shares for the different SITC groups (*t*-values in parentheses).

(*a*) From all countries to the developed market economies:

| *Explanatory variable*[a] | *XN.WD.01* | *XN.WD.24* | *XN.WD.3* | *XN.WD.5* | *XN.WD.68* | *XN.WD.7* | *XN.WD.9* |
|---|---|---|---|---|---|---|---|
| *P.XWD*.01 | 0.027 | – | – | – | – | – | – |
| | (1.00) | | | | | | |
| *P.XWD*.24 | 0.012 | 0.13 | – | – | – | – | – |
| | (3.21) | (4.43) | | | | | |
| *P.XWD*.3 | –0.004 | –0.017 | 0.11 | – | – | – | – |
| | (0.91) | (3.06) | (30.42) | | | | |
| *P.XWD*.5 | –0.036 | –0.005 | –0.015 | 0.092 | – | – | – |
| | (2.31) | (0.25) | (4.85) | (4.38) | | | |
| *P.XWD*.68 | –0.14 | 0.004 | 0.044 | 0.013 | 0.035 | – | – |
| | (4.95) | (0.12) | (11.36) | (0.52) | (0.68) | | |
| *P.XWD*.7 | –0.06 | 0.017 | –0.06 | 0.006 | –0.1 | 0.059 | – |
| | (2.44) | (0.84) | (9.99) | (0.35) | (3.36) | (1.53) | |
| *P.XWD*.9 | – | – | – | – | – | – | – |
| Const | 0.24 | 0.13 | 0.21 | –0.003 | 0.16 | 0.07 | – |
| | (20.23) | (3.66) | (28.73) | (0.38) | (14.01) | (4.77) | |
| *XN.WD/PXNDW* | 0.066 | –0.083 | 0.003 | 0.04 | 0.05 | 0.092 | – |
| | (10.7) | (15.11) | (0.68) | (8.1) | (8.27) | (11.15) | |
| $R^2$ | 0.97 | 0.98 | 0.99 | 0.85 | 0.9 | 0.88 | – |

(*b*) From all countries to the developing market economies:

| *Explanatory variable*[a] | *XN.WG.01* | *XN.WG.24* | *XN.WG.3* | *XN.WG.5* | *XN.WG.68* | *XN.WG.7* | *XN.WG.9* |
|---|---|---|---|---|---|---|---|
| *P.XWG.01* | 0.079 | – | – | – | – | – | – |
| | (3.53) | | | | | | |
| *P.XWG.24* | –0.002 | 0.006 | – | – | – | – | – |
| | (0.36) | (0.25) | | | | | |
| *P.XWG.3* | 0.01 | 0.002 | 0.091 | – | – | – | – |
| | (1.95) | (0.21) | (20.69) | | | | |
| *P.XWG.5* | –0.027 | –0.042 | –0.03 | 0.094 | – | – | – |
| | (1.54) | (1.93) | (6.77) | (2.77) | | | |
| *P.XWG.68* | –0.1 | 0.043 | –0.029 | 0.026 | 0.064 | – | – |
| | (3.75) | (2.16) | (4.54) | (1.05) | (1.32) | | |
| *P.XWG.7* | 0.002 | –0.021 | –0.056 | –0.012 | –0.084 | 0.062 | – |
| | (0.06) | (1.1) | (5.01) | (0.3) | (1.31) | (0.57) | |
| *P.XWG.9* | – | – | – | – | – | – | – |
| Const | 0.16 | 0.084 | 0.19 | 0.05 | 0.21 | 0.28 | – |
| | (23.1) | (1.54) | (45.53) | (8.77) | (25.78) | (21.58) | |
| *XN.WG/PXNWG* | –0.066 | –0.017 | –0.04 | 0.04 | 0.024 | 0.1 | – |
| | (7.87) | (2.78) | (6.86) | (5.89) | (2.31) | (6.16) | |
| $R^2$ | 0.93 | 0.81 | 0.98 | 0.62 | 0.86 | 0.81 | – |

[a]The remaining coefficients can be computed from the parameter restrictions. $P.X$ = price index for the corresponding goods and country group. Const = constant term of the equation. $PXNWG$ = aggregate Stone price index. $R^2$ = coefficient of determination.

*Table 24.3.* Estimation results of the second stage AIDS models determining the import shares for the different SITC groups ($t$-values in parentheses).

| *SITC goods classification* | | $a_i$ | $b_i$ | $c_i$ | $d_i$ | $R^2$ |
|---|---|---|---|---|---|---|
| | | *Parameter* | | | | |
| (*a*) From developing to developed market economies: | | | | | | |
| 0 + 1 | Food, etc. | 0.47 | 0.25 | –0.067 | 0.91 | 0.93 |
| | | (8.41) | (7.49) | (7.66) | (8.43) | |
| 2 + 4 | Raw materials excl. fuels | 0.25 | 0.11 | –0.028 | 0.63 | 0.88 |
| | | (2.05) | (2.04) | (1.71) | (2.20) | |
| 3 | Fuels, etc. | 0.51 | 0.16 | 0.003 | 0.70 | 0.67 |
| | | (6.06) | (3.49) | (0.19) | (4.98) | |
| 5 | Chemicals | 0.031 | 0.007 | 0.003 | 0.99 | 0.72 |
| | | (5.50) | (2.18) | (3.72) | (4.98) | |
| 6 + 8 | Other manufactures | 0.015 | 0.012 | 0.025 | 0.78 | 0.88 |
| | | (1.19) | (0.49) | (3.29) | (4.12) | |
| 7 | Machinery | 0.008 | 0.007 | 0.002 | 0.52 | 0.78 |
| | | (2.04) | (1.16) | (2.37) | (2.58) | |
| 9 | Commodities not classified | 0.1 | –0.084 | 0.047 | – | 0.02 |
| | | (1.83) | (0.21) | (1.35) | | |
| (*b*) From developed to developing market economies: | | | | | | |
| 0 + 1 | Food, etc. | 0.65 | 0.06 | 0.026 | – | 0.49 |
| | | (65.52) | (2.04) | (4.69) | | |
| 2 + 4 | Raw materials excl. fuels | 0.53 | 0.31 | 0.062 | – | 0.62 |
| | | (36.31) | (3.23) | (6.34) | | |
| 3 | Fuels, etc. | –0.018 | 0.1 | 0.024 | 0.47 | 0.89 |
| | | (0.67) | (3.48) | (2.26) | (3.52) | |
| 5 | Chemicals | 0.92 | 0.029 | –0.023 | – | 0.91 |
| | | (364.47) | (2.07) | (14.19) | | |
| 6 + 8 | Other manufactures | 0.91 | 0.019 | –0.051 | – | 0.87 |
| | | (93.12) | (0.35) | (13.05) | | |
| 7 | Machinery | 0.97 | 0.031 | –0.012 | – | 0.62 |
| | | (128.92) | (1.74) | (5.61) | | |
| 9 | Commodities not classified | 0.92 | 0.17 | –0.023 | – | 0.11 |
| | | (31.87) | (0.78) | (1.38) | | |

*Table 24.4.* Assumptions about the world market development of the Bonn–IIASA Project.

| *Scenario* | *Percentage growth of the volume index of world trade* | *Percentage growth of the price index of world trade* |
|---|---|---|
| I | 2 | 5 |
| II | 4 | 5 |
| III | 6 | 5 |
| IV | 2 | 8 |
| V | 4 | 8 |
| VI | 6 | 8 |

*Table 24.5.* Exports from all countries to all market economies in current prices (billion US$) for scenarios II, IV, and V.

| *SITC* | *1970* | *1980* | *1985* | *1990* | *1995* | *2000* |
|---|---|---|---|---|---|---|
| Scenario II: | | | | | | |
| 0 + 1 | 35.16 | 164.11 | 153.01 | 211.15 | 289.22 | 396.08 |
| 2 + 4 | 27.66 | 113.12 | 94.33 | 118.13 | 144.75 | 172.25 |
| 3 | 24.82 | 421.87 | 380.30 | 655.80 | 1,114.49 | 1,880.08 |
| 5 | 19.06 | 127.61 | 128.32 | 203.41 | 315.80 | 486.29 |
| 6 + 8 | 77.16 | 431.06 | 420.34 | 639.80 | 959.64 | 1,433.88 |
| 7 | 77.90 | 440.47 | 461.84 | 719.90 | 1,102.55 | 1,677.95 |
| 9 | 4.16 | 21.31 | 21.82 | 26.23 | 34.63 | 48.09 |
| Total | 265.93 | 1,719.27 | 1,659.96 | 2,574.43 | 3,961.08 | 6,094.6 |
| Scenario IV: | | | | | | |
| 0 + 1 | 35.16 | 164.11 | 153.01 | 263.62 | 384.28 | 644.36 |
| 2 + 4 | 27.66 | 113.12 | 94.33 | 139.94 | 226.25 | 384.95 |
| 3 | 24.82 | 421.57 | 380.30 | 714.90 | 1,341.84 | 2,480.50 |
| 5 | 19.06 | 127.61 | 128.32 | 195.20 | 280.72 | 384.23 |
| 6 + 8 | 77.16 | 431.06 | 420.34 | 637.42 | 941.86 | 1,366.26 |
| 7 | 77.90 | 440.47 | 461.84 | 706.79 | 1,040.20 | 1,483.53 |
| 9 | 4.16 | 21.31 | 21.82 | 39.36 | 85.26 | 182.03 |
| Total | 265.93 | 1,719.27 | 1,659.96 | 2,670.21 | 4,300.41 | 6,925.86 |
| Scenario V: | | | | | | |
| 0 + 1 | 35.16 | 164.11 | 153.01 | 240.55 | 396.98 | 677.54 |
| 2 + 4 | 27.66 | 113.12 | 94.33 | 136.36 | 210.55 | 343.28 |
| 3 | 24.82 | 421.57 | 380.30 | 766.80 | 1,571.56 | 3,172.64 |
| 5 | 19.06 | 127.61 | 128.32 | 182.01 | 362.98 | 585.34 |
| 6 + 8 | 77.16 | 431.06 | 420.34 | 694.25 | 1,144.20 | 1,859.17 |
| 7 | 77.90 | 440.57 | 461.84 | 778.72 | 1,299.40 | 2,119.17 |
| 9 | 4.16 | 21.31 | 21.82 | 35.07 | 71.84 | 155.88 |
| Total | 265.93 | 1,719.27 | 1,659.96 | 2,869.76 | 5,057.50 | 8,913.02 |

1970, and 1985 are added. *Table 24.5* shows the trade volume in billions of US dollars for the seven commodity groups traded between all market economies.

If we define as primary products the SITC groups 1, 2, 3, and 4, and as secondary products the SITC groups 5, 6, 7, and 8, we observe that low real growth of world trade and high inflation in world trade prices induce a share shift from the secondary products to the primary products. The share of energy goods, especially, increases from 24.5% in 1980 to about 35.6% or 35.8% for an inflation rate of 8% and 2% or 4% real growth of world trade. Only if inflation is comparably low at 5% and real growth of world trade is at 4% does the share of energy commodities increase less.

For scenario II this increase in world trade of energy goods is completely compensated by the decrease in the trade shares of food and other raw materials. In scenario II the share of raw materials is nearly identical in the year 2000 to the share in the year 1980. The decline of inflation of world market prices, especially the decline of oil prices, helps the industrialized countries to participate successfully in the world markets. A similar effect results from higher real growth of world trade by constant inflation in prices.

Comparing the outcomes of scenarios IV and V, one sees that the increase in real growth induces higher export shares of industry products balanced by a decrease in exports of food and raw materials. The world trade share of fuels remains nearly constant in both scenarios.

For industrial products, the share of other manufactures (SITC 6+8) decreases for the year 2000 compared with 1980 in scenario II, while the shares of chemical products and machinery increase. Therefore, for the growth of the world economy and the related growth in world trade, machinery and chemicals are more important than other manufactures. This kind of reaction can be observed for the other two scenarios, IV and V.

*Table 24.6* gives the corresponding percentage shares for the three scenarios.

*Table 24.6.* Percentage shares of growth in world trade for scenario II, IV and V.

| SITC | *1970* | *1980* | *1985* | *1990* | *1995* | *2000* |
|---|---|---|---|---|---|---|
| Scenario II: | | | | | | |
| 0 + 1 | 13.22 | 9.55 | 9.22 | 8.20 | 7.30 | 6.50 |
| 2 + 4 | 10.40 | 6.58 | 5.68 | 4.59 | 3.65 | 2.83 |
| 3 | 9.33 | 24.52 | 22.91 | 25.47 | 28.14 | 30.85 |
| 5 | 7.17 | 7.42 | 7.73 | 7.90 | 7.97 | 7.98 |
| 6 + 8 | 29.02 | 25.07 | 25.32 | 24.85 | 24.23 | 23.53 |
| 7 | 29.29 | 25.62 | 27.82 | 27.96 | 27.83 | 27.53 |
| 9 | 1.56 | 1.24 | 1.31 | 1.02 | 0.87 | 0.79 |
| Total | 100.0 | 100.0 | 100.0 | 100.0 | 100.0 | 100.0 |
| Scenario IV: | | | | | | |
| 0 + 1 | 13.22 | 9.55 | 9.22 | 9.87 | 8.94 | 9.30 |
| 2 + 4 | 10.40 | 6.58 | 5.68 | 5.24 | 5.26 | 5.56 |
| 3 | 9.33 | 24.52 | 22.91 | 26.77 | 31.20 | 35.82 |
| 5 | 7.17 | 7.42 | 7.73 | 7.31 | 6.53 | 5.55 |
| 6 + 8 | 29.02 | 25.07 | 25.32 | 23.87 | 21.90 | 19.73 |
| 7 | 29.29 | 25.62 | 27.82 | 26.47 | 24.19 | 21.42 |
| 9 | 1.56 | 1.24 | 1.31 | 1.47 | 1.98 | 2.63 |
| Total | 100.0 | 100.0 | 100.0 | 100.0 | 100.0 | 100.0 |
| Scenario V: | | | | | | |
| 0 + 1 | 13.22 | 9.55 | 9.22 | 8.38 | 7.85 | 7.60 |
| 2 + 4 | 10.40 | 6.58 | 5.68 | 4.75 | 4.16 | 3.85 |
| 3 | 9.33 | 24.52 | 22.91 | 26.72 | 31.07 | 35.60 |
| 5 | 7.17 | 7.42 | 7.73 | 7.60 | 7.18 | 6.57 |
| 6 + 8 | 29.02 | 25.07 | 25.32 | 24.19 | 22.62 | 20.86 |
| 7 | 29.29 | 25.62 | 27.82 | 27.14 | 25.69 | 23.78 |
| 9 | 1.56 | 1.24 | 1.31 | 1.22 | 1.42 | 1.75 |
| Total | 100.0 | 100.0 | 100.0 | 100.0 | 100.0 | 100.0 |

Additionally we present in *Table 24.7* the percentage shares of the inter- and intraregional trade development of the two country groups for scenario II, which seems the most likely under the current circumstances.

*Table 24.7.* Percentage shares of exports from ... to ... market economies for scenario II.

| *SITC* | *1970* | *1980* | *1985* | *1990* | *1995* | *2000* |
|---|---|---|---|---|---|---|
| From developed to developed countries: | | | | | | |
| 0 + 1 | 10.84 | 9.59 | 8.91 | 8.27 | 7.73 | 7.26 |
| 2 + 4 | 9.72 | 7.31 | 6.00 | 4.82 | 3.79 | 2.82 |
| 3 | 3.75 | 8.42 | 7.37 | 7.60 | 7.64 | 7.49 |
| 5 | 7.90 | 9.77 | 10.43 | 11.15 | 11.82 | 12.48 |
| 6 + 8 | 32.60 | 30.98 | 30.86 | 30.83 | 30.78 | 30.77 |
| 7 | 33.70 | 32.50 | 35.05 | 36.29 | 37.32 | 38.30 |
| 9 | 1.50 | 1.42 | 1.37 | 1.04 | 0.92 | 0.88 |
| Total | 100.0 | 100.0 | 100.0 | 100.0 | 100.0 | 100.0 |
| From developed to developing countries: | | | | | | |
| 0 + 1 | 10.83 | 10.66 | 10.12 | 9.62 | 9.13 | 8.68 |
| 2 + 4 | 4.63 | 4.42 | 4.51 | 4.46 | 4.41 | 4.38 |
| 3 | 1.48 | 2.40 | 1.46 | 1.46 | 1.32 | 1.13 |
| 5 | 10.81 | 10.93 | 9.83 | 9.71 | 9.55 | 9.36 |
| 6 + 8 | 28.03 | 26.61 | 24.73 | 24.31 | 23.82 | 23.32 |
| 7 | 41.15 | 43.53 | 46.79 | 48.22 | 49.82 | 51.43 |
| 9 | 3.08 | 1.46 | 2.55 | 2.22 | 1.95 | 1.71 |
| Total | 100.0 | 100.0 | 100.0 | 100.0 | 100.0 | 100.0 |
| From developing to developing countries: | | | | | | |
| 0 + 1 | 18.36 | 9.27 | 9.37 | 8.06 | 6.86 | 5.84 |
| 2 + 4 | 14.03 | 6.42 | 6.21 | 5.44 | 4.70 | 4.06 |
| 3 | 34.69 | 54.73 | 52.65 | 55.04 | 57.23 | 59.31 |
| 5 | 3.62 | 3.25 | 3.14 | 3.10 | 3.03 | 2.94 |
| 6 + 8 | 19.24 | 17.07 | 19.03 | 18.77 | 18.60 | 18.35 |
| 7 | 9.27 | 8.24 | 9.08 | 9.15 | 9.18 | 9.17 |
| 9 | 0.08 | 1.02 | 0.53 | 0.45 | 0.39 | 0.33 |
| Total | 100.0 | 100.0 | 100.0 | 100.0 | 100.0 | 100.0 |
| From developing to developed countries: | | | | | | |
| 0 + 1 | 24.55 | 8.70 | 9.31 | 7.23 | 5.60 | 4.33 |
| 2 + 4 | 18.34 | 6.59 | 5.51 | 3.89 | 2.65 | 1.72 |
| 3 | 34.38 | 66.98 | 66.30 | 70.50 | 74.13 | 77.20 |
| 5 | 1.22 | 0.94 | 1.11 | 1.10 | 1.07 | 1.02 |
| 6 + 8 | 17.42 | 13.31 | 14.08 | 13.83 | 13.29 | 12.66 |
| 7 | 3.59 | 2.75 | 3.11 | 3.05 | 2.91 | 2.76 |
| 9 | 0.47 | 0.74 | 0.58 | 0.41 | 0.34 | 0.32 |
| Total | 100.0 | 100.0 | 100.0 | 100.0 | 100.0 | 100.0 |

Obviously, the intraregional trade of the developed country group will increase the shares of the manufactured, chemical, machinery and transport equipment commodities and reduce the trade share for agricultural products and raw materials except fuels. The trade share for fuels is relatively stable over the forecasting time horizon.

The intraregional trade from the developed to the developing countries shows a share shift that increases only for machinery and transport equipment. This commodity group has the greatest growth potential for the developed market economies. It is interesting to notice that the trade share of the commodity group of SITC 6+8 decreases moderately in the exports from the developed to the developing countries. This shows the growing potential of at least a number of developing countries, e.g., the NICs, to substitute trade with developed countries for trade with developing economies. The reverse effect does not appear: developing countries cannot export an increasing share of industrial goods to the developed countries. Because the demand for agricultural products and raw materials (except fuels) loses some importance, only fuel exports are increasing in the intraregional trade relationship between the developing and developed economies. This finding offers little hope for overcoming the debt problem of the majority of the developing countries caused by long-lasting trade deficits.

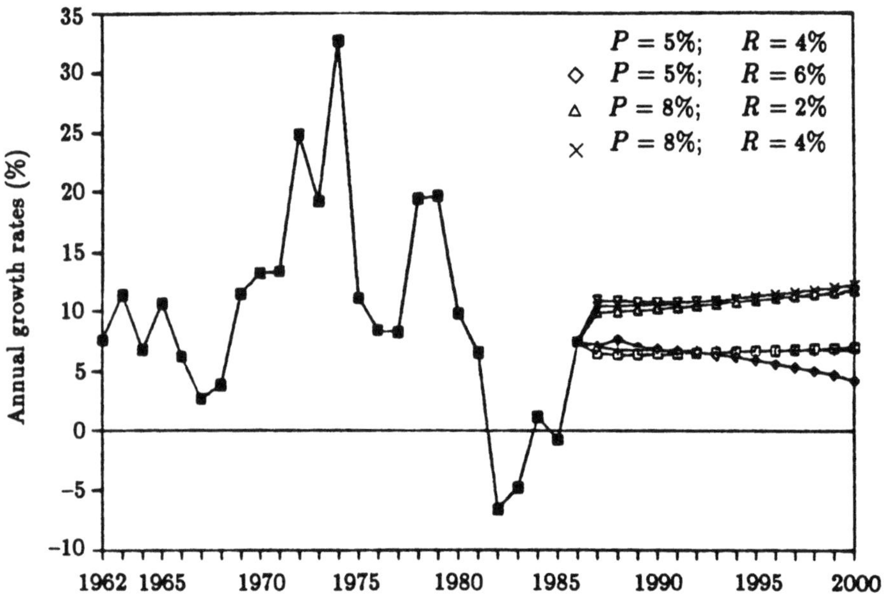


*Figure 24.2.* Exports of crude materials, fats, etc. (except fuels) from developed to developed market economies: historical observations, 1962–1980; forecasted values, 1981–2000.

*Figures 24.2, 24.3,* and *24.4* show the development of three variables from 1962 until 2000 (forecasted from 1981 to 2000) for the six scenarios. As we can see, the model adjusts after a couple of years to a stable path. This is reasonable since the aggregate variables driving the system were set to a constant growth trajectory. As soon as there is less steady variation in the aggregate variables,

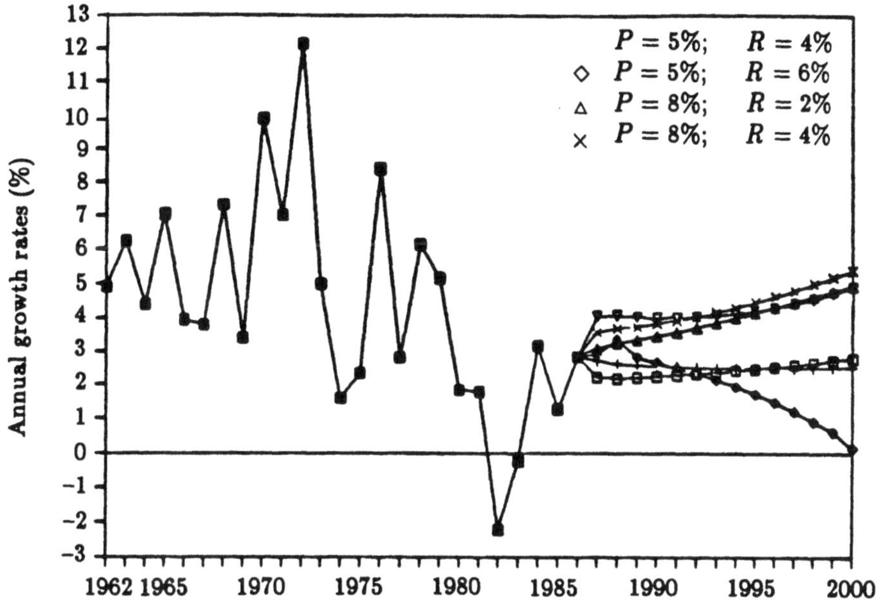


*Figure 24.3.* Volume index of exports of crude materials, fats, etc. (except fuels) from developed to developed market economies: historical observations, 1962–1980; forecasted values, 1981–2000.

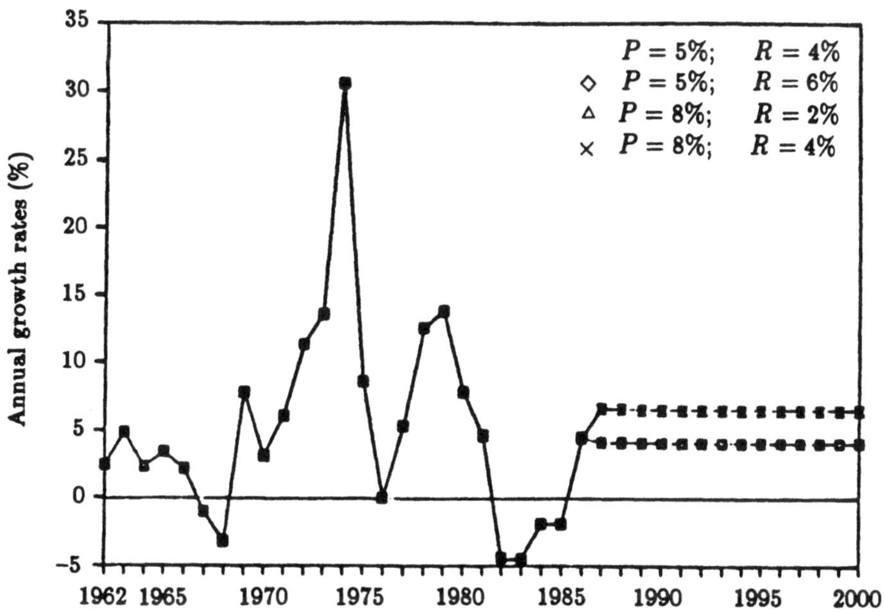


*Figure 24.4.* Price index of exports of crude materials, fats, etc. (except fuels) from developed to developed market economies: 1962–1980; forecasted values, 1981–2000.

the shape of the forecast of the endogenous variables is much more similar to the historical observations. This is obvious from the development of the variables in the years 1981 to 1986, where IMF data were used. The model is able to follow up-and-down swings of world trade and can adjust to the two oil price shocks, as *ex post* simulation of the model has shown.

## 24.4. Summary

This chapter presented some results of six different *ex ante* forecast scenarios with the SITC submodel of the FIND project. The two-stage, almost ideal demand systems used to explain import demand of the developed and developing countries produced reasonable results. They showed that changes in the rate of inflation and real growth of world trade had considerable impacts on the structure of commodities traded and, thereby, on the growth potential of the two country groups.

Slow growth and high inflation favor the producers of raw materials, whereas faster growth increases disproportionately the demand for industrial goods. As long as the supply conditions of the developing countries do not change in the direction of more industrial goods, which could be traded with the developed countries, North–South trade is trapped in an unbalanced state.

Second, energy prices largely determine how the shares between North–South world market exports split up. The presented scenarios suggest that the share of the commodity group of fuel (SITC 3) would increase under all assumptions – a finding that should be reconsidered in light of the current decline of energy prices. The effects of energy substitution through conservation and increased use of nuclear and alternative energies would have an impact, which is not fully accounted by the model. A reestimation with more recent data from 1981 to 1985 may correct for some biases that result from using a data set ending in 1980.

Further investigation should check this outcome with respect to the sensitivity of the transfer of world price inflation to the energy prices. Since the estimation period of the model included the two oil price crises, this could bias the forecast. Additional scenarios that investigate unequal growth between the two country groups should give more insights into different development strategies.

### References

Armington, P.S. (1969), A Theory of Demand for Products Distinguished by Place of Production. *International Monetary Fund Staff Papers* (Washington, DC), pp. 159–178.

Artis, M. and Ostry, S. (1986), *International Economic Policy Coordination*, Chatham House Papers No. 30. (London and New York: Royal Institute for International Affairs).

Barney, O. (1980), *The Global 2000 Report to the President.* Sponsored by the Council on Environment Quality and the US Department of State. (Washington, DC: US Government Printing Office).

Blazejczak, J. (1986), Simulation gesamtwirtschaftlicher Perspektiven mit einem ökonometrischen Modell für die Bundesrepublik Deutschland, Berlin (unpublished).

Deaton, A.S. and Muellbauer, J. (1980), An Almost Ideal Demand System, *American Economic Review*, pp. 312–326

DIW (1983), Erhöhter Handlungsbedarf im Strukturwandel – Analyse der strukturellen Entwicklung der deutschen Wirtschaft – Strukturberichterstattung 1983, Gutachten im Auftrage des Bundesministers für Wirtschaft, Berlin.

Erber, G. (1986a), The FIND-Model – The Theoretical Framework of a new Econometric Model for the Federal Republic of Germany, Berlin VH 3/86, pp. 108–109.

Erber, G. (1986b), The FIND-Model – Foreign Trade Modeling for the Federal Republic of Germany, Berlin VH 3/86, pp. 135–155.

Erber, G. and Hass, H. (1983), Fortführung der Arbeiten mit dem disaggregierten Modell, pp. 1–31: in: *Methodenband Strukturberichterstattung 1983.* (Berlin: DIW).

Erber, G., Haas, H. and Kiy, M. (1983), Das FIND-Modell – Zur Konzeption eines neuen disaggregierten Modells für die Bundesrepublik Deutschland, Berlin VH 4/83, pp. 273–313.

Erber, G., Haas, H., and Horn, G.A. (1986), An Almost Ideal Demand System for a Dynamic Multisectoral Model of Private Consumption, Berlin VH 3/86, pp. 123–132.

Horn, G.A. (1986), FIND-Modell – Sektorale Vorratsinvestitionen in der Bundesrepublik Deutschland 1961–1980, Eine empirische Analyse, Berlin VH 4/86.

Hugger, W. (1974), *Weltmodelle auf dem Prüfstand – Anspruch und Leistungen der Weltmodelle von J.W. Forrester und D. Meadows.* (Basel und Stuttgart: Birkhäuser).

International Money Fund (1985), *World Economic Outlook, August 1985.* (Washington, DC: IMF).

Krelle, W., Erber, G. and Kiy, M. (1982), Forschungsplan für die Entwicklung eines disaggregierten Modells 2, Bonn und Berlin (unpublished).

Meadows, D.L. *et al.* (1972), *The Limits to Growth* (A Report for the Club of Rome's Project on the Predicament of Mankind). (New York: Universe Books).

Meier, K. (1986), Weiterführung der Arbeiten mit dem FIND-Modell: Implementation der sektoralen Vermögens- und Potentialrechnung, Berlin, DIW (internes Diskussionspapier).

Nakamura, S. (1986), A Dynamic Multisectoral Model of Production, Investment and Prices based on Flexible Cost Functions, VH 3/86, pp. 110–122.

Ross, H. (1986), On Econometric Work based on Growth Theory, Bonn-IIASA Discussion Paper, Sofia.

UN (various years), *Yearbooks of International Trade Statistics.* (New York: United Nations).

CHAPTER 25

# Foreign Debt and the Structure of World Trade

*Peter Pauly*

## Summary

The dramatic increase in the level of external debt of a large number of Third World countries has no doubt significantly affected the volume of world trade in the early 1980s. In this chapter, we examine whether, in addition, the *structure* of trade, i.e., the direction of trade flows, has been altered through financial constraints. The aim of the chapter is to examine the causal relationships between various global measures of structural change in bilateral trade flows and developing countries' debt. The results are largely affirmative, and suggest that medium- and long-term projections of world trade structures will have to be based on an assessment of monetary conditions as well as relative prices and technological trends. The analysis indicates the need for an in-depth analysis of these mechanisms based on the dynamics of bilateral trade shares.

## 25.1. Introduction

The dramatic increase in the net external debt position for a significant number of developing countries has been one of the most important disruptions of the global economy during the first half of the 1980s. Beyond the immediate impact of both the inflow of foreign capital into these countries, and the subsequent adjustment policies imposed by private and official creditors on the debtor countries themselves, the world economy as a whole has been affected severely. It is well understood that the observed variability in growth rates of world trade over the period 1977–1985 is, among other things, related to the sharp swings in trade activities in debtor countries. Little attention has so far been paid, however, to the effects of the debt crisis on the structure of world trade flows. It is the purpose of this chapter to address that issue.

Based on a highly aggregated trade matrix analysis, we shall examine the trade flows between the largest debtor countries and other regions. The primary intention is to identify interregional reallocations of trade that can be traced to the accumulations of foreign debt in these countries. Clearly, an aggregate analysis cannot immediately be translated into a forecast of structural change. Here an examination of the dynamics of trade shares based on detailed matrices, disaggregated by country and commodity, is in order. Unfortunately, however, such an econometric study is, at present, infeasible due to the short history of available data. It is therefore most appropriate to look at the present exercise as a pilot study toward that ultimate goal.

The chapter is organized as follows: in Section 25.2 we review briefly the history of the debt crisis, with special reference to the 10 major countries under investigation in this study. Section 25.3 contains a discussion of the evolution of trade shares between major regions in the world. In Section 25.4, we propose various measures for structural change based on trade share analysis as well as bilateral flows. In Section 25.5 we report the empirical performance of those measures over the period of investigation, and attempt to test for a significant impact of the surge in foreign debt on various measures of structural change. Section 25.6 contains a short summary of our results.

## 25.2. The Accumulation of Debt since 1974

For the purposes of this analysis we concentrate on the 10 largest debtor countries, based on data available from 1985. In *Table 25.1* we summarize the level of outstanding debt, current interest payments, and the 1985 trade balances for those countries. With regard to the stock of debt, these countries account for about $450 billion of the total developing country debt of $600 billion; other significant debtor countries not included here are Colombia, Peru, Nigeria, Thailand and Algeria. With two exceptions, the current interest burden for these countries equals or exceeds available foreign exchange receipts from merchandise transactions; the fall in oil prices in 1986 has worsened the situation for Mexico and Venezuela significantly.

The rapid buildup of external debt by these countries has certainly augmented their purchasing power in world markets in the late 1970s and early 1980. But it would be a mistake to assume that this borrowing was associated with large merchandise trade deficits in all cases. While it is impossible to trace how the purchasing power provided by external creditors was utilized, the five Latin American countries as a group matched all of their imports with exports of merchandise.

As shown in *Table 25.2*, the cumulated trade deficits of Brazil, Chile and Mexico were less than one-fifth as large as their buildup in external debts. Argentina, Venezuela and Indonesia ran substantial cumulative trade surpluses. Korea and the Philippines did show cumulated trade deficits equal to about two-thirds and one-half, respectively, of their buildup in external debt – a

*Table 25.1.* Debt and debt service in 1985 (billion US$).

| *Country* | *Total debt* | *Interest payments* | *Trade balance* |
|---|---|---|---|
| Mexico | 96.7 | 8.8 | 8.7 |
| Brazil | 100.1 | 9.5 | 10.9 |
| Argentina | 47.8 | 5.9 | 4.3 |
| Venezuela | 32.4 | 3.2 | 7.4 |
| Chile | 20.4 | 1.8 | 0.7 |
| Korea | 46.7 | 4.1 | 0.1 |
| Indonesia | 34.3 | 2.4 | 5.4 |
| Philippines | 26.5 | 2.3 | −0.3 |
| Egypt | 32.0 | 2.2 | −6.1 |
| India | 21.7 | 0.7 | −5.4 |

*Table 25.2.* Cumulated stock of debt since 1973 and its sources in 1982 (billion US$).

| *Country* | *Change in gross external debt* | *Increase in official reserves* | *Net service and other current account payments* | *Trade deficit* | *Increase in private claims on nonresidents* |
|---|---|---|---|---|---|
| Mexico | 82.6 | −0.1 | 36.4 | 9.0 | 36.3 |
| Brazil | 93.5 | −1.3 | 68.0 | 15.6 | 11.4 |
| Argentina | 32.6 | 2.3 | 20.1 | −9.8 | 20.2 |
| Venezuela | 27.0 | 9.1 | 26.0 | −33.3 | 25.5 |
| Chile | 15.4 | 2.6 | 10.8 | 3.2 | 0.1 |
| Korea | 33.6 | 5.8 | −1.1 | 21.3 | 5.9 |
| Indonesia | 28.1 | 6.6 | 3.1 | −14.8 | 33.2 |
| Philippines | 19.9 | 2.4 | 0 | 13.4 | 3.9 |
| Egypt | 26.1 | 2.1 | 0.8 | 19.8 | 3.4 |
| India | 18.0 | 3.0 | 1.2 | 12.0 | 1.8 |

Source: Dooley *et al.* (1986) and own calculations.

pattern more typical of developing countries as a group. Hence, one cannot assume that the buildup in external debt was related to net imports of goods, or growth in productive capacity, in any simple manner [Dooley *et al.* (1986)].

The most important net use of foreign exchange for a number of countries was net service payments, largely in the form of interest payments on existing debt. The buildup in official reserve assets was an important use of funds for only three countries: Chile, Korea and Venezuela. In contrast, the accumulation of private claims on nonresidents accounted for more than half of the buildup of external debt for Venezuela and Argentina and accounted for 20% or more of the buildup for four other countries.

The estimates for private claims on nonresidents are based on a residual calculation, and as such, subject to potential errors. For example, if trade values are consistently misreported, perhaps to conceal capital flight, this estimate would be biased downward. It should also be noted that gross capital inflows and outflows are not in themselves the cause of debt problems. Many countries have gross external debts which are roughly offset by gross external assets. This situation is a problem only if the external receipts are not available to those who have to make payments on external debt.

There is enough evidence to show that the source of external debt has varied considerably among the countries examined here. Nevertheless, it is our contention that as far as the effects on trade structures are concerned, an aggregate analysis will still be sufficient; for a detailed analysis, see Cline (1984) and Dornbusch and Fischer (1985).

## 25.3. The Structure of Aggregate World Trade

Before we turn to the formal analysis of intertemporal changes in the structure of world trade, a brief historical overview is in order. We base our analysis on aggregate trade for SITC categories 0 to 9. The data base was compiled by the United Nations for the period 1965 to 1983. Extrapolations for 1984 to 1986 were obtained as preliminary estimates based on the Direction of Trade Statistics. The definition of regional groupings for the purposes of this analysis is as follows:

| | |
|---|---|
| Group 1: | 10 largest debtor countries (Mexico, Argentina, Brazil, Venezuela, Chile, Korea, Indonesia, Philippines, Egypt, India) [DEBT] |
| Group 2: | United States [USA] |
| Group 3: | Japan [JAPAN] |
| Group 4: | EEC [EEC] |
| Group 5: | Rest of OECD [OECD] |
| Group 6: | OPEC (excl. Venezuela, Indonesia) [OPEC] |
| Group 7: | Other LDCs (non-oil, non-debt, incl. China) [LDC] |
| Group 8: | Centrally planned economies [CMEA] |

The historical structure of world trade is summarized in *Table 25.3*, where we report export shares for all 8 regional groups for three representative years: 1965, 1975, and 1985. In comparison with 1965, the structure for 1975 is already affected by the first oil shock, while the table for 1985 also reflects the effects of the debt crisis. Generally, the EEC countries account for a little more than one-third of world trade, with the USA and the rest of OECD contributing about 15% each. The share of CMEA countries is about 10%, but declining – about the same size as the non-debtor developing countries. The debtor countries and Japan account for about 6% of merchandise trade, respectively. Note that export shares add up to unity row-wise.

Elements on the main diagonal indicate the extent of intraregional trade which is substantial for both the EEC and the CMEA countries. However, while the EEC share is slightly increasing, the CMEA share shows a significant downward trend after 1975.

The particular interest of this chapter is centered around the behavior of trade flows between debtor countries and other regions over time. In *Figures 25.1* and *25.2*, we summarize trade between group 1 and groups 2–8. *Figure 25.1* reports shares of exports to groups 2–8 in total exports of the group of major

*Table 25.3.* World trade matrices (export shares).

| | | Regional group code | | | | | | | |
|---|---|---|---|---|---|---|---|---|---|
| *Year* | *Code* | *DEBT* | *USA* | *JAPAN* | *EEC* | *OECD* | *LDC* | *OPEC* | *CMEA* |
| 1965 | DEBT | 0.060640 | 0.302151 | 0.064988 | 0.271310 | 0.083732 | 0.132809 | 0.0080073 | 0.076362 |
| | USA | 0.156116 | 0.000000 | 0.082119 | 0.288238 | 0.326862 | 0.116103 | 0.0252929 | 0.005270 |
| | JAPAN | 0.118310 | 0.317153 | 0.000000 | 0.100149 | 0.131052 | 0.272586 | 0.0339904 | 0.026762 |
| | EEC | 0.041927 | 0.083201 | 0.008187 | 0.500142 | 0.212578 | 0.090910 | 0.0320048 | 0.031052 |
| | OECD | 0.030301 | 0.235882 | 0.040658 | 0.425583 | 0.123157 | 0.069331 | 0.0090378 | 0.066051 |
| | LDC | 0.028327 | 0.065714 | 0.108811 | 0.650677 | 0.073549 | 0.053695 | 0.0101029 | 0.009125 |
| | OPEC | 0.041626 | 0.173536 | 0.067108 | 0.340882 | 0.077861 | 0.178887 | 0.0196201 | 0.100481 |
| | CMEA | 0.028157 | 0.006963 | 0.012156 | 0.119530 | 0.068665 | 0.093914 | 0.0096797 | 0.660936 |
| 1975 | DEBT | 0.055483 | 0.284331 | 0.153379 | 0.188740 | 0.070552 | 0.144836 | 0.0361663 | 0.066512 |
| | USA | 0.149294 | 0.000000 | 0.091542 | 0.239456 | 0.308268 | 0.125798 | 0.0582433 | 0.027401 |
| | JAPAN | 0.134562 | 0.226632 | 0.000000 | 0.116907 | 0.112560 | 0.279239 | 0.0888325 | 0.041269 |
| | EEC | 0.038881 | 0.064746 | 0.009903 | 0.533455 | 0.170519 | 0.067973 | 0.0625679 | 0.051954 |
| | OECD | 0.036760 | 0.251712 | 0.069664 | 0.334307 | 0.144838 | 0.064777 | 0.0254895 | 0.072454 |
| | LDC | 0.072421 | 0.107590 | 0.169878 | 0.437341 | 0.069031 | 0.097443 | 0.0145939 | 0.031703 |
| | OPEC | 0.046666 | 0.247372 | 0.098176 | 0.260746 | 0.070567 | 0.165843 | 0.0358476 | 0.074782 |
| | CMEA | 0.023415 | 0.010473 | 0.017472 | 0.145944 | 0.086177 | 0.063213 | 0.0140637 | 0.639244 |
| 1985 | DEBT | 0.056378 | 0.348563 | 0.152485 | 0.177001 | 0.073028 | 0.126322 | 0.0308277 | 0.035397 |
| | USA | 0.134649 | 0.000000 | 0.112674 | 0.236074 | 0.326873 | 0.127373 | 0.0450492 | 0.017308 |
| | JAPAN | 0.092114 | 0.372141 | 0.000000 | 0.110136 | 0.098177 | 0.219632 | 0.0709985 | 0.036803 |
| | EEC | 0.030967 | 0.110623 | 0.012976 | 0.533091 | 0.146129 | 0.067431 | 0.0657696 | 0.033012 |
| | OECD | 0.032467 | 0.331499 | 0.062291 | 0.302628 | 0.100546 | 0.065900 | 0.0386914 | 0.065978 |
| | LDC | 0.098475 | 0.104579 | 0.190626 | 0.347474 | 0.063951 | 0.124418 | 0.0388215 | 0.031655 |
| | OPEC | 0.051283 | 0.283408 | 0.105899 | 0.187847 | 0.062173 | 0.232371 | 0.0474404 | 0.029579 |
| | CMEA | 0.020186 | 0.015550 | 0.021077 | 0.253274 | 0.123474 | 0.068616 | 0.0293432 | 0.468479 |

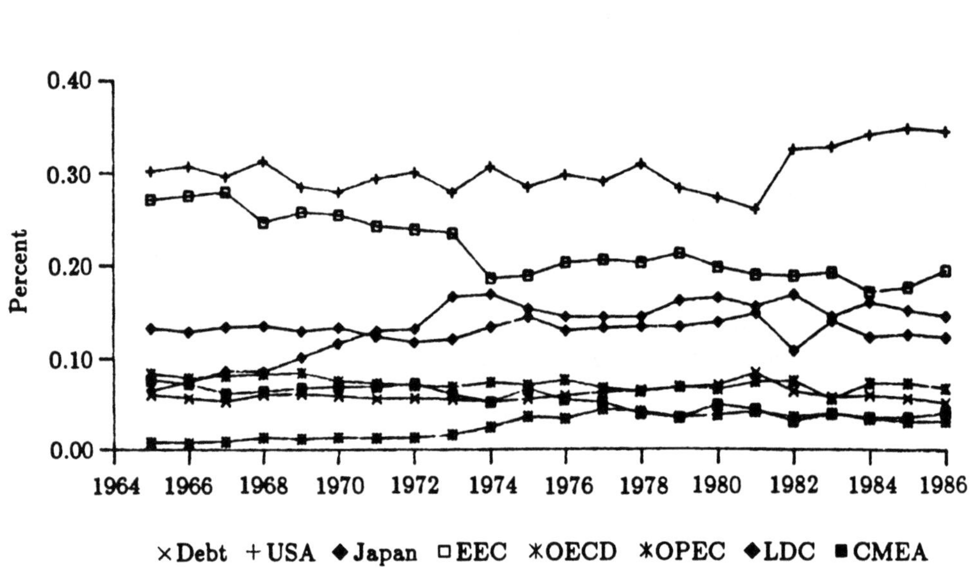


*Figure 25.1.* Export shares of debtor countries 1965–1986.

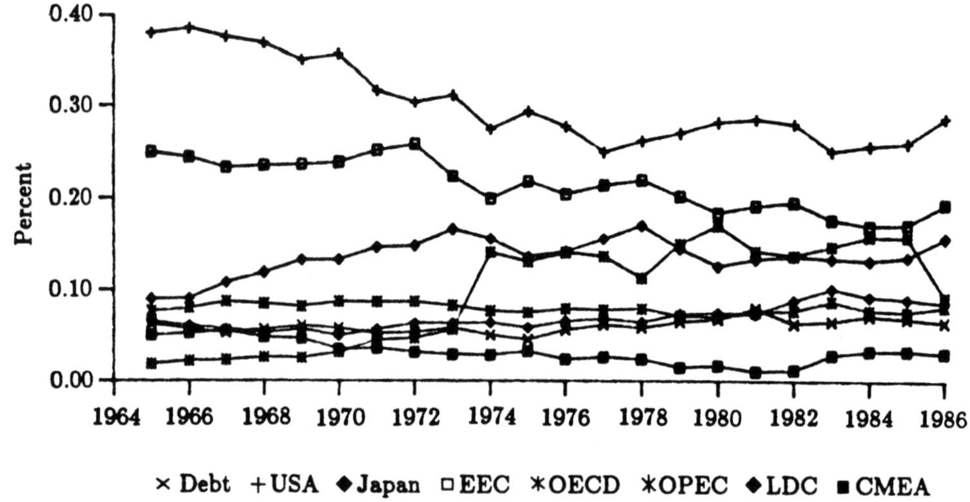


*Figure 25.2.* Import shares of debtor countries 1965–1986.

debtor countries, while *Figure 25.2* reports import shares defined analogously. The data cover the period 1965 to 1986; of particular interest in the present context are the years 1977 to 1986, corresponding to the debt buildup. On the export side, it appears as if, after the initial demand stimulus following the inflow of funds, the share of intradebtor countries trade has fallen sharply. Minor reductions seem to have occurred for exports to other LDCs and to Japan. Upon first inspection, the USA seems to have absorbed the bulk of the adjustment. On the import side the important question is whether the reduction in real imports, caused by domestic stabilization programs in the debtor countries, has been distributed evenly across regions [see also Dornbusch (1985), Marquez and McNeilly (1986)]. It appears that initially imports from the USA and the EEC have been affected more than proportionally, while the intradebtor share and the share of other LDCs remained unaffected. Imports from Japan have not been affected noticeably, while the CMEA share has improved somewhat. A more formal analysis of these developments is deferred to Section 25.5.

## 25.4. Measures of Structural Change

A natural way of analyzing the structure of world trade over time focuses on the dynamics of trade shares. Generally, a structure can be characterized by an $n$-dimensional vector of trade shares. Two structures, at different points in time, $(t_1, t_2)$ can be compared through the angle that these two vectors form with each other. A convenient measure of structural change is then

$$\Lambda = \arccos \frac{\sum_{i=1}^{n} s_{i,t_1} \cdot s_{i,t_2}}{\sqrt{\left(\sum_{i=1}^{n} s_{i,t_1}^2\right)\left(\sum_{i=1}^{n} s_{i,t_2}^2\right)}} \tag{25.1}$$

where $s_{i,t_j}$ $(j = 1, 2)$ denotes the trade shares in both periods. This measure has the following useful properties:

$$\Lambda(S_{t_1}, S_{t_2}) = 0 \quad \text{if } s_{j,t_1} = s_{j,t_2} \, \forall j \tag{25.1a}$$

$$\Lambda(S_{t_1}, S_{t_2}) = \Lambda\,(S_{t_2}, S_{t_1}) \tag{25.1b}$$

$$\Lambda(S_{t_1}, S_{t_3}) \leq \Lambda\,(S_{t_1}, S_{t_2}) + \Lambda\,(S_{t_2}, S_{t_3}) \tag{25.1c}$$

Alternatively, one can attempt to focus on bilateral balances directly. Arrange the cumulative shares of exports and imports of countries in world trade in monotonic order of their export/import ratio $(X_i/M_i)$. These shares thus are

$$\sum_{j=1}^{t} X_j / \sum_{j=1}^{n} X_j = SX_i \text{ for exports} \tag{25.2}$$

$$\sum_{j=1}^{t} M_j / \sum_{j=1}^{n} M_j = SM_i \text{ for imports} \tag{25.3}$$

The locus of each point $(SX_i, SM_i)$ in *Figure 25.3* resembles the Lorenz curve in income distribution analysis. The shaded area in the diagram is equal to

$$A = \frac{1}{2} - \frac{1}{2} \sum_{i=1}^{n} (SX_i - SX_{i-1})(SM_i - SM_{i-1}) \tag{25.4}$$

and the area relative to the triangle is

$$C = 1 - \sum_{i=1}^{n} (SX_i - SX_{i-1})(SM_i - SM_{i-1}) \tag{25.5}$$

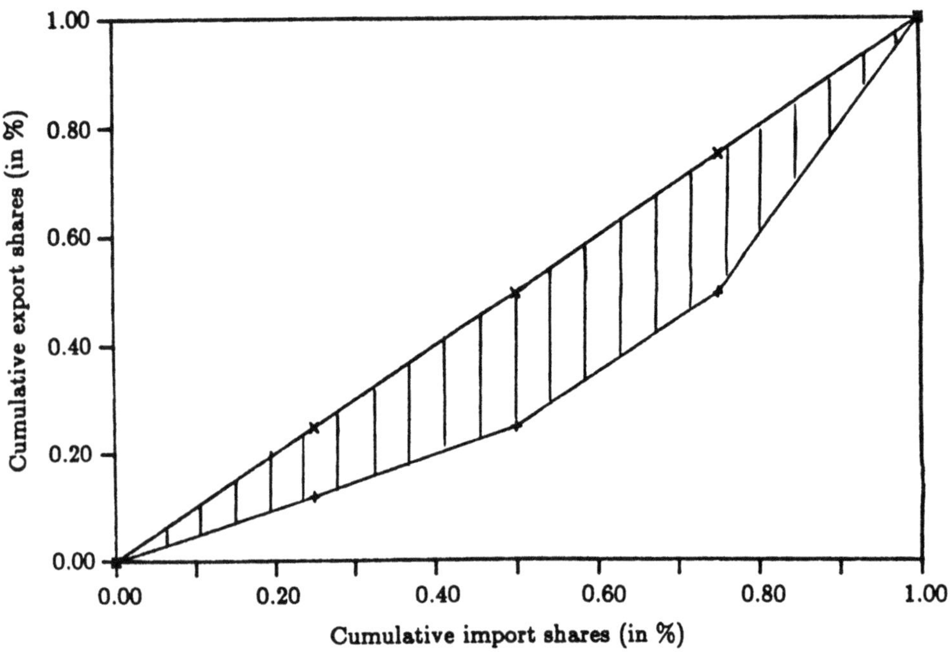


*Figure 25.3.* Cumulative trade shares.

This measures the extent of global bilateral trade imbalance. If $SX_i = SM_i$ for all $i$, then $C = 0$. Hence, $0 \leq C \leq 1$. An intertemporal change in the value of $C$ is proportional to the shaded area, which allows us to evaluate the extent of global maladjustment numerically [Nakamura and Nakamura (1983)].

## 25.5. Empirical Results

Based on the various measures introduced in the previous section, we now evaluate the dynamics of world trade structures numerically. In *Table 25.4* we report the time profile for three indicators for intertemporal comparison:

$\Lambda_{XS}$, based on comparison of export shares in subsequent periods.
$\Lambda_{MS}$, based on comparison of import shares in subsequent periods.
$\Delta_C$, based on comparison of bilateral trade balances in subsequent periods.

For the purposes of this study, we concentrate on trade of group 1 (the 10 largest debtor countries) with all other regions.

*Table 25.4.* Measures of structural change.

| *Year* | $\Lambda_{XS}$ | $\Lambda_{XM}$ | $C$ |
|---|---|---|---|
| 1966 | 0.029994 | 0.030084 | 0.071 |
| 1967 | 0.046854 | 0.046995 | 0.002 |
| 1968 | 0.084184 | 0.084436 | –0.013 |
| 1969 | 0.072258 | 0.072475 | –0.017 |
| 1970 | 0.045710 | 0.045847 | –0.025 |
| 1971 | 0.059537 | 0.059716 | –0.083 |
| 1972 | 0.024062 | 0.024134 | –0.091 |
| 1973 | 0.098764 | 0.099060 | 0.041 |
| 1974 | 0.135160 | 0.135565 | 0.009 |
| 1975 | 0.075653 | 0.075880 | –0.023 |
| 1976 | 0.067549 | 0.067751 | –0.042 |
| 1977 | 0.037334 | 0.037446 | 0.016 |
| 1978 | 0.049396 | 0.049545 | 0.033 |
| 1979 | 0.079895 | 0.080135 | 0.042 |
| 1980 | 0.052769 | 0.052928 | 0.158 |
| 1981 | 0.061565 | 0.061750 | 0.132 |
| 1982 | 0.177499 | 0.178032 | 0.062 |
| 1983 | 0.105099 | 0.105414 | 0.081 |
| 1984 | 0.087808 | 0.088071 | 0.144 |
| 1985 | 0.028183 | 0.028268 | 0.116 |
| 1986 | 0.047773 | 0.047916 | 0.102 |

Upon first inspection, there does indeed seem to be an increase in the order of magnitude of structural change beginning in the second half of the 1970s. This holds in particular for bilateral balances. A comparison of the coefficient of variation for the periods 1965–1976 and 1977–1986 confirms this shift to be statistically significant.

The question of interest here is to what extent that shift can be attributed, at least in part, to the emergence of the debt cycle. We base our formal analysis on the estimation of a simple transfer function model of the form

$$A(L)\ Y_t + B(L)\ D_t = C(L)\ u_t \tag{25.6}$$

with

$$Y = \{\Lambda_{XS,t}, \Lambda_{MS,t}, \Delta C_t\}$$
$$A(L) = (1 - a_1)$$
$$B(L) = (b_0 + b_1)$$
$$C(L) = (1 + c_1)$$

$\Delta D_t$ denotes the change in the stock of debt, and $u_t$ is a white noise residual. The idea behind this specification is similar to the standard Granger non-causality tests: effects of the change of debt on the structure of trade will only be identified if they contribute to a time-series behavior that cannot be ascribed to an ARMA representation for the measure of structural change.

The estimation results are summarized in *Table 25.5.* It is clear that over the entire sample period – based on an $F$-test of the joint hypothesis $b_0 = b_1 = 0$ – the null of no effect of the debt cannot be rejected. This holds for all three measures. In the case of trade share measures, a pooled estimation of both import and export share equations allows us to reject the null at the 5% level. Furthermore, focusing on the relevant time period by eliminating the early observations in the sample tends to generate statistically significant rejections of the null. We therefore conclude that there is some evidence for an impact of increased debt on trade structures, even though the relatively short experience with this phenomenon makes a statistically sound statement rather difficult.

*Table 25.5.* Estimation results ($t$-values in parentheses).

| *Dependent variable* | *Period* | $a_1$ | $b_0$ | $b_1$ | $c_1$ | $R^2/F$ |
|---|---|---|---|---|---|---|
| $\Lambda_{XS}$ | 1966–1986 | 0.920 | 0.002 | 0.016 | 0.122 | 0.985 |
| | | (7.91) | (1.18) | (0.98) | (2.22) | 1.05 |
| | 1976–1986 | 0.895 | 0.024 | 0.008 | – | 0.992 |
| | | (6.05) | (1.86) | (1.65) | | 2.48 |
| $\Lambda_{MS}$ | 1966–1986 | 0.903 | 0.007 | 0.030 | 0.267 | 0.970 |
| | | (12.1) | (1.12) | (0.48) | (3.32) | 1.16 |
| | 1976–1986 | 0.914 | 0.035 | 0.011 | – | 0.981 |
| | | (10.2) | (2.18) | (1.37) | | 2.56 |
| $[\Lambda_{XS}, \Lambda_{MS}]$ | 1966–1986 | 0.908 | 0.005 | 0.018 | 0.205 | 0.955 |
| | | (14.7) | (1.99) | (1.54) | (2.67) | 2.40 |
| | 1976–1986 | 0.907 | 0.029 | 0.010 | – | 0.960 |
| | | (12.0) | (2.65) | (2.05) | | 3.45 |
| $\Delta C$ | 1966–1986 | 0.985 | 0.123 | 0.156 | – | 0.912 |
| | | (15.2) | (1.04) | (1.62) | | 1.85 |
| | 1976–1986 | 0.975 | 0.407 | 0.284 | – | 0.963 |
| | | (18.1) | (2.34) | (3.07 | | 4.02 |

## 25.6. Conclusion

It has been the purpose of this study to examine the effects of the surge in foreign indebtedness for a significant number of developing countries on the structure of world trade. Based on various measures of structural change, we performed statistical tests for bilateral relations between debtor countries and seven other world regions, using aggregate trade data. The statistical evidence is somewhat supportive of the original hypothesis, even though the data set appears to be too short to provide conclusive answers. It is hoped that further analysis on a pooled set of disaggregated trade measures will overcome these difficulties. However, even these preliminary results suggest that projections of world trade structures will have to be based on an assessment of financial conditions, in addition to more standard relative price mechanisms.

## References

Cline, W. (1984), *International Debt: Systematic Risk and Policy Response.* (Washington, DC: Institute for International Economics).

Dooley, M., Helkie, W., Tryon, R., and Underwood, J. (1986), An Analysis of External Debt Positions of Eight Developing Countries Through 1990, *Journal of Development Economics*, **21**, pp. 283–318.

Dornbusch, R. (1985), Policy and Performance Links between LDC Debtors and Industrial Nations, *Brookings Papers on Economic Activity*, No. 2, pp. 303–366.

Dornbusch, R. and Fischer, S. (1985), The World Debt Problem: Origin and Prospects, *Journal of Development Planning*, **16**, pp. 168–189.

Marquez, J. and McNeilly, C. (1986), *Can Debtor Countries Service their Debts? Income and Price Statistics for Exports of Developing Countries.* Federal Reserve Board, International Finance Discussion Paper No. 277, April.

Nakamura, T and Nakamura, Y. (1983), Structural Analysis of Global Trade, *Economic Studies Quarterly*, **34**, pp. 259–275.

Pinies, J. de (1986), *Debt Sustainability and Overadjustment.* United Nations DRPA/DIESA, Working Paper No. 1, December.

Sachs, J. (1985), External Debt and Macroeconomic Performance in Latin America and East Asia, *Brookings Papers on Economic Activity*, No. 2, pp. 523–573.

CHAPTER 26

# The Debt Situation of the Developing Countries: A Medium-Term Scenario Analysis

*Jürgen Richtering*


## Summary

The present chapter attempts to evaluate quantitatively the growth prospects of the different regions of developing countries up to the year 2000 with special consideration paid to the evolution of their external financial situations. Starting with a baseline scenario, which is an extrapolation of current trends, two alternative scenarios are presented to assess the impacts of changes in key assumptions. The analysis shows that external financial obligations will continue to be a critical constraint on growth of most developing countries in the medium term. Exporters of manufactures are likely to benefit more from additional import demand in the developed market economies than those of primary commodities. The analytical framework used for this exercise is UNCTAD's System for Interlinked Global Modeling and Analysis (SIGMA).


## 26.1. Introduction

Since the beginning of the 1980s a large number of developing countries have been faced with severe problems in meeting their external financial obligations. In consequence, many of them carried out strenuous adjustment measures, often in conjunction with debt rescheduling operations and in compliance with IMF programs. The adjustment efforts made by developing countries in many cases have resulted in a very significant strengthening of their trade and current account balances. For example, from 1982 to 1985, the current account deficit/exports ratios for the major eurocurrency borrowers improved from -23% to -1% while their trade balance has been positive by a substantial margin since 1983. There has, however, so far not been any improvement in their aggregate debt/exports ratio. The adjustment efforts have put a heavy burden on many developing countries; GDP per capita declined in Africa and in Latin America in the first half of this decade. The outlook for the medium term is not very promising as the accumulated debt and its associated debt-servicing obligations continue to be a critical constraint on growth in many developing countries. Non-oil primary commodity prices have declined dramatically in the past three years, reaching a 30-year low, while real interest rates continue at historically high levels; and there is not much hope that OECD growth rates will pick up significantly in the medium term.

The present chapter attempts to evaluate quantitatively the growth prospects of the different regions of developing countries up to the year 2000 with special consideration paid to the evolution of their external financial situations. Starting with a baseline scenario, which is an extrapolation of current trends under broadly realistic assumptions for the major variables of the world economy, two alternative scenarios are presented to assess the impacts of changes in key assumptions. The analytical framework used for this exercise is UNCTAD's System for Interlinked Global Modeling and Analysis (SIGMA). Section 26.2 gives a brief description of the SIGMA system, followed by an analysis of the various scenarios in Section 26.3 and conclusions in Section 26.4.

## 26.2. The Analytical Framework

UNCTAD's SIGMA system has been developed as an aid to the analysis of the medium- and long-term prospects of developing countries, taking explicit account of the implications of current account deficits for capital flows and consequent debt accumulation. The system may be used to project the capital requirements of the different regions of developing countries for a given set of assumptions concerning the domestic and the international environment. Alternatively, the availability of foreign capital may be made subject to behavioral assumptions and a search carried out for a growth path consistent with the given financial constraints.

The SIGMA system consists of 15 regional macroeconomic models linked together by bilateral trade and financial flows. (Appendix 26A contains a listing of country groupings.) The specification of each regional model, which consists of five blocks of equations, is standardized. In the first block, *aggregate demand relations* are determined. GDP is either set exogenously (capital requirements mode) or solved iteratively to satisfy imposed constraints on the growth of private debt (debt constraint mode). Consumption functions are estimated on a cross-regional basis taking into account income per capita and dependency ratios. Gross fixed capital formation is explained by the availability of imported capital goods and the level of GDP. Total imports and exports are aggregates of SITC groups derived in the trade block.

In the second block of equations, total output is disaggregated into seven *production sectors, viz.* food, non-food agriculture, mining and quarrying, manufacturing, construction, utilities, and services, using sectoral share functions estimated on a cross-regional sample with GDP per capita as the main explanatory variable.

In the *trade* block, imports for five categories of goods are determined using income elasticities estimated on a regional basis. The five trade sectors are food, non-food primary commodities, fuels, heavy equipment, and other manufacturing. For each commodity group exports are derived by applying the corresponding bilateral trade matrix to the relevant import margin and are then passed to the regional models. Imports and exports of non-factor services are determined by their historical relationships to imports and exports of goods, respectively, but with the latter adjusted to be equal to imports of non-factor services at the global level.

In the *balance of payments* block, the current account balance and the net flow of foreign capital are determined. Factor payments, consisting mainly of interest charges and direct investment income are calculated from estimated stock figures. Other factor payments, mainly labor income, are linked to the GDP of the region where the payments originate in the case of debits, or to the GDP of the developed market economies or the oil-dominant countries in the case of credits. Net private transfers are related to the GDP of donor regions. The stock of reserves is adjusted so as to cover, on average, three months of imports of goods in the simulation period. For the developing regions, the current account balance is financed through six types of international capital flows. These flows are assigned in the financial flow module which is described below.

Regional *GDP deflators* are based on the weighted average of their import and export prices and the regional exchange rate index. Import prices are calculated as trade share-weighted averages of the export prices of the exporting regions. Export prices are assumed to follow world prices. At present, the exchange rate index is kept constant at its 1983 level since a regional subsystem of prices has yet to be developed. The differences in regional inflation rates therefore reflect the differences in structures of regional trade.

The international dollar prices of capital goods and other manufactures are provided by an estimated function which includes, as explanatory variables, the US inflation rate, change of the US dollar/SDR index as well as the real growth rate of both world trade and GDP in the developed market economies. In turn, the prices for primary commodities (including fuels) are determined by the respective growth rates of world trade in each of these commodity groups, GDP growth in the developed market economies and the price index for manufactured goods. The nominal interest rate (one year LIBOR) is assumed to be sensitive to inflation and real GDP growth in the developed market economies.

For each of the five categories of traded goods, time series of $15 \times 15$ bilateral trade share matrices have been estimated up to 1983. Although the model allows for trade share adjustments over time, trade shares in the present exercise are kept constant at their 1983 levels. Regional export volumes and import prices for each of the commodity groups are determined by applying regional import volumes and the export prices to the relevant trade share matrices.

The financial-cum-debt module keeps track of six types of international capital flows, *viz.* public grants, official development assistance (ODA) loans, other official flows (OOF), private long-term financial flows, direct investment and short-term flows. Separate submodels are used to determine the terms structure of bilateral and multilateral ODA, OOF and private flows as well as the level and distribution of new gross flows. In contrast to the relatively comprehensive treatment of commodity trade, the financial flow model contains full details only for the developing regions. For each bilateral ODA, OOF and private flows, the module keeps account of debt outstanding, commitments, disbursements, amortization, interest payments, interest rates, grace period and repayment period. Short-term flows and direct investment are treated only on a pooled basis. The allocation of new flows to a recipient region in a given year depends upon the type of flows. Bilateral and multilateral ODA grants and ODA loans are related to the GDP of donor regions. Other official flows depend on the imports of recipient regions. Short-term flows are determined on the basis of debt outstanding and imports. Direct investment is tied to the export performance of the recipient region. Flows are allocated in a predetermined order of priority, depending on the current account deficit of a recipient region as well as previously commitmented flows. As a rule, OOF and private flows are scaled to fit the financing gap.

For each year of the projections period, the system is solved by means of a double iterative loop. First, each module is iterated until convergence is achieved, and linkage variables passed to other relevant modules. When all modules have converged, a new pass is started with the updated linkage variables. The process is repeated until each individual module requires only one iteration to converge.

## 26.3. Debt Simulations

The scenarios presented below are intended to illustrate the magnitude of the debt problem of the developing countries and its potential repercussions on their economic performance. These projections (capital requirements exercises) provide indications of the situations that developing countries may face under a given set of assumptions and are not intended as forecasts. It is not the intention here to analyze how the different developing regions might adjust their domestic economies to potential external constraints, for example, via import substitution or export expansion. The fact that certain results may seem unrealistic, especially in the year 2000, serves to indicate the need for policy changes.

### 26.3.1. The baseline scenario

The baseline scenario assumes that variables of central importance to the international economic environment are in line with current trends (see *Table 26.1*).

The fairly low growth of GDP of the developed market economies in the medium term is a reflection of the expectations of a modest reduction in unemployment rates and a decline in the growth of the labor force. (The growth rates for GDP for the developed market economies and the socialist countries up to 1990 are taken in large part from the October 1986 forecast of Project Link.) The GDP deflator of the United States is assumed to maintain a level of around 4% growth throughout the projection period. Nominal interest rates are set to decline slowly to about 7% in the year 2000, leaving real rates at a level of 3% to 4%, which is relatively high in view of historical trends.

*Table 26.1.* Development of key variables in the baseline scenario, 1985–2000 (in %).

| *Variable* | *1985–1990* | *1990–1995* | *1995–2000* |
|---|---|---|---|
| GDP of developed market economies | 2.5 | 2.4 | 2.3 |
| US dollar inflation | 4.0 | 4.0 | 4.0 |
| LIBOR | 7.5 | 7.3 | 7.1 |
| International prices | | | |
| Food | 1.4 | 2.4 | 2.4 |
| Raw materials | 0.8 | 3.7 | 3.6 |
| Energy | -8.1 | 4.4 | 4.3 |
| Heavy equipment | 6.2 | 2.9 | 2.9 |
| Other manufactures | 7.0 | 4.4 | 4.3 |
| Import income elasticity | | | |
| Developed market economies | 1.7 | 1.6 | 1.6 |
| Developing countries | | | |
| Africa | 0.9 | 1.3 | 1.4 |
| Asia | 0.9 | 1.5 | 1.5 |
| Latin America | 0.7 | 1.3 | 1.3 |
| of which | | | |
| Least developed countries | 1.0 | 1.7 | 1.7 |

Source: UNCTAD Secretariat calculations based on SIGMA projections.
Note: Africa does not include Algeria, Libya and Nigeria; Asia does not include the oil-dominant countries of West Asia. The reported growth rates are period averages.

The projected rather low rate of inflation as well as the projected slow real growth in the developed market economies cannot be expected to create enough demand to raise the depressingly low non-oil primary commodity prices significantly. This observation is supported by the fact that competition among debt-ridden developing countries tends to become more intensive as they try to improve their foreign exchange earnings through export expansion. The price of fuels, after having declined in dollar terms by 45% in 1986, is assumed to rise in accordance with the prices of manufactures other than heavy equipment. The US dollar/SDR rate is kept constant from 1987 onward.

Assumptions regarding import income elasticities are of crucial importance to the growth prospects of developing countries at a time when the access to external capital is rather restricted. At the beginning of this decade, many developing countries were forced to compress their imports radically owing to mounting external financial obligations. In Latin America, import volume fell by more than 20% from 1980 to 1986, while in Africa it increased only marginally.

In contrast, Asia, which had less accumulated debt, was able to maintain moderate rates of import growth. The observed import compression should be seen as a short-term reaction to the debt crisis rather than as a fundamental shift in policy. Therefore, in the various scenarios one should not expect the extremely low and even negative import income elasticities to be maintained in the medium or longer term, if growth prospects are not to be jeopardized. Accordingly, import income elasticities in 1987 are assumed to return to their historical levels, estimated on the basis of data for 1970–1983. Consequently, import/GDP ratios, especially in Latin America, remain for a considerable number of years far below the levels reached before the debt crisis in 1982. Allowance for import/GDP ratios to reach the precrisis levels in the next few years would imply a very substantial increase in external financing.

Import income elasticities of the developed market economies play a key role in the simulation exercise as they affect the export earnings of the developing countries, their main source of foreign exchange. The estimated import income elasticities are not adjusted to allow for the current rise in protectionist tendencies nor for increased market penetration of developing countries' exports in the developed market economies.

Within the context of the above mentioned assumptions and in view of the restricted availability of additional foreign capital, growth prospects for developing countries are rather limited. (The term "developing countries", as used in the discussion of the scenario analysis, is meant to comprise all developing countries other than the following: Algeria, Iraq, Kuwait, Libyan Arab Jamahiriya, Nigeria, Oman, Qatar, Saudi Arabia, and United Arab Emirates. Because of the complete dependence on oil exports for their foreign exchange earnings, the development in these countries is not comparable with that of other developing countries.)

The results of the baseline scenario, presented in *Table 26.2*, show that the average growth rate of GDP in the developing regions is assumed to surpass that of the developed market economies only by about one percentage point. In spite of this modest GDP growth of the developing countries, their capital requirements are expected to increase, with debt growing at an average of 1.4% in 1986–1990 to 5.9% in 1996–2000. The marked resurgence of debt growth and the subsequent reduction of output growth toward the end of the projection period are the results of an ever-increasing current account deficit. The situation is attributable, in part, to the deceleration of GDP growth in the developed market economies due to demographic patterns. Inasmuch as the growth rates of debt are lower than those of nominal exports and GDP for the developing countries as a whole, the ratios of interest payments and debt to exports and GDP in the year 2000 are substantially lower than present levels. Partly responsible for the initial decline of the interest payments/exports ratio from 1985 to 1990 is a fall of about two percentage points in LIBOR rate.

At the regional level, Latin America and Asia manage to increase their per capita GDP and lower their debt ratios simultaneously. Africa and the least

*Table 26.2.* Baseline scenario, 1985–2000 (%).

| *Regional group/ variable* | *1985* | *1985–1990* | *1990* | *1990–1995* | *1995* | *1995–2000* | *2000* |
|---|---|---|---|---|---|---|---|
| ***Total of regional groupings*** | | | | | | | |
| Growth of GDP | | 3.8 | | 3.4 | | 3.3 | |
| Growth of debt | | 1.4 | | 3.2 | | 5.9 | |
| Growth of GDP per capita | | 1.6 | | 1.3 | | 1.3 | |
| Current account balance/exports | –5 | | –6 | | –8 | | –12 |
| Interest payments/exports | 11 | | 8 | | 6 | | 5 |
| Debt/exports | 145 | | 115 | | 92 | | 84 |
| Debt/GDP | 38 | | 29 | | 24 | | 22 |
| ***Africa*** | | | | | | | |
| Growth of GDP | | 2.5 | | 2.6 | | 2.5 | |
| Growth of debt | | 11.9 | | 9.9 | | 9.6 | |
| Growth of GDP per capita | | –0.5 | | –0.4 | | –0.4 | |
| Current account balance/exports | –42 | | –54 | | –57 | | –60 |
| Interest payments/exports | 14 | | 20 | | 24 | | 27 |
| Debt/exports | 271 | | 378 | | 434 | | 494 |
| Debt/GDP | 67 | | 91 | | 108 | | 127 |
| ***Asia*** | | | | | | | |
| Growth of GDP | | 4.9 | | 3.7 | | 3.5 | |
| Growth of debt | | –0.8 | | 3.0 | | 9.1 | |
| Growth of GDP per capita | | 2.9 | | 1.9 | | 1.8 | |
| Current account balance/exports | –0 | | –0 | | –5 | | –10 |
| Interest payments/exports | 5 | | 3 | | 2 | | 2 |
| Debt/exports | 78 | | 52 | | 40 | | 42 |
| Debt/GDP | 23 | | 15 | | 12 | | 13 |
| ***Latin America*** | | | | | | | |
| Growth of GDP | | 3.1 | | 3.2 | | 3.2 | |
| Growth of debt | | –1.7 | | –2.1 | | –2.7 | |
| Growth of GDP per capita | | 0.9 | | 1.2 | | 1.4 | |
| Current account balance/exports | –3 | | –5 | | –2 | | –1 |
| Interest payments/exports | 22 | | 16 | | 9 | | 5 |
| Debt/exports | 231 | | 177 | | 112 | | 69 |
| Debt/GDP | 49 | | 35 | | 22 | | 14 |
| ***Least developed countries*** | | | | | | | |
| Growth of GDP | | 2.8 | | 2.7 | | 2.7 | |
| Growth of debt | | 7.9 | | 9.8 | | 12.0 | |
| Growth of GDP per capita | | –0.2 | | –0.1 | | –0.1 | |
| Current account balance/exports | –75 | | –67 | | –79 | | –97 |
| Interest payments/exports | 11 | | 9 | | 11 | | 16 |
| Debt/exports | 360 | | 372 | | 423 | | 534 |
| Debt/GDP | 60 | | 64 | | 74 | | 96 |

Source: UNCTAD Secretariat calculations based on SIGMA projections. More detailed results are available from the authors upon request.

Note: Africa does not include Algeria, Libya and Nigeria; Asia does not include the oil-dominant countries of West Asia. The reported growth rates are period averages. Debt refers to medium- and long-term debt only. Exports include goods and services.

developed countries, on the other hand, see their living standards deteriorate further, together with an increasing debt burden. The regional results reveal a strong inverse relationship between the debt/GDP ratio and the associated growth performance. Economic reality, however, is more complex, and one should not try to reduce the development process to a simple bivariate relationship. Certainly, external financial obligations will continue to be one, if not the main, constraint on growth in many developing countries, but there are also other key factors. For instance, the export structure of Latin America and Asia is largely responsible for their export performance, given the relatively high world import elasticity for manufactured goods. On the other hand, Western Europe, Africa's main trading partner, has the lowest growth rates among the three regions of developed market economies in the baseline scenario. Nevertheless, one can agree with the latest report of the Committee for Development Planning (1986), which stated (referring to the accumulated debt burdens) that "the debris of the past will continue to impede economic progress in many developing countries" (p 9).

### 26.3.2. Alternative scenarios

Starting from the baseline scenario, two alternative scenarios are presented to show the sensitivity of the projections to changes in the main variables. The first, scenario A, shows how the debt ratios would change if the developing countries were to achieve a 1% higher GDP growth in the years 1988–2000. In the second, scenario B, it is further assumed that the developed market economies are able to improve their growth performance by one percentage point in the same period. The results are summarized in *Table 26.3*.

In the first scenario, the current account balances worsen dramatically in all regions, leading eventually to increasing debt/export ratios. It is very unlikely that the external capital required to finance this extra growth would be available. In comparison with the baseline scenario, the difference in accumulated debt for the total of all three regional groups amounted to about 20 billion US dollars in 1990, and 10 times and 100 times this amount in 1995 and 2000, respectively.

In contrast to scenario A, the external balances in scenario B are very close to those of the baseline scenario. In all regional groupings, with the exception of the least developed countries, there are improvements in the external financial indicators throughout the project period. Evidently, the higher import demand of the developed market economies plays a significant part in improving the trade balances of the developing countries. In this scenario, exporters of manufactures would profit the most as the import elasticities for manufactures are much higher than those for primary commodities. This explains why Asia and Latin America could improve their external position much more than Africa and the least developed countries, relative to scenario A. In addition, the increased demand of the developed market economies generates higher inflation

*Table 26.3.* Results of alternative scenarios of GDP growth in the period 1988–2000 (in %).

| *Regional group/ variable* | *1990* | *1995* | *2000* |
|---|---|---|---|
| A. Developing countries: + 1% | | | |
| *Total of regional groupings* | | | |
| Current account balance/exports | –10 | –20 | –35 |
| Interest payments/exports | 8 | 7 | 10 |
| Debt/exports | 119 | 119 | 183 |
| *Africa* | | | |
| Current account balance/exports | –59 | –73 | –93 |
| Interest payments/exports | 21 | 27 | 37 |
| Debt/exports | 388 | 493 | 655 |
| *Asia* | | | |
| Current account balance/exports | –4 | –17 | –33 |
| Interest payments/exports | 3 | 3 | 7 |
| Debt/exports | 54 | 60 | 140 |
| *Latin America* | | | |
| Current account balance/exports | –8 | –13 | –23 |
| Interest payments/exports | 16 | 12 | 11 |
| Debt/exports | 184 | 148 | 155 |
| *Least developed countries* | | | |
| Current account balance/exports | –75 | –106 | –154 |
| Interest payments/exports | 9 | 16 | 34 |
| Debt/exports | 384 | 515 | 804 |
| B. Developing and developed market economies: + 1% | | | |
| *Total of regional groupings* | | | |
| Current account balance/exports | –5 | –7 | –9 |
| Interest payments/exports | 8 | 5 | 4 |
| Debt/exports | 108 | 78 | 66 |
| *Africa* | | | |
| Current account balance/exports | –53 | –55 | –58 |
| Interest payments/exports | 20 | 23 | 25 |
| Debt/exports | 360 | 389 | 422 |
| *Asia* | | | |
| Current account balance/exports | 1 | –3 | –7 |
| Interest payments/exports | 3 | 1 | 1 |
| Debt/exports | 49 | 33 | 33 |
| *Latin America* | | | |
| Current account balance/exports | –4 | –1 | –0 |
| Interest payments/exports | 16 | 8 | 4 |
| Debt/exports | 165 | 93 | 49 |
| *Least developed countries* | | | |
| Current account balance/exports | –68 | –84 | –110 |
| Interest payments/exports | 9 | 12 | 21 |
| Debt/exports | 356 | 404 | 537 |

Source: UNCTAD Secretariat calculations based on SIGMA projections.

rates in world trade without affecting the price structure of commodity groups. This leads to a devaluation of the nominal debt stock. The real rate of interest, however, is kept constant as the nominal LIBOR rate is adjusted for the increase in inflation. Taken together, one may conclude that under the given assumptions, 1% additional GDP growth in the developed market economies would allow for somewhat more than 1% additional growth in the developing countries without leading to a deterioration in the latter's external accounts. The impact would be higher for exporters of manufactures and lower for exporters of primary commodities.

## 26.4. Conclusions

The results of the various scenarios presented above show that the external financial sector is likely to continue to be a critical constraint on the growth of most developing countries in the medium term. Import demand growth in the developed market economies seems to be of paramount importance to the growth prospects of all developing regions, in terms of the latter's ability to earn foreign exchange to finance their development and to service their debt. The impact of GDP growth in the developed market economies on the developing countries is felt more significantly in countries which are exporters of manufactures than those of primary commodities which face low demand elasticities. The analysis suggests than an increase in GDP growth of the developed market economies of 1% would enable developing regions to increase their growth by about the same amount without risking a deterioration in their debt ratios.

It is, however, rather unlikely that GDP growth in the industrialized countries will rise much above the projected trends, given present demographic patterns. Nevertheless, developing countries will have to earn enough foreign exchange to maintain adequate rates of output growth while paying their import bills and servicing their debts. If no growth impulses can be expected from the international environment, domestic adjustment efforts in developing countries will have to be continued. Exports promotion may, in some cases, succeed in expanding trade shares of developing countries. This requires, however, that the present protectionist trends can be reversed. But, more exports are perhaps not possible for all countries because of limited demand for their goods facing low demand elasticities, as is the case with most primary commodities, or because too much competition would bring down prices. This constellation may force some countries to replace part of their imports by domestically produced goods. Import substitution may be a viable strategy for the larger developing countries; in other cases it may require (regional) economic cooperation among developing countries. For a number of the poorest developing countries, however, the outflow of resources due to the accumulated debt burden will have to be reduced, because it simply can no longer be financed without ruining these countries. In those cases debt relief and debt write-off may be the only solutions to more promising growth prospects.

## Acknowledgment

The scenario analysis is based on UNCTAD's System for Interlinked Global Modeling and Analysis, which was designed and built by the Review and Appraisal, Interdependence Section, Money, Finance and Development Division, of which the author is a staff member. The author is grateful to staff members of his division for their comments. The opinions expressed in this chapter are those of the author and do not necessarily reflect those of the UNCTAD Secretariat.

## Reference

Committee for Development Planning (1986), *Doubling Development Finance: Meeting a Global Challenge. Views and Recommendations of the Committee for Development Planning* (ST/ESA/187). (New York: United Nations, Department of International Economic and Social Affairs).

## Appendix 26A. SIGMA Country Groupings

| Code | Regional name | Countries |
|---|---|---|
| R01 | North America | Canada; Puerto Rico; USA |
| R02 | Western Europe | Austria; Belgium; Cyprus; Denmark; Faeroe Islands; Finland; France; Germany, Federal Republic of; Gibraltar; Greece; Greenland; Iceland; Ireland; Israel; Italy; Liechtenstein; Luxembourg; Malta; Netherlands; Norway; Portugal; Spain; Svalbard and Jan Mayen Islands; Sweden; Switzerland; Turkey; United Kingdom; Yugoslavia |
| R03 | Pacific and South Africa developed market economies | Australia; Japan; New Zealand; South Africa |
| R04 | Socialist countries of Eastern Europe | Albania; Bulgaria; Czechoslovakia; German Democratic Republic; Hungary; Poland; Romania; USSR |
| R05 | Socialist countries of Asia | China, People's Republic of; Democratic Kampuchea; Korea, Democratic People's Republic of; Mongolia; Viet Nam |
| R06 | Latin America: Eurocurrency countries | Argentina; Brazil; Chile; Colombia; Peru; Venezuela |
| R07 | Latin America: Others | Antigua and Barbuda; Bahamas; Barbados; Belize; Bermuda; Bolivia; British Antarctic Territory; British Virgin Islands; Cayman Islands; Costa Rica; Cuba; Dominica; Dominican Republic; Ecuador; El Salvador; Falkland Islands; French Guiana; Grenada; Guadaloupe; Guatemala; Guyana; Honduras; Jamaica; Martinique; Montserrat; Netherlands Antilles; Nicaragua; Panama; Paraguay; St. Kitts-Nevis; St. Lucia; St. Pierre and Miquelon; St. Vincent and the Grenadines; Suriname; Trinidad and Tobago; Turks and Caicos Islands; Uruguay |
| R08 | Africa: Eurocurrency countries | Algeria; Nigeria. |
| R09 | Africa: Least developed countries | Benin; Botswana: Burkina Faso; Burundi Cape Verde; Central African Republic; Chad; Comoros; Equatorial Guinea; Ethiopia; Djibouti; Gambia; Guinea; Guinea-Bissau; Lesotho; Malawi; Mali; Mauritania, Niger; Rwanda; Sao Tome and Principe; Sierra Leone; Somalia; Sudan; Togo; Uganda; United Republic of Tanzania |

**Appendix 26A.** *Continued*

| *Code* | *Regional name* | *Countries* |
|---|---|---|
| R10 | Africa: Others | Angola; British Indian Ocean Territory; Cameroon; Congo; Cote d'Ivoire; Egypt; French Southern and Antarctic Territories; Gabon; Ghana; Kenya; Liberia: Madagascar; Mauritius; Morocco; Mozambique; Namibia; Reunion; St. Helena; Senegal; Seychelles; Swaziland; Tunisia; Western Sahara; Zaire; Zambia; Zimbabwe |
| R11 | Asia: Oil-dominant economies | Iraq; Kuwait; Libyan Arab Jamahiriya; Oman; Qatar; Saudi Arabia; United Arab Emirates |
| R12 | Asia: Eurocurrency countries | Taiwan Province of China; Hong Kong; Indonesia; Korea, Republic of; Malaysia; Philippines |
| R13 | Asia: Least developed countries | Afghanistan; Bangladesh; Bhutan; Democratic Yemen; Kiribati; Haiti; Lao People's Democratic Republic; Maldives; Nepal; Samoa; Tuvalu; Vanuatu; Yemen |
| R14 | Asia: Others | Bahrain; Brunei Darussalam; Burma; East Timor; India; Iran; Jordan; Lebanon; Macau; Pakistan; Singapore; Sri Lanka; Syrian Arab Republic; Thailand |
| R15 | Oceania | American Samoa; Canton and Endenbury Islands; Christmas Island; Cocos (Keeling) Islands; Cook Islands; Fiji; French Polynesia; Guam; Johnston Island; Nauru; New Caladonia; Niue; Norfolk Island; Pacific Islands; Papua New Guinea; Pitcairn; Solomon Islands Tokelau; Tonga; Wallis and Futuna Islands |

# PART IV

# Some General and Methodological Problems Related to Economic Growth and Structural Change

CHAPTER 27

# Structural Change and the Long Wave

*S. Menshikov and L. Klimenko*


## Summary

The structural crisis is seen as a phase in the long wave when structural change is so intense that a substantial part of previously well-established industries undergoes a prolonged decline or stagnation, which leads to an overall macroeconomic slowdown. The intensity of structural change in the US economy appears to fluctuate periodically with peaks every 25–30 years. This is demonstrated by a latent variable estimation technique developed by Wold. It is modified in order to deal with disaggregated total fixed capital investment. It is split into three components (extensive, productivity-increasing investment and investment to produce new products) with differing impacts on structural change and on the cyclical behavior of the economy.


## 27.1. Periodicity in Structural Change

In a previous paper (Menshikov, 1985) we have indicated that the downward phase of the long wave coincides with a structural crisis in the economy. The crisis manifests itself in widespread stagnation of large groups of relatively old industries which have exhausted their growth potentials, while other, newer groups continue to flourish and still other industries make their first appearance on the economic scene. Structural change accompanies all phases of the long wave; however, it is during the crisis phase that there is not just uneven development of different sectors, but long-term stagnation and downturn in many industries, which in the previous long upturn had served as the backbone and mainstay of general economic growth, and still account for a major part of overall output.

We have also indicated that the principal motive force behind the long upturns is the opening up of new directions in technology, i.e., technological revolutions. In the course of such revolutions new sectors and industries are created, and the spread of new technologies and products brings about substantial change in the structure of the economy. During the period when a new economic structure is being created, there is a boost to the development of most industries, though some old industries may be wiped out and disappear completely in spite of the general boom. In this phase of the long wave, dynamism is characteristic of most industries, resulting in overall economic acceleration.

As the further development of new technological directions becomes less radical and more evolutionary in character, and the benefits from such development become more marginal, conditions are set for a new long downturn which ends up in a structural crisis including stagnation of some of the previously buoyant industries. At the same time the search for new areas of efficiency and profitability is renewed.

The economic mechanism, by which changes in technology are translated into long waves and structural change, operates through the rate of profit. As shown in Menshikov and Klimenko (1985), technological change is reflected in long-term fluctuations of labor productivity, capital intensity and the output/capital ratio. The latter tends to rise, starting with the later part of the structural crisis, and reaches a maximum at the peak of a technological revolution. The ratio tends to fall when changes in technology become evolutionary and reaches a trough early in a structural crisis. The average profit rate in the economy tends to follow the output/capital ratio with a lag. Since the profit rate is a major determinant of business investment, its fluctuations set off different patterns of behavior of firms *vis-à-vis* new technology.

When the profit rate is rising, conditions are favorable for the spread of new technology. But as the profit rate reaches high levels, the predominant attitude is to give priority to minor improvements in existing technology, rather than to basic innovations. When the profit rate starts falling from its peak, the climate for innovations becomes unfavorable. However, when the profit rate reaches historically low levels, conditions become ripe for a new take-off in basic innovations. These become widespread as the profit rate resumes the upward movement from its long-term trough.

Thus, there is an interplay between technological change and the rate of profit, the first determining the course of the latter through changes in the output/capital ratio, and the profit rate determining changes in investment which in turn lead to fluctuations in the output/capital ratio.

In the present chapter we shall: (1) consider statistical evidence of long-term fluctuations in structural change; and (2) attempt to disaggregate capital investment into components with different effect on structural change.

## 27.2. Statistical Evidence of Fluctuations

A number of indices are used to measure the intensity of structural change in the economy. One of them, recommended by the UN Economic Commission for

Europe (ECE), is the weighted average annual increment of shares of industries which, during a given period of time, have expanded faster than the total. This index is defined as

$$I = \sum_{i=1}^{q} [s\,(i_T) - s\,(i_0)]/q * t \tag{27.1}$$

where $S(i)$ are shares of $i$th industry in total output (or employment, or capital stock, etc.); 0 and $T$ indicate the beginning and the end of the period which is being measured; $t = T - 0$ is the length of the period; and $q$ is the number of industries which have increased their share.

We have computed the ECE index for US manufacturing divided into 20 large industries. *Figure 27.1* shows changes in the shares in total value added and employment for 1899–1982. Subperiods roughly correspond to actually observed business cycles of average duration: 1899–1909, 1909–1921, 1921–1929, 1929–1937, 1937–1947, 1947–1957, 1957–1967, 1967–1973, 1973–1982 (*Historical Statistics...*, 1975).

The value-added index has two definite peaks (in 1921–1929 and 1947–1957), and troughs (in 1937–1947 and 1967–1973). The frequency of oscillation in both cases is about 30 years. However, if one considers 1899–1909 as another trough, the periodicity increases to 40 years.

The index of employment has the same peaks in 1921–1929, 1937–1947, 1957–1967, and troughs in 1929–1937, 1947–1957, 1967–1973. In this case fluctuation frequency is 20, not 30 years, except for the starting cycle, which is at least 30 years.

From a methodological viewpoint the ECE index raises a number of questions. The larger the number of industries with a rising share, the smaller – other things being equal – the indicator of intensity. The broken lines in *Figure 27.1* show the change in the number of such industries. It is obvious that the number of industries is negatively correlated to the index.

Let us consider, for example, 1899–1909. According to both indices (for value added and employment), this was a period of relatively little structural change. However, in that decade a majority of industries (13 out of 20) increased their shares – a record repeated only in 1909–1921 (another low-intensity period according to the indices). Among the rising industries there were machinery and equipment, instruments, chemicals, oil products, rubber and plastics. Receding were food, tobacco, textiles, leather, and wood. With a few exceptions there was a widespread move in favor of the relatively new branches of heavy industry – quite an important structural change by any criterion.

Similar results were obtained by calculating the angle between $q$-dimensional vectors of industry shares for different periods of time, as suggested by Pauly (Chapter 25 in this volume):

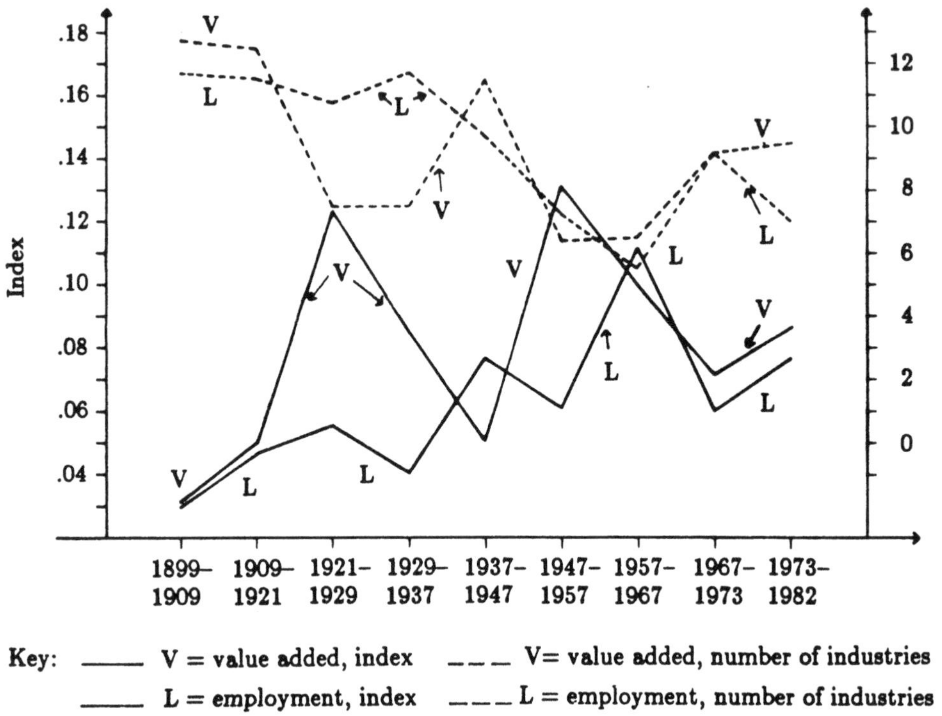


*Figure 27.1.* Index of intensity of structural change.

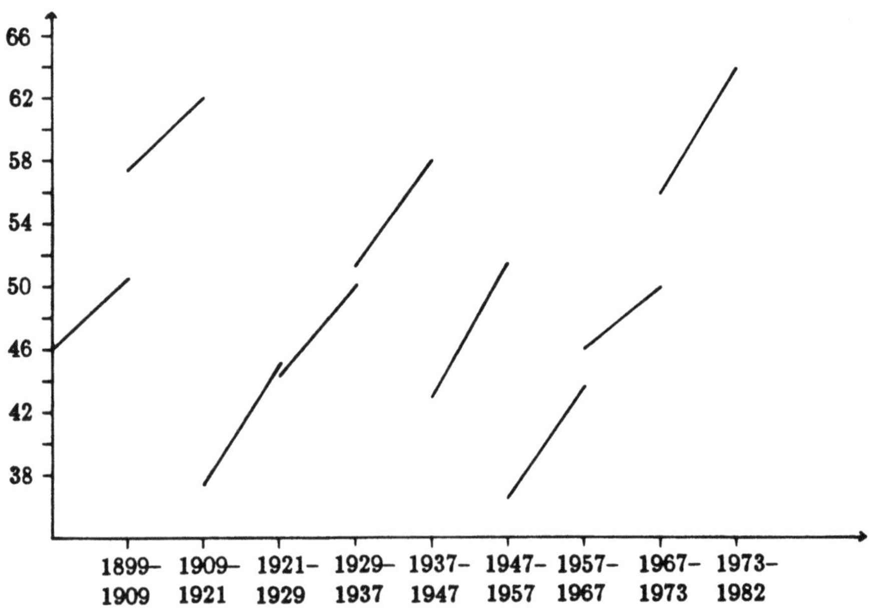


*Figure 27.2.* Intensity of structural change, according to the increase in total share of rising industries (value added).

$$L = \arccos \frac{\sum_{i=1}^{q} s_{i,}t_1 * s_{i,}t_2}{\sqrt{\left(\sum_{i=1}^{q} s_{i,}t_1^2\right) * \left(\sum_{i=1}^{q} s_{i,}t_2^2\right)}} \tag{27.2}$$

where $s_{i,t_j}$ denotes industry shares in the $j$th period.

In this calculation peaks are reached in 1921–1929 and 1947–1957, as in the ECE index, but troughs occur in both cases immediately following the peaks, i.e., in 1929–1937 and 1957–1967.

To take care of these shortcomings of the EEC index, we suggest another indicator: the total shares of rising industries at the beginning and the end of each period. These are shown in *Figure 27.2*. The position of the line in each period shows the relative weight of industries undergoing expansion, while its incline shows the relative speed of their expansion.

In 1899–1909 the expanding industries started with a total share of 46.1% in value added, and ended with 50.8%. This was closer to the average for all periods, rather than to the minimum. According to this measure, 1899–1909 was all but a trough in intensity of structural change. Compare this to 1921–1929, when only 8 out of 20 industries expanded their shares, increasing the total from 37.1% to 45.4%. The speed of the increase was twice that of 1899–1909 (22.3% as against 10.0%); but in terms of the total weight of rising industries, structural change in 1921–1929 was much smaller.

The indicator which measures structural change by total weight of expanding industries in value added singles out 1909–1921, 1937–1947 and 1973–1982 as peak periods, whereas 1921–1929 and 1957–2967 were troughs. Periodicity varies between 30 and 40 years.

*Figures 27.1* and *27.2* point to qualitatively heterogeneous periods of structural change. Widespread, but relatively slow expansion of a majority of industries is characteristic of the late upturn phase in the long wave. The slowness points to a predominantly evolutionary rise of new industries. This is supported by some old industries, which are being stimulated by overall growth in the economy. Contrary to this, fast expansion of a small range of new industries characterizes the start of a new technological upsurge in the midst of a long recession.

Thus, there seems to be a double interchange of intensive and extensive expansion during one long wave of approximately 50 to 60 years. The interchange occurs twice within the long wave and points to two qualitatively different phenomena.

Let us observe the various types of long-term movement of individual industries. *Figure 27.3* shows changes in the share (in value added) of six industries, which have been expanding throughout the century. These are: electrical equipment, instruments, nonelectrical machinery, fabricated metal products, chemicals, rubber and plastics. The most interesting is the case of electrical machinery, which has also undergone intensive internal structural change in

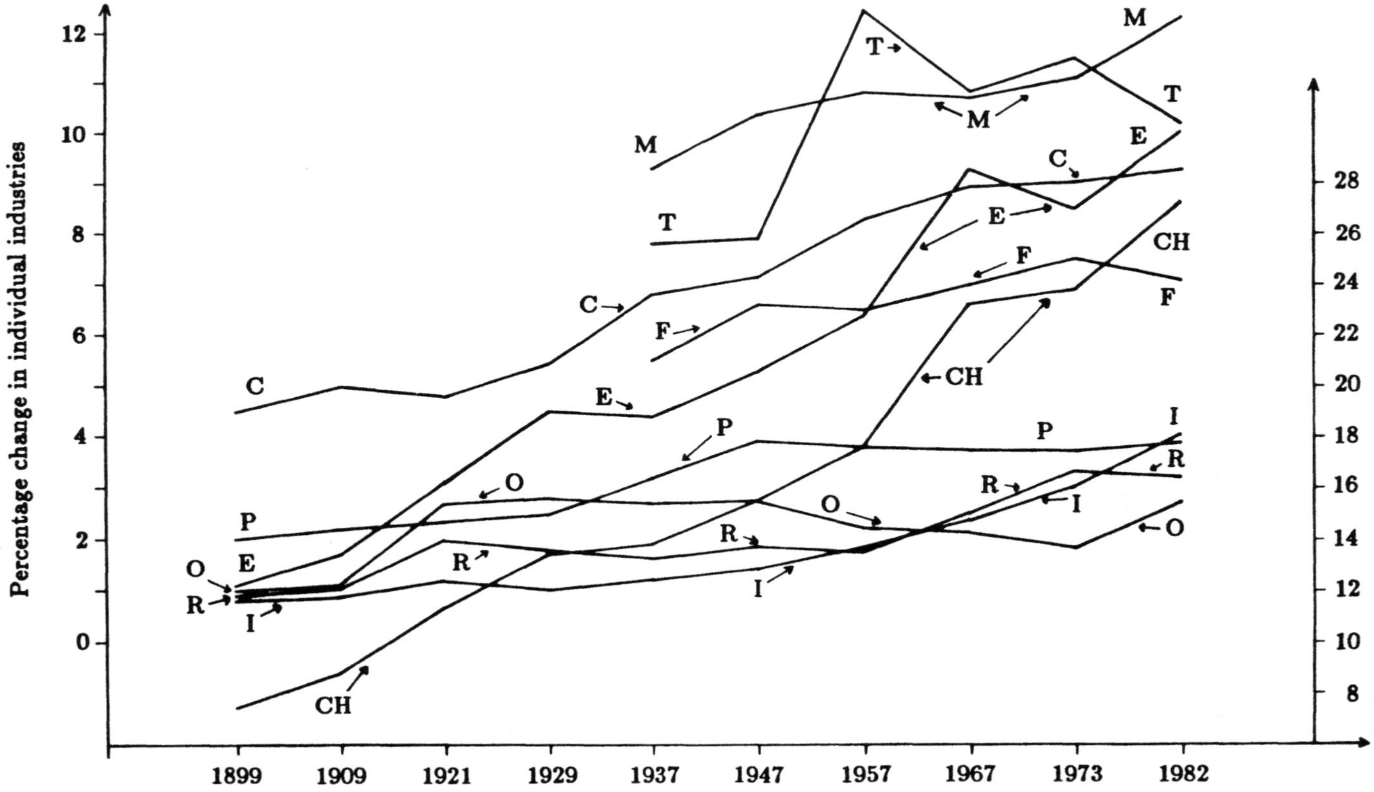


Key: E = electric equipment
I = instruments
M = machinery, except electric
F = fabricated metal products
CH = electric equipment, instruments, plastics and chemicals, combined
T = transportation equipment
O = petroleum products
P = paper
R = rubber, plastics
C = chemicals

*Figure 27.3.* Changes in shares of rising industries (value added).

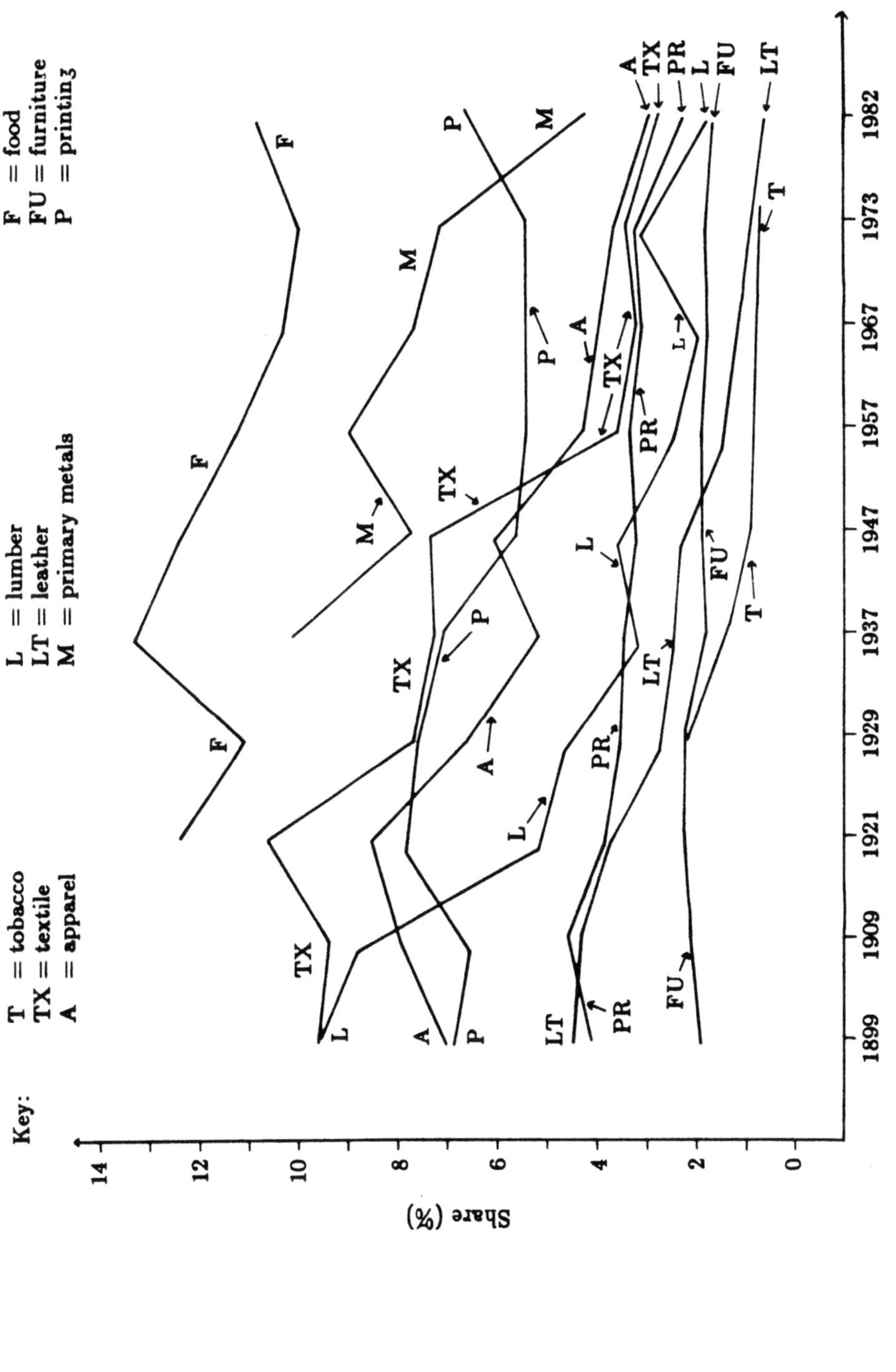


*Figure 27.4.* Changes in shares of receding industries.

output and technology. The 30- to 40-year cycle is apparent here. The same is true of chemicals, fabricated metal products, and nonelectrical machinery. The total share of four rising industries (electrical equipment, instruments, plastics and chemicals, combined), for which statistics exist throughout the whole period, shows the same pattern.

Another group of industries represented in *Figure 27.3* are those which, after rising for a number of decades, finally became stagnant. Prominent among these are transport equipment, petroleum products, and paper. There is evidence of a 50-year wave in some of these, but it is not clear-cut.

*Figure 27.4* shows long-term dynamics of receding industries. Here textiles and apparel present the most apparent cases of 30-year periodicity. Lumber and primary metals are closer to a 20-year fluctuation, possibly associated with the Kuznets construction cycle.

The general conclusion is the existence of 30- to 40-year periodicities of structural intensity within the framework of a longer 50- to 60-year wave. This conclusion is supplemented by the analysis of structural components of business investment.

## 27.3. Disaggregation of Capital Investment by Structural Components

All technological and structural change in the economy has to pass through the phase of capital investment. Whatever new technology is applied or new products are introduced in the markets, capital has to be invested to achieve this result. Investment is essential in creating new plants and industries and in expanding existing production units or infrastructure facilities. In a sense any given economic structure is based on capital investment which has been physically materialized and accumulated over time.

Traditionally, total capital investment has been statistically disaggregated by industry, functional use (plant or equipment), or reproduction role (depreciation or expansion). To our knowledge, there are no statistical series which differentiate between investment used to expand existing technologies and products, and investment going into new technologies and products. Statistical data on research and development expenditure serve as a useful indicator, but they are not identical with investment in new technology and new products.

An indirect way of disaggregating investment according to this criterion involves the estimation of latent variables, as developed by Wold (1985). Our addition to the Wold technique is to use latent variables to break down aggregates into new time series which add up to existing statistical series. Thus, it is possible to disaggregate statistical series into components which are not to be found in statistical publications.

For our purposes we have used this technique to break down net capital investment in the USA from 1889 to 1982 into three parts:

(1) "Extensive investment" – used to increase the stock of existing technologies.
(2) "Intensive investment, type 1" – used to introduce technologies which increase factor productivity in existing plants.
(3) "Intensive investment type 2" – used to create new products or spheres of production.

All three types of investment serve to bring about structural changes, albeit in a different way. The most direct result is produced by intensive investment, type 2, which serves to create new products and industries. The role of intensive investment, type 1, is first and foremost to accelerate factor productivity growth and to implement new technologies which promote a general upswing in the economy. Its eventual result is the spread of new technology across the board, and the creation of a new structure in the economy, as a whole. Extensive investment serves to expand existing technology, and in this way supports the further development of a new economic structure. The most active and revolutionary case is intensive investment of both types, while extensive investment is more conservative and evolutionary in nature.

In order to estimate latent variables it is necessary to explicitly define their causal relationships with other variables for which statistical series do exist. The outcome of the exercise is largely dependent on the hypothesis or criteria used.

We define extensive investment ($EK$) as investment which is necessary to expand output:

$$EK = f\,(GPDP,\ GPDPD) \tag{27.3}$$

where $GPDP$ is current gross private domestic product and $GPDPD$ is the desired $GPDP$. All variables are measured in constant prices.

Intensive investment, type 1 ($IEK$), is investment necessary to expand factor productivity:

$$IEK = f\,(YL,\ YLD) \tag{27.4}$$

where $YL$ is existing factor productivity, and $YLD$ is the desired productivity.

Intensive investment, type 2 ($IIK$), is investment in new products or spheres of production and is determined by the need to raise the individual rate of profit of enterprises above the prevailing average:

$$IIK = f\,(P,\ PD) \tag{27.5}$$

where $P$ is current profit rate, and $PD$ is the desired profit rate.

We assume that the sum of the three types of investment equals total net capital investment in the economy:

$$dPCS = EK + IEK + IIK \tag{27.6}$$

where *PCS* stands for private capital stock, and *d* for first differences.

The computation of variables *EK*, *IEK* and *IIK* is made in four steps:

(1) Compute latent variables $L_1$ and $L_2$, representing, respectively, accumulated extensive and accumulated total intensive investment (of both types) based on Wold's partial least squares iteration technique and postulated functional relationships (see below).

Because latent variables are not directly observable, it is, perhaps, necessary to explain their economic meaning. $L_1$ can be interpreted as fixed capital stock which is necessary to maintain a desired (rising or falling) output level of existing products with existing technology. $L_2$ is the fixed capital stock needed to maintain a desired output level of new products and also to introduce new production technologies.

(2) Compute variables *EK* and *IK* (*IK* = *IEK* + *IIK*) by estimating OLS function:

$$dPCS = a_1 dL_1 + a_2 dL_2 + EPS\ (dPCS) \quad a_1 > 0, \quad a_2 > 0 \tag{27.7}$$

where *EPS* (*dPCS*) is the random component of *dPCS*; and assuming that:

$$EK = a_1 dL_1 + EPS(dPCS) * a_1/(a_1 + a_2) \tag{27.8}$$

$$IK = a_2 dL_2 + EPS(dPCS) * a_2/(a_1 + a_2) \tag{27.9}$$

$dL_1$ and $dL_2$ are desired investments needed to achieve $L_1$ and $L_2$, respectively. Their sum may differ from actual net investments of the extensive and intensive types (*EK* and *IK*), which together are equal to total net investment into fixed capital:

$$dPCS = EK + IK \tag{27.10}$$

The difference between desired and actual net investment is accounted for by the stochastic element *EPS*(*dPCS*), which is distributed between *EK* and *IK* according to the estimated relative shares $a_1$ and $a_2$ of extensive and intensive investment.

(3) Compute latent variables $L_3$ and $L_4$, representing intensive investment of two different types by the same technique as in step (1). We interpret $L_3$ as the fixed capital stock necessary to maintain a desired rate of introducing new production technologies, and $L_4$ as the capital stock needed to maintain desired output levels of new products and rates of expansion into new spheres of production.

(4) Compute variables *EIK* and *IIK* by estimating function:

$$IK = a_3 dL_3 + a_4 dL_4 + EPS\,(IK), \quad a_3 > 0, \quad a_4 > 0 \tag{27.11}$$

assuming that:

$$EIK = a_3 dL_3 + EPS(dIK) * a_3/(a_3 + a_4) \tag{27.12}$$

$$IIK = a_4 dL_4 + EPS(dIK) * a_4/(a_3 + a_4) \tag{27.13}$$

$dL_3$ and $dL_4$ are desired levels of net investments needed to achieve $L_3$ and $L_4$, respectively. They may be different from actual productivity and new spheres investment (*IEK*, *IIK*) which together equal net intensive investment (*IK*):

$$IK = IEK + IIK \tag{27.14}$$

To compute the latent variables, various functional criteria were tested. The most significant estimation at step (1) was found to combine:

$$L_1(T) = f[GPDP(T), PCS(T-1)] \tag{27.15}$$

and

$$L_2(T) = f[YL(T), P(T)] \tag{27.16}$$

where $YL$ = labor productivity in period $T$ and $P$ = profit rate in period $T$.

The estimated equations are [after six iterations in the Wold procedure; six iterations were necessary to achieve practical convergence of coefficient, values and measures of statistical significance ($R^2$, $S$, $DW$, $F$, $t$-statistics). The criteria were set at no-change in the first 5 digits of the coefficients and measures. They were found to converge approximately simultaneously]:

$$L_1(T) = \underset{(23.58)}{1.2295\ 10^{-5}}\ GPDP - \underset{(8.16)}{1.9758^{-6}}\ PCS(T-1) \tag{27.17}$$

$$R^2 = 0.989575; \quad F = 4414; \quad DW = 0.336$$

$$L_2(T) = \underset{(47.69)}{0.01191}\ YL(T) + \underset{(4.69)}{0.0306}\ P(T) \tag{27.18}$$

$$R^2 = 0.98957; \quad F = 4415.1; \quad DW = 0.339$$

At step (2) the best results were represented by equation:

$$\underset{}{dPCS(T)} = \underset{(2.6)}{70887.72\ dL_1(T)} + \underset{(1.03)}{34008.82\ dL_2(T)} \tag{27.19}$$

$$R^2 = 0.22; \quad F = 13.9; \quad DW = 0.56$$

At step (3) the most significant functions were found to be:

$$L_3(T) = f\,[YL(T-1),\ YL(T-2)] \tag{27.20}$$

$$L_4(T) = f\,[P(T),\ P(T-2)] \tag{27.21}$$

The estimated equations (after six iterations) are:

$$L_3(T) = \underset{(2.156)}{0.0376\ \ YL(T-1)} - \underset{(1.511)}{0.0268\ \ YL(T-2)} \tag{27.22}$$

$$R^2 = 0.728; \quad F = 124; \quad DW = 0.352$$

$$L_4(T) = \underset{(4.576)}{0.1749\ \ P(T)} + \underset{(3.279)}{0.1264\ \ P(T-2)} \tag{27.23}$$

$$R^2 = 0.728; \quad F = 124; \quad DW = 0.317$$

At step (4) the best results were represented by equation:

$$IK(T) = \underset{(4.70)}{2576.3\ dL_3} + \underset{(9.47)}{4552.0\ dL_4} \tag{27.24}$$

$$R^2 = 0.713; \quad F = 115.2; \quad DW = 0.350$$

A few comments are in order on the signs of the estimated parameters. In the equation for $L_1$, we assumed that capital invested in existing technology would correlate positively with current levels of output and negatively with total capital stock lagged. The results of the estimation supported this expectation.

In the equation for $L_2$, one could expect productivity to have a negative sign on the assumption that low labor productivity would induce firms to invest in productivity-improving capital goods, while profits, with a positive sign, would give the financial means to do that. However, in the estimation both signs were positive. One has to be reminded that $L_2$ represents *total* intensive investment, which does not necessarily follow this logic. Productivity investment proper ($L_3$) was found to have both a positive correlation (with productivity lagged one period) and a negative (with productivity lagged two periods). This shows that the effects of growing productivity on productivity investment are contradictory. On the one hand, high productivity achieved by the same firm in the past acts as a constraint on new investment into capital goods improving productivity. On the other hand, one can argue that, in a highly competitive environment, high levels of productivity achieved by many firms would induce them to strive for even better results, so as to gain additional advantages.

Another contradiction is found in the equation for new spheres investment ($L_4$), which showed positive correlation with both current and lagged profit rates. This conforms with the assumption that adequate financing and improving business serve as conditions for large-scale expansion into new products and new branches of industry. However, it can also be argued that low or falling profit rates should induce at least some firms to risk undertaking innovations with the expectation of higher-than-average returns. In this case, as in the previous one, further investigation is necessary.

Time series were calculated for $EK$, $IEK$ and $IIK$. On the average, extensive investment leading to quantitative growth in existing technologies accounts for two-thirds of total net investment; investment meant to raise factor productivity accounts for 12%; and intensive investment in new commodities and spheres of output, 21%.

The last share may appear to be larger than expected. This is true only if such investment is limited to the introduction of new products. However, when one considers also capital flowing into new branches of industry, where new plants have to be built, not just old ones modernized, then the figure looks fairly reasonable. This would also include new technologies used for producing new, rather than old, commodities. These are, of course, intuitive judgments which depend on how one defines structural components of investment and their causal relationships with directly observed variables. This chapter is meant to suggest a basic principle of disaggregation and a method of estimation, rather than to reach definite conclusions as to actual values or proportions of investment components.

*Figures 27.5, 27.6* and *27.7* show deviations of these variables from their respective time trends. In general, they tend to follow the cyclical pattern of total net investment, with frequencies of approximately 30–40 years. In $EK$ long-term peaks occur in 1927 and 1971 (span of 44 years), but shorter-term fluctuations are also indicated by intermediate peaks in 1913 and 1953, and by

*Figure 27.5.* Extensive investment: deviation from second-order trend (moving average).

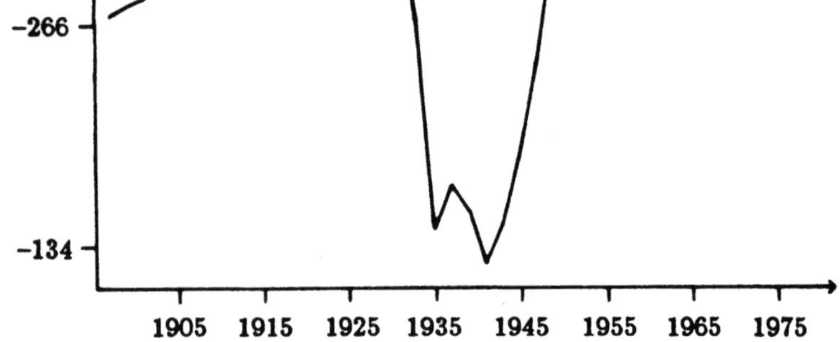


*Figure 27.6.* Intensive investment, type 1: deviation from second-order trend (moving average).

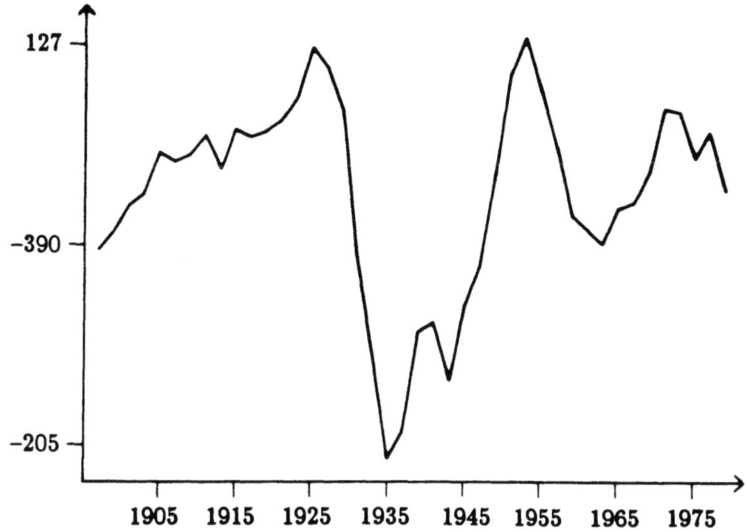


*Figure 27.7.* Intensive investment, type 2: deviation from second-order trend (moving average).

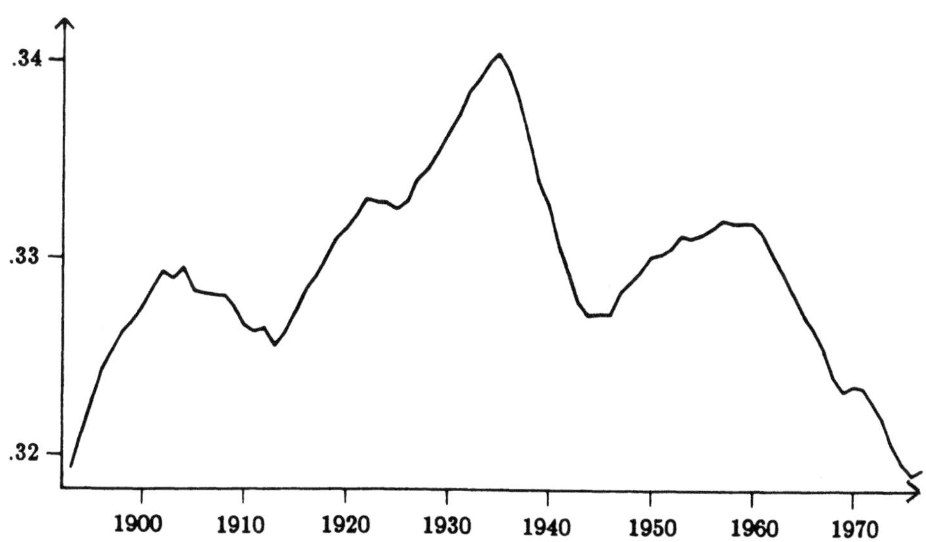


*Figure 27.8.* Intensive investment: share in total accumulated investment.

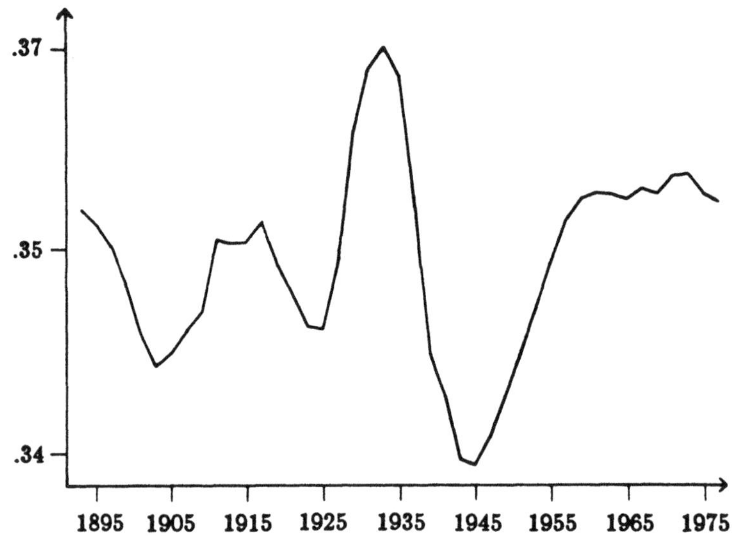


*Figure 27.9.* Productivity investment: share in intensive accumulated investment.

troughs in 1935 and 1965 (span of 30 years). Productivity investment (*IEK*) follows this general pattern with a lag of two to six years. Innovation investment (*IIK*), on the contrary, has a tendency to lead both in upswings and downswings.

It is difficult to directly assess the movement of the relative shares of the components in total net investment, since in downswings they tend to become negative – either all at once or one at a time. Therefore, at critical points the shares of *EK*, *IEK* and *IIK* tend to fluctuate wildly from positive to negative values and vice versa. A better indication of their long-term oscillations is given by *Figures 27.8, 27.9, 27.10*, showing the movement of relative shares of accumulated investment in their totals. Accumulated investment is defined as

$$EKK\ (T) = 0.676\ PCS\ (o) + \sum_{x=1}^{T} EK\ (x) \tag{27.25}$$

and, accordingly,

$$IEKK\ (T) = 0.117\ PCS(o) + \sum_{x=1}^{T} IEK\ (x) \tag{27.26}$$

$$IIKK\ (T) = 0.207\ PCS\ (o) + \sum_{x=1}^{T} IIK\ (x) \tag{27.27}$$

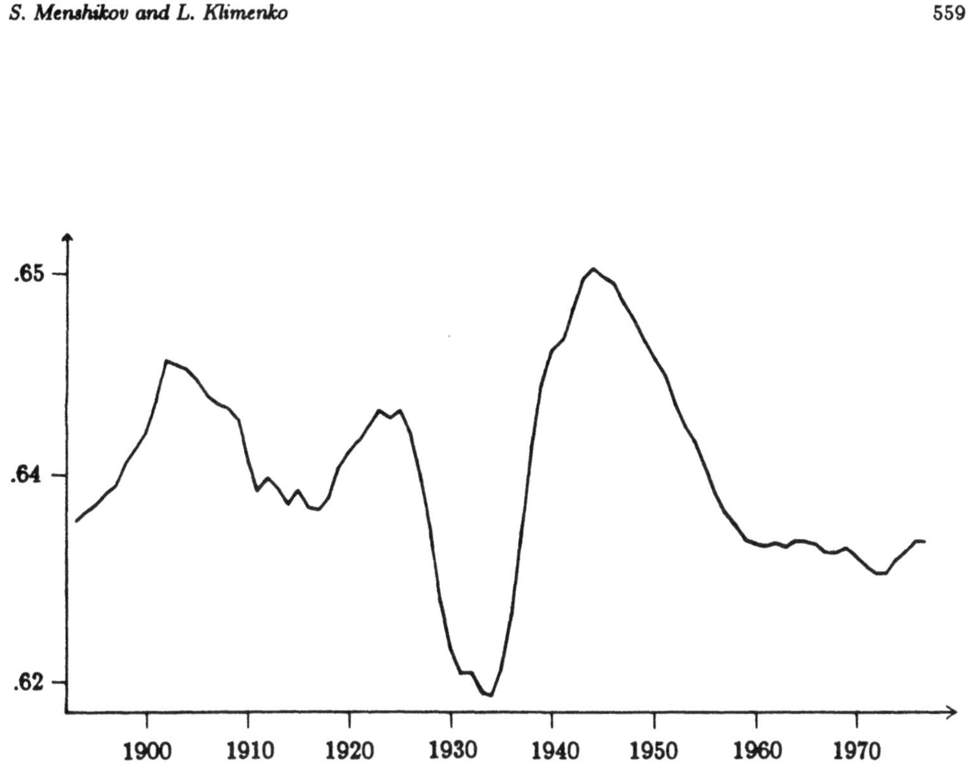


*Figure 27.10.* New spheres investment: share in intensive accumulated investment.

Shares in the starting period are assumed to be equal to their long-time averages.

The share of total accumulated intensive investment reaches local maxima in 1902, 1935, 1959, and minima in 1913, 1944, 1976. The average periodicity of these fluctuations is about 30 years. The share of *IK* grew in the long boom period which started in the mid-1890s, then fell in the later part of the upswing, and resumed its long rise from 1913 to 1935 as the long depression developed and gained force. Falling in the late 1930s and early 1940s, while extensive and overall investment expanded rapidly, the share of intensive investment then went on to rise in the late 1940s and early 1950s, as the new upswing gained momentum. It fell again in the 1960s and early 1970s at the late stage of the upswing. A new turnabout may have occurred around 1976, in the midst of the latest long recession.

This movement is similar to the 30-year interchange of intensive and extensive growth evident in indices of structural change in US manufacturing. The interchange is apparently caused by two upswings in intensive investment – one characteristic of the recessionary phase and serving to bring about its reversal, and the second associated with intensive growth at a certain stage of the long upswing. There are clear differences, though, between the behavior of intensive investment in the 1920–1930 and the 1970–1980 recessions. The upswing of intensive investment started much earlier in the 1920–1930 period.

The pattern of the two types of accumulated intensive investment within their total is very different. It is the share of productivity (not new spheres) investment that reached a maximum in 1934. During the whole period between

1903 and 1934, except for a temporary decline in 1917–1923, there was a sharp upturn associated with the advent of new factor-saving technology – the assembly line, Taylorism, Fordism. Productivity investment continued straight into the Great Depression, adding to mass unemployment of those years. In the late 1930s and during the war years the reintroduction of relatively cheap labor reduced the importance of labor-saving technology. It regained its importance in the postwar years as the new production technology spread across the board. The pause in the 1960s and early 1970s occurred at the later stage of the upswing, and did not yet reflect the new era of automation based on the microprocessor and robotics. As to the frequency of oscillations in productivity investment, there is a 42-year interval between the two major troughs of 1903 and 1945, and, similarly, a 40-year distance between the two major peaks of 1933 and 1973. Thus, productivity investment brings us closer to the long-term Kondratiev wave.

The share of new spheres investment tends to rise and fall with the general long-term upswings and downswings. This is not unexpected considering long-term oscillations in the profit rate. Massive expansion into new markets and spheres of production has to wait for any improvement in general business conditions in order to take off. This pattern is entirely different from productivity investment where long periods of low profits tend to generate active interest in new technology. Thus, one can single out productivity investment as the most active and leading investment component of structural change in the long wave.

## References

*Historical Statistics of the United States* (1975), Part 2, pp. 666, 669–680. (Washington, DC: US Bureau of the Census).

Menshikov, S. (1987), Structural Crisis as a Phase in Long-term Economic Fluctuations, pp. 66–75 in: T. Vasko (ed.), *The Long-Wave Debate*. (Berlin *et al.*: Springer-Verlag).

Menshikov, S. and Klimenko, L. (1985), On Long Waves in the Economy, in: G. Bianchi, G. Bruckmann, Y. Delbeke, T. Vasko (eds.) *Long Waves, Depression, and Innovation*. Proceedings of the Siena/Florence Meeting (26–30 October 1983). Collaborative Paper CP-85-9. (Laxenburg, Austria: International Institute for Applied Systems Analysis).

*Statistical Abstract of the United States* (1971), pp. 690–695; (1976), pp. 768–773; (1983), pp. 772–777; (1986), pp. 745–749.

Wold, H. (1985), Systems Analysis by Partial Least Squares, in: P. Nijkamp *et al.* (eds.), *Measuring the Unmeasurable*. (Dordrecht, Netherlands: Martinus Nijhoff).

CHAPTER 28

# The Emergence of Service Economics

*G. d'Alcantara*

## Summary

Structural change can be conceived as a process of quantitative economic growth or as a process of qualitative transformation. Services can be studied under both perspectives: service industries and standardized service processes, on the one hand, and creative and regulatory services, on the other hand. Economic regulatory mechanisms can be seen as the provision of services. Price regulations, contractual mechanisms, planning procedures or hierarchical organizations are regulatory systems with specific properties and fields of implementation. Beyond economic regulatory mechanisms, other mechanisms, such as psychosomatic, cultural or political regulation, are required to control the socialization process. The socialization process, which is a prerequisite for economic regulation, is produced by services. Services produce the cultural consensus and integrate the ambivalent, both violent and convivial, tendencies of individuals into competitive and cooperative social equilibria. Economics has to consider these regulatory functions in a theory of services.

## 28.1. Structural Change through Service

In economics, services used to be regarded as the appendages of goods. They were regarded as secondary functions ensuring appropriation, reproduction and exchange of material objects: distribution and transport of goods, services to goods-producing companies, education for the purpose of the production process. The very archaic way in which the quantity and price of services are identified in national accounting clearly illustrates this view. It reflects a lack of understanding of the nature of services and leads to numerous errors in economic analysis.

The fact that services now account for a large proportion (55% to 65%) of economic activity, as measured by employment or by the value of gross domestic product, has led to increased interest in basic issues of service economics. New technologies, mainly the combination of computer technologies and telecommunications, have helped to bring about deep structural changes in the industrialized economies and to increase the functions of services. New insights and a theory about services are now necessary.

Strictly speaking, growth is characterized by the increased production of industrial goods. The division of labor has increased the scale of elementary tasks and made possible the mass production of identical goods. The tasks and goods produced are standardized and reproduced in a repetitive way. This has led to quantitative and materialistic growth.

The increased production of services cannot be measured just in terms of growth, but rather as the generator of the transformation process of society itself. Services produce access to goods or other services; the regulation of personal, cultural, social, economic and political relationships; and the foundation and development of technologies, preference structures and modes of distribution of access. Service activity determines a qualitative transformation process of the nature of both objects and relationships.

So far, services have produced the behavioral patterns which enable industrial economies to reach high levels of standardized material growth. These behavioral patterns are produced not only by education or professional training, but also by services, such as religion or legal repression, which produce public morality and public order, and banking services which determine the criteria for credibility and worthiness and which issue the securities and financial assets necessary to get access to resources. Cultural conformity is necessary for the standardization of social and economic roles implied by the division of labor. It selectively represses individual creativity and differentiation and controls deviating behavior through social exclusion. The efficiency of many services is due to the fact that these cultural patterns result from the agents' self-regulation according to their belief systems and neuro-linguistic programs.

In industrial economies, cultural values are dominated by economic values symbolized by the criterion of quantitative growth. But these cultural values are produced by services. In service economies, instead of dividing labor into fixed, standardized and repetitive tasks, new technologies gradually transform the division of labor into an infinite differentiation of work. Telematics, robotics and new materials allow an infinite differentiation of product qualities and characteristics. It becomes possible to adapt them to each individual culture and person. Previously, most cultures had to adapt themselves to the constraints of mechanization. Now, technologies and production instruments can be reset so as to differentiate production according to very varied cultural patterns.

Two opposite developments coexist: on the one hand, standardization due to increased communication and accelerated diffusion of goods and services and, on the other hand, diversification due to the increased specialization and accelerated rhythm of innovation, invention and creation. When the number of specialized elements in a system increases, the number of relationships between these elements increases more than proportionally. The production of relationships requires services and has to be regulated through services.

The shift of employment to the service sector therefore follows two paths at one and the same time: on the one hand, the growth of service industries, and, on the other hand, the accelerated differentiation of tasks and their regulation, bringing about an overall qualitative transformation of society.

The first class of services is industrialized services. Traditional services are turned into standardized reproducible processes. Good examples here are the fast-food industries or communication network industries. The second class of services, such as guidance and advisory services for individuals, companies or public administrations, is infinitely diversified and made-to-measure in such a way that these services can be adapted to the preservation of specific identities and relational transformation. When these services have predominantly regulatory and fundamental functions, rather than access functions, they also have a dominant qualitative nature: the capacity to discover new fields, to explore society and the universe, to perceive and interpret unknown events and apparently random phenomena, to face uncertainty, to confront the unreliability of nature and human beings themselves. The added value of creativity is necessary for survival and can be achieved thanks to an increasingly subtle knowledge of society and the universe - both the exterior and the interior universe, the macro-cosmos and the micro-cosmos.

In this field of creativity, diversity guarantees cultural wealth. In the domain of cultural wealth, what is important is not what is useful and available in limited quantity, reproducible, exchangeable and appropriable, but rather what is unique in each human being, in each personal experience, in each culture, in each combination of an infinite number of characteristics. This can only be discovered in a climate of attention which is favorable to the free expression of differences between individuals, firms, regions and so on. The growing number of soft services present in a service economy help produce this type of climate. Creative processes, artistic production or research and development, go beyond the law of competition resulting from mimetic relationships. Their fundamental function precedes the need for regulation.

## 28.2. Economic Regulatory Mechanisms Seen as Services

Economic theory is mainly based on an objective conception of preferences and activities. The essence of services, whether seen as production or consumption, domestic or marketed, concerns processes of relationships, socialization, organization and information. In fact, the general theory of competitive equilibrium in economics implies underlying service systems. They have to be made explicit in the theory.

This dominant economic paradigm, founded on an accurate mathematical formula and applicable to market economies as well as planned economies, does not propose any distinction between goods and services. Both are commodities, produced by firms or consumed by consumers. Under optimizing behavioral assumptions it is shown that a Pareto optimal equilibrium exists which defines prices for produced, traded and consumed quantities. This model can be extended to include intertemporal behavior and uncertainty, without modifying the basic concepts. It is useful to realize that economics portrays an extremely powerful regulatory mechanism. This model, however, also needs to be seen as the description of the provision of a regulatory service. It has been steadily extended to solve regulatory problems at an economic level.

T. Anderson (1985), for example, distinguishes two kinds of regulatory mechanism: market-clearing, flexible-price regulation and nonmarket-clearing, rigid-price, rationing regulation. Beyond these price regulatory mechanisms we find hierarchical organizations.

### 28.2.1. Market-clearing or flexible-price regulations

When prices are sufficiently flexible, the interchange of access to one another's goods and services is agreed upon; in other words, demand and supply meet at *ex post* equilibrium prices. Auction theory, for example, studies various alternative processes through which these equilibrium prices can be reached. As such, the auction process constitutes a service activity. This process has to be seen as a regulatory mechanism. This means it regulates competitive access to private goods. A regulatory mechanism is equivalent to an information exchange procedure or a revelation mechanism. This is a fundamental theorem of information economics from which one must start in order to understand economic regulatory mechanisms.

Within the class of market-clearing regulation, several alternatives exist with respect to function and structures of the information streams which deviate from the perfect information model:

(1) The agents may be separated from one another in space: communication services can be formalized if one starts from information islands where different Walrasian auctioneers shift prices toward market equilibrium without having integrated the different spatial markets from the informational point of view.

(2) The agents are separated from one another over time: numerous information services are related to the fact that information from the past gets lost and information about the future is subject to the formulation of expectations. Many markets for future transactions do not exist, in such a way that future prices are lacking and one has to work with temporary equilibria. The flexible price mechanism is a remarkably efficient regulatory mechanism since, in the purely theoretical case, complete market information is reflected by prices. It does require a service to be implemented and additional services are required to regulate such cases as production economies of scale or externalities of consumption and production.

### 28.2.2. Nonmarket-clearing or fixed-price and rationing regulations

When prices are not flexible, because of contractual agreements or planning procedures, for example, market incentives and sanctions are insufficiently or indirectly perceived. Another regulatory mechanism is needed, since economic behavior is subjected only to weak external pressure, and markets do not tend automatically toward equilibrium. This is also the case with public goods and nonmarket activities, such as services, which are financed by subscriptions, contributions, donations, taxes or budgets, as in large organizations or public administrations.

When prices are fixed or preset, the explicit formalization of the information sets of the agents and information streams between the agents provides various regulatory mechanisms to solve the coordination problems between agents. We distinguish:

(1) *Anticipatory price setting*, which allows for small deviations with respect to the perfect information model; contract theory deals with asymmetric information structures; a principal agent requires a service from a firm without knowing the cost structure of the service, which is known to the firm. In such cases, for example, research and development services in new fields, the regulatory mechanisms have to deal with such problems as incentives, commitments, errors and uncertainty (adverse selection and moral hazard).

(2) *Centralized price setting*, which allows for public goods or natural monopolies. In the case of public transport or communication network monopolies, the tariffs fixed are not a free reflection of marginal costs or supply-and-demand conditions, but rather reflect a political consensus about the networks supplied, the distribution among users of access to these networks, the rules which determine priorities in waiting lists, etc. Various public goods allocation mechanisms have been developed in the literature which guarantee the desired distribution of access, the information processing required and the compatibility of economic agents' natural incentives. In practice, many public goods are regulated on the basis of their private complements. The state can impose something private (an identity card or a ticket, for example) as a condition of access to the public good or to the national territory. The nonexcludability principle characterizing public goods can also be compensated by regulations, threats and controls produced by the state.

### 28.2.3. Hierarchical organizations

Incentives, commitments and economic behavior are the results of a socialization process which requires a large number of services. The properties of price regulatory mechanisms rely on such social behavior, which cannot be taken for granted. Direct control through collection of information and supervision may be needed to solve the coordination problem when a price system is not sufficient to induce a competitive equilibrium. The theory of hierarchical organizations provides a regulatory mechanism or strategies to deal with human fallibility, random errors of the first and the second type, and various forms of moral hazard, such as shirking, threats to specific asset or free-riding. Shirking consists in a team member's failure to keep agreements about working programs. This endangers the team's income without causing a comparable loss to the shirker. Similarly, specific assets have a value which depends on collaboration or cooperation with another person (agreements on technological standards, quality norms and reputation linked to labels or brands). An interruption in programmed cooperation causes considerable marginal costs. Free-riding is a rational strategy for an agent who can use a public good without bearing its costs.

These three cases require regulatory devices, such as centralized property, so that all parties share the risks of joint profits or losses, and direct supervision, which can enforce cooperative behavior because deviations from agreed transactions and programs will be easily detected.

Beyond these mechanisms, which are of an economic nature, there are other regulatory mechanisms which are part of the socialization process and which are necessary for the functioning of any economic system: psychosomatic regulatory modes, cultural regulation, including "habitus" and linguistic differentiation [see Bourdieu (1982)], political regulation, including the legal and repressive order of the state [see Enriques (1983)], which has the legitimate monopoly of and over violence. Psychosomatic, cultural and political regulation are necessary conditions for the existence of economic regulation.

## 28.3. A Starting Point for a Systems Theory of Services

The competitive equilibrium paradigm has not yet integrated (assuming it can do so) those activities which are prerequisites for establishing the existence of the competitive equilibrium itself. They simply cannot be included as goods of the utility functions and production functions. To identify a competitive equilibrium, one has to presuppose and therefore establish the existence, identity and conditions of objects and agents and to guarantee a socialization process – in other words, a cultural consensus.

By cultural consensus we mean the fundamental mechanism that establishes the existence, stability and consistency of society, from the local to the global level. This consensus may be conscious or unconscious. It exists whenever a social group is able to survive. Survival, however, requires not only that technologies and resources produce primary necessities, but also the control of self-destructive processes which, as the anthropologist R. Girard (1978) has pointed out, threaten any society. The control of self-destructive processes on the individual and the collective level, is an essential function of services. This service function is not regulated by the competitive equilibrium mechanism, since the former is a prerequisite for the latter.

Economic theory has to be made compatible with the fundamental anthropological findings of R. Girard. Authors such as M. Aglietta and A. Orlean (1982) or P. Dumouchel and J.P. Dupuy (1979, 1983) have already embarked on this task. We propose to use these ideas to construct a theory of services, by increasing the integration of economics and of human sciences through a systems approach.

Thomas More (1516), who, long before Adam Smith, perceived the nation as an economic order rather than as the moral order of the Platonic tradition, reminded us that "servire" is related to two different Latin words:

(1) "Servire", which means to be someone's slave.
(2) "Inservire", which means to be useful to someone.

What is said here is that a service is not an object, but a relationship. A relationship is the involvement of the lives of two persons, or groups of persons, with one another. Like life itself, which is $\beta\iota\alpha$ (violence) and $\beta\iota o\varsigma$ (animation), service carries the ambivalence of violence and conviviality, competition and cooperation. This is due to the structure of relationships and desire, defined in a Freudian perspective.

As shown by R. Girard, this structure is inherently two-sided because it is generated by the dominant mimetic patterns of human behavior and evolution. The mimetic relationship turns the mimetic model into a mimetic rival. This happens when the mimesis – the mimetic interaction – is triggered by the model's move to appropriate or have access to an object. This configuration contains the essential violence of any relationship. Any desire is generated as the desire for the object of someone else's desire and is therefore violent by nature. This is the fundamental socialization problem which has to be solved by a regulatory mechanism. The origin and founding moment of this mechanism provides the basis for a social consensus against the auto-destructive process of reciprocal mimetic violence. This founding moment is the starting point of a self-enforcing regulatory mechanism. The origin of collective survival, as argued by R. Girard, is generated by a violent consensus against a projected cause of violence: a scapegoat. The external projection and exclusion of this cause restores unanimity and peace. At the pivot of the mechanism, brute violence is transformed qualitatively into more subtle forms of control, such as mythical stories, ritual practices or modern institutions and ideologies.

The term "violence" is used here in a generic sense. Violence can take different forms, such as military violence; terrorist or criminal violence; the political struggle between states, social classes and interest groups; economic competition and cultural segregation. This violence is transformed according to the regulatory systems. Culture operates the most refined forms of regulation through, for example, sports championships, beauty competitions, artistic prizes or academic debates between scientists, etc. Language is the most powerful cultural regulator. Through language, services extend their regulatory function within the human unconscious. Services have such a cultural dimension and are therefore fundamental for the survival of human communities. They recall the "violent" nature of unanimity and the permanent risk that human rivalry will degenerate into brute violence. They are also cultural because of their conventional and arbitrary characteristics. These characteristics, however, become essential to the cultural identity of a human community which defines itself in a fuzzy way as the reference group of its members.

The socialization process comprises, on the one hand, the foundation of a regulatory mechanism which controls access to objects and, on the other hand, the development of the life-cycle of this regulatory mechanism. After such a regulatory mechanism has proved its qualities, it diffuses itself and then progressively loses its power to regulate relationships because it itself becomes an object of desire, competition and violence.

The mimetic complex hypothesis formulated by R. Girard leads to a formalization of the inner structure of the human psyche and the socialization process required for any economic order. It determines the structure of preferences and the selection of goods included in its scope. It provides the basis for a theory connecting individual and collective behavior. It provides a fundamental structure for the essence of services. This approach to services, within and on top of the competitive equilibrium framework, provides a methodology for the solution of problems of social cohesion, such as those, left open by Harsanyi (1977), of dominant loyalties, reference group boundary or anti-social attitudes.

In service industries, services have become increasingly tradeable on the national and international level because of economies of scale and cost-reducing new technologies in the transmission, storage and treatment of information. This trend explains the pressure for the liberalization of international service activities, as reflected by GATT's trade negotiation agenda.

For regulating services, however, especially services with dominant regulatory functions, tradeability is linked to numerous externalities and spill-overs. One regulatory system may be neutralizing, contradictory and destructive for another system of rules. The external effects of such service systems have cultural consequences. Market-service imports can make traditional regulatory modes obsolete and result in indirect costs, such as increased criminality or political violence. This justifies various levels of regulatory protection, up to the protection of community languages as instruments for the preservation of identity.

## 28.4. Conclusion

The economic system as described by the competitive equilibrium paradigm is characterized by an increasing fraction of service activities which have dominant regulatory and creative functions. These functions cannot be seen as inert arguments of utility functions and production functions. They contribute to producing the economic system itself: the identity of the agents, the form of the utility and production functions, the nature of the relationships in this system.

Construction of a theory of services is far beyond the ability of an isolated researcher. However, concepts from various disciplines can be gathered to form a framework and to deal with the issue. Recent developments in economic theory aiming at the integration of micro- and macroeconomics can be combined with the new bridges between psychology and sociology contained in Girardian theory. The balance of both the inner individual psyches and social groups is regulated by interrelated "mechanisms". The study of the interactions between such mechanisms and the dynamic systems in which they are embedded would identify both the field and the method for a theory of services.

## References

Aglietta, M. and Orlean, A. (1982), *La violence de la monnaie.* (Paris: Presses Universitaires de France).

Andersen, T.M. (1985), Allocation under Differential Information – Flexible and Fixed Prices, doctoral thesis, Catholic University of Louvain.

Arrow, K.J. and Debreu, G. (1954), Existence of an Equilibrium for a Competitive Economy, *Econometrica*, **22**, pp. 265–290.

Baudrillard, J. (1968), *Le système des objects.* (Paris: Gallimard).

Bourdieu, P. (1980), *Le sens pratique.* (Paris: Editions de Minuit).

Bourdieu, P. (1982), *Ce que parler veut dire* [L'économie des échanges linguistiques]. (Paris: Fayard).

Dumouchel, P. (1985), *Violence et vérité autour de René Girard* (Colloque de Cérisy). (Paris: Bernard Grasset).

Dumouchel, P. and Dupuy, J.P. (1983), *L'auto-organisation de la physique au politique.* (Colloque de Cérisy). (Paris: Editions du Seuil).

Dupuy, J.P. (1982), *Ordres et desordres. Enquête sur un nouveau paradigme.* (Paris: Editions du Seuil).

Dupuy, J.P. and Dumouchel, P. (1979), *L'enfer des choses. René Girard et la logique de l'économie.* (Paris: Editions du Seuil).

Enriques, E. (1983), *De la horde à l'état. Essai de psychanalyse du lien social.* (Paris: Gallimard).

Girard, R. (1978), *Des choses cachées depuis la fondation du monde* (recherches avec J.M. Oughourlian et G. Lefort). (Paris: Bernard Grasset).

Harsanyi, J. (1977), *Rational Behavior and Bargaining Equilibrium in Games and Social Situations.* (New York and London: Cambridge University Press).

Laffont, J.J. (1977), *Effets externes et théorie économique* (monographies du seminaire d'économétrie). (Paris: CNRS).

Laffont, J.J. (1985), *Procurement of Services with Incentives.* (Leuven: European Seminar on Productivity in the Services).

Perroux, F. (1975), *Unités actives et mathémathiques nouvelles. Revision de la théorie de l'équilibre économique général.* (Paris–Bruxelles–Montreal: Bordas).

Prigogine, I. and Stengers, I. (1984), *Order out of Chaos. Man's New Dialogue with Nature.* (Toronto–New York–London–Sydney: Bantam Books).

Selten, R. (1985), *Bureaucratic Budget Bargaining as a Game of Incomplete Information.* (Leuven: European Seminar on Productivity in the Services).

CHAPTER 29

# Economic Structural Changes: The Problems of Forecasting

*Iouri Tchijov*


## Summary

An accuracy analysis of several long-term forecasts of the US economy was performed from the structural point of view. More disaggrated industrial levels usually provide less correct forecasts of structural changes. The main reason for this is the lack of explicit accounting for technological progress in models used for prediction. Thus, the problems of structural change forecasting plague the technological as well as the industrial level.


## 29.1. Introduction

There are two general approaches to the modeling and forecasting of structural changes which are known at the present time. The first one is based on moving down from the macrolevel to the industrial level [Levitsky *et al.* (1975); Preston (1972)]. The problems investigators come across here are connected with the mechanism of transition – the development of the disaggregation matrix moving from the macroeconomic level to the industrial one. Practically nobody reaches a technological level in this way.

The second approach is based on moving up – from the technological level through the industrial level to the macroeconomic level [Leontief and Duchin (1986)]. The main problems in this procedure are connected with the estimation of economic parameters of technological processes such as profitability, based on competitive advantages, the rate of risk, potential penetration, etc. Another problem is the estimation of economic system elasticities with respect to technological changes, such as the influence of a new technology on input–output coefficients or labor demand. Practically, the direct impact of any technology on macroeconomic variables is much lower than in the model used for accuracy estimation.

The first approach has a longer history; a lot of structural forecasts are available now, and some of them are verifiable. That is why we shall try to analyze forecasting accuracy from the viewpoint of prediction of economic structural changes.

## 29.2. GNP Structure

The main widespread approach to structural disaggregation at the macrolevel is the division of GNP into its main components – consumption, investment, government purchases and foreign trade balance. The accuracy analysis of several long-term forecasts of the US GNP and its components for the 1960s and 1970s [Tchijov (1981)] showed an underestimation of average rates of growth for the first period as well as an overestimation for the second period. The main reason for the relatively low accuracy was that real technological progress was not predicted correctly.

The results of a long-term forecast, developed at Syracuse University (as shown in *Table 29.1*), demonstrate that in spite of a very wide range of alternatives, the real dynamics were usually close to either a highest variant (in the case of the labor force) or to a most pessimistic one (in the case of population and GNP).

*Table 29.1.* Forecasted versus real changes in long-term economic growth, USA, 1960–1980 (in %).[a]

| | *Forecast variant* | | | |
|---|---|---|---|---|
| *Variable* | *Low* | *Medium* | *High* | *Real*[b] |
| Population | +26 | +36 | +55 | +26 |
| Participation rate | –0.8 | –0.2 | –0.5 | +4.4 |
| Labor force | +34 | +40 | +49 | +52 |
| Labor productivity | +42 | +51 | +67 | +26 |
| GNP in constant prices | +92 | +110 | +125 | +91 |

[a]Syracuse University Research Corporation (1964).
[b]Council of Economic Advisers (1986).

The fact that reality follows the lowest variant in the population case and the highest one in the labor force case is explained by the completely wrong forecast of the participation rate. Similarly, the comparison of the GNP forecast with that of the labor force reflects a wrong prediction (overestimation) of labor productivity dynamics – the main variable reflecting technological progress.

From the structural viewpoint (see *Table 29.2*), incorrect directions of share changes were predicted in the cases of personal consumption, government purchases, and construction. This means that in the cases of such long-term forecasts, the macrostructure of GNP is not quite accurately predicted in spite of a rather high stability of the macroshares. This is especially true for forecasts developed without the use of macroeconomic models, such as input–output models.

*Table 29.2.* Forecasted versus real GNP structure (in constant prices), USA, 1960–1980 (in %).[a]

| *Component of GNP* | *1960 (real)* | *1980 (forecast)* | *1980 (real)*[b] |
|---|---|---|---|
| Personal consumption | 65.3 | 62.3 | 66.8 |
| Investments | 14.3 | 16.0 | 15.2 |
| Government purchases | 19.8 | 22.6 | 15.6 |
| Durable goods | 19.0 | 21.6 | 22.9 |
| Nondurable goods | 32.1 | 27.6 | 26.2 |
| Construction | 11.3 | 12.3 | 8.9 |
| Services | 37.5 | 39.2 | 42.6 |

[a]Syracuse University Research Corporation (1964).
[b]Council of Economic Advisers (1986).

The forecast made by C. Almon *et al.* (1985) was almost perfect as far as the direction of changes was concerned. He predicted the directions correctly for all GNP components by using his famous input–output model. But the real dimension of the changes was underestimated in all cases. It reflects the inertia features of the models of such a type (see *Table 29.3*)

*Table 29.3.* Almon's forecasted versus real GNP structure, USA, 1971–1985 (in %).[a]

| *Component of GNP* | *1971*[b] *(real)* | *1985*[b] *(forecast)* | *1985*[c] *(real)* |
|---|---|---|---|
| Personal consumption: | 63.3 | 64.6 | 65.5 |
| durables | 9.9 | 10.5 | 11.5 |
| nondurables | 26.4 | 25.9 | 24.0 |
| services | 27.0 | 28.0 | 30.0 |
| Gross private domestic investments: | 14.5 | 15.9 | 17.0 |
| equipment | 6.4 | 8.5 | 9.6 |
| structures | 7.7 | 6.8 | 7.4 |
| Exports | 6.3 | 7.4 | 8.5 |
| Imports | –6.2 | –6.7 | –9.9 |
| Government purchases | 22.2 | 18.8 | 19.1 |

[a]Almon *et al.* (1985, p. E6).
[b]1971 prices.
[c]1972 prices; see Council of Economic Advisers (1986).

The examples can be continued, and the analysis of a wider range of input–output forecasts permits the conclusion that the forecasts of GNP structure were closer to its previous values (inertial forecasts) than to its real changes.

## 29.3. Production Structure

Moving down from the macroeconomic level, we come upon the production structure which relates to the industrial disaggregation of the total product (or GNP), and in the first place to two-digit industries.

The smooth curves in *Figure 29.1* show the long-term dynamics in the manufacturing structure and, moreover, the tendency toward structural stabilization in consumer-oriented industries (SIC 20–27). Even in high-tech machinery (SIC 35–37) such a stabilization becomes evident. This means that technological progress is hidden more deeply than at the two-digit level. Though the share of electrical machinery was stable in US manufacturing, inside the industry there was a dramatic shift from traditional electrical equipment to electronic components. The share of industrial electrical equipment and lighting fixtures decreased from 21% of this industry's (SIC 36) shipments in 1972 to 13% in 1982. At the same time the share of semiconductor devices and electronic components increased from 19% up to 53% [US Department of Commerce (1986)]. A dramatic decrease in metal-cutting equipment took place with a simultaneous growth of NC – metal-cutting machines.

One of the detailed forecasts of the production structure for the US economy was developed by C. Almon. We have chosen this forecast for our analysis not because of its accuracy (though it may have possibly been one of the best among such disaggregated forecasts), but because of its complete availability in book form [Almon *et al.* (1985)].

The macroindustry forecast (see *Table 29.4*) shows, as in the case of GNP structure, the right directions of the changes, but it is completely wrong from the viewpoint of the size of change. The huge decrease of the 33rd and 34th industries' shares was not predicted nor was the dramatic increase of the 36th industry's share (which was growing due to its "electronic part"). The most accurate forecast was made for the 35th industry – nonelectrical machinery, but inside that industry the real structural changes were forecasted incorrectly (see *Table 29.5*).

For the traditional four-digit industries, the directions of changes were forecasted incorrectly. Instead of the predicted growth, the shares of the mentioned industries (turbines, metal-cutting and metal-forming machines, etc.) decreased markedly. These industries were replaced by computing equipment production within the nonelectrical machinery two-digit industry.

This review of structural forecasts permits us to draw certain common conclusions:

(1) Macroeconomic structures are rather stable; that is why their forecasts are usually close to reality, though cyclical oscillations may sometimes cause disturbances around a trend.
(2) It is extremely difficult to follow up the impact of modeled technological progress on macroeconomic structures by using macrolevel structural models because of a lack of explicit technological parameters in such models.

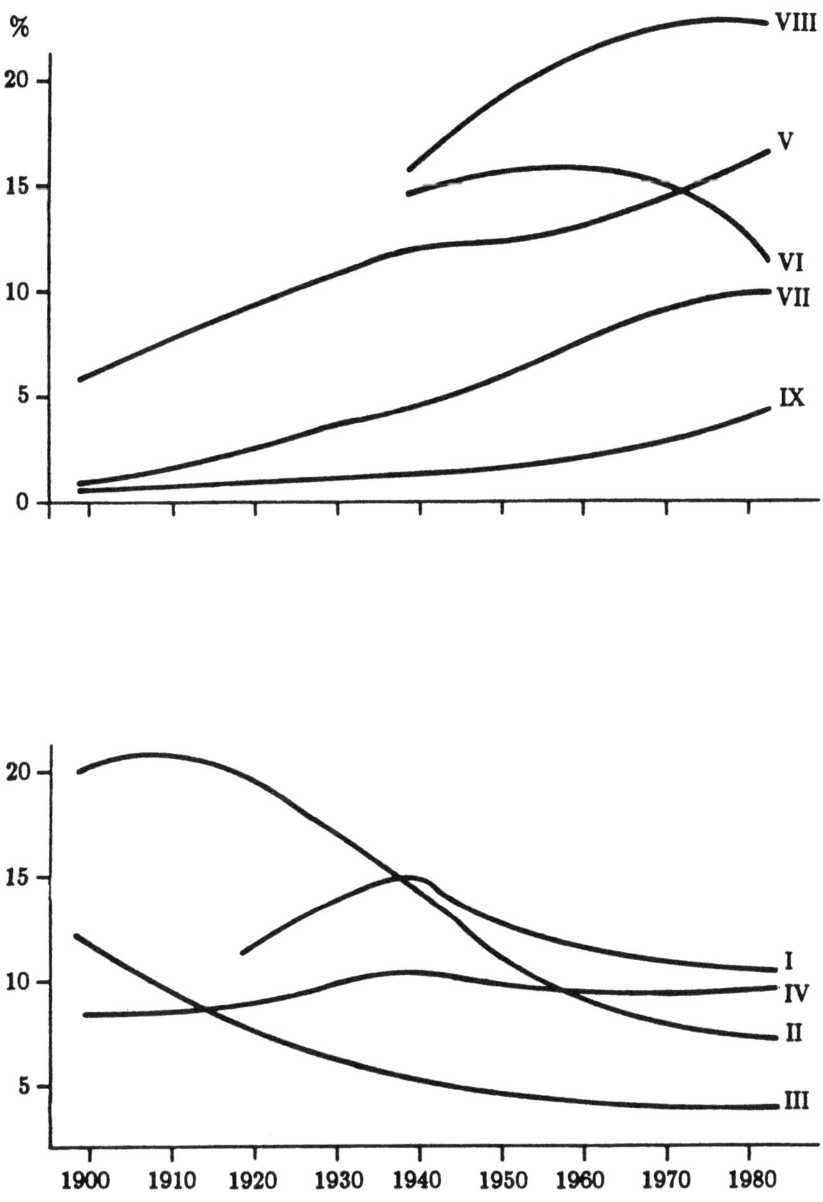


*Figure 29.1.* The long-term dynamics of the shares of two-digit US manufacturing industries (value added). I. food and tobacco products (SIC 20+21); II. textile, apparel and leather products (22+23+31); III. lumber products and furniture (24+25); IV. paper products and printing (26+27); V. chemical, petroleum and rubber products (28+29+30); VI. primary metals and fabricated metal products (33+34); VII. nonelectrical machinery and transportation equipment (35+37); VIII. electrical machinery (36); IX. instruments (38).

*Table 29.4.* Almon's forecasted versus real structure of the US metal working industry, shipments, 1971–1985 (in %).[a]

| SIC | Industry | 1971[b] (real) | 1985[b] (forecast) | 1985[c] (real) |
|---|---|---|---|---|
| 33 | Primary metals | 17.2 | 13.2 | 5.6 |
| 34 | Fabricated metal products | 13.8 | 13.0 | 4.7 |
| 35 | Nonelectrical machinery | 18.5 | 21.6 | 21.2 |
| 36 | Electrical machinery | 15.9 | 17.8 | 38.8 |
| 37 | Transportation equipment | 30.3 | 28.9 | 23.4 |
| 38 | Instruments | 4.3 | 5.5 | 6.2 |
| | Total shipments, billion $ | 300 | 524 | 516 |

[a] Almon *et al.* (1985, pp. E26–E27).
[b] 1971 prices.
[c] 1972 prices; see US Department of Commerce (1986).

*Table 29.5.* Almon's forecasted versus real structure of the US nonelectrical machinery industry, shipments, 1971–1985 (in %).[a]

| SIC | Industry | 1971[b] (real) | 1985[b] (forecast) | 1985[c] (real) |
|---|---|---|---|---|
| 3511, 3519 | Turbine generator sets, diesel/other engines | 1.72 | 2.13 | 0.96 |
| 3531–3533 | Construction, mining and oilfield machinery | 2.19 | 2.19 | 1.43 |
| 3541 | Metal-cutting machines | 0.40 | 0.53 | 0.20 |
| 3542 | Metal-forming machines | 0.21 | 0.22 | 0.08 |
| 3561, 3564 | Pumps, blowers and fans | 1.01 | 1.09 | 0.64 |
| 3562 | Ball and roller bearings | 0.43 | 0.52 | 0.25 |
| | Total | 5.96 | 6.68 | 3.56 |

[a] Almon *et al.* (1985, pp. E27–E28).
[b] 1971 prices.
[c] 1972 prices; US Department of Commerce (1986).

(3) The attempts to forecast industrial structures at a four-digit level were not successful either, because of the absence of endogenous technological progress in the industrial relationships even at this level.

## 29.4. From the Technological Level to the National Economy

Now there is only one known way for incorporating modern technological progress into model systems, i.e., to describe the latest technologies in economic terms. The first-order difficulties in this process are connected with the lack of necessary data on production, consumption, profitability, capital/labor savings, and unit costs of new technologies. The second-order problems are due to the low reliability of the estimation of a potential penetration rate, though some results have been published for such a case as robotics [Ayres and Miller (1983)]. Moreover, even the statistical information published in several public and special issues is contradictory because of different definitions and calculation methods.

It is quite obvious that there are a lot of bottlenecks in technological progress forecasting, based on estimates of concrete new technologies (NT). The first problem is the preliminary estimation of NT economic parameters and their comparison with the parameters of traditional technologies. The second one is the methodology of the development and penetration (or diffusion rate) forecasting. Well-known estimates for long-term tendencies of technological substitution, usually based on Mansfield's approach (1986), do not take the different economic aspects into account. The latter are very important, especially for the expansion phase, and sometimes also for the embryonic phase.

For example, during the embryonic phase [up to 8–10% of the new technology's share, according to our estimates; see Tchijov (1987)], the NT share usually declines in recession periods and increases during the expansion phase. The technology share after saturation usually does not decrease to zero because of the differences in the products' specific features provided by different technologies.

The importance of taking economic conditions into account is further underlined when we estimate the saving possibilities to provide the adequate investments. Practically, a majority of amortization funds, as well as part of net investments, are used for traditional technologies reproduction. This means that only a small share of savings can be used for investments into new technologies; and this is why the investment climate is an important determinant for NT diffusion. It is well known that NT demand much heavier expenditures at the initial stages of their development than the traditional technologies. This means that two economic conditions are necessary to accept NT: a strong financial position of the innovative firm and future growth of demand for the product produced when using NT.

But the most important thing in NT penetration is the relative NT costs compared with the recognized costs of traditional technologies. In this calculation the relative importance of different cost elements (capital, labor, material, including materials, and energy) has to be taken into consideration.

When we deal with new technologies, it is necessary to provide different approaches to the investigation of their diffusion, depending on the phase of their life cycle. During the embryonic (or childhood) phase, the traditional determinants of NT share dynamics, such as the rate of profitability or relative cost, are not as important as during the expansion (or adolescence) phase. For example, robot production is not yet profitable in the United States [US Department of Commerce (1986]: their price equals approximately 50% of the production cost. Nevertheless, an accelerating diffusion of robots is taking place. Negative profitability of the embryonic technologies is rather typical, and this is why alternative explanatory variables should be used for embryonic phase forecasting.

From our viewpoint the most productive way of structural change forecasting, taking into account technological progress, is a combination of the two approaches mentioned above. The principal scheme of such a combination is shown in *Figure 29.2*. Of course, some important details, such as industrial

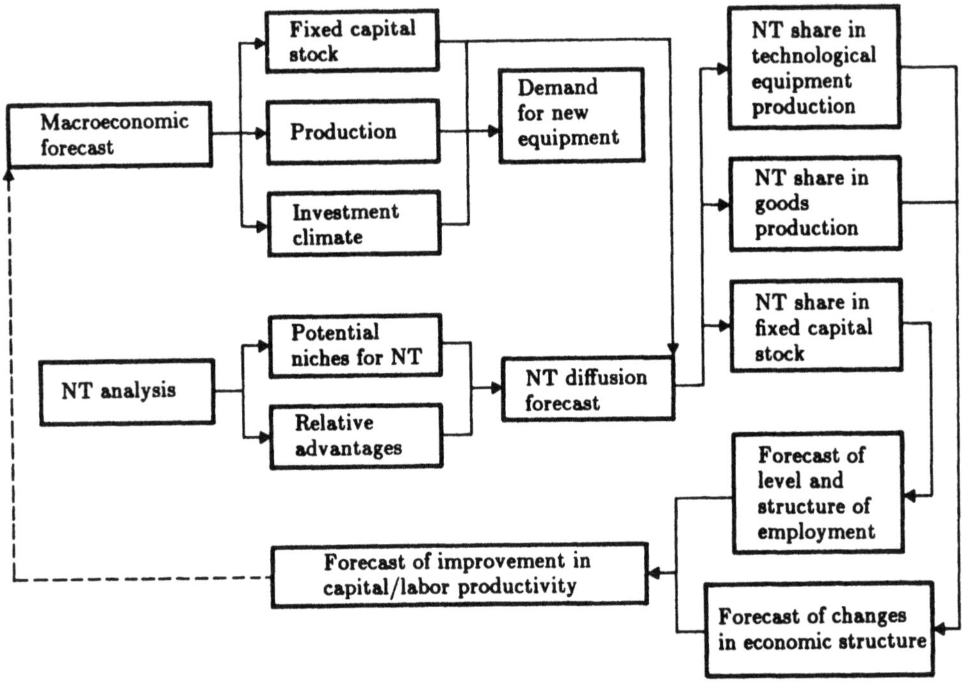


*Figure 29.2.* Scheme of combination of macroeconomic and technological approaches.

disaggregation, capital/labor ratio, etc., are missing here. But this combined approach provides the possibility of describing technological progress explicitly and taking into account interrelationships between technological progress, economic growth and structural changes.

## References

Almon, C. *et al.* (1985), *Interindustry Forecasts of the American Economy*, 2nd edition. (Lexington, MA: Heath–Lexington).

Ayres, R. and Miller, S. (1983), *Robotics: Applications and Social Implications.* (Cambridge, MA: MIT Press).

Council of Economic Advisors (1986), *Economic Report of the President.* (Washington, DC: US Government Printing Office).

Levitsky, E., Menshikov, S., and Tchijov, I. (1975), *American Economy Modeling.* Novosibirsk [in Russian].

Leontief, W. and Duchin, F. (1986), *The Future Impact of Automation on Workers.* (New York: Oxford University Press).

Mansfield, E. (1986), *The Economics of Technological Change.* (New York: Norton & Co.).

Preston, R. (1972), *The Wharton Annual and Industry Forecasting Model.* (Philadelphia, PA: Wharton School of Business, University of Pennsylvania).

Tchijov, I. (1981), The Problems of Long-Term Forecasting of a Market Economy by Using Macroeconomic Models. In: A. Bakaev and I. Tchijov (eds.), *The Problems of Capitalist Economy Modeling*. Novosibirsk [in Russian].

Tchijov, I. (1987), *The Cyclical Dynamics of Diffusion Rates*. IIASA Working Paper WP-87-14. (Laxenburg, Austria: International Institute for Applied Systems Analysis).

US Department of Commerce (1986), *US Industrial Outlook*. (Washington, DC: US Government Printing Office).

Syracuse University Research Corporation (1964), *The United States and the World in the 1985 Era*. (Syracuse, NY: SURC).

CHAPTER 30

# Comparison of Gross Domestic Product and Net Material Product

*Jaroslav Češka*
*(with an appendix by Rumen Dobrinsky)*

## Summary

This study deals with the relationship between the corresponding important production aggregates used in the System of National Accounts (for market economies) and the System of Balances of National Economy (for nonmarket economies). This comparison of the gross domestic product (GDP) and the national income (NMP) is based on the concept of the economic production sphere and definitions of the relevant aggregates of both systems of national accounting. The results of illustrative calculations and other matters of relevance to the GDP and NMP comparison are discussed.

## 30.1. Introduction

The study of the relationship between the System of National Accounts (SNA) used in countries with market economies and the System of Balances of National Economy (MPS) used in countries with centrally planned economies is needed for any authoritative comparative analysis of the corresponding national aggregates of these two systems, for the adjustments that need to be made in their comparison and mutual conversion and for many other specific purposes.

Both SNA and MPS have in the course of their development served many purposes, among which functioning as a program for international reporting of comparable national aggregated data is obviously the most important. As the differences between SNA and MPS are numerous and diverse, this intersystem comparison is limited to the relevant main production aggregates and related issues. This is preceded by a short description of both systems.

## 30.2. Structure and Main Characteristics of SNA

### 30.2.1. Matrix representation

The structure, main characteristics and the relationship of the aggregates of SNA, which provides a comprehensive framework for a systematic and integrated recording of transaction flows in an economy, is usually described by reference to a matrix representing the system as a whole, to a set of standard accounts and tables, and to the applied concepts, definitions and classifications. The following short description follows the United Nations System of National Accounts. The matrix provides a symbolic presentation of the SNA system. An example of the simplified matrix is set out in *Table 30.1*.

In the above-mentioned matrix the headings of the columns (not shown) are the same as the headings for the rows. Only the symbol $T$ is used. The first subscript relates to the row; the second, to the column. The matrix may be viewed as a set of double entry accounts presenting the different flows in the economy in a comprehensive fashion. Each row and the corresponding column in the matrix constitute a separate account: the row represents the receipts or incomings of the accounts; and the columns, the expenditure or outgoings. In the system itself each of the headings of the rows and columns is further subdivided according to appropriate classification.

The accounting structure of the United Nations System of National Accounts covers the production accounts, which are further subdivided into the commodity accounts, the accounts for other goods and services and the activity accounts, the income and outlays accounts, the capital finance accounts and the external transactions account.

For each group of domestic transactions, a production, income and outlay and capital finance account is distinguished. The production accounts are presented for transaction groups separately and in consolidated form for the domestic economy as a whole. In consolidated form they are called the accounts for gross domestic product and expenditure, the account for national disposable income and its appropriation and the capital finance account, respectively. In addition to the accounts for the domestic sector, there is a single account covering transactions with the rest of the world – the external transactions account.

The contents of the accounts can be specified in terms of matrix elements. For the *commodity accounts*, which correspond to rows and columns 1 and 2, the supply, shown in the columns, consists of commodities produced by domestic units ($T_{3.1}$, $T_{3.2}$, $T_{4.1}$, $T_{4.2}$, $T_{5.1}$, $T_{5.2}$, $T_{6.1}$ and $T_{6.2}$) and commodities imported ($T_{20.1}$, $T_{20.2}$). Import duties ($T_{10.1}$ and $T_{10.2}$) presented as a component of value are added to this.

### 30.2.2. The main aggregates and their definitions

The definitions are limited to gross output, intermediate consumption and gross domestic product.

*Table 30.1.* SNA matrix. (Note: For technical reasons, columns 8–20 are omitted.)

| *Accounts* | *T* | 1 | 2 | 3 | 4 | 5 | 6 | 7 | ... | 20 |
|---|---|---|---|---|---|---|---|---|---|---|
| PRODUCTION | | | | | | | | | | |
| Commodities (incl. taxes): | | | | | | | | | | |
| goods and material services | 1 | | | $T_{1.3}$ | $T_{1.4}$ | $T_{1.5}$ | $T_{1.6}$ | $T_{1.7}$ | | |
| nonmaterial services | 2 | | | $T_{2.3}$ | $T_{2.4}$ | $T_{2.5}$ | $T_{2.6}$ | $T_{2.7}$ | | |
| Activities: | | | | | | | | | | |
| industries (material | 3 | $T_{3.1}$ | $T_{3.2}$ | | | | | | | |
| industries (nonmaterial) | 4 | $T_{4.1}$ | $T_{4.2}$ | | | | | | | |
| producers of government services | 5 | $T_{5.1}$ | $T_{5.2}$ | | | | | $T_{5.7}$ | | |
| private services | 6 | $T_{6.1}$ | $T_{6.2}$ | | | | | $T_{6.7}$ | | |
| CONSUMPTION | | | | | | | | | | |
| Expenditures: | | | | | | | | | | |
| household goods and services | 7 | | | | | | | | | |
| government purposes | 8 | | | | | | | | | |
| purposes of private nonprofit bodies | 9 | | | | | | | | | |
| Income and outlays: | | | | | | | | | | |
| value added | 10 | $T_{10.1}$ | $T_{10.2}$ | $T_{10.3}$ | $T_{10.4}$ | $T_{10.5}$ | $T_{10.6}$ | | | |
| institutional sector of origin | 11 | | | | | | | | | |
| form of income | 12 | | | | | | | | | |
| institutional sector of receipt | 13 | | | | | | | | | |
| ACCUMULATION | | | | | | | | | | |
| Capital formation: | | | | | | | | | | |
| increase in stocks | 14 | | | | | | | | | |
| fixed capital formation | 15 | | | | | | | | | |
| Capital finance: | | | | | | | | | | |
| net purchases of land and intangible assets | 16 | | | | | | | | | |
| capital transfers | 17 | | | | | | | | | |
| financial assets | 18 | | | | | | | | | |
| institutional sectors | 19 | | | | | | | | | |
| REST OF THE WORLD | | | | | | | | | | |
| Current and capital transactions | 20 | $T_{20.1}$ | $T_{20.2}$ | | | $T_{20.5}$ | | $T_{20.7}$ | | |

*Gross output.* Gross output of goods and services covers both the value of goods and services produced for sale and the value of goods and services produced for own use. It includes:

(1) The domestic production of goods and services which are either for sale or for transfer to others.
(2) Net additions to work in progress valued at cost, and to stocks of finished goods valued in producers prices.
(3) Products made on own account for government or private consumption or for gross capital formation.
(4) Rents received on structures, machinery and equipment and imputed rent for owner-occupied dwellings.

Gross output at producers' value is the initial point of measurement of production and can be expressed in the relevant terms of the matrix entries as the sum of:

(1) Gross output of commodities.
(2) Gross output of noncommodity goods and services.
(3) Import duties.

*Intermediate consumption.* Intermediate consumption covers nondurable goods and services used up in production, including repair and maintenance, research and development and exploration costs. It also includes indirect outlays on financing capital formation such as flotation costs involved in the purchase and sale of intangible assets and financial claims. (Intermediate consumption is, whenever possible, valued at purchasers' prices at the moment of use).

*Gross domestic product and value added.* The gross domestic product is equal to the value added for the total of all domestic producers plus import duties and value added tax. The value added of industries at producers' prices is equal to the gross output of the industries at producers' prices less the value of their intermediate consumption at purchasers prices. The gross domestic product may alternatively be defined as the sum of final expenditure in the domestic economy, or as the sum of incomes received in the domestic economy.

## 30.3. Structure and Main Characteristics of MPS

### 30.3.1. Tabular representation

The System of Balances of the National Economy is based on a set of mutually linked tables, each of which covers a certain part (category, field) of the whole reproduction process in the national economy. The system is presented in the form of basic source and use tables (referred to as balances) and a certain number of supplementary tables providing further necessary details.

The system set out in *Basic Principles of the System of Balances of the National Economy*, published by the United Nations, contains the following five basic tables:

(1) The balance of production, consumption and accumulation of the global product (the material balance).
(2) The balance of production, distribution, redistribution and final utilization of the global product and the national income (the financial balance).
(3) The balance of manpower resources.
(4) The balance of fixed capital.
(5) The indicators (balance) of national wealth.

The first two balances, i.e., the material and financial balances, are most appropriate for the purpose of this survey.

The material balance is a presentation of the volume of the supply of goods and material services originating in domestically produced global product and imports and their disposition to consumption, capital formation and exports, classified by different production activity categories.

The financial balance is a presentation of income flows generated in production in the material sphere, their redistribution through transactions in the non-material sphere and through other transfer flows and finally their disbursement to consumption and capital formation.

The structure and main characteristics of MPS can be described with reference to the basic and, if need be, supplementary tables, specifying the contents of the individual rows and columns and the derived elements.

Although in the MPS some accountancy rules are applied and followed, source and use approaches to the main aggregates have to be fully balanced and the system itself makes great use of statistical and accountancy data, there is no standard system of accounts similar to that of SNA to cover the relevant flows of transactions.

As these accounts can be derived from MPS balances – namely, from the financial balance – they are sometimes constructed together with the MPS matrix. The derived MPS conceptual matrix is then comparable with that of SNA, and is used together with the description of entries for comparing the structure of MPS with that of SNA in the relevant comparison and reconciliation systems studies.

### 30.3.2. Main production aggregates and their definitions

The definitions below are limited to the MPS main production aggregates parallel to those of SNA shown in Section 30.2.2.

*Global product.* Global product covers the value of goods and material services produced. Deliveries of goods and material services within the same enterprise are excluded. Included are the value of own-account constructed capital goods and capital repairs of fixed assets, the value of work in progress and the value of finished goods added to stocks. Also covered is the value of goods and material services provided free to employees (the services are valued at the material costs involved). In the contribution to global product by agriculture are included seeds and feed produced and consumed at the same farm, agricultural and other goods produced on private plots for individual consumption or for sale, including the cost of their processing.

*Intermediate material consumption including depreciation.* Intermediate material consumption consists of the value of goods and material services used up in production during the period under review by production units of the material sphere, including the consumption of fixed assets during the same period. Consistent with the scope of the gross output of goods and material services certain items are included in intermediate consumption, e.g., in agriculture seeds and animal feed, which are produced and used by the same unit.

The intermediate outputs of raw materials, etc. are valued net of the value of scrap and wastes originating in the process of production. Purchased items are valued at purchasers' values, items produced on own account are valued at cost in the case of state and cooperative enterprises and at average purchasers' prices in the case of personal plots of households.

Depreciation and consumption of fixed assets include an allowance for normal wear and tear and foreseen obsolescence of fixed assets used in the material sphere based on standard rates of depreciation and, furthermore, the difference between the book value of scrapped fixed assets and their scrap or actual value. The allowances for depreciation are often based on the original cost of the assets, which may be periodically adjusted toward replacement costs.

*National income (net material product).* National income (NMP) can be defined differently according to the method used. Following the production approach, net material product is equal to the difference between global product (gross output) of goods and material services and intermediate material consumption including consumption of fixed assets.

National income from the income side is the sum of primary incomes of the population (comparable to compensation of employees in the SNA) and primary incomes of enterprises (comparable to operating surplus in the SNA).

Using the expenditure approach, national income is defined as the sum of the final uses of goods and material services, i.e., personal consumption, material consumption of units in the nonmaterial sphere serving individuals and that of similar units serving the community as a whole, net capital formation, replacement for losses and the balance between exports and imports of goods and material services.

## 30.4. Relationship between the SNA and the MPS Main Production Aggregates

The following survey of the relationship (similarities and differences) between the main production aggregates used in SNA and MPS is based on the relevant concepts and definitions. Its goal is not to provide a basis for complete and exhaustive comparison of all factors which may influence the actual comparisons, but which may not be of the same qualitative significance. Some of them are also dealt with in other parts of this review.

### 30.4.1. Fundamental conceptual differences

The most important and well-known difference between the SNA and the MPS concerns the definition of economic production. While in the SNA with a few exceptions practically all activities relating to the production of goods and the provision of services are treated as "productive" and embraced by the concept of economic production, in MPS the production concept is limited to productive activities in the material production sphere, where national income originates, i.e., to the industries producing material goods and material services such as repairs, transportation, communication and distribution of goods. Such activities as general government services, finance, research and scientific services, housing, medical and other health services, educational and cultural services are considered nonmaterial services; they are outside the production sphere and are therefore included in those branches of the economy where the processes of redistribution and final consumption occur.

Certain activities, e.g., passenger transport, though not strictly falling under the definition of material services, are for practical reasons also treated as belonging to material services in some MPS countries. The difference in the concept of economic production has a great impact on comparability of not only SNA and MPS production aggregates, but also on the categories of consumption, distribution and redistribution of income. The breakdown of other categories by type of sphere (productive, nonproductive) and all relevant categories by industrial origin are affected as well.

The activities which are treated differently in SNA and in MPS have to be specified in terms of the relevant classification when adjusting data of one system to the concepts of another system, and comparability requirements in this respect are needed.

The above-mentioned activities are specified in terms of groups of the International Standard Industrial Classification of All Economic Activities (ISIC). They should, in principle, be excluded from the scope of economic production in converting SNA data to MPS concepts and vice versa. In addition, there are some other activities falling under ISIC group being treated as nonmaterial activities in MPS. Such activities (e.g., tourist agencies and tourist development services) should, in principle, be treated in the same way as those previously mentioned. However, as these activities are usually of relatively little importance, in practical calculations of the relevant aggregates they are not separately specified along with the corresponding data.

It should be pointed out that the above-mentioned description of the differences in the concepts of economic production used in SNA and MPS, which are of considerable importance for actual comparisons of the aggregates, is based on the relevant documents approved by the UN Statistical Commission and published as internationally recommended guidelines and manuals as well as on other relevant studies.

The description then does not cover the additional differences which could occur within the individual systems themselves, e.g., between the United Nations international standards and the recommendations of other international bodies, or even between these standards and the national systems in the relevant fields. Although it is not likely that there is a significant difference in the concept of the production sphere, differences may occur mainly for practical reasons.

The different treatment of transport and communication activities can be mentioned as an example of the current practice in MPS countries. While statistical data for Czechoslovakia in the relevant national material production aggregates include only part of the transport branch (industry), i.e., goods transport, in some other East European countries, owing to practical difficulties in distinguishing between activities and assets, the whole transport branch is treated as productive. The quantitative significance of the different practical treatment of activities in transport and communication services can, to a certain extent, be judged from the composition of GDP and NMP aggregates by branches (industries) of national economy shown in the relevant statistical publications.

From the above-mentioned description of the concepts of production sphere used in SNA and MPS countries can be deduced that the differences are of real significance and have to be taken into account in direct comparison of the corresponding SNA and MPS aggregates.

### 30.4.2. Gross output compared with global product

The MPS global product defined in 30.3.2 is more restrictive as compared with the gross output in SNA due to the fact that it includes only goods and material services. Besides this, a certain quantitative difference between these two production aggregates appears in the treatment of the output of the catering trades, waste materials sold by producing units, second-hand goods bought from households and in the treatment of tips (gratuities).

Without giving a detailed description of the individual points mentioned above, a conclusion can be made that the differences relating to the treatment of the catering trades and to the sales of waste materials by producing units affect only the gross output or the global product, but not the GDP and the NMP (national income). The differences in the treatment of other issues mentioned affect both the "gross" and the "net" production aggregates, i.e., also the comparison between GDP and NMP (national income). However, as they are likely to be of little quantitative significance, they can be disregarded.

### 30.4.3. Intermediate consumption aggregates in SNA and MPS

As to the relationship between intermediate consumption and production aggregates, the purpose of the specification of intermediate consumption aggregates is generally to derive unduplicated production aggregates and their differences rather than to serve direct comparisons.

In addition to the main difference concerning the sphere of production and some other differences already mentioned (the catering trades, sales of waste material), the SNA and MPS intermediate consumption is differently defined with regard to enterprise expenditures on cultural, sport and similar facilities for their employees; travel expenses in connection with business; uniforms; fixed assets used for special purposes and losses of stocks. The first two additional differences mentioned, i.e., different treatment of expenditures on cultural and other facilities and travel expenses, are probably of some quantitative importance and should be taken into account in the intersystem comparisons.

Another significant difference results from the treatment of fixed capital consumption in relation to the scope of the three main aggregates under review. While in SNA consumption of fixed capital is an element of value added aggregate and gross domestic product, the MPS national income (NMP) is defined as net, i.e., without the consumption of fixed capital which is treated as part of intermediate consumption.

Concerning capital consumption itself, it is, in principle, treated in the same way as far as it relates to fixed capital used in the material production sphere. In MPS the capital consumption in the nonmaterial production sphere is treated as final consumption. In some specific points there are differences between SNA and MPS in this respect (e.g., other than normal losses in fixed assets, differences between the actual and the written-off value of scrapped assets, valuation of fixed capital consumption). These differences may be of some quantitative significance, but are usually disregarded in the conceptual framework for intersystem comparisons.

### 30.4.4. Gross domestic product and national income

As has already been shown, the gross domestic product (total value added) and national income (NMP) are defined as gross output or global product minus the corresponding intermediate consumption which for MPS includes also capital consumption. The differences between these two most important production aggregates are therefore those that have already been described in connection with the concept of the material production sphere and also with the definitions and comparisons of the previous aggregates (30.3.2 and 30.3.3).

The main difference is apparently that MPS national income (NMP) is confined to activities and income originating in the material production sphere whereas gross domestic product (total value added) includes activities and income which are of material and also nonmaterial character. Another relatively significant difference arises from the treatment of capital consumption. While capital consumption is a component of SNA gross domestic product or value added, in MPS it is considered to be intermediate consumption and therefore it is not included in the national income (NMP) aggregate. In order to allow for the adjustments needed for the conversion of SNA national aggregates into the relevant MPS aggregates and vice versa, the components of various flows which

*Table 30.2.* Conversion of the selected aggregates from MPS to SNA and *vice versa*.

| *MPS aggregates* | *Elementary flows* | | | | | | | | | | | *SNA aggregates* |
|---|---|---|---|---|---|---|---|---|---|---|---|---|
| | *Gross output of the nonmaterial sphere* | | | *Intermediate consumption of the nonmaterial sphere* | | | | | | | | |
| | Gross output of services of nonbudgetary units, other than providing housing services | Cost of housing services: (a) provided by units in the nonmaterial sphere (b) owner-occupied dwellings | Cost of services of budgetary units | Material consumption of units in the nonmaterial sphere (excluding depreciation) | Nonmaterial services bought by units in the nonmaterial sphere | Expenditure in connection with owner-occupied dwellings: (a) material (excluding depreciation) (b) nonmaterial | Nonmaterial services bought by units in the material sphere | Consumption of fixed capital in the material sphere | Expenditure on travel for official business | Purchases by enterprises for recreation and culture for employees | Losses in stocks other than those due to normal events in production | |
| 1 | 2 | 3 | 4 | 5 | 6 | 7 | 8 | 9 | 10 | 11 | 12 | 13 |
| Global product $A = (13-2-3-4)$ | × | × | × | | | | | | | × | | Gross output $A = (1+2+3+4-11)$ |
| Intermediate material consumption incl. depreciation $B = (13-5-6-7-8+9-10-12)$ | | | | × | × | × | × | × | × | | × | Intermediate consumption $B = (1+5+6+7+8-9+10+12)$ |
| National income $C = A - B = (13-2-3-4+5+6+7+8-9+10+12)$ | × | × | × | × | × | × | × | × | | × | × | Gross domestic product $C = A - B = (1+2+3+4-5-6-7-8+9-10-11-12)$ |

are treated differently in both systems must be separated. The construction of such elementary flows and their groupings or regroupings make it possible to move from an aggregate of one system to its counterpart of the other system. A table presenting the adjustments needed for the conversion of MPS global product, intermediate material consumption and national income (NMP) into SNA gross output, intermediate consumption and gross domestic product and vice versa, based on the differently treated components and elementary flows, is shown in *Table 30.2.*

The components or elementary flows shown in adjustment tables are also identified in terms of the classification of activities, categories of transactions and/or other instruments adopted in SNA and MPS.

### 30.4.5. Other issues involved in the systems comparison

The relationship between the SNA and MPS production aggregates discussed in Subsections 30.4.2, 30.4.3 and 30.4.4 display the main conceptual and methodological differences which can be encountered in their comparisons. In addition to these differences and problems involved in the availability of the appropriate statistical data there are also other issues of some importance which should be taken into account in an actual intersystems comparison.

There is no identity between the statistical (transaction) units used in SNA and MPS. While in SNA such a unit is the establishment or equivalent unit, in MPS a statistical unit used is the enterprise. Enterprises are organized on a kind-of-activity basis. In some cases an enterprise is engaged in a single kind of activity, in other cases an enterprise comprises several establishments, for which separate statistics and accounting records are kept. For enterprises engaged in more than one activity, the relevant statistical data have to be reported separately (through special statistical and accountancy questionnaires) to meet the needs of MPS statistics.

Another problem may arise in connection with territorial or national concepts, if applied differently. In SNA only production taking place in the domestic territory of a given country is encompassed. In MPS the concept actually used corresponds to the SNA concept but there are some minor differences.

Both in SNA and MPS the relevant economic activity classifications are used. However, in SNA the activities are classified in terms of the International Classification of All Economic Activities (ISIC); in MPS countries the classification of Branches of National Economy (CBNE) is used for the relevant classification of activities and data. The significance of differences, resulting from the applications of these two classifications in the corresponding SNA and MPS production aggregates, can be judged from the conversion keys and their testing. It is, of course, likely that in the case of subaggregates and similar categories the differences will be relatively more important.

## 30.5. Calculations of Gross Domestic Product and National Income for Selected Countries

The calculations of GDP and national income (NMP) in terms of actual figures undoubtedly represent one of the most positive contributions to the studies of comparisons and links between the SNA and MPS and their relevant aggregates.

The objectives of the actual calculations are to serve not only as a basis for analyzing ratios between the corresponding SNA and MPS aggregates, but also to test the recommended conversion methodology together with the quantitative significance of selected items.

The results of experimental illustrative calculations of the GDP and NDP for selected countries with centrally planned economies and the national income (NMP) for selected countries with market economies are presented in *Table 30.3*.

*Table 30.3.* Relationship between domestic product and net material product (NMP = 100.0) aggregates for selected countries (in %).

| *Country* | *Year* | *Gross domestic product and net material product* | *Net domestic product and net material product* |
|---|---|---|---|
| USA | 1963[a] | 147.0 | 133.9 |
| | 1967[a] | 147.3 | 134.9 |
| | 1972 | 152.5 | 139.0 |
| United Kingdom | 1963[a] | 136.1 | 125.7 |
| | 1971[a] | 143.0 | 130.1 |
| | 1972[a] | 144.4 | 131.1 |
| | 1974 | 147.5 | 132.1 |
| Japan | 1965[a] | 129.5 | 112.9 |
| | 1970[a] | 126.1 | 109.7 |
| | 1975 | 139.0 | 120.1 |
| France | 1976 | 145.2 | – |
| Canada | 1976[b] | 151.2 | 136.1 |
| Hungary | 1976 | 124.9 | – |

[a]Previous calculations (from first round).
[b]Preliminary figures.

The experimental illustrative calculations of those and other aggregates were carried out by the United Nations Statistical Office in close cooperation with national statistical offices which also provided additional data and assistance in further statistical operations. The illustrative calculations are presented in the form of ratios referring to the relationships between corresponding production aggregates of SNA and MPS. The derivation of GDP and national income (NMP) has been carried out on the basis of several types of conversion tables aiming at double-checking estimates obtained through the calculations.

As can be seen from *Table 30.3*, there are significant differences between countries regarding their ratios of GDP or NDP to national income (NMP). The ratio is generally higher in countries with a higher level of economic development measured in terms of per capita GDP. It should be noted that the calculations have been made in national currencies and current prices.

Year-to-year fluctuations in the ratio of gross and net domestic product to national income are relatively small, but fluctuations over longer periods are significant and cannot be disregarded. The ratio of gross and net domestic product to national income also tends to increase over the years.

In view of significant differences in ratios of the corresponding aggregates and, of course, of a very limited sample of calculations carried out as yet, we do not recommend applying the ratio calculations for one country to calculations for another country, especially if the countries are different in terms of the structure of their national economies and the level of their economic development.

## 30.6. Other Projects in the Field of Gross Domestic Product and National Income Comparison

There are other important projects which should be mentioned in relation to the comparison of gross domestic product and national income (NMP) aggregates.

### 30.6.1. International comparison project

Following the consideration by the UN Statistical Commission at it fifteenth session on issues relating to international comparison of production, income and expenditure aggregates, the United Nations International Comparison Project (ICP) was established in 1968. The long-term objective of ICP was to establish a worldwide system for detailed comparisons of the purchasing power of currencies and of *gross domestic product* in terms of these purchasing powers.

The basic methodological approach used in ICP is that the quantity comparisons between countries are obtained by means of price comparisons of a number of carefully specified goods and services covering all categories of final expenditures on GDP. Final expenditures on GDP are divided into approximately 150 detailed categories and price relatives are computed and averaged for each category. These category price relatives show the purchasing power parities between the countries compared, and they are used to convert the category expenditures of a given country to the currency units of a country (numeraire country) undertaking the role of base country. Thus, all expenditure categories and aggregated summary categories and GDP are directly comparable.

After the development of a methodology for such a comprehensive system of comparisons, the ICP started its experimental implementation in several successive phases.

Following the requested involvement of the UN Regional Commission, the Conference of European Statisticians included in its work program a project on "Comparison within the ICP Framework for the European Region" with 1980 as a reference year. Eighteen European countries divided in two groups agreed to participate in the project. Twelve countries (Group I) carried out a comparison under the auspices of the Statistical Office of the European Communities. Group II is composed of six countries: Austria, Finland, Hungary, Poland, Romania and Yugoslavia with Austria undertaking the role of a base country.

According to the accepted program of work in that field by the UN Statistical Commission, phase IV of ICP extended further the country coverage to 60 countries in a full-scale comparison; 1980 was chosen as reference year. The report on phase IV containing the results of the group of countries (regions) comparisons was published in 1985.

In phase V, 16 countries from all of the UN regional commissions and other major international groups have been asked to assume the role of core countries for 1985 comparisons. The core countries representing various types of economies, regions and memberships in international organizations are linked to groups of countries encompassing all countries participating in the comparisons. With the UN Statistical Office being responsible for the core country comparisons, in phase V, its results will be issued in greater detail when more countries have completed their data collection.

### 30.6.2. Bilateral comparisons in SNA–MPS aggregates

Bilateral comparisons or comparative studies of basic economic aggregates and other main indicators (structures, prices) between European countries have a long tradition. They have been carried out either on the basis of cooperation agreements between the national statistical offices of the countries concerned or under the auspices of the Conference of European Statisticians or other international bodies.

As these comparisons also involve countries with different economic systems, the numerical results together with the methodological or analytical conclusions have made a valuable contribution to the development of intersystem and intercountry comparisons.

In 1979, France and Hungary agreed to test the conceptual framework for intersystem (SNA–MPS) comparisons. This original experiment enriched the recommended methodology for intersystem comparisons. It showed that the systematic use of input–output tables considerably improved comparison, revealing the different stages in the conversion of one system into the other and guaranteeing that adjustments were made in a balanced manner, both in the general equation of goods and services and in the equation for primary income distribution.

Another bilateral undertaking was the comparison of prices and gross domestic expenditure agreed on between Austria and Poland for the years 1975 and 1978. The results of this study, containing the numerical calculations together with the relevant analytical part and methodological descriptions, were published by the central statistical offices of Austria and Poland in 1982.

In 1983, the Central Statistical Offices of Bulgaria and Finland agreed to carry out a bilateral comparison and verification of the conceptual links between SNA and MPS with 1982 as a base year. According to the agreed approach Bulgaria converted its aggregates from MPS into SNA categories on the basis of its national data and, accordingly, Finland converted its national aggregates into relevant MPS aggregates using its national data. The results of calculations of the corresponding SNA–MPS production aggregates are shown in *Tables 30.4* and *30.5*.

*Table 30.4.* Bulgaria, 1982 (in %).

| *MPS categories* | | | *SNA categories* |
|---|---|---|---|
| Global product | 100.0 | 109.7 | Gross output |
| Intermediate material consumption including depreciation | 100.0 | 100.0 | Intermediate consumption |
| Net material product | 100.0 | 128.5 | Gross domestic product |

*Table 30.5.* Finland, 1982 (in %).

| *SNA categories* | | | *MPS categories* |
|---|---|---|---|
| Gross output | 100.0 | 78.4 | Global product |
| Intermediate consumption | 100.0 | 89.0 | Intermediate material consumption including depreciation |
| Gross domestic product | 100.0 | 67.1 | Net material product |

The results of the experimental calculations together with the relevant methodological conclusions were published by the Central Statistical Offices of Bulgaria and Finland in November 1985.

There are other important SNA–MPS aggregates, which are also a subject of bilateral comparisons. For example in 1986, Czechoslovakia and Finland agreed to carry out a comparative study on total consumption by population, which will contribute to the building up of so-called "common aggregates" in SNA and MPS systems which play an important role in international statistics.

## 30.7. Further Development of the SNA–MPS Comparisons

Together with the development of the existing accounting systems and their revisions, great attention is focused on the further improvement of the SNA–MPS comparisons and links – namely, in their conceptual relationship together with the revision of the relevant international recommendation documents.

At its twenty-fourth session in 1987, the UN Statistical Commission considered the report on major results of the study on SNA–MPS links and discussed the preparatory work of the UN Statistical Office on the revision of the Comparisons of the System of National Accounts and the System of Balances of the National Economy in the field of conceptual relationship. The specific proposals cover the construction of an improved version of the conceptual framework for the SNA–MPS comparison, including the modification of its structure. They include the drafts of the revised conversion tables designed for the derivation of GDP for countries with centrally planned economies. The Commission will also consider the results of experimental calculation of GDP and NMP, which are to be carried out by both the UN Statistical Office (for the Federal Republic of Germany, the Netherlands and other countries) and by countries on a bilateral basis. This is in relation to the recommendation of the Commission that the work on the revision of the methodological document on SNA–MPS comparisons should reflect the experience gained through the illustrative calculations.

## References

United Nations Statistical Office (1977), ***Comparisons of the System of National Accounts and the System of Balances of the National Economy***, Studies in Methods, Series F, No. 20. (United Nations, New York).

Central Statistical Office of Bulgaria, Central Statistical Office of Finland (1985), ***Comparison of the Balance Sheets and the National Accounts of the PR Bulgaria and the Republic of Finland.*** (Sofia – Helsinki).

# Appendix 30A. GDP Estimates for the CMEA countries in the Bonn–IIASA Project

*Rumen Dobrinsky*

In accordance with the available statistical data (Dobrinsky, 1986), the CMEA country models within the Bonn–IIASA Research Project on Economic Growth and Structural Change were developed in accordance with the MPS concept (see Chapters 4 and 5). However, the nature of the empirical work within the project made it necessary to have some means of comparing the results for the CMEA countries with those for the developed marked economies and the developing countries whose models are based on the SNA concept. As it is pointed out in Chapter 30, this is by no means a trivial task. On the other hand, due to the limited resources of the project, we could not put too much effort into an in-depth treatment of the problem. Our approach was a rather simplified one, following the practical goals of the project. Accordingly, the GDP figures which were constructed and used for the CMEA countries within the project should be regarded only as some rough approximation.

In principle, we tried to base our figures on estimations which were available in the literature. However, since there do not exist generally accepted data on this subject, some additional assumptions and adjustments had to be made.

One of the most comprehensive empirical studies in this field is the World Bank-sponsored project reported in Marer (1985). It served as a basic source for our estimations. Other sources which were used for checks and comparisons were the study of the Economic Commission for Europe (ECE, 1980); a comparison of the level of per capita income in six CMEA countries reported by a Hungarian economist (Szilágyi, 1978); an estimation of GDP levels in the CMEA countries in the period 1970–1980 performed by two Bulgarian economists (Kostov and Videnov, 1985) as well as the comparisons of the levels of the national income in USSR and USA published in the Statistical Yearbooks of the USSR (*Narodnoye Khozyaistvo SSSR*, various years).

Two main problems had to be tackled with respect to the needs of the project: the problem of static conversion (conversion of NMP figures of certain years into internationally comparable GDP figures) and the problem of dynamic conversion (conversion of MNP growth rates into GDP growth rates).

Since the "international" measurement unit for real figures in the project was the 1975 US$, it was decided to perform the basic static conversion for the year 1975. The actual GDP figures for this year were based on the GNP per capita estimations reported in Marer (1985). Using the official population figures we arrived at the following approximate GDP levels for the seven European CMEA countries for 1975, in US$ of the same year: Bulgaria – 29.13 billion; CSSR – 57.54 billion; GDR – 69.25 billion; Hungary – 36.47 billion; Poland – 119.0 billion; Romania – 49.34 billion; USSR – 875.81 billion. In most cases these figures compare quite favorably with other estimates quoted earlier.

Marer (1985) also reports on the analysis of the implied GNP:NMP ratios, measured in domestic prices and suggests the following figures for 1980 (pp. 18–19): Bulgaria – 1.32; CSSR – 1.19; GDR – 1.26; Hungary – 1.24; Poland – 1.28; Romania – 1.20; USSR – 1.29. From here, under the assumption that these ratios do not change substantially over time, we calculated approximate GDP figures for 1975 in constant domestic prices. The ratio of GDP in domestic prices to GDP in US$ gives the conversion factor reported in Chapter 2.

The next and much more complex problem was the dynamic conversion. It is widely argued that NMP growth rates are not directly comparable to GDP growth rates. One obvious reason for this is the different composition of the two aggregates. The team of P. Marer claims also that the CMEA methodology of calculating NMP by the "double deflation method" (the values of gross output and secondary inputs are both deflated separately) leads to an upward bias in the estimated rates of growth of NMP as compared to the relevant GDP growth rates. Instead, they propose an "adjusted factor cost approach" by which they reestimate the NMP growth rates in a way which, they claim, makes them comparable to the GDP growth rates. On the basis of their recalculation for the period 1970–1980 they suggest the following ratios between NMP growth rates and comparable GDP growth rates for this period (Marer, 1985, pp. 184–185): Bulgaria – 2.1; CSSR – 1.6; GDR – 1.5; Hungary – 1.5; Poland – 1.4; Romania – 1.6; USSR – 1.5. This approach is not indisputable with respect to the economies of the CMEA countries (the adjusted factor cost approach is based on the assumption that the factor prices are proportional to the marginal factor productivities – an assumption which is usually rejected on methodological grounds; see Chapter 5). However, the quoted ratios are not very different from the ones which can be derived from other estimations.

Due to these statistical and methodological problems, we decided to use within the project two different estimates for the GDP dynamics in the case of the CMEA countries. The first one (which we refer to as GDP type 1) is calculated from the quoted 1975 US$ GDP figures, by multiplying the latter by the NMP index of growth. Thus the rate of growth of GDP type 1 is the same as the rates of growth of the corresponding NMP. The second one (which we refer to as GDP type 2) is calculated in a similar way from the 1975 figures but using an adjusted index of growth. The adjustment factor was the approximate ratio linking NMP growth rates to GDP growth rates. The actual ratios which we used were: USSR and Poland – 1.3; CSSR, GDR and Hungary – 1.5; Bulgaria and Romania – 1.8. They were taken on the basis of the ratios quoted above with some corrections stemming from other available estimations.

The two estimates, GDP type 1 and type 2, can be regarded as upper and lower limits of the corresponding GDP figures in dynamics.

Finally, we would like to repeat that, due to the very crude (and probably mechanical) nature of this approach, the resulting figures should be treated rather cautiously.

## Appendix References

Dobrinsky, R. (1986), *The Macroeconomic Data Base for the CMEA Countries*. Working Paper WP-86-54. (Laxenburg, Austria: International Institute for Applied Systems Analysis).

ECE (1980), Comparative GDP Levels, in: *Economic Bulletin for Europe*. (New York: UN, ECE).

Kostov, I. and Videnov, A. (1985), International comparison by physical indicators. *Ikonomicheska Misal*, 2. pp. 46–57.

Marer, P. (1985), *Dollar GNPs of the USSR and Eastern Europe*. A World Bank Publication. (Baltimore, MD: Johns Hopkins University Press).

*Narodnoye Khozyaistvo SSSR* [Statistical Yearbook of the USSR]. (various years). Moscow.

Szilágyi, G. (1978), *The Comparison of Economic Level and Structure via Factor Analysis*. Budapest.

# Annex 1

# Coding for Countries and Country Groups

| Codes | | Names of Countries or Groups | IMF |
|---|---|---|---|
| | | Main industrialized countries | |
| 01 | | United States of America | 111 |
| 02 | | Federal Republic of Germany | 134 |
| 03 | | Japan | 158 |
| 04 | | France | 132 |
| 05 | | United Kingdom | 112 |
| 06 | | Italy | 136 |
| 07 | | Netherlands | 138 |
| 08 | | Belgium/Luxembourg, i.e. | 124 |
| | 81 | Belgium | 124 |
| | 82 | Luxembourg | 137 |
| 09 | | Canada | 156 |
| 10 | | Other developed market economies (= other DMEs) | |
| | 80 | Sweden | 144 |
| | 83 | Norway | 142 |
| | 84 | Finland | 172 |
| | 85 | Denmark | 128 |
| | 86 | Ireland | 178 |
| | 87 | Spain | 184 |
| | 88 | South Africa | 199 |
| | 89 | Switzerland | 146 |
| | 90 | Austria | 122 |
| | 91 | Australia | 193 |
| | 92 | New Zealand | 196 |
| 11 | | Oil-exporting developing countries | |
| | 20 | Saudi Arabia | 456 |
| | 21 | Indonesia | 536 |
| | 22 | Venezuela | 299 |
| | 23 | Iran | 429 |
| | 24 | Nigeria | 694 |
| | 25 | Libya | 672 |
| | 26 | Algeria | 612 |
| | 27 | Iraq | 433 |
| | 28 | Kuwait | 443 |
| 12I | | Developing Asian countries without India | |
| | 30 | Korea | 542 |
| | 31 | Hong Kong | 532 |
| | 32 | Singapore | 576 |
| | 33 | Malaysia | 548 |
| | 35 | Thailand | 578 |
| | 36 | Philippines | 566 |
| 34 | - | India | 534 |
| 13 | | Developing African countries | |
| | 40 | Cameroon | 622 |
| | 41 | Gabon | 646 |
| | 42 | Tanzania | 738 |

| Codes | | Names of Countries or Groups | IMF |
|---|---|---|---|
| | 43 | Zambia | 754 |
| | 44 | Togo | 742 |
| | 45 | Ivory Coast | 662 |
| | 46 | Zaire | 636 |
| | 47 | Kenya | 664 |
| 14 | | Developing Latin American countries | |
| | 50 | Chile | 228 |
| | 51 | Colombia | 233 |
| | 52 | Ecuador | 248 |
| | 53 | Peru | 293 |
| | 54 | Uruguay | 298 |
| | 55 | Trinidad & Tobago | 369 |
| 15 | | Further developed Latin American countries | |
| | 60 | Mexico | 273 |
| | 61 | Brazil | 223 |
| | 62 | Argentina | 213 |
| 16 | | Union of Soviet Socialist Republics | 974 |
| (17) | | USSR plus the other European CMEA countries | |
| (70) | | European CMEA countries without USSR, i.e. | |
| 71 | | Bulgaria | 918 |
| 72 | | Czechoslovakia | 934 |
| 73 | | German Democratic Republic | 938 |
| 74 | | Hungary | 944 |
| 75 | | Poland | 964 |
| 76 | | Romania | 968 |
| 18 | | Other Mid-East & North African developing countries | |
| | 38 | Egypt | 469 |
| | 39 | Syrian Arab Republic | 463 |
| | 48 | Tunisia | 744 |
| | 49 | Morocco | 686 |
| (00) | | Sum of the countries and groups 01 - 18 | |
| (97) | | Rest of the world with respect to 01 - 18 (including the P.R. of China | |
| (99) | | World total | 001 |

*Note*: Separate models were built for all countries or country groups whose codes are put in the first column with the exception of the last three groups (00, 97, 99).

# Annex 2

# Outline of the Bonn–IIASA World Model

*Editor's note*:

**We present here the general approach of the model, using somewhat simpler notation than that found in some chapters of the book. See specific chapters for details.**

1. We subdivide the world economy into

9 OECD countries + remainder (= other industrialized market economies),

7 CMEA countries + remainder (= other centrally planned economies),

7 developing countries (or groups of developing countries) + remainder[1)].

Details: see Annex 1.

2. We assume a production function

for each country or group of countries:

$$Y^* = \tau L^{\alpha_1} \cdot K^{\alpha_2} \cdot (IM_R)^{\alpha_3} \quad , \quad \alpha_i > 0 \quad , \quad \sum_i \alpha_i = 1 \quad ,$$

or

$$Y^* = (L\tau^*)^{\alpha_1} \cdot K^{\alpha_2} \cdot (IM_R)^{\alpha_3} \quad , \quad \tau^* = \tau^{1/\alpha} \quad ,$$

where

$$Y^* = Y + IM_R \cdot P'IM_R/P'Y \quad ,$$

and $Y^*$ = total production minus secondary domestic inputs (for developing countries $Y^*$ is defined as total production minus domestic secondary inputs minus production Y'MIN of the oil sector), Y = real GDP for OECD countries, Y = NMP + AD = net material product NMP plus

---

1) The P.R. of China is still among this remainder. It is a task for the future to deal with this nation separately.

depreciation AD of real fixed assets for CMEA countries, $Y = GDP - M'MIN$ for developing countries, $IM_R$ = real imports of secondary inputs, $P'IM_R$ = price of these imports in local currency, $P'Y$ = price level of Y, L = labor employed (for CMEA countries: employed in the material sphere), K = capital, $\tau$ = index of the state of technology (Hicks-neutral technical progress), $\tau^*$ = index of the state of technology (Harrod-neutral technical progress). All prices are in domestic currency if not otherwise indicated. In fact, as an approximation we assumed $P'IM_R = P'Y$. Therefore the last equation above can be substituted by

$$Y^* = Y + IM_R \ .$$

The imports $IM_R$ of secondary inputs are related to total imports IM by

$$IM_R = \mu_R \cdot IM \ ,$$

where $\mu_R$ = const. for CMEA and developing countries and a function for OECD countries (see below). We estimated $\alpha_1, \ldots, \alpha_3$ by assuming a linear trend for the rate $w_\tau$ of technical progress[1) ] in the reference period, usually 1960 - 1983, i.e.

$$\tau = \tau_o e^{a_1 (a_2 t + \frac{a_3}{2} t^2)}$$

and (for OECD countries) by assuming that $\alpha_1, \ldots, \alpha_3$ is related to the average of the relative shares of labor, capital and imports in the value of production Y*; for details see chapter 3. For CMEA countries a special estimating procedure is used, see chapter 5.

---

1) We use the notation $w_x$ for the growth rate of a variable x, i.e. $w_x = \dot{x}/x$ if x is continuous in time, and $w_x = (x - x_{-1})/x_{-1}$ if x is discrete in time.

3. For OECD and developing countries capital accumulation is determined by

$$K = K_{-1}(1-d) + I'F_{-1} \quad ,$$

where d = rate of depreciation (exogenous, see below), I'F = gross fixed investment. I'F is determined by the investment ratio s:

$$I'F = s \cdot Y \quad ,$$

s exogenous (see below).

Since real exports EX and real imports IM are explained by export and import functions (see below), real consumption C follows from the GDP identity:

$$C = Y - I - EX + IM \quad ,$$

where I = gross investment and is related to I'F by I = I'IF · I'F, I'IF exogenous.

4. For CMEA countries capital accumulation is basically explained in the same way. But the above concepts have to be related to the official national accounting figures based on the NMP concept. This yields:

$$D = D'K \cdot K \quad ,$$

where D = "depreciation" (= sorting out) of fixed assets, D'K is the rate of sorting out of fixed assets (given exogenously), and

$$ACD = ACD'I \cdot I \quad ,$$

where ACD = gross accumulation (according to the NMP concept) and ACD'I = factor relating gross accumulation to gross investment in the NMP concept (exogenous). Consumption is determined by the NMP identity:

$$C = Y - ACD - EX + IM \ .$$

5. <u>For OECD countries, the domestic price level P'Y</u> is determined by Fisher's equation:

$$P'Y = v \cdot M/Y \quad ,$$

where v = velocity of money, M = the money supply M2 (exogenous, see below). The velocity of money is explained for each country by a variant of the function

$$v = v_{-1}^{a_1} \left( \frac{K/Y}{K_{-3}/Y_{-3}} \right)^{a_2} \cdot \left( \frac{Y/L}{Y_{-3}/L_{-3}} \right)^{a_3} \cdot e^{a_4 (CCA/Y)} \cdot e^{a_5} \quad ,$$

where CCA = accumulated current account deficits or surpluses. Thus CCA/Y is the relative foreign debt or surplus position of the country.

The <u>nominal rate of interest</u> r is explained by a variant of the function

$$r = a_1 r_{-1} + a_2 w_{P'Y} + a_3 r_{USA} + a_4 r_{USA,-1} + a_5 (s/s_{-4})$$

$$+ a_6 CCA/Y + a_7 w_{FX} + a_8 \quad ,$$

where $r_{USA}$ = rate of interest in the USA and FX = "foreign exchange" = exchange rate to the $ = price of US $ in domestic currency.

The <u>exchange rate</u> FX of domestic currency to the $ is derived from an exchange rate index I'FX by

$$FX = FX'C \cdot I'FX \quad ,$$

where FX'C = const. = exchange rate in the base year 1975.

The exchange rate index (1975 = 1) is forecast on the basis of purchasing power parity theory:

$$I'FX = I'FX_{-1} \cdot \left[ \frac{1 + w_{P'Y}}{1 + w_{P'Y(USA)}} \right] ,$$

where $w_{P'Y(USA)}$ is the rate of inflation in the USA.[1)]

From this follows the nominal GDP in domestic currency:

$$Y'N = Y \cdot P'Y ,$$

in current US $:

$$Y'\$N = Y'N/FX$$

and real GDP in $ of the base year:

$$Y'\$ = Y/FX'C .$$

6. For CMEA countries

The domestic price index P'Y (which is identified with the price level of final material product Y) is determined by the ratio of the nominal wage rate W'N to the real wage rate W

$$P'Y = W'N/W .$$

The nominal wage rate is determined by political authorities and is exogenous (see below). The real wage rate is estimated by

$$W = a_1 W_{-1} + a_2 \cdot Y/L + a_3 \cdot R_{-1} \cdot K/L + a_4 ,$$

1) Unfortunately we could not use the originally planned and much more refined approach to exchange rate determination put forward in chapter 8 by Dr. Welsch, where all exchange rates are determined simultaneously. In the time available for our research we were unable to get a solution for this much more complicated system. Thus we had to simplify the exchange rate equations.

where R is the primary income of enterprises per unit of fixed assets (which may be considered as an analog to the rate of interest):

$$R = (Y - W \cdot L)/K \quad .$$

The "<u>GDP-exchange rate</u>" FX to the \$ is derived from an exchange rate index in the same way as for the OECD countries:

$$FX = FX'C \cdot I'FX \quad , \quad \text{see above.}$$

The exchange rate index is estimated by a function which can be understood by a variant of purchasing power parity theory:

$$I'FX = (I'FX_{-1})^{a_1} \left[I'FX_{-1} \frac{1 + w_{P'Y}}{1 + w_{\$FT}}\right]^{a_2} \cdot a_3 \quad ,$$

where

$$\$FT = (EX\$N + IM\$N)/(EX\$ + IM\$)$$

is the ratio of nominal exports and imports in \$ to the real exports and imports in \$. This approach has been chosen because CMEA countries use different exchange rates for imports and exports. We determine the export exchange rate index I'FXDE by

$$I'FXDE = I'FXDE_{-1}(1 + w_{FX})$$

and the import exchange rate index I'FXDI by

$$I'FXDI = I'FXDI_{-1}(1 + w_{FX}) \quad .$$

The export and import exchange rates themselves are explained by

$$FXDE = FXDE'C \cdot I'FXDE \quad \text{and} \quad FXDI = FXDI'C \cdot I'FXDI \quad ,$$

where FXDE'C and FXDI'C are the constant export and import exchange rates, respectively, in the base year.

From this follows Y or NMP in domestic prices:

$$Y'N = Y \cdot P'Y \quad \text{and} \quad NMP'N = NMP \cdot P'Y \quad ,$$

as well as "GDP type 1" in constant \$ prices:[1)]

$$Y'\$ = NMP \cdot P'NMP75/FX'C \quad ,$$

where P'NMP75 is the price level in the base year 1975 and FX'C is the exchange rate in the same year. "GDP type 1" in current \$ is defined by

$$Y'\$N = GDP\$ \; \frac{P'Y/P'NMP75}{I'FX} \quad .$$

7. <u>For developing countries</u>

we first have to estimate the <u>production Y'MIN of the oil sector</u> in order to get GDP of the oil-exporting countries from the equation GDP = Y* + Y'MIN - $IM_R$, see section 2 above. We explain this in a preliminary way as a function of the exports:

$$Y'MIN = a_1 EX^{a_2} \quad .$$

We derive the <u>price level</u> P'Y for some groups of countries by export and import prices:

$$P'Y = a_1 (P'IM)^{a_2} \cdot (P'Y_{-1})^{a_3} \cdot (P'EX)^{a_4} \quad ,^{2)}$$

for other countries in the same way as for the OECD countries by money supply and the velocity of money:

---

1) GDP "type 1" refers to the official figures which appear in the statistical yearbooks of the CMEA countries and are based on NMP national accounting. These figures are not directly comparable to the GDP figures of the OECD and developing countries. For comparison one has to modify them. This yields GDP "type 2" for CMEA countries, see chapters 2 and 5.

2) P'IM and P'EX are explained below.

$$P'Y = v \cdot M/Y \quad .$$

The velocity of money is estimated as

$$v = \exp[a_1 v_{-1} + a_2 w_{P'Y,-1} + a_3 \frac{1}{Y/L} + a_4] \; .$$

For investment goods a special price index has been estimated by:

$$P'I = a_1 (P'Y)^{a_2} \quad .$$

Thus we get for nominal GDP and nominal investment

$$Y'N = Y \cdot P'Y \quad \text{and} \quad I'N = I \cdot P'Y \quad .$$

For the groups of developing countries we use an artificial currency the exchange rate of which, with respect to the $, is one in 1975. Therefore:

$$Y = Y'\$ \quad \text{in 1975.}$$

For the same reason we only need an exchange rate index I'FX, i.e. I'FX = FX. It is explained by

$$I'FX = (I'FX_{-1})^{a_1} \cdot \left(\frac{P'\$Y_{-1}/P'Y_{-1}}{P'\$Y/P'Y}\right)^{a_2} \cdot a_3 \quad ,$$

where

$$P'\$Y = \frac{IM'\$N + EX'\$N}{IM + EX} = \text{world market price level,}$$

IM'$N and EX'$N = import and export values, respectively, in $ (see below).

The nominal GDP in $ is defined by

$$Y'\$N = Y/I'FX \quad .$$

8. Aggregation over countries yields:

a) for OECD countries (= "World 1"):
GDP of OECD countries in current ∅:

$$Y'\text{∅}NW1 = \sum_i Y'N_i/FX_i \quad , \; i = \text{OECD country } i.$$

GDP of OECD countries in constant ∅:

$$Y'\text{∅}W1 = \sum_i Y_i/FX'C_i \quad .$$

Price level in OECD countries:

$$P'\text{∅}W1 = Y'\text{∅}NW1/Y'\text{∅}W1 \quad ;$$

b) for CMEA countries (= "World 2"):

$$Y'\text{∅}NW2 = \sum_i Y'N_i/FX_i \quad , \quad i = \text{CMEA country } i,$$

$$Y'\text{∅}W2 = \sum_i Y'\text{∅}_i \quad ,$$

$$P'\text{∅}W2 = Y'\text{∅}NW2/Y'\text{∅}W2 \quad ;$$

c) for developing countries (= "World 3"):
an equivalent formula is used as for the OECD countries.

d) for the whole world (= "World 0"):

$$Y'\text{∅}NWO = \sum_{i=1}^{3} Y'\text{∅}NWi \quad ,$$

$$Y'\text{∅}WO = \sum_{i=1}^{3} Y'\text{∅}Wi \quad ,$$

$$P'\text{∅}WO = Y'\text{∅}NWO/Y'\text{∅}WO \quad .$$

We now turn to foreign trade. This links the economies of the different countries to an interdependent system. The foreign trade part of the model is constructed in such a way that the adding up constraints are fully preserved with respect to real as well as nominal imports and exports, i.e.: total real imports = total real exports and total nominal imports = total nominal exports on the world level.

9. Real imports
(in domestic currency) are explained by a dynamic version of the linear expenditure system. We use variants of the equation:

$$IM = a_1 IM_{-1} + a_2 \frac{Y'N + IM'N - EX'N}{P'IM} + a_4 \frac{EX'N}{P'IM} - a_5 \frac{P'Y}{P'IM} Y_{-1} \quad {}^{1)},$$

where P'IM = price level of imports (in domestic currency), I'N = nominal investment, EX'N = nominal exports.
Real imports in \$ are defined by

$$IM'\$ = IM/FX'C \quad {}^{2)},$$

where FX'C = the \$ exchange rate in the base year 1975.

The price level of imports P'IM in domestic currency is explained by the import price level P'\$IM in \$ and the exchange rate index I'FX:

---

1) For developing countries a simplified version was used: $IM = a_1 IM_{-1} + a_2 \frac{I'N}{P'IM} + a_3 \frac{EX'N}{P'IM} + a_4 \frac{Y'N_{-1}}{P'IM}$ .

2) For CMEA countries the index FXDI'C of the \$ exchange rate used for imports in the base year has to be applied. Therefore another scaling factor P'IMD must be introduced. The above equation then becomes: $IM'\$ = IM \cdot P'IMD/FXDI'C$.

$$P'IM = P'\$IM \cdot I'FX$$ [1) .]

The price level P'$IM is related to the estimated price index P'$MM of imports in $ by a scaling factor D'P$IM which is equal for all countries and very near to one and is exactly one if the foreign trade statistics are consistent in real terms. In order to make the model consistent, i.e. in order to assure that total exports equal total imports also in real terms, adapt all import prices as they are forecast in the import price equations by the common factor D'P$IM (see below). Thus we have

$$P'\$IM = P'\$MM \cdot D'P\$IM \ .$$

The price index P'$MM is basically explained by the average $ export price index P'M$EX abroad. A variant of the following equation is used:

$$P'\$MM = a_1 \frac{P'IM_{-1}}{I'FX} + a_2 \; P'M\$EX + a_3 \; P'M\$EX_{-1}$$

$$+ \; a_4 \frac{I'FX_{-1}}{I'FX} + a_5 \quad ,$$

where

$$P'M\$EX = \sum_{i=1}^{3} SMW_i \cdot P'\$EXW_i \quad ,$$

$SMW_i$ = the appropriate weight for "World i", $P'\$EXW_i$ = price index of $ exports of "World i", see below. Thus we get for the imports in current $:

1) For CMEA countries this equation must be modified to P'IM = P'$IM · P'IMD · I'FXDI for the reason indicated in footnote (1).

$$IM'\$N = IM'\$ \cdot P'\$IM$$ [1]

and for the current value of imports in domestic currency

$$IM'N = IM \cdot P'IM \quad .$$

10. Import aggregation
over countries yields:

a) Imports of "World i" in current \$ values:

$$IM'\$NWi = \sum_{j \in J(i)} IM'\$N_j \quad , \qquad i = 1,2,3 \; ,$$

where J(i) = set of all countries belonging to "World i".

b) Imports of "World i" in constant \$ values:

$$IM'\$Wi = \sum_{j \in J(i)} IM'\$_j \quad , \qquad i = 1,2,3 \; .$$

c) Price index of imports of "World i":

$$P'\$IMWi = IM'\$NWi/IM'\$Wi \quad , \qquad i = 1,2,3 \; .$$

d) World Totals (= "World 0"):

$$IM'\$NW0 = \sum_{i=1}^{3} IM'\$NWi \quad ,$$

$$IM'\$W0 = \sum_{i=1}^{3} IM'\$Wi \quad ,$$

$$P'\$IMW0 = IM'\$NW0/IM'\$W0 \quad .$$

Of course, total world imports in real as well as in nominal terms must be equal to exports in real or nominal terms, respectively. This implies that the total world import price level equals the total world export price level.

---

1) For developing countries IM = IM\$ since for these countries we calculate in \$.

11. For OECD countries nominal exports
EX'\$N in current \$ are explained by a dynamic version of the linear expenditure system. Because of their large role in world trade, the USA (= country 1), the FRG (= country 2) and Japan (= country 3) will be treated separately. For these countries we estimate:

$$EX'\$N_i = a_1\, P'\$EX_i \cdot EX'\$_{i,-1} + a_2(IM'\$NWO + IM'\$NWR) - \sum_{\substack{j=1\\ j\neq i}}^{3} b_j\, P'\$EX_j \cdot EX'\$_{j,-1} - a_3\, P'\$EXW1A \cdot EX'\$W1A_{-1} - \sum_{j=2}^{3} c_j\, P'\$EXWj \cdot EX'\$Wj_{-1} ,$$

where $P'\$EX_j$ is the \$ export price of country j, $EX'\$_j$ is the export volume (in \$), P'\$EXWj is the export price of "World j", and "World 1A" includes all OECD countries other than the USA, FRG and Japan. $IM'\$NW_R$ denotes those imports (in current \$) which are not explained by total exports. If the statistics were consistent IM'\$NWR would be zero. We put it to zero in the forecasts.

For the other OECD countries which are treated as individual countries we relate the exports to the exports of the rest of the OECD :

$$EX'\$N_i = a_1\, P'EX_i \cdot EX'\$_{i,-1} + a_2\, EX'\$NW1a - \sum_{j\neq i} b_j\, P'\$EX_j \cdot EX'\$_j \quad , \quad j = 4,\dots,10 .$$

The nominal exports of the rest of the OECD countries which are not considered individually (= country 10) are estimated as a residual:

$$EX'\$N_{10} = EX'\$NW1A - \sum_{i=4}^{9} EX'\$N_i \quad .$$

The aggregated exports (in current $ value) of the OECD countries except the USA, FRG and Japan are estimated by

$$EX'\$NW1A = a_1\ P'\$EXW1A \cdot EX'\$W1A_{-1}$$
$$+ a_2(IM'\$NWO + IM'\$NWR)$$
$$- \sum_{j=1}^{3} b_j\ P'\$EX_j \cdot EX'\$_j - \sum_{i=2}^{3} P'\$EXWi \cdot EX'\$Wi \ .$$

Of course in real terms

$$EX'\$W1A = \sum_{i=4}^{10} EX'\$_i \quad .$$

Therefore

$$P'\$EXW1A = EX'\$NW1A/EX'\$W1A \ .$$

The export prices in domestic currency are determined by

$$P'EX = (P'Y)^{a_1} \cdot (P'IM)^{a_2} \cdot (P'IM_{-1})^{a_3} \cdot (P'EX_{-1})^{a_4} \cdot a_5 \ .$$

The $ export prices are

$$P'\$EX = P'EX/I'FX \quad .$$

Thus we get the nominal and real exports in domestic currency by

$EX'N = EX'\$N \cdot FX$ and $EX = EX'N/P'EX$ , respectively.

The balance of trade in current $ value is:

$$EX'\$IM = EX'\$N - IM'\$N \quad .$$

The export totals for OECD countries (= "World 1") follow as:

$$EX'\$NW1 = \sum_{i=1}^{3} EX'\$N_i + EX'\$NW1 ,$$

$EX'\$W1$ similarly, and $P'\$EXW1 = EX'\$NW1/EX'\$W1$ .

12. For the exports of CMEA countries
(in current $ value) basically the same approach is used. The exports of CMEA country i (with the exception of Romania) are explained as a function of the total exports of all CMEA countries:

$$EX'\$N_i = a_1 P'\$EX_i \cdot EX'\$_{i,-1} + a_2 EX'\$NW2 - \sum_{j \neq i} P'\$EX_j \cdot EX'\$_j ,$$

where j refers to CMEA countries and $EX'\$NW2$ means the $ value of the exports of all CMEA countries (= "World 2"). It is estimated as a function of total world imports, of the exports of the US (= country 1), the FRG (= country 2) and Japan (= country 3), of the exports of all other OECD countries (W1A) and of all developing countries (W3):

$$EX'\$NW2 = a_1 P'\$EXW2 \cdot EX'W2_{-1} + a_2 (IM'\$NWO + IM'\$NWR) - \sum_{i=1}^{3} b_i P'\$EX_i \cdot EX'\$_i - a_3 P'\$EXW1A \cdot EX'\$W1A_{-1} - a_3 P'\$EXW3 \cdot EX'\$W3_{-1} .$$

The exports $EX'\$_R$ of Romania are explained as a residual:

$$EX'\$_R = EX'\$NW2 - \sum_{i \neq R} EX'\$N_i ,$$

where i refers to all CMEA countries.

The <u>export prices</u> are estimated in current $\not\$$ value basically as a function of the $\not\$$ export prices of OECD countries:

$$P'\not\$EX = (P'\not\$EX_{-1})^{a_1} \cdot (P'\not\$EXW1)^{a_2} \cdot a_3 \quad .$$

From this we get the real exports in $\not\$$:

$$EX'\not\$ = EX'\not\$N/P'\not\$EX$$

and the nominal and real exports in domestic currency:

$$EX'N = EX'\not\$N \cdot FXDE \quad , \quad EX = EX'N/P'EX \quad .$$

The export price in domestic currency and in $\not\$$ are related by:

$$P'EX = P'\not\$EX \cdot I'FXDE \quad .$$

The <u>balance of trade</u> is defined as for OECD countries. The <u>total real exports</u> of all CMEA countries (in constant $\not\$$) are

$$EX'\not\$W2 = \sum_j EX'\not\$_j \quad ,$$

where j refers to CMEA countries.

The <u>price level</u> of these exports is

$$P'\not\$EXW2 = EX'\not\$NW2/EX'\not\$W2 \quad .$$

13. <u>The exports of developing countries</u>
(in current $\not\$$) with the exception of the exports of North Africa and the Middle East (= country 18) are explained similarly as the exports of OECD and CMEA countries, but by a simpler equation. They are related to total world 3 exports by:

$$EX'\not\$N_i = a_1\ P'\not\$EX_i \cdot EX_{i,-1} + a_2\ EX'\not\$NW3 \quad ,$$

where EX = EX'\$ and i refers to all developing countries or groups of countries without the group of countries 18. For this group the exports are explained as a residual:

$$EX'\$N_{18} = EX'\$NW3 - \sum_{j \neq 18} EX'\$N_i \quad ,$$

where j refers to all developing countries.

The nominal exports of "World 3" are also determined as a residual from total world exports minus the exports of "World 1" and "World 2":

$$EX'\$NW3 = EX'\$NWO - EX'\$NW1 - EX'\$NW2 \quad .$$

For total world exports EX'\$NWO, see the next section.

The export prices (in \$) of developing countries i (with the exception of the oil-exporting countries i = group 11) are basically explained by the import prices of OECD countries:

$$P'\$EX_i = (P'\$EX_{i,-1})^{a_1} \cdot (P'\$IMW1)^{a_2} \cdot a_3 \quad , \quad i \neq 11 \ .$$

For the oil-exporting countries (group 11), the export price level is related to the import price level P'\$MFLW1 of minerals and fuels of OECD countries which is estimated separately for the forecasts. This price is exogenous in this model. We estimate it by:

$$P'\$EX_{11} = (P'\$EX_{11,-1})^{a_1} \cdot (P'\$MFLW1)^{a_2} \cdot (P'\$MFLW1_{-1})^{a_3} \cdot a_4 \ .$$

The nominal exports EX'N in domestic prices are for each country

$$EX'N = EX'\$N \cdot I'FXDA$$

and the real exports:

$$EX = EX'\$N/P'\$EX \ .$$

The trade balance in \$ is, as usual, defined by

$$EX'\$IM = EX'\$N - IM'\$N \quad .$$

From this we get the totals for "World 3":

$$EX'\$W3 = \sum_i EX_i \quad ,$$

where i refers to developing countries and

$$P'EXW3 = EX'\$NW3/EX\$W3 \quad .$$

14. <u>The grand totals of foreign trade are:</u>

$$EX'\$NWO = IM'\$NWO + IM'\$NWR \quad .$$

Total world exports (in current \$) equal total imports (in current \$), but there is a small statistical discrepancy IM'\$NWR which we put to zero in the forecasts. This guarantees that the foreign trade model is consistent in value terms.

Total real exports (in \$) are

$$EX'\$WO = \sum_{i=1}^{3} EX'\$Wi \quad .$$

The world export price level (in \$) follows from

$$P'\$EXWO = EX'\$NWO/EX'\$WO \quad .$$

Now we have to define the scaling factor D'P\$IM for all \$ import prices which will guarantee consistency also in real terms. We put

$$D'P\$IM = \sum_i \frac{IM'\$N_i}{P'\$MM_i} \Big/ D'P\$M \cdot EX'\$WO \quad ,$$

where D'P$M is a correction factor for statistical discrepancies defined by

$$D'P\$M \cdot EX'\$WO = IM'\$WO \quad .$$

In the reference period this correction factor is around .97. We put it to one for forecasting. This guarantees consistency[1] and completes the model.

15. <u>The total aggregated model</u>

consists of 1069 equations and as many <u>endogenous variables</u>. It is controlled by relatively few <u>exogenous variables</u>. They have to be estimated separately for the forecasting period and are the <u>driving forces</u> of economic development. For each country (or group of countries) we have to specify for the forecasting period:

a. the labor force L[2],

b. the state of technology $\tau$ or the rate of technical progress $w_\tau$[3],

---

1) This is sufficient to guarantee consistency also in real terms. By construction of the model we have for the forecasting periods:

$$EX'\$NWO = IM'\$NWO \; .$$

Therefore $EX'\$WO = \frac{EX'\$NWO}{P'\$EXWO}$ will be equal to

$IM'\$WO = \frac{IM'\$NWO}{P'\$IMWO}$ if $P'\$EXWO = P'\$IMWO$. In fact there is a discrepancy if we apply our price forecasting formulae without adaptation, i.e. we get

$P'\$EXWO = A \cdot P'\$IMWO \quad , \quad A \neq 1$ in general (though not very different from one). But if we adapt all import prices for forecasting by the common factor

$$A = \frac{P'\$EXWO}{P'\$IMWO} = \frac{EX'\$NWO/EX'\$WO}{IM'\$NWO/IM'\$WO} = \frac{IM'\$WO}{EX'\$WO} = \frac{\sum_i IM'\$N_i/P'\$MM_i}{EX'\$WO}$$

we get consistency also in real terms. For forecasting: $A = D'P\$IM$.

2) For countries where no statistical forecasts of the labor force were available we put $L = L_{-1}(1 + w_L)$ and forecast $w_L$ from the past observations.

3) We put $\tau = \tau_{-1}(1 + w_\tau)$ and forecast $w_\tau$ (see chapter 2-6).

c. the investment ratio s[1],
d. the rate of depreciation d[2],
e. the ratio $\mu_R$ of secondary factor imports to total imports[3],
f. the money supply M or (for CMEA countries) the nominal wage rate (see chapter 2).

In addition, we have to forecast exogenously the $ import price P'$MFLW1 of fuels and related products. The oil price is subject to political decisions which should be taken as exogenous forces. The forecasts are given in chapter 2.

16. The commodity structure
of GDP or NMP and of exports and imports was estimated by relating the structure to the variables estimated in the aggregated model. For details see chapters 2-6.

---

1) For OECD countries we put $s = \lambda s_{-1} + (1-\lambda)\bar{s}$, $\lambda = .9$. For $\bar{s}$ and the medium scenario we took the average value of s from 1975-1984:

| | USA | FRG | Jap | Fr | GB | Ita | NL | B/L | Can | Rest OECD |
|---|---|---|---|---|---|---|---|---|---|---|
| $\bar{s}$ in % | 18.0 | 20.5 | 31.7 | 21.8 | 18.0 | 19.4 | 19.2 | 20.0 | 22.5 | 23.0 |

In the pessimistic scenario, $\bar{s}$ is 25% lower, in the optimistic scenario 25% higher.
For CMEA and developing countries, s is exogenous, see chapter 2.

2) For OECD countries we put $d = \lambda d_{-1} + (1-\lambda)\bar{d}$, $\lambda = .9$, $\bar{d}$ = average value of d for 1975-1984:

| | USA | FRG | Jap | Fr | GB | Ita | NL | B/L | Can | Rest OECD |
|---|---|---|---|---|---|---|---|---|---|---|
| $\bar{d}$ in % | 5.6 | 3.9 | 5.9 | 4.5 | 4.7 | 3.4 | 2.7 | 2.8 | 5.0 | 2.9 |

These figures were used for all three scenarios. For CMEA and developing countries, d is exogenous , see chapter 2.

3) For OECD countries we put $\mu_R = \lambda \mu_{R,-1} + (1-\lambda)\bar{\mu}_R$, $\lambda = .9$, $\bar{\mu}_R$ estimated on the basis of past observations as:

| | USA | FRG | Jap | Fr | GB | Ita | NL | B/L | Can | Rest OECD |
|---|---|---|---|---|---|---|---|---|---|---|
| $\bar{\mu}_R$ in % | 14.0 | 16.0 | 42.0 | 13.0 | 11.0 | 22.0 | 17.0 | 15.0 | 7.0 | 13.0 |

These figures were used for all scenarios. For CMEA and developing countries $\mu_R$ is exogenous, see chapter 2.

# Annex 3

# Basic Data from Other Sources

*Annex Table 8.1.* Total world population (in millions).[a]

| | 1960 | | | | | | | | | 2000 | |
|---|---|---|---|---|---|---|---|---|---|---|---|
| | Mill. | % | 1965 | 1970 | 1975 | 1980 | 1985 | 1990 | 1995 | Mill. | % |
| Industrialized Market Economies | 610 | 20.9 | 651 | 684 | 718 | 745 | 772 | 799 | 827 | 857 | 14.7 |
| European CMEA countries | 311 | 10.7 | 331 | 346 | 361 | 375 | 388 | 399 | 407 | 413 | 7.1 |
| Developing countries without P.R. of China | 1317 | 45.1 | 1492 | 1692 | 1915 | 2160 | 2432 | 2724 | 3039 | 3365 | 57.7 |
| (Thereof: included in the model) | (990) | / | (1121) | (1271) | (1437) | (1615) | (1807) | (2011) | (2229) | (2451) | / |
| P.R. of China | 681 | 23.3 | 721 | 816 | 912 | 975 | 1014<br>1033 | 1036<br>1090 | 1050<br>1145 | 1055<br>1196 | /<br>20.5 |
| World Total | 2919 | 100.0 | 3195 | 3538 | 3906 | 4255 | 4609<br>4628 | 4958<br>5012 | 5323<br>5418 | 5690<br>5831 | /<br>100.0 |

[a]Sources: Office of Development Research and Policy Analysis (DRPA) of the UN. These figures differ from those of the World Development Reports of the World Bank because the classification and coverage of countries differ. Population figures for the P.R. of China are borrowed, with gratitude, from the data bank of Professor Lawrence Lau, Stanford University. There are upper and lower figures for the future population of the P.R. of China.

*Annex Table 3.2.* Employed labor force (in millions).[a,b]

| | | 1960 Mill. | 1960 % | 1965 | 1970 | 1975 | 1980 | 1985 | 1990 | 1995 | 2000 Mill. | 2000 % |
|---|---|---|---|---|---|---|---|---|---|---|---|---|
| Industrialized Market economies | total | 249 | 19.8 | 265 | 280 | 293 | 314 | 328 | 352 | 380 | 411 | 15.9 |
| | (thereof: 9 OECD countries in the model) | (212) | / | (224) | (236) | (246) | (266) | (273) | (294) | (314) | (355) | / |
| European CMEA countries | total | 138 | 11.0 | 145 | 158 | 171 | 180 | 186 | 191 | 195 | 198 | 7.7 |
| | (thereof: employed in the material sphere) | (117) | / | (119) | (127) | (134) | (138) | (142) | (142) | (143) | (143) | / |
| Developing countries without P.R. of China | total | 519 | 41.4 | 568 | 630 | 699 | 775 | 864 | 963 | 1092 | 1221 | 47.3 |
| | (thereof: included in the model) | (388) | / | (424) | (470) | (522) | (577) | (641) | (711) | (802) | (892) | / |
| P.R. of China | | 348 | 27.8 | 372 | 420 | 488 | 543 | {593<br>604 | {640<br>674 | {653<br>712 | {661<br>750 | /<br>29.1 |
| World Total | | 1254 | 100.0 | 1350 | 1488 | 1651 | 1812 | {1971<br>1982 | {2146<br>2180 | {2320<br>2379 | {2491<br>2580 | /<br>100.0 |

[a]Sources: see *Annex Table 3.1.*
[b]For developing countries, the numbers reflect "economically active population".

*Annex Table 3.3.* Average growth rate of employed labor force.[a,b,c]

| | 1960–65 | 1965–70 | 1970–75 | 1975–80 | 1980–85 | 1985–90 | 1990–95 | 1995–2000 |
|---|---|---|---|---|---|---|---|---|
| 9 OECD countries | 1.10 | 1.05 | .083 | 1.57 | .74 | 1.27 | 1.32 | 1.30 |
| European CMEA countries | .34 | 1.73 | 1.08 | .59 | .57 | .00 | .14 | .00 |
| Developing countries (without P.R. of China) | 1.84 | 2.06 | 2.12 | 2.09 | 2.19 | 2.20 | 2.54 | 2.26 |
| Developing countries included in the model | 1.82 | 2.06 | 2.12 | 2.04 | 2.12 | 2.09 | 2.44 | 2.16 |
| P.R. of China | 1.34 | 2.46 | 3.05 | 2.16 | { 1.78<br>2.15 | { 1.54<br>2.22 | { .40<br>1.10 | { .24<br>1.05 |

[a] Sources: see *Annex Table 3.1.*
[b] For CMEA countries: growth rate of persons employed in the material sphere; for developing countries: growth rate of the economically active population.
[c] For comparison, we get from the figures in Appendix 2A (Chapter 2) for the average growth rate, 1985–2000, of employed labor force for the nine OECD countries:

$$W_L{}^* = W_{POP} + W_{LPR} + W_{ER}$$

| | USA | FRG | Japan | Fra | GB | Ita | NL | B/L | Can |
|---|---|---|---|---|---|---|---|---|---|
| $W_L$* | 1.7 | .0 | .6 | .1 | -.3 | .5 | .5 | -.2 | 1.9 |

# Annex 4

# Main Results of the Bonn–IIASA Research Project

*Annex Table 4.1.* Growth rates of GDP, OECD countries, medium scenario (to 1981: observations; 1982–1984: adjusted forecasts; 1985–1999: trend forecasts).

| Period | USA | FRG | Japan | France | UK |
|---|---|---|---|---|---|
| 62 | 5.542 | 4.501 | 7.054 | 6.614 | 1.051 |
| 63 | 4.048 | 3.163 | 10.504 | 5.399 | 4.051 |
| 64 | 5.272 | 6.673 | 13.196 | 6.309 | 5.265 |
| 65 | 6.075 | 5.488 | 5.132 | 4.882 | 2.401 |
| 66 | 5.828 | 2.735 | 10.587 | 5.063 | 1.898 |
| 67 | 2.764 | 0.020 | 10.769 | 4.725 | 2.806 |
| 68 | 4.047 | 5.730 | 12.752 | 4.221 | 4.035 |
| 69 | 2.831 | 7.400 | 12.253 | 6.704 | 1.153 |
| 70 | −0.283 | 5.234 | 9.805 | 5.657 | 2.111 |
| 71 | 3.060 | 3.269 | 4.605 | 5.526 | 2.427 |
| 72 | 5.404 | 4.157 | 8.812 | 5.701 | 1.837 |
| 73 | 5.510 | 4.468 | 8.801 | 5.185 | 7.462 |
| 74 | −0.656 | 0.632 | −1.034 | 3.125 | −0.944 |
| 75 | −0.694 | −1.581 | 2.344 | 0.710 | −0.866 |
| 76 | 4.886 | 5.327 | 5.296 | 4.817 | 3.689 |
| 77 | 5.241 | 3.095 | 5.251 | 3.003 | 1.227 |
| 78 | 4.681 | 3.113 | 5.032 | 3.774 | 3.508 |
| 79 | 2.350 | 4.074 | 5.134 | 3.088 | 1.780 |
| 80 | −0.223 | 1.917 | 4.931 | 0.991 | −2.571 |
| 81 | 2.901 | 0.095 | 4.199 | 0.640 | −2.185 |
| 82 | −2.194 | 0.998 | 0.546 | 1.816 | 1.852 |
| 83 | 1.774 | 1.327 | 4.045 | 1.167 | 0.144 |
| 84 | 4.371 | 2.108 | 3.518 | 2.521 | 2.194 |
| 85 | 2.211 | 2.850 | 4.214 | 2.803 | 1.145 |
| 86 | 2.218 | 2.898 | 4.431 | 2.808 | 1.147 |
| 87 | 2.221 | 2.920 | 4.407 | 2.792 | 1.140 |
| 88 | 2.223 | 2.931 | 4.362 | 2.783 | 1.136 |
| 89 | 2.225 | 2.939 | 4.323 | 2.786 | 1.132 |
| 90 | 2.228 | 2.947 | 4.292 | 2.798 | 1.129 |
| 91 | 2.230 | 2.950 | 4.260 | 2.808 | 1.125 |
| 92 | 2.232 | 2.952 | 4.229 | 2.817 | 1.121 |
| 93 | 2.234 | 2.952 | 4.200 | 2.824 | 1.115 |
| 94 | 2.236 | 2.952 | 4.175 | 2.831 | 1.109 |
| 95 | 2.237 | 2.951 | 4.153 | 2.836 | 1.102 |
| 96 | 2.238 | 2.951 | 4.134 | 2.841 | 1.095 |
| 97 | 2.239 | 2.951 | 4.118 | 2.844 | 1.088 |
| 98 | 2.240 | 2.951 | 4.104 | 2.846 | 1.080 |
| 99 | 2.241 | 2.951 | 4.093 | 2.848 | 1.072 |

For comparison. Optimistic scenario:

| | | | | | |
|---|---|---|---|---|---|
| 1990 | 2.768 | 3.885 | 5.579 | 3.719 | 1.634 |
| 1999 | 2.862 | 3.989 | 5.374 | 3.823 | 1.590 |

Pessimistic scenario:

| | | | | | |
|---|---|---|---|---|---|
| 1990 | 1.687 | 2.672 | 3.086 | 1.995 | .613 |
| 1999 | 1.560 | 1.860 | 2.833 | 1.821 | .457 |

*Annex Table 4.1.* Continued.

| Period | Italy | Netherlands | Belgium/ Luxembourg | Canada | Other DMEs |
|---|---|---|---|---|---|
| 62 | 5.856 | 4.146 | 5.082 | 5.330 | 6.276 |
| 63 | 4.900 | 3.753 | 4.333 | 5.709 | 5.374 |
| 64 | 2.691 | 8.066 | 6.901 | 6.937 | 6.214 |
| 65 | 3.141 | 5.445 | 3.524 | 6.494 | 3.853 |
| 66 | 5.739 | 2.895 | 3.102 | 7.375 | 4.676 |
| 67 | 7.135 | 5.371 | 3.832 | 4.124 | 3.766 |
| 68 | 6.315 | 6.508 | 4.035 | 5.381 | 5.511 |
| 69 | 5.915 | 6.120 | 6.551 | 4.200 | 6.003 |
| 70 | 5.171 | 6.763 | 6.105 | 4.945 | 4.917 |
| 71 | 1.102 | 4.337 | 2.813 | 7.196 | 3.612 |
| 72 | 2.822 | 3.415 | 5.285 | 4.875 | 5.063 |
| 73 | 7.818 | 5.422 | 6.348 | 7.066 | 4.635 |
| 74 | 4.260 | 3.340 | 4.433 | 2.420 | 3.132 |
| 75 | −4.611 | −0.520 | −2.084 | 1.965 | 1.439 |
| 76 | 6.455 | 5.033 | 5.391 | 6.153 | 2.629 |
| 77 | 1.508 | 2.624 | 0.627 | 3.068 | 1.351 |
| 78 | 2.282 | 2.494 | 3.202 | 3.564 | 2.751 |
| 79 | 4.774 | 2.501 | 2.528 | 2.735 | 2.523 |
| 80 | 4.458 | 0.877 | 3.154 | 1.512 | 3.174 |
| 81 | 3.002 | −0.580 | −0.910 | 3.634 | 1.073 |
| 82 | −2.689 | 0.541 | 1.627 | −2.914 | 0.779 |
| 83 | 0.360 | −0.975 | 2.011 | 1.460 | 2.940 |
| 84 | −0.589 | 2.022 | 2.516 | 1.733 | 3.068 |
| 85 | 2.967 | 2.057 | 2.649 | 2.444 | 3.035 |
| 86 | 2.954 | 2.085 | 2.696 | 2.480 | 3.029 |
| 87 | 2.901 | 2.096 | 2.731 | 2.507 | 2.969 |
| 88 | 2.881 | 2.104 | 2.753 | 2.527 | 2.919 |
| 89 | 2.876 | 2.112 | 2.780 | 2.544 | 2.889 |
| 90 | 2.879 | 2.123 | 2.820 | 2.559 | 2.877 |
| 91 | 2.879 | 2.126 | 2.845 | 2.570 | 2.869 |
| 92 | 2.877 | 2.127 | 2.867 | 2.580 | 2.865 |
| 93 | 2.875 | 2.125 | 2.882 | 2.587 | 2.863 |
| 94 | 2.874 | 2.123 | 2.899 | 2.593 | 2.864 |
| 95 | 2.873 | 2.120 | 2.910 | 2.598 | 2.865 |
| 96 | 2.872 | 2.117 | 2.922 | 2.601 | 2.868 |
| 97 | 2.871 | 2.114 | 2.931 | 2.602 | 2.871 |
| 98 | 2.870 | 2.110 | 2.941 | 2.603 | 2.874 |
| 99 | 2.870 | 2.106 | 2.948 | 2.602 | 2.877 |

For comparison. Optimistic scenario:

| | | | | | |
|---|---|---|---|---|---|
| 1990 | 4.385 | 3.407 | 3.716 | 3.440 | 3.948 |
| 1999 | 4.421 | 3.552 | 3.873 | 3.624 | 4.058 |

Pessimistic scenario:

| | | | | | |
|---|---|---|---|---|---|
| 1990 | 1.812 | 1.018 | 1.995 | 1.749 | 1.820 |
| 1999 | 1.715 | 0.820 | 1.862 | 1.610 | 1.636 |

*Annex Table 4.2.* Growth rates of NMP (GDP type "1"), CMEA countries, medium scenario (to 1981: observations; 1982–1985: adjusted forecasts; 1986–1999: trend forecasts).

| Period | Bulgaria | CSSR | GDR | Hungary | Poland | Romania | USSR |
|---|---|---|---|---|---|---|---|
| 62 | 6.209 | 1.402 | 2.695 | 5.854 | 2.105 | 9.765 | 5.781 |
| 63 | 7.518 | −2.415 | 3.550 | 5.581 | 6.948 | 9.239 | 3.643 |
| 64 | 9.848 | 0.855 | 4.937 | 4.219 | 6.728 | 8.692 | 8.914 |
| 65 | 7.035 | 3.427 | 4.647 | 0.558 | 7.009 | 9.438 | 7.378 |
| 66 | 11.066 | 9.150 | 4.880 | 8.006 | 7.107 | 7.571 | 9.447 |
| 67 | 9.374 | 5.276 | 5.397 | 8.141 | 5.690 | 7.650 | 8.877 |
| 68 | 6.189 | 7.186 | 5.101 | 5.071 | 9.498 | 6.595 | 8.243 |
| 69 | 10.033 | 7.311 | 5.203 | 7.805 | 2.438 | 7.147 | 4.827 |
| 70 | 7.038 | 5.678 | 5.603 | 5.282 | 5.198 | 6.969 | 8.694 |
| 71 | 6.796 | 5.496 | 4.437 | 5.914 | 8.101 | 12.983 | 5.625 |
| 72 | 7.770 | 5.749 | 5.651 | 6.619 | 10.568 | 10.626 | 2.593 |
| 73 | 8.108 | 5.199 | 5.603 | 7.355 | 10.816 | 10.387 | 9.403 |
| 74 | 7.615 | 5.909 | 6.468 | 6.166 | 10.447 | 13.052 | 5.330 |
| 75 | 8.806 | 6.237 | 4.874 | 6.531 | 7.195 | 10.263 | 2.954 |
| 76 | 6.523 | 3.656 | 3.456 | 2.980 | 6.846 | 9.280 | 5.199 |
| 77 | 6.315 | 4.196 | 5.049 | 8.235 | 5.009 | 9.062 | 5.320 |
| 78 | 5.575 | 4.109 | 3.698 | 4.500 | 3.002 | 7.310 | 4.744 |
| 79 | 6.599 | 3.053 | 4.048 | 2.330 | −2.286 | 6.368 | 2.521 |
| 80 | 5.711 | 2.916 | 4.415 | −0.651 | −6.000 | 2.983 | 3.438 |
| 81 | 4.993 | −0.109 | 4.817 | 2.517 | −11.994 | 2.491 | 2.516 |
| 82 | 4.143 | 1.820 | 3.218 | 1.663 | 3.398 | 3.378 | 5.073 |
| 83 | 3.957 | 2.202 | 5.093 | 1.648 | −1.226 | 5.687 | 2.278 |
| 84 | 4.082 | 2.147 | 4.333 | 2.109 | 4.107 | 5.416 | 3.962 |
| 85 | 2.079 | 3.070 | 4.731 | −0.081 | 3.108 | 5.698 | 3.479 |
| 86 | 4.485 | 2.741 | 4.342 | 2.480 | 2.561 | 5.093 | 5.061 |
| 87 | 5.045 | 3.117 | 4.510 | 2.932 | 2.646 | 5.589 | 4.137 |
| 88 | 4.956 | 3.034 | 4.416 | 2.828 | 2.541 | 5.389 | 4.070 |
| 89 | 4.943 | 3.018 | 4.407 | 2.809 | 2.502 | 5.314 | 4.023 |
| 90 | 4.917 | 3.002 | 4.409 | 2.790 | 2.473 | 5.252 | 3.981 |
| 91 | 4.891 | 3.023 | 4.415 | 2.779 | 3.398 | 5.199 | 3.941 |
| 92 | 4.865 | 2.968 | 4.423 | 2.772 | 3.406 | 5.157 | 3.904 |
| 93 | 4.839 | 2.949 | 4.432 | 2.761 | 3.429 | 5.121 | 3.867 |
| 94 | 4.815 | 2.930 | 4.441 | 2.744 | 3.449 | 5.090 | 3.833 |
| 95 | 4.793 | 2.912 | 4.449 | 2.730 | 3.459 | 5.068 | 3.799 |
| 96 | 4.770 | 2.892 | 4.454 | 2.719 | 3.483 | 5.047 | 3.766 |
| 97 | 4.751 | 2.874 | 4.460 | 2.698 | 3.500 | 5.033 | 3.735 |
| 98 | 4.730 | 2.893 | 4.464 | 2.684 | 3.513 | 5.017 | 3.704 |
| 99 | 4.711 | 2.838 | 4.469 | 2.659 | 3.522 | 5.002 | 3.674 |

For comparison. Optimistic scenario:

| | | | | | | | |
|---|---|---|---|---|---|---|---|
| 1990 | 6.223 | 4.394 | 5.371 | 4.358 | 3.704 | 7.021 | 5.217 |
| 1999 | 6.107 | 4.339 | 5.513 | 4.421 | 4.733 | 6.967 | 5.005 |

Pessimistic scenario:

| | | | | | | | |
|---|---|---|---|---|---|---|---|
| 1990 | 3.612 | 1.773 | 3.602 | 1.513 | 1.220 | 3.285 | 2.757 |
| 1999 | 3.317 | 1.543 | 3.581 | 1.315 | 2.200 | 2.769 | 2.345 |

*Annex Table 4.3.* Growth rates of GDP type "2", CMEA countries, medium scenario (observations and forecasts as in *Table 4.2*).

| Period | Bulgaria | CSSR | GDR | Hungary | Poland | Romania | USSR |
|---|---|---|---|---|---|---|---|
| 62 | 3.449 | 0.935 | 1.797 | 3.902 | 1.619 | 5.425 | 4.447 |
| 63 | 4.177 | -1.610 | 2.366 | 3.721 | 5.345 | 5.133 | 2.803 |
| 64 | 5.471 | 0.570 | 3.291 | 2.813 | 5.176 | 4.829 | 6.857 |
| 65 | 3.908 | 2.285 | 3.098 | 0.372 | 5.391 | 5.243 | 5.675 |
| 66 | 6.148 | 6.100 | 3.253 | 5.337 | 5.467 | 4.206 | 7.267 |
| 67 | 5.208 | 3.517 | 3.598 | 5.427 | 4.377 | 4.250 | 6.828 |
| 68 | 3.438 | 4.790 | 3.401 | 3.381 | 7.306 | 3.664 | 6.341 |
| 69 | 5.574 | 4.874 | 3.469 | 5.204 | 1.876 | 3.970 | 3.713 |
| 70 | 3.910 | 3.785 | 3.735 | 3.521 | 3.999 | 3.871 | 6.688 |
| 71 | 3.775 | 3.664 | 2.958 | 3.942 | 6.231 | 7.213 | 4.327 |
| 72 | 4.317 | 3.832 | 3.767 | 4.412 | 8.129 | 5.903 | 1.995 |
| 73 | 4.505 | 3.466 | 3.735 | 4.903 | 8.320 | 5.771 | 7.233 |
| 74 | 4.231 | 3.940 | 4.312 | 4.111 | 8.036 | 7.251 | 4.100 |
| 75 | 4.892 | 4.158 | 3.249 | 4.354 | 5.535 | 5.701 | 2.272 |
| 76 | 3.624 | 2.437 | 2.304 | 1.987 | 5.267 | 5.156 | 3.999 |
| 77 | 3.508 | 2.797 | 3.366 | 5.490 | 3.853 | 5.034 | 4.092 |
| 78 | 3.097 | 2.739 | 2.465 | 3.000 | 2.309 | 4.061 | 3.649 |
| 79 | 3.666 | 2.035 | 2.699 | 1.553 | -1.758 | 3.538 | 1.940 |
| 80 | 3.173 | 1.944 | 2.944 | -0.434 | -4.615 | 1.657 | 2.645 |
| 81 | 2.774 | -0.072 | 3.211 | 1.678 | -9.226 | 1.384 | 1.936 |
| 82 | 2.302 | 1.214 | 2.145 | 1.108 | 2.614 | 1.877 | 3.903 |
| 83 | 2.198 | 1.468 | 3.395 | 1.098 | -0.943 | 3.160 | 1.752 |
| 84 | 2.268 | 1.431 | 2.889 | 1.406 | 3.159 | 3.009 | 3.048 |
| 85 | 1.155 | 2.047 | 3.154 | -0.054 | 2.391 | 3.165 | 2.676 |
| 86 | 2.492 | 1.828 | 2.895 | 1.654 | 1.970 | 2.829 | 3.893 |
| 87 | 2.802 | 2.078 | 3.007 | 1.954 | 2.035 | 3.105 | 3.182 |
| 88 | 2.753 | 2.023 | 2.944 | 1.885 | 1.955 | 2.994 | 3.131 |
| 89 | 2.746 | 2.012 | 2.938 | 1.873 | 1.925 | 2.952 | 3.095 |
| 90 | 2.732 | 2.001 | 2.939 | 1.860 | 1.902 | 2.918 | 3.063 |
| 91 | 2.717 | 2.015 | 2.943 | 1.852 | 2.614 | 2.889 | 3.032 |
| 92 | 2.703 | 1.979 | 2.949 | 1.848 | 2.620 | 2.865 | 3.003 |
| 93 | 2.688 | 1.966 | 2.955 | 1.841 | 2.638 | 2.845 | 2.975 |
| 94 | 2.675 | 1.954 | 2.960 | 1.829 | 2.653 | 2.828 | 2.948 |
| 95 | 2.663 | 1.941 | 2.966 | 1.820 | 2.661 | 2.815 | 2.922 |
| 96 | 2.650 | 1.928 | 2.969 | 1.813 | 2.679 | 2.804 | 2.897 |
| 97 | 2.639 | 1.916 | 2.973 | 1.799 | 2.692 | 2.796 | 2.873 |
| 98 | 2.628 | 1.928 | 2.976 | 1.789 | 2.703 | 2.787 | 2.849 |
| 99 | 2.617 | 1.892 | 2.979 | 1.772 | 2.710 | 2.779 | 2.826 |

For comparison. Optimistic scenario:

| | | | | | | | |
|---|---|---|---|---|---|---|---|
| 1990 | 3.457 | 2.929 | 3.580 | 2.905 | 2.849 | 3.900 | 4.013 |
| 1999 | 3.393 | 2.893 | 3.675 | 2.947 | 3.641 | 3.870 | 3.850 |

Pessimistic scenario:

| | | | | | | | |
|---|---|---|---|---|---|---|---|
| 1990 | 2.006 | 1.182 | 2.402 | 1.008 | 0.939 | 1.825 | 2.121 |
| 1999 | 1.843 | 1.029 | 2.387 | 0.877 | 1.692 | 1.538 | 1.804 |

*Annex Table 4.4.* Growth rates of GDP, developing countries, medium scenario (to 1981: observations; 1982–1999: trend forecasts).

| Period | Oil-Exporting countries | Asian countries without India | India | Black Africa | Latin Amer without M.,B.,A. | Mexico Brazil Argentina | Mid-East & North Africa |
|---|---|---|---|---|---|---|---|
| 62 | 5.457 | 7.345 | 3.868 | 9.563 | 5.335 | 3.425 | 13.231 |
| 63 | 7.405 | 10.087 | 7.301 | 4.737 | 3.517 | 2.576 | 10.220 |
| 64 | 9.712 | 6.797 | 7.196 | 6.005 | 6.077 | 7.562 | 9.685 |
| 65 | 8.103 | 6.728 | -4.127 | 5.866 | 3.956 | 5.217 | -2.742 |
| 66 | 4.507 | 8.683 | -2.254 | 4.507 | 6.475 | 4.027 | 6.334 |
| 67 | 1.564 | 6.547 | 8.416 | 5.117 | 3.036 | 5.107 | 5.983 |
| 68 | 11.085 | 7.670 | 3.394 | 8.324 | 3.619 | 8.657 | 5.822 |
| 69 | 9.386 | 9.540 | 6.756 | 6.199 | 3.865 | 8.226 | 5.716 |
| 70 | 9.206 | 6.956 | 5.743 | 7.404 | 5.624 | 7.566 | 5.997 |
| 71 | 8.454 | 7.172 | 2.448 | 4.684 | 5.153 | 7.501 | 6.076 |
| 72 | 6.523 | 7.557 | -0.756 | 4.151 | 4.055 | 8.492 | 5.941 |
| 73 | 9.408 | 11.589 | 3.931 | 6.607 | 4.434 | 10.034 | 5.397 |
| 74 | 6.021 | 6.042 | 0.337 | 7.234 | 4.370 | 7.231 | 4.934 |
| 75 | -0.129 | 5.774 | 9.694 | 2.031 | 1.382 | 4.866 | 6.567 |
| 76 | 10.398 | 12.037 | 1.442 | 5.434 | 4.723 | 6.502 | 7.813 |
| 77 | 5.423 | 9.396 | 8.011 | 0.358 | 3.515 | 5.277 | 7.529 |
| 78 | -0.448 | 8.531 | 6.533 | 4.980 | 5.019 | 4.674 | 7.750 |
| 79 | 7.583 | 7.638 | -4.584 | 0.742 | 5.075 | 7.651 | 7.326 |
| 80 | -2.373 | 4.537 | 7.115 | 2.895 | 3.861 | 7.348 | 7.454 |
| 81 | -6.928 | 7.880 | 5.360 | 2.185 | 3.107 | 1.175 | 8.235 |
| 82 | -0.757 | 3.840 | 1.021 | 3.609 | 6.212 | 4.250 | 10.111 |
| 83 | -0.647 | 4.244 | 1.770 | 1.470 | 3.093 | 2.852 | 4.870 |
| 84 | 3.213 | 3.869 | 3.214 | 1.025 | 2.812 | 2.683 | 4.609 |
| 85 | -0.214 | 3.867 | 3.651 | 1.283 | 3.273 | 5.624 | 4.445 |
| 86 | 5.687 | 4.059 | 4.604 | 1.788 | 4.766 | 5.680 | 5.244 |
| 87 | 8.539 | 4.053 | 3.408 | 2.510 | 5.349 | 5.791 | 5.598 |
| 88 | 5.975 | 4.107 | 3.525 | 2.405 | 5.298 | 6.106 | 6.000 |
| 89 | 5.940 | 4.385 | 3.501 | 2.847 | 5.106 | 5.920 | 6.122 |
| 90 | 6.880 | 4.238 | 3.543 | 2.919 | 5.096 | 7.564 | 6.864 |
| 91 | 7.034 | 4.710 | 3.707 | 3.611 | 5.508 | 6.432 | 6.520 |
| 92 | 6.906 | 4.662 | 3.674 | 3.732 | 5.256 | 7.376 | 7.019 |
| 93 | 6.919 | 4.683 | 3.635 | 3.955 | 5.357 | 6.046 | 6.988 |
| 94 | 6.750 | 4.592 | 3.584 | 3.853 | 5.301 | 7.303 | 6.470 |
| 95 | 6.581 | 4.433 | 3.546 | 3.839 | 5.254 | 6.251 | 6.931 |
| 96 | 6.303 | 4.390 | 3.510 | 3.824 | 4.902 | 6.831 | 6.595 |
| 97 | 6.170 | 4.271 | 3.489 | 3.807 | 5.167 | 6.346 | 6.721 |
| 98 | 6.152 | 4.140 | 3.439 | 3.709 | 4.833 | 6.621 | 6.546 |
| 99 | 5.996 | 3.995 | 3.404 | 3.773 | 5.087 | 6.407 | 6.663 |

For comparison. Optimistic scenario:

| | | | | | | | |
|---|---|---|---|---|---|---|---|
| 1990 | 9.738 | 5.819 | 4.387 | 4.132 | 5.971 | 8.152 | 7.892 |
| 1999 | 7.405 | 5.257 | 4.316 | 4.539 | 5.896 | 7.342 | 7.568 |

Pessimistic scenario:

| | | | | | | | |
|---|---|---|---|---|---|---|---|
| 1990 | 3.462 | 3.082 | 3.368 | 1.542 | 3.844 | 5.853 | 5.934 |
| 1999 | 3.620 | 2.673 | 3.195 | 2.763 | 4.050 | 5.289 | 2.968 |

*Annex Table 4.5.* Growth rates of GDP, world regions, medium scenario (observations and forecasts as in *Table.4.4*).

| | | GDP type"1" | | with GDP type"1" for CMEA countr. | GDP type"2" | with GDP type"2" for CMEA countr. |
|---|---|---|---|---|---|---|
| Period | OECD countries total | CMEA countries total | Less developed countries total | World total | CMEA countries total | World total |
| 62 | 5.351 | 5.105 | 5.120 | 5.285 | 3.805 | 5.021 |
| 63 | 4.798 | 3.877 | 5.798 | 4.753 | 2.897 | 4.528 |
| 64 | 6.260 | 7.882 | 7.996 | 6.713 | 5.879 | 6.367 |
| 65 | 5.134 | 6.812 | 4.477 | 5.340 | 5.074 | 5.053 |
| 66 | 5.373 | 8.866 | 3.571 | 5.764 | 6.585 | 5.417 |
| 67 | 3.771 | 8.163 | 4.299 | 4.586 | 6.067 | 4.272 |
| 68 | 5.436 | 7.924 | 8.031 | 6.155 | 5.909 | 5.797 |
| 69 | 5.149 | 5.024 | 8.177 | 5.452 | 3.693 | 5.180 |
| 70 | 3.328 | 7.886 | 7.471 | 4.609 | 5.876 | 4.273 |
| 71 | 3.567 | 6.013 | 6.785 | 4.388 | 4.456 | 4.102 |
| 72 | 5.241 | 4.069 | 6.007 | 5.108 | 2.977 | 4.879 |
| 73 | 5.949 | 9.057 | 8.435 | 6.822 | 6.759 | 6.395 |
| 74 | 0.705 | 6.228 | 5.718 | 2.357 | 4.585 | 2.056 |
| 75 | -0.155 | 4.117 | 3.925 | 1.194 | 2.975 | 0.971 |
| 76 | 4.756 | 5.317 | 7.623 | 5.232 | 3.931 | 4.948 |
| 77 | 3.719 | 5.486 | 5.803 | 4.349 | 4.034 | 4.052 |
| 78 | 3.959 | 4.611 | 3.877 | 4.083 | 3.397 | 3.835 |
| 79 | 3.064 | 2.430 | 5.845 | 3.294 | 1.724 | 3.159 |
| 80 | 1.411 | 2.523 | 3.637 | 1.938 | 1.836 | 1.796 |
| 81 | 2.011 | 1.375 | 0.303 | 1.647 | 0.933 | 1.562 |
| 82 | -0.545 | 4.496 | 2.861 | 0.954 | 3.377 | 0.692 |
| 83 | 1.917 | 2.368 | 2.064 | 2.034 | 1.673 | 1.888 |
| 84 | 3.179 | 3.935 | 3.095 | 3.330 | 2.910 | 3.113 |
| 85 | 2.704 | 3.480 | 3.358 | 2.961 | 2.563 | 2.766 |
| 86 | 2.751 | 4.663 | 5.178 | 3.498 | 3.463 | 3.233 |
| 87 | 2.747 | 4.080 | 5.879 | 3.475 | 2.995 | 3.242 |
| 88 | 2.742 | 4.003 | 5.406 | 3.399 | 2.940 | 3.170 |
| 89 | 2.741 | 3.964 | 5.373 | 3.394 | 2.910 | 3.167 |
| 90 | 2.745 | 3.930 | 6.272 | 3.531 | 2.883 | 3.309 |
| 91 | 2.748 | 3.961 | 5.997 | 3.512 | 2.912 | 3.288 |
| 92 | 2.750 | 3.931 | 6.342 | 3.574 | 2.889 | 3.354 |
| 93 | 2.752 | 3.905 | 5.856 | 3.507 | 2.869 | 3.288 |
| 94 | 2.755 | 3.880 | 6.257 | 3.581 | 2.849 | 3.366 |
| 95 | 2.758 | 3.856 | 5.819 | 3.519 | 2.830 | 3.304 |
| 96 | 2.761 | 3.833 | 5.932 | 3.548 | 2.812 | 3.336 |
| 97 | 2.764 | 3.811 | 5.725 | 3.522 | 2.795 | 3.311 |
| 98 | 2.768 | 3.790 | 5.789 | 3.542 | 2.778 | 3.335 |
| 99 | 2.772 | 3.768 | 5.678 | 3.531 | 2.760 | 3.325 |
| For comparison. Optimistic scenario: | | | | | | |
| 1990 | 3.629 | 5.194 | 7.738 | 4.629 | 3.813 | 4.340 |
| 1999 | 3.762 | 5.132 | 6.805 | 4.684 | 3.758 | 4.401 |
| Pessimistic scenario: | | | | | | |
| 1990 | 1.928 | 2.688 | 4.444 | 2.453 | 1.968 | 2.301 |
| 1999 | 1.787 | 2.416 | 4.343 | 2.365 | 1.769 | 2.236 |

*Annex Table 4.6.* GDP per capita (in 1,000 $, 1975 value), OECD countries, medium scenario (observations and forecasts as in *Table 4.2*).

| Period | USA | FRG | Japan | France | UK |
|---|---|---|---|---|---|
| 62 | 5.412 | 4.619 | 1.853 | 3.894 | 3.141 |
| 63 | 5.550 | 4.711 | 2.027 | 4.034 | 3.252 |
| 64 | 5.762 | 4.967 | 2.271 | 4.245 | 3.403 |
| 65 | 6.037 | 5.174 | 2.362 | 4.411 | 3.464 |
| 66 | 6.315 | 5.272 | 2.589 | 4.597 | 3.509 |
| 67 | 6.419 | 5.240 | 2.838 | 4.776 | 3.588 |
| 68 | 6.612 | 5.513 | 3.164 | 4.942 | 3.715 |
| 69 | 6.734 | 5.894 | 3.510 | 5.230 | 3.743 |
| 70 | 6.637 | 6.175 | 3.811 | 5.477 | 3.812 |
| 71 | 6.754 | 6.317 | 3.935 | 5.725 | 3.891 |
| 72 | 7.043 | 6.539 | 4.223 | 5.999 | 3.951 |
| 73 | 7.361 | 6.798 | 4.530 | 6.258 | 4.236 |
| 74 | 7.246 | 6.833 | 4.424 | 6.409 | 4.195 |
| 75 | 7.125 | 6.748 | 4.471 | 6.418 | 4.160 |
| 76 | 7.402 | 7.145 | 4.657 | 6.715 | 4.314 |
| 77 | 7.712 | 7.379 | 4.855 | 6.892 | 4.370 |
| 78 | 7.988 | 7.620 | 5.053 | 7.111 | 4.524 |
| 79 | 8.086 | 7.913 | 5.268 | 7.300 | 4.602 |
| 80 | 7.973 | 8.049 | 5.485 | 7.335 | 4.478 |
| 81 | 8.123 | 8.043 | 5.673 | 7.341 | 4.349 |
| 82 | 7.865 | 8.127 | 5.665 | 7.433 | 4.430 |
| 83 | 7.930 | 8.264 | 5.854 | 7.486 | 4.433 |
| 84 | 8.202 | 8.472 | 6.022 | 7.644 | 4.522 |
| 85 | 8.300 | 8.730 | 6.238 | 7.834 | 4.578 |
| 86 | 8.400 | 9.001 | 6.476 | 8.030 | 4.635 |
| 87 | 8.502 | 9.283 | 6.721 | 8.230 | 4.693 |
| 88 | 8.605 | 9.574 | 6.972 | 8.434 | 4.751 |
| 89 | 8.709 | 9.875 | 7.230 | 8.643 | 4.809 |
| 90 | 8.815 | 10.187 | 7.496 | 8.858 | 4.869 |
| 91 | 8.923 | 10.508 | 7.768 | 9.079 | 4.928 |
| 92 | 9.032 | 10.840 | 8.049 | 9.307 | 4.988 |
| 93 | 9.142 | 11.183 | 8.337 | 9.541 | 5.049 |
| 94 | 9.254 | 11.536 | 8.633 | 9.782 | 5.110 |
| 95 | 9.367 | 11.900 | 8.938 | 10.030 | 5.172 |
| 96 | 9.482 | 12.276 | 9.252 | 10.284 | 5.234 |
| 97 | 9.598 | 12.663 | 9.575 | 10.544 | 5.296 |
| 98 | 9.716 | 13.063 | 9.909 | 10.812 | 5.358 |
| 99 | 9.836 | 13.475 | 10.253 | 11.087 | 5.421 |

For comparison. Optimistic scenario:

| | | | | | |
|---|---|---|---|---|---|
| 1990 | 9.045 | 10.704 | 8.039 | 9.312 | 4.994 |
| 1999 | 10.644 | 15.462 | 12.292 | 12.684 | 5.831 |

Pessimistic scenario:

| | | | | | |
|---|---|---|---|---|---|
| 1990 | 8.592 | 9.738 | 7.025 | 8.497 | 4.745 |
| 1999 | 9.062 | 11.794 | 8.621 | 9.802 | 5.014 |

*Annex Table 4.6.* Continued.

| Period | Italy | Netherlands | Belgium/ Luxembourg | Canada | Other DMEs |
|---|---|---|---|---|---|
| 62 | 2.301 | 4.082 | 3.926 | 4.511 | 3.017 |
| 63 | 2.369 | 4.175 | 4.065 | 4.680 | 3.134 |
| 64 | 2.414 | 4.456 | 4.304 | 4.909 | 3.275 |
| 65 | 2.470 | 4.633 | 4.417 | 5.135 | 3.322 |
| 66 | 2.595 | 4.706 | 4.521 | 5.412 | 3.431 |
| 67 | 2.762 | 4.900 | 4.670 | 5.536 | 3.511 |
| 68 | 2.919 | 5.169 | 4.838 | 5.744 | 3.656 |
| 69 | 3.072 | 5.422 | 5.138 | 5.900 | 3.824 |
| 70 | 3.211 | 5.718 | 5.445 | 6.107 | 3.958 |
| 71 | 3.225 | 5.893 | 5.592 | 6.465 | 4.042 |
| 72 | 3.292 | 6.030 | 5.862 | 6.705 | 4.186 |
| 73 | 3.517 | 6.305 | 6.213 | 7.101 | 4.323 |
| 74 | 3.633 | 6.468 | 6.466 | 7.166 | 4.400 |
| 75 | 3.440 | 6.382 | 6.311 | 7.201 | 4.415 |
| 76 | 3.640 | 6.645 | 6.637 | 7.544 | 4.483 |
| 77 | 3.676 | 6.780 | 6.671 | 7.692 | 4.488 |
| 78 | 3.782 | 6.904 | 6.877 | 7.895 | 4.555 |
| 79 | 3.951 | 7.032 | 7.044 | 8.039 | 4.615 |
| 80 | 4.118 | 7.038 | 7.258 | 8.072 | 4.709 |
| 81 | 4.235 | 6.943 | 7.198 | 8.235 | 4.705 |
| 82 | 4.111 | 6.952 | 7.315 | 7.901 | 4.681 |
| 83 | 4.111 | 6.860 | 7.455 | 7.932 | 4.761 |
| 84 | 4.077 | 6.969 | 7.627 | 7.993 | 4.849 |
| 85 | 4.186 | 7.084 | 7.821 | 8.075 | 4.937 |
| 86 | 4.296 | 7.203 | 8.024 | 8.161 | 5.027 |
| 87 | 4.408 | 7.325 | 8.235 | 8.250 | 5.114 |
| 88 | 4.521 | 7.449 | 8.453 | 8.342 | 5.201 |
| 89 | 4.637 | 7.576 | 8.680 | 8.436 | 5.288 |
| 90 | 4.757 | 7.706 | 8.916 | 8.532 | 5.376 |
| 91 | 4.879 | 7.839 | 9.160 | 8.631 | 5.464 |
| 92 | 5.004 | 7.974 | 9.413 | 8.731 | 5.554 |
| 93 | 5.133 | 8.111 | 9.675 | 8.834 | 5.646 |
| 94 | 5.264 | 8.250 | 9.945 | 8.937 | 5.738 |
| 95 | 5.399 | 8.391 | 10.225 | 9.043 | 5.833 |
| 96 | 5.538 | 8.535 | 10.513 | 9.150 | 5.929 |
| 97 | 5.680 | 8.680 | 10.810 | 9.259 | 6.027 |
| 98 | 5.825 | 8.828 | 11.117 | 9.368 | 6.127 |
| 99 | 5.975 | 8.978 | 11.433 | 9.479 | 6.228 |

For comparison. Optimistic scenario:

| | | | | | |
|---|---|---|---|---|---|
| 1990 | 5.185 | 8.272 | 9.384 | 8.914 | 5.783 |
| 1999 | 7.443 | 10.884 | 13.046 | 10.796 | 7.404 |

Pessimistic scenario:

| | | | | | |
|---|---|---|---|---|---|
| 1990 | 4.477 | 7.258 | 8.535 | 8.200 | 5.001 |
| 1999 | 5.098 | 7.593 | 10.059 | 8.390 | 5.229 |

*Annex Table 4.7.* GDP type "1" per capita (in 1,000 $, 1975 value), CMEA countries, medium scenario (observations and forecasts as in *Table 4.2*).

| Period | Bulgaria | CSSR | GDR | Hungary | Poland | Romania | USSR |
|---|---|---|---|---|---|---|---|
| 62 | 1.301 | 2.213 | 2.115 | 1.717 | 1.534 | 0.837 | 1.732 |
| 63 | 1.388 | 2.144 | 2.195 | 1.806 | 1.620 | 0.908 | 1.768 |
| 64 | 1.512 | 2.145 | 2.310 | 1.877 | 1.707 | 0.980 | 1.897 |
| 65 | 1.605 | 2.203 | 2.423 | 1.881 | 1.804 | 1.066 | 2.007 |
| 66 | 1.770 | 2.400 | 2.540 | 2.024 | 1.919 | 1.132 | 2.175 |
| 67 | 1.923 | 2.519 | 2.675 | 2.180 | 2.016 | 1.204 | 2.344 |
| 68 | 2.028 | 2.695 | 2.812 | 2.284 | 2.193 | 1.267 | 2.513 |
| 69 | 2.216 | 2.883 | 2.956 | 2.453 | 2.232 | 1.341 | 2.608 |
| 70 | 2.355 | 3.040 | 3.120 | 2.573 | 2.333 | 1.417 | 2.806 |
| 71 | 2.502 | 3.186 | 3.266 | 2.714 | 2.499 | 1.585 | 2.936 |
| 72 | 2.681 | 3.348 | 3.461 | 2.883 | 2.739 | 1.737 | 2.984 |
| 73 | 2.883 | 3.500 | 3.663 | 3.083 | 3.008 | 1.899 | 3.234 |
| 74 | 3.087 | 3.682 | 3.910 | 3.260 | 3.292 | 2.127 | 3.375 |
| 75 | 3.340 | 3.887 | 4.110 | 3.460 | 3.498 | 2.322 | 3.442 |
| 76 | 3.543 | 3.998 | 4.268 | 3.543 | 3.700 | 2.514 | 3.588 |
| 77 | 3.747 | 4.135 | 4.489 | 3.818 | 3.848 | 2.715 | 3.746 |
| 78 | 3.952 | 4.274 | 4.658 | 3.976 | 3.928 | 2.887 | 3.890 |
| 79 | 4.201 | 4.381 | 4.849 | 4.062 | 3.808 | 3.047 | 3.956 |
| 80 | 4.429 | 4.482 | 5.066 | 4.032 | 3.550 | 3.114 | 4.059 |
| 81 | 4.635 | 4.474 | 5.310 | 4.133 | 3.096 | 3.170 | 4.127 |
| 82 | 4.813 | 4.541 | 5.494 | 4.204 | 3.173 | 3.258 | 4.299 |
| 83 | 4.992 | 4.626 | 5.770 | 4.278 | 3.114 | 3.416 | 4.371 |
| 84 | 5.186 | 4.710 | 6.016 | 4.373 | 3.220 | 3.579 | 4.507 |
| 85 | 5.285 | 4.836 | 6.297 | 4.374 | 3.299 | 3.760 | 4.626 |
| 86 | 5.515 | 4.952 | 6.566 | 4.487 | 3.362 | 3.929 | 4.824 |
| 87 | 5.787 | 5.090 | 6.857 | 4.624 | 3.429 | 4.126 | 4.988 |
| 88 | 6.070 | 5.228 | 7.155 | 4.760 | 3.494 | 4.326 | 5.157 |
| 89 | 6.367 | 5.369 | 7.466 | 4.899 | 3.559 | 4.534 | 5.332 |
| 90 | 6.671 | 5.512 | 7.790 | 5.041 | 3.624 | 4.749 | 5.514 |
| 91 | 6.987 | 5.657 | 8.128 | 5.187 | 3.724 | 4.974 | 5.702 |
| 92 | 7.315 | 5.807 | 8.482 | 5.336 | 3.828 | 5.208 | 5.897 |
| 93 | 7.658 | 5.960 | 8.853 | 5.489 | 3.934 | 5.453 | 6.100 |
| 94 | 8.014 | 6.115 | 9.240 | 5.646 | 4.045 | 5.708 | 6.309 |
| 95 | 8.385 | 6.273 | 9.644 | 5.807 | 4.160 | 5.976 | 6.526 |
| 96 | 8.771 | 6.435 | 10.067 | 5.971 | 4.279 | 6.255 | 6.750 |
| 97 | 9.174 | 6.599 | 10.509 | 6.139 | 4.403 | 6.548 | 6.982 |
| 98 | 9.593 | 6.765 | 10.971 | 6.311 | 4.530 | 6.854 | 7.222 |
| 99 | 10.030 | 6.936 | 11.454 | 6.486 | 4.662 | 7.174 | 7.469 |

For comparison. Optimistic scenario:

| | | | | | | | |
|---|---|---|---|---|---|---|---|
| 1990 | 7.063 | 5.879 | 8.131 | 5.426 | 3.838 | 5.136 | 5.830 |
| 1999 | 11.926 | 8.395 | 13.039 | 8.082 | 5.489 | 9.110 | 8.829 |

Pessimistic scenario:

| | | | | | | | |
|---|---|---|---|---|---|---|---|
| 1990 | 6.292 | 5.203 | 7.513 | 4.743 | 3.419 | 4.349 | 5.213 |
| 1999 | 8.410 | 5.853 | 10.265 | 5.429 | 3.928 | 5.467 | 6.312 |

*Annex Table 4.8.* GDP type "2" per capita (in 1,000 $, 1975 value), CMEA countries, medium scenario (observations and forecasts as in *Table 4.2*).

| Period | Bulgaria | CSSR | GDR | Hungary | Poland | Romania | USSR |
|---|---|---|---|---|---|---|---|
| 62 | 2.035 | 2.718 | 2.614 | 2.192 | 1.893 | 1.378 | 2.086 |
| 63 | 2.102 | 2.655 | 2.682 | 2.265 | 1.969 | 1.438 | 2.113 |
| 64 | 2.199 | 2.649 | 2.779 | 2.322 | 2.045 | 1.497 | 2.224 |
| 65 | 2.266 | 2.691 | 2.871 | 2.323 | 2.128 | 1.566 | 2.315 |
| 66 | 2.389 | 2.849 | 2.963 | 2.438 | 2.230 | 1.611 | 2.459 |
| 67 | 2.496 | 2.940 | 3.068 | 2.560 | 2.313 | 1.659 | 2.601 |
| 68 | 2.564 | 3.075 | 3.172 | 2.639 | 2.466 | 1.699 | 2.738 |
| 69 | 2.688 | 3.216 | 3.280 | 2.765 | 2.496 | 1.745 | 2.812 |
| 70 | 2.774 | 3.329 | 3.401 | 2.852 | 2.579 | 1.790 | 2.970 |
| 71 | 2.863 | 3.428 | 3.510 | 2.953 | 2.715 | 1.900 | 3.069 |
| 72 | 2.971 | 3.538 | 3.653 | 3.071 | 2.909 | 1.994 | 3.101 |
| 73 | 3.088 | 3.638 | 3.798 | 3.209 | 3.123 | 2.088 | 3.294 |
| 74 | 3.202 | 3.755 | 3.971 | 3.328 | 3.344 | 2.219 | 3.397 |
| 75 | 3.340 | 3.887 | 4.110 | 3.460 | 3.498 | 2.322 | 3.442 |
| 76 | 3.446 | 3.951 | 4.221 | 3.509 | 3.645 | 2.419 | 3.547 |
| 77 | 3.549 | 4.031 | 4.368 | 3.685 | 3.749 | 2.516 | 3.660 |
| 78 | 3.655 | 4.112 | 4.478 | 3.782 | 3.802 | 2.595 | 3.761 |
| 79 | 3.778 | 4.173 | 4.602 | 3.836 | 3.705 | 2.665 | 3.803 |
| 80 | 3.888 | 4.229 | 4.740 | 3.815 | 3.506 | 2.689 | 3.872 |
| 81 | 3.983 | 4.224 | 4.892 | 3.879 | 3.153 | 2.707 | 3.915 |
| 82 | 4.062 | 4.261 | 5.009 | 3.924 | 3.207 | 2.743 | 4.033 |
| 83 | 4.143 | 4.309 | 5.176 | 3.971 | 3.156 | 2.807 | 4.079 |
| 84 | 4.228 | 4.357 | 5.322 | 4.031 | 3.234 | 2.873 | 4.169 |
| 85 | 4.270 | 4.429 | 5.486 | 4.033 | 3.290 | 2.946 | 4.246 |
| 86 | 4.371 | 4.495 | 5.641 | 4.105 | 3.334 | 3.013 | 4.378 |
| 87 | 4.488 | 4.574 | 5.807 | 4.190 | 3.380 | 3.089 | 4.485 |
| 88 | 4.609 | 4.652 | 5.974 | 4.273 | 3.425 | 3.165 | 4.595 |
| 89 | 4.734 | 4.730 | 6.145 | 4.358 | 3.468 | 3.243 | 4.709 |
| 90 | 4.857 | 4.810 | 6.322 | 4.444 | 3.513 | 3.322 | 4.827 |
| 91 | 4.981 | 4.888 | 6.503 | 4.531 | 3.582 | 3.403 | 4.948 |
| 92 | 5.107 | 4.969 | 6.691 | 4.620 | 3.653 | 3.485 | 5.073 |
| 93 | 5.237 | 5.051 | 6.884 | 4.710 | 3.727 | 3.570 | 5.202 |
| 94 | 5.368 | 5.134 | 7.083 | 4.801 | 3.802 | 3.657 | 5.335 |
| 95 | 5.503 | 5.217 | 7.288 | 4.894 | 3.880 | 3.746 | 5.472 |
| 96 | 5.640 | 5.301 | 7.500 | 4.988 | 3.960 | 3.837 | 5.612 |
| 97 | 5.780 | 5.386 | 7.718 | 5.084 | 4.042 | 3.931 | 5.757 |
| 98 | 5.923 | 5.469 | 7.942 | 5.180 | 4.127 | 4.028 | 5.905 |
| 99 | 6.068 | 5.556 | 8.174 | 5.278 | 4.214 | 4.127 | 6.057 |

For comparison. Optimistic scenario:

| | | | | | | | |
|---|---|---|---|---|---|---|---|
| 1990 | 5.017 | 5.023 | 6.508 | 4.670 | 3.672 | 3.473 | 5.041 |
| 1999 | 6.697 | 6.319 | 8.923 | 6.122 | 4.783 | 4.729 | 6.898 |
| Pessimistic scenario: | | | | | | | |
| 1990 | 4.698 | 4.627 | 6.169 | 4.265 | 3.358 | 3.160 | 4.621 |
| 1999 | 5.493 | 4.957 | 7.590 | 4.684 | 3.691 | 3.539 | 5.317 |

*Annex Table 4.9.* GDP per capita (in 1,000 $, 1975 value), developing countries, medium scenario (observations and forecasts as in *Table 4.4*).

| Period | Oil-Exporting countries | Asian countries without India | India | Black Africa | Latin Amer without M.,B.,A. | Mexico Brazil Argentina | Mid-East & North Africa |
|---|---|---|---|---|---|---|---|
| 62 | 0.492 | 0.284 | 0.122 | 0.206 | 0.645 | 0.812 | 0.317 |
| 63 | 0.514 | 0.304 | 0.128 | 0.210 | 0.648 | 0.810 | 0.340 |
| 64 | 0.551 | 0.315 | 0.134 | 0.217 | 0.670 | 0.847 | 0.365 |
| 65 | 0.581 | 0.328 | 0.126 | 0.225 | 0.678 | 0.868 | 0.364 |
| 66 | 0.591 | 0.346 | 0.120 | 0.229 | 0.703 | 0.878 | 0.349 |
| 67 | 0.584 | 0.359 | 0.127 | 0.233 | 0.706 | 0.896 | 0.346 |
| 68 | 0.632 | 0.376 | 0.128 | 0.246 | 0.711 | 0.948 | 0.359 |
| 69 | 0.674 | 0.402 | 0.134 | 0.254 | 0.723 | 0.999 | 0.385 |
| 70 | 0.718 | 0.419 | 0.138 | 0.266 | 0.743 | 1.047 | 0.391 |
| 71 | 0.757 | 0.437 | 0.138 | 0.269 | 0.764 | 1.096 | 0.404 |
| 72 | 0.784 | 0.458 | 0.134 | 0.272 | 0.778 | 1.158 | 0.415 |
| 73 | 0.836 | 0.499 | 0.136 | 0.282 | 0.794 | 1.241 | 0.413 |
| 74 | 0.864 | 0.516 | 0.134 | 0.294 | 0.810 | 1.297 | 0.430 |
| 75 | 0.840 | 0.534 | 0.144 | 0.291 | 0.809 | 1.327 | 0.481 |
| 76 | 0.903 | 0.583 | 0.143 | 0.297 | 0.825 | 1.377 | 0.508 |
| 77 | 0.927 | 0.623 | 0.151 | 0.289 | 0.835 | 1.415 | 0.520 |
| 78 | 0.900 | 0.660 | 0.158 | 0.294 | 0.857 | 1.445 | 0.554 |
| 79 | 0.944 | 0.695 | 0.147 | 0.288 | 0.881 | 1.518 | 0.576 |
| 80 | 0.899 | 0.711 | 0.155 | 0.287 | 0.895 | 1.592 | 0.601 |
| 81 | 0.814 | 0.749 | 0.160 | 0.284 | 0.904 | 1.572 | 0.611 |
| 82 | 0.787 | 0.761 | 0.158 | 0.285 | 0.936 | 1.601 | 0.654 |
| 83 | 0.762 | 0.777 | 0.158 | 0.279 | 0.942 | 1.608 | 0.667 |
| 84 | 0.767 | 0.789 | 0.160 | 0.274 | 0.946 | 1.613 | 0.680 |
| 85 | 0.747 | 0.803 | 0.163 | 0.269 | 0.957 | 1.667 | 0.691 |
| 86 | 0.768 | 0.818 | 0.168 | 0.264 | 0.980 | 1.720 | 0.707 |
| 87 | 0.813 | 0.834 | 0.170 | 0.262 | 1.008 | 1.779 | 0.727 |
| 88 | 0.840 | 0.851 | 0.173 | 0.260 | 1.036 | 1.846 | 0.751 |
| 89 | 0.868 | 0.872 | 0.176 | 0.259 | 1.066 | 1.914 | 0.776 |
| 90 | 0.906 | 0.892 | 0.180 | 0.259 | 1.096 | 2.015 | 0.808 |
| 91 | 0.945 | 0.917 | 0.183 | 0.259 | 1.132 | 2.098 | 0.840 |
| 92 | 0.986 | 0.941 | 0.187 | 0.260 | 1.164 | 2.206 | 0.876 |
| 93 | 1.029 | 0.968 | 0.190 | 0.262 | 1.201 | 2.291 | 0.914 |
| 94 | 1.073 | 0.995 | 0.194 | 0.264 | 1.238 | 2.409 | 0.950 |
| 95 | 1.117 | 1.021 | 0.198 | 0.266 | 1.274 | 2.510 | 0.993 |
| 96 | 1.160 | 1.050 | 0.201 | 0.267 | 1.309 | 2.628 | 1.032 |
| 97 | 1.203 | 1.077 | 0.205 | 0.268 | 1.351 | 2.740 | 1.077 |
| 98 | 1.248 | 1.103 | 0.209 | 0.270 | 1.388 | 2.865 | 1.120 |
| 99 | 1.294 | 1.130 | 0.213 | 0.271 | 1.428 | 2.993 | 1.167 |

For comparison. Optimistic scenario:

| | | | | | | | |
|---|---|---|---|---|---|---|---|
| 1990 | 1.168 | 1.005 | 0.192 | 0.286 | 1.176 | 2.256 | 0.879 |
| 1999 | 1.937 | 1.423 | 0.246 | 0.321 | 1.648 | 3.619 | 1.374 |

Pessimistic scenario:

| | | | | | | | |
|---|---|---|---|---|---|---|---|
| 1990 | 0.735 | 0.835 | 0.174 | 0.238 | 1.012 | 1.841 | 0.761 |
| 1999 | 0.837 | 0.942 | 0.203 | 0.222 | 1.196 | 2.519 | 1.037 |

*Annex Table 4.10.* GDP per capita (in 1,000 $, 1975 value), world regions, medium scenario (to 1981: observations; 1982–1983: forecasts; 1984–1999: trend forecasts).

| | GDP type "1" | | with GDP type "1" for CMEA Countr. | | GDP type "2" | with GDP type "2" for CMEA Countr. |
|---|---|---|---|---|---|---|
| Period | OECD countries total | CMEA countries total | Less developed countries total | World total | CMEA countries total | World total |
| 62 | 3.794 | 1.691 | 0.334 | 1.643 | 2.084 | 1.706 |
| 63 | 3.924 | 1.735 | 0.345 | 1.688 | 2.118 | 1.749 |
| 64 | 4.120 | 1.848 | 0.364 | 1.768 | 2.215 | 1.826 |
| 65 | 4.275 | 1.950 | 0.371 | 1.828 | 2.298 | 1.883 |
| 66 | 4.459 | 2.104 | 0.374 | 1.898 | 2.428 | 1.948 |
| 67 | 4.581 | 2.256 | 0.380 | 1.949 | 2.553 | 1.995 |
| 68 | 4.785 | 2.414 | 0.401 | 2.033 | 2.681 | 2.073 |
| 69 | 4.985 | 2.513 | 0.423 | 2.106 | 2.755 | 2.143 |
| 70 | 5.100 | 2.687 | 0.444 | 2.165 | 2.892 | 2.196 |
| 71 | 5.224 | 2.825 | 0.462 | 2.218 | 2.995 | 2.243 |
| 72 | 5.441 | 2.915 | 0.478 | 2.289 | 3.059 | 2.310 |
| 73 | 5.710 | 3.153 | 0.505 | 2.402 | 3.238 | 2.414 |
| 74 | 5.699 | 3.321 | 0.522 | 2.416 | 3.358 | 2.422 |
| 75 | 5.644 | 3.429 | 0.530 | 2.404 | 3.429 | 2.404 |
| 76 | 5.871 | 3.581 | 0.556 | 2.487 | 3.534 | 2.480 |
| 77 | 6.044 | 3.747 | 0.575 | 2.551 | 3.647 | 2.537 |
| 78 | 6.240 | 3.889 | 0.583 | 2.612 | 3.741 | 2.591 |
| 79 | 6.381 | 3.955 | 0.604 | 2.654 | 3.778 | 2.630 |
| 80 | 6.420 | 4.025 | 0.612 | 2.662 | 3.819 | 2.634 |
| 81 | 6.497 | 4.051 | 0.599 | 2.660 | 3.828 | 2.629 |
| 82 | 6.415 | 4.202 | 0.602 | 2.641 | 3.928 | 2.604 |
| 83 | 6.494 | 4.278 | 0.601 | 2.651 | 3.971 | 2.610 |
| 84 | 6.656 | 4.415 | 0.606 | 2.696 | 4.058 | 2.649 |
| 85 | 6.790 | 4.537 | 0.613 | 2.732 | 4.134 | 2.679 |
| 86 | 6.930 | 4.718 | 0.630 | 2.782 | 4.249 | 2.721 |
| 87 | 7.071 | 4.881 | 0.653 | 2.833 | 4.350 | 2.765 |
| 88 | 7.216 | 5.047 | 0.674 | 2.884 | 4.452 | 2.809 |
| 89 | 7.362 | 5.219 | 0.695 | 2.936 | 4.557 | 2.853 |
| 90 | 7.513 | 5.397 | 0.724 | 2.994 | 4.665 | 2.903 |
| 91 | 7.666 | 5.585 | 0.751 | 3.051 | 4.779 | 2.952 |
| 92 | 7.822 | 5.780 | 0.782 | 3.112 | 4.896 | 3.004 |
| 93 | 7.981 | 5.981 | 0.811 | 3.172 | 5.016 | 3.056 |
| 94 | 8.144 | 6.191 | 0.845 | 3.237 | 5.140 | 3.112 |
| 95 | 8.310 | 6.407 | 0.877 | 3.302 | 5.267 | 3.168 |
| 96 | 8.480 | 6.631 | 0.911 | 3.369 | 5.398 | 3.226 |
| 97 | 8.653 | 6.863 | 0.944 | 3.437 | 5.532 | 3.284 |
| 98 | 8.829 | 7.103 | 0.980 | 3.509 | 5.669 | 3.346 |
| 99 | 9.010 | 7.351 | 1.017 | 3.581 | 5.810 | 3.408 |

For comparison. Optimistic scenario:

| | | | | | | |
|---|---|---|---|---|---|---|
| 1990 | 7.881 | 5.715 | 0.836 | 3.195 | 4.868 | 3.090 |
| 1999 | 10.279 | 8.724 | 1.305 | 4.218 | 6.519 | 3.980 |

Pessimistic scenario:

| | | | | | | |
|---|---|---|---|---|---|---|
| 1990 | 7.190 | 5.099 | 0.650 | 2.830 | 4.472 | 2.753 |
| 1999 | 7.951 | 6.197 | 0.812 | 3.072 | 5.120 | 2.951 |

***Annex Table 4.11.*** GDP per employed person ≈ labor productivity (in 1,000 \$, 1975 value), OECD countries, medium scenario (observations and forecasts as in *Table 4.2*).

| Period | USA | FRG | Japan | France | UK |
|---|---|---|---|---|---|
| 62 | 15.134 | 10.004 | 3.899 | 9.725 | 6.919 |
| 63 | 15.501 | 10.309 | 4.272 | 10.086 | 7.186 |
| 64 | 15.954 | 11.006 | 4.773 | 10.547 | 7.474 |
| 65 | 16.499 | 11.557 | 4.938 | 10.996 | 7.575 |
| 66 | 17.028 | 11.917 | 5.351 | 11.459 | 7.670 |
| 67 | 17.151 | 12.322 | 5.816 | 11.963 | 8.002 |
| 68 | 17.481 | 13.013 | 6.450 | 12.501 | 8.370 |
| 69 | 17.519 | 13.770 | 7.185 | 13.117 | 8.454 |
| 70 | 17.297 | 14.326 | 7.806 | 13.669 | 8.665 |
| 71 | 17.672 | 14.711 | 8.123 | 14.357 | 8.963 |
| 72 | 17.995 | 15.383 | 8.830 | 15.091 | 9.137 |
| 73 | 18.337 | 15.951 | 9.364 | 15.674 | 9.589 |
| 74 | 17.853 | 16.281 | 9.306 | 16.052 | 9.464 |
| 75 | 17.925 | 16.501 | 9.550 | 16.357 | 9.413 |
| 76 | 18.185 | 17.537 | 9.964 | 17.028 | 9.845 |
| 77 | 18.459 | 18.112 | 10.348 | 17.389 | 9.949 |
| 78 | 18.513 | 18.561 | 10.736 | 17.980 | 10.236 |
| 79 | 18.415 | 19.061 | 11.141 | 18.531 | 10.281 |
| 80 | 18.286 | 19.228 | 11.570 | 18.707 | 10.075 |
| 81 | 18.611 | 19.384 | 11.958 | 18.985 | 10.288 |
| 82 | 18.350 | 19.942 | 11.905 | 19.283 | 10.649 |
| 83 | 18.435 | 20.532 | 12.179 | 19.607 | 10.761 |
| 84 | 18.487 | 21.011 | 12.535 | 20.057 | 10.797 |
| 85 | 18.597 | 21.610 | 12.985 | 20.620 | 10.954 |
| 86 | 18.710 | 22.236 | 13.480 | 21.199 | 11.113 |
| 87 | 18.823 | 22.886 | 13.990 | 21.792 | 11.274 |
| 88 | 18.938 | 23.557 | 14.513 | 22.399 | 11.437 |
| 89 | 19.054 | 24.250 | 15.050 | 23.023 | 11.601 |
| 90 | 19.171 | 24.965 | 15.603 | 23.668 | 11.768 |
| 91 | 19.289 | 25.702 | 16.170 | 24.333 | 11.936 |
| 92 | 19.408 | 26.461 | 16.754 | 25.019 | 12.107 |
| 93 | 19.529 | 27.242 | 17.353 | 25.726 | 12.279 |
| 94 | 19.650 | 28.046 | 17.970 | 26.454 | 12.452 |
| 95 | 19.772 | 28.874 | 18.604 | 27.205 | 12.628 |
| 96 | 19.896 | 29.727 | 19.258 | 27.979 | 12.805 |
| 97 | 20.020 | 30.604 | 19.931 | 28.775 | 12.983 |
| 98 | 20.145 | 31.508 | 20.626 | 29.594 | 13.163 |
| 99 | 20.272 | 32.438 | 21.342 | 30.438 | 13.345 |

For comparison. Optimistic scenario:

| | | | | | |
|---|---|---|---|---|---|
| 1990 | 19.670 | 26.232 | 16.733 | 24.880 | 12.071 |
| 1999 | 21.937 | 37.220 | 25.586 | 34.823 | 14.354 |

Pessimistic scenario:

| | | | | | |
|---|---|---|---|---|---|
| 1990 | 18.685 | 23.866 | 14.623 | 22.703 | 11.470 |
| 1999 | 18.677 | 28.392 | 17.944 | 26.909 | 12.343 |

*Annex Table 4.11.* Continued.

| Period | Italy | Netherlands | Belgium/ Luxembourg | Canada | Other DMEs |
|---|---|---|---|---|---|
| 62 | 5.702 | 11.128 | 10.207 | 13.486 | 7.347 |
| 63 | 6.068 | 11.391 | 10.574 | 13.920 | 7.640 |
| 64 | 6.256 | 12.097 | 11.150 | 14.359 | 7.993 |
| 65 | 6.609 | 12.648 | 11.503 | 14.728 | 8.190 |
| 66 | 7.111 | 12.914 | 11.813 | 15.130 | 8.482 |
| 67 | 7.532 | 13.650 | 12.327 | 15.230 | 8.733 |
| 68 | 8.016 | 14.404 | 13.150 | 15.681 | 9.142 |
| 69 | 8.571 | 15.036 | 13.753 | 15.841 | 9.542 |
| 70 | 8.965 | 15.864 | 14.309 | 16.442 | 9.833 |
| 71 | 9.084 | 16.454 | 14.554 | 17.223 | 10.056 |
| 72 | 9.489 | 17.165 | 15.331 | 17.543 | 10.440 |
| 73 | 10.133 | 18.088 | 16.091 | 17.889 | 10.711 |
| 74 | 10.341 | 18.684 | 16.554 | 17.591 | 10.902 |
| 75 | 9.801 | 18.775 | 16.417 | 17.629 | 11.075 |
| 76 | 10.356 | 19.661 | 17.440 | 18.329 | 11.336 |
| 77 | 10.403 | 19.937 | 17.581 | 18.560 | 11.421 |
| 78 | 10.590 | 20.232 | 18.144 | 18.597 | 11.744 |
| 79 | 10.974 | 20.463 | 18.389 | 18.374 | 11.936 |
| 80 | 11.304 | 20.178 | 18.979 | 18.152 | 12.253 |
| 81 | 11.598 | 20.102 | 19.204 | 18.333 | 12.388 |
| 82 | 11.320 | 20.293 | 19.743 | 18.387 | 12.273 |
| 83 | 11.361 | 20.484 | 20.306 | 18.507 | 12.534 |
| 84 | 11.376 | 20.241 | 20.912 | 18.260 | 12.817 |
| 85 | 11.655 | 20.576 | 21.510 | 18.357 | 13.102 |
| 86 | 11.940 | 20.922 | 22.135 | 18.460 | 13.392 |
| 87 | 12.225 | 21.277 | 22.785 | 18.570 | 13.681 |
| 88 | 12.515 | 21.638 | 23.460 | 18.684 | 13.969 |
| 89 | 12.811 | 22.008 | 24.162 | 18.801 | 14.259 |
| 90 | 13.115 | 22.387 | 24.893 | 18.922 | 14.554 |
| 91 | 13.426 | 22.772 | 25.654 | 19.046 | 14.853 |
| 92 | 13.744 | 23.165 | 26.443 | 19.172 | 15.158 |
| 93 | 14.069 | 23.564 | 27.261 | 19.301 | 15.469 |
| 94 | 14.401 | 23.969 | 28.108 | 19.432 | 15.787 |
| 95 | 14.742 | 24.381 | 28.985 | 19.564 | 16.111 |
| 96 | 15.090 | 24.799 | 29.893 | 19.698 | 16.442 |
| 97 | 15.446 | 25.223 | 30.832 | 19.833 | 16.781 |
| 98 | 15.811 | 25.653 | 31.803 | 19.969 | 17.127 |
| 99 | 16.184 | 26.090 | 32.807 | 20.106 | 17.481 |

For comparison. Optimistic scenario:

| | | | | | |
|---|---|---|---|---|---|
| 1990 | 14.296 | 24.029 | 26.202 | 19.768 | 15.658 |
| 1999 | 20.162 | 31.628 | 37.435 | 22.898 | 20.782 |

Pessimistic scenario:

| | | | | | |
|---|---|---|---|---|---|
| 1990 | 12.344 | 21.084 | 23.831 | 18.185 | 13.540 |
| 1999 | 13.809 | 22.064 | 28.864 | 17.796 | 14.678 |

*Annex Table 4.12.* GDP type "1" per employed person ≈ labor productivity type "1" (in 1,000 \$, 1975 value), CMEA countries, medium scenario (observations and forecasts as in *Table 4.2*).

| Period | Bulgaria | CSSR | GDR | Hungary | Poland | Romania | USSR |
|---|---|---|---|---|---|---|---|
| 62 | 3.310 | 4.939 | 4.721 | 3.798 | 3.454 | 1.629 | 3.975 |
| 63 | 3.503 | 4.791 | 4.912 | 3.988 | 3.630 | 1.770 | 4.078 |
| 64 | 3.795 | 4.803 | 5.146 | 4.097 | 3.842 | 1.915 | 4.387 |
| 65 | 4.006 | 4.898 | 5.372 | 4.107 | 4.013 | 2.094 | 4.635 |
| 66 | 4.389 | 5.239 | 5.629 | 4.421 | 4.209 | 2.221 | 4.988 |
| 67 | 4.744 | 5.450 | 5.909 | 4.735 | 4.281 | 2.380 | 5.327 |
| 68 | 4.978 | 5.752 | 6.212 | 4.916 | 4.589 | 2.527 | 5.651 |
| 69 | 5.408 | 6.067 | 6.507 | 5.170 | 4.611 | 2.713 | 5.803 |
| 70 | 5.727 | 6.342 | 6.851 | 5.341 | 4.810 | 2.905 | 6.182 |
| 71 | 5.975 | 6.669 | 7.131 | 5.623 | 5.112 | 3.261 | 6.402 |
| 72 | 6.308 | 7.044 | 7.519 | 5.961 | 5.526 | 3.596 | 6.447 |
| 73 | 6.722 | 7.370 | 7.907 | 6.371 | 6.006 | 3.950 | 6.933 |
| 74 | 7.100 | 7.729 | 8.355 | 6.748 | 6.528 | 4.444 | 7.167 |
| 75 | 7.594 | 8.150 | 8.713 | 7.171 | 6.957 | 4.861 | 7.274 |
| 76 | 7.983 | 8.408 | 8.935 | 7.374 | 7.452 | 5.272 | 7.552 |
| 77 | 8.522 | 8.693 | 9.340 | 8.000 | 7.793 | 5.729 | 7.832 |
| 78 | 8.940 | 8.970 | 9.614 | 8.380 | 8.038 | 6.133 | 8.151 |
| 79 | 9.407 | 9.154 | 9.922 | 8.555 | 7.904 | 6.504 | 8.237 |
| 80 | 9.817 | 9.326 | 10.309 | 8.512 | 7.292 | 6.679 | 8.414 |
| 81 | 10.175 | 9.255 | 10.713 | 8.829 | 6.382 | 6.828 | 8.559 |
| 82 | 10.544 | 9.387 | 10.962 | 8.998 | 6.763 | 7.024 | 8.924 |
| 83 | 10.905 | 9.553 | 11.479 | 9.161 | 6.424 | 7.379 | 9.086 |
| 84 | 11.294 | 9.717 | 11.932 | 9.369 | 6.651 | 7.745 | 9.382 |
| 85 | 11.475 | 9.966 | 12.451 | 9.376 | 6.820 | 8.154 | 9.646 |
| 86 | 11.938 | 10.196 | 12.945 | 9.624 | 6.957 | 8.538 | 10.072 |
| 87 | 12.490 | 10.470 | 13.479 | 9.922 | 7.103 | 8.983 | 10.431 |
| 88 | 13.061 | 10.743 | 14.023 | 10.218 | 7.245 | 9.438 | 10.800 |
| 89 | 13.661 | 11.021 | 14.588 | 10.522 | 7.387 | 9.910 | 11.184 |
| 90 | 14.270 | 11.304 | 15.175 | 10.833 | 7.530 | 10.403 | 11.582 |
| 91 | 14.900 | 11.590 | 15.788 | 11.152 | 7.746 | 10.917 | 11.996 |
| 92 | 15.554 | 11.885 | 16.426 | 11.479 | 7.968 | 11.454 | 12.426 |
| 93 | 16.233 | 12.185 | 17.091 | 11.814 | 8.199 | 12.015 | 12.872 |
| 94 | 16.938 | 12.490 | 17.785 | 12.158 | 8.438 | 12.604 | 13.334 |
| 95 | 17.669 | 12.801 | 18.509 | 12.510 | 8.687 | 13.221 | 13.813 |
| 96 | 18.428 | 13.117 | 19.263 | 12.870 | 8.944 | 13.867 | 14.309 |
| 97 | 19.216 | 13.439 | 20.049 | 13.239 | 9.212 | 14.544 | 14.823 |
| 98 | 20.034 | 13.763 | 20.867 | 13.616 | 9.488 | 15.255 | 15.354 |
| 99 | 20.884 | 14.096 | 21.721 | 14.001 | 9.774 | 16.000 | 15.903 |

For comparison. Optimistic scenario:

| | | | | | | | |
|---|---|---|---|---|---|---|---|
| 1990 | 15.107 | 12.056 | 15.841 | 11.659 | 7.974 | 11.250 | 12.247 |
| 1999 | 24.832 | 17.061 | 24.725 | 17.445 | 11.507 | 20.318 | 18.799 |

Pessimistic scenario:

| | | | | | | | |
|---|---|---|---|---|---|---|---|
| 1990 | 13.459 | 10.671 | 14.636 | 10.192 | 7.104 | 9.525 | 10.951 |
| 1999 | 17.512 | 11.896 | 19.465 | 11.719 | 8.235 | 12.193 | 13.441 |

*Annex Table 4.13.* GDP type "2" per employed person ≈ labor productivity type "2" (in 1,000 $, 1975 value), CMEA countries, medium scenario (observations and forecasts as in *Table 4.2*).

| Period | Bulgaria | CSSR | GDR | Hungary | Poland | Romania | USSR |
|---|---|---|---|---|---|---|---|
| 62 | 5.176 | 6.066 | 5.836 | 4.849 | 4.262 | 2.682 | 4.788 |
| 63 | 5.308 | 5.933 | 6.002 | 5.001 | 4.413 | 2.805 | 4.871 |
| 64 | 5.520 | 5.930 | 6.190 | 5.069 | 4.603 | 2.926 | 5.142 |
| 65 | 5.658 | 5.981 | 6.367 | 5.072 | 4.735 | 3.077 | 5.347 |
| 66 | 5.924 | 6.219 | 6.567 | 5.324 | 4.890 | 3.162 | 5.640 |
| 67 | 6.159 | 6.362 | 6.777 | 5.560 | 4.911 | 3.280 | 5.908 |
| 68 | 6.296 | 6.563 | 7.009 | 5.679 | 5.160 | 3.388 | 6.158 |
| 69 | 6.562 | 6.766 | 7.220 | 5.828 | 5.156 | 3.529 | 6.257 |
| 70 | 6.746 | 6.946 | 7.468 | 5.920 | 5.317 | 3.670 | 6.542 |
| 71 | 6.839 | 7.177 | 7.663 | 6.117 | 5.553 | 3.909 | 6.692 |
| 72 | 6.989 | 7.444 | 7.935 | 6.351 | 5.871 | 4.127 | 6.700 |
| 73 | 7.199 | 7.660 | 8.197 | 6.632 | 6.237 | 4.343 | 7.061 |
| 74 | 7.365 | 7.883 | 8.487 | 6.888 | 6.631 | 4.636 | 7.214 |
| 75 | 7.594 | 8.150 | 8.713 | 7.171 | 6.957 | 4.861 | 7.274 |
| 76 | 7.766 | 8.309 | 8.836 | 7.303 | 7.342 | 5.073 | 7.466 |
| 77 | 8.072 | 8.476 | 9.088 | 7.722 | 7.593 | 5.309 | 7.652 |
| 78 | 8.268 | 8.631 | 9.243 | 7.973 | 7.780 | 5.511 | 7.881 |
| 79 | 8.461 | 8.720 | 9.416 | 8.077 | 7.691 | 5.690 | 7.919 |
| 80 | 8.618 | 8.801 | 9.645 | 8.054 | 7.199 | 5.767 | 8.027 |
| 81 | 8.743 | 8.736 | 9.869 | 8.286 | 6.499 | 5.833 | 8.119 |
| 82 | 8.900 | 8.809 | 9.994 | 8.399 | 6.835 | 5.912 | 8.371 |
| 83 | 9.049 | 8.900 | 10.296 | 8.504 | 6.511 | 6.063 | 8.479 |
| 84 | 9.208 | 8.989 | 10.555 | 8.637 | 6.680 | 6.218 | 8.678 |
| 85 | 9.271 | 9.128 | 10.848 | 8.646 | 6.802 | 6.389 | 8.853 |
| 86 | 9.462 | 9.256 | 11.121 | 8.803 | 6.899 | 6.546 | 9.142 |
| 87 | 9.688 | 9.409 | 11.414 | 8.990 | 7.002 | 6.726 | 9.380 |
| 88 | 9.918 | 9.559 | 11.707 | 9.173 | 7.101 | 6.906 | 9.625 |
| 89 | 10.156 | 9.710 | 12.007 | 9.360 | 7.199 | 7.089 | 9.878 |
| 90 | 10.388 | 9.864 | 12.315 | 9.549 | 7.298 | 7.276 | 10.139 |
| 91 | 10.622 | 10.014 | 12.631 | 9.742 | 7.450 | 7.468 | 10.410 |
| 92 | 10.860 | 10.170 | 12.956 | 9.937 | 7.606 | 7.664 | 10.690 |
| 93 | 11.101 | 10.327 | 13.291 | 10.136 | 7.766 | 7.866 | 10.978 |
| 94 | 11.347 | 10.485 | 13.634 | 10.338 | 7.931 | 8.074 | 11.275 |
| 95 | 11.596 | 10.645 | 13.988 | 10.544 | 8.102 | 8.287 | 11.581 |
| 96 | 11.849 | 10.806 | 14.350 | 10.751 | 8.277 | 8.507 | 11.897 |
| 97 | 12.107 | 10.968 | 14.723 | 10.963 | 8.458 | 8.732 | 12.221 |
| 98 | 12.369 | 11.127 | 15.106 | 11.176 | 8.644 | 8.964 | 12.555 |
| 99 | 12.635 | 11.291 | 15.500 | 11.393 | 8.834 | 9.203 | 12.898 |

For comparison. Optimistic scenario:

| | | | | | | | |
|---|---|---|---|---|---|---|---|
| 1990 | 10.731 | 10.301 | 12.678 | 10.035 | 7.629 | 7.608 | 10.589 |
| 1999 | 13.943 | 12.843 | 16.921 | 13.215 | 10.027 | 10.546 | 14.689 |

Pessimistic scenario:

| | | | | | | | |
|---|---|---|---|---|---|---|---|
| 1990 | 10.050 | 9.489 | 12.098 | 9.166 | 6.976 | 6.922 | 9.708 |
| 1999 | 11.438 | 10.075 | 14.393 | 10.111 | 7.737 | 7.892 | 11.321 |

*Annex Table 4.14.* GDP per economically active population ≈ labor productivity (in 1,000 \$, 1975 value), developing countries, medium scenario (observations and forecasts as in *Table 4.4*).

| Period | Oil-Exporting countries | Asian countries without India | India | Black Africa | Latin Amer without M.,B.,A. | Mexico Brazil Argentina | Mid-East & North Africa |
|---|---|---|---|---|---|---|---|
| 62 | 1.385 | 0.703 | 0.287 | 0.443 | 2.058 | 2.516 | 1.115 |
| 63 | 1.459 | 0.756 | 0.304 | 0.454 | 2.099 | 2.518 | 1.206 |
| 64 | 1.571 | 0.790 | 0.321 | 0.472 | 2.164 | 2.650 | 1.298 |
| 65 | 1.667 | 0.825 | 0.304 | 0.490 | 2.202 | 2.724 | 1.307 |
| 66 | 1.702 | 0.875 | 0.291 | 0.500 | 2.297 | 2.763 | 1.259 |
| 67 | 1.689 | 0.910 | 0.310 | 0.513 | 2.319 | 2.823 | 1.256 |
| 68 | 1.835 | 0.957 | 0.315 | 0.543 | 2.356 | 2.989 | 1.311 |
| 69 | 1.964 | 1.024 | 0.331 | 0.564 | 2.385 | 3.149 | 1.413 |
| 70 | 2.100 | 1.071 | 0.344 | 0.592 | 2.487 | 3.306 | 1.441 |
| 71 | 2.226 | 1.118 | 0.346 | 0.605 | 2.536 | 3.457 | 1.492 |
| 72 | 2.319 | 1.173 | 0.337 | 0.615 | 2.576 | 3.652 | 1.532 |
| 73 | 2.483 | 1.277 | 0.345 | 0.640 | 2.658 | 3.915 | 1.530 |
| 74 | 2.577 | 1.322 | 0.340 | 0.671 | 2.695 | 4.086 | 1.592 |
| 75 | 2.521 | 1.366 | 0.366 | 0.669 | 2.687 | 4.174 | 1.784 |
| 76 | 2.721 | 1.494 | 0.366 | 0.688 | 2.707 | 4.333 | 1.879 |
| 77 | 2.805 | 1.596 | 0.389 | 0.674 | 2.714 | 4.435 | 1.923 |
| 78 | 2.733 | 1.692 | 0.408 | 0.691 | 2.778 | 4.531 | 2.045 |
| 79 | 2.878 | 1.781 | 0.383 | 0.680 | 2.875 | 4.749 | 2.125 |
| 80 | 2.752 | 1.821 | 0.404 | 0.684 | 2.927 | 4.975 | 2.215 |
| 81 | 2.501 | 1.919 | 0.418 | 0.681 | 2.931 | 4.900 | 2.253 |
| 82 | 2.424 | 1.952 | 0.416 | 0.688 | 3.012 | 4.970 | 2.402 |
| 83 | 2.352 | 1.988 | 0.416 | 0.680 | 3.021 | 4.984 | 2.453 |
| 84 | 2.373 | 2.021 | 0.423 | 0.672 | 3.025 | 4.980 | 2.500 |
| 85 | 2.318 | 2.050 | 0.431 | 0.664 | 3.044 | 5.129 | 2.535 |
| 86 | 2.387 | 2.093 | 0.443 | 0.660 | 3.109 | 5.275 | 2.592 |
| 87 | 2.531 | 2.128 | 0.451 | 0.656 | 3.182 | 5.435 | 2.651 |
| 88 | 2.618 | 2.173 | 0.460 | 0.655 | 3.259 | 5.614 | 2.735 |
| 89 | 2.714 | 2.219 | 0.468 | 0.656 | 3.346 | 5.799 | 2.838 |
| 90 | 2.835 | 2.273 | 0.477 | 0.660 | 3.437 | 6.080 | 2.934 |
| 91 | 2.946 | 2.321 | 0.485 | 0.661 | 3.508 | 6.285 | 3.048 |
| 92 | 3.071 | 2.374 | 0.493 | 0.665 | 3.588 | 6.547 | 3.151 |
| 93 | 3.197 | 2.430 | 0.501 | 0.670 | 3.664 | 6.754 | 3.260 |
| 94 | 3.323 | 2.486 | 0.509 | 0.676 | 3.742 | 7.035 | 3.393 |
| 95 | 3.446 | 2.543 | 0.517 | 0.683 | 3.825 | 7.283 | 3.515 |
| 96 | 3.578 | 2.599 | 0.525 | 0.690 | 3.924 | 7.572 | 3.656 |
| 97 | 3.711 | 2.655 | 0.533 | 0.698 | 4.012 | 7.856 | 3.798 |
| 98 | 3.840 | 2.709 | 0.542 | 0.707 | 4.118 | 8.163 | 3.953 |
| 99 | 3.978 | 2.762 | 0.551 | 0.715 | 4.214 | 8.477 | 4.109 |

For comparison. Optimistic scenario:

| | | | | | | | |
|---|---|---|---|---|---|---|---|
| 1990 | 3.655 | 2.562 | 0.511 | 0.728 | 3.687 | 6.807 | 3.189 |
| 1999 | 5.953 | 3.479 | 0.636 | 0.847 | 4.862 | 10.253 | 4.836 |

Pessimistic scenario:

| | | | | | | | |
|---|---|---|---|---|---|---|---|
| 1990 | 2.299 | 2.129 | 0.464 | 0.605 | 3.172 | 5.555 | 2.760 |
| 1999 | 2.571 | 2.303 | 0.524 | 0.586 | 3.528 | 7.136 | 3.652 |

*Annex Table 4.15.* GDP per employed person or economically active population ≈ labor productivity (in 1,000 \$, 1975 value), world regions, medium scenario (observations and forecasts as in *Table 4.10*).

| Period | OECD countries total | CMEA countries total (GDP type"1") | Less developed countries total | World total (with GDP type"1" for CMEA Countr.) | CMEA countries total (GDP type"2") | World total (with GDP type"2" for CMEA Countr.) |
|---|---|---|---|---|---|---|
| 62 | 9.294 | 3.828 | 0.866 | 4.088 | 4.718 | 4.244 |
| 63 | 9.653 | 3.939 | 0.900 | 4.224 | 4.809 | 4.377 |
| 64 | 10.115 | 4.204 | 0.955 | 4.439 | 5.037 | 4.585 |
| 65 | 10.498 | 4.425 | 0.981 | 4.604 | 5.216 | 4.742 |
| 66 | 10.919 | 4.742 | 0.995 | 4.785 | 5.472 | 4.912 |
| 67 | 11.247 | 5.034 | 1.016 | 4.923 | 5.696 | 5.039 |
| 68 | 11.743 | 5.338 | 1.075 | 5.141 | 5.928 | 5.244 |
| 69 | 12.171 | 5.506 | 1.140 | 5.324 | 6.038 | 5.417 |
| 70 | 12.445 | 5.845 | 1.202 | 5.481 | 6.290 | 5.558 |
| 71 | 12.809 | 6.094 | 1.255 | 5.629 | 6.460 | 5.693 |
| 72 | 13.326 | 6.240 | 1.302 | 5.814 | 6.547 | 5.867 |
| 73 | 13.789 | 6.703 | 1.384 | 6.084 | 6.884 | 6.116 |
| 74 | 13.754 | 7.005 | 1.432 | 6.124 | 7.083 | 6.138 |
| 75 | 13.828 | 7.207 | 1.459 | 6.132 | 7.207 | 6.132 |
| 76 | 14.313 | 7.511 | 1.537 | 6.347 | 7.412 | 6.330 |
| 77 | 14.606 | 7.829 | 1.593 | 6.505 | 7.620 | 6.469 |
| 78 | 14.912 | 8.147 | 1.622 | 6.657 | 7.837 | 6.604 |
| 79 | 15.108 | 8.255 | 1.683 | 6.757 | 7.886 | 6.695 |
| 80 | 15.219 | 8.364 | 1.711 | 6.791 | 7.937 | 6.719 |
| 81 | 15.506 | 8.422 | 1.680 | 6.812 | 7.957 | 6.734 |
| 82 | 15.456 | 8.765 | 1.691 | 6.797 | 8.192 | 6.702 |
| 83 | 15.658 | 8.904 | 1.690 | 6.835 | 8.266 | 6.730 |
| 84 | 15.875 | 9.198 | 1.707 | 6.942 | 8.455 | 6.821 |
| 85 | 16.170 | 9.463 | 1.729 | 7.044 | 8.621 | 6.908 |
| 86 | 16.477 | 9.851 | 1.780 | 7.181 | 8.872 | 7.025 |
| 87 | 16.788 | 10.201 | 1.845 | 7.318 | 9.091 | 7.143 |
| 88 | 17.104 | 10.560 | 1.904 | 7.456 | 9.315 | 7.261 |
| 89 | 17.424 | 10.931 | 1.967 | 7.599 | 9.545 | 7.383 |
| 90 | 17.751 | 11.317 | 2.049 | 7.755 | 9.781 | 7.519 |
| 91 | 18.083 | 11.723 | 2.118 | 7.888 | 10.030 | 7.632 |
| 92 | 18.422 | 12.144 | 2.198 | 8.033 | 10.287 | 7.755 |
| 93 | 18.766 | 12.581 | 2.271 | 8.175 | 10.551 | 7.876 |
| 94 | 19.116 | 13.035 | 2.357 | 8.329 | 10.823 | 8.007 |
| 95 | 19.472 | 13.505 | 2.437 | 8.482 | 11.102 | 8.138 |
| 96 | 19.835 | 13.992 | 2.525 | 8.646 | 11.390 | 8.278 |
| 97 | 20.204 | 14.497 | 2.612 | 8.812 | 11.685 | 8.420 |
| 98 | 20.580 | 15.020 | 2.705 | 8.984 | 11.988 | 8.567 |
| 99 | 20.963 | 15.560 | 2.799 | 9.160 | 12.299 | 8.718 |

For comparison. Optimistic scenario:

| | | | | | | |
|---|---|---|---|---|---|---|
| 1990 | 18.622 | 11.982 | 2.365 | 8.276 | 10.207 | 8.003 |
| 1999 | 23.996 | 18.467 | 3.591 | 10.790 | 13.969 | 10.180 |

Pessimistic scenario:

| | | | | | | |
|---|---|---|---|---|---|---|
| 1990 | 16.990 | 10.691 | 1.839 | 7.331 | 9.377 | 7.129 |
| 1999 | 18.500 | 13.118 | 2.236 | 7.858 | 10.837 | 7.548 |

*Annex Table 4.16.* Imports (in billions $, 1975 value), OECD countries, medium scenario (to 1981: observations; 1982–1984: adjusted forecasts; 1985–1999: trend forecasts).

| Period | USA | FRG | Japan | France | UK |
|---|---|---|---|---|---|
| 62 | 60.362 | 36.749 | 15.318 | 19.340 | 36.893 |
| 63 | 62.121 | 38.560 | 18.082 | 22.063 | 38.404 |
| 64 | 65.525 | 42.136 | 20.634 | 25.399 | 42.366 |
| 65 | 72.780 | 48.205 | 22.085 | 25.970 | 42.686 |
| 66 | 83.908 | 49.469 | 24.825 | 28.712 | 43.573 |
| 67 | 89.806 | 48.876 | 30.904 | 31.091 | 46.808 |
| 68 | 103.664 | 55.484 | 34.229 | 35.101 | 50.210 |
| 69 | 109.770 | 64.593 | 38.594 | 41.955 | 51.517 |
| 70 | 113.988 | 74.323 | 47.041 | 44.583 | 53.951 |
| 71 | 121.276 | 81.613 | 49.241 | 48.652 | 56.798 |
| 72 | 134.683 | 86.800 | 54.130 | 56.422 | 62.092 |
| 73 | 140.982 | 90.391 | 67.074 | 64.977 | 69.387 |
| 74 | 136.579 | 90.896 | 70.462 | 68.329 | 70.036 |
| 75 | 119.754 | 93.274 | 63.745 | 63.847 | 65.096 |
| 76 | 143.866 | 103.574 | 67.893 | 76.813 | 67.781 |
| 77 | 156.277 | 106.995 | 71.285 | 79.027 | 68.370 |
| 78 | 173.587 | 113.457 | 76.457 | 83.258 | 70.867 |
| 79 | 176.447 | 124.192 | 86.435 | 92.475 | 78.265 |
| 80 | 170.797 | 129.071 | 80.524 | 98.931 | 75.283 |
| 81 | 181.171 | 125.328 | 81.912 | 100.509 | 72.696 |
| 82 | 184.021 | 126.038 | 79.188 | 100.796 | 76.414 |
| 83 | 204.469 | 127.815 | 81.056 | 101.277 | 77.072 |
| 84 | 259.600 | 132.353 | 83.935 | 108.430 | 80.842 |
| 85 | 265.457 | 138.233 | 88.737 | 114.973 | 83.501 |
| 86 | 272.564 | 144.981 | 95.940 | 121.478 | 86.466 |
| 87 | 280.230 | 152.228 | 103.551 | 127.311 | 89.497 |
| 88 | 288.335 | 159.820 | 111.338 | 132.667 | 92.653 |
| 89 | 296.940 | 167.715 | 119.319 | 137.901 | 95.993 |
| 90 | 306.162 | 175.977 | 127.565 | 143.417 | 99.588 |
| 91 | 315.987 | 184.524 | 136.015 | 149.203 | 103.425 |
| 92 | 326.436 | 193.337 | 144.666 | 155.310 | 107.508 |
| 93 | 337.509 | 202.384 | 153.504 | 161.722 | 111.824 |
| 94 | 349.250 | 211.694 | 162.572 | 168.512 | 116.385 |
| 95 | 361.670 | 221.263 | 171.884 | 175.652 | 121.182 |
| 96 | 374.810 | 231.122 | 181.488 | 183.183 | 126.225 |
| 97 | 388.695 | 241.287 | 191.422 | 191.094 | 131.515 |
| 98 | 403.367 | 251.794 | 201.742 | 199.411 | 137.063 |
| 99 | 418.859 | 262.667 | 212.491 | 208.132 | 142.874 |

For comparison. Optimistic scenario:

| | | | | | |
|---|---|---|---|---|---|
| 1990 | 109.472 | 76.522 | 70.956 | 79.807 | 286.381 |
| 1999 | 206.158 | 138.013 | 124.450 | 146.821 | 484.601 |

Pessimistic scenario:

| | | | | | |
|---|---|---|---|---|---|
| 1990 | 75.107 | 53.336 | 49.541 | 56.558 | 219.548 |
| 1999 | 84.614 | 59.219 | 50.283 | 63.167 | 240.075 |

*Annex Table 4.16.* Continued.

| Period | Italy | Netherlands | Belgium/ Luxembourg | Canada | Other DMEs |
|---|---|---|---|---|---|
| 62 | 17.113 | 15.267 | 14.180 | 14.211 | 57.438 |
| 63 | 20.970 | 16.766 | 15.353 | 14.619 | 62.191 |
| 64 | 19.681 | 19.257 | 16.773 | 16.508 | 70.424 |
| 65 | 20.074 | 20.435 | 17.846 | 18.598 | 77.050 |
| 66 | 22.891 | 21.874 | 19.453 | 21.156 | 80.926 |
| 67 | 25.978 | 23.262 | 19.680 | 22.260 | 84.442 |
| 68 | 27.505 | 26.291 | 21.955 | 24.380 | 88.938 |
| 69 | 32.812 | 30.008 | 25.299 | 27.677 | 99.076 |
| 70 | 38.065 | 34.369 | 27.337 | 27.101 | 109.540 |
| 71 | 39.045 | 36.473 | 28.365 | 29.117 | 111.633 |
| 72 | 43.330 | 38.363 | 30.909 | 32.937 | 117.089 |
| 73 | 47.772 | 42.555 | 36.474 | 37.759 | 133.269 |
| 74 | 48.362 | 42.096 | 38.166 | 42.076 | 140.960 |
| 75 | 43.667 | 40.372 | 34.935 | 40.612 | 133.819 |
| 76 | 50.090 | 44.524 | 39.097 | 43.963 | 145.985 |
| 77 | 50.073 | 45.805 | 40.807 | 44.179 | 145.587 |
| 78 | 54.139 | 48.685 | 42.125 | 45.391 | 145.680 |
| 79 | 61.469 | 51.620 | 46.039 | 48.527 | 156.912 |
| 80 | 66.624 | 51.392 | 46.800 | 47.056 | 163.495 |
| 81 | 63.221 | 48.383 | 46.103 | 48.146 | 165.052 |
| 82 | 60.655 | 48.020 | 45.966 | 42.753 | 165.365 |
| 83 | 61.703 | 46.897 | 46.599 | 44.557 | 173.445 |
| 84 | 63.606 | 49.342 | 48.899 | 47.639 | 185.351 |
| 85 | 68.367 | 51.341 | 50.329 | 50.640 | 197.167 |
| 86 | 73.300 | 53.570 | 51.900 | 53.871 | 209.597 |
| 87 | 77.556 | 55.909 | 53.574 | 57.217 | 220.835 |
| 88 | 81.605 | 58.348 | 55.305 | 60.531 | 230.931 |
| 89 | 85.711 | 60.916 | 57.134 | 63.788 | 240.381 |
| 90 | 90.046 | 63.675 | 59.153 | 67.072 | 249.738 |
| 91 | 94.545 | 66.567 | 61.305 | 70.364 | 259.131 |
| 92 | 99.204 | 69.573 | 63.583 | 73.696 | 268.716 |
| 93 | 103.996 | 72.668 | 65.966 | 77.084 | 278.557 |
| 94 | 108.959 | 75.865 | 68.478 | 80.574 | 288.786 |
| 95 | 114.087 | 79.154 | 71.101 | 84.172 | 299.430 |
| 96 | 119.414 | 82.550 | 73.853 | 87.908 | 310.568 |
| 97 | 124.949 | 86.055 | 76.727 | 91.788 | 322.228 |
| 98 | 130.720 | 89.685 | 79.735 | 95.831 | 334.463 |
| 99 | 136.740 | 93.447 | 82.878 | 100.042 | 347.299 |

For comparison. Optimistic scenario:

| | | | | | |
|---|---|---|---|---|---|
| 1990 | 109.472 | 76.522 | 70.956 | 79.807 | 286.381 |
| 1999 | 206.158 | 138.450 | 124.450 | 146.821 | 484.601 |

Pessimistic scenario:

| | | | | | |
|---|---|---|---|---|---|
| 1990 | 75.107 | 53.336 | 49.541 | 56.558 | 219.548 |
| 1999 | 84.614 | 59.219 | 50.283 | 63.167 | 240.075 |

*Annex Table 4.17.* Imports (in billions $, 1975 value), CMEA countries, medium scenario (observations and forecasts as in *Table 4.2*).

| Period | Bulgaria | CSSR | GDR | Hungary | Poland | Romania | USSR |
|---|---|---|---|---|---|---|---|
| 62 | 1.559 | 3.487 | 5.418 | 1.724 | 3.197 | 1.902 | 11.496 |
| 63 | 1.813 | 3.612 | 5.213 | 1.959 | 3.376 | 2.053 | 12.579 |
| 64 | 2.045 | 3.996 | 5.658 | 2.230 | 3.483 | 2.286 | 13.622 |
| 65 | 2.222 | 4.475 | 6.098 | 2.279 | 4.011 | 2.153 | 14.519 |
| 66 | 2.629 | 4.663 | 6.550 | 2.304 | 4.333 | 2.429 | 14.071 |
| 67 | 2.930 | 4.599 | 6.976 | 2.593 | 4.659 | 3.134 | 15.764 |
| 68 | 3.308 | 5.321 | 7.149 | 2.703 | 5.050 | 3.313 | 17.519 |
| 69 | 3.311 | 5.613 | 8.296 | 2.906 | 5.501 | 3.550 | 19.390 |
| 70 | 3.488 | 6.138 | 9.150 | 3.971 | 6.182 | 3.870 | 21.238 |
| 71 | 3.410 | 6.688 | 9.615 | 4.722 | 7.026 | 3.984 | 22.397 |
| 72 | 3.888 | 6.959 | 10.329 | 4.484 | 8.558 | 4.420 | 25.614 |
| 73 | 4.035 | 7.635 | 11.005 | 4.625 | 10.486 | 4.753 | 26.595 |
| 74 | 4.719 | 8.415 | 11.479 | 5.436 | 11.997 | 5.630 | 27.794 |
| 75 | 5.393 | 8.488 | 11.265 | 5.716 | 12.548 | 5.330 | 35.718 |
| 76 | 5.202 | 9.243 | 12.039 | 5.870 | 13.763 | 5.658 | 36.430 |
| 77 | 5.262 | 9.427 | 12.324 | 6.459 | 13.752 | 6.221 | 36.731 |
| 78 | 5.674 | 9.555 | 11.807 | 7.296 | 13.925 | 6.724 | 39.027 |
| 79 | 5.735 | 9.585 | 11.397 | 6.990 | 13.759 | 6.838 | 38.556 |
| 80 | 5.722 | 9.604 | 11.502 | 6.909 | 13.478 | 6.893 | 40.411 |
| 81 | 6.607 | 8.939 | 11.406 | 6.922 | 11.142 | 6.683 | 45.653 |
| 82 | 6.502 | 9.022 | 11.595 | 7.023 | 11.391 | 5.593 | 47.704 |
| 83 | 6.556 | 9.141 | 11.862 | 7.006 | 11.064 | 6.149 | 49.373 |
| 84 | 6.733 | 9.332 | 12.089 | 7.228 | 12.100 | 6.575 | 51.039 |
| 85 | 6.748 | 9.641 | 12.397 | 7.098 | 12.937 | 7.001 | 52.725 |
| 86 | 7.318 | 9.951 | 12.417 | 7.554 | 13.766 | 7.214 | 54.497 |
| 87 | 7.803 | 10.292 | 12.513 | 8.052 | 14.540 | 7.403 | 56.007 |
| 88 | 8.218 | 10.662 | 12.659 | 8.531 | 15.156 | 7.554 | 57.545 |
| 89 | 8.617 | 11.068 | 12.879 | 8.994 | 15.667 | 7.692 | 59.234 |
| 90 | 9.032 | 11.514 | 13.180 | 9.448 | 16.132 | 7.834 | 61.134 |
| 91 | 9.470 | 12.003 | 13.558 | 9.901 | 16.616 | 7.987 | 63.250 |
| 92 | 9.935 | 12.527 | 14.004 | 10.359 | 17.148 | 8.159 | 65.589 |
| 93 | 10.429 | 13.090 | 14.514 | 10.823 | 17.749 | 8.352 | 68.148 |
| 94 | 10.954 | 13.689 | 15.081 | 11.297 | 18.435 | 8.570 | 70.930 |
| 95 | 11.509 | 14.324 | 15.702 | 11.786 | 19.213 | 8.814 | 73.927 |
| 96 | 12.096 | 14.994 | 16.374 | 12.295 | 20.097 | 9.085 | 77.141 |
| 97 | 12.716 | 15.698 | 17.094 | 12.825 | 21.093 | 9.385 | 80.568 |
| 98 | 13.370 | 16.439 | 17.862 | 13.381 | 22.205 | 9.714 | 84.209 |
| 99 | 14.060 | 17.212 | 18.676 | 13.963 | 23.438 | 10.073 | 88.066 |

For comparison. Optimistic scenario:

| | | | | | | | |
|---|---|---|---|---|---|---|---|
| 1990 | 10.071 | 12.431 | 14.128 | 11.098 | 18.397 | 9.014 | 66.594 |
| 1999 | 17.867 | 21.042 | 22.744 | 19.420 | 32.638 | 14.104 | 113.679 |

Pessimistic scenario:

| | | | | | | | |
|---|---|---|---|---|---|---|---|
| 1990 | 8.148 | 10.763 | 12.402 | 8.128 | 14.178 | 6.718 | 56.592 |
| 1999 | 11.107 | 14.251 | 15.424 | 10.023 | 16.244 | 6.767 | 67.812 |

*Annex Table 4.18.* Imports (in billions $, 1975 value), developing countries, medium scenario (observations and forecasts as in *Table 4.4*).

| Period | Oil-Ex- porting countries | Asian countries without India | India | Black Africa | Latin Amer without M.,B.,A. | Mexico Brazil Argentina | Mid-East & North Africa |
|---|---|---|---|---|---|---|---|
| 62 | 12.940 | 11.662 | 3.343 | 3.789 | 5.096 | 9.311 | 4.940 |
| 63 | 12.594 | 12.520 | 3.606 | 3.833 | 5.113 | 8.940 | 5.148 |
| 64 | 14.689 | 12.063 | 3.922 | 4.331 | 5.701 | 9.137 | 4.964 |
| 65 | 15.795 | 13.710 | 3.632 | 4.415 | 5.670 | 9.197 | 5.155 |
| 66 | 16.292 | 15.698 | 3.790 | 5.096 | 6.676 | 10.045 | 5.030 |
| 67 | 17.712 | 17.166 | 3.816 | 5.536 | 6.652 | 10.663 | 5.183 |
| 68 | 19.895 | 19.707 | 3.369 | 5.909 | 6.824 | 12.104 | 5.301 |
| 69 | 22.607 | 22.204 | 3.079 | 6.174 | 7.252 | 13.373 | 6.212 |
| 70 | 24.550 | 24.118 | 3.316 | 6.913 | 8.007 | 14.879 | 6.673 |
| 71 | 27.557 | 26.261 | 4.592 | 7.698 | 8.372 | 16.432 | 6.763 |
| 72 | 30.217 | 27.422 | 4.592 | 7.729 | 8.162 | 18.378 | 7.365 |
| 73 | 35.939 | 32.123 | 3.553 | 7.669 | 8.535 | 21.126 | 7.541 |
| 74 | 51.167 | 34.865 | 6.214 | 8.965 | 10.419 | 26.218 | 9.449 |
| 75 | 67.610 | 34.231 | 6.778 | 8.532 | 9.925 | 25.793 | 11.225 |
| 76 | 78.008 | 39.207 | 5.995 | 9.412 | 9.504 | 24.863 | 11.500 |
| 77 | 93.690 | 44.725 | 4.993 | 11.039 | 10.913 | 23.631 | 12.984 |
| 78 | 94.183 | 52.754 | 5.466 | 10.103 | 11.133 | 25.587 | 12.578 |
| 79 | 92.290 | 60.445 | 5.831 | 9.068 | 12.441 | 30.922 | 13.679 |
| 80 | 100.908 | 64.532 | 6.377 | 10.160 | 14.899 | 36.816 | 13.730 |
| 81 | 111.790 | 69.332 | 6.924 | 9.600 | 16.262 | 37.066 | 15.301 |
| 82 | 109.615 | 71.293 | 6.182 | 9.636 | 14.783 | 31.948 | 13.834 |
| 83 | 104.546 | 74.006 | 5.925 | 9.453 | 12.999 | 28.706 | 12.467 |
| 84 | 101.697 | 77.003 | 5.744 | 9.181 | 11.918 | 25.835 | 11.554 |
| 85 | 95.214 | 81.138 | 6.306 | 9.395 | 11.475 | 28.465 | 10.486 |
| 86 | 86.599 | 85.903 | 7.211 | 9.672 | 11.880 | 30.917 | 9.933 |
| 87 | 84.039 | 90.334 | 8.060 | 9.997 | 13.006 | 34.290 | 9.737 |
| 88 | 83.351 | 94.731 | 8.975 | 10.258 | 14.359 | 37.797 | 9.933 |
| 89 | 85.454 | 99.321 | 9.863 | 10.668 | 15.990 | 40.961 | 10.372 |
| 90 | 90.262 | 104.040 | 10.752 | 10.996 | 17.472 | 46.223 | 10.992 |
| 91 | 95.811 | 108.977 | 11.571 | 11.346 | 18.871 | 49.863 | 11.641 |
| 92 | 101.596 | 114.168 | 12.331 | 11.720 | 20.211 | 53.550 | 12.346 |
| 93 | 107.514 | 119.630 | 13.060 | 12.125 | 21.557 | 56.440 | 13.107 |
| 94 | 113.488 | 125.385 | 13.778 | 12.543 | 22.927 | 60.123 | 13.912 |
| 95 | 119.478 | 131.419 | 14.501 | 12.977 | 24.339 | 63.236 | 14.779 |
| 96 | 125.444 | 137.754 | 15.240 | 13.429 | 25.769 | 66.933 | 15.691 |
| 97 | 131.405 | 144.382 | 16.003 | 13.899 | 27.275 | 70.444 | 16.651 |
| 98 | 137.436 | 151.306 | 16.790 | 14.381 | 28.813 | 74.363 | 17.651 |
| 99 | 143.516 | 158.521 | 17.605 | 14.886 | 30.443 | 78.294 | 18.696 |
| For comparison. Optimistic scenario: | | | | | | | |
| 1990 | 131.053 | 118.704 | 13.849 | 12.728 | 22.048 | 63.075 | 14.269 |
| 1999 | 256.262 | 214.779 | 22.877 | 19.541 | 34.965 | 116.462 | 28.581 |
| Pessimistic scenario: | | | | | | | |
| 1990 | 61.626 | 92.890 | 8.431 | 8.605 | 11.612 | 32.628 | 8.177 |
| 1999 | 74.713 | 116.229 | 14.220 | 10.107 | 17.230 | 50.558 | 11.477 |

*Annex Table 4.19.* Imports (in billions $, 1975 value), world regions, medium scenario (observations and forecasts as in *Table 4.10*).

| Period | OECD countries total | CMEA countries total | Less developed countries total | World total |
|---|---|---|---|---|
| 62 | 286.870 | 28.783 | 51.082 | 366.735 |
| 63 | 309.127 | 30.605 | 51.753 | 391.485 |
| 64 | 338.702 | 33.320 | 54.807 | 426.828 |
| 65 | 365.729 | 35.758 | 57.574 | 459.061 |
| 66 | 396.787 | 36.979 | 62.627 | 496.393 |
| 67 | 423.105 | 40.655 | 66.728 | 530.488 |
| 68 | 467.756 | 44.363 | 73.109 | 585.229 |
| 69 | 521.301 | 48.567 | 80.903 | 650.771 |
| 70 | 570.298 | 54.036 | 88.457 | 712.791 |
| 71 | 602.212 | 57.842 | 97.676 | 757.730 |
| 72 | 656.754 | 64.252 | 103.865 | 824.870 |
| 73 | 730.639 | 69.133 | 116.486 | 916.258 |
| 74 | 747.961 | 75.471 | 147.297 | 970.729 |
| 75 | 699.119 | 84.459 | 164.094 | 947.672 |
| 76 | 783.584 | 88.205 | 178.489 | 1050.279 |
| 77 | 808.404 | 90.175 | 201.976 | 1100.554 |
| 78 | 853.646 | 94.008 | 211.805 | 1159.458 |
| 79 | 922.380 | 92.860 | 224.675 | 1239.914 |
| 80 | 929.972 | 94.518 | 247.423 | 1271.912 |
| 81 | 932.520 | 97.353 | 266.276 | 1296.149 |
| 82 | 929.216 | 98.831 | 257.291 | 1285.338 |
| 83 | 964.890 | 101.152 | 248.104 | 1314.145 |
| 84 | 1059.996 | 105.097 | 242.931 | 1408.023 |
| 85 | 1108.745 | 108.547 | 242.478 | 1459.770 |
| 86 | 1163.669 | 112.716 | 242.115 | 1518.499 |
| 87 | 1217.907 | 116.610 | 249.464 | 1583.981 |
| 88 | 1271.531 | 120.325 | 259.403 | 1651.259 |
| 89 | 1325.798 | 124.151 | 272.629 | 1722.577 |
| 90 | 1382.394 | 128.274 | 290.738 | 1801.406 |
| 91 | 1441.066 | 132.786 | 308.081 | 1881.932 |
| 92 | 1502.030 | 137.722 | 325.922 | 1965.674 |
| 93 | 1565.214 | 143.105 | 343.432 | 2051.750 |
| 94 | 1631.075 | 148.956 | 362.157 | 2142.188 |
| 95 | 1699.597 | 155.275 | 380.729 | 2235.601 |
| 96 | 1771.121 | 162.083 | 400.260 | 2333.464 |
| 97 | 1845.760 | 169.379 | 420.059 | 2435.198 |
| 98 | 1923.811 | 177.181 | 440.740 | 2541.732 |
| 99 | 2005.429 | 185.487 | 461.959 | 2652.876 |

For comparison. Optimistic scenario:

| | | | | |
|---|---|---|---|---|
| 1990 | 1598.296 | 141.734 | 375.726 | 2115.755 |
| 1999 | 2847.664 | 241.494 | 701.233 | 3790.391 |

Pessimistic scenario:

| | | | | |
|---|---|---|---|---|
| 1990 | 1204.414 | 116.930 | 223.969 | 1545.313 |
| 1999 | 1333.204 | 141.628 | 294.533 | 1769.365 |

*Annex Table 4.20.* Exports (in billions $, 1975 value), CMEA countries, medium scenario (observations and forecasts as in *Table 4.2*).

| Period | Bulgaria | CSSR | GDR | Hungary | Poland | Romania | USSR |
|---|---|---|---|---|---|---|---|
| 62 | 1.229 | 3.434 | 4.318 | 1.407 | 2.949 | 1.674 | 12.413 |
| 63 | 1.328 | 3.779 | 4.793 | 1.541 | 3.086 | 1.853 | 13.242 |
| 64 | 1.561 | 3.975 | 5.115 | 1.678 | 3.644 | 1.964 | 13.271 |
| 65 | 1.827 | 4.234 | 5.395 | 1.856 | 3.934 | 2.084 | 15.280 |
| 66 | 2.092 | 4.494 | 5.627 | 1.997 | 4.097 | 2.188 | 16.982 |
| 67 | 2.424 | 4.623 | 6.132 | 2.023 | 4.571 | 2.733 | 20.184 |
| 68 | 2.790 | 4.861 | 6.719 | 2.193 | 5.273 | 2.925 | 22.512 |
| 69 | 2.979 | 5.310 | 7.296 | 2.633 | 5.665 | 3.205 | 25.539 |
| 70 | 3.010 | 5.974 | 7.643 | 3.038 | 6.156 | 3.482 | 26.921 |
| 71 | 3.318 | 6.454 | 8.320 | 3.259 | 6.565 | 3.792 | 28.507 |
| 72 | 3.642 | 6.840 | 9.142 | 3.884 | 7.581 | 4.299 | 28.566 |
| 73 | 3.974 | 7.086 | 9.425 | 4.451 | 8.468 | 4.937 | 30.850 |
| 74 | 4.185 | 7.388 | 9.865 | 4.606 | 9.510 | 5.425 | 31.752 |
| 75 | 4.679 | 7.809 | 10.066 | 4.807 | 10.288 | 5.330 | 32.189 |
| 76 | 4.910 | 8.396 | 10.140 | 5.132 | 10.444 | 5.688 | 34.183 |
| 77 | 5.605 | 9.145 | 10.385 | 5.864 | 11.278 | 6.402 | 37.748 |
| 78 | 6.022 | 9.575 | 10.467 | 5.969 | 11.927 | 6.407 | 38.446 |
| 79 | 6.809 | 10.015 | 10.703 | 6.635 | 12.733 | 6.335 | 38.694 |
| 80 | 7.421 | 10.123 | 10.542 | 6.686 | 12.196 | 6.510 | 38.155 |
| 81 | 7.565 | 10.215 | 11.586 | 6.866 | 9.876 | 7.636 | 43.874 |
| 82 | 7.987 | 10.607 | 12.087 | 7.175 | 10.313 | 7.383 | 43.238 |
| 83 | 8.501 | 11.100 | 12.701 | 7.544 | 10.879 | 7.280 | 43.654 |
| 84 | 9.139 | 11.744 | 13.469 | 7.997 | 11.608 | 7.346 | 45.399 |
| 85 | 9.578 | 12.140 | 14.035 | 8.328 | 12.106 | 7.251 | 45.373 |
| 86 | 10.088 | 12.626 | 14.760 | 8.707 | 12.697 | 7.321 | 46.283 |
| 87 | 10.605 | 13.121 | 15.546 | 9.100 | 13.312 | 7.426 | 47.575 |
| 88 | 11.135 | 13.633 | 16.386 | 9.514 | 13.959 | 7.558 | 49.022 |
| 89 | 11.688 | 14.174 | 17.289 | 9.959 | 14.653 | 7.725 | 50.614 |
| 90 | 12.275 | 14.758 | 18.264 | 10.444 | 15.407 | 7.932 | 52.390 |
| 91 | 12.891 | 15.377 | 19.302 | 10.969 | 16.215 | 8.168 | 54.242 |
| 92 | 13.540 | 16.036 | 20.407 | 11.538 | 17.083 | 8.433 | 56.176 |
| 93 | 14.225 | 16.736 | 21.582 | 12.154 | 18.011 | 8.725 | 58.172 |
| 94 | 14.950 | 17.485 | 22.835 | 12.827 | 19.010 | 9.047 | 60.266 |
| 95 | 15.716 | 18.282 | 24.165 | 13.556 | 20.078 | 9.395 | 62.415 |
| 96 | 16.529 | 19.133 | 25.583 | 14.350 | 21.225 | 9.773 | 64.662 |
| 97 | 17.389 | 20.038 | 27.089 | 15.212 | 22.450 | 10.177 | 66.980 |
| 98 | 18.302 | 21.004 | 28.694 | 16.150 | 23.763 | 10.612 | 69.394 |
| 99 | 19.268 | 22.030 | 30.400 | 17.167 | 25.164 | 11.074 | 71.890 |

For comparison. Optimistic scenario:

| | | | | | | | |
|---|---|---|---|---|---|---|---|
| 1990 | 13.236 | 15.923 | 19.572 | 11.074 | 16.575 | 8.794 | 59.329 |
| 1999 | 23.979 | 27.614 | 27.792 | 16.437 | 28.877 | 14.702 | 98.999 |

Pessimistic scenario:

| | | | | | | | |
|---|---|---|---|---|---|---|---|
| 1990 | 11.490 | 13.806 | 17.195 | 9.928 | 14.451 | 7.227 | 46.730 |
| 1999 | 15.510 | 17.582 | 24.901 | 14.407 | 20.379 | 8.197 | 50.808 |

*Annex Table 4.21.* Exports (in billions $, 1975 value), OECD countries, medium scenario (to 1981: observations: 1982–1985: adjusted forecasts; 1986–1999: trend forecasts).

| Period | USA | FRG | Japan | France | UK |
|---|---|---|---|---|---|
| 62 | 54.638 | 41.040 | 11.386 | 20.283 | 32.139 |
| 63 | 58.318 | 44.304 | 12.220 | 21.714 | 33.590 |
| 64 | 65.372 | 48.009 | 14.790 | 23.159 | 34.875 |
| 65 | 66.921 | 51.140 | 18.126 | 25.826 | 36.452 |
| 66 | 71.756 | 56.317 | 20.932 | 27.532 | 37.917 |
| 67 | 74.844 | 60.631 | 22.131 | 29.538 | 38.516 |
| 68 | 80.672 | 68.428 | 27.248 | 32.320 | 43.149 |
| 69 | 84.583 | 74.651 | 32.584 | 37.404 | 47.029 |
| 70 | 92.644 | 79.035 | 38.057 | 43.434 | 49.319 |
| 71 | 93.516 | 83.761 | 44.467 | 48.216 | 52.626 |
| 72 | 102.301 | 89.630 | 46.782 | 54.446 | 52.989 |
| 73 | 122.223 | 99.136 | 50.139 | 60.890 | 59.072 |
| 74 | 133.155 | 111.232 | 61.502 | 67.198 | 63.094 |
| 75 | 131.429 | 104.463 | 63.958 | 66.221 | 61.382 |
| 76 | 136.608 | 115.679 | 76.234 | 73.216 | 66.689 |
| 77 | 139.411 | 120.302 | 85.608 | 79.786 | 70.702 |
| 78 | 152.550 | 124.341 | 85.608 | 85.021 | 71.874 |
| 79 | 165.895 | 130.143 | 89.079 | 90.908 | 74.551 |
| 80 | 182.709 | 137.187 | 105.660 | 93.068 | 74.363 |
| 81 | 180.697 | 147.805 | 120.670 | 97.861 | 73.024 |
| 82 | 178.899 | 145.053 | 121.960 | 94.698 | 71.893 |
| 83 | 181.909 | 151.913 | 128.314 | 97.289 | 73.178 |
| 84 | 190.275 | 171.415 | 140.827 | 105.840 | 77.492 |
| 85 | 189.733 | 171.995 | 144.309 | 104.344 | 78.183 |
| 86 | 193.223 | 179.362 | 153.205 | 106.804 | 80.244 |
| 87 | 197.410 | 186.834 | 162.939 | 109.762 | 82.597 |
| 88 | 201.893 | 193.816 | 172.972 | 112.952 | 85.091 |
| 89 | 206.825 | 200.744 | 183.519 | 116.574 | 87.780 |
| 90 | 212.478 | 208.190 | 194.961 | 120.889 | 90.775 |
| 91 | 218.368 | 215.307 | 206.701 | 125.383 | 93.887 |
| 92 | 224.597 | 222.389 | 218.922 | 130.164 | 97.147 |
| 93 | 231.080 | 229.349 | 231.552 | 135.138 | 100.516 |
| 94 | 237.958 | 236.506 | 244.838 | 140.447 | 104.049 |
| 95 | 245.107 | 243.669 | 258.651 | 145.967 | 107.698 |
| 96 | 252.622 | 251.048 | 273.176 | 151.783 | 111.500 |
| 97 | 260.443 | 258.555 | 288.373 | 157.841 | 115.436 |
| 98 | 268.623 | 266.301 | 304.364 | 164.189 | 119.529 |
| 99 | 277.131 | 274.246 | 321.149 | 170.801 | 123.772 |

For comparison. Optimistic scenario:

| | | | | | |
|---|---|---|---|---|---|
| 1990 | 242.979 | 251.760 | 228.373 | 144.082 | 104.892 |
| 1999 | 390.454 | 411.301 | 446.589 | 253.076 | 178.254 |

Pessimistic scenario:

| | | | | | |
|---|---|---|---|---|---|
| 1990 | 187.823 | 172.850 | 167.434 | 101.726 | 79.421 |
| 1999 | 192.323 | 171.460 | 218.921 | 108.131 | 83.190 |

*Annex Table 4.21.* Continued.

| Period | Italy | Netherlands | Belgium/ Luxembourg | Canada | Other DMEs |
|---|---|---|---|---|---|
| 62 | 13.977 | 15.212 | 13.990 | 15.690 | 58.331 |
| 63 | 14.884 | 16.121 | 15.094 | 17.126 | 62.914 |
| 64 | 16.492 | 17.944 | 16.560 | 19.283 | 67.744 |
| 65 | 19.787 | 19.300 | 17.567 | 20.162 | 71.409 |
| 66 | 22.000 | 20.308 | 18.815 | 22.988 | 76.398 |
| 67 | 23.577 | 21.653 | 19.590 | 25.483 | 79.805 |
| 68 | 26.847 | 24.421 | 21.962 | 28.518 | 87.069 |
| 69 | 30.003 | 28.070 | 25.305 | 30.973 | 96.523 |
| 70 | 31.750 | 31.522 | 27.852 | 33.703 | 103.883 |
| 71 | 33.963 | 34.974 | 29.113 | 35.392 | 109.374 |
| 72 | 37.566 | 38.747 | 32.202 | 37.840 | 118.045 |
| 73 | 38.813 | 43.555 | 36.770 | 41.788 | 125.289 |
| 74 | 42.094 | 44.642 | 38.403 | 40.971 | 128.875 |
| 75 | 43.699 | 43.290 | 35.254 | 38.183 | 125.232 |
| 76 | 49.131 | 47.663 | 39.588 | 42.184 | 135.680 |
| 77 | 53.324 | 46.809 | 40.510 | 45.420 | 143.326 |
| 78 | 58.739 | 48.354 | 41.602 | 50.000 | 152.501 |
| 79 | 64.296 | 51.935 | 44.903 | 51.551 | 161.309 |
| 80 | 61.350 | 52.719 | 46.610 | 52.051 | 165.067 |
| 81 | 63.858 | 53.533 | 47.951 | 53.753 | 169.989 |
| 82 | 61.991 | 51.519 | 45.980 | 53.642 | 172.555 |
| 83 | 63.546 | 52.334 | 48.073 | 55.090 | 178.358 |
| 84 | 69.161 | 56.669 | 54.198 | 59.006 | 188.343 |
| 85 | 69.447 | 56.577 | 52.864 | 59.868 | 191.620 |
| 86 | 73.217 | 58.990 | 54.539 | 62.019 | 197.348 |
| 87 | 77.605 | 61.810 | 56.322 | 64.303 | 203.026 |
| 88 | 82.035 | 64.759 | 58.082 | 66.589 | 208.667 |
| 89 | 86.570 | 67.900 | 60.016 | 68.966 | 214.569 |
| 90 | 91.445 | 71.371 | 62.325 | 71.579 | 221.055 |
| 91 | 96.343 | 74.874 | 64.627 | 74.245 | 227.797 |
| 92 | 101.399 | 78.476 | 67.035 | 77.020 | 234.891 |
| 93 | 106.591 | 82.130 | 69.496 | 79.875 | 242.280 |
| 94 | 112.068 | 85.932 | 72.124 | 82.874 | 250.084 |
| 95 | 117.762 | 89.823 | 74.826 | 85.973 | 258.223 |
| 96 | 123.766 | 93.867 | 77.673 | 89.212 | 266.771 |
| 97 | 130.058 | 98.046 | 80.625 | 92.571 | 275.697 |
| 98 | 136.693 | 102.402 | 83.722 | 96.074 | 285.043 |
| 99 | 143.666 | 106.932 | 86.943 | 99.712 | 294.800 |

For comparison. Optimistic scenario:

| | | | | | |
|---|---|---|---|---|---|
| 1990 | 109.845 | 85.685 | 76.294 | 84.356 | 251.731 |
| 1999 | 209.760 | 156.874 | 132.188 | 146.022 | 414.305 |

Pessimistic scenario:

| | | | | | |
|---|---|---|---|---|---|
| 1990 | 76.428 | 59.708 | 50.959 | 61.250 | 196.303 |
| 1999 | 91.838 | 68.118 | 52.976 | 64.857 | 205.078 |

*Annex Table 4.22.* Exports (in billions $, 1975 value), developing countries, medium scenario (to 1981: observations; 1982–1999: trend forecasts).

| Period | Oil-Exporting countries | Asian countries without India | India | Black Africa | Latin Amer without M.,B.,A. | Mexico Brazil Argentina | Mid-East & North Africa |
|---|---|---|---|---|---|---|---|
| 62 | 48.090 | 9.811 | 3.545 | 3.635 | 5.866 | 8.624 | 4.482 |
| 63 | 51.705 | 10.905 | 4.182 | 3.752 | 5.820 | 9.266 | 4.776 |
| 64 | 58.683 | 10.931 | 4.262 | 4.333 | 6.274 | 9.235 | 4.974 |
| 65 | 65.698 | 12.279 | 3.710 | 4.145 | 6.597 | 10.102 | 4.893 |
| 66 | 70.452 | 13.979 | 3.791 | 4.270 | 6.764 | 10.966 | 5.087 |
| 67 | 72.199 | 15.322 | 4.000 | 4.547 | 7.088 | 10.606 | 4.290 |
| 68 | 81.450 | 17.075 | 4.355 | 5.037 | 7.930 | 11.644 | 4.793 |
| 69 | 90.186 | 18.968 | 4.202 | 5.538 | 7.738 | 13.624 | 5.382 |
| 70 | 99.438 | 20.568 | 3.952 | 5.647 | 7.854 | 14.240 | 6.060 |
| 71 | 106.419 | 22.278 | 3.907 | 5.743 | 7.697 | 14.091 | 6.199 |
| 72 | 110.905 | 25.078 | 4.398 | 6.506 | 8.134 | 16.957 | 7.239 |
| 73 | 122.328 | 29.453 | 4.577 | 6.479 | 8.010 | 20.265 | 7.011 |
| 74 | 120.029 | 29.191 | 4.979 | 7.051 | 8.144 | 18.626 | 6.959 |
| 75 | 107.948 | 30.426 | 5.760 | 6.784 | 8.423 | 18.357 | 7.095 |
| 76 | 118.611 | 37.450 | 6.921 | 8.178 | 8.848 | 21.433 | 7.827 |
| 77 | 122.918 | 42.575 | 6.653 | 8.036 | 8.972 | 25.243 | 9.286 |
| 78 | 116.472 | 47.255 | 5.760 | 8.160 | 10.039 | 26.257 | 9.561 |
| 79 | 124.854 | 52.131 | 7.078 | 7.644 | 11.055 | 27.251 | 10.630 |
| 80 | 114.872 | 58.377 | 8.462 | 7.956 | 11.203 | 30.285 | 10.423 |
| 81 | 97.616 | 65.537 | 8.529 | 8.118 | 10.856 | 35.008 | 10.663 |
| 82 | 81.006 | 68.569 | 7.821 | 8.144 | 10.703 | 36.403 | 10.277 |
| 83 | 75.597 | 71.959 | 7.587 | 8.222 | 10.687 | 37.982 | 10.121 |
| 84 | 79.471 | 76.585 | 8.051 | 8.443 | 10.988 | 40.178 | 10.376 |
| 85 | 74.913 | 79.117 | 7.310 | 8.441 | 10.851 | 41.320 | 10.071 |
| 86 | 86.967 | 81.636 | 6.923 | 8.447 | 10.768 | 42.453 | 9.874 |
| 87 | 100.541 | 84.402 | 6.925 | 8.494 | 10.777 | 43.717 | 9.834 |
| 88 | 114.094 | 87.436 | 7.171 | 8.583 | 10.868 | 45.126 | 9.915 |
| 89 | 128.070 | 90.788 | 7.585 | 8.718 | 11.041 | 46.708 | 10.104 |
| 90 | 143.039 | 94.552 | 8.148 | 8.910 | 11.305 | 48.512 | 10.404 |
| 91 | 157.773 | 98.644 | 8.771 | 9.147 | 11.628 | 50.498 | 10.774 |
| 92 | 172.570 | 103.079 | 9.450 | 9.432 | 12.007 | 52.678 | 11.204 |
| 93 | 187.225 | 107.836 | 10.168 | 9.762 | 12.430 | 55.041 | 11.677 |
| 94 | 202.178 | 112.942 | 10.935 | 10.138 | 12.896 | 57.604 | 12.194 |
| 95 | 217.160 | 118.366 | 11.733 | 10.557 | 13.393 | 60.355 | 12.738 |
| 96 | 232.531 | 124.130 | 12.575 | 11.022 | 13.922 | 63.304 | 13.311 |
| 97 | 248.199 | 130.214 | 13.449 | 11.529 | 14.476 | 66.445 | 13.904 |
| 98 | 264.429 | 136.635 | 14.367 | 12.080 | 15.054 | 69.788 | 14.517 |
| 99 | 281.208 | 143.389 | 15.327 | 12.675 | 15.654 | 73.333 | 15.148 |

For comparison. Optimistic scenario:

| | | | | | | | |
|---|---|---|---|---|---|---|---|
| 1990 | 184.715 | 105.826 | 10.573 | 10.019 | 13.374 | 54.086 | 12.662 |
| 1999 | 413.888 | 195.479 | 23.925 | 17.663 | 23.608 | 99.442 | 23.448 |

Pessimistic scenario:

| | | | | | | | |
|---|---|---|---|---|---|---|---|
| 1990 | 108.713 | 85.454 | 6.173 | 8.013 | 9.641 | 44.008 | 8.582 |
| 1999 | 168.621 | 103.161 | 8.550 | 8.783 | 9.656 | 53.041 | 8.878 |

***Annex Table 4.23.*** Exports (in billions $, 1975 value), world regions, medium scenario (observations and forecasts as in *Table 4.16*).

| Period | OECD countries total | CMEA countries total | Less developed countries total | World total |
|---|---|---|---|---|
| 62 | 276.685 | 27.425 | 84.052 | 388.162 |
| 63 | 296.285 | 29.622 | 90.407 | 416.314 |
| 64 | 324.227 | 31.209 | 98.693 | 454.128 |
| 65 | 346.690 | 34.610 | 107.425 | 488.724 |
| 66 | 374.962 | 37.476 | 115.308 | 527.746 |
| 67 | 395.767 | 42.689 | 118.052 | 556.509 |
| 68 | 440.634 | 47.272 | 132.284 | 620.190 |
| 69 | 487.124 | 52.627 | 145.638 | 685.389 |
| 70 | 531.197 | 56.224 | 157.759 | 745.180 |
| 71 | 565.402 | 60.216 | 166.335 | 791.952 |
| 72 | 610.547 | 63.954 | 179.217 | 853.718 |
| 73 | 677.675 | 69.191 | 198.122 | 944.987 |
| 74 | 731.165 | 72.731 | 194.978 | 998.875 |
| 75 | 713.110 | 75.168 | 184.795 | 973.073 |
| 76 | 782.672 | 78.893 | 209.270 | 1070.834 |
| 77 | 825.198 | 86.428 | 223.684 | 1135.310 |
| 78 | 870.589 | 88.811 | 223.504 | 1182.904 |
| 79 | 924.568 | 91.925 | 240.641 | 1257.134 |
| 80 | 970.781 | 91.634 | 241.577 | 1303.991 |
| 81 | 1009.140 | 97.618 | 236.328 | 1343.085 |
| 82 | 998.188 | 98.787 | 222.921 | 1319.896 |
| 83 | 1030.023 | 101.651 | 222.161 | 1353.834 |
| 84 | 1113.243 | 106.694 | 234.099 | 1454.036 |
| 85 | 1118.940 | 108.810 | 232.022 | 1459.772 |
| 86 | 1158.950 | 112.482 | 247.068 | 1518.500 |
| 87 | 1202.607 | 116.684 | 264.689 | 1583.980 |
| 88 | 1246.855 | 121.209 | 283.194 | 1651.258 |
| 89 | 1293.463 | 126.100 | 303.014 | 1722.576 |
| 90 | 1345.061 | 131.476 | 324.871 | 1801.408 |
| 91 | 1397.533 | 137.162 | 347.236 | 1881.931 |
| 92 | 1452.042 | 143.211 | 370.422 | 1965.675 |
| 93 | 1508.006 | 149.605 | 394.141 | 2051.752 |
| 94 | 1566.881 | 156.419 | 418.886 | 2142.186 |
| 95 | 1627.688 | 163.609 | 444.321 | 2235.619 |
| 96 | 1691.414 | 171.256 | 470.795 | 2333.466 |
| 97 | 1757.638 | 179.340 | 498.228 | 2435.207 |
| 98 | 1826.942 | 187.915 | 526.872 | 2541.730 |
| 99 | 1899.152 | 196.992 | 556.730 | 2652.874 |

For comparison. Optimistic scenario:

| | | | | |
|---|---|---|---|---|
| 1990 | 1579.998 | 144.503 | 391.254 | 2115.756 |
| 1999 | 2738.824 | 254.111 | 797.454 | 3790.389 |

Pessimistic scenario:

| | | | | |
|---|---|---|---|---|
| 1990 | 1153.901 | 120.828 | 270.584 | 1545.314 |
| 1999 | 1256.889 | 151.786 | 360.692 | 1769.367 |

*Annex Table 4.24.* Current trade balance (in billion $), including services, OECD countries, medium scenario (observations and forecasts as in *Table 4.21*).

| Period | USA | FRG | Japan | France | UK |
|---|---|---|---|---|---|
| 62 | 3.348 | 1.145 | 0.111 | 0.692 | -0.092 |
| 63 | 4.181 | 1.470 | -0.561 | 0.280 | -0.280 |
| 64 | 5.943 | 1.620 | -0.150 | -0.194 | -1.526 |
| 65 | 4.550 | 0.325 | 1.278 | 0.872 | -0.697 |
| 66 | 2.761 | 2.063 | 1.658 | 0.277 | -0.067 |
| 67 | 2.419 | 4.497 | 0.278 | 0.298 | -1.047 |
| 68 | 0.070 | 4.830 | 1.642 | -0.094 | -0.701 |
| 69 | -0.095 | 4.243 | 2.753 | -0.779 | 0.674 |
| 70 | 2.061 | 3.751 | 2.614 | 0.718 | 1.250 |
| 71 | -1.514 | 3.922 | 6.292 | 1.540 | 2.180 |
| 72 | -5.670 | 5.187 | 7.039 | 1.854 | 0.070 |
| 73 | 1.010 | 10.099 | 0.110 | 1.493 | -4.177 |
| 74 | -4.573 | 16.586 | -3.420 | -3.820 | -9.711 |
| 75 | 11.675 | 11.190 | 0.212 | 2.374 | -3.714 |
| 76 | -6.327 | 10.199 | 4.502 | -4.049 | -2.221 |
| 77 | -27.247 | 12.312 | 11.325 | -1.080 | 2.208 |
| 78 | -30.483 | 16.215 | 16.893 | 4.642 | 4.771 |
| 79 | -28.894 | 4.523 | -9.136 | 0.811 | 1.754 |
| 80 | -20.605 | -4.099 | -9.478 | -11.544 | 13.200 |
| 81 | -21.586 | 6.082 | 9.295 | -9.053 | 15.695 |
| 82 | -6.371 | -1.173 | 5.413 | -14.277 | 5.851 |
| 83 | -24.739 | 1.562 | 13.218 | -12.574 | 3.334 |
| 84 | -84.369 | 10.134 | 24.224 | -11.767 | 0.634 |
| 85 | -62.714 | 8.745 | 24.468 | -13.381 | 3.422 |
| 86 | -47.895 | 13.427 | 29.586 | -13.295 | 5.050 |
| 87 | -43.907 | 15.386 | 29.782 | -15.692 | 5.254 |
| 88 | -42.381 | 16.297 | 29.773 | -17.835 | 5.700 |
| 89 | -40.996 | 16.976 | 30.112 | -19.383 | 6.424 |
| 90 | -39.104 | 18.191 | 31.044 | -20.353 | 7.495 |
| 91 | -37.778 | 18.773 | 31.754 | -21.515 | 8.526 |
| 92 | -36.605 | 19.169 | 32.449 | -22.738 | 9.595 |
| 93 | -35.765 | 19.238 | 32.974 | -24.171 | 10.627 |
| 94 | -34.783 | 19.511 | 33.614 | -25.644 | 11.743 |
| 95 | -33.967 | 19.650 | 34.111 | -27.364 | 12.842 |
| 96 | -32.981 | 20.048 | 34.656 | -29.216 | 14.008 |
| 97 | -31.947 | 20.537 | 35.109 | -31.293 | 15.202 |
| 98 | -30.677 | 21.358 | 35.552 | -33.541 | 16.474 |
| 99 | -29.276 | 22.419 | 35.851 | -36.026 | 17.799 |

For comparison. Optimistic scenario:

| | | | | | |
|---|---|---|---|---|---|
| 1990 | -23.244 | 37.672 | 44.919 | -31.289 | 12.352 |
| 1999 | -14.256 | 66.726 | 88.473 | -100.068 | 27.622 |

Pessimistic scenario:

| | | | | | |
|---|---|---|---|---|---|
| 1990 | -50.809 | 2.402 | 20.102 | -11.496 | 4.056 |
| 1999 | -28.322 | -9.449 | -9.276 | 29.824 | 13.940 |

*Annex Table 4.24.* Continued.

| Period | Italy | Netherlands | Belgium/ Luxembourg | Canada | Other DMEs |
|---|---|---|---|---|---|
| 62 | -0.181 | 0.051 | -0.021 | -0.037 | -0.265 |
| 63 | -1.181 | -0.106 | -0.159 | 0.327 | 0.063 |
| 64 | 0.186 | -0.366 | -0.079 | 0.461 | -1.312 |
| 65 | 1.643 | -0.132 | -0.025 | -0.129 | -2.463 |
| 66 | 1.430 | -0.301 | -0.143 | -0.061 | -1.770 |
| 67 | 0.971 | -0.212 | 0.108 | 0.646 | -1.852 |
| 68 | 1.963 | -0.004 | 0.140 | 0.888 | -0.222 |
| 69 | 1.512 | -0.059 | 0.347 | 0.168 | -0.455 |
| 70 | 0.539 | -0.576 | 0.750 | 2.233 | -2.620 |
| 71 | 1.224 | -0.117 | 0.650 | 1.732 | -1.127 |
| 72 | 1.164 | 1.324 | 1.288 | 0.890 | 3.200 |
| 73 | -3.256 | 2.018 | 1.216 | 1.588 | 0.079 |
| 74 | -7.678 | 2.035 | 0.783 | 0.346 | -11.725 |
| 75 | 0.032 | 2.918 | 0.320 | -2.429 | -8.587 |
| 76 | -2.361 | 3.230 | 0.310 | -0.971 | -12.043 |
| 77 | 2.757 | 1.361 | -0.548 | 0.005 | -8.564 |
| 78 | 6.773 | 0.051 | -0.893 | 1.043 | 4.958 |
| 79 | 4.345 | -0.813 | -1.798 | 1.650 | 1.862 |
| 80 | -11.430 | -0.910 | -3.326 | 4.978 | -9.736 |
| 81 | -6.702 | 4.962 | -2.450 | 3.427 | -11.410 |
| 82 | -3.085 | 2.883 | -5.195 | 10.802 | -8.491 |
| 83 | -0.628 | 5.580 | -4.248 | 10.534 | -5.767 |
| 84 | 3.623 | 7.229 | -0.712 | 12.091 | -0.371 |
| 85 | 2.006 | 5.177 | -3.942 | 9.392 | -0.859 |
| 86 | 2.146 | 5.570 | -3.744 | 9.857 | -0.719 |
| 87 | 0.101 | 5.846 | -3.811 | 10.249 | -3.617 |
| 88 | -0.938 | 6.197 | -4.054 | 10.559 | -6.478 |
| 89 | -1.397 | 6.712 | -4.221 | 10.949 | -8.627 |
| 90 | -1.491 | 7.509 | -4.140 | 11.592 | -9.989 |
| 91 | -1.787 | 8.214 | -4.288 | 12.311 | -11.263 |
| 92 | -2.144 | 8.951 | -4.491 | 13.201 | -12.389 |
| 93 | -2.625 | 9.680 | -4.814 | 14.239 | -13.531 |
| 94 | -3.057 | 10.543 | -5.096 | 15.502 | -14.557 |
| 95 | -3.569 | 11.453 | -5.481 | 16.930 | -15.658 |
| 96 | -4.056 | 12.508 | -5.863 | 18.576 | -16.748 |
| 97 | -4.577 | 13.675 | -6.309 | 20.419 | -17.909 |
| 98 | -5.086 | 15.017 | -6.764 | 22.498 | -19.103 |
| 99 | -5.635 | 16.518 | -7.269 | 24.811 | -20.394 |

For comparison. Optimistic scenario:

| | | | | | |
|---|---|---|---|---|---|
| 1990 | -2.272 | 9.133 | -2.576 | 11.432 | -13.074 |
| 1999 | -14.940 | 21.784 | -7.873 | 24.510 | -35.716 |

Pessimistic scenario:

| | | | | | |
|---|---|---|---|---|---|
| 1990 | -1.794 | 6.019 | -5.487 | 11.979 | -6.849 |
| 1999 | 1.175 | 11.976 | -3.659 | 27.792 | -3.471 |

*Annex Table 4.25.* Current trade balance (in billion $), including services, CMEA countries, medium scenario (observations and forecasts as in *Table 4.21*).

| Period | Bulgaria | CSSR | GDR | Hungary | Poland | Romania | USSR |
|---|---|---|---|---|---|---|---|
| 62 | -0.012 | 0.112 | -0.012 | -0.109 | -0.240 | -0.123 | 0.466 |
| 63 | -0.098 | 0.302 | 0.396 | -0.119 | -0.210 | -0.107 | 0.212 |
| 64 | -0.081 | 0.239 | 0.310 | -0.198 | 0.023 | -0.167 | -0.054 |
| 65 | -0.003 | 0.016 | 0.260 | -0.025 | -0.112 | 0.024 | 0.118 |
| 66 | -0.173 | 0.007 | -0.008 | -0.005 | -0.221 | -0.026 | 0.941 |
| 67 | -0.114 | 0.185 | 0.174 | -0.116 | -0.118 | -0.153 | 1.119 |
| 68 | -0.166 | -0.072 | 0.403 | -0.011 | 0.007 | -0.140 | 1.222 |
| 69 | 0.045 | 0.025 | 0.047 | 0.134 | -0.066 | -0.108 | 1.335 |
| 70 | 0.173 | 0.097 | -0.265 | -0.198 | -0.060 | -0.109 | 1.068 |
| 71 | 0.060 | 0.168 | 0.096 | -0.553 | -0.170 | -0.004 | 1.334 |
| 72 | 0.061 | 0.254 | 0.283 | 0.114 | -0.399 | -0.017 | -0.687 |
| 73 | 0.030 | -0.082 | -0.333 | 0.375 | -1.431 | 0.230 | 0.345 |
| 74 | -0.490 | -0.465 | -0.895 | -0.377 | -2.171 | -0.268 | 2.552 |
| 75 | -0.714 | -0.679 | -1.199 | -0.909 | -2.260 | -0.000 | -3.529 |
| 76 | -0.246 | -0.643 | -1.825 | -0.596 | -2.849 | 0.039 | -0.942 |
| 77 | -0.017 | -0.828 | -2.302 | -0.690 | -2.357 | 0.014 | 4.224 |
| 78 | -0.166 | -0.765 | -1.364 | -1.556 | -1.975 | -0.656 | 1.545 |
| 79 | 0.349 | -1.019 | -1.230 | -0.737 | -1.338 | -1.148 | 6.591 |
| 80 | 0.725 | -0.260 | -1.769 | -0.559 | -2.090 | -0.973 | 7.932 |
| 81 | 0.216 | -1.021 | 0.974 | -0.417 | -2.226 | -0.394 | 4.351 |
| 82 | 0.447 | -0.730 | 1.194 | 0.065 | -0.907 | 2.232 | 1.275 |
| 83 | 0.650 | -0.643 | 1.077 | 0.618 | 0.597 | 1.579 | -0.671 |
| 84 | 0.848 | -0.580 | 1.213 | 0.933 | 0.544 | 0.975 | -1.048 |
| 85 | 1.526 | -0.443 | 1.173 | 1.906 | 1.126 | 0.719 | -0.987 |
| 86 | 1.167 | -0.480 | 0.718 | 1.862 | 1.200 | 0.333 | -1.439 |
| 87 | 1.010 | -0.438 | 0.672 | 1.737 | 0.960 | 0.081 | -3.020 |
| 88 | 0.974 | -0.373 | 0.857 | 1.607 | 0.665 | -0.021 | -3.140 |
| 89 | 1.007 | -0.282 | 1.156 | 1.510 | 0.512 | -0.028 | -2.686 |
| 90 | 1.080 | -0.165 | 1.524 | 1.458 | 0.525 | 0.032 | -1.949 |
| 91 | 1.167 | -0.052 | 1.927 | 1.438 | 0.624 | 0.122 | -1.190 |
| 92 | 1.264 | 0.069 | 2.363 | 1.449 | 0.784 | 0.237 | -0.405 |
| 93 | 1.365 | 0.187 | 2.833 | 1.489 | 0.971 | 0.369 | 0.347 |
| 94 | 1.472 | 0.307 | 3.350 | 1.564 | 1.173 | 0.520 | 1.130 |
| 95 | 1.582 | 0.424 | 3.914 | 1.669 | 1.368 | 0.680 | 1.851 |
| 96 | 1.697 | 0.545 | 4.538 | 1.806 | 1.536 | 0.854 | 2.585 |
| 97 | 1.816 | 0.669 | 5.228 | 1.980 | 1.666 | 1.037 | 3.282 |
| 98 | 1.939 | 0.794 | 5.994 | 2.189 | 1.745 | 1.232 | 3.980 |
| 99 | 2.066 | 0.937 | 6.845 | 2.440 | 1.759 | 1.437 | 4.648 |

For comparison. Optimistic scenario:

| | | | | | | | |
|---|---|---|---|---|---|---|---|
| 1990 | 0.554 | -0.228 | 1.715 | 0.283 | -0.420 | -0.421 | 1.892 |
| 1999 | 1.121 | 1.578 | 9.093 | -0.571 | -2.051 | 0.345 | 17.607 |

Pessimistic scenario:

| | | | | | | | |
|---|---|---|---|---|---|---|---|
| 1990 | 1.598 | -0.096 | 1.380 | 2.399 | 1.430 | 0.652 | -5.035 |
| 1999 | 2.859 | 0.236 | 4.961 | 4.343 | 5.009 | 2.817 | -7.429 |

*Annex Table 4.26.* Current trade balance (in billion $), including services, developing countries, medium scenario (observations and forecasts as in *Table 4.22*).

| Period | Oil-Ex-porting countries | Asian countries without India | India | Black Africa | Latin Amer without M.,B.,A. | Mexico Brazil Argentina | Mid-East & North Africa |
|---|---|---|---|---|---|---|---|
| 62 | 2.340 | -0.401 | -0.785 | 0.163 | -0.256 | -0.122 | -0.404 |
| 63 | 3.581 | -0.618 | -0.794 | 0.277 | -0.115 | 0.500 | -0.447 |
| 64 | 3.102 | -0.562 | -1.073 | 0.317 | 0.036 | 0.364 | -0.234 |
| 65 | 3.219 | -0.374 | -1.117 | 0.156 | 0.175 | 0.834 | -0.362 |
| 66 | 3.205 | -0.383 | -1.242 | 0.192 | -0.095 | 0.788 | -0.223 |
| 67 | 2.939 | -0.706 | -0.924 | 0.197 | 0.001 | 0.457 | -0.491 |
| 68 | 3.490 | -1.139 | -0.400 | 0.273 | 0.089 | 0.069 | -0.270 |
| 69 | 3.346 | -1.120 | -0.163 | 0.882 | 0.235 | 0.370 | -0.435 |
| 70 | 4.051 | -1.249 | -0.067 | 0.233 | 0.090 | -0.177 | -0.612 |
| 71 | 7.234 | -1.972 | -0.227 | -0.581 | -0.556 | -0.649 | -0.627 |
| 72 | 8.049 | -0.964 | 0.224 | -0.566 | -0.515 | -0.471 | -0.599 |
| 73 | 15.783 | 0.053 | -0.452 | 0.062 | 0.137 | 0.443 | -0.651 |
| 74 | 59.902 | -3.678 | -1.173 | -0.225 | -0.142 | -6.190 | -1.594 |
| 75 | 40.339 | -3.805 | -1.018 | -1.748 | -1.502 | -7.435 | -4.130 |
| 76 | 43.900 | -0.514 | 0.592 | -0.334 | 0.206 | -2.528 | -4.822 |
| 77 | 33.762 | -0.526 | 0.137 | -1.397 | 0.004 | 0.509 | -6.195 |
| 78 | 13.683 | -3.682 | -0.391 | -1.835 | 0.216 | -0.398 | -7.042 |
| 79 | 53.423 | -6.589 | -1.822 | -0.920 | 1.733 | -6.057 | -6.763 |
| 80 | 84.512 | -9.222 | -6.157 | -2.099 | -0.625 | -10.438 | -6.630 |
| 81 | 53.117 | -10.597 | -5.335 | -2.371 | -5.824 | -7.335 | -9.078 |
| 82 | 27.109 | -7.207 | -3.068 | -0.433 | -4.551 | 6.558 | -8.344 |
| 83 | 11.043 | -4.222 | -2.075 | 0.036 | -1.595 | 14.594 | -7.259 |
| 84 | 21.329 | -1.622 | -0.701 | 0.685 | 0.660 | 22.217 | -6.171 |
| 85 | 9.744 | -1.426 | -1.945 | 0.680 | 0.817 | 20.025 | -5.230 |
| 86 | -10.218 | -2.123 | -3.544 | 0.260 | -0.063 | 17.560 | -5.216 |
| 87 | 1.995 | -3.996 | -5.418 | -0.430 | -1.393 | 13.985 | -5.336 |
| 88 | 14.210 | -5.531 | -7.249 | -0.906 | -2.969 | 10.916 | -5.881 |
| 89 | 22.763 | -6.969 | -8.906 | -1.487 | -4.963 | 8.543 | -6.719 |
| 90 | 28.721 | -8.129 | -10.490 | -1.893 | -6.711 | 3.026 | -7.784 |
| 91 | 33.912 | -9.318 | -12.008 | -2.299 | -8.352 | -0.070 | -8.849 |
| 92 | 39.154 | -10.539 | -13.472 | -2.709 | -9.934 | -3.263 | -9.998 |
| 93 | 44.536 | -11.878 | -14.978 | -3.145 | -11.580 | -5.092 | -11.276 |
| 94 | 50.453 | -13.304 | -16.542 | -3.573 | -13.331 | -8.316 | -12.678 |
| 95 | 56.702 | -14.868 | -18.251 | -4.007 | -15.248 | -10.470 | -14.296 |
| 96 | 63.788 | -16.561 | -20.108 | -4.445 | -17.290 | -13.764 | -16.117 |
| 97 | 71.371 | -18.414 | -22.171 | -4.893 | -19.610 | -16.678 | -18.190 |
| 98 | 79.721 | -20.403 | -24.438 | -5.333 | -22.130 | -20.504 | -20.521 |
| 99 | 88.927 | -22.540 | -26.947 | -5.791 | -25.014 | -24.395 | -23.170 |

For comparison. Optimistic scenario:

| | | | | | | | |
|---|---|---|---|---|---|---|---|
| 1990 | 16.780 | -12.142 | -12.933 | -2.469 | -10.163 | -14.188 | -11.312 |
| 1999 | 80.650 | -25.722 | -23.015 | -3.635 | -28.888 | -49.386 | -33.391 |

Pessimistic scenario:

| | | | | | | | |
|---|---|---|---|---|---|---|---|
| 1990 | 31.574 | -5.909 | -8.853 | 0.010 | -0.274 | 17.503 | -4.503 |
| 1999 | 54.287 | -24.343 | -34.448 | -5.047 | -11.857 | -6.000 | -15.920 |

*Annex Table 4.27.* Current trade balance (in billion $), including services, world regions, medium scenario. (For details, see *Tables 4.19* and *4.28*).

| Period | OECD countries total | CMEA countries total | Less developed countries total |
|---|---|---|---|
| 62 | 4.752 | 0.082 | 1.311 |
| 63 | 4.034 | 0.375 | 2.384 |
| 64 | 4.582 | 0.072 | 1.950 |
| 65 | 5.223 | 0.278 | 2.531 |
| 66 | 5.846 | 0.514 | 2.241 |
| 67 | 6.107 | 0.977 | 1.473 |
| 68 | 8.511 | 1.242 | 2.111 |
| 69 | 8.310 | 1.412 | 3.116 |
| 70 | 10.720 | 0.706 | 2.269 |
| 71 | 14.782 | 0.932 | 2.622 |
| 72 | 16.346 | -0.391 | 5.158 |
| 73 | 10.180 | -0.866 | 15.374 |
| 74 | -21.178 | -2.113 | 46.899 |
| 75 | 13.991 | -9.291 | 20.701 |
| 76 | -9.731 | -7.062 | 36.498 |
| 77 | -7.471 | -1.956 | 26.293 |
| 78 | 23.970 | -4.938 | 0.551 |
| 79 | -25.696 | 1.469 | 33.004 |
| 80 | -52.949 | 3.006 | 49.340 |
| 81 | -11.741 | 1.483 | 12.577 |
| 82 | -13.643 | 3.577 | 10.066 |
| 83 | -13.728 | 3.207 | 10.521 |
| 84 | -39.283 | 2.887 | 36.396 |
| 85 | -27.686 | 5.019 | 22.667 |
| 86 | -0.016 | 3.361 | -3.345 |
| 87 | -0.409 | 1.002 | -0.593 |
| 88 | -3.160 | 0.568 | 2.591 |
| 89 | -3.451 | 1.188 | 2.262 |
| 90 | 0.755 | 2.505 | -3.260 |
| 91 | 2.947 | 4.037 | -6.984 |
| 92 | 4.998 | 5.762 | -10.760 |
| 93 | 5.851 | 7.561 | -13.412 |
| 94 | 7.776 | 9.516 | -17.292 |
| 95 | 8.947 | 11.488 | -20.438 |
| 96 | 10.934 | 13.562 | -24.497 |
| 97 | 12.906 | 15.676 | -28.585 |
| 98 | 15.730 | 17.874 | -33.607 |
| 99 | 18.797 | 20.130 | -38.930 |

For comparison. Optimistic scenario:

| | | | |
|---|---|---|---|
| 1990 | 43.053 | 3.374 | -46.427 |
| 1999 | 56.262 | 27.122 | -83.387 |

Pessimistic scenario:

| | | | |
|---|---|---|---|
| 1990 | -31.876 | 2.329 | 29.547 |
| 1999 | 30.529 | 12.796 | -43.328 |

*Annex Table 4.28.* Index of the dollar exchange rate (1975 = 1), CMEA countries, medium scenario. (For details, see *Tables 4.17* and *4.20*).

| Period | Bulgaria | CSSR | GDR | Hungary | Poland | Romania | USSR |
|---|---|---|---|---|---|---|---|
| 62 | 1.455 | 1.127 | 1.421 | 0.935 | 1.110 | 1.205 | 1.075 |
| 63 | 1.452 | 1.137 | 1.411 | 0.929 | 1.106 | 1.204 | 1.097 |
| 64 | 1.440 | 1.130 | 1.396 | 0.951 | 1.109 | 1.206 | 1.098 |
| 65 | 1.431 | 1.158 | 1.362 | 0.968 | 1.109 | 1.205 | 1.127 |
| 66 | 1.413 | 1.159 | 1.342 | 0.968 | 1.141 | 1.205 | 1.143 |
| 67 | 1.405 | 1.127 | 1.328 | 0.959 | 1.151 | 1.205 | 1.199 |
| 68 | 1.401 | 1.130 | 1.315 | 0.930 | 1.176 | 1.206 | 1.196 |
| 69 | 1.385 | 1.213 | 1.306 | 0.952 | 1.124 | 1.205 | 1.229 |
| 70 | 1.336 | 1.240 | 1.288 | 0.966 | 1.101 | 1.205 | 1.218 |
| 71 | 1.309 | 1.248 | 1.280 | 0.971 | 1.141 | 1.205 | 1.246 |
| 72 | 1.181 | 1.138 | 1.189 | 0.991 | 1.060 | 1.108 | 1.146 |
| 73 | 1.043 | 1.025 | 1.057 | 1.023 | 0.983 | 1.000 | 1.015 |
| 74 | 1.021 | 1.022 | 1.049 | 1.067 | 1.016 | 1.000 | 1.007 |
| 75 | 1.000 | 1.000 | 1.000 | 1.000 | 1.000 | 1.000 | 1.000 |
| 76 | 0.979 | 1.003 | 0.987 | 0.991 | 0.915 | 1.000 | 1.011 |
| 77 | 0.962 | 0.991 | 0.996 | 0.991 | 0.908 | 0.999 | 1.007 |
| 78 | 0.892 | 0.937 | 0.929 | 0.922 | 0.864 | 0.962 | 0.955 |
| 79 | 0.849 | 0.913 | 0.906 | 0.854 | 0.845 | 0.931 | 0.935 |
| 80 | 0.795 | 0.880 | 0.922 | 0.779 | 0.892 | 0.848 | 0.945 |
| 81 | 0.851 | 1.002 | 0.926 | 0.837 | 0.986 | 0.648 | 0.959 |
| 82 | 0.857 | 1.038 | 0.951 | 0.870 | 1.754 | 0.704 | 0.979 |
| 83 | 0.862 | 1.059 | 0.968 | 0.898 | 2.011 | 0.726 | 0.987 |
| 84 | 0.870 | 1.092 | 0.984 | 0.930 | 2.273 | 0.741 | 0.999 |
| 85 | 0.886 | 1.123 | 1.017 | 0.980 | 2.633 | 0.768 | 1.018 |
| 86 | 0.895 | 1.131 | 1.053 | 1.007 | 2.847 | 0.793 | 1.037 |
| 87 | 0.901 | 1.134 | 1.078 | 1.034 | 3.001 | 0.804 | 1.050 |
| 88 | 0.902 | 1.132 | 1.093 | 1.059 | 3.136 | 0.812 | 1.053 |
| 89 | 0.901 | 1.127 | 1.103 | 1.084 | 3.266 | 0.817 | 1.053 |
| 90 | 0.898 | 1.122 | 1.109 | 1.108 | 3.397 | 0.821 | 1.052 |
| 91 | 0.894 | 1.116 | 1.112 | 1.131 | 3.485 | 0.823 | 1.049 |
| 92 | 0.890 | 1.110 | 1.113 | 1.154 | 3.572 | 0.825 | 1.045 |
| 93 | 0.885 | 1.104 | 1.113 | 1.177 | 3.657 | 0.826 | 1.040 |
| 94 | 0.879 | 1.099 | 1.112 | 1.201 | 3.743 | 0.827 | 1.035 |
| 95 | 0.873 | 1.094 | 1.109 | 1.224 | 3.827 | 0.827 | 1.030 |
| 96 | 0.867 | 1.089 | 1.107 | 1.247 | 3.912 | 0.827 | 1.024 |
| 97 | 0.861 | 1.085 | 1.104 | 1.271 | 3.997 | 0.826 | 1.019 |
| 98 | 0.855 | 1.082 | 1.101 | 1.295 | 4.083 | 0.826 | 1.014 |
| 99 | 0.850 | 1.079 | 1.098 | 1.320 | 4.170 | 0.826 | 1.009 |

For comparison. Optimistic scenario:

| | | | | | | | |
|---|---|---|---|---|---|---|---|
| 1990 | 0.893 | 1.120 | 1.094 | 1.081 | 3.325 | 0.810 | 1.047 |
| 1999 | 0.851 | 1.081 | 1.079 | 1.178 | 3.969 | 0.807 | 1.008 |

Pessimistic scenario:

| | | | | | | | |
|---|---|---|---|---|---|---|---|
| 1990 | 0.904 | 1.123 | 1.120 | 1.130 | 3.472 | 0.834 | 1.056 |
| 1999 | 0.842 | 1.069 | 1.104 | 1.386 | 4.347 | 0.845 | 1.004 |

*Annex Table 4.29.* Index of the dollar exchange rate, OECD countries, medium scenario. (For details, see *Tables 4.16* and *4.21*).

| | | DM/$ | 100 ¥/$ | FFR/$ | £/$ |
|---|---|---|---|---|---|
| Period | USA | FRG | Japan | France | UK |
| 62 | 1.000 | 4.000 | 0.360 | 4.937 | 0.357 |
| 63 | 1.000 | 4.000 | 0.360 | 4.937 | 0.357 |
| 64 | 1.000 | 4.000 | 0.360 | 4.937 | 0.357 |
| 65 | 1.000 | 4.000 | 0.360 | 4.937 | 0.357 |
| 66 | 1.000 | 4.000 | 0.360 | 4.937 | 0.357 |
| 67 | 1.000 | 4.000 | 0.360 | 4.937 | 0.362 |
| 68 | 1.000 | 4.000 | 0.360 | 4.937 | 0.417 |
| 69 | 1.000 | 3.943 | 0.360 | 5.194 | 0.417 |
| 70 | 1.000 | 3.660 | 0.360 | 5.554 | 0.417 |
| 71 | 1.000 | 3.491 | 0.349 | 5.543 | 0.411 |
| 72 | 1.000 | 3.189 | 0.303 | 5.044 | 0.400 |
| 73 | 1.000 | 2.673 | 0.272 | 4.454 | 0.408 |
| 74 | 1.000 | 2.588 | 0.292 | 4.810 | 0.428 |
| 75 | 1.000 | 2.460 | 0.297 | 4.286 | 0.452 |
| 76 | 1.000 | 2.518 | 0.297 | 4.780 | 0.557 |
| 77 | 1.000 | 2.322 | 0.269 | 4.913 | 0.573 |
| 78 | 1.000 | 2.009 | 0.210 | 4.513 | 0.522 |
| 79 | 1.000 | 1.833 | 0.219 | 4.254 | 0.472 |
| 80 | 1.000 | 1.818 | 0.227 | 4.226 | 0.430 |
| 81 | 1.000 | 2.260 | 0.221 | 5.435 | 0.498 |
| 82 | 1.000 | 2.427 | 0.249 | 6.572 | 0.572 |
| 83 | 1.000 | 2.553 | 0.238 | 7.621 | 0.659 |
| 84 | 1.000 | 2.846 | 0.238 | 8.739 | 0.748 |
| 85 | 1.000 | 2.868 | 0.247 | 9.239 | 0.782 |
| 86 | 1.000 | 2.865 | 0.256 | 9.603 | 0.818 |
| 87 | 1.000 | 2.860 | 0.267 | 9.932 | 0.862 |
| 88 | 1.000 | 2.822 | 0.272 | 10.201 | 0.907 |
| 89 | 1.000 | 2.758 | 0.274 | 10.416 | 0.952 |
| 90 | 1.000 | 2.676 | 0.273 | 10.586 | 0.996 |
| 91 | 1.000 | 2.581 | 0.270 | 10.717 | 1.040 |
| 92 | 1.000 | 2.478 | 0.266 | 10.817 | 1.081 |
| 93 | 1.000 | 2.370 | 0.261 | 10.891 | 1.122 |
| 94 | 1.000 | 2.260 | 0.255 | 10.947 | 1.161 |
| 95 | 1.000 | 2.151 | 0.249 | 10.988 | 1.198 |
| 96 | 1.000 | 2.043 | 0.242 | 11.019 | 1.235 |
| 97 | 1.000 | 1.939 | 0.236 | 11.043 | 1.270 |
| 98 | 1.000 | 1.839 | 0.229 | 11.063 | 1.305 |
| 99 | 1.000 | 1.743 | 0.223 | 11.080 | 1.340 |

For comparison. Optimistic scenario:

| | | | | | |
|---|---|---|---|---|---|
| 1990 | 1.000 | 2.653 | 0.267 | 10.020 | 1.001 |
| 1999 | 1.000 | 1.750 | 0.224 | 10.797 | 1.476 |

Pessimistic scenario:

| | | | | | |
|---|---|---|---|---|---|
| 1990 | 1.000 | 2.698 | 0.278 | 11.051 | 0.994 |
| 1999 | 1.000 | 1.683 | 0.215 | 10.955 | 1.193 |

*Annex Table 4.29.* Continued.

| Period | 1000 LIR/$<br>Italy | FL/$<br>Netherlands | BFR/$<br>Belgium/ Luxembourg | Can $/$<br>Canada | ϕ currency Rest/$<br>Other DMEs |
|---|---|---|---|---|---|
| 62 | 0.625 | 3.620 | 50.000 | 1.070 | 1.200 |
| 63 | 0.625 | 3.620 | 50.000 | 1.081 | 1.197 |
| 64 | 0.625 | 3.620 | 50.000 | 1.081 | 1.196 |
| 65 | 0.625 | 3.620 | 50.000 | 1.081 | 1.194 |
| 66 | 0.625 | 3.620 | 50.000 | 1.081 | 1.193 |
| 67 | 0.625 | 3.620 | 50.000 | 1.081 | 1.204 |
| 68 | 0.625 | 3.620 | 50.000 | 1.081 | 1.260 |
| 69 | 0.625 | 3.620 | 50.000 | 1.081 | 1.258 |
| 70 | 0.625 | 3.620 | 50.000 | 1.048 | 1.259 |
| 71 | 0.620 | 3.502 | 48.870 | 1.010 | 1.239 |
| 72 | 0.583 | 3.210 | 44.015 | 0.990 | 1.174 |
| 73 | 0.583 | 2.796 | 38.977 | 1.000 | 1.029 |
| 74 | 0.650 | 2.688 | 38.952 | 0.978 | 1.011 |
| 75 | 0.653 | 2.529 | 36.779 | 1.017 | 1.000 |
| 76 | 0.832 | 2.644 | 38.605 | 0.986 | 1.081 |
| 77 | 0.882 | 2.454 | 35.843 | 1.063 | 1.128 |
| 78 | 0.849 | 2.164 | 31.492 | 1.141 | 1.075 |
| 79 | 0.831 | 2.006 | 29.319 | 1.171 | 1.020 |
| 80 | 0.856 | 1.988 | 29.243 | 1.169 | 1.030 |
| 81 | 1.137 | 2.495 | 37.131 | 1.199 | 1.211 |
| 82 | 1.353 | 2.670 | 45.691 | 1.234 | 1.401 |
| 83 | 1.519 | 2.854 | 51.132 | 1.232 | 1.446 |
| 84 | 1.757 | 3.209 | 57.784 | 1.295 | 1.495 |
| 85 | 1.910 | 3.284 | 58.849 | 1.378 | 1.609 |
| 86 | 2.071 | 3.377 | 59.072 | 1.447 | 1.694 |
| 87 | 2.273 | 3.488 | 59.204 | 1.518 | 1.775 |
| 88 | 2.464 | 3.589 | 58.871 | 1.580 | 1.843 |
| 89 | 2.646 | 3.679 | 58.163 | 1.634 | 1.901 |
| 90 | 2.822 | 3.758 | 57.161 | 1.682 | 1.954 |
| 91 | 2.995 | 3.826 | 55.943 | 1.724 | 2.003 |
| 92 | 3.165 | 3.883 | 54.573 | 1.762 | 2.049 |
| 93 | 3.336 | 3.929 | 53.101 | 1.797 | 2.094 |
| 94 | 3.509 | 3.965 | 51.567 | 1.829 | 2.138 |
| 95 | 3.686 | 3.992 | 50.002 | 1.860 | 2.182 |
| 96 | 3.868 | 4.010 | 48.430 | 1.888 | 2.227 |
| 97 | 4.056 | 4.022 | 46.871 | 1.917 | 2.271 |
| 98 | 4.253 | 4.028 | 45.336 | 1.944 | 2.317 |
| 99 | 4.458 | 4.028 | 43.837 | 1.972 | 2.363 |

For comparison. Optimistic scenario:

| | | | | | |
|---|---|---|---|---|---|
| 1990 | 2.760 | 3.696 | 56.489 | 1.662 | 1.892 |
| 1999 | 4.233 | 3.997 | 46.142 | 2.039 | 2.334 |

Pessimistic scenario:

| | | | | | |
|---|---|---|---|---|---|
| 1990 | 2.855 | 3.806 | 57.625 | 1.700 | 2.019 |
| 1999 | 4.346 | 3.875 | 40.472 | 1.850 | 2.334 |

*Annex Table 4.30.* Index of the dollar exchange rate (1975 = 1), developing countries, average, medium scenario. (For details, see *Tables 4.18* and *4.22*).

| Period | Oil-Exporting countries | Asian countries without India | India | Black Africa | Latin Amer without M.,B.,A. | Mexico Brazil Argentina | Mid-East & North Africa |
|---|---|---|---|---|---|---|---|
| 62 | 0.288 | 0.897 | 0.569 | 0.584 | 0.023 | 0.100 | 1.089 |
| 63 | 0.357 | 0.884 | 0.569 | 0.646 | 0.038 | 0.125 | 1.130 |
| 64 | 0.431 | 0.926 | 0.569 | 0.848 | 0.045 | 0.200 | 1.136 |
| 65 | 0.548 | 0.940 | 0.569 | 0.810 | 0.046 | 0.245 | 1.154 |
| 66 | 0.946 | 0.926 | 0.759 | 0.811 | 0.063 | 0.283 | 1.153 |
| 67 | 0.984 | 0.914 | 0.895 | 0.990 | 0.073 | 0.326 | 1.154 |
| 68 | 1.076 | 0.909 | 0.895 | 1.061 | 0.091 | 0.372 | 1.155 |
| 69 | 1.086 | 0.909 | 0.895 | 1.079 | 0.098 | 0.401 | 1.153 |
| 70 | 1.108 | 0.963 | 0.895 | 1.099 | 0.114 | 0.445 | 1.155 |
| 71 | 1.123 | 0.982 | 0.896 | 1.097 | 0.098 | 0.489 | 1.154 |
| 72 | 1.084 | 0.992 | 0.907 | 1.066 | 0.109 | 0.626 | 1.116 |
| 73 | 1.023 | 0.979 | 0.924 | 1.002 | 0.288 | 0.632 | 1.021 |
| 74 | 1.009 | 0.973 | 0.967 | 1.037 | 0.573 | 0.647 | 1.034 |
| 75 | 1.000 | 1.000 | 1.000 | 1.000 | 1.000 | 1.000 | 1.000 |
| 76 | 1.016 | 1.007 | 1.070 | 1.188 | 1.446 | 1.559 | 1.037 |
| 77 | 1.020 | 1.001 | 1.043 | 1.222 | 1.863 | 2.380 | 1.043 |
| 78 | 1.014 | 0.994 | 0.978 | 1.163 | 2.293 | 3.124 | 1.020 |
| 79 | 1.040 | 0.993 | 0.970 | 1.264 | 2.743 | 4.758 | 1.227 |
| 80 | 1.025 | 1.034 | 0.939 | 1.300 | 3.148 | 7.392 | 1.248 |
| 81 | 1.082 | 1.080 | 1.034 | 1.615 | 3.655 | 9.332 | 1.406 |
| 82 | 1.152 | 1.262 | 1.201 | 1.872 | 4.775 | 12.203 | 1.533 |
| 83 | 1.194 | 1.437 | 1.398 | 2.272 | 6.263 | 16.921 | 1.736 |
| 84 | 1.188 | 1.642 | 1.595 | 2.751 | 8.515 | 23.909 | 1.985 |
| 85 | 1.205 | 1.998 | 1.815 | 3.520 | 12.463 | 35.268 | 2.391 |
| 86 | 1.272 | 2.345 | 1.991 | 4.273 | 17.446 | 50.231 | 2.788 |
| 87 | 1.241 | 2.624 | 2.096 | 4.918 | 23.437 | 69.453 | 3.156 |
| 88 | 1.220 | 2.898 | 2.211 | 5.612 | 31.006 | 95.345 | 3.506 |
| 89 | 1.210 | 3.169 | 2.347 | 6.340 | 40.758 | 130.123 | 3.858 |
| 90 | 1.204 | 3.452 | 2.494 | 7.111 | 53.259 | 175.668 | 4.196 |
| 91 | 1.199 | 3.725 | 2.641 | 7.894 | 69.102 | 236.195 | 4.561 |
| 92 | 1.197 | 4.007 | 2.787 | 8.713 | 89.534 | 315.603 | 4.923 |
| 93 | 1.195 | 4.296 | 2.934 | 9.561 | 115.725 | 421.883 | 5.304 |
| 94 | 1.194 | 4.602 | 3.086 | 10.470 | 149.398 | 561.444 | 5.730 |
| 95 | 1.194 | 4.929 | 3.245 | 11.442 | 192.688 | 747.833 | 6.155 |
| 96 | 1.194 | 5.276 | 3.412 | 12.490 | 248.866 | 994.455 | 6.625 |
| 97 | 1.194 | 5.652 | 3.589 | 13.623 | 320.684 | 1323.522 | 7.117 |
| 98 | 1.194 | 6.060 | 3.775 | 14.857 | 413.883 | 1760.604 | 7.653 |
| 99 | 1.195 | 6.506 | 3.972 | 16.192 | 533.362 | 2343.403 | 8.218 |
| For comparison. Optimistic scenario: | | | | | | | |
| 1990 | 1.193 | 3.074 | 2.247 | 6.110 | 48.348 | 161.277 | 3.846 |
| 1999 | 1.178 | 5.561 | 3.514 | 13.488 | 478.690 | 2109.562 | 7.533 |
| Pessimistic scenario: | | | | | | | |
| 1990 | 1.215 | 3.643 | 2.576 | 8.143 | 59.548 | 186.794 | 4.472 |
| 1999 | 1.217 | 7.017 | 3.653 | 20.652 | 613.846 | 2496.158 | 8.313 |

***Annex Table 4.31.*** Change rate of the general price level = inflation rate (%), CMEA countries, medium scenario. (For details, see *Table 4.17*).

| Period | Bulgaria | CSSR | GDR | Hungary | Poland | Romania | USSR |
|---|---|---|---|---|---|---|---|
| 62 | 2.314 | 0.417 | −0.592 | −0.501 | 0.843 | −3.522 | 1.930 |
| 63 | 2.087 | 0.597 | −0.338 | −0.377 | 1.151 | 0.203 | −1.509 |
| 64 | −0.352 | −1.970 | −0.134 | 0.767 | 1.685 | 0.990 | −1.503 |
| 65 | −0.295 | −1.017 | −1.708 | −1.954 | −0.263 | 1.704 | −0.393 |
| 66 | −1.119 | −0.371 | 0.440 | 2.805 | −0.520 | 2.095 | −0.875 |
| 67 | −1.134 | 15.350 | 0.401 | 0.685 | 0.173 | −0.399 | 0.139 |
| 68 | 2.081 | 2.635 | 0.653 | 5.171 | 0.866 | 0.241 | −0.069 |
| 69 | −0.363 | 5.342 | 0.857 | 4.071 | 0.532 | 0.743 | 2.224 |
| 70 | 4.294 | 1.278 | 1.492 | 3.427 | 2.283 | 0.887 | 1.568 |
| 71 | −6.526 | −0.372 | −1.370 | 1.602 | 5.402 | −0.461 | −0.417 |
| 72 | 0.124 | −0.574 | 0.255 | 1.613 | 0.786 | −0.738 | −0.551 |
| 73 | 0.083 | −0.334 | 0.639 | 2.584 | 0.678 | 0.162 | −0.998 |
| 74 | 0.169 | 1.161 | −0.886 | −1.547 | 2.284 | −0.451 | −0.893 |
| 75 | 0.235 | −0.943 | 0.166 | 0.810 | 5.610 | 1.071 | −1.111 |
| 76 | −0.481 | −0.992 | 1.665 | 6.253 | 9.491 | 1.143 | −0.198 |
| 77 | −3.264 | −5.029 | −0.249 | 2.003 | 3.982 | −0.954 | 0.273 |
| 78 | −0.047 | 1.000 | 0.935 | 3.448 | 7.096 | 0.293 | 0.032 |
| 79 | 1.178 | 1.795 | 2.180 | 4.937 | 8.402 | 0.978 | 0.192 |
| 80 | 9.986 | 2.352 | 3.161 | 5.175 | 11.692 | −0.529 | −0.391 |
| 81 | 1.361 | −2.174 | 2.725 | 6.429 | 22.381 | 1.598 | 1.005 |
| 82 | −0.097 | 4.203 | 1.892 | 5.482 | 111.514 | 14.780 | 3.891 |
| 83 | 0.752 | 1.885 | 0.191 | 5.009 | 17.426 | 1.371 | 1.058 |
| 84 | 1.400 | 4.547 | 0.197 | 5.957 | 16.076 | 0.583 | 1.501 |
| 85 | 1.200 | 2.682 | 0.198 | 7.490 | 17.680 | 0.482 | 1.214 |
| 86 | 1.345 | 1.152 | 1.282 | 4.353 | 9.775 | 0.980 | 3.436 |
| 87 | 1.357 | 1.168 | 1.389 | 4.775 | 8.269 | 1.075 | 2.475 |
| 88 | 1.325 | 1.169 | 1.400 | 5.003 | 8.495 | 1.173 | 1.322 |
| 89 | 1.306 | 1.170 | 1.394 | 5.130 | 8.644 | 1.255 | 1.279 |
| 90 | 1.294 | 1.172 | 1.375 | 5.202 | 8.735 | 1.320 | 1.239 |
| 91 | 1.284 | 1.176 | 1.348 | 5.245 | 6.841 | 1.375 | 1.204 |
| 92 | 1.275 | 1.182 | 1.318 | 5.277 | 6.863 | 1.420 | 1.173 |
| 93 | 1.265 | 1.187 | 1.287 | 5.300 | 6.870 | 1.457 | 1.146 |
| 94 | 1.255 | 1.192 | 1.258 | 5.316 | 6.865 | 1.488 | 1.123 |
| 95 | 1.244 | 1.199 | 1.230 | 5.332 | 6.851 | 1.514 | 1.104 |
| 96 | 1.234 | 1.206 | 1.203 | 5.351 | 6.837 | 1.538 | 1.089 |
| 97 | 1.223 | 1.214 | 1.178 | 5.367 | 6.821 | 1.558 | 1.078 |
| 98 | 1.212 | 1.222 | 1.155 | 5.387 | 6.807 | 1.578 | 1.070 |
| 99 | 1.202 | 1.233 | 1.132 | 5.406 | 6.795 | 1.597 | 1.065 |

For comparison. Optimistic scenario.

| | | | | | | | |
|---|---|---|---|---|---|---|---|
| 1990 | 0.765 | 0.749 | 0.446 | 3.848 | 7.657 | 0.015 | 0.476 |
| 1999 | 0.636 | 0.434 | 0.181 | 3.799 | 5.629 | 0.146 | 0.166 |

Pessimistic scenario.

| | | | | | | | |
|---|---|---|---|---|---|---|---|
| 1990 | 1.818 | 1.534 | 2.132 | 6.329 | 9.842 | 2.817 | 2.001 |
| 1999 | 1.740 | 1.886 | 1.961 | 6.646 | 8.069 | 3.247 | 1.959 |

*Annex Table 4.32.* Change rate of the general price level = inflation rate (%), OECD countries, medium scenario. (For details, see *Table 4.16*).

| Period | USA | FRG | Japan | France | UK |
|---|---|---|---|---|---|
| 62 | 2.022 | 4.079 | 3.602 | 4.730 | 3.535 |
| 63 | 1.439 | 2.734 | 4.501 | 6.362 | 2.186 |
| 64 | 1.513 | 3.014 | 4.443 | 4.352 | 3.605 |
| 65 | 2.142 | 3.594 | 5.077 | 2.633 | 4.899 |
| 66 | 3.412 | 3.498 | 5.002 | 3.041 | 4.544 |
| 67 | 2.934 | 1.233 | 5.799 | 3.145 | 2.891 |
| 68 | 4.924 | 2.028 | 5.180 | 4.287 | 4.258 |
| 69 | 5.099 | 4.226 | 4.805 | 6.859 | 5.591 |
| 70 | 5.420 | 7.498 | 7.333 | 5.704 | 7.418 |
| 71 | 5.324 | 7.626 | 5.182 | 5.647 | 9.584 |
| 72 | 4.313 | 5.370 | 5.210 | 6.391 | 8.643 |
| 73 | 5.645 | 6.592 | 11.936 | 7.967 | 7.290 |
| 74 | 8.741 | 6.664 | 20.595 | 11.252 | 14.901 |
| 75 | 9.056 | 5.934 | 7.806 | 12.812 | 27.148 |
| 76 | 5.695 | 3.560 | 6.402 | 10.228 | 14.927 |
| 77 | 6.002 | 3.618 | 5.672 | 9.039 | 13.856 |
| 78 | 7.444 | 4.200 | 4.591 | 9.479 | 11.104 |
| 79 | 8.741 | 4.109 | 2.616 | 10.652 | 14.758 |
| 80 | 9.326 | 4.383 | 2.841 | 12.112 | 19.743 |
| 81 | 9.422 | 3.776 | 2.864 | 11.658 | 12.998 |
| 82 | 6.070 | 2.696 | 4.179 | 12.632 | 7.079 |
| 83 | 5.691 | 3.190 | -0.169 | 9.055 | 8.434 |
| 84 | 6.317 | 2.333 | 2.834 | 6.029 | 3.900 |
| 85 | 3.144 | 3.929 | 7.212 | 8.869 | 7.688 |
| 86 | 4.454 | 4.339 | 8.172 | 8.395 | 9.009 |
| 87 | 4.536 | 4.348 | 8.493 | 7.961 | 9.918 |
| 88 | 5.279 | 3.958 | 7.330 | 7.982 | 10.499 |
| 89 | 5.876 | 3.633 | 6.502 | 7.990 | 10.838 |
| 90 | 6.347 | 3.368 | 5.996 | 7.975 | 11.004 |
| 91 | 6.718 | 3.161 | 5.680 | 7.956 | 11.055 |
| 92 | 7.006 | 2.997 | 5.467 | 7.935 | 11.031 |
| 93 | 7.229 | 2.868 | 5.314 | 7.921 | 10.961 |
| 94 | 7.398 | 2.767 | 5.201 | 7.909 | 10.868 |
| 95 | 7.524 | 2.689 | 5.117 | 7.902 | 10.764 |
| 96 | 7.614 | 2.630 | 5.054 | 7.896 | 10.658 |
| 97 | 7.676 | 2.585 | 5.004 | 7.894 | 10.558 |
| 98 | 7.715 | 2.552 | 4.965 | 7.893 | 10.466 |
| 99 | 7.736 | 2.528 | 4.934 | 7.894 | 10.383 |

For comparison. Optimistic scenario:

| | | | | | |
|---|---|---|---|---|---|
| 1990 | 5.345 | 2.175 | 4.597 | 6.438 | 10.453 |
| 1999 | 5.792 | 0.990 | 3.673 | 6.725 | 9.819 |

Pessimistic scenario:

| | | | | | |
|---|---|---|---|---|---|
| 1990 | 7.319 | 4.496 | 7.322 | 9.352 | 11.574 |
| 1999 | 10.373 | 4.105 | 6.240 | 9.247 | 11.058 |

*Annex Table 4.32.* Continued.

| Period | Italy | Netherlands | Belgium/ Luxembourg | Canada | Other DMEs |
|---|---|---|---|---|---|
| 62 | 6.136 | 3.557 | 1.517 | 2.133 | 2.826 |
| 63 | 9.192 | 4.685 | 3.052 | 1.581 | 4.434 |
| 64 | 6.600 | 8.545 | 4.751 | 2.424 | 4.390 |
| 65 | 4.325 | 6.055 | 4.990 | 3.693 | 5.822 |
| 66 | 2.469 | 5.973 | 4.146 | 3.951 | 4.882 |
| 67 | 2.817 | 4.142 | 2.966 | 2.942 | 4.764 |
| 68 | 1.954 | 4.113 | 2.954 | 3.417 | 3.378 |
| 69 | 4.231 | 6.383 | 4.265 | 4.757 | 4.453 |
| 70 | 7.007 | 5.548 | 4.966 | 3.322 | 6.493 |
| 71 | 7.761 | 8.572 | 6.268 | 3.450 | 7.878 |
| 72 | 6.645 | 9.379 | 6.196 | 5.197 | 8.210 |
| 73 | 10.801 | 8.450 | 7.109 | 9.466 | 10.719 |
| 74 | 18.328 | 9.302 | 12.414 | 16.308 | 13.318 |
| 75 | 18.714 | 11.196 | 11.995 | 10.106 | 12.674 |
| 76 | 17.371 | 8.891 | 7.527 | 10.149 | 11.008 |
| 77 | 19.535 | 6.316 | 6.997 | 7.005 | 11.796 |
| 78 | 14.316 | 5.402 | 4.202 | 6.432 | 10.510 |
| 79 | 16.033 | 3.785 | 4.206 | 10.855 | 10.955 |
| 80 | 20.018 | 5.650 | 4.452 | 10.353 | 11.037 |
| 81 | 15.094 | 5.395 | 5.267 | 11.600 | 11.334 |
| 82 | 20.396 | 3.974 | 7.048 | 9.250 | 10.368 |
| 83 | 14.151 | 3.608 | 4.383 | 6.209 | 8.862 |
| 84 | 14.243 | 2.271 | 4.389 | 6.408 | 9.732 |
| 85 | 11.845 | 5.506 | 4.988 | 9.564 | 10.741 |
| 86 | 12.908 | 7.273 | 4.832 | 9.479 | 9.789 |
| 87 | 14.258 | 7.816 | 4.760 | 9.409 | 9.292 |
| 88 | 13.688 | 8.182 | 4.716 | 9.353 | 9.103 |
| 89 | 13.276 | 8.395 | 4.674 | 9.308 | 9.051 |
| 90 | 13.003 | 8.495 | 4.624 | 9.271 | 9.107 |
| 91 | 12.826 | 8.518 | 4.588 | 9.242 | 9.214 |
| 92 | 12.708 | 8.485 | 4.557 | 9.218 | 9.328 |
| 93 | 12.632 | 8.413 | 4.533 | 9.200 | 9.429 |
| 94 | 12.584 | 8.315 | 4.508 | 9.186 | 9.513 |
| 95 | 12.559 | 8.203 | 4.489 | 9.177 | 9.582 |
| 96 | 12.548 | 8.086 | 4.471 | 9.170 | 9.637 |
| 97 | 12.548 | 7.968 | 4.456 | 9.166 | 9.680 |
| 98 | 12.554 | 7.854 | 4.442 | 9.164 | 9.712 |
| 99 | 12.565 | 7.746 | 4.430 | 9.164 | 9.737 |

For comparison. Optimistic scenario:

| | | | | | |
|---|---|---|---|---|---|
| 1990 | 11.466 | 7.341 | 3.760 | 8.279 | 7.683 |
| 1999 | 10.522 | 6.061 | 3.426 | 8.001 | 8.325 |

Pessimistic scenario.

| | | | | | |
|---|---|---|---|---|---|
| 1990 | 14.186 | 9.515 | 5.434 | 10.206 | 10.502 |
| 1999 | 14.120 | 9.253 | 5.574 | 10.324 | 11.169 |

*Annex Table 4.33.* Change rate of the general price level = inflation rate (%), developing countries, medium scenario. (For details, see *Table 4.18*).

| Period | Oil-Exporting countries | Asian countries without India | India | Black Africa | Latin Amer without M.,B.,A. | Mexico Brazil Argentina | Mid-East & North Africa |
|---|---|---|---|---|---|---|---|
| 62 | 22.522 | -7.756 | 3.037 | -0.054 | 7.329 | 30.097 | 0.978 |
| 63 | 24.271 | 4.418 | 7.132 | 12.752 | 53.154 | 39.594 | 1.573 |
| 64 | 12.622 | 1.146 | 9.366 | 29.496 | 29.776 | 44.133 | 1.708 |
| 65 | 26.725 | 1.925 | 9.138 | 3.618 | 6.437 | 33.714 | 4.767 |
| 66 | 68.203 | 3.537 | 17.373 | 8.565 | 33.473 | 27.067 | 3.230 |
| 67 | 9.394 | 4.443 | 7.676 | 14.126 | 17.356 | 19.657 | 2.310 |
| 68 | 10.832 | 2.537 | -0.332 | 12.456 | 19.883 | 15.146 | 2.009 |
| 69 | 4.090 | 4.533 | 3.725 | 7.613 | 16.144 | 12.803 | -0.937 |
| 70 | 5.263 | 1.934 | 3.318 | -0.683 | 21.796 | 17.957 | 2.333 |
| 71 | 10.210 | 4.129 | 5.128 | 1.046 | -6.379 | 15.329 | 3.367 |
| 72 | 7.734 | 8.310 | 11.425 | 7.044 | 21.768 | 30.592 | 2.166 |
| 73 | 18.588 | 19.371 | 18.385 | 15.400 | 173.558 | 21.434 | 9.955 |
| 74 | 66.397 | 22.273 | 17.636 | 18.250 | 132.675 | 27.984 | 18.540 |
| 75 | 14.782 | 4.390 | -2.609 | 6.771 | 76.449 | 56.880 | 1.638 |
| 76 | 12.301 | 8.868 | 6.020 | 26.018 | 52.407 | 70.906 | 10.264 |
| 77 | 12.592 | 9.524 | 3.755 | 20.574 | 47.406 | 50.412 | 11.076 |
| 78 | 7.298 | 12.781 | 2.138 | 9.833 | 25.774 | 50.688 | 7.925 |
| 79 | 15.888 | 15.440 | 15.035 | 21.416 | 45.287 | 78.343 | 9.048 |
| 80 | 26.999 | 10.979 | 11.786 | 10.967 | 40.830 | 80.506 | 18.970 |
| 81 | 16.871 | 5.788 | 9.066 | 8.898 | 29.083 | 37.747 | 16.153 |
| 82 | 8.715 | 12.598 | 13.846 | 19.985 | 25.882 | 35.060 | 9.499 |
| 83 | 2.560 | 12.462 | 15.182 | 20.623 | 31.225 | 37.511 | 15.210 |
| 84 | -1.148 | 13.048 | 12.441 | 20.493 | 35.151 | 40.056 | 15.683 |
| 85 | -5.580 | 13.173 | 9.301 | 19.892 | 36.884 | 37.822 | 16.052 |
| 86 | -8.691 | 12.964 | 6.303 | 18.863 | 35.491 | 38.998 | 15.185 |
| 87 | -5.578 | 13.014 | 7.044 | 17.802 | 34.445 | 40.674 | 14.813 |
| 88 | -2.261 | 12.945 | 8.348 | 17.576 | 34.446 | 40.691 | 14.292 |
| 89 | 0.178 | 12.641 | 9.450 | 16.761 | 34.820 | 40.688 | 14.062 |
| 90 | 1.790 | 12.774 | 9.875 | 16.341 | 35.045 | 39.851 | 13.268 |
| 91 | 2.933 | 12.286 | 9.796 | 15.591 | 34.937 | 39.924 | 13.597 |
| 92 | 3.779 | 12.340 | 9.666 | 15.229 | 35.402 | 39.500 | 13.117 |
| 93 | 4.402 | 12.318 | 9.630 | 14.822 | 35.578 | 39.925 | 13.184 |
| 94 | 4.868 | 12.425 | 9.700 | 14.766 | 35.785 | 39.560 | 13.716 |
| 95 | 5.201 | 12.604 | 9.814 | 14.679 | 35.943 | 39.881 | 13.273 |
| 96 | 5.445 | 12.669 | 9.924 | 14.626 | 36.341 | 39.790 | 13.639 |
| 97 | 5.603 | 12.816 | 9.999 | 14.594 | 36.159 | 39.990 | 13.537 |
| 98 | 5.700 | 12.978 | 10.072 | 14.615 | 36.470 | 39.978 | 13.740 |
| 99 | 5.752 | 13.156 | 10.132 | 14.544 | 36.311 | 40.087 | 13.649 |

For comparison. Optimistic scenario.

| | | | | | | | |
|---|---|---|---|---|---|---|---|
| 1990 | 1.355 | 10.863 | 8.473 | 13.910 | 33.467 | 38.270 | 11.999 |
| 1999 | 4.780 | 11.719 | 8.797 | 13.398 | 35.144 | 38.523 | 12.676 |

Pessimistic scenario.

| | | | | | | | |
|---|---|---|---|---|---|---|---|
| 1990 | 2.133 | 14.227 | 10.047 | 19.465 | 37.880 | 41.643 | 14.478 |
| 1999 | 3.569 | 14.790 | 10.409 | 16.589 | 38.052 | 42.062 | 14.434 |

*Annex Table 4.34.* Relative size of economies.

**(a) OECD countries.**

| Ratio of GDP in % / Year | FRG/USA | Jap/USA | Fra/USA | UK/USA | Ita/USA | Can/USA | Ratio 8 OECD countries/USA |
|---|---|---|---|---|---|---|---|
| 1962 | 26.0 | 17.5 | 18.1 | 16.6 | 11.5 | 8.3 | 1.0 |
| 1984 | 26.7 | 37.4 | 21.7 | 13.2 | 12.0 | 10.4 | 1.3 |
| 1999 optimistic scenario | 31.4 | 55.2 | 24.9 | 11.1 | 15.2 | 11.4 | 1.6 |
| 1999 medium scenario | 29.6 | 49.8 | 23.6 | 11.2 | 13.2 | 10.9 | 1.5 |
| 1999 pessimistic scenario | 28.1 | 45.5 | 22.6 | 11.2 | 12.2 | 10.4 | 1.4 |

**(b) CMEA countries (GDP type "1" or NMP).**

| Ratio of GDP in % / Year | Bulg/USSR | CSSR/USSR | GDR/USSR | Hung/USSR | Pol/USSR | Rom/USSR | Ratio 6 CMEA countries/USSR |
|---|---|---|---|---|---|---|---|
| 1962 | 2.8 | 8.4 | 10.3 | 4.6 | 12.5 | 3.8 | .42 |
| 1984 | 3.6 | 5.8 | 8.1 | 3.6 | 9.4 | 6.7 | .37 |
| 1999 optimistic scenario | 4.2 | 5.2 | 8.5 | 3.3 | 8.5 | 8.5 | .34 |
| 1999 medium scenario | 4.2 | 5.1 | 8.8 | 3.1 | 8.5 | 7.9 | .34 |
| 1999 pessimistic scenario | 4.1 | 5.1 | 9.3 | 3.1 | 8.5 | 7.2 | .35 |

*Annex Table 4.34.* Continued.

**(*c*) CMEA countries (GDP type "2").**

| Ratio of GDP in % / Year | Bulg/USSR | CSSR/USSR | GDR/USSR | Hung/USSR | Pol/USSR | Rom/USSR | Ratio 6 CMEA countries/USSR |
|---|---|---|---|---|---|---|---|
| 1962 | 3.5 | 8.2 | 9.7 | 4.8 | 12.5 | 5.6 | .44 |
| | | 5 | | | | | .37 |
| 1984 | 3.3 | 5.9 | 7.8 | 3.8 | 10.4 | 5.7 | .37 |
| 1999 optimistic scenario | 3.0 | 5.1 | 7.4 | 3.2 | 9.5 | 5.7 | .34 |
| 1999 medium scenario | 3.1 | 5.1 | 7.7 | 3.1 | 9.5 | 5.6 | .34 |
| 1999 pessimistic scenario | 3.2 | 5.1 | 8.2 | 3.1 | 9.5 | 5.5 | .35 |

**(*d*) Ratio of GDP type "2" of the USSR to GDP of the USA (%).**

| 1984 | 1999 optimistic scenario | 1999 medium scenario | 1999 pessimistic scenario |
|---|---|---|---|
| 58.8 | 69.5 | 66.1 | 62.9 |

*Annex Table 4.35.* Sectoral composition (%) of GDP, OECD countries (1962 and 1984: actual figures for share of value added; 1999: forecasts based on the individual country models).

| | | Agric. | Min. & Quar. | Manuf. | Public Services[a] | Construct. | Services |
|---|---|---|---|---|---|---|---|
| USA | 1962 | 4.3 | 3.0 | 23.9 | 2.2 | 6.6 | 59.9 |
| | 1984 | 3.0 | 2.5 | 24.0 | 2.7 | 4.9 | 63.1 |
| 1999 | medium scenario | 2.1 | 2.1 | 25.9 | 2.6 | 3.3 | 64.5 |
| | pessimistic scenario | 2.3 | 2.2 | 24.9 | 2.0 | 3.3 | 65.8 |
| FRG | 1962 | 4.0 | 3.0 | 36.3 | 1.8 | 8.4 | 46.6 |
| | 1984 | 2.7 | .9 | 36.4 | 2.9 | 5.9 | 52.2 |
| 1999 | medium scenario | 1.2 | .5 | 35.9 | 3.6 | 5.8 | 54.1 |
| | pessimistic scenario | 1.5 | .6 | 36.5 | 3.5 | 4.8 | 53.1 |
| Japan | 1962 | 11.9 | 1.1 | 22.9 | 1.9 | 7.9 | 54.3 |
| | 1984 | 4.8 | .7 | 30.4 | 2.2 | 9.7 | 52.3 |
| 1999 | medium scenario | 4.3 | .5 | 34.3 | 2.3 | 10.5 | 49.7 |
| | pessimistic scenario | 5.4 | .5 | 34.3 | 2.1 | 9.9 | 49.6 |
| France | 1962 | 8.9 | 2.1 | 26.8 | 1.3 | 7.7 | 53.2 |
| | 1984 | 4.5 | .7 | 28.7 | 2.1 | 5.5 | 58.4 |
| 1999 | medium scenario | 3.2 | .4 | 30.0 | 2.2 | 6.1 | 57.4 |
| | pessimistic scenario | 3.5 | .5 | 26.9 | 2.2 | 4.2 | 62.6 |
| UK | 1962 | 2.7 | 3.1 | 27.2 | 2.2 | 8.4 | 56.4 |
| | 1984 | 2.9 | 3.0 | 25.3 | 3.3 | 6.3 | 59.1 |
| 1999 | medium scenario | 2.9 | 3.3 | 25.4 | 3.7 | 5.4 | 55.2 |
| | pessimistic scenario | 3.2 | 7.9 | 22.6 | 3.5 | 4.6 | 54.3 |
| Italy | 1962 | 10.6 | 2.5 | 25.2 | 3.9 | 12.0 | 45.7 |
| | 1984 | 7.4 | 2.4 | 31.2 | 5.0 | 7.3 | 46.9 |
| 1999 | medium scenario | 6.1 | 2.2 | 31.1 | 5.9 | 6.6 | 47.0 |
| | pessimistic scenario | 6.8 | 2.4 | 34.5 | 4.8 | 5.9 | 46.6 |
| Canada | 1962 | 7.6 | 4.3 | 21.3 | 1.8 | 8.6 | 56.4 |
| | 1984 | 4.4 | 3.0 | 19.5 | 3.5 | 6.6 | 62.3 |
| 1999 | medium scenario | 3.8 | 3.1 | 18.6 | 4.2 | 6.5 | 64.8 |
| | pessimistic scenario | 4.2 | 3.0 | 19.2 | 3.3 | 6.1 | 63.7 |

[a]Electricity, water, and gas.

***Annex Table 4.36.*** **Sectoral composition (%) of NMP, CMEA countries (1960 and 1985: observed figures; 1999: forecasts based on the individual country models).**

| | | Agric. | Industr. | Construct. | Trade | Traffic Commun. | Others |
|---|---|---|---|---|---|---|---|
| USSR | 1960 | 38.3 | 37.4 | 11.5 | 7.8 | 5.0 | / |
| | 1985 | 14.5 | 60.3 | 10.4 | 8.3 | 6.4 | / |
| 1999 | medium scenario | 6.9 | 67.7 | 10.3 | 8.3 | 6.8 | / |
| | pessimistic scenario | 7.9 | 68.3 | 9.3 | 7.9 | 6.6 | / |
| Bulg. | 1960 | 57.5 | 24.3 | 6.3 | 3.1 | 3.9 | 5.0 |
| | 1985 | 19.9 | 48.4 | 9.1 | 10.7 | 8.4 | 3.5 |
| 1999 | medium scenario | 14.3 | 52.6 | 9.8 | 11.2 | 9.0 | 3.0 |
| | pessimistic scenario | 16.2 | 51.1 | 9.3 | 11.4 | 8.9 | 3.0 |
| CSSR | 1960 | 16.9 | 51.4 | 11.0 | 15.2 | 4.9 | .6 |
| | 1985 | 9.8 | 55.4 | 12.5 | 16.8 | 4.7 | .7 |
| 1999 | medium scenario | 8.6 | 55.5 | 13.0 | 17.6 | 4.6 | .7 |
| | pessimistic scenario | 9.0 | 56.7 | 13.0 | 15.6 | 4.7 | 1.0 |
| GDR | 1960 | 17.6 | 58.2 | 5.9 | 10.5 | 5.1 | 2.7 |
| | 1985 | 7.4 | 69.7 | 5.4 | 10.1 | 4.5 | 3.0 |
| 1999 | medium scenario | 5.2 | 71.0 | 5.9 | 10.1 | 4.3 | 3.4 |
| | pessimistic scenario | 5.3 | 71.3 | 5.7 | 10.0 | 4.3 | 3.3 |
| Hung. | 1960 | 35.5 | 36.7 | 10.5 | 12.3 | 4.8 | .4 |
| | 1985 | 17.9 | 52.2 | 10.3 | 13.0 | 5.7 | 1.0 |
| 1999 | medium scenario | 15.4 | 52.8 | 11.3 | 12.3 | 6.6 | 1.6 |
| | pessimistic scenario | 17.4 | 54.2 | 11.0 | 9.2 | 6.9 | 1.4 |
| Poland | 1960 | 38.2 | 31.9 | 13.2 | 13.1 | 3.0 | .7 |
| | 1985 | 18.2 | 48.9 | 12.3 | 15.0 | 4.4 | 1.3 |
| 1999 | medium scenario | 11.9 | 53.1 | 12.7 | 16.0 | 4.8 | 1.5 |
| | pessimistic scenario | 9.5 | 54.4 | 13.1 | 16.5 | 4.8 | 1.7 |
| Rom. | 1960 | 53.2 | 22.9 | 8.7 | 8.2 | 4.0 | 3.0 |
| | 1985 | 22.0 | 54.3 | 7.8 | 6.0 | 7.7 | 2.2 |
| 1999 | medium scenario | 16.0 | 60.1 | 7.5 | 6.2 | 7.8 | 2.3 |
| | pessimistic scenario | 14.0 | 61.4 | 6.5 | 6.4 | 8.6 | 3.1 |

***Annex Table 4.37.*** **Sectoral composition (%) of GDP, developing countries (1960 and 1982: observed figures; 2000: medium scenario forecasts based on the individual country models).**

| | | Agric. | Min. & Quarr. | Manuf. | Public Services | Construct. | Services |
|---|---|---|---|---|---|---|---|
| Group 11 | 1960 | 41.9 | / | 9.9 | .6 | 7.3 | 40.4 |
| (oil-exporting | 1982 | 19.0 | / | 12.5 | 1.3 | 13.9 | 53.3 |
| developing countries) | 2000 | 20.5 | / | 11.7 | 1.5 | 15.0 | 51.2 |
| Group 12I | 1960 | 34.6 | 2.2 | 14.5 | .7 | 4.0 | 43.9 |
| (Asian dev. countries, | 1982 | 18.2 | 1.7 | 27.4 | 1.5 | 6.0 | 45.2 |
| without India) | 2000 | 12.8 | 1.1 | 33.8 | 1.9 | 5.2 | 45.3 |
| | 1960 | 50.8 | 1.1 | 13.2 | .5 | 4.7 | 29.7 |
| India | 1982 | 40.4 | 1.3 | 16.7 | 1.3 | 4.7 | 35.3 |
| | 2000 | 34.8 | 1.3 | 18.0 | 2.1 | 3.5 | 40.2 |
| Group 13 | 1960 | 36.5 | 10.3 | 7.2 | 1.3 | 4.8 | 40.0 |
| (Black Africa) | 1982 | 27.1 | 6.9 | 11.5 | 2.7 | 6.9 | 44.9 |
| | 2000 | 27.1 | 8.8 | 11.0 | 1.6 | 7.4 | 44.0 |
| Group 14 | 1960 | 18.9 | 6.3 | 21.0 | .8 | 4.1 | 48.9 |
| (Latin America | 1982 | 14.6 | 6.3 | 23.0 | 1.4 | 3.6 | 51.1 |
| without M.,B.,A., | 2000 | 11.7 | 7.4 | 23.2 | 1.8 | 2.7 | 53.2 |
| Group 15 | 1960 | 16.3 | 1.3 | 23.7 | 1.1 | 6.3 | 51.3 |
| (Mexico, Brazil | 1982 | 10.4 | 1.8 | 29.3 | 2.2 | 4.9 | 51.4 |
| Argentina) | 2000 | 6.6 | 1.7 | 28.5 | 3.8 | 6.0 | 53.4 |
| Group 18 | 1960 | 30.2 | 5.3 | 15.6 | .7 | 3.7 | 44.5 |
| (Middle East, | 1982 | 18.4 | 8.5 | 14.8 | 1.6 | 5.3 | 51.5 |
| North Africa) | 2000 | 19.4 | 6.8 | 16.1 | 2.4 | 5.2 | 50.0 |

# The Authors

BRILLET, Jean-Louis
Institut National de la Statistique
et des Études Economiques
18 Blvd. Adolphe Pinard
75014 Paris
France

ČEŠKA, Jaroslav
Federal Statistical Office
Sokolovská 142
186 13 Prague
Czechoslovakia

CZYŻEWSKI, Adam B.
University of Łódź
Institute of Econometrics
and Statistics
41 Rewolucji 1905 r. 90
214 Łódź
Poland

D'ALCANTARA, G.
European Cooperation Fund
Hoornstraat 34
1040 Brussels
Belgium

DJAROVA, J.
Institute of Social Management
ul. Pionerski put, 21
Sofia 1635
Bulgaria

DOBRINSKY, Rumen
Institute of Social Management
ul. Pionerski put, 21
Sofia 1635
Bulgaria

DONGES, Jürgen B.
Kiel Institute of World Economics
Düsternbroker Weg 120
2300 Kiel 1
FRG

DUBOVSKY, S.V.
USSR Academy of Sciences
Institute for Systems Studies
Prospect 60 Let Octyabria 9
117312 Moscow
USSR

EISMONT, O.A.
USSR Academy of Sciences
Institute for Systems Studies
Prospect 60 Let Octyabria 9
117312 Moscow
USSR

ERBER, Georg
Deutsches Institut für
Wirtschaftsforschung
Königin-Luise-Straße 5
1000 Berlin 33
FRG

GAJDA, Jan B.
University of Łódź
Institute of Econometrics
and Statistics
41 Rewolucji 1905 r. 90
214 Łódź
Poland

HICKMAN, Bert G.
Stanford University
Department of Economics
Stanford, CA 94305–6072
USA

KLIMENKO, L.
World Marxist Review
Thakurova 3
Prague 6
Czechoslovakia

KLODT, Henning
Kiel Institute of World Economics
Düsternbroker Weg 120
2300 Kiel 1
FRG

KRELLE, Wilhelm
University of Bonn
Institut für Gesellschafts- und
Wirtschaftswissenschaften
Adenauerallee 24–42
5300 Bonn
FRG

LANDESMANN, Michael
University of Cambridge
Department of Applied Economics
Sidgwick Avenue
Cambridge CB3 9DE
UK

MENSHIKOV, S.
World Marxist Review
Thakurova 3
Prague 6
Czechoslovakia

MORIGUCHI, Chikashi
Osaka University
Institute of Social and
Economic Research
6–1 Mihogaoka, Ibaraki
Osaka 567
Japan

PANOV, O.
Institute of Social Management
ul. Pionerski put, 21
Sofia 1635
Bulgaria

PAULY, Peter
University of Pennsylvania
Department of Economics
3718 Locust Walk CR
Philadelphia, PA 19104–6297
USA

RICHTERING, Jürgen
UNCTAD
Money, Finance and Development
Division
Palais des Nations
1211 Geneva 10
Switzerland

ROSS, Hermann
University of Bonn
Institut für Gesellschafts- und
Wirtschaftswissenschaften
Adenauerallee 24–42
5300 Bonn
FRG

SARRAZIN, Hermann
University of Bonn
Institut für Gesellschafts- und
Wirtschaftswissenschaften
Adenauerallee 24–42
5300 Bonn
FRG

SCHLEICHER, Stefan
University of Graz
Institute of Economics
Schubertstraße 6A
8010 Graz
Austria

SCHMIDT, Klaus-Dieter
Kiel Institute of World Economics
Düsternbroker Weg 120
2300 Kiel 1
FRG

SIMON, Andras
Karl-Marx-University of Economics
Department of Economic Planning and Modeling
Dimitrov tér 8
Budapest 1093
Hungary

SNELL, Andrew
Department of Economics
University of Edinburgh
Edinburgh
Scotland

SUJAN, Ivan
Institute of Socio-Economic Information and Automation in Management, Bratislava
Dúbravská 3
84221 Bratislava
Czechoslovakia

SZÉKELY, István
Karl-Marx-University of Economics
Department of Economic Planning and Modeling
Dimitrov tér 8
Budapest 1093
Hungary

TCHIJOV, Iouri
International Institute for Applied Systems Analysis
2361 Laxenburg
Austria

WELFE, Wladyslaw
University of Łódź
Institute of Econometrics and Statistics
41 Rewolucji 1905 r. 90
214 Łódź
Poland

WELSCH, Heinz
University of Bonn
Institut für Gesellschafts- und Wirtschaftswissenschaften
Adenauerallee 24–42
5300 Bonn
FRG

ZEMAN, Karl
Central Institute of Economic Research
nám. Hrdinu 4/1635
14096 Prague 4 – Nusle
Czechoslovakia

**THE INTERNATIONAL INSTITUTE FOR APPLIED SYSTEMS ANALYSIS**

is a nongovernmental research institution, bringing together scientists from around the world to work on problems of common concern. Situated in Laxenburg, Austria, IIASA was founded in October 1972 by the academies of science and equivalent organizations of twelve countries. Its founders gave IIASA a unique position outside national, disciplinary, and institutional boundaries so that it might take the broadest possible view in pursuing its objectives:

*To promote international cooperation* in solving problems from social, economic, technological, and environmental change

*To create a network of institutions* in the national member organization countries and elsewhere for joint scientific research

*To develop and formalize systems analysis* and the sciences contributing to it, and promote the use of analytical techniques needed to evaluate and address complex problems

*To inform policy advisors and decision makers* about the potential application of the Institute's work to such problems

The Institute now has national member organizations in the following countries:

**Austria**
The Austrian Academy of Sciences

**Bulgaria**
The National Committee for Applied Systems Analysis and Management

**Canada**
The Canadian Committee for IIASA

**Czechoslovakia**
The Committee for IIASA of the Czechoslovak Socialist Republic

**Finland**
The Finnish Committee for IIASA

**France**
The French Association for the Development of Systems Analysis

**German Democratic Republic**
The Academy of Sciences of the German Democratic Republic

**Federal Republic of Germany**
Association for the Advancement of IIASA

**Hungary**
The Hungarian Committee for Applied Systems Analysis

**Italy**
The National Research Council

**Japan**
The Japan Committee for IIASA

**Netherlands**
The Foundation IIASA–Netherlands

**Poland**
The Polish Academy of Sciences

**Sweden**
The Swedish Council for Planning and Coordination of Research

**Union of Soviet Socialist Republics**
The Academy of Sciences of the Union of Soviet Socialist Republics

**United States of America**
The American Academy of Arts and Sciences

MIX
Papier aus verantwortungsvollen Quellen
Paper from responsible sources
FSC® C105338

If you have any concerns about our products,
you can contact us on
**ProductSafety@springernature.com**

In case Publisher is established outside the EU,
the EU authorized representative is:
**Springer Nature Customer Service Center GmbH**
**Europaplatz 3, 69115 Heidelberg, Germany**

Printed by Libri Plureos GmbH
in Hamburg, Germany